Advanced Studies on High-Performance Metal-Ion Capacitors: Technologies, Systems and Applications

Advanced Studies on High-Performance Metal-Ion Capacitors: Technologies, Systems and Applications

Guest Editors

Kangyu Zou
Tianjing Wu
Jiangmin Jiang

MDPI

Basel • Beijing • Wuhan • Barcelona • Belgrade • Novi Sad • Cluj • Manchester

Guest Editors
Kangyu Zou
School of Materials Science and Engineering
Changsha University of Science and Technology
Changsha
China

Tianjing Wu
School of Chemistry
Xiangtan University
Xiangtan
China

Jiangmin Jiang
School of Materials Science and Physics
China University of Mining and Technology
Xuzhou
China

Editorial Office
MDPI AG
Grosspeteranlage 5
4052 Basel, Switzerland

This is a reprint of the Special Issue, published open access by the journal *Batteries* (ISSN 2313-0105), freely accessible at: https://www.mdpi.com/journal/batteries/special_issues/1P3BU89I8G.

For citation purposes, cite each article independently as indicated on the article page online and as indicated below:

Lastname, A.A.; Lastname, B.B. Article Title. *Journal Name* **Year**, *Volume Number*, Page Range.

ISBN 978-3-7258-6195-8 (Hbk)
ISBN 978-3-7258-6196-5 (PDF)
https://doi.org/10.3390/books978-3-7258-6196-5

Contents

batteries

MDPI

Article

Biomass-Derived Carbon Materials for Advanced Metal-Ion Hybrid Supercapacitors: A Step Towards More Sustainable Energy

Syed Shaheen Shah

Department of Material Chemistry, Graduate School of Engineering, Kyoto University, Nishikyo-ku, Kyoto 615-8520, Japan; shah.syedshaheen.78r@st.kyoto-u.ac.jp

Abstract: Modern research has made the search for high-performance, sustainable, and efficient energy storage technologies a main focus, especially in light of the growing environmental and energy-demanding issues. This review paper focuses on the pivotal role of biomass-derived carbon (BDC) materials in the development of high-performance metal-ion hybrid supercapacitors (MIHSCs), specifically targeting sodium (Na)-, potassium (K)-, aluminium (Al)-, and zinc (Zn)-ion-based systems. Due to their widespread availability, renewable nature, and exceptional physicochemical properties, BDC materials are ideal for supercapacitor electrodes, which perfectly balance environmental sustainability and technological advancement. This paper delves into the synthesis, functionalization, and structural engineering of advanced biomass-based carbon materials, highlighting the strategies to enhance their electrochemical performance. It elaborates on the unique characteristics of these carbons, such as high specific surface area, tuneable porosity, and heteroatom doping, which are pivotal in achieving superior capacitance, energy density, and cycling stability in Na-, K-, Al-, and Zn-ion hybrid supercapacitors. Furthermore, the compatibility of BDCs with metal-ion electrolytes and their role in facilitating ion transport and charge storage mechanisms are critically analysed. Novelty arises from a comprehensive comparison of these carbon materials across metal-ion systems, unveiling the synergistic effects of BDCs' structural attributes on the performance of each supercapacitor type. This review also casts light on the current challenges, such as scalability, cost-effectiveness, and performance consistency, offering insightful perspectives for future research. This review underscores the transformative potential of BDC materials in MIHSCs and paves the way for next-generation energy storage technologies that are both high-performing and ecologically friendly. It calls for continued innovation and interdisciplinary collaboration to explore these sustainable materials, thereby contributing to advancing green energy technologies.

Keywords: biomass; activated carbon; zinc-ion hybrid supercapacitor; sodium-ion hybrid supercapacitor; potassium-ion hybrid supercapacitor; aluminium-ion hybrid supercapacitor

1. Introduction

The efficient utilization of renewable energy sources, such as solar and wind, relies heavily on sustainable energy storage. Hybrid supercapacitors (HSCs) have been created to store energy using electric double-layer capacitance and Faradaic pseudocapacitance modes [1]. Even though they are used in various applications, HSCs offer a larger specific capacitance, a longer service life, and a greater power density than general batteries and double-layer capacitors [2]. Hence, their exceptional performance makes them an outstanding choice for advanced energy storage devices in the next generation [3]. Supercapacitors, sustainable energy storage options, are gaining popularity in fulfilling energy requirements and decreasing dependence on traditional energy sources [4]. Electrochemical capacitors, also known as supercapacitors, are potential alternatives or additions to batteries because of their high specific power, durability, and rapid charging and discharging capabilities [5].

These devices have many applications, from portable electronics to hybrid electric cars and power plants. When combined with fuel cells or batteries in hybrid electric vehicles, HSCs can provide the necessary energy for rapid acceleration [6]. Over the years, supercapacitors have demonstrated an existence ranging from low-scale to high-power energy storage systems. Supercapacitors are heavily influenced by the elements incorporated in preparing HSCs due to their high capacitance values [7–9]. Overall, it is essential that sustainable energy storage technologies, such as HSCs, be used to produce renewable energy because they provide good proficiency and long durability.

Metal-ion HSCs (MIHSCs) are an unequalled venture in sustainable energy storage technologies. They incorporate two distinctive types of devices: supercapacitors and batteries. MIHSCs are a hybrid of supercapacitors and batteries, with both the high power density of supercapacitors and the high energy density of batteries. Li et al. [10] explained that by incorporating supercapacitor-type and battery-type electrodes, MIHSCs inherit two charge storage mechanisms, delivering a high level of power per unit volume from the supercapacitor and a high level of energy per unit volume from the battery. In other words, MIHSCs are vital in maintaining modern devices due to their acutely growing energy demands. Moreover, it could be inferred that the hybrid device type is already quite promising in energy storage technologies that can satiate the rapid increase in energy needs [11]. They are essential in maintaining energy storage because batteries and supercapacitors can be used in a hybrid configuration since MIHSCs combine supercapacitors' long-cycle stability with batteries' inherent disadvantages [12,13]. Furthermore, these devices are instrumental in refining energy storage. They pave the way for achieving equilibrium in energy utilization in density and power. Their use in energy storage allows for wide operating potential windows (OPWs), high energy, safer application, and high power density; thus, MIHSCs show promising potential as a leading contender for the future generation of storage devices [14,15].

MIHSCs, comprising zinc-, sodium-, potassium-, and aluminium-ion HSCs, are differentiated based on efficiency and performance. Zinc-ion HSCs (ZIHSCs) are a potential form of energy storage which possess both the large storage capacity of zinc-ion batteries and the high power of a supercapacitor; high energy/power density is suitable for effective and consistent energy storage. ZIHSCs were presented as having better initial and first-cycle capacity retention [16,17]. Sodium-ion HSCs (NIHSCs) are new alloying-based MIHSCs fabricated by overcoming sodium-ion and lithium-ion devices' explosive and toxic drawbacks. The NIHSCs are structurally proven to occupy the neutralizing position with the best energy and power density ratio. However, high-power alkalis have recently gained much attention since the breakthrough of ion batteries with sodium and potassium [18,19]. Potassium-ion HSCs (KIHSCs) exhibit a significant ability to maintain their capacity over multiple cycles. KIHSCs have the most reasonable energy density of 58.2 Wh/kg with exceptional cycling durability [19]. Aluminium-ion HSCs (AIHSCs) made entirely from nitrogen-doped carbon spheres with a hierarchical porous hollow structure exhibit much better performance [20]. AIHSCs were thus developed to have drastically superior energy and power densities and show great potential for electrochemical performance [21]. Therefore, these MIHSCs are optimal for energy storage systems fulfilling requirements.

Biomass-derived carbon (BDC) materials are essential for improving the efficiency and performance of MIHSCs. Carbon materials possess numerous benefits, such as cost-effectiveness, environmental friendliness, and sustainability, making them highly appealing for energy conversion and storage. BDC materials provide multiple characteristics that render them highly suitable for enhancing the electrochemical efficiency of supercapacitors [22–24]. The inherent characteristics of these carbon compounds encompass a substantial specific surface area (SSA), commendable electrical conductivity, and adjustable porosity. The efficiency of devices, especially MIHSCs, can be enhanced by utilizing BDC materials. Specifically, in a ZIHSC [25], BDC materials could function as high-performance electrode materials. Such materials offer a high SSA that ensures sufficient charge storage and an overall improvement in the capacitive behaviour of such devices. Utilizing BDC materials in a NIHSC [24] could also realize high energy and power density. For KIHSCs, BDC materials could enhance capacity retention and cycling stability [26]. Lastly, these materials could improve ion adsorption and mobility in AIHSCs, increasing their energy and power density [27]. BDC materials, therefore, present a sustainable way of integrating high performance into MIHSCs. The unique natural properties of this type of carbon make it suitable for enhancing energy storage in these supercapacitors, which would, therefore, find an extensive range of applications in the current shift to sustainability.

This review study thoroughly examines using carbon materials produced from biomass to develop high-performance MIHSCs. The focus is primarily on sodium-, potassium-, aluminium-, and zinc-ion-based systems. BDC compounds are highlighted for their abundant availability, renewable nature, and exceptional physicochemical properties. These materials are considered excellent choices for supercapacitor electrodes, as they effectively combine environmental sustainability with technical progress. This paper explores creating, modifying, and designing the structure of these materials, focusing on their significant SSA, adjustable porosity, and the introduction of diverse elements, which are crucial for gaining excellent electrochemical performance in energy storage systems. An important innovation is the thorough comparison of these carbon materials in various metal-ion systems, revealing the combined effects of their structural characteristics on each type of supercapacitor. This review also discusses the difficulties of scalability and maintaining consistent performance and provides valuable insights for improving sustainable energy storage technologies. It highlights the potential of using BDC materials to develop advanced MIHSCs.

2. Basic Principles of MIHSCs

Capacitors, batteries, and HSCs are three energy storage technologies with unique attributes. Capacitors, specifically electrolytic capacitors, are efficient energy storage devices with notably high power capability and cyclic life [28]. They offer unprecedented ultrahigh power density and charge–discharge speed advantages, outperforming chemical batteries and electrochemical supercapacitors [29]. In contrast, batteries have notably high energy densities and are known for storing large amounts of energy. Various types of batteries are utilized in several industries, including portable electronic devices, electric cars, and large-scale electrical storage at power stations [30]. They use chemical reactions to store significantly more energy than capacitors, while their power capabilities are considerably lower than those of capacitance devices. Supercapacitors, alternatively called electrical double-layer capacitors [31], are rechargeable energy storage approaches with capacitors' and batteries' characteristics. Such power and energy delivery within traditional capacitors and batteries lend supercapacitors unique characteristics that make them promising energy storage device candidates [32]. Supercapacitors have a significantly greater power density than lithium-ion batteries but are behind these in energy density [12,33]. HSCs, specifically MIHSCs [16], combine the best of both worlds by providing energy and power density superior to capacitors or batteries. The devices consist of a battery-type electrode and a capacitive electrode, giving potential avenues for enhancing energy storage systems in the future [14]. Energy storage devices can be categorized based on the area in which they excel, with capacitors having superior power density and cyclic life, batteries having high energy density, supercapacitors offering power and many cycles, and hybrid capacitors having both exceptional power and energy density.

MIHSCs represent a significant advancement in energy storage technology since they effectively merge the advantages of batteries and supercapacitors for efficient energy storage applications. These devices are believed to enhance both unique power and energy densities [12]. MIHSCs leverage the advantageous features of both battery-type and capacitive electrodes. This allows them to achieve a high energy density similar to batteries and a high power density comparable to supercapacitors [16]. A diverse array of metal oxide/hydroxide materials can unite to undermine the integrity of the material, exhibiting a shift in valence at different potentials, which is incorporated to enhance the efficiency of HSCs [34]. Different MIHSCs, including LIC and NHC, have distinct properties and performances [35]. Coating capacitor-type and active battery-type material for pseudocapacitive reactions is the standard format in which HSCs use more of what works well in the performance [36]. Batteries and supercapacitors are then integrated to achieve optimal performance in pseudocapacitive reactions, such as the features of battery-type material [37]. Bunker and his colleague's work on materials such as ZHSCs and PIHCs attempts to understand a battery that delivers optimal functionality [21,38]. Furthermore, researchers have created multivalent-ion HSCs to attain elevated energy and power densities [39]. MIHSCs constitute a significant advancement in energy storage technology that combines the benefits of batteries and supercapacitors. These devices utilize a range of metal ions and electrode materials to enhance power and energy density, making them a promising alternative for the future of energy storage.

MIHSCs integrate battery-type and capacitive electrodes to achieve high energy density, fast energy discharge, and extended lifespan [40]. The process of charging and discharging between the two electrodes exhibits significant differences. During the charging process, ions are stored in the electrolyte on the electrode's surface, forming an electric double layer in electrodes similar to supercapacitors [41]. For a battery-type electrode, ions are propelled into an electrode material via a redox process. The charging process of MIHSCs commences by applying a voltage to the device, enabling the ions in the electrolyte to propel themselves towards the electrode's surface through different methods. Electrons are generated via intermediary electrodes [42] or redox reactions [43]. Charging refers to the input of electric current retained within the device and availed during discharge. In this regard, discharging involves the removal of stored charge as the ions migrate back into the composition from the negatrode and positrode.

Consequently, a current is produced and can drive electrically driven devices [44]. The working principles of MIHSCs are the primary determinants of efficiency when measuring their performance. The device can produce a high power output and store a significant amount of charge throughout multiple life cycles due to the integration of supercapacitors and battery technologies [45]. The choice of electrodes, the electrolyte composition, and the charging approach determine the overall efficiency [12]. Highly appropriate charge–discharge operations are crucial to fully realizing sophisticated energy storage devices' energy storage and distribution functionalities. MIHSCs exploit capacitive and battery-type securities as electrodes; hence, they avail the merged properties of supercapacitors and batteries. Battery-type securities undertake redox responses, with capacitive securities acting as an electrochemical double layer or pseudocapacitors during charging and discharging, as demonstrated in Figure 1a,b. Therefore, when designing MIHSCs, ensuring security, including the battery, is essential [46,47].

Carbon-based zinc HSCs' energy storage and release mechanism involves multiple processes. The charge and discharge processes are enabled by the reversible deposition and removal of Zn/Zn^{2+} on the zinc metal electrode during charge and discharge, respectively (Equation (1)). However, the mechanism of the carbon negatrode is more complex. Some researchers believe that the Zn^{2+} ions' adsorption/desorption on the carbon surface is a critical component of the capacity (Equation (2)) [48,49]. In contrast, others propose that anions, such as sulfate ($SO_4{}^{2-}$), also play a substantial role in the energy storage process (Equation (3)) [50]. The behaviour of these negatively charged ions in the electric double layer is contingent upon the carbon material's operating voltage and zero charge potential. In addition to physical adsorption/desorption, chemical adsorption/desorption can occur, especially in carbon materials doped with heteroatoms such as oxygen and nitrogen. This leads to the generation of extra pseudocapacitance (Equations (4) and (5)) [51].

Zn positrode:

$$Zn^{2+} + 2e^{-} \leftrightarrow Zn^{2+} \tag{1}$$

Carbon-based negatrode:

i. Process of physical adsorption or desorption

$$C + X^- \leftrightarrow C \parallel X^- \quad (2)$$

$$C + Zn^{2+} \leftrightarrow C \parallel Zn^{2+} \quad (3)$$

Here, C stands for the carbon, while X^- symbolizes the electrolyte anion.

ii. Process of chemical adsorption or desorption

$$C \cdots O + Zn^{2+}/H^+ + 2e^- \leftrightarrow C \cdots O \cdots Zn/C \cdots O \cdots H \quad (4)$$

$$N \cdots O + Zn^{2+}/H^+ + 2e^- \leftrightarrow N \cdots O \cdots Zn/C \cdots O \cdots H \quad (5)$$

Here, $C \cdots O$ represents oxygen-containing functional groups on the carbon material, while $N \cdots O$ denotes nitrogen-containing functional groups. These groups play roles in the chemical processes occurring on carbon-based electrodes.

Carbon-based MIHSCs have demonstrated exceptional performance, which promises to transform future energy storage. This technology exhibits a strong combination of high energy density, rapid charge–discharge rate, and excellent cycle life. It is a significant competitor in multiple industries, such as portable devices, electric cars, and renewable energy systems.

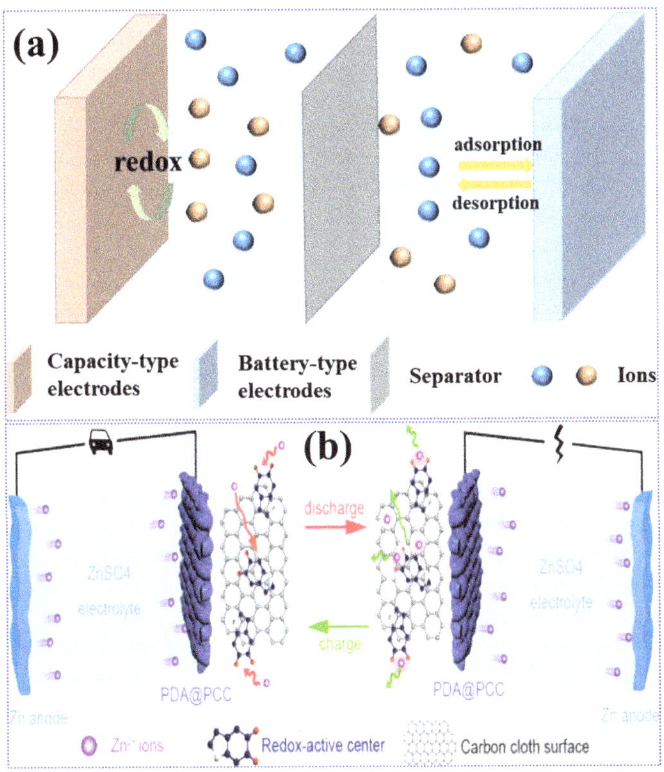

Figure 1. (**a**) A schematic representation for an MIHSC consisting of battery-type and capacitor-type electrodes. Reproduced with permission from Ref. [46]. Reproduced under the term CC BY 4.0. Copyright 2024, Fan et al., Springer Nature. (**b**) A specific charge–discharge demonstration of a ZIHSC using a carbon-based negatrode. Reproduced with permission from Ref. [47]. Copyright 2020, American Chemical Society.

3. Synthesis of BDC Materials for MIHSCs

Several essential techniques and aspects are involved in synthesizing BDC materials for MIHSCs to improve their electrochemical performance. Pyrolysis is the most commonly used technique for converting biomass into carbon materials at high temperatures without oxygen to produce porous carbon structures suitable for energy storage applications [52]. Hydrothermal carbonization is another critical technique for converting biomass into carbon at high temperatures and pressure in water, resulting in unique properties of carbon materials [53]. Template methods are also used to control the morphology and porosity of BDC materials, which involves sacrificing templates to form specific structuring in the synthesized materials [54]. Activation techniques, such as chemical or physical methods, are mainly used to increase the SSA and porosity of carbon compounds derived from biomass sources. While these alterations directly enhance the SSA and porosity, they also indirectly improve the electrochemical performance. This is because a larger surface area and enhanced porosity enable more efficient movement of ions and easier access to the electroactive sites. This can result in a noticeable improvement in conductivity during operation [55]. Alloying or doping bio-based carbons with heteroatoms such as nitrogen or phosphorus enhances conductivity and produces the appropriate materials for supercapacitors [56]. The porosity development of BDC materials is critical to maximizing energy storage capacity. Surface functionalization methods include surface chemistry modification for improved interaction with various electrolytes that contribute generously to the supercapacitors' electrochemical performance [57]. The performance qualities of bio-based carbons are influenced by their electrochemical properties, including capacitance and cyclic stability. These features are considered while choosing appropriate synthesis methods [58]. Figure 2a depicts the essential electrode materials for MIHSCs, with the electrolyte playing a critical role in charging the contact between the electrode and the electrolyte [59]. Research on electrode materials aims to enhance long-term charge propagation and stability in MIHSCs by optimizing pore size and distribution. Similarly, Figure 2b showcases a range of carbon-based materials employed in supercapacitive storage systems, highlighting their distinct attributes [60]. The synthesis of BDC materials for MIHSCs involves critical techniques, pyrolysis, hydrothermal carbonization, template methods, activation techniques, heteroatom doping, porosity development, surface functionalization, and other processes to ensure proper suitability for supercapacitors.

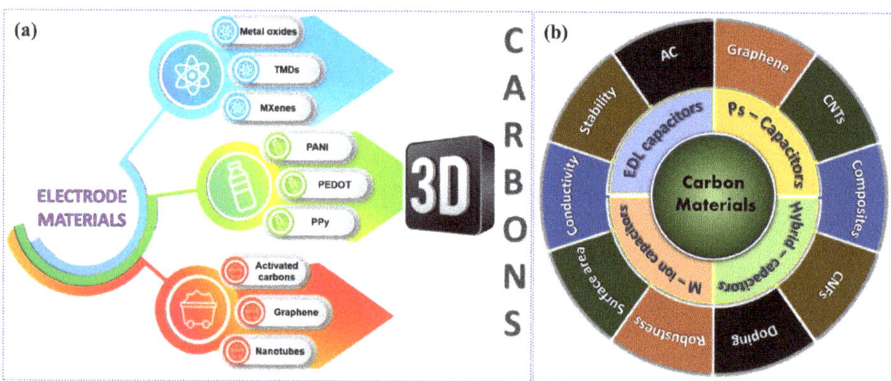

Figure 2. (**a**) Segmentation of electrode materials for MIHSCs. Reproduced with permission from Ref. [59]. Reproduced under the term CC BY 4.0. Copyright 2020, Galek et al., Frontiers. (**b**) Schematic representation for various carbon-based materials used as electrodes in different supercapacitive storage systems. Reproduced with permission from Ref. [60]. Reproduced under the term CC BY 4.0. Copyright 2022, Olabi et al., MDPI.

3.1. Synthesis of BDC Materials for ZIHSCs

BDC materials have attracted considerable interest because of their varied structural characteristics and possible uses in energy storage systems, particularly in synthesizing ZIHSCs. These materials offer renewability, sustainability, and cost-effectiveness advantages, making them attractive for various electrochemical applications [58,61]. BDC materials' structural diversity, porosity, and surface heteroatom doping have significantly impacted their performance in supercapacitors and other energy storage systems [61]. Additionally, synthesizing activated carbon (AC) from lignocellulosic biomass via chemical activation is promising for obtaining suitable precursors for energy storage applications [62]. Furthermore, incorporating self-doped heteroatoms in biomass-derived ACs has enhanced their electrochemical performance, laying a foundation for their supercapacitor application [63]. The potential of BDC materials in synthesizing ZIHSCs has been further supported by developing novel electrode materials and electrolytes. For instance, combining a zinc-activated positrode and a BDC negatrode has been explored, leading to the assembly of an aqueous ZIHSC with promising performance [16,64,65]

Moreover, using hydrogel electrolytes based on naturally abundant cellulose has been identified as an ideal approach for constructing flexible solid-state ZIHSCs, aligning with the principles of resource renewability and environmental sustainability [66]. Furthermore, the study of hollow mesoporous carbon spheres as electrode materials for symmetrical and aqueous ZIHSCs has shown promise for developing high-performance energy storage devices [67]. BDC materials provide a flexible foundation for creating electrode materials and electrolytes, which allows for the advancement of high-performance ZIHSCs. Due to their varied structure, ability to hold substances, the introduction of different elements on their surface, ability to be replenished, and ability to be maintained over time, these materials are appealing options for meeting the increasing need for effective energy storage solutions [68,69].

Samage et al. [70] explored the transformation of *Solanum melongena* (SM), an agro-waste, into a self-doped carbon material suitable for symmetric supercapacitors and ZIHSCs through a one-step pyrolysis process. Initially, the SM was washed to remove impurities, chopped into small pieces, and sun-dried before being crushed into a fine powder, as shown in Figure 3a. The pyrolysis process was carried out in a nitrogen atmosphere that does not react with the materials. The temperature was increased gradually at a rate of 5 °C per minute, reaching 700 °C, 800 °C, and 900 °C successively. The pyrolysis process lasted for a total of 3 h. After the pyrolysis process, the carbonized samples were subjected to a washing step using 0.5 M HCl. They were then rinsed with distilled water until the pH became neutral and dried at 80 °C for 12 h. The resulting SM-derived carbon (SMC) exhibited a three-dimensional porous structure with uneven flake-like shapes, with SMC-700 showing a more hydrophilic nature than SMC-800 and SMC-900. Surface morphology examined through FESEM revealed the formation of porous structures at 700 °C and 800 °C, while 900 °C samples showed pore structure collapsing due to aggregated carbon, which undergoes a process where it transforms into thick layers resembling flakes. The nitrogen sorption isotherms revealed that SMC-800 had a significant SSA of 686.29 m^2/g, mainly attributed to micropores and mesopores, resulting in a high porosity. The iron content in the precursor was used strategically for graphitization, converting amorphous carbon into graphitic nanostructures during pyrolysis.

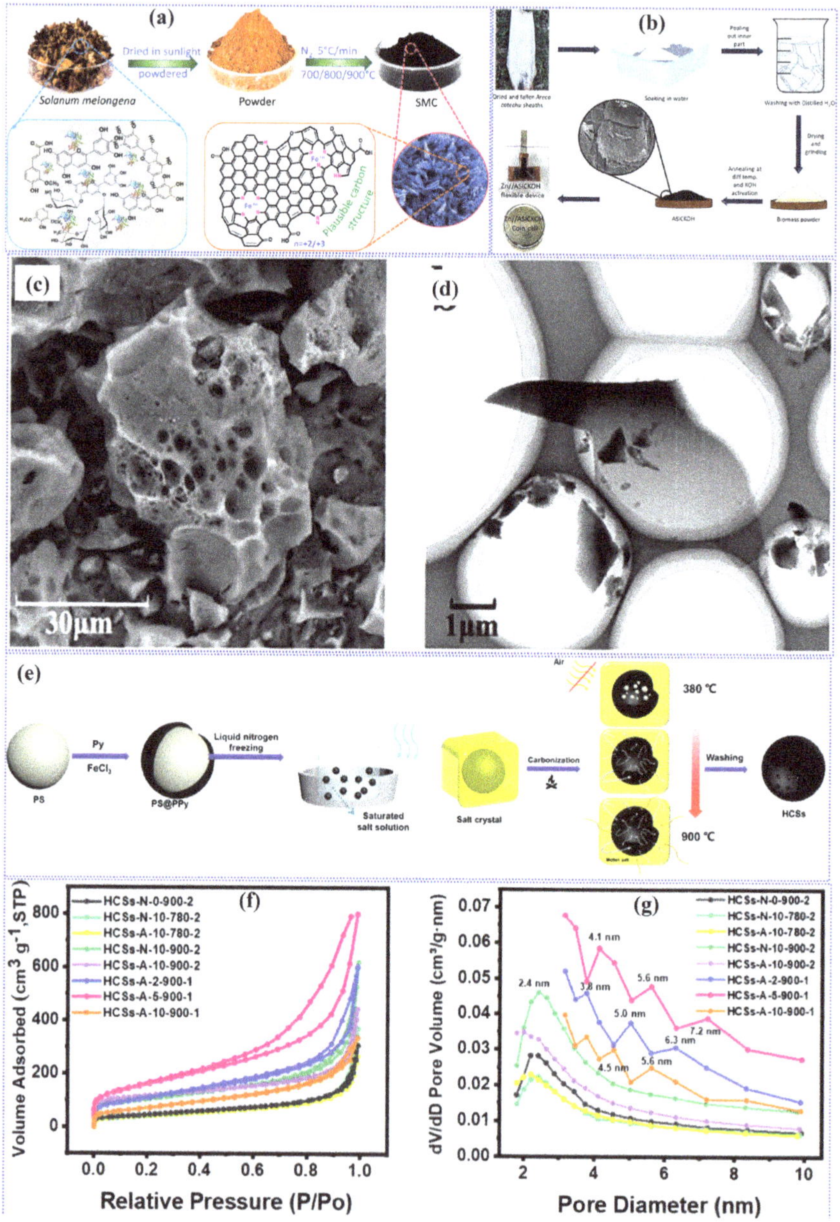

Figure 3. (**a**) A diagram illustrating the process of preparing the self-doped SMC. Reproduced with permission from Ref. [70]. Copyright 2024, Elsevier. (**b**) Diagram showing the process for synthesizing an AC from Areca Catechu sheaths, along with its application in ZIHSCs in coin and pouch cell formats. Reproduced with permission from Ref. [71]. Copyright 2022, American Chemical Society. (**c**) FESEM and (**d**) TEM micrographs of the N/S-AC. Reproduced with permission from Ref. [72]. Copyright 2023, American Chemical Society. (**e**) Diagram illustrating the process of preparing HCS. (**f**) Nitrogen adsorption/desorption isotherms and (**g**) pore size distribution of the HCSs under different circumstances. Reproduced with permission from Ref. [73]. Copyright 2024, American Chemical Society.

Naik et al. [71] investigated the production of high-performance flexible ZIHSCs using carbon sheets produced from agricultural waste. The carbon material is derived from *Areca Catechu sheaths*, an abundant agricultural waste, through a multi-step process. As demonstrated in Figure 3b, the synthesis begins with collecting dried sheaths from a local farm, washing them, and soaking them in water to remove the inner part. After peeling, the inner section is dehydrated, pulverized into a fine powder, and subjected to annealing in a tubular furnace at temperatures between 700 and 900 °C for 3 h while exposed to a nitrogen flow. Post-carbonization, the carbon materials undergo a washing procedure using a solution of 1 M HCl and a mixture of ethanol and water. Subsequently, they are dried at 80 °C for one night. The carbon obtained at a temperature of 900 °C exhibited superior performance and was subsequently subjected to KOH activation, leading to KOH-AC formation. The carbon compounds demonstrate a significant SSA, with KOH-AC boasting a value of 2760 m^2/g. The enhanced SSA and reduced pore diameter in KOH-AC contribute to its improved electrochemical performance.

Lignin, a significant biomass constituent, produces nitrogen and sulfur co doped hierarchical porous carbon (N/S-AC) materials for high-performance ZIHSCs [72]. The preparation entails a dual-phase procedure: charring and activation using co-doping. The process involves combining lignin with urea and thiourea to provide nitrogen and sulfur, subjecting the mixture to carbonization at a temperature of 750 °C. As a result, nitrogen and sulfur co-doped activated carbon is formed. The materials possess a hierarchical pore structure, which is essential for offering active sites with a high SSA and an appropriate distribution of pore sizes. The FESEM image (Figure 3c) reveals that the N/S-AC has micrometre-sized particles with a porous structure, while TEM analysis (Figure 3d) confirms an interconnected porous nature. The BET SSA for the N/S-AC material is estimated at 2773.02 m^2/g, demonstrating significant porosity. Additionally, the doped heteroatoms contribute to adjusting the pore structure and increasing surface wettability, providing enhanced conductivity and pseudocapacitance, making them ideal for high-performance ZIHSCs.

Zhao et al. [73] proposed a sustainable technique for producing hierarchically porous hollow carbon spheres (HCSs) as negatrode materials in ZIHSCs. The procedure entails the pyrolysis of polystyrene–polypyrrole nanoparticles within a sodium chloride confinement, as shown in Figure 3e. This synthesis is performed in a single vessel, in the presence of air, eliminating the requirement to use inert gas to protect the reaction. The HCSs obtained display a hierarchical pore structure with a significant SSA of 578.2 m^2/g. They are also doped with 2.6 at.% oxygen and 5.6 at.% of nitrogen. Figure 3f,g display the N_2 adsorption/desorption isotherms and the related pore size distribution of the HCSs under different circumstances. The carbon spheres' form and structure remain intact throughout the pyrolysis process, and their porous nature enhances their electrochemical properties. Du et al. [74] proposed a technique for creating a two-dimensional (2D) porous carbon matrix co-doped with nitrogen and oxygen. This matrix is synthesized using biomass generated from yeast cell walls. The synthesis involves a molten salt method, where yeast cell walls are mixed with a urea solution, freeze-dried to form precursors, and heated at 800 °C in an air atmosphere with NaCl and KCl. The resulting material exhibits a 2D lamellar structure, with its porous nature further enhanced by the urea decomposition, which introduces micro- and mesopores into the carbon matrix. The material, known as NAC-20, possesses a significant SSA of 368.63 m^2/g. It is composed of a carbon structure that has been co-doped with nitrogen (6.03%) and oxygen (9.96%). This co-doping enhances the material's electrical conductivity and wettability. Several BDCs have been synthesized and utilized in ZIHSC configurations. Biomass-based nitrogen–oxygen co-doped carbons have exhibited high SSAs, heteroatom doping, and porous morphologies contributing to the performance of these electrodes. Their characteristic properties and performance capabilities, including morphology and SSA, are listed in Table 1. Thus, it is clear from the literature that biomass-based carbons with improved morphologies, porous structures, and enhanced SSAs were addressed to permit effective ZIHSCs. The hierarchical pores and substantial heteroatom doping of catalytic zinc ions may improve these devices' specific capacitance, energy, cyclability, and power.

3.2. Synthesis of BDC Materials for NIHSCs

BDC materials have attracted considerable interest for their possible use in NIHSCs due to their capacity to be renewed, their sustainability, and their cost-effectiveness [23,75,76]. These materials have been investigated as potential electrode materials for high-energy NIHSCs, showing promising electrochemical performance [23]. BDC materials' inherent structural benefits, cost-effectiveness, and renewability make them highly appealing for sodium-ion energy storage systems [76]. Moreover, introducing heteroatoms into BDC materials increases the interlayer distance, enhancing the efficiency of transporting and storing sodium ions. As a result, sodium-ion batteries exhibit exceptional performance at high rates [61]. The synthesis of BDC materials has been a subject of interest, with research focusing on tailoring their structure to enhance their performance in NIHSCs. For instance, hierarchically porous carbon derived from shiitake mushrooms has been investigated for capacitive energy storage, demonstrating the potential for BDC materials in energy storage applications [77].

Furthermore, the favourable tubular composition of BDC has been identified as a possible positrode option for sodium-ion batteries, thanks to its abundant availability, affordable price, and sustainable nature [78]. In addition, there have been new developments in energy storage using NIHSCs and ZIHSCs, revealing the potential of biomass-derived materials to meet energy storage requirements [18]. BDC materials show significant potential for NIHSCs, providing a viable and economical option for energy storage applications. These materials' diverse structure and electrochemical activity, combined with continuing research on their synthesis and performance improvement, make them viable contenders for meeting the increasing need for effective sodium-ion energy storage solutions.

The research works by Niu et al. [79] mainly focused on synthesizing two types of BDC materials using gelatine and phytic acid as precursors, as demonstrated in Figure 4a. The synthesis was carried out using sustainable concepts, including non-toxic precursors, which are readily available, low in cost, and ecologically friendly processes. The synthesis also included the utilization of physisorption, nanostructure engineering, and heteroatom doping, particularly phosphorus and nitrogen, to enhance the efficiency of carbon materials. P- and N-doped networked carbon nanosheets with hierarchical porosity were formed. They exhibited porous nanosheet networks synthesized directly by KCl/ice as a dual template or KOH for microporosity. The materials illustrated several morphologic features, such as nanosheet networks and a high SSA, essential for their application as electrodes in NIHSCs. Guo et al. [80] investigated the sustainable fabrication of dual (N, O)-doped biocarbon nanosheets from marine biomass. As described in Figure 4b, the direct pyrolysis of cuttlebones, along with the air oxidation activation approach, allows for a precise decrease in the thickness of carbon sheets from 35 to 5 nm. Consequently, this leads to the formation of a distinct hierarchical porous structure characterized by enhanced SSA and porosity and an increase in N-doping content from 7.5% to 13.9% at the atomic level. The controlled N-doping content of the nanosheets was independently monitored by the new role of air oxidation activation in regulating heteroatom content in the carbon matrix, ensuring novel biocarbon nanosheets with significant nitrogen and oxygen doping in the carbon matrix. The dual-doped structure provides improved conductivity and additional charge storage.

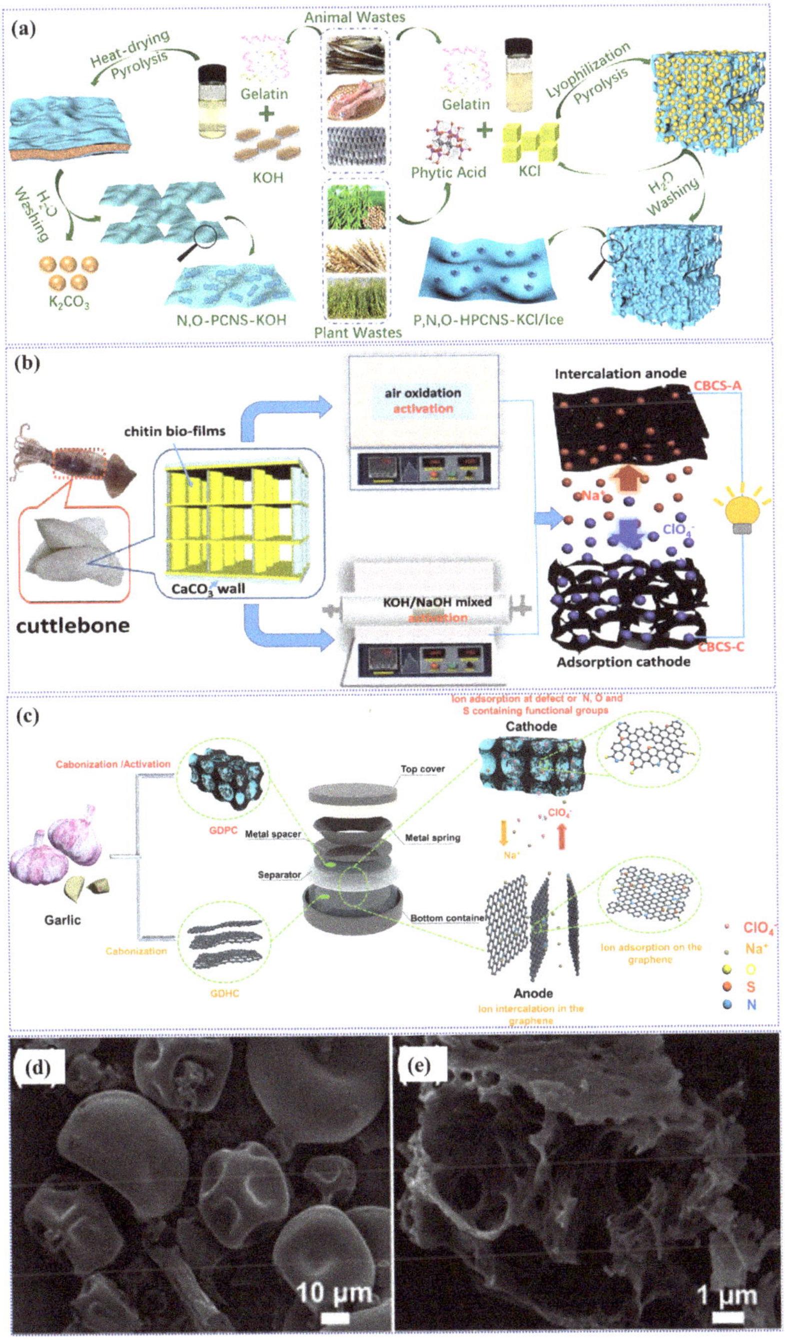

Figure 4. (**a**) A diagram illustrating the synthesis of interconnected carbon nanosheets with hierarchical porosity, incorporating phosphorus (P) and nitrogen (N) dopants. This synthesis involves dual KCl templates and ice (P, N-HPCNS-KCl/Ice). Reproduced with permission from Ref. [79]. Copyright 2020, American Chemical Society. (**b**) The diagram illustrates the synthesis process of the material and the corresponding charge storage processes at the positrode and negatrode in a NIHSC. Reproduced

with permission from Ref. [80]. Copyright 2018, American Chemical Society. (**c**) A scheme outlining the GDHC and GDPC preparation process and the mechanisms underlying sodium-ion storage in NIHSCs. Reproduced with permission from Ref. [81]. Copyright 2019, American Chemical Society. (**d**,**e**) FESEM micrographs of the PCNS at different magnifications. Reproduced with permission from Ref. [82]. Copyright 2020, Elsevier.

Liu et al. [81] examined a NIHSC made from carbon compounds obtained from garlic. The synthesis (Figure 4c) involves two methods: high-temperature pyrolysis at 1100 °C to create a durable carbon positrode, referred to as GDHC, and carbonization tailored by KOH activation to generate a porous carbon negatrode, referred to as GDPC. The GDHC positrode demonstrates a pseudo-graphitic arrangement characterized by an interlayer distance ranging from 0.365 to 0.376 nm. Conversely, GDPC possesses a highly permeable structure, with an SSA measuring 1682 m^2/g. The carbon material structure is assessed using scanning electron microscopy, which identifies the block-like morphology of GDHC and the porous structure of GDPC. In addition, XRD and Raman spectra provide evidence of a well-organized structure and the presence of graphite. Zhang et al. [82] studied fabricating a three-dimensional porous carbon nanosheet (PCNS) utilizing soy protein and zinc nitrate. The synthesis involves using a chemical gas expansion approach followed by calcination to produce a three-dimensional PCNS material doped with nitrogen and sulfur. The resulting PCNS has a hierarchical porous structure characterized by a significant abundance of micropores and mesopores, as revealed in Figure 4d,e. The PCNS exhibits an SSA of ~516.6 m^2/g. The carbon compounds exhibit an interlayer spacing of 0.385 nm, surpassing that of graphite. This higher spacing aids in the process of Na^+ intercalation and deintercalation. The result showcases a distinctive linked structure of carbon nanosheets, which offer abundant pathways for ion transport and improve its ability to store salt. Zhang et al. [83] examined the production of rigid carbon material from lotus stems. The synthesis entails subjecting the lotus stems to carbonization at several temperatures, namely 1200, 1400, and 1600 °C. The resultant materials, LS1200, LS1400, and LS1600, maintain a porous structure with a hierarchical channel array inherited from the lotus stems. FESEM and TEM investigations demonstrate that greater carbonization temperatures result in an elevated graphitization degree and a decreased tube wall thickness. The materials exhibit d_{002} spacings of 3.69 Å, 3.68 Å, and 3.61 Å, corresponding to LS1200, LS1400, and LS1600, respectively. These values indicate different levels of graphitic ordering. The carbons have a relatively low SSA ranging from 23.73 to 25.81 m^2/g. This SSA gradually decreases as the carbonization temperature increases. Several more carbon materials generated from biomass, characterized by their unique shapes, porous architectures, and large SSA, have greatly improved the efficiency of NIHSCs. The presence of a distinct microstructure, which includes both hierarchical porosity and doped structures, enables enhanced ion transport and charge storage capabilities. These supercapacitors have remarkable electrochemical properties, including a high capacitance, energy density, power density, and great cycle stability. All of these factors are thoroughly presented in Table 1.

3.3. *Synthesis of BDC Materials for AIHSCs*

BDC materials are considered promising options for developing AIHSCs because of their capacity to be renewed, their sustainability, and their environmentally beneficial characteristics [61,84]. These materials provide a viable and long-lasting option for storing energy, in line with the ideals of using renewable resources and promoting environmental sustainability. BDC materials' structural diversities and synthesis methods have been demonstrated to significantly enhance their performance in supercapacitors [61,84]. An example is using a nitrogen-doped carbon nanofiber network obtained from biomass as a template to decorate ultrathin nickel–cobalt-layered double hydroxide nanosheets [52]. This demonstrates the high performance of the nanosheets as an electrode for an asymmetric supercapacitor [25]. Furthermore, using biomass straw-derived porous carbon to produce supercapacitors through ball milling has attracted interest because of its renewable nature, abundant availability, affordability, and positive environmental impact [85]. Moreover, researchers have investigated the various structural variations in BDC materials for their potential uses in energy storage, focusing on the significance of their porous nature and the introduction of heteroatoms on their surfaces [61]. The synthesis methods of BDC materials are a subject of interest, with research focusing on tailoring their structure to enhance their performance in AIHSCs. These materials' renewable and eco-friendly characteristics and their wide range of structures make them appealing for meeting energy storage requirements, especially in AIHSCs. Using BDC materials in AIHSCs has significant potential for creating long-lasting and effective energy storage solutions. BDC materials provide a sustainable and cost-effective framework for creating AIHSCs. Their structural diversity and capacity to be renewed make them excellent options for meeting the increasing need for efficient energy storage solutions.

The research reported by Feng et al. [86] presents a sustainable and solvent-free technique for producing polymer spheres using m-phenylenediamine and hexamethylenetetramine (HMTA), with Pluronic F127 serving as a guiding agent for the structure, as demonstrated in Figure 5a. The combination is subjected to a temperature of 160 °C, initiating the polymerization process of MPF resin spheres. Subsequently, these spheres are subjected to a temperature of 600 °C, converting them into porous carbon spheres (PCSs). The PCSs possess a significant SSA of 3115 m^2/g and a 3.05 cm^3/g pore volume. The FESEM study (Figure 5b,c) reveals that the resin spheres maintain a spherical shape throughout the activation and carbonization techniques. Figure 5d displays the solid-state 13C NMR spectrum, which reveals a peak at 18 ppm for methyl and methylene. The 70, 73, and 75 ppm peaks show bonding between carbon and oxygen. A series of distinct peaks (104, 113, 129, and 147 ppm) associated with different carbon environments in the benzene ring are produced by the chemical shift over 100 ppm. Figure 5e displays the FTIR spectra of MPF resin spheres, revealing aromatic, methylene, amino, and hydroxyl functional groups. The FTIR data corroborate the atomic structure of the MPF resin. According to NMR and FTIR investigations, the m-phenylenediamine and formaldehyde derived from the breakdown of HMTA polymerize almost entirely; full polymerization results in volumetric productivity and ultrahigh yield. The carbon material obtained has a hierarchical structure containing micropores and mesopores, making it well suited for energy storage.

Seon et al. [87] examined the production of AIHSCs that incorporate a battery-like aluminium positrode with a capacitor-like graphene negatrode. They employed an ionic liquid, namely 1-ethyl-3-methylimidazolium chloride (EMImCl), and aluminium chloride ($AlCl_3$) as the electrolyte. The investigation is focused on two surface treatment approaches, electropolishing and electrodeposition, to improve the electrochemical performance of the aluminium positrode. The aluminium positrode, formed by electrodeposition on a molybdenum foil substrate, has a dendritic structure and a significant SSA. The presence of roughness and dendritic development improves the electrochemical activity of the material, resulting in higher efficiency in storing Al^{3+} ions. The XPS and GIXRD analysis reveals a reduction in the thickness of oxide layers and the presence of Al–OH species, which contribute to the formation of redox-active sites.

Lei et al. [88] reported the development and production of AIHSCs with a nitrogen-doped graphene negatrode, aluminium positrode, and [EMIm]Cl and $AlCl_3$ electrolyte with high capacitance. A layer of nitrogen-doped graphene was applied to a tantalum foil, creating a flexible negatrode with a nitrogen concentration of 3–5% by weight. Doping alters the arrangement of energy levels in the band structure, amplifies the ability of graphene to conduct electricity, and boosts its ability to store electrical charge. The FESEM study revealed crumpled graphene layers stacked on each other, resulting in a large SSA that facilitates electrochemical reactions. Sun et al. [89] presented using a nitrogen-doped micro-mesoporous carbon sphere (NCS) as a negatrode material in aluminium-based HSCs. The NCS was produced by combining phenolic resin with melamine in a 1:1 mass ratio, followed by carbonization at 600 °C in nitrogen. Subsequently, the carbon precursor was mixed with KOH and heated at 800 °C, forming a porous configuration. The SESEM study indicated the presence of round particles with sizes ranging from 0.1 to 1 μm. TEM demonstrated that these particles have an amorphous structure. The NCS has a BET SSA of 3078 m^2/g, considerably greater than the 506 m^2/g of the NC precursor containing nitrogen. This difference allows for more adsorption sites and facilitates more efficient ion diffusion.

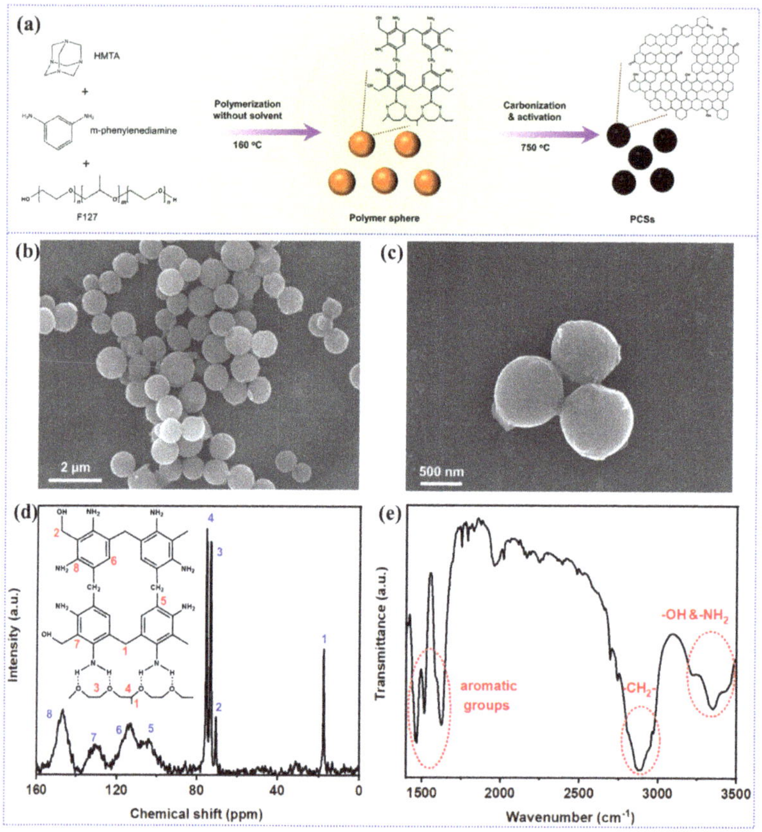

Figure 5. (**a**) The solvent-free synthesis method for MPF polymer spheres and PCSs is illustrated schematically. Morphological and structural characterization of MPF resin spheres; (**b**) low- and (**c**) high-magnification FESEM micrographs; (**d**) solid-state 13C NMR spectrum; and (**e**) FTIR spectrum. Reproduced with permission from Ref. [86]. Copyright 2023, Wiley.

Optimizing the synthesis parameters during the carbon generation from biomass is essential for improving the efficiency of AIHSCs. BDC materials provide a sustainable and environmentally friendly option for producing electrodes, including distinctive structures that may be fine-tuned to achieve exceptional electrochemical capabilities. Research emphasizes the importance of various parameters, including SSA, porosity, nitrogen doping, and the existence of electrochemically active metals in carbons generated from biomass [90–92]. These properties can all be affected by the procedure used for synthesis. Researchers may customize the characteristics of BDCs for supercapacitor applications by precisely controlling culture conditions, nutrient concentrations, and activation methods. This enables them to attain a high specific capacitance, exceptional cycle stability, and enhanced energy storage capacities. Various essential parameters for producing BDCs for AIHSCs are tabulated in Table 1.

3.4. Synthesis of BDC Materials for KIHSCs

BDC materials have garnered considerable attention in the field of KIHSCs due to their renewable nature, sustainability, and promise for efficient energy storage applications [93,94]. These materials offer a sustainable and cost-effective energy storage alternative, aligning with resource renewability and environmental sustainability principles. BDC materials' synthesis processes and structural design have been demonstrated to be significantly improved [93–95]. An example is the deliberate arrangement of well-distributed, extremely thin MoS_2 nanosheets within a hollow carbon structure, showing rapid potassium storage. This emphasizes the promise of BDC materials in KIHSCs [96]. Furthermore, the recent advancements and prospects of biomass-derived hierarchically porous carbon have been examined, focusing on the significance of developing and fabricating these materials for supercapacitor applications, specifically KIHSCs [94]. Using BDC materials in KIHSCs has significant potential for creating sustainable and highly effective energy storage solutions. The SSA, distribution of pore sizes, and ultimate electrochemical performance of the carbon materials are determined by the type of biomass precursor used and the activation processes employed. This emphasizes the significance of synthesis methods in customizing the properties of these materials for KIHSCs. Moreover, the recent advancements in creating and using hybrid electrodes made from biomass have given us valuable knowledge about the electrochemical properties of these materials and the ways they store ions in various types of batteries, such as potassium-ion batteries. This highlights the potential of BDC materials in KIHSCs [97]. BDC materials offer a sustainable and cost-effective platform for developing KIHSCs, with their structural diversity and renewability positioning them as valuable candidates for addressing the increasing need for effective energy storage technologies.

Graphene-like AC nanosheets and hard carbon (HC), two different carbon compounds made from almond shells, were synthesized according to Pham et al. [98]. The production of the HC involves a straightforward carbonization procedure at various temperatures, whereas the AC is generated by chemically activating it with KOH as the activating agent. An analysis of the structural properties of the materials shows that the hard carbon has a notable distance between its layers, which is advantageous for the process of potassium-ion (de)intercalation. The AC nanosheets possess a high SSA and an intricate porous structure. These observations, along with the analysis of the structure and chemistry, suggest that both HC and AC possess qualities that make them appropriate for storing potassium ions. A schematic representation for the dual-carbon KIHSCs using carbon nanosheet negatrodes derived from biomass and similar to graphene is shown in Figure 6a. Chen et al. [99] presented an innovative "plains–hills" model of carbon structure for producing electrode materials inspired by a lotus leaf's surface structure. This design comprises ultrathin carbon nanosheets, also known as "plains", which are formed with multiple lumps, referred to as "hills", (Figure 6b) using a one-step pyrolysis method. The synthesis entails utilizing $CaCl_2$ as a dual-purpose compound, complexing agent, and oxygen scavenger in a Ca^{2+}-bio gel, as demonstrated in Figure 6c. This structure integrates an unstacked nanosheet with different levels of imperfections, achieving a harmonious balance between pseudocapacitance and conductivity. The carbon materials obtained exhibit a balanced combination of high pseudocapacitance and conductivity. They have an SSA of 581 m^2/g and a drop in oxygen doping from 24.9% to 12.3%. This reduction contributes to the presence of both ordered and amorphous carbon structures.

Huang et al. [100] outlined a single-step calcination technique for converting mango seed husk into BDC doped with nitrogen and oxygen (NOHC). This material is intended to be a positrode KIHSC. The NOHC materials that have been synthesized exhibit a significant level of graphitization, with interlayer spacings that vary between 0.39 and 0.38 nm in correlation with increasing temperature. The materials demonstrate a BET SSA of up to

489.4 m^2/g, suggesting a porous structure advantageous for capacitive charge storage. A combination of rod- and sheet-like structures characterizes the shape of the NOHC. Doped nitrogen and oxygen are included to enhance the electronic conductivity and create active spots.

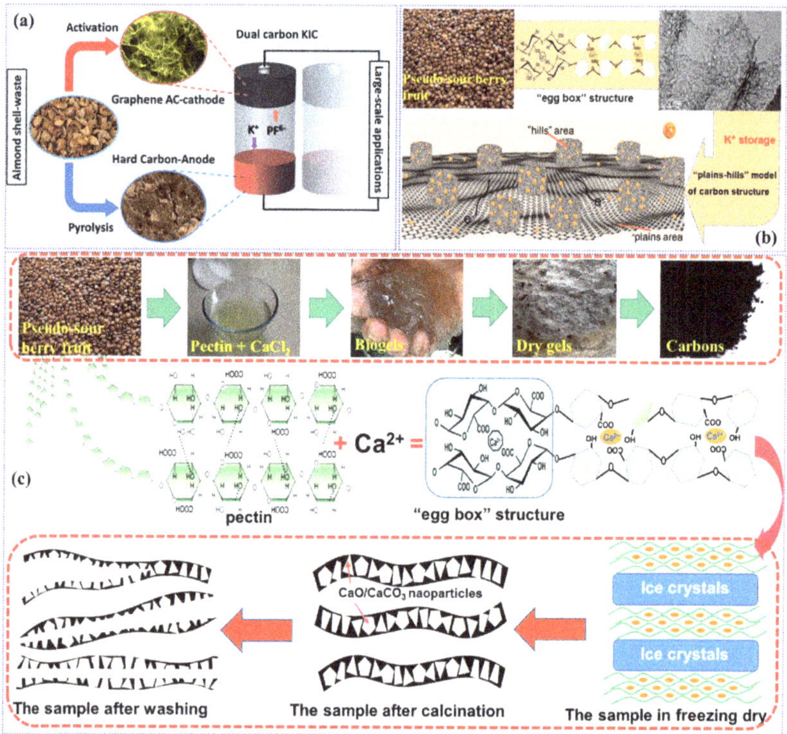

Figure 6. (**a**) Dual-carbon KIHSCs using carbon nanosheet negatrodes derived from biomass and similar to graphene. Reproduced with permission from Ref. [98]. Copyright 2020, American Chemical Society. (**b**) A graphical representation for the K^+-ion storage in BDC materials. (**c**) A graphical depiction of the CACF samples' synthesis protocol. Reproduced with permission from Ref. [99]. Copyright 2021, American Chemical Society.

Developing efficient synthesis techniques for carbon produced from biomass is crucial in determining the optimal performance of HSCs that use potassium ions. Supercapacitors have attracted significant attention because of their remarkable power density, long cycle life, and ecologically sustainable characteristics. However, the properties of the carbon compounds used as the electrode material significantly impact the device's performance. BDC is an ideal substitute for carbon since it is naturally abundant, inexpensive, and environmentally friendly. The synthesis techniques used to produce carbon, such as carbonization and activation processes, significantly impact its structure and other properties, thus affecting the performance of supercapacitors. The optimization of KIHSCs can be influenced by various parameters, as indicated in Table 1. The criteria encompass SSA, pore size distribution, surface functional groups, electrical conductivity, and other aspects, such as element content, which will be elaborated on later. Scientists can customize the carbon material by optimizing the synthesis technique to improve its capacitive behaviour, ion transport kinetics, and other energy-related characteristics. Thus, to enhance the above procedures, researchers can utilize several precursors and employ varied carbonization temperatures, activation methods, and post-treatment strategies.

Table 1. Synthetic protocols for preparing BDC materials in high-performance MIHSCs.

Biomass Source	Carbon Preparation Protocols	Morphology	SSA (m^2/g)	Ref.
Solanum melongena	Pyrolysis; 700, 800, and 900 °C; N_2 gas	Irregular flake-like 3D porous structure	574.34, 686.29, and 239.16	[70]
Areca Catechu sheaths	Pyrolysis; KOH activation; 900 °C; N_2 gas	Porous nanosheet	2760	[71]
Lignin	Pyrolysis; KOH activation; 750 °C; N_2 gas	Particles with a porous structure	2773.02	[72]
Salt crystals	Carbonization; $FeCl_3$ activation; 900 °C; N_2 gas	Hierarchically porous hollow spheres	578.2	[73]
Recycled yeast cell wall	Carbonization, NaCl/KCl activation, 800 °C; air	2D lamellar structure	368.63	[74]
Ipomoea carnea leaves	Carbonized; KOH activation; 900 °C; N_2 gas	Transparent spheres	736.7	[101]
Resorcinol–furfural resin	Carbonization, KOH activation, 900 °C, N_2 gas	3D hierarchical porous structure	1350	[102]
Anhydrous glucose	Pyrolysis; magnesium oxide and potassium bicarbonate activation; 750 °C; Ar gas	3D honeycomb-like structure with interconnected pores	2265	[103]
Sodium lignosulfonate	Carbonization; KOH activation; 900 °C; N_2 gas	Hierarchical porous structure	2460	[104]
Tannin	Pyrolysis; 800 °C; N_2 gas	Hierarchical micro- and mesoporous structure	967	[105]
Pine needles	Carbonization; KOH activation; 800 °C; Ar gas	Microporous structures with interconnected pores and channels	2493.4	[106]
Zeolitic imidazolate framework	Carbonization; thiourea mixing; 800 °C; Ar gas	Cubic porous structure	677.2	[107]
Fallen autumn leaves	Carbonization; KOH activation; 950 °C; N_2 gas	Amorphous and porous structures, with minor graphitic flakes	2114	[108]
Osmanthus flowers	Freeze drying; carbonization; 800 °C; Ar gas	Foamy-like structure with hierarchical pores	690.3	[109]
Aerial roots of Ficus macrocarpa	Pyrolysis; $KHCO_3$ activation; 900 °C; N_2 gas	Hierarchical porous nanosheet structure	1454.7	[110]
Gelatine/phytic acid	Pyrolysis; KOH/$KHCO_3$ activation; 900 °C; N_2 gas	Ultrathin and small nanosheets	1911	[79]
Marine biomass	Carbonization; 600 °C; N_2 gas	Porous nanosheets	1489	[80]
Garlic	Pyrolysis; KOH activation; 1100 °C; N_2 gas	Blocky morphology	1682	[81]
Soy protein	Pyrolysis; zinc nitrate activation; 700 °C; Ar gas	Hierarchical porous structure with abundant micropores and mesopores	516.6	[82]
Lotus stems	Carbonization; 1200, 1400, and 1600 °C; Ar gas	Turbostratic graphitic structure	25.81, 24.37, and 23.73	[83]
Apricot shell	Pyrolysis; 900, 1100, 1300, and 1500 °C; H_2/Ar atmosphere	Porous structure with well-connected porous network	2.7 to 56.7	[111]
Mango peels	Carbonization; sulfur and hexamethylenetetramine mixing; 1000 °C/800 °C; Ar gas	Porous flake-like morphology	1079.88	[112]

Table 1. *Cont.*

Biomass Source	Carbon Preparation Protocols	Morphology	SSA (m^2/g)	Ref.
Agricultural wastes, specifically peanut shell, wheat straw, rice straw, corn stalk, cotton stalk, and soybean stalk	Hydrothermal reaction with H_2SO_4 aqueous solution and subsequent chemical activation with KOH at 800 °C; Ar atmosphere	Porous sheet structure with interconnected macropores, mesopores, and micropores	534.3 to 1994.2	[113]
Cassava peel	Pre-carbonization 350 °C; KOH activation at 600 °C; Ar atmosphere	Slit-type pores and a highly disordered structure	1032	[114]
Corn silk	Pre-carbonization at 350 °C; KOH activation; carbonization at 800 °C; Ar atmosphere	Microporous nanosheets	2550	[115]
Sugarcane bagasse	$NaH_2PO_2{\cdot}2H_2O$ and melamine dispersion; hydrothermal reaction at 180 °C for 24 h; KOH activation; carbonization; 700 °C; Ar atmosphere	3D interconnected hierarchical porous structure	2803.2	[116]
m-phenylenediamine-formaldehyde resin	Carbonization; KOH activation; 750 °C; Ar atmosphere	Spherical morphology	3115	[86]
Graphene	...	Nanosheets	2675	[87]
Nitrogen-doped graphene derived from a synthetic approach	...	Wrinkled and overlapping morphology with a few layers	...	[88]
Phenol resin and melamine	Pre-carbonized at 600 °C; KOH activation at 800 °C; N_2 environment	Spherical particles	3078	[89]
Almond shells	Carbonization; KOH activation; 1100 °C; N_2 environment	Sheetlike nanostructures	127	[98]
Sour berry fruits	Freeze drying, $CaCl_2$ curing; bio gel and dry gel production; acid treatment	Nanoflakes	581	[99]
Mango seed shuck	Carbonization; 900, 1100, and 1300 °C; Ar atmosphere	Rod- or sheet-shaped feature	489.4, 383.2, and 319.1	[100]
Petroleum pitch	Pyrolysis; $KHCO_3$ activation; 800 °C; inert atmosphere	Cross-linked nanosheets	48.8	[117]
Gpositroderma lucidum spores	Carbonization at 850 °C; mixing with diethylene triamine and ammonium molybdate tetrahydrate; hydrothermal reaction at 200 °C; pyrolysis at 850 °C; N_2 atmosphere	Porous cage-like structure with double walls and a 3D network of interconnected carbon structures	217	[118]
Potato	Two-step carbonization; pre-pyrolysed at 500 °C; carbonization at various temperatures (900 °C, 1000 °C, and 1100 °C); Ar atmosphere	Mesoporous	620.77, 531.67, and 325.73	[119]
Hemp stalk core	Carbonization; 800 °C; N_2 atmosphere	Honeycomb-like structure	1185.3	[120]
Walnut septum	Pyrolysis; urea activation, 800 °C; N_2 atmosphere	Tubular cellular structure	176.9	[121]

4. Application of BDCs in MIHSCs

Carbon materials derived from biomass have emerged as potential candidates for MIHSCs, especially in the case of zinc and other moieties, namely sodium, potassium, and AIHSCs. This approach is attractive, as using such precursors would be sustainable and relatively cheaper than other biomass-based energy storage materials. Carbon materials can be prepared from biomass using pyrolysis, hydrothermal carbonization, or activation methods to obtain porous carbons with desired properties as a supercapacitor electrode [105,122,123]. Since the initiation of research on MIHSCs, scientists have investigated many types of electrode materials [124]. Figure 7a illustrates the configuration of these capacitors, emphasizing the primary materials employed for electrodes. The positrode materials are classified into various groups, including carbon-based materials, titanium-based compounds, niobium pentoxide (Nb_2O_5), and other transition metal compounds such as molybdenum, tungsten, and vanadium. Scientists have also employed various methods to customize and optimize the active substances to improve the rate capability, prolong the lifespan, and increase the capacity of the capacitors, as shown in Figure 7b. The following discussion will explore the progress in developing practical positrode materials, organized by their respective types. Among various MIHSCs, biomass-based carbon materials are most commonly used in ZIHSCs. Prior research has demonstrated that using carbon derived from biomass can enhance the energy storage capacity of ZIHSCs, resulting in improved gravimetric and volumetric energy density [21,76,125]. Hierarchical carbon structures prepared from biomass have mainly been effective in increasing the energy storage of ZIHSCs. Moreover, the nitrogen doping of biomass-originated carbon can facilitate the performance of potassium supercapacitors, thus suggesting new possibilities for the applications of advanced energy storage solutions. Additionally, sodium and AIHSC applications have been investigated using BDC materials. These materials enabled relatively high energy storage for the sodium battery to develop a sustainable energy storage solution. Biomass-originated carbons are still being studied in AIHSCs to improve energy retention performance and energy density without volume expansion, as previously reported in the field. The possible designs and study suggest a sustainable and flexible approach to MIHSCs using biomass-based carbon material.

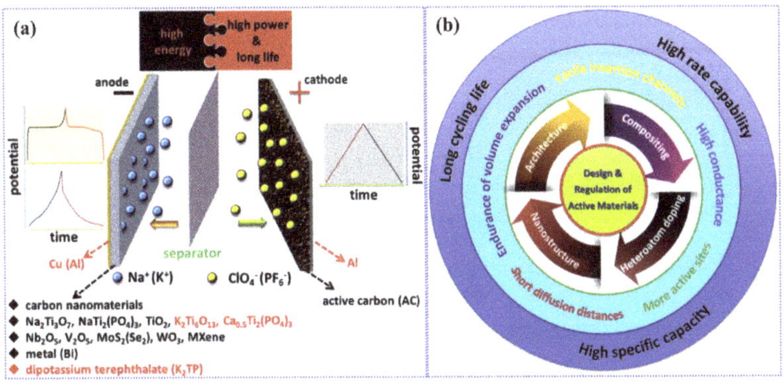

Figure 7. (**a**) Structural diagram of an MIHSC, illustrating the key components and materials. (**b**) Illustration of various strategies employed in designing and optimizing electrode materials for enhanced performance. Reproduced with permission from Ref. [124]. Copyright 2019, Elsevier.

The overall performance of MIHSCs is substantially influenced by the interaction of electrolyte volume, electrode mass, and surface area, which are essential considerations. The electrolyte volume impacts the quantity of charge that may be stored, where a smaller volume may result in a greater energy density [126]. MIHSCs can increase the device's energy density by reducing the electrolyte's total volume. In addition, selecting electrolytes such as water-in-salt electrolytes can significantly impact the performance of these supercapacitors [127]. The mass of the electrode is a crucial factor affecting MIHSCs' performance. High capacitance, cycle life, and operating voltage require optimal electrode mass ratios [128]. The electrode material's mass directly impacts the MIHSCs' charge storage capacity, affecting energy density [128]. In addition, the electrode material should possess a high surface area relative to its volume or mass to improve its performance [129]. The electrochemical performance of MIHSCs is significantly influenced by surface area. Maximizing the electrochemical double-layer capacitance and pseudocapacitance requires a large surface area for the electrodes, which leads to enhanced performance [130]. BDCs with mesopores are preferred for MIHSCs because of their expansive specific surface area and appropriate distribution of pore sizes, which enhance their overall performance. Additionally, it is essential to develop electrode materials that possess significant active surface areas and facilitate rapid ion transfer to attain superior energy density and rate capability in supercapacitors. Overall, the performance of MIHSCs is influenced by the synergistic interaction of electrolyte volume, electrode mass, and surface area. By optimizing these characteristics, it is possible to improve the energy density, power density, and cycling stability of these sophisticated energy storage devices.

4.1. ZIHSCs

BDC materials are considered suitable candidates for optimizing the performance of ZIHSCs. BDC materials are recognized for their ability to have a regulated microstructure with a large SSA. Due to their environmentally benign nature, they are considered a desirable choice for creating energy storage systems [131]. In this case, the application of BDC materials in ZIHSCs is addressed to assess the opportunity to improve the power capability of these supercapacitors. Porous carbon compounds derived from biomass are commonly recognized as superior electrodes for charge storage devices, such as supercapacitors, owing to their affordability and resilience. These materials have proven to be highly efficient in tackling energy storage and conversion challenges [23,132].

Consequently, implementing BDC materials in ZHSCs is aligned with the sustainable strategy. It provides a direction for one of the various commercial implementations to develop high-efficiency energy storage systems. Additionally, the negatrode material for ZIHSCs is critical to the carbon-based material's structural characteristics and electrochemical performance. Recent research analysed the influence of carbon materials, including AC, N-doped AC, graphene oxide, and carbon nanotube, on the development of ZIHSCs, focusing on the role of carbon materials in high-efficiency and -power functionality [133,134]. By incorporating the unique qualities offered by BDC materials, particularly the large SSA and regulated porosity, enhancing the energy density and power density of ZIHSCs is necessary to assure the practicality of high-power applications. This type of development collaborates with high-performance ZIHSCs for the marketable implementation of zinc-ion batteries due to the opportunity to provide high-efficiency, sustainable energy storage systems.

The ZIHSCs were created using the SMC as the negatrode, commercial zinc foil as the positrode, and a 2 M $ZnSO_4 \cdot 7H_2O$ solution as the electrolyte, with a Whatman filter paper as the separator, as shown in Figure 8a [70]. The electrochemical studies indicated the potential for energy storage applications. The specific capacitance, energy, and power for SMC-800-based ZIHSCs were 313.08 F/g, 141.35 Wh/kg, and 6935.38 W/kg, respectively. The ZIHSC devices exhibited excellent cyclic stability, losing only 1.92% over 20,000 GCD cycles at 4 A/g. The symmetric supercapacitors fabricated using produced carbon samples with 6 M KOH delivered high specific capacitance, with SMC-800 achieving the best performance. The symmetric supercapacitor with SMC-800 demonstrated a high specific capacitance of 154.64 F/g at 0.1 A/g and excellent stability with over 99% retention after 50,000 GCD cycles at 10 A/g. These results suggest that the developed BDC materials suit high-performance energy storage applications in ZIHSCs. The flexible ZIHSC was fabricated with KOH-AC as the positrode electrode and metallic zinc as the negatrode [71]. The positrode in the coin cell format was formed by depositing the active material slurry onto a pre-cut titanium foil using drop-casting, followed by drying in an oven. The electrolyte employed was a zinc sulfate solution with a concentration of 1 Molar. The ZIHSC with the KOH-AC positrode exhibited a notable specific capacitance, reaching 208 F/g at a current density of 0.1 A/g. Additionally, it demonstrated exceptional rate capability, maintaining 84.5% of its capacitance after undergoing 10,000 cycles at a 5 A/g current density. Figure 8b displays the CV curves of ZIHSC utilizing Areca catechu sheath-derived AC, together with the background information, which includes a digital image of the sandwich-type pouch cell architecture. A laser-scribed carbon pattern on a polyimide sheet was used as the current collector for both the positrode and negatrode in order to create a flexible pouch cell. This design showcased excellent flexibility and stability when subjected to mechanical stress. Under bending conditions, the flexible device exhibits an energy density of 32.6 $\mu Wh/cm^2$ and a high power density of ~2 W/cm^2. The electrochemical tests suggest that BDC sheets derived from agricultural waste show great potential for high-performance ZIHSCs, providing both sustainability and cost-effectiveness in energy storage applications.

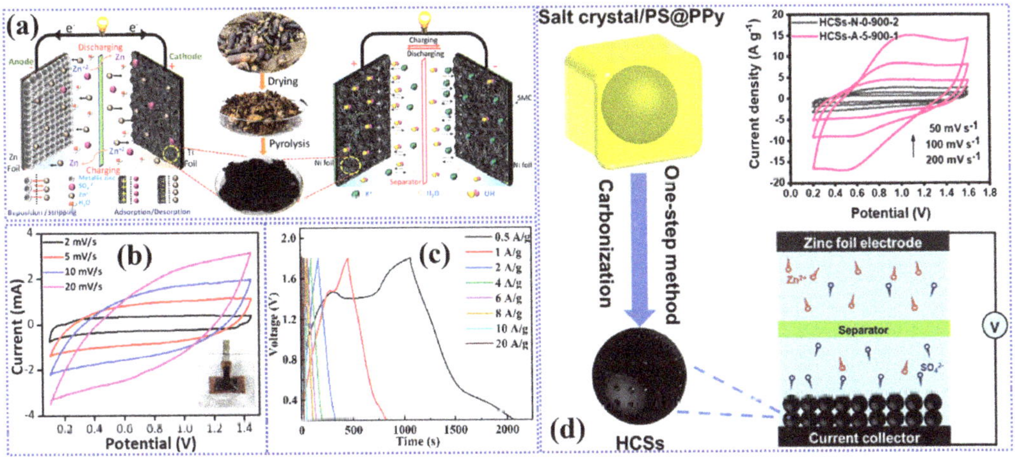

Figure 8. (**a**) The diagram illustrates the process of creating a ZIHSC utilizing SMC as the positrode. Reproduced with permission from Ref. [70]. Copyright 2024, Elsevier. (**b**) CV curves of ZIHSC using Areca catechu sheath-derived AC; the background includes a digital image of the sandwich-type pouch cell configuration. Reproduced with permission from Ref. [71]. Copyright 2022, American Chemical Society. (**c**) The GCD profiles of the N/S-AC-based ZIHSC at various current densities. Reproduced with permission from Ref. [72]. Copyright 2023, American Chemical Society. (**d**) Schematic representation of the HCSs' synthesis and their electrochemical performance in ZIHSCs. Reproduced with permission from Ref. [73]. Copyright 2024, American Chemical Society.

The synthesized N/S-AC materials were used to fabricate the ZIHSC with a carbon negatrode and zinc foil positrode, employing 2 M $ZnSO_4$ aqueous electrolyte [72]. The electrochemical results (Figure 8c) designate that the specific capacitance for the N/S-AC-based ZIHSCs reached 307 F/g at 0.5 A/g, demonstrating an excellent capacity retention of 99.72% after 20,000 cycles at 10 A/g. The achieved energy density was 108.8 Wh/kg, whereas the achieved power density was 115,200 W/kg. The exceptional electrochemical capabilities of this material can be attributed to the combined effects of nitrogen and sulfur doping, as well as the hierarchical porous structure. Moreover, the manufacturing procedure enables the creation of a significant quantity of BDC materials, demonstrating the capacity of these materials for creating high-performance ZIHSCs on a wide scale.

Zhao et al. [73] created ZIHSCs by using the manufactured HCSs as negative electrodes, zinc foil as the positrode, and a 2 M $ZnSO_4$ electrolyte, as shown in Figure 8d. The ZISCs based on HCSs-A-5-900-1 had a specific capacitance of 108 F/g at a current density of 0.2 A/g. They also had an energy density of 26.2 Wh/kg and a power density of 140.1 W/kg. In addition, the supercapacitors demonstrated remarkable cycling stability, maintaining 65% of their capacity after undergoing 5000 cycles. The ZIHSCs exhibited a significant capacitive-controlled storage level, contributing to their exceptional electrochemical performance. The hierarchical porosity and doping levels of the HCSs are essential factors that significantly contribute to their performance, rendering these carbon materials highly favourable for energy storage applications. Du et al. [74] demonstrated the electrochemical performance of the synthesized carbon matrix by constructing ZIHSCs using NAC-20 as the negatrode, zinc flakes as the positrode, and a 2 M $ZnSO_4$ electrolyte. The ZIHSCs showed a capacitance of 191.1 F/g at 0.5 A/g, a stable cycle life of over 10,000 cycles, and a 103% capacity retention. Additionally, these supercapacitors achieved an energy density of 37 Wh/kg at a power density of 91 W/kg. Combining a high SSA, porous structure, and heteroatom doping contributes to this remarkable performance, allowing ZIHSCs assembled with NAC-20 to store energy effectively and light up a red LED for nearly 5 min, demonstrating their practical application. Several different BDCs have been used as electrodes in ZIHSCs, significantly increasing their performance. Nitrogen–oxygen co-doped carbon obtained from yeast cell walls and other feedstock demonstrated exceptional electrochemical characteristics. There has been a notable increase in interest in electrochemical energy storage in recent years, with a specific emphasis on ZIHSCs. These investigations have been conducted to explore the use of BDC materials. These materials are valued for their sustainability, cost-effectiveness, and ecologically favourable properties. These materials have shown great potential in improving the electrochemical performance of ZIHSCs through several methods, such as maximizing surface area, porosity, and electrical conductivity [134–141]. Various studies have examined ZIHSCs using BDCs derived from hemp biomass [142], glutinous rice [143], platanus achene fruits [144], pyrolytic carbon black and critical PANI [145], metal organic frameworks (MOFs) [146,147], carp scales [148], green cokes [149], bamboo [150,151], rice husk [152], heavy fraction of bio-oil [153], yeast [154], by-product of traditional Chinese medicine [155], wood [156], lignin [157,158], corncob cellulose [159], heavy bio-oil [160], and sugarcane [161]. The wide variety of BDCs, each with their own unique structure and electrochemical properties, has played a significant role in the progress of ZIHSCs. Their utilization has resulted in significant enhancements in performance indicators, including specific capacitance, energy density, power density, and cycling stability, establishing their position in the future of sustainable energy storage technologies. Several important performance metrics from the experimental results of these carbonaceous materials are presented in Table 2, revealing their ability to drive ZIHSCs to the next level.

4.2. NIHSCs

BDC materials have emerged as potential materials for enhancing the efficiency of NIHSCs in recent years. BDC materials are affordable and eco-friendly alternatives that are extensively used to improve the efficiency of SIHSCs. Moreover, these materials possess other distinctive characteristics compared to conventional carbon, such as large interlayer spaces, disorderly carbon structures, and more active functional groups that ensure better charge storage than conventional carbon [94,162]. The commercial potential of BDC materials in supercapacitors arises from the diverse structure and superior properties of these materials. They fit the supercapacitor performance property and tailored structure. It is feasible to enhance the energy density, power density, and cycling stability of BDC materials used in supercapacitors. Therefore, their commercial potential in the commercial sector plays a natural role in producing sodium-ion batteries for industrial applications. Their eco-friendly nature and sources, coupled with their versatile high-performance structure moulded by nature, make BDCs promising candidates for sustainable energy storage in the sodium-ion battery industry [76]. The progress in research on BDC materials for sodium-ion energy storage devices indicates their potential for innovation and collaborative development in supercapacitors.

Table 2. A summary of electrochemical performance for the BDC materials in high-performance MIHSCs.

Positrode	Negatrode	Electrolyte	HSC Type	Specific Capacitance (F/g) @ Current Density (A/g)	Energy Density (Wh/kg) @ Power Density (W/kg)	Capacitance Retention (% @ No. of Cycles)	Ref.
SMC	Zn foil	2 M $ZnSO_4 \cdot 7H_2O$	ZIHSC	313.08 @ 0.1	141.35 @ 6935.38	99 @ 50,000	[70]
KOH-AC	Zn foil	1 M $ZnSO_4$	ZIHSC	208 @ 0.1	52 @ 1906.4	84.5 @ 10,000	[71]
N/S-AC	Zn foil	2 M $ZnSO_4$	ZIHSC	307 @ 0.5	108.8 @ 115,200	99.72 @ 20,000	[72]
HCSs	Zn foil	2 M $ZnSO_4$	ZIHSC	108.7 @ 0.2	26.2 @ 140.1	65 @ 5000	[73]
NAC	Zin sheet	2 M $ZnSO_4$	ZIHSC	191.1 @ 0.5	37 @ 91	103 @ 10,000	[74]
ICF	Zn foil	1 M $ZnSO_4/MnSO_4$	ZIHSC	168 @ 1	39.7 @ 9000	79.5 @ 5000	[101]
Porous carbon	Zn foil	2 M $ZnSO_4$	ZIHSC	291 @ 0.1	79 @ 391	96 @ 50,000	[102]
HHPC	Zn foil	2 M $ZnSO_4$	ZIHSC	147 mAh/g @ 0.2	117.6 @ 160	88 @ 20,000	[103]
LHPC	Zinc metal	1 M $ZnSO_4$	ZIHSC	298 @ 0.1	135 @ 101	97 @ 8000	[104]
TOMC	Zn foil	2 M $ZnSO_4$	ZIHSC	331.2 @ 0.2	148.9 @ 180	85.7 @ 10,000	[105]
PNMNC	Zn foil	1 M $ZnSO_4$	ZIHSC	472.5 @ 0.1	39.7 @ 9000	97.1 @ 10,000	[106]
S, N-CNC	Zn foil	2 M $ZnSO_4$	ZIHSC	331 @ 1	148.9 @ 900	70 @ 10,000	[107]
AC650	Zn foil	2 M $ZnSO_4$	ZIHSC	95 @ 0.1	86 @ 210	73 @ 10,000	[108]
NPFC700	Zn foil	1 M $Zn(CF_3SO_3)_2$	ZIHSC	207.9 @ 0.1	85.7 @ 15.4	97.4 @ 20,000	[109]
FHPCNSs-800	Zn foil	2 M $ZnSO_4$	ZIHSC	220.1 mAh/g @ 0.2	181.6 @ 165	100 @ 30,000	[110]
P,N-HPCNS-KCl/Ice	HPCB-KOH	1 M $NaClO_4$	NIHSC	119 mAh/g @ 0.035	135.3 @ 30	88.6 @ 8000	[79]
CBCS-A3	CBCS-C	1 M $NaClO_4$	NIHSC	640 mAh/g @ 0.1	36,000 @ 53,000	90.5 @ 8000	[80]
GDHC	GDPC	1 M $NaClO_4$	NIHSC	260 mAh/g @ 0.05	156 @ 355	73 @ 10,000	[81]
PCN	HPC	1 M $NaClO_4$	NIHSC	205 mAh/g @ 0.5	119 @ 200	82 @ 8000	[82]
LS1400	Sodium foil	1 M $NaClO_4$	NIHSC	350 mAh/g @ 0.1	...	94 @ 450	[83]
HASH	HASH	1 M $NaClO_4$	NIHSC	400 mAh/g @ 0.1	...	91.7, 500	[111]
NS-MPC	Na-metal foil	1 M $NaClO_4$	NIHSC	765 mAh/g @ 0.1	...	52 @ 2500	[112]
BDC	$Na_2Ti_{2.97}Nb_{0.03}O_7$	1 M $NaClO_4$	NIHSC	60.5 @ 0.2	169.4 @ 120.5	83.8 @ 500	[113]
CAC	CAC	1 M $NaClO_4$	NIHSC	247 @ 0.5	160 @ 425	100 @ 10,000	[114]
CSC	CSC	1 M $NaClO_4$	NIHSC	126 @ 0.3	116 @ 1560	68 @ 300	[115]
SBNPK	SBNPK	1 M $NaClO_4$	NIHSC	356.4 @ 1	6.5 @ 251.9	96.5 @ 20,000	[116]
PCSs	Al foil	$EMIMAlCl_4$ (1-ethyl-3-methylimidazolium tetrachloroaluminate)	AIHSC	200 mAh/g @ 0.100	220 @ 900	80 @ 1500	[86]
Graphene	Al foil	EMImCl and $AlCl_3$	AIHSC	211 @ 3	51 @ 3390	87 @ 5000	[87]
Nitrogen-doped graphene	Al foil	EMImCl and $AlCl_3$	AIHSC	240 @ 0.1	...	90 @ 2000	[88]
NCS	Al foil	EMImCl and $AlCl_3$	AIHSC	224 mAh/g @ 0.3	...	100 @ 35,000	[89]
AS-AC	AS-HC	1 M KPF_6	KIHSC	88 @ 0.053	105 @ 250	96 @ 10,000	[98]
CACF	CACF	1 M potassium bis(fluorosulfonyl)imide (KFSI)	KIHSC	350 mAh/g @ 0.1	178.4 @ 1115	75.2 @ 10,000	[99]
NOHC	Hard carbon	0.8 M KPF_6	KIHSC	186.3 mAh/g @ 0.1	68 @ 2000	80 @ 1000	[100]
ACNs	SCNs	0.8 M KPF_6	KIHSC	89.1 @ 0.1	124.0 @ 198.9	89.4 2 @ 9000	[117]
NOBC	Potassium disk	1 M KPF_6	KIHSC	251.2 mAh/g @ 0.5	...	77.74 @ 2000	[118]
PBPC	Potassium metal	3 M potassium bis(fluoro-slufonyl)imide (KFSI)	KIHSC	248 mAh/g @ 0.1	...	87.5 @ 400	[119]
N-CHC	N-CHC	1 M KPF_6	KIHSC	42.4 mAh/g @ 0.3	27.36 @ 203.4	91 @ 600	[120]
NHPC	NHPC	0.8 M KPF_6	KIHSC	263.6 mAh/g @ 0.1	...	84 @ 1000	[121]

Marine-biomass-derived porous carbon nanosheets (CBCS-A3) served as the positrodes in NIHSCs [80]. As demonstrated in Figure 9a, the CBCS-A3 positrode showed an outstanding capacity of 640 mAh/g at 0.1 A/g, and it showed state-of-the-art Na^+ storage performance, which is very close to the same performance of lithium-ion storage in similar materials. Furthermore, devices showed high cycling retention, retaining 270 mAh/g and 138 mAh/g after ten thousand cycles at 2 A/g and 10 A/g, respectively. The NIHSCs which used a CBCS-A3 positrode and a CBC-C negatrode displayed an excellent combination of energy–power densities at 5 A/g, 36 Wh/kg at 53,000 W/kg, and retained 90.5% capacitance retention after 8000 cycles. This research demonstrates the applicability of BDC materials for enhanced Na-ion storage and extensive energy. Niu et al. [79] obtained carbon materials from biomass and implemented them as electrode materials in NIHSCs. Specifically, the P- and N-doped and N-doped BDCs acted as the negatrode and positrode, respectively. The supercapacitors demonstrated excellent electrochemical performance, as demonstrated by the GCD profiles in Figure 9b,c, combining a high energy density (135.3 Wh/kg) with high power density (16.1 kW/kg), and outstanding long-term stability, retaining 88.6% of the initial capacity even after 8000 cycles. These results can be explained by the unique structural features of carbon materials, comprising hierarchical porosity and heteroatom doping. These features enabled the supercapacitors to maintain excellent kinetic compatibility between the negatrode and positrode. Therefore, utilizing carbon materials obtained from renewable biomass sources in sophisticated energy systems is extremely advantageous because of their sustainable production and exceptional electrochemical capabilities. A NIHSC was made using a GDHC positrode and a GDPC negatrode, with the electrodes having a 1:1 mass ratio [81]. The developed NIHSC performed excellently in a wide (4 V) operating potential window, as shown in Figure 9d. The capacitor has an energy density range of 31 to 156 Wh/kg and a power density range of 355 to 38,910 W/kg. In addition, the capacitor exhibits remarkable cycling stability (Figure 9e), maintaining 73% of its original capacity after undergoing 10,000 cycles at a 30 A/g current density. The capacitance decline shown in Figure 9e is mainly ascribed to the interconnected phenomena that affect the electrochemical performance and durability of the electrode materials in NIHSCs. One of the most important factors is the creation and continuous development of the Solid Electrolyte Interface (SEI). Although initially advantageous in stabilizing electrochemical reactions at the interface, the passivating layer can gradually thicken and lose conductivity, hindering ion movement and causing capacity degradation. Simultaneously, the electrode material undergoes mechanical degradation due to the stress caused by volumetric changes during ion insertion and extraction. This degradation leads to the loss of electrical contact and active material from the structure of the electrode. In addition, the frequent insertion of ions can lead to the blockage of pores and the formation of a protective layer on the electrodes, resulting in a significant decrease in the accessible surface area for ion storage and a reduction in the overall capacitance. The decline is worsened by the decomposition of electrolytes, which results in the creation of non-conductive deposits on the surfaces of electrodes, further impeding the movement of ions. The combined impacts lead to a gradual decrease in the capacitor's capacity to hold charge. This requires sophisticated material designs and engineering techniques to improve the strength of the structure, optimize the interaction between the electrolyte and electrode, and maintain consistent electrochemical performance over long periods of use. It is essential to tackle these problems to enhance the longevity and effectiveness of sodium-ion capacitors in real-world use. The electrochemical performance of the NIHSC showcases the potential of carbon materials derived from biomass in the advancement of sustainable and high-performance energy storage systems. This study focuses on the synthesis technique of carbon materials and their efficient utilization in developing high-performance NIHSCs.

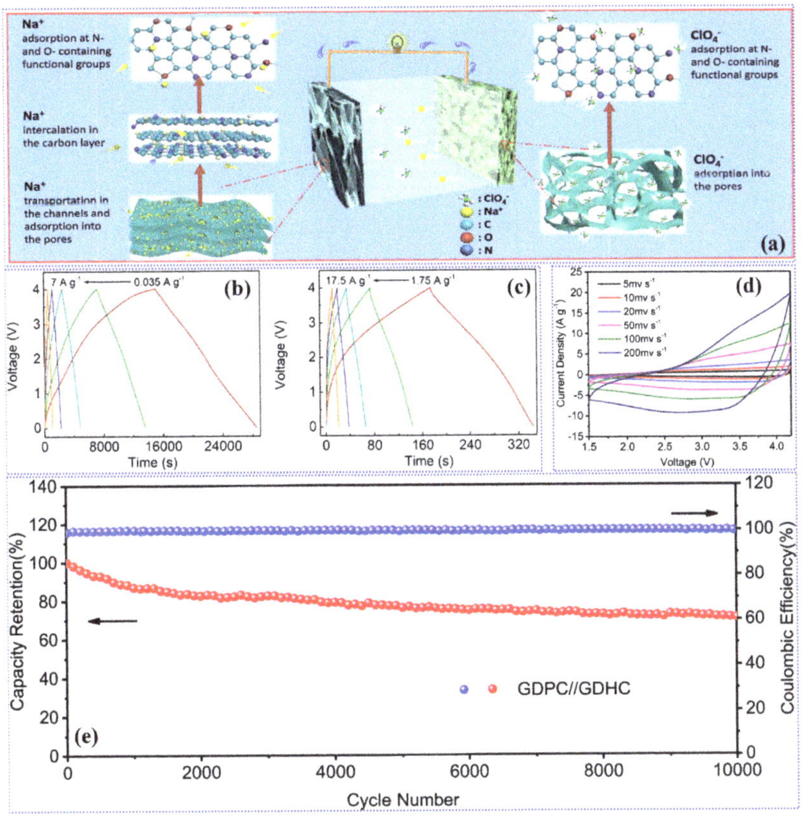

Figure 9. (**a**) Schematic depiction of the charge storage techniques employed in the NIHSC. Reproduced with permission from Ref. [80]. Copyright 2018, American Chemical Society. (**b**,**c**) GCD profiles at different current densities for the BDC-based NIHSC. Reproduced with permission from Ref. [79]. Copyright 2020, American Chemical Society. (**d**) CV curves at various scan rates and (**e**) cyclic stability and Coulombic efficiency for 10,000 GCD cycles of the GDPC//GDHC NIHSC. Reproduced with permission from Ref. [81]. Copyright 2019, American Chemical Society.

The PCNS functions as the positrode in NIHSCs in conjunction with a negatrode made of hierarchically porous carbon with capacitive properties [82]. The supercapacitor exhibits an energy density of 119 Wh/kg when subjected to a power density of 200 W/kg and a lower energy density of 53 Wh/kg when subjected to a higher power density of 20 kW/kg. Furthermore, it maintains 82% of its capacity after undergoing 8000 cycles at a current density of 5 A/g, demonstrating exceptional cycling stability. The outstanding performance of these devices may be due to the synergistic coupling of a battery-type positrode and a capacitive-type negatrode, which leads to a device with high energy and power density. This work highlights the capacity of carbon materials obtained from biomass to be used in sodium-ion storage systems that are both sustainable and high-performing. The created carbon compounds were evaluated as positrodes in NIHSCs [83]. When exposed to a current density of 100 mA/g, the LS1400 material exhibits a substantial reversible 350 mAh/g capacity. Additionally, it exhibits a plateau capacity of 250 mAh/g, indicating its possible suitability as a material for positrodes. The material has remarkable cycling stability, retaining 94% of its initial capacity after surviving 450 cycles at a current density of 100 mA/g. The EIS analysis showed that LS1400 has a lower charge transfer resistance and improved Na^+ diffusion kinetics compared to LS1200 and LS1600. The sodium storage method entails the adsorption of Na^+ ions in regions with imperfections, the insertion of Na^+ ions between graphite layers, and the deposition of Na metal in confined pores [163]. The LS1400 version exhibits a significant proportion of closed pores, considerably enhancing its plateau capacity. This study highlights the ability of hard carbons derived from lotus stems to serve as high-performance positrode materials for nonaqueous NIHSCs. Several alternative carbon compounds from biomass have been used as electrodes in NIHSCs, showcasing their capability in advanced energy storage technologies [164–168]. These materials, obtained from natural sources including lotus stems, marine biomass, and other plant-based precursors, display various shapes, layered porosity, and doping characteristics that enhance their exceptional electrochemical capabilities. The progress in SIHSCs has received significant attention in recent years due to their promise as affordable and environmentally friendly energy storage options [78,79,169–172]. As a result of this interest, several studies have been conducted on utilizing BDC materials obtained from various biomasses. These studies aim to exploit these materials' renewable, environmentally benign, and economically advantageous properties. These materials have demonstrated considerable potential in enhancing the electrochemical efficiency of SIHSCs by optimizing variables such as the specific surface area, porosity, and electrical conductivity [173,174]. Several types of biomasses, including carrageenan [175], jute [176], cork [177], sodium lauryl sulfate/PANI [178], peanut shell [179], MOFs [180], sucrose [181], spring onion [182], rice husk [183], wheat bran [184], and recycled olive pits [185], have been documented as suitable sources for synthesizing BDCs for SIHSCs. The wide variety of carbon compounds obtained from biomass and their specific structural and electrochemical properties have played a crucial role in the progress in SIHSCs. Their utilization has resulted in significant enhancements in performance indicators, including the specific capacitance, energy density, power density, and cycling stability, consequently reinforcing their position in the future of sustainable energy storage solutions. Their performance, including important factors such as capacitance, cycle stability, and energy and power density, is outlined in Table 2.

4.3. AIHSCs

BDC materials have proven to be a capable solution to boost the AIHSCs' performance. These materials, which are well known for their sustainability and low cost, have remarkably supported the efficiency of AIHSCs. The unique characteristics of BDC, including hierarchical porous structures, nitrogen doping, and high SSAs, facilitate excellent charge storage and improved electrochemical performance of AIHSCs [186]. The materials used in BDC have shown exceptional energy density, power density, and cycling stability. This makes them very appropriate for commercial application in aluminium-ion batteries for AIHSCs [187]. The application of BDC materials in AIHSCs is commercially applicable due to their structural flexibility and excellent electrochemical characteristics. The technology, therefore, comes with customization design and performance qualities required for high-performance supercapacitors. BDC materials help to improve the overall performance and efficiency of AIHSCs, making them commercially viable in the aluminium-ion battery industry. The materials' contributions are based on their renewable, abundant, and non-toxic nature and adjustable features. These qualities make them highly suitable for sustainable energy storage solutions in the aluminium-ion battery industry [187]. The existing research and applications regarding BDC for aluminium-ion energy storage solutions can potentially drive real-world applications and innovation in high-performance energy systems.

Feng et al. [86] reported an AIHSC composed of a PCS negatrode, an aluminium positrode, and an ionic liquid electrolyte called $EMIMAlCl_4$. The AIHSC device and its accompanying electrochemical performances are depicted Figure 10a–h. The PCS negatrode, utilizing a double-layer capacitive process, demonstrates a consistent and reliable charge storage mechanism. It has a specific capacity of around 200 milliampere-hours per gram at a current density of 100 milliamperes per gram. The device's electrochemical performance is characterized by quasi-rectangular cyclic voltammetry profiles, which indicate its capacitive nature. The PCS negatrode exhibits remarkable rate capabilities, maintaining approximately 80% capacity retention even at a current density of 2500 mA/g. Extended cycling investigations have demonstrated that the material's capacity reaches 150 mAh/g after undergoing 1500 cycles at a current density of 2000 mA/g. At a temperature of 50 °C, the supercapacitor achieves an energy density of 220 Wh/kg and a power density of 9 kW/kg, outperforming other aluminium-based energy storage devices in terms of both energy and power capabilities. Additionally, it has a lifespan of over 1000 cycles.

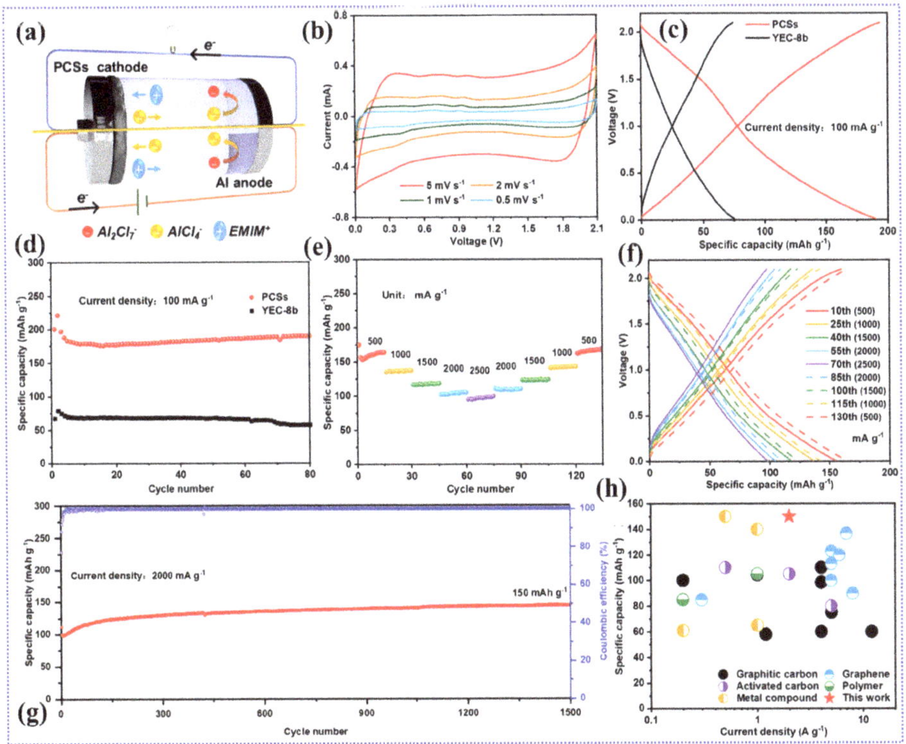

Figure 10. Electrochemical performance evaluation of AIHSC based on PCSs at 25 °C. (**a**) Schematic depiction of the rechargeable AIHSC working mechanism, (**b**) CV profiles observed at diverse scan rates, (**c**) GCD curves for PCSs and YEC-8b at 100 mA/g, (**d**) cycling stability comparison between PCSs and YEC-8b at 100 mA/g, (**e**) rate performance analysis and (**f**) corresponding GCD curves for PCSs at numerous current densities, (**g**) long-term cycling stability and Coulombic efficiency observed at 2000 mA/g, and (**h**) specific capacity comparison of Al-based AIHSCs with a cycle life exceeding 1000 cycles. Reproduced with permission from Ref. [86]. Copyright 2023, Wiley.

Soen et al. [87] demonstrated an AIHSC that has favourable electrochemical characteristics. The AIHSC produced using electrodeposition has a specific capacitance of 211 F/g, surpassing the electropolished counterpart by 20%. The ASC also exhibits an energy density of 151 Wh/kg at a power density of 3390 W/kg and maintains 70% of its capacitance even at a current density of 20 A/g, demonstrating exceptional rate capability. The device retains 87% of its capacity after completing 5000 cycles. The extraordinary performance may be attributed to the reduced resistance to charge transfer, the irregular and branching structure, and areas with elevated surface energy. This study demonstrates that using a surface treatment for aluminium electrodes can significantly enhance the performance of AIHSCs, rendering them a viable option for advanced energy storage devices in the future. Lie et al. [88] reported an HSC composed of an aluminium positrode, a graphene negatrode, and an ionic liquid electrolyte. It has a specific discharge capacitance of 240 F/g at a current density of 0.1 A/g and a Coulombic efficiency of 90%. After undergoing 1000 cycles at a current density of 0.3 A/g, the device exhibits a capacitance of 254 F/g. Furthermore, after 2000 cycles at a current density of 2.0 A/g, the capacitance decreases to 130 F/g. These results indicate that the device has excellent cyclic stability and achieves almost 100% Coulombic efficiency. The electrochemical impedance spectrum indicates a charge transfer resistance of 13.15 Ω, a series resistance of 3.82 Ω, and a dispersed resistance of 6.36 Ω. The primary energy storage mechanism in this case comprises the combination of electric double-layer capacitance resulting from the adsorption and desorption of $AlCl^{4-}$ ions, as well as weak intercalation and deintercalation in nitrogen-doped graphene. This makes it a desirable option for high-performance supercapacitors. Sun et al. [89] developed an AIHSC composed of a positrode made of aluminium foil, a negatrode made of NCS, and an electrolyte made of 1-ethyl-3-methylimidazolium chloride ([EMIm]Cl) and $AlCl_3$. The device demonstrates non-ideal rectangular CV curves and achieves a high specific discharge capacitance of 224 mAh/g at a current density of 0.300 A/g, while maintaining a Coulombic efficiency of 98%. The NCS negatrode exhibits outstanding cycling performance, with a 114 mAh/g capacity at 10 A/g even after undergoing 35,000 cycles. Electrochemical impedance spectroscopy revealed a progressive reduction in charge transfer resistance throughout cycling, resulting in consistent and reliable performance. The Al-HSC exhibits a blend of battery-like and capacitor-like properties, making it a very promising energy storage device. It provides fast charge–discharge cycles and maintains long-term stability.

Using the most suitable carbon materials obtained from biomass can significantly improve the efficiency of AIHSCs, enhancing their electrochemical performance. BDC materials, such as corn wastes, corn cob, jute sticks, corn silk, sulfur-doped carbon, nitrogen-doped carbon, and porous carbon sheets, possess distinctive characteristics such as a high SSA, hierarchical porous structures, and diverse elemental distributions [188–192]. These characteristics render them very well suited for utilization as electrode materials. The materials exhibit outstanding electrochemical performance, characterized by a high specific capacitance, enhanced energy densities, and excellent cycling stability. The qualities listed in Table 2 are essential for improving the overall efficiency of supercapacitors. By carefully selecting and optimising carbon materials derived from biomass, it is possible to tailor the electrochemical properties of AIHSCs, leading to improved performance and long-lasting durability for energy storage.

4.4. KIHSCs

BDC materials significantly improve the efficiency of KIHSCs. Biomass-derived materials have advantages such as hierarchical porous structures with nitrogen doping and high SSA. Rich nitrogen doping and defects with BDC gain due to these parameters exhibit high SSA and other features favourable to electrochemical energy storage. This means that a KIHSC has a larger charge storage capability [193]. According to the mentioned benefits, BDC materials can significantly enhance the energy and power density and the cyclic stability of KIHSCs. They possess great commercial potential in the potassium-ion battery field. BDC materials' structural and electrochemical properties prove their high commercial acumen. Strong structural characteristics and additional electrochemical properties are available. Two advantages are critical to the success of supercapacitors with high performance: they are highly effective and environmentally friendly. Unique materials are appreciated because of their design characteristics and possibilities for developing high-performance supercapacitors [94], a good choice for scaling KIHSCs for commercial application in the potassium-ion battery market.

The unique structural characteristics and cost-effectiveness of porous carbons derived from biomass make them promising electrode materials for KIHSCs. BDCs possess many advantageous properties that make them suitable for KIHSC applications. A substantial SSA and an intricate pore structure comprising micropores, mesopores, and macropores facilitate the efficient migration of ions and provide abundant active sites for storing potassium ions. Heteroatom doping involves the introduction of heteroatoms, such as nitrogen, oxygen, and sulfur, into carbon materials derived from biomass [194]. This process can improve the capacity of the carbons to be wetted and conduct electricity, resulting in enhanced electrochemical performance. Tuneable surface chemistry refers to the ability to modify the chemical properties of BDCs by choosing specific biomass precursors and activation procedures. This allows for the enhancement of the potassium-ion storage capacity. KIHSCs, utilizing porous carbon electrodes generated from biomass, have exhibited encouraging electrochemical characteristics [99]. The high specific capacitance values, often exceeding 300 F/g, are attributed to the hierarchical pore structure and the incorporation of heteroatom doping. The exceptional rate capacity of the material is attributed to its linked pore network and strong electrical conductivity, which enable fast ion transport and consequently lead to outstanding rate capability and high power density. The robust carbon structure and uniform surface chemistry of carbons derived from biomass ensure exceptional stability after several charge–discharge cycles [195]. Although biomass-derived porous carbons have demonstrated encouraging outcomes in KIHSCs, some obstacles still need to be addressed. Additional investigation is required to enhance the pores' configuration and the surface's chemical composition to achieve the highest possible efficiency in storing and transporting potassium ions. To improve the energy density of KIHSCs, one can utilize strategies like incorporating battery-type materials or integrating pseudocapacitive materials. In order to make biomass-derived porous carbons commercially feasible, it is crucial to develop synthesis methods that are both scalable and cost-effective. Porous carbons generated from biomass provide a sustainable and cost-effective alternative for KIHSCs. There is potential for enhancing their performance by optimizing material design and synthesis.

Using AS-HC as the negatrode and AS-AC as the positrode in a potassium-based electrolyte, a KIHSC was created, as shown in Figure 11a [98]. The device has a remarkable electrochemical performance, as demonstrated in Figure 11b–d, with a specific energy of 105 Wh/kg and no capacitance degradation after 10,000 cycles, indicating exceptional cycling stability. The AC electrode exhibits a dual electrical layer capacitance characteristic and supplementary pseudocapacitive peaks, which augment the device's overall performance. The specific discharge capacity of the AS-HC electrode reaches a stable state after a few cycles, suggesting that potassium storage is reversible and plays a role in the device's high Coulombic efficiency and structural stability.

Chen et al. [99] constructed a KIHSC using plains–hills carbon as the positrode and negatrode. The device exhibits a reversible capacity of 147.2 mAh/g at a current density of 10 A/g after undergoing 5000 cycles, indicating excellent stability. CV and GCD experiments were used to evaluate the electrode's performance, demonstrating an energy density of 178.4 Wh/kg at a power density of 1115 W/kg. The KIHSC exhibits remarkable cycling performance, maintaining 75.2% of its capacity after 10,000 cycles. Its ideal blend of conductivity and pseudocapacitance accounts for the device's outstanding performance, making it an extremely advanced electrode material for potassium ion storage. A KIHSC device was constructed using the NOHC as the positrode and AC as the negatrode [100]. The device exhibits an initial reversible capacity of 186.3 mAh/g and maintains 93.5% of its capacity after undergoing 500 cycles at a current density of 0.1 A/g, thus displaying exceptional cycle stability. Furthermore, the device demonstrates an energy density of 68 Wh/kg at a power density of 2000 W/kh and sustains an 80% capacity retention after 1000 cycles at a current of 1 ampere per gram. The NOHC's structure and the development of a stable SEI layer during cycling enable a combination of K-ion intercalation and capacitive processes that contribute to the system's performance. Ultimately, while improving the synthesis methods, it is equally important to conduct electrochemical analysis on biomass-generated carbon to achieve optimal performance in KIHSCs. Electrochemical characterization techniques are crucial for comprehending the carbon material's charge storage mechanisms, kinetics, and overall electrochemical behaviour when exposed to potassium ions.

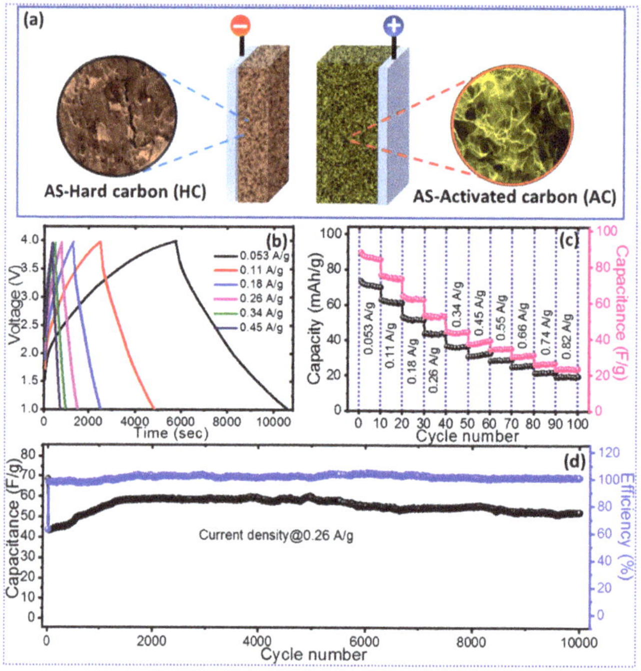

Figure 11. The complete cell KIHSC electrochemical performances of the HC-1000/AC-800: (**a**) a schematic representation of the KIHSC complete cell, (**b**) GCD profiles recorded at various current densities, (**c**) specific capacity and specific capacitance variation with several cycles at various current densities, and (**d**) cyclic stability of the full cell with corresponding Coulombic efficiency over 10,000 cycles. Reproduced with permission from Ref. [98]. Copyright 2020, American Chemical Society.

The field of energy storage has seen significant progress in the development of KIHSCs in recent years. The increasing interest in BDC compounds has led to several studies exploring their use, taking advantage of their sustainable, cost-effective, and environmentally beneficial properties. These materials have shown significant promise in improving the electrochemical performance of KIHSCs by optimizing variables such as specific surface area, porosity, and electrical conductivity [196–200]. Multiple prominent research studies have shown the utilization of different carbon materials derived from agaric [201], ganoderma lucidum [202], coffee grounds [203], rapeseed meal [204], lignin [205], and 2D tremella [206] for KIHSCs. The wide variety of carbon compounds obtained from biomass, each with its unique structure and electrochemical properties, has significantly contributed to the progress in potassium-ion hybrid supercapacitors (KIHSCs). Their utilization has resulted in significant enhancements in performance measures, including the specific capacitance, energy density, power density, and cycling stability, consequently confirming their position in the future of sustainable energy storage technologies. Table 2 presents a concise overview of the different electrochemical parameters required to evaluate the effectiveness of BDC in KIHSCs. Specific capacitance, rate capability, cycling stability, charge transfer resistance, and diffusion coefficients are crucial in optimizing electrochemical characterization techniques. The optimization of electrochemical characterization processes involves the utilization of many methods, including CV, GCD experiments, and EIS. Surface characterization techniques offer additional information about electrochemical behaviour, particularly in situ approaches. This information allows researchers to discover limitations and optimize the carbon material. Through meticulous analysis of the electrochemical data, researchers can gain insights into the ion transport pathways, kinetics of charge storage, and the impact of functional surface groups on the electrochemical performance. This comprehension can improve the synthesis processes, electrode manufacturing procedures, and electrolyte compositions, boosting the energy storage capacity and lifespan of KIHSCs. Optimizing electrochemical characterization techniques is crucial for establishing correlations between structure and properties and constructing prediction models for carbon derived from biomass. Predictive models can direct and expedite the logical creation of carbon materials for certain uses, advancing high-efficiency and financially feasible energy storage technologies.

5. Various Factors Affecting the Electrochemical Behaviour of BDC-Based MIHSCs

BDC electrodes possess numerous advantages compared to alternative carbon-based electrodes for MIHSCs. The advantages encompass heightened electrical conductivity, enhanced surface wettability, the induction of pseudocapacitance, expedited charge transfer, and facilitated reactions at the electrode/electrolyte interface [207]. BDC materials provide both sustainability and cost-effectiveness, rendering them a compelling choice for supercapacitor electrodes. They have a structure that allows for the passage of substances through small holes, a high ability to store electrical charge per unit mass, and the ability to remain stable over time. These properties make them better than many other carbon-based materials used in MIHSCs [208]. Lignin/cellulose nanofibers from biomass exhibit potential as electrode materials because of their exceptional durability, favourable electronic conductivity, and porous structure [209]. Moreover, carbon compounds obtained from biomass, which are renewable, have shown considerable promise in improving the performance of electrodes for MIHSCs [84]. Porous BDCs, which can be tailored to have specific nano/microstructures, are attracting interest due to their high effectiveness and cost-effectiveness as materials for supercapacitor electrodes [210]. The use of biomass-generated carbon can result in various electrochemical characteristics, including specific capacitance and surface area, due to its varied internal structure [211]. BDC materials have distinctive characteristics, such as a highly developed porous structure, a substantial specific surface area, and excellent electrical conductivity. These attributes render them highly favourable as electrode materials for supercapacitors with exceptional performance capabilities [212]. Electrodes made from BDCs provide a sustainable, economically viable, and highly effective alternative for supercapacitors. Their distinctive qualities and benefits have led to their growing popularity in energy storage.

Several vital parameters that control the electrochemical behaviour of BDC-based MIHSCs for energy storage are tabulated in Table 3. Electrolyte choice, OPW range, and electrode material qualities are essential. Supercapacitors' ion transport efficiency depends on electrolyte choice. The electrolyte must promote fast ion movement and chemical stability within the device's working voltage range. BDCs, which frequently incorporate inherent functional groups derived from natural doping elements such as nitrogen, sulfur, and phosphorus, exhibit distinct interactions with electrolyte ions, influencing charge storage. Given the complex surface chemistry of BDCs, electrolyte stability and conductivity are crucial. These electrodes could use electrolytes with higher ionic conductivity to maximize their porous structure. The OPW determines the voltage range, which affects supercapacitor energy and power density. BDCs can withstand higher voltages due to their surface chemistries and structural stability. This trait may improve energy storage. However, deviating from the constant voltage range may cause rapid material degradation. Biomass carbons may have inherent heteroatoms that affect voltage range stability, supercapacitor lifetime, and efficiency. This depends on how heteroatoms interact with the electrolyte. BDC electrodes have many advantages over carbon-based electrodes. BDC electrodes are more sustainable since they use renewable resources. This reduces the environmental effect of electrode material manufacture and promotes sustainability. Bio-wastes minimize material costs compared to conventional electrodes that use expensive processed carbons or synthetic precursors. High porosity in BDCs helps MIHSCs store more ions. This structural benefit is achieved without activation or templating, which traditional carbon composites require. BDCs store energy using pseudocapacitance mechanisms due to their heteroatoms. This increases energy storage capacities beyond double-layer capacitance. Supercapacitors can be customized and function better by changing electrode properties like conductivity and wettability using physical or chemical approaches. A number of biomass types can be used to make carbons with customized supercapacitor characteristics. Customization is harder with standard carbon sources. In addition, BDCs are thermally stable, ensuring constant performance across all operational temperatures. This is essential for energy storage applications that need durability. Carbon electrodes made from biomass are a sustainable energy storage solution. They have enhanced performance and are cost-effective and adaptable for supercapacitor applications. These advantages make them attractive alternatives to carbon-based electrodes, which are often limited by their environmental impact, high cost, and inflexibility.

Table 3. Merits of BDC electrodes compared with other types of carbon-based electrodes used in MIHSCs.

Feature	BDC Electrodes	Traditional Carbon-Based Electrodes
Sustainability	Produced from renewable materials, rendering them ecologically sound and enduring.	Frequently obtained from finite resources or artificial methods, which might result in a greater ecological footprint.
Cost	The cost is typically cheaper because waste or natural by-products are utilized, decreasing material prices.	Usually, the higher cost is a result of utilizing new or artificial materials.
Porous structure	The material has inherent high porosity, which results in a diverse range of pore diameters. This characteristic significantly improves its ability to store ions.	Porosity must be deliberately designed and controlled, often necessitating the use of supplementary processing procedures.
Natural doping	Includes inherent heteroatoms like nitrogen, sulfur, and phosphorus, improving electrochemical performance by increasing pseudocapacitance.	Doping necessitates the use of further chemical processing to insert heteroatoms, hence augmenting the intricacy and expense.
Modification flexibility	Can be readily adjusted through physical or chemical activation to customize characteristics such as surface area and conductivity.	Modifying synthetic or processed carbon materials is feasible, but their inherent qualities frequently limit flexibility.
Unique chemistries	The electrode characteristics can be modified to suit individual applications, as they vary greatly depending on the type of biomass used.	Typically uniform, with minimal variation in electrochemical characteristics until intentionally modified.
Thermal stability	Excellent thermal stability is advantageous for preserving performance over various temperature conditions.	The thermal stability of a material can vary and is influenced by the precursor and the procedure used to create it. Achieving consistent thermal stability may necessitate strict management.
Renewable production	The production cycle can be included in a sustainable bio-waste usage loop, fostering a circular economy.	Typically, it depends on industrial operations not included in a renewable cycle, with a greater emphasis on extracting resources.

5.1. Effect of Electrolytes on the Electrochemical Behaviour of BDC Electrodes

Electrolytes have a crucial role in determining the electrochemical characteristics of BDCs in MIHSCs. The selection of electrolytes substantially influences the performance of the supercapacitor system. Yoo et al. [126] showed that newly created nonaqueous electrolytes containing bis(trifluoromethylsulfonyl)imide salts allowed reversible deposition on metal positrodes and reversible adsorption on an AC negatrode. This led to the development of MIHSCs. Momodu et al. [213] highlighted the significance of electrolytes by demonstrating that gel-based electrolytes had exceptional stability and distinctive recuperative properties. This suggests that gel electrolytes formed from plant biomass can be effectively utilized in supercapacitor devices. In their study, Zhu et al. [214] emphasized that carbon-based supercapacitors demonstrated significant gravimetric capacitance and energy density while using organic and ionic liquid electrolytes. This highlights the importance of choosing the proper electrolyte to improve the electrochemical performance of carbon electrodes generated from biomass. In addition, Li et al. [215] examined progress in electrode materials and electrolyte innovation for lithium-ion capacitors, highlighting the pivotal significance of electrolytes in energy storage devices. In addition, Biswal et al. [216] reported that carbon electrodes displayed almost perfect capacitive behaviour with minimal ionic resistance, highlighting the significance of the rapid diffusion of electrolyte ions into the porous network for effective supercapacitor functioning. Figure 12a depicts the intricate ecology of electrode materials and electrolytes in MIHSCs, highlighting the importance of selecting the proper electrolyte to enhance performance. These findings emphasize the crucial importance of electrolytes in controlling the electrochemical properties of BDCs for MIHSCs.

Han et al. [127] showed that water-in-salt electrolytes based on acetate successfully enhanced the performance of ZIHSCs. These electrolytes exhibited excellent reversibility and stability during zinc plating and stripping. Choosing the appropriate electrolyte composition is crucial for maintaining the durability and effectiveness of the supercapacitor device. In addition, Zhang et al. [217] emphasized the significance of the electrolyte mixture in improving the electrochemical efficiency of ZIHSCs. The study demonstrated that using a particular electrolyte mixture enhanced electrochemical performance by substituting the bound water within the solvation structure. This substitution led to an improvement in the solvation structure, ultimately resulting in superior overall performance. This highlights the electrolyte composition's significance in enhancing the supercapacitor device's electrochemical performance. In addition, Eskusson et al. [218] examined the electrochemical characteristics of different aqueous electrolytes in ZIHSCs. The study compared several electrolytes and their effects on the electrochemical properties of the supercapacitor device. A comprehensive understanding of the interactions between electrolytes and electrodes is essential for optimal performance and durability in ZIHSCs. Ultimately, selecting electrolytes plays a crucial role in influencing the electrochemical characteristics and overall efficiency of BDC electrodes in ZIHSCs. Optimizing the electrolyte composition can improve the supercapacitor system's efficiency, stability, and durability, enhancing energy storage capacity.

Shao et al. [219] highlighted the significance of asymmetric supercapacitors in extending the voltage range beyond the point when electrolytes break down due to thermodynamic factors. This offers a solution to the energy storage constraints faced by symmetric supercapacitors. This emphasizes the importance of considering the compatibility of electrolytes and electrode materials to improve the overall performance of supercapacitors. Moreover, Simon and Gogotsi [220] emphasized that combining appropriate electrode materials with the proper electrolyte can augment the supercapacitor system's energy storage capacity and power. It is crucial to recognize that there is no flawless active substance, and the choice of electrolyte should align with specific performance goals. Comprehending the interplay between electrode materials and electrolytes is essential for attaining optimal electrochemical performance in NIHSCs. In addition, Maurya et al. [221] examined the configuration of NIHSCs, highlighting the importance of the three constituents, i.e., positrode, negatrode, and electrolyte, which resemble those found in batteries. This highlights the need for electrolytes that can fulfil the specific demands of NIHSCs. By carefully choosing and designing electrolytes, it is possible to improve the electrochemical efficiency and longevity of carbon electrodes generated from biomass in NIHSCs. Ultimately, the selection of electrolyte composition has a substantial impact on the electrochemical characteristics and overall efficacy of BDC electrodes in NIHSCs. By integrating findings from studies on asymmetric supercapacitors, electrode–electrolyte compatibility, and NIHSCs design, it is possible to enhance the efficiency and performance of supercapacitor systems for energy storage purposes. MIHSC applications make use of different types of aqueous electrolytes, including acidic, basic, and neutral solutions, as shown in Figure 12b.

The electrolyte's composition is vital for maximizing the performance and efficiency of the supercapacitor system. Dubal et al. [222] stressed the significance of creating a streamlined pathway for electron transport and ensuring optimal electrolyte access to the electrochemically active materials in carbon fibres. They underscored the importance of electrolyte permeability and connectivity within the electrode structure. In addition, Tomboc et al. [223] examined the energy storage mechanisms in supercapacitors, specifically emphasizing ion adsorption at the interface between the electrode and electrolyte (electric double-layer capacitance) and faradaic processes that involve electron transfer through oxidation/reduction reactions (pseudocapacitance). Comprehending these principles is essential for choosing electrolytes that can efficiently facilitate both forms of energy storage processes in AIHSCs. In addition, Kim et al. [224] showed that electric double-layer capacitors store charges by physically adsorbing electrolyte ions onto porous carbon electrodes, highlighting the significance of the electrode–electrolyte interface in the performance of supercapacitors. This emphasizes the need to ensure that the electrolyte is compatible with the electrode material in order to achieve efficient charge storage and transfer in AIHSCs. Ultimately, the choice of the correct electrolyte composition is crucial for controlling the electrochemical characteristics and overall efficiency of BDC electrodes in AIHSCs. By incorporating findings from studies on electron transport pathways, energy storage processes, and electrode–electrolyte interactions, it is feasible to improve the efficiency and functionality of supercapacitor devices used for energy storage purposes.

Dubal et al. [222] stressed the significance of creating a streamlined pathway for electron transport and ensuring that the electrolyte can effectively reach the electrochemically active materials in carbon fibres. They highlighted the importance of having good electrolyte permeability and connectivity within the electrode structure to improve the overall electrochemical performance. In addition, Kim et al. [224] explained how electric double-layer capacitors store charges utilizing the physisorption of electrolyte ions onto BDC electrodes. This highlights the importance of the interface between the electrode and electrolyte in determining the performance of supercapacitors. It emphasizes the need to use electrolytes compatible with the electrode material to achieve efficient storage and charge transfer in KIHSCs. Figure 12c illustrates the many energy storage technologies, including MIHSCs, and emphasizes the strategic integration of specific electrolytes to enhance performance. Stressing the importance of electrolyte selection [225], this study demonstrates how particular electrolytes utilized in BDC electrodes directly influence the efficiency and efficacy of MIHSCs in different applications. Choosing electrolyte composition is crucial for controlling BDC electrodes' electrochemical characteristics and overall efficiency in KIHSCs. By incorporating findings from studies on electron transport pathways, energy storage processes, and electrode–electrolyte interactions, it is feasible to enhance the efficiency and usefulness of supercapacitor systems for energy storage purposes.

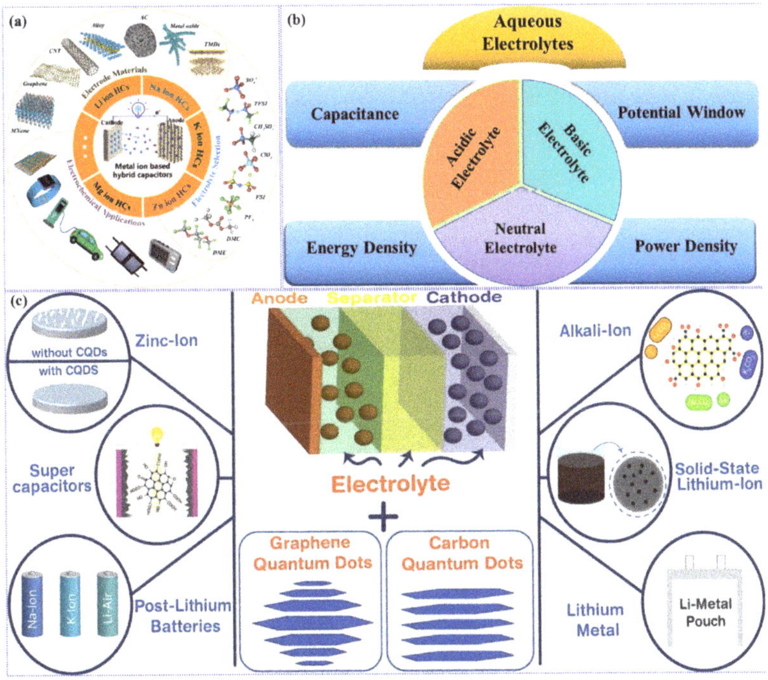

Figure 12. (**a**) Exploration of materials and electrolytes for MIHSCs. Reproduced with permission from Ref. [226]. Copyright 2021, Elsevier. (**b**) The impact of aqueous electrolytes on the efficiency of MIHSCs. Reproduced with permission from Ref. [227]. Copyright 2020, Elsevier. (**c**) Schematic representation of the effect of electrolytes on carbon-based electrodes in MIHSCs. Reproduced with permission from Ref. [225]. Reproduced under the term CC BY 4.0. Copyright 2024, Shaker et al., Elsevier.

5.2. Advancements in High-OPW MIHSCs

The energy density and overall performance of MIHSCs are greatly affected by a high OPW, making it a crucial aspect of these energy storage devices. Maximizing the OPW in such systems is essential for augmenting their capabilities. A practical approach to increase the OPWs in MIHSCs is utilizing electrode materials that can function efficiently at higher potentials. For instance, using materials such as BDCs and MoO_2 with a more significant work function can aid in advancing high-voltage supercapacitors [228]. By incorporating materials like silicene into MIHSCs, it is possible to create devices with a broad range of OPWs, reaching up to 3 V [229]. This allows for higher energy and power densities in the MIHSCs. Moreover, the configuration of asymmetrical MIHSCs using BDCs can also enhance the OPWs. By manipulating electrode work functions and utilising specific electrode combinations, it is possible to fabricate high-voltage asymmetric MIHSCs that exhibit improved performance [228]. Furthermore, sophisticated electrode materials such as highly porous BDCs and metallic nanosheets can provide exceptional energy storage capabilities at increased OPWs, highlighting the need to select appropriate materials to achieve high supercapacitor OPWs [230]. Ultimately, high OPWs in MIHSCs are of utmost importance, as they are directly correlated with these devices' energy density and overall efficiency. Employing suitable electrode materials, implementing asymmetric configurations, and integrating novel substances such as silicene and metallic nanosheets makes it feasible to enhance the voltage in MIHSCs, thus unleashing their complete energy storage capacity.

High-voltage aqueous ZIHSCs are a noteworthy advancement in energy storage. They offer improved safety features and higher energy densities compared to conventional systems. The ability to attain elevated operational voltages is dependent on the inventive restructuring of the hydrogen bond network in the electrolyte, which is assisted by the use of co-solvents such as polyethylene glycol and N, N-dimethylformamide [231]. These co-solvents alter the structure of the hydrogen bonds, which decreases the movement and reactivity of water molecules. As a result, the electrolyte can withstand greater OPWs up to 4.27 V without breaking down. This method avoids the usual problems with water-based systems and takes advantage of the natural safety and environmental advantages of electrolytes based on water. This opens up the possibility of developing ZIHSCs in the future which have far better energy storage capacity. The advancement of high-voltage KIHSCs represents a crucial breakthrough in energy storage technology, providing a combination of high power and energy density essential for various applications [232]. These KIHSCs achieve high OPWs by exploiting the distinctive characteristics of potassium ions, such as their larger size and lower solvation energy than lithium and sodium. This enables easier removal of solvent molecules and faster movement of ions during ion insertion and removal processes. The lower typical electrode potential of potassium in electrolytes such as propylene carbonate is also a contributing factor to the high voltage capability of these devices. The reduced potential and the creation of robust electrochemical interfaces enable potassium-ion supercapacitors to function effectively at higher voltages. This maximizes energy density and improves the overall performance and efficiency of the MIHSCs.

5.3. Theoretical Calculations: Insights into Energy Storage Mechanisms in MIHSCs

Theoretical calculations are crucial for comprehending the energy storage mechanisms of MIHSCs. By employing density functional theory (DFT) calculations, researchers can obtain valuable information regarding charge separation mechanisms and the efficient confinement of precursors inside these systems [233,234]. The characteristics of metal-based hydroxides have been investigated using DFT calculations. These calculations have also been used to examine how the incorporation of transition metal ions and lattice defects affects the electronic properties of these hydroxides [235]. In addition, DFT has played a crucial role in comparing the characteristics of several polypyridyl dyes based on metal ions for dye-sensitized solar cells [236]. MIHSCs are a form of energy storage device that combines battery-like electrodes with capacitive electrodes. They have shown great potential in high energy and remarkable power densities [16]. Hybrid supercapacitors exhibit superior energy density compared to conventional electric double-layer capacitors and effectively surpass the power density constraints of batteries [52,237,238]. MIHSCs, specifically ZIHSCs, are recognized for their exceptional safety, impressive energy and power outputs, long-lasting performance, and cost-efficiency [106]. These devices combine the benefits of zinc-ion batteries, which have high energy density, with supercapacitors, which have excellent power density and cycle stability [14]. Zinc metal positrodes have generated considerable interest in energy storage systems due to their inherent advantages, leading to the development of aqueous ZIHSCs [38]. DFT simulations provide valuable insights into the energy storage mechanisms in MIHSCs, which contribute to developing high-performance energy storage devices with improved energy and power densities.

Recent advancements in DFT calculations have shed light on the energy barriers involved in the adsorption and desorption of Zn^{2+} ions on doped carbon electrodes, which is crucial for optimizing the chemisorption kinetics in ZIHSCs [239]. The addition of nitrogen and sulfur heteroatoms to the carbon structure considerably impacts the energy barriers, leading to improved electrochemical performance. DFT investigations reveal that the adsorption of Zn^{2+} ions commences by their contact with the oxygen sites of hydroxyl groups. This interaction leads to the breaking of O-H bonds and the subsequent creation of C-O-Zn bonds, accompanied by the release of H^+ ions. The significance of this sequence is in its ability to determine the chemisorption kinetics, which directly impact the performance of ZIHSCs. The computational analysis in Figure 13a,b reveals that the energy barrier for this process is significantly reduced when N/S co-doping is present, measuring at 1.17 eV, in contrast to 1.91 eV for N-doped alone and 3.21 eV for undoped electrodes. Incorporating N/S co-doping decreases the energy barriers, resulting in faster adsorption and improved ion exchange reversibility and efficiency at the electrode interface during charge–discharge cycles. The findings have significant ramifications, indicating that by strategically adding N and S heteroatoms to the electrode material of ZIHSCs, these capacitors' energy storage capacity and cycling stability can be significantly improved. Reducing energy barriers enables faster and more effective ion adsorption and desorption, which is crucial for high-performance supercapacitors that achieve rapid energy delivery and high power densities. Therefore, combining N/S co-doping is essential to improve future supercapacitors' design and functional capabilities, making them more efficient and long-lasting.

DFT computations are essential in studying and improving ion diffusion pathways within positrode materials in NIHSCs. These calculations play a crucial role in determining the energy barriers related to the movement of sodium ions. These barriers are essential factors that affect the electrochemical performance of supercapacitors. DFT is used to analyse and understand how changes in the structure and composition of the positrode might improve ion transport and increase the device's overall efficiency [240]. This is conducted by comparing the diffusion barriers in different materials or modified structures. A detailed analysis of the obstacles that hinder the movement of sodium ions in two different situations involving electrodes made of manganese oxide (MnO) is depicted in Figure 13c–g. The initial scenario investigates the diffusion of sodium ions across the main body of cubic MnO. In this process, ions move by directly jumping between interstitial sites. This process does not entail the displacement of lattice atoms, but it does exhibit a relatively high energy barrier, which suggests that ion transport occurs at a slower rate. The second scenario examines the diffusion of sodium ions at the interface between Mn and Na_2O, a crucial region for enhancing ion transport. The interface reduces the energy needed for Na ions to move, making them more mobile than their movement through the more significant MnO material. The decrease in energy barriers at the interface can be ascribed to the distinctive structural and electrical interactions that occur, facilitating the migration of Na ions across the electrode. The insights obtained from DFT calculations are crucial and show that by strategically manipulating the interfaces in positrode materials, we may significantly influence the performance of NIHSCs. Reducing the obstacles to the movement of ions at essential points, like the interface between manganese and sodium oxide, makes it feasible to raise the speed at which ions are transported. This, in turn, can lead to improved performance regarding the rate at which the supercapacitor can charge and discharge and its overall efficiency. This research enhances our comprehension of material behaviours at the atomic scale and guides the future development and creation of advanced electrode materials for improved energy storage technologies.

DFT simulations have played a crucial role in understanding the adsorption process on heteroatom-doped carbon negatrodes in AIHSCs [241]. These theoretical insights are essential for developing negatrodes with the most efficient energy storage capacities. The DFT simulations were used to study the interaction between the $[AlCl_4]^-$ moieties and the nitrogen and sulfur co-doped carbon negatrode (N,S-C). The DFT analysis demonstrates that adding nitrogen and sulfur to the carbon structure dramatically improves the adsorption potency of $[AlCl_4]^-$ ions. The improvement is due to the distinctive electronic characteristics provided by the N and S dopants, which alter the electronic surroundings of the carbon structure, thereby enabling more robust and more enduring interactions with the $[AlCl_4]^-$ ions. Specifically, graphitic and pyridinic nitrogen sites were discovered to enhance adsorption energies, suggesting a more advantageous adsorption situation than carbon that is not doped. The computational findings provide a crucial explanation for the experimental data on the enhanced capacitance and cycling stability of the N,S-doped carbon negatrodes in AIHSCs. DFT's predictive and explanatory power concerning the impact of heteroatom doping on adsorption characteristics highlights its usefulness in advancing energy storage materials. The combination of theoretical predictions and experimental advancements offers a strong foundation for improving the efficiency of aluminium-ion batteries and supercapacitors.

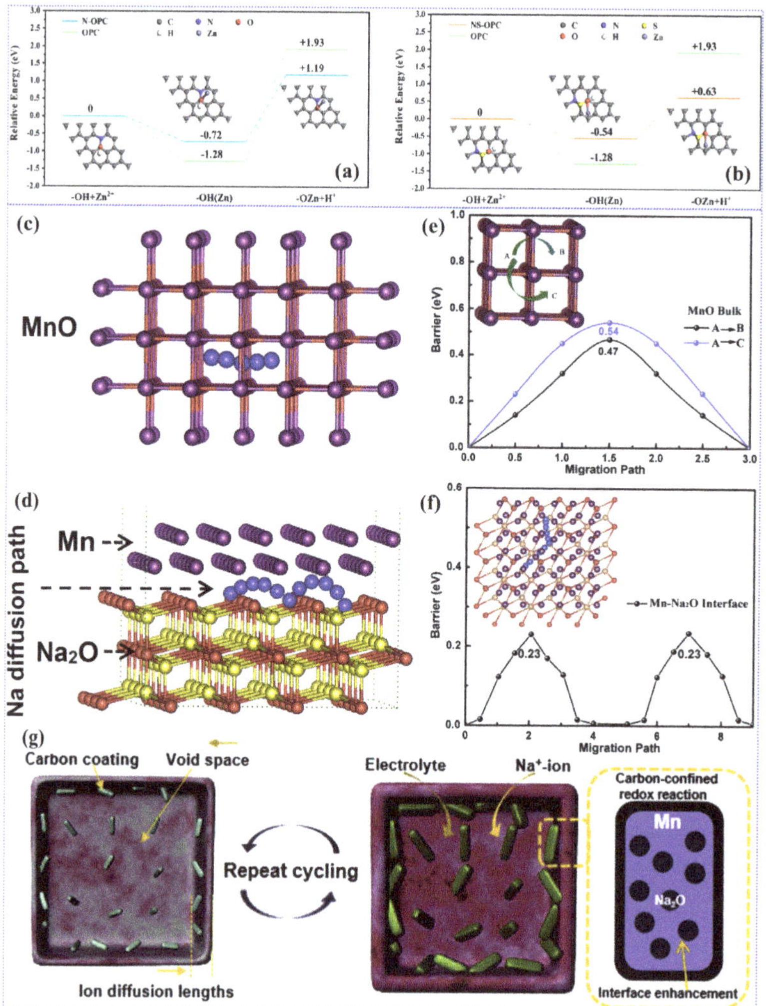

Figure 13. (**a**,**b**) Variations in the relative energy levels during the chemisorption process for ZIHSCs. Reproduced with permission from Ref. [239]. Copyright 2023, American Chemical Society. Atomic structures of various models and distinct Na diffusion pathways in (**c**) MnO and (**d**) Mn/Na_2O interface. Diffusion energy barriers for Na atoms are shown for (**e**) MnO bulk and (**f**) Mn/Na_2O interface. (**g**) Diagram illustrating the energy storage mechanisms for the MnO@HCNb electrode. Reproduced with permission from Ref. [240]. Copyright 2020, Elsevier.

DFT calculations provide valuable insights into the interactions at the interface and the mechanisms of charge transport in the electrode materials of KIHSCs [242]. These calculations have a substantial impact on the electrochemical performance of the supercapacitors. More precisely, bonding polyaniline (PANI) with molybdenum disulfide (MoS_2) promotes improved reaction speed, essential for adequate energy storage. DFT investigations demonstrate that incorporating PANI into the MoS_2 layers substantially reduces the barrier for K^+ diffusion. This indicates that the molecular-level hybridization reinforces the material's structural integrity and improves its electrical conductivity. This alteration leads to an electronic structure that is easier to access and facilitates faster and more efficient ion transport. The formation of covalent connections between PANI and MoS_2 establishes novel routes for electron transmission, reducing the energy obstacles linked to the insertion and removal of potassium ions in the charge–discharge procedures. Theoretical findings highlight the potential of employing organic–inorganic hybrid materials in supercapacitors to overcome the constraints imposed by conventional electrode materials. These hybrids can lower energy barriers and improve the movement of ions. As a result, they can attain a high capacitance and maintain excellent cycling stability. This makes them well suited for energy storage applications that require high performance. The findings obtained from DFT calculations play a vital role in directing the manufacture and enhancement of electrode materials to use these favourable characteristics fully.

6. Challenges and Future Research Perspectives

BDC materials are promising for high-performance MIHSCs, primarily due to their sustainable nature, widespread availability, and affordability. These materials exhibit significant attributes, including structural diversity, porousness, and heteroatom doping, greatly enhancing their electrochemical performance. This review thoroughly examines these characteristics and the obstacles that must be overcome for practical supercapacitor use. A major challenge lies in synthesizing BDC compounds with consistent and optimal structures for energy storage. The intricate chemical composition of biomasses hinders the continuous production of carbon nanotube fibres. Additionally, generating high-performance shape-memory carbon aerogels from biomass is a difficult task. To tackle these problems, several techniques have been suggested, such as altering the distance between layers in carbon materials, which can improve ion movement and storage. Incorporating nitrogen, phosphorus, and sulfur directly into the carbon structure enhances conductivity and electrochemical activity. Introducing multiple heteroatoms simultaneously generates a synergistic effect, improving overall supercapacitor performance.

BDC compounds also show promise as raw materials for metal-ion batteries, contributing to developing materials for such batteries. However, obstacles such as insufficient electrochemical storage capacity render BDCs unsuitable for battery applications. Efforts to enhance specific capacity and lifespan are necessary to make them viable for battery use. The complex structure of biomass precursors also hinders the continuous production of carbon nanotube fibres and shape-memory carbon aerogels, which are essential for battery applications. Several methodologies have been proposed to address these issues, such as biomass-derived nitrogen-doped porous carbons, which can replace noble metal catalysts in metal-ion batteries, improving battery performance. Improving electrode design, refining electrolyte composition, and innovating new energy storage mechanisms are crucial for progressing BDCs in MIHSCs. A deeper understanding of the correlation between the composition and characteristics of carbon compounds obtained from biomass will guide research focused on specific synthesis techniques, such as interlayer modulation and self-doping, to produce tailor-made materials with improved performance.

Advances in BDC materials for high-performance MIHSCs include electrode materials with high energy densities and mechanical stability, which are crucial for large-scale production. Surface engineering is vital for enhancing charge–discharge capabilities and maintaining consistent performance across multiple cycles, which is essential for industrial-scale applications. Encasing metal nanoparticles with metal oxides within zeolitic imidazolate frameworks can improve the performance of carbon materials for energy storage. Further enhancing capacitance is possible by using high-mass-loading biomass-based porous carbon electrodes.

BDCs have garnered considerable interest in advanced MIHSCs because of their notable characteristics, such as large surface area, conductivity, and cost-effectiveness [243].

Recent studies have investigated different types of carbon materials, such as nanoporous carbon, graphene, carbon nanosheets, and carbon aerogels obtained from sustainable sources like bagasse. These studies have shown that these materials can be inexpensive and effective electrode materials for supercapacitors [244,245]. Moreover, combining MOFs and conductive materials such as polypyrrole has been recognized as a method to improve the capacitance properties of supercapacitors, effectively resolving concerns regarding inadequate electrical conductivity [246–248]. In addition, the progress in creating all-carbon hybrids, specifically rGO/CNF nanostructures, has demonstrated encouraging outcomes in attaining superior charging and discharging abilities for supercapacitor electrodes [249]. Researchers have explored using carbon materials and transition metal compounds to improve electrode materials for KIHSCs to achieve higher power and energy density in advanced metal-ion hybrid supercapacitors [12]. In addition, the use of yolk–shell-structured nickel cobalt sulfide and CNT composites has demonstrated excellent performance in hybrid supercapacitors, emphasizing the significance of carbon materials as key electrode components [250]. The development of MIHSCs has enabled the creation of energy storage devices that combine the advantages of batteries and supercapacitors. These devices offer large energy storage capacity, fast charge–discharge rates, and an extended cycle life [18,41]. These advancements highlight the potential of carbon materials sourced from biomass to influence the future of MIHSCs. They offer a sustainable and efficient alternative for energy storage applications.

The ongoing exploration of BDC compounds for energy storage indicates a promising future. Techniques such as co-incorporating phosphorus and boron can significantly prolong supercapacitors' lifespan, making them more suitable for practical applications. These materials can become ideal for commercial use by creating scalable and economical synthesis methods and advancing material design and performance. The challenges and outlook for using BDC materials in high-performance MIHSCs must address structural, functional, and material difficulties. Advanced synthesis methods, innovative design strategies, and a thorough study of structure–property relationships are essential for the progress of these materials in energy storage. Tackling these challenges will allow BDC materials to substantially contribute to advancing sustainable and high-performance energy storage systems.

7. Conclusions

This review study thoroughly examines the crucial significance of BDC materials in advancing the exceptional performance of MIHSCs. This paper concentrates explicitly on systems based on sodium, potassium, and zinc ions. These materials are highlighted for their plentiful supply, capacity to be renewed, and remarkable physical and chemical qualities, making them ideal for energy storage applications. This review extensively examines the process of creating, modifying, and structurally enhancing carbon materials obtained from biomass. It highlights the significance of their large SSA, adjustable porosity, and introduction of heteroatoms in enhancing their electrochemical capabilities. This study explores many methods of creating carbon materials from biomass, such as pyrolysis, hydrothermal carbonization, and activation procedures. These processes make materials with various structural properties and regulated shapes. These techniques facilitate the production of carbon materials with hierarchical porous architectures and tailored surface chemistry, rendering them appropriate for MIHSCs. This review analyses the electrochemical characteristics of carbon materials obtained from biomass in various MIHSCs. It showcases their capacity to attain high capacitance, energy density, and cycling stability. This research emphasizes the suitability of BDCs with metal-ion electrolytes, their function in aiding ion transport, and the mechanisms underlying their ability to store charges. This study discusses the practical uses of BDC materials in sodium, potassium, aluminium, and ZIHSCs. It highlights how these materials can improve the performance of these systems. The analysis of several metal-ion systems demonstrates the benefits of BDCs' structural characteristics for each form of supercapacitor. This review examines the present obstacles in developing carbon materials produced from biomass for supercapacitors, encompassing scalability, cost-effectiveness, and performance consistency.

Additionally, it provides valuable insights for future studies, including enhancing the synthesis procedures, optimizing the structural configuration, and investigating novel energy storage mechanisms. This research highlights the significant impact of BDC materials in MIHSCs, which can lead to the development of environmentally friendly and efficient energy storage systems. This review is unusual because it comprehensively compares carbon compounds obtained from biomass in various metal-ion systems. It uncovers the distinct contributions of these materials to different types of supercapacitors. This comparison offers vital insights into the structural characteristics that improve electrochemical performance, influencing the advancement of sustainable energy storage solutions. This paper incorporates the latest developments in the field, emphasizing the breakthroughs in creating carbon materials from biomass and their usage in MIHSCs. The statement underscores the importance of ongoing innovation and interdisciplinary cooperation in investigating these sustainable materials better. This will help advance green energy technologies and harmoniously balance environmental sustainability and technical advancement. Overall, this review paper offers an in-depth examination of carbon materials produced from biomass in MIHSCs with exceptional performance. This study emphasizes the capacity of these materials to bring about significant changes, tackles existing obstacles, and provides insights for future investigations, thereby promoting the progress of sustainable energy storage technologies.

Funding: This work is supported by the Japan Society for the Promotion of Science (JSPS) through KAKENHI Grant Number 22F22336.

Institutional Review Board Statement: Not applicable.

Informed Consent Statement: Not applicable.

Data Availability Statement: Data are contained within this article.

Acknowledgments: The author acknowledges the support of the Japan Society for the Promotion of Science (JSPS) for the postdoctoral fellowship.

Conflicts of Interest: The author declares no conflicts of interest.

References

1. Guo, H.; Qiao, M.; Yan, J.; Jiang, L.; Yu, J.; Li, J.; Deng, S.; Qu, L. Fabrication of Hybrid Supercapacitor by $MoCl_5$ Precursor-Assisted Carbonization with Ultrafast Laser for Improved Capacitance Performance. *Adv. Funct. Mater.* **2023**, *33*, 2213514. [CrossRef]
2. Han, X.; Wang, B.; Yang, C.; Meng, G.; Zhao, R.; Hu, Q.; Triana, O.; Iqbal, M.; Li, Y.; Han, A.; et al. Inductive Effect in Mn-Doped NiO Nanosheet Arrays for Enhanced Capacitive and Highly Stable Hybrid Supercapacitor. *ACS Appl. Energy Mater.* **2019**, *2*, 2072–2079. [CrossRef]
3. Lee, S.-H.; Kim, J.H.; Yoon, J.-R. Laser Scribed Graphene Cathode for Next Generation of High Performance Hybrid Supercapacitors. *Sci. Rep.* **2018**, *8*, 8179. [CrossRef]
4. Larcher, D.; Tarascon, J.-M. Towards Greener and More Sustainable Batteries for Electrical Energy Storage. *Nat. Chem.* **2015**, *7*, 19–29. [CrossRef]
5. Zhai, Y.; Dou, Y.; Zhao, D.; Fulvio, P.F.; Mayes, R.T.; Dai, S. Carbon Materials for Chemical Capacitive Energy Storage. *Adv. Mater.* **2011**, *23*, 4828–4850. [CrossRef] [PubMed]
6. Zhang, X.; Hou, L.; Ciesielski, A.; Samorì, P. 2D Materials Beyond Graphene for High-Performance Energy Storage Applications. *Adv. Energy Mater.* **2016**, *6*, 1600671. [CrossRef]
7. Mondal, M.; Goswami, D.K.; Bhattacharyya, T.K. High-Performing Asymmetric 2 V Supercapacitor Assembled with Leucine-Capped RGO-α-Fe_2O_3 as Anode and PANI Decorated MwCNT-V_2O_5 as Cathode. *J. Electrochem. Soc.* **2023**, *170*, 110521. [CrossRef]
8. Li, J.; Zhu, J.; Dong, Z.; Wu, Q. Nanomaterials Derived from a Template Method for Supercapacitor Applications. *ChemistrySelect* **2023**, *8*, e202204487. [CrossRef]
9. Genc, R.; Alas, M.O.; Harputlu, E.; Repp, S.; Kremer, N.; Castellano, M.; Colak, S.G.; Ocakoglu, K.; Erdem, E. High-Capacitance Hybrid Supercapacitor Based on Multi-Colored Fluorescent Carbon-Dots. *Sci. Rep.* **2017**, *7*, 11222. [CrossRef]

10. Li, S.; Chen, W.; Huang, X.; Ding, L.; Ren, Y.; Xu, M.; Zhu, J.; Miao, Z.; Liu, H. Enabling Wasted A4 Papers as a Promising Carbon Source to Construct Partially Graphitic Hierarchical Porous Carbon for High-Performance Aqueous Zn-Ion Storage. *ACS Appl. Mater. Interfaces* **2024**, *16*, 10126–10137. [CrossRef]
11. Díez, N.; Sevilla, M. Hybrid Metal-Ion Capacitors Based on Carbon Nanospheres. *ChemElectroChem* **2024**, *11*, e202300475. [CrossRef]
12. Wang, Z.; Hong, P.; Zhao, H.; Lei, Y. Recent Developments and Future Prospects of Transition Metal Compounds as Electrode Materials for Potassium-Ion Hybrid Capacitors. *Adv. Mater. Technol.* **2023**, *8*, 2200515. [CrossRef]
13. Gao, Q.; Li, P.; Ding, S.; He, H.; Cai, M.; Ning, X.; Cai, Y.; Zhang, M. Cu2Se-ZnSe Heterojunction Encapsulated in Carbon Fibers for High-Capacity Anodes of Sodium-Ion Batteries. *Ionics* **2020**, *26*, 5525–5533. [CrossRef]
14. Tang, H.; Yao, J.; Zhu, Y. Recent Developments and Future Prospects for Zinc-Ion Hybrid Capacitors: A Review. *Adv. Energy Mater.* **2021**, *11*, 2003994. [CrossRef]
15. Zeng, F.; Gong, X.; Xu, Z.; Du, Z.; Xu, J.; Deng, T.; Wang, D.; Zeng, Y.; Yu, S.; Meng, Z.; et al. A New Selection Criterion for Voltage Windows of Aqueous Zinc Ion Hybrid Capacitors: Achieving a Balance between Energy Density and Cycle Stability. *J. Mater. Chem. A* **2023**, *11*, 26698–26706. [CrossRef]
16. Shang, K.; Liu, Y.; Cai, P.; Li, K.; Wen, Z. N, P, and S Co-Doped 3D Porous Carbon-Architectured Cathode for High-Performance Zn-Ion Hybrid Capacitors. *J. Mater. Chem. A* **2022**, *10*, 6489–6498. [CrossRef]
17. Zheng, H.; Zhou, H.; Zheng, B.; Wei, C.; Ma, A.; Jin, X.; Chen, W.; Liu, H. Stable Flexible Electronic Devices under Harsh Conditions Enabled by Double-Network Hydrogels Containing Binary Cations. *ACS Appl. Mater. Interfaces* **2024**, *16*, 7768–7779. [CrossRef]
18. Shaikh, N.S.; Lokhande, V.C.; Praserthdam, S.; Lokhande, C.D.; Ezema, F.I.; Salunkhe, D.J.; Shaikh, J.S.; Kanjanaboos, P. Recent Advancements in Energy Storage Based on Sodium Ion and Zinc Ion Hybrid Supercapacitors. *Energy Fuels* **2021**, *35*, 14241–14264. [CrossRef]
19. Dong, S.; Li, Z.; Xing, Z.; Wu, X.; Ji, X.; Zhang, X. Novel Potassium-Ion Hybrid Capacitor Based on an Anode of $K_2Ti_6O_{13}$ Microscaffolds. *ACS Appl. Mater. Interfaces* **2018**, *10*, 15542–15547. [CrossRef]
20. Qiu, D.; Guan, J.; Li, M.; Kang, C.; Wei, J.; Li, Y.; Xie, Z.; Wang, F.; Yang, R. Kinetics Enhanced Nitrogen-Doped Hierarchical Porous Hollow Carbon Spheres Boosting Advanced Potassium-Ion Hybrid Capacitors. *Adv. Funct. Mater.* **2019**, *29*, 1903496. [CrossRef]
21. Ruan, J.; Mo, F.; Chen, Z.; Liu, M.; Zheng, S.; Wu, R.; Fang, F.; Song, Y.; Sun, D. Rational Construction of Nitrogen-Doped Hierarchical Dual-Carbon for Advanced Potassium-Ion Hybrid Capacitors. *Adv. Energy Mater.* **2020**, *10*, 1904045. [CrossRef]
22. Buliyaminu, I.A.; Aziz, M.A.; Shah, S.S.; Mohamedkhair, A.K.; Yamani, Z.H. Preparation of Nano-Co_3O_4-Coated Albizia Procera-Derived Carbon by Direct Thermal Decomposition Method for Electrochemical Water Oxidation. *Arab. J. Chem.* **2020**, *13*, 4785–4796. [CrossRef]
23. Natarajan, S.; Lee, Y.; Aravindan, V. Biomass-Derived Carbon Materials as Prospective Electrodes for High-Energy Lithium- and Sodium-Ion Capacitors. *Chem. Asian J.* **2019**, *14*, 936–951. [CrossRef] [PubMed]
24. Shah, S.S.; Qasem, M.A.A.; Berni, R.; Del Casino, C.; Cai, G.; Contal, S.; Ahmad, I.; Siddiqui, K.S.; Gatti, E.; Predieri, S.; et al. Physico-Chemical Properties and Toxicological Effects on Plant and Algal Models of Carbon Nanosheets from a Nettle Fibre Clone. *Sci. Rep.* **2021**, *11*, 6945. [CrossRef] [PubMed]
25. Lai, F.; Miao, Y.-E.; Zuo, L.; Lu, H.; Huang, Y.; Liu, T. Biomass-Derived Nitrogen-Doped Carbon Nanofiber Network: A Facile Template for Decoration of Ultrathin Nickel-Cobalt Layered Double Hydroxide Nanosheets as High-Performance Asymmetric Supercapacitor Electrode. *Small* **2016**, *12*, 3235–3244. [CrossRef] [PubMed]
26. Wang, Q.; Qin, B.; Qu, C.; Wang, B.; Duan, H.; Cao, Q.; Li, H.; Qi, J. Synthesis of Hierarchical Porous Carbon from Bio-Oil for Supercapacitor Application. *Energy Fuels* **2023**, *37*, 16970–16978. [CrossRef]
27. Wang, T.; Hu, S.; Yu, W.; Hu, Y.; Yan, S.; Wang, M.; Zhao, W.; Xu, J.; Zhang, J. Biologically Inspired Small Herbal Biomolecules and Biomass Carbon for High-Performance Supercapacitors. *ACS Appl. Energy Mater.* **2023**, *6*, 2347–2357. [CrossRef]
28. Zhou, Y.; Parker, C.B.; Joshi, P.; Naskar, A.K.; Glass, J.T.; Cao, C. 4D Printing of Stretchable Supercapacitors via Hybrid Composite Materials. *Adv. Mater. Technol.* **2021**, *6*, 2001055. [CrossRef]
29. Liu, H.; Yao, M.; Yao, X. Improving Energy Density of Crystalline–Amorphous Multilayer Films Deposited on Ti Foils by Structural Modulation. *J. Am. Ceram. Soc.* **2021**, *104*, 1379–1390. [CrossRef]
30. Choi, N.; Chen, Z.; Freunberger, S.A.; Ji, X.; Sun, Y.; Amine, K.; Yushin, G.; Nazar, L.F.; Cho, J.; Bruce, P.G. Challenges Facing Lithium Batteries and Electrical Double-Layer Capacitors. *Angew. Chem. Int. Ed.* **2012**, *51*, 9994–10024. [CrossRef]
31. He, M.; Fic, K.; Frąckowiak, E.; Novák, P.; Berg, E.J. Towards More Durable Electrochemical Capacitors by Elucidating the Ageing Mechanisms under Different Testing Procedures. *ChemElectroChem* **2019**, *6*, 566–573. [CrossRef]
32. Wu, Z.; Parvez, K.; Feng, X.; Müllen, K. Graphene-Based in-Plane Micro-Supercapacitors with High Power and Energy Densities. *Nat. Commun.* **2013**, *4*, 2487. [CrossRef]
33. Kandambeth, S.; Jia, J.; Wu, H.; Kale, V.S.; Parvatkar, P.T.; Czaban-Jóźwiak, J.; Zhou, S.; Xu, X.; Ameur, Z.O.; Abou-Hamad, E.; et al. Covalent Organic Frameworks as Negative Electrodes for High-Performance Asymmetric Supercapacitors. *Adv. Energy Mater.* **2020**, *10*, 2001673. [CrossRef]
34. Nguyen, T.; Montemor, M.d.F. Metal Oxide and Hydroxide–Based Aqueous Supercapacitors: From Charge Storage Mechanisms and Functional Electrode Engineering to Need-Tailored Devices. *Adv. Sci.* **2019**, *6*, 1801797. [CrossRef] [PubMed]

35. Naoi, K.; Ishimoto, S.; Miyamoto, J.; Naoi, W. Second Generation 'Nanohybrid Supercapacitor': Evolution of Capacitive Energy Storage Devices. *Energy Environ. Sci.* **2012**, *5*, 9363–9373. [CrossRef]
36. Guo, Q.; Liu, J.; Bai, C.; Chen, N.; Qu, L. 2D Silicene Nanosheets for High-Performance Zinc-Ion Hybrid Capacitor Application. *ACS Nano* **2021**, *15*, 16533–16541. [CrossRef]
37. Xia, L.; Tang, B.; Wei, J.; Zhou, Z. Recent Advances in Alkali Metal-Ion Hybrid Supercapacitors. *Batter. Supercaps* **2021**, *4*, 1108–1121. [CrossRef]
38. Wang, Q.; Wang, S.; Li, J.; Ruan, L.; Wei, N.; Huang, L.; Dong, Z.; Cheng, Q.; Xiong, Y.; Zeng, W. A Novel Aqueous Zinc-Ion Hybrid Supercapacitor Based on TiS_2 (De)Intercalation Battery-Type Anode. *Adv. Electron. Mater.* **2020**, *6*, 2000388. [CrossRef]
39. Liu, Y.; Zhang, Y.; Sun, Z.; Cheng, S.; Cui, P.; Wu, Y.; Zhang, J.; Fu, J.; Xie, E. New Insight into the Mechanism of Multivalent Ion Hybrid Supercapacitor: From the Effect of Potential Window Viewpoint. *Small* **2020**, *16*, 2003403. [CrossRef] [PubMed]
40. Wang, H.; Zhu, C.; Chao, D.; Yan, Q.; Fan, H.J. Nonaqueous Hybrid Lithium-Ion and Sodium-Ion Capacitors. *Adv. Mater.* **2017**, *29*, 1702093. [CrossRef]
41. Ma, Y.; Chang, H.; Zhang, M.; Chen, Y. Graphene-Based Materials for Lithium-Ion Hybrid Supercapacitors. *Adv. Mater.* **2015**, *27*, 5296–5308. [CrossRef]
42. Zuo, W.; Li, R.; Zhou, C.; Li, Y.; Xia, J.; Liu, J. Battery-Supercapacitor Hybrid Devices: Recent Progress and Future Prospects. *Adv. Sci.* **2017**, *4*, 1600539. [CrossRef] [PubMed]
43. Peng, L.; Peng, X.; Liu, B.; Wu, C.; Xie, Y.; Yu, G. Ultrathin Two-Dimensional MnO_2/Graphene Hybrid Nanostructures for High-Performance, Flexible Planar Supercapacitors. *Nano Lett.* **2013**, *13*, 2151–2157. [CrossRef]
44. Karri, S.N.; Ega, S.P.; Perupogu, V.; Srinivasan, P. Enhancing the Electrochemical Performance of Polyaniline Using Fly Ash of Coal Waste for Supercapacitor Application. *ChemistrySelect* **2021**, *6*, 2576–2589. [CrossRef]
45. Unnikrishnan, B.; Wu, C.-W.; Chen, I.-W.P.; Chang, H.-T.; Lin, C.-H.; Huang, C.-C. Carbon Dot-Mediated Synthesis of Manganese Oxide Decorated Graphene Nanosheets for Supercapacitor Application. *ACS Sustain. Chem. Eng.* **2016**, *4*, 3008–3016. [CrossRef]
46. Fan, W.; Wang, F.; Xiong, X.; Song, B.; Wang, T.; Cheng, X.; Zhu, Z.; He, J.; Liu, Y.; Wu, Y. Recent Advances in Functional Materials and Devices for Zn-Ion Hybrid Supercapacitors. *NPG Asia Mater.* **2024**, *16*, 18. [CrossRef]
47. Huang, C.; Zhao, X.; Xu, Y.; Zhang, Y.; Yang, Y.; Hu, A.; Tang, Q.; Song, X.; Jiang, C.; Chen, X. Sewable and Cuttable Flexible Zinc-Ion Hybrid Supercapacitor Using a Polydopamine/Carbon Cloth-Based Cathode. *ACS Sustain. Chem. Eng.* **2020**, *8*, 16028–16036. [CrossRef]
48. Zhang, H.; Liu, Q.; Fang, Y.; Teng, C.; Liu, X.; Fang, P.; Tong, Y.; Lu, X. Boosting Zn-Ion Energy Storage Capability of Hierarchically Porous Carbon by Promoting Chemical Adsorption. *Adv. Mater.* **2019**, *31*, 1904948. [CrossRef] [PubMed]
49. Zheng, Y.; Zhao, W.; Jia, D.; Liu, Y.; Cui, L.; Wei, D.; Zheng, R.; Liu, J. Porous Carbon Prepared via Combustion and Acid Treatment as Flexible Zinc-Ion Capacitor Electrode Material. *Chem. Eng. J.* **2020**, *387*, 124161. [CrossRef]
50. Li, Z.; Chen, D.; An, Y.; Chen, C.; Wu, L.; Chen, Z.; Sun, Y.; Zhang, X. Flexible and Anti-Freezing Quasi-Solid-State Zinc Ion Hybrid Supercapacitors Based on Pencil Shavings Derived Porous Carbon. *Energy Storage Mater.* **2020**, *28*, 307–314. [CrossRef]
51. Yin, J.; Zhang, W.; Wang, W.; Alhebshi, N.A.; Salah, N.; Alshareef, H.N. Electrochemical Zinc Ion Capacitors Enhanced by Redox Reactions of Porous Carbon Cathodes. *Adv. Energy Mater.* **2020**, *10*, 2001705. [CrossRef]
52. Shah, S.S.; Aziz, M.A.; Ali, M.; Hakeem, A.S.; Yamani, Z.H. Advanced High-Energy All-Solid-State Hybrid Supercapacitor with Nickel-Cobalt-Layered Double Hydroxide Nanoflowers Supported on Jute Stick-Derived Activated Carbon Nanosheets. *Small* **2023**, 2306665. [CrossRef]
53. Wang, R.; Han, M.; Zhao, Q.; Ren, Z.; Guo, X.; Xu, C.; Hu, N.; Lu, L. Hydrothermal Synthesis of Nanostructured Graphene/Polyaniline Composites as High-Capacitance Electrode Materials for Supercapacitors. *Sci. Rep.* **2017**, *7*, 44562. [CrossRef]
54. Young, C.; Park, T.; Yi, J.W.; Kim, J.; Hossain, M.S.A.; Kaneti, Y.V.; Yamauchi, Y. Advanced Functional Carbons and Their Hybrid Nanoarchitectures towards Supercapacitor Applications. *ChemSusChem* **2018**, *11*, 3546–3558. [CrossRef]
55. Abouelamaiem, D.I.; He, G.; Parkin, I.; Neville, T.P.; Jorge, A.B.; Ji, S.; Wang, R.; Titirici, M.-M.; Shearing, P.R.; Brett, D.J.L. Synergistic Relationship between the Three-Dimensional Nanostructure and Electrochemical Performance in Biocarbon Supercapacitor Electrode Materials. *Sustain. Energy Fuels* **2018**, *2*, 772–785. [CrossRef]
56. Wang, K.; Zhang, Z.; Sun, Q.; Wang, P.; Li, Y. Durian Shell-Derived N, O, P-Doped Activated Porous Carbon Materials and Their Electrochemical Performance in Supercapacitor. *J. Mater. Sci.* **2020**, *55*, 10142–10154. [CrossRef]
57. Liu, B.; Zhang, X.; Tian, D.; Li, Q.; Zhong, M.; Chen, S.; Hu, C.; Ji, H. In Situ Growth of Oriented Polyaniline Nanorod Arrays on the Graphite Flake for High-Performance Supercapacitors. *ACS Omega* **2020**, *5*, 32395–32402. [CrossRef]
58. Mehmandoust, M.; Li, G.; Erk, N. Biomass-Derived Carbon Materials as an Emerging Platform for Advanced Electrochemical Sensors: Recent Advances and Future Perspectives. *Ind. Eng. Chem. Res.* **2023**, *62*, 4628–4635. [CrossRef]
59. Galek, P.; Mackowiak, A.; Bujewska, P.; Fic, K. Three-Dimensional Architectures in Electrochemical Capacitor Applications—Insights, Opinions, and Perspectives. *Front. Energy Res.* **2020**, *8*, 139. [CrossRef]
60. Olabi, A.G.; Abbas, Q.; Abdelkareem, M.A.; Alami, A.H.; Mirzaeian, M.; Sayed, E.T. Carbon-Based Materials for Supercapacitors: Recent Progress, Challenges and Barriers. *Batteries* **2022**, *9*, 19. [CrossRef]
61. Jiang, L.; Sheng, L.; Fan, Z. Biomass-Derived Carbon Materials with Structural Diversities and Their Applications in Energy Storage. *Sci. China Mater.* **2018**, *61*, 133–158. [CrossRef]

62. Falco, C.; Marco-Lozar, J.P.; Salinas-Torres, D.; Morallón, E.; Cazorla-Amorós, D.; Titirici, M.M.; Lozano-Castelló, D. Tailoring the Porosity of Chemically Activated Hydrothermal Carbons: Influence of the Precursor and Hydrothermal Carbonization Temperature. *Carbon* **2013**, *62*, 346–355. [CrossRef]
63. Chen, D.; Yang, L.; Li, J.; Wu, Q. Effect of Self-Doped Heteroatoms in Biomass-Derived Activated Carbon for Supercapacitor Applications. *ChemistrySelect* **2019**, *4*, 1586–1595. [CrossRef]
64. Guo, D.; Li, Z.; Wang, D.; Sun, M.; Wang, H. Design and Synthesis of Zinc-Activated $Co_xNi_{2-x}P$/Graphene Anode for High-Performance Zinc Ion Storage Device. *ChemSusChem* **2021**, *14*, 2205–2215. [CrossRef]
65. Yang, J.; Bissett, M.A.; Dryfe, R.A.W. Investigation of Voltage Range and Self-Discharge in Aqueous Zinc-Ion Hybrid Supercapacitors. *ChemSusChem* **2021**, *14*, 1700–1709. [CrossRef]
66. Lu, J.; Lin, X.; Wang, S.; Xu, X.; Zhou, Y.; Zhang, Y.; Li, Q.; Liu, H. High Ionic Conductivity and Toughness Hydrogel Electrolyte for High-Performance Flexible Solid-State Zinc-Ion Hybrid Supercapacitors Enabled by Cellulose-Bentonite Coordination Interactions. *Green Chem.* **2023**, *25*, 1635–1646. [CrossRef]
67. Chen, S.; Yang, G.; Zhao, X.; Wang, N.; Luo, T.; Chen, X.; Wu, T.; Jiang, S.; van Aken, P.A.; Qu, S.; et al. Hollow Mesoporous Carbon Spheres for High Performance Symmetrical and Aqueous Zinc-Ion Hybrid Supercapacitor. *Front. Chem.* **2020**, *8*, 663. [CrossRef]
68. Chang, H.; Khan, I.; Yuan, A.; Khan, S.; Sadiq, S.; Khan, A.; Shah, S.A.; Chen, L.; Humayun, M.; Usman, M. Polyarylimide-Based COF/MOF Nanoparticle Hybrids for CO_2 Conversion, Hydrogen Generation, and Organic Pollutant Degradation. *ACS Appl. Nano Mater.* **2024**, *7*, 10451–10465. [CrossRef]
69. Shah, S.A.; Khan, I.; Yuan, A. MoS_2 as a Co-Catalyst for Photocatalytic Hydrogen Production: A Mini Review. *Molecules* **2022**, *27*, 3289. [CrossRef]
70. Samage, A.; Halakarni, M.; Yoon, H.; Sanna Kotrappanavar, N. Sustainable Conversion of Agricultural Biomass Waste into Electrode Materials with Enhanced Energy Density for Aqueous Zinc-Ion Hybrid Capacitors. *Carbon* **2024**, *219*, 118774. [CrossRef]
71. Naik, P.B.; Yadav, P.; Nagaraj, R.; Puttaswamy, R.; Beere, H.K.; Maiti, U.N.; Mondal, C.; Sanna Kotrappanavar, N.; Ghosh, D. Developing High-Performance Flexible Zinc Ion Capacitors from Agricultural Waste-Derived Carbon Sheets. *ACS Sustain. Chem. Eng.* **2022**, *10*, 1471–1481. [CrossRef]
72. Li, S.; Luo, X.; Xiao, H.; Li, D.; Chen, Y. Nitrogen and Sulfur Codoped Hierarchical Porous Carbon Derived from Lignin for High-Performance Zinc Ion Capacitors. *ACS Appl. Energy Mater.* **2023**, *6*, 6700–6711. [CrossRef]
73. Zhao, X.; Hong, R.; Lu, R.; Chen, Y.; Yang, X. Sustainable Synthesis of Hierarchically Porous Hollow Carbon Spheres for Enhanced Zinc-Ion Hybrid Supercapacitors. *ACS Appl. Energy Mater.* **2024**, *7*, 931–940. [CrossRef]
74. Du, X.; Ma, Y.; Xie, X.; Jiang, H.; Sun, X.; Yang, X.; Zhang, Y.; Hou, C.; Du, W. Preparation of Two-Dimensional Porous Nitrogen-oxygen Co-Doped Recycled Yeast Cell Wall Derived-carbon Matrix for High-Performance Zinc Ion Supercapacitors. *J. Energy Storage* **2024**, *82*, 110428. [CrossRef]
75. Wang, J.; Nie, P.; Ding, B.; Dong, S.; Hao, X.; Dou, H.; Zhang, X. Biomass Derived Carbon for Energy Storage Devices. *J. Mater. Chem. A* **2017**, *5*, 2411–2428. [CrossRef]
76. Yan, M.; Qin, Y.; Wang, L.; Song, M.; Han, D.; Jin, Q.; Zhao, S.; Zhao, M.; Li, Z.; Wang, X.; et al. Recent Advances in Biomass-Derived Carbon Materials for Sodium-Ion Energy Storage Devices. *Nanomaterials* **2022**, *12*, 930. [CrossRef]
77. Cheng, P.; Gao, S.; Zang, P.; Yang, X.; Bai, Y.; Xu, H.; Liu, Z.; Lei, Z. Hierarchically Porous Carbon by Activation of Shiitake Mushroom for Capacitive Energy Storage. *Carbon* **2015**, *93*, 315–324. [CrossRef]
78. Liu, H.; Liu, H.; Di, S.; Zhai, B.; Li, L.; Wang, S. Advantageous Tubular Structure of Biomass-Derived Carbon for High-Performance Sodium Storage. *ACS Appl. Energy Mater.* **2021**, *4*, 4955–4965. [CrossRef]
79. Niu, J.; Guan, J.; Dou, M.; Zhang, Z.; Kong, J.; Wang, F. Sustainable Synthesis of Biomass-Derived Carbon Electrodes with Hybrid Energy-Storage Behaviors for Use in High-Performance Na-Ion Capacitors. *ACS Appl. Energy Mater.* **2020**, *3*, 2478–2489. [CrossRef]
80. Guo, Y.; Liu, W.; Wu, R.; Sun, L.; Zhang, Y.; Cui, Y.; Liu, S.; Wang, H.; Shan, B. Marine-Biomass-Derived Porous Carbon Sheets with a Tunable N-Doping Content for Superior Sodium-Ion Storage. *ACS Appl. Mater. Interfaces* **2018**, *10*, 38376–38386. [CrossRef]
81. Liu, H.; Liu, X.; Wang, H.; Zheng, Y.; Zhang, H.; Shi, J.; Liu, W.; Huang, M.; Kan, J.; Zhao, X.; et al. High-Performance Sodium-Ion Capacitor Constructed by Well-Matched Dual-Carbon Electrodes from a Single Biomass. *ACS Sustain. Chem. Eng.* **2019**, *7*, 12188–12199. [CrossRef]
82. Zhang, L.; Sun, J.; Zhao, H.; Sun, Y.; Dai, L.; Yao, F.; Fu, Y.; Zhu, J. Gas Expansion-Assisted Preparation of 3D Porous Carbon Nanosheet for High-Performance Sodium Ion Hybrid Capacitor. *J. Power Sources* **2020**, *475*, 228679. [CrossRef]
83. Zhang, N.; Liu, Q.; Chen, W.; Wan, M.; Li, X.; Wang, L.; Xue, L.; Zhang, W. High Capacity Hard Carbon Derived from Lotus Stem as Anode for Sodium Ion Batteries. *J. Power Sources* **2018**, *378*, 331–337. [CrossRef]
84. Gao, Z.; Zhang, Y.; Song, N.; Li, X. Biomass-Derived Renewable Carbon Materials for Electrochemical Energy Storage. *Mater. Res. Lett.* **2017**, *5*, 69–88. [CrossRef]
85. Jiang, B.; Cao, L.; Yuan, Q.; Ma, Z.; Huang, Z.; Lin, Z.; Zhang, P. Biomass Straw-Derived Porous Carbon Synthesized for Supercapacitor by Ball Milling. *Materials* **2022**, *15*, 924. [CrossRef] [PubMed]
86. Feng, S.; Xing, L.; Li, K.; Wang, H.; An, Q.; Zhou, L.; Mai, L. Solvent-Free Synthesis of Polymer Spheres and the Activation to Porous Carbon Spheres for Advanced Aluminum-Ion Hybrid Capacitors. *Small Methods* **2023**, *7*, 2300150. [CrossRef]

87. Seon, E.; Jang, S.; Raj, M.R.; Tak, Y.; Lee, G. Ultrahigh Energy Density and Long-Life Cyclic Stability of Surface-Treated Aluminum-Ion Supercapacitors. *ACS Appl. Mater. Interfaces* **2022**, *14*, 45059–45072. [CrossRef] [PubMed]
88. Lei, H.; Tu, J.; Tian, D.; Jiao, S. A Nitrogen-Doped Graphene Cathode for High-Capacitance Aluminum-Ion Hybrid Supercapacitors. *New J. Chem.* **2018**, *42*, 15684–15691. [CrossRef]
89. Sun, W.; Xing, L.; Zhang, B.; Shi, W.; Ren, J.; Zhang, X.; Xiong, F.; An, Q. Ultra-High-Performance Aluminum-Based Hybrid Supercapacitors Prepared for Nitrogen-Doped Micro-Mesoporous Carbon Sphere. *J. Power Sources* **2023**, *571*, 233052. [CrossRef]
90. Zhang, K.; Liu, M.; Si, M.; Wang, Z.; Zhuo, S.; Chai, L.; Shi, Y. Polyhydroxyalkanoate-Modified Bacterium Regulates Biomass Structure and Promotes Synthesis of Carbon Materials for High-Performance Supercapacitors. *ChemSusChem* **2019**, *12*, 1732–1742. [CrossRef]
91. Feng, T.; Wang, S.; Hua, Y.; Zhou, P.; Liu, G.; Ji, K.; Lin, Z.; Shi, S.; Jiang, X.; Zhang, R. Synthesis of Biomass-Derived N,O-Codoped Hierarchical Porous Carbon with Large Surface Area for High-Performance Supercapacitor. *J. Energy Storage* **2021**, *44*, 103286. [CrossRef]
92. Shah, S.S.; Aziz, M.A.; Yamani, Z.H. Recent Progress in Carbonaceous and Redox-Active Nanoarchitectures for Hybrid Supercapacitors: Performance Evaluation, Challenges, and Future Prospects. *Chem. Rec.* **2022**, *22*, e202200018. [CrossRef] [PubMed]
93. Sudhan, N.; Subramani, K.; Karnan, M.; Ilayaraja, N.; Sathish, M. Biomass-Derived Activated Porous Carbon from Rice Straw for a High-Energy Symmetric Supercapacitor in Aqueous and Non-Aqueous Electrolytes. *Energy Fuels* **2017**, *31*, 977–985. [CrossRef]
94. Yin, Y.; Liu, Q.; Zhao, Y.; Chen, T.; Wang, J.; Gui, L.; Lu, C. Recent Progress and Future Directions of Biomass-Derived Hierarchical Porous Carbon: Designing, Preparation, and Supercapacitor Applications. *Energy Fuels* **2023**, *37*, 3523–3554. [CrossRef]
95. Zhang, W.; Yin, J.; Wang, C.; Zhao, L.; Jian, W.; Lu, K.; Lin, H.; Qiu, X.; Alshareef, H.N. Lignin Derived Porous Carbons: Synthesis Methods and Supercapacitor Applications. *Small Methods* **2021**, *5*, 2100896. [CrossRef]
96. Cui, Y.; Liu, W.; Feng, W.; Zhang, Y.; Du, Y.; Liu, S.; Wang, H.; Chen, M.; Zhou, J. Controlled Design of Well-Dispersed Ultrathin MoS 2 Nanosheets inside Hollow Carbon Skeleton: Toward Fast Potassium Storage by Constructing Spacious "Houses" for K Ions. *Adv. Funct. Mater.* **2020**, *30*, 1908755. [CrossRef]
97. Baskar, A.V.; Singh, G.; Ruban, A.M.; Davidraj, J.M.; Bahadur, R.; Sooriyakumar, P.; Kumar, P.; Karakoti, A.; Yi, J.; Vinu, A. Recent Progress in Synthesis and Application of Biomass-Based Hybrid Electrodes for Rechargeable Batteries. *Adv. Funct. Mater.* **2023**, *33*, 2208349. [CrossRef]
98. Pham, H.D.; Mahale, K.; Hoang, T.M.L.; Mundree, S.G.; Gomez-Romero, P.; Dubal, D.P. Dual Carbon Potassium-Ion Capacitors: Biomass-Derived Graphene-like Carbon Nanosheet Cathodes. *ACS Appl. Mater. Interfaces* **2020**, *12*, 48518–48525. [CrossRef]
99. Chen, M.; Liu, W.; Du, Y.; Cui, Y.; Feng, W.; Zhou, J.; Gao, X.; Wang, T.; Liu, S.; Jin, Y. "Plains–Hills": A New Model to Design Biomass-Derived Carbon Electrode Materials for High-Performance Potassium Ion Hybrid Supercapacitors. *ACS Sustain. Chem. Eng.* **2021**, *9*, 3931–3941. [CrossRef]
100. Huang, F.; Liu, W.; Wang, Q.; Wang, F.; Yao, Q.; Yan, D.; Xu, H.; Xia, B.Y.; Deng, J. Natural N/O-Doped Hard Carbon for High Performance K-Ion Hybrid Capacitors. *Electrochim. Acta* **2020**, *354*, 136701. [CrossRef]
101. Rajkumar, P.; Thirumal, V.; Radhika, G.; Gnanamuthu, R.M.; Subadevi, R.; Sivakumar, M.; Yoo, K.; Kim, J. Eco-Friendly Production of Carbon Electrode from Biomass for High Performance Lithium and Zinc Ion Capacitors with Hybrid Energy Storage Characteristics. *Mater. Lett.* **2024**, *354*, 135320. [CrossRef]
102. Tian, Z.; Yang, C.; Zhang, C.; Han, X.; Han, J.; Liu, K.; He, S.; Duan, G.; Jian, S.; Hu, J.; et al. In-Situ Activation of Resorcinol-Furfural Resin Derived Hierarchical Porous Carbon for Supercapacitors and Zinc-Ion Hybrid Capacitors. *J. Energy Storage* **2024**, *85*, 111130. [CrossRef]
103. Chen, G.; Hu, Z.; Pan, Z.; Wang, D. Design of Honeycomb-like Hierarchically Porous Carbons with Engineered Mesoporosity for Aqueous Zinc-Ion Hybrid Supercapacitors Applications. *J. Energy Storage* **2021**, *38*, 102534. [CrossRef]
104. Zhao, L.; Jian, W.; Zhang, X.; Wen, F.; Zhu, J.; Huang, S.; Yin, J.; Lu, K.; Zhou, M.; Zhang, W.; et al. Multi-Scale Self-Templating Synthesis Strategy of Lignin-Derived Hierarchical Porous Carbons toward High-Performance Zinc Ion Hybrid Supercapacitors. *J. Energy Storage* **2022**, *53*, 105095. [CrossRef]
105. Yu, H.; Chen, X.; Zhou, J.; Wang, H. Tannin-Derived Ordered Mesoporous Carbon Cathode for Zn-Ion Hybrid Supercapacitor with Remarkable Energy Density. *Ind. Eng. Chem. Res.* **2023**, *62*, 62–71. [CrossRef]
106. Zheng, J.; Song, Q.; Qi, Y.; Leng, H.; Zhou, W.; Li, S.; Qiu, J. N, O Co-Doped Porous Carbon Derived from Pine Needles for Zinc-Ion Hybrid Supercapacitors. *New J. Chem.* **2023**, *47*, 9692–9700. [CrossRef]
107. Gupta, H.; Dahiya, Y.; Rathore, H.K.; Awasthi, K.; Kumar, M.; Sarkar, D. Energy-Dense Zinc Ion Hybrid Supercapacitors with S, N Dual-Doped Porous Carbon Nanocube Based Cathodes. *ACS Appl. Mater. Interfaces* **2023**, *15*, 42685–42696. [CrossRef] [PubMed]
108. Sielicki, K.; Maślana, K.; Mijowska, E. Fallen Autumn Leaves—The Source of Highly Porous Carbon for Zn-Ion Hybrid Supercapacitors. *Diam. Relat. Mater.* **2024**, *144*, 111021. [CrossRef]
109. Wei, F.; Wei, Y.; Wang, J.; Han, M.; Lv, Y. N, P Dual Doped Foamy-like Carbons with Abundant Defect Sites for Zinc Ion Hybrid Capacitors. *Chem. Eng. J.* **2022**, *450*, 137919. [CrossRef]
110. Wang, H.; Chen, X.; Zhang, J.; Yuan, Z.; Ye, P.; Shen, J.; Zhong, Y.; Hu, Y. Unveiling the Cooperative Roles of Pyrrolic-N and Carboxyl Groups in Biomass-Derived Hierarchical Porous Carbon Nanosheets for High Energy-Power Zn-Ion Hybrid Supercapacitors. *Appl. Surf. Sci.* **2022**, *598*, 153819. [CrossRef]
111. Zhu, Y.; Chen, M.; Li, Q.; Yuan, C.; Wang, C. A Porous Biomass-Derived Anode for High-Performance Sodium-Ion Batteries. *Carbon* **2018**, *129*, 695–701. [CrossRef]

112. Muruganantham, R.; Wang, F.-M.; Liu, W.-R. A Green Route N, S-Doped Hard Carbon Derived from Fruit-Peel Biomass Waste as an Anode Material for Rechargeable Sodium-Ion Storage Applications. *Electrochim. Acta* **2022**, *424*, 140573. [CrossRef]
113. Chen, J.; Zhou, X.; Mei, C.; Xu, J.; Zhou, S.; Wong, C.-P. Evaluating Biomass-Derived Hierarchically Porous Carbon as the Positive Electrode Material for Hybrid Na-Ion Capacitors. *J. Power Sources* **2017**, *342*, 48–55. [CrossRef]
114. Ragul, S.; Sujithkrishnan, E.; Elumalai, P. Inherent Heteroatom-Enriched Amorphous Carbon as High-Performance Electrode for Sodium-Ion Battery and Sodium-Ion Ultracapacitor. *Energy Fuels* **2022**, *36*, 15221–15233. [CrossRef]
115. Vadivazhagan, M.; Parameswaran, P.; Mani, U.; Nallathamby, K. Waste-Driven Bio-Carbon Electrode Material for Na-Ion Storage Applications. *ACS Sustain. Chem. Eng.* **2018**, *6*, 13915–13923. [CrossRef]
116. Zou, X.; Dong, C.; Jin, Y.; Wang, D.; Li, L.; Wu, S.; Xu, Z.; Chen, Y.; Li, Z.; Yang, H. Engineering of N, P Co-Doped Hierarchical Porous Carbon from Sugarcane Bagasse for High-Performance Supercapacitors and Sodium Ion Batteries. *Colloids Surf. A Physicochem. Eng. Asp.* **2023**, *672*, 131715. [CrossRef]
117. Wu, C.; Xu, Q.; Ning, H.; Zhao, Y.; Guo, S.; Sun, X.; Wang, Y.; Hu, H.; Wu, M. Petroleum Pitch Derived Carbon as Both Cathode and Anode Materials for Advanced Potassium-Ion Hybrid Capacitors. *Carbon* **2022**, *196*, 727–735. [CrossRef]
118. Yang, M.; Kong, Q.; Feng, W.; Yao, W. N/O Double-Doped Biomass Hard Carbon Material Realizes Fast and Stable Potassium Ion Storage. *Carbon* **2021**, *176*, 71–82. [CrossRef]
119. Cao, W.; Zhang, E.; Wang, J.; Liu, Z.; Ge, J.; Yu, X.; Yang, H.; Lu, B. Potato Derived Biomass Porous Carbon as Anode for Potassium Ion Batteries. *Electrochim. Acta* **2019**, *293*, 364–370. [CrossRef]
120. Wang, P.; Gong, Z.; Ye, K.; Gao, Y.; Zhu, K.; Yan, J.; Wang, G.; Cao, D. N-Rich Biomass Carbon Derived from Hemp as a Full Carbon-Based Potassium Ion Hybrid Capacitor Anode. *Appl. Surf. Sci.* **2021**, *553*, 149569. [CrossRef]
121. Gao, C.; Wang, Q.; Luo, S.; Wang, Z.; Zhang, Y.; Liu, Y.; Hao, A.; Guo, R. High Performance Potassium-Ion Battery Anode Based on Biomorphic N-Doped Carbon Derived from Walnut Septum. *J. Power Sources* **2019**, *415*, 165–171. [CrossRef]
122. Gong, Y.; Fu, D.; Fan, M.; Zheng, S.; Xue, Y. Multilayer Core–Sheath Wires with Radially Aligned N-Doped Carbon Nanohole Arrays for Boosting Energy Storage in Zinc-Ion Hybrid Supercapacitors. *ACS Appl. Mater. Interfaces* **2024**, *16*, 4793–4802. [CrossRef] [PubMed]
123. Sun, Z.; Jiao, X.; Chu, S.; Li, Z. Low-cost Porous Carbon Materials Prepared from Peanut Red Peels for Novel Zinc-ion Hybrid Capacitors. *ChemistrySelect* **2023**, *8*, e202304071. [CrossRef]
124. Chen, J.; Yang, B.; Liu, B.; Lang, J.; Yan, X. Recent Advances in Anode Materials for Sodium- and Potassium-Ion Hybrid Capacitors. *Curr. Opin. Electrochem.* **2019**, *18*, 1–8. [CrossRef]
125. Zeng, Z.; Mao, Y.; Hu, Z.; Chen, K.; Huang, Q.; Song, Y.; Wu, Z.; Zhang, P.; Chen, T.; Guo, X. Research Progress and Commercialization of Biologically Derived Hard Carbon Anode Materials for Sodium-Ion Batteries. *Ind. Eng. Chem. Res.* **2023**, *62*, 15343–15359. [CrossRef]
126. Yoo, H.D.; Han, S.-D.; Bayliss, R.D.; Gewirth, A.A.; Genorio, B.; Rajput, N.N.; Persson, K.A.; Burrell, A.K.; Cabana, J. "Rocking-Chair"—Type Metal Hybrid Supercapacitors. *ACS Appl. Mater. Interfaces* **2016**, *8*, 30853–30862. [CrossRef]
127. Han, J.; Mariani, A.; Zarrabeitia, M.; Jusys, Z.; Behm, R.J.; Varzi, A.; Passerini, S. Zinc-Ion Hybrid Supercapacitors Employing Acetate-Based Water-in-Salt Electrolytes. *Small* **2022**, *18*, 2201563. [CrossRef]
128. Varanasi, S.R.; Bhatia, S.K. Optimal Electrode Mass Ratio in Nanoporous Carbon Electrochemical Supercapacitors. *J. Phys. Chem. C* **2016**, *120*, 27925–27933. [CrossRef]
129. Cho, S.; Lim, J.; Seo, Y. Flexible Solid Supercapacitors of Novel Nanostructured Electrodes Outperform Most Supercapacitors. *ACS Omega* **2022**, *7*, 37825–37833. [CrossRef]
130. Cherusseri, J.; Sambath Kumar, K.; Choudhary, N.; Nagaiah, N.; Jung, Y.; Roy, T.; Thomas, J. Novel Mesoporous Electrode Materials for Symmetric, Asymmetric and Hybrid Supercapacitors. *Nanotechnology* **2019**, *30*, 202001. [CrossRef]
131. Wang, J.; Kaskel, S. KOH Activation of Carbon-Based Materials for Energy Storage. *J. Mater. Chem.* **2012**, *22*, 23710–23725. [CrossRef]
132. Ahmad, S.; Hussain, A.; Mian, S.A.; Rahman, G.; Ali, S.; Jang, J. Sensing and Conversion of Carbon Dioxide to Methanol Using Ag-Decorated Zinc Oxide Nanocatalyst. *Mater. Adv.* **2024**, *5*, 1119–1129. [CrossRef]
133. Wang, Z.; Zhang, M.; Ma, W.; Zhu, J.; Song, W. Application of Carbon Materials in Aqueous Zinc Ion Energy Storage Devices. *Small* **2021**, *17*, 2100219. [CrossRef]
134. Javed, M.S.; Asim, S.; Najam, T.; Khalid, M.; Hussain, I.; Ahmad, A.; Assiri, M.A.; Han, W. Recent Progress in Flexible Zn-ion Hybrid Supercapacitors: Fundamentals, Fabrication Designs, and Applications. *Carbon Energy* **2023**, *5*, e271. [CrossRef]
135. Yu, J.; Wang, L.; Peng, J.; Jia, X.; Zhou, L.; Yang, N.; Li, L. O-Doped Porous Carbon Derived from Biomass Waste for High-Performance Zinc-Ion Hybrid Supercapacitors. *Ionics (Kiel)* **2021**, *27*, 4495–4505. [CrossRef]
136. Amiri, A.; Swart, E.N.; Polycarpou, A.A. Recent Advances in Electrochemically-Efficient Materials for Zinc-Ion Hybrid Supercapacitors. *Renew. Sustain. Energy Rev.* **2021**, *148*, 111288. [CrossRef]
137. Hou, X.; Ren, P.; Tian, W.; Xue, R.; Fan, B.; Ren, F.; Jin, Y. High-Performance Zn-Ion Hybrid Supercapacitors Based on Biomass-Derived Hierarchical Porous Carbon through Template-Activated Bifunctional Induced and Ice-Crystal Assisted Strategy. *J. Power Sources* **2024**, *603*, 234408. [CrossRef]
138. Zhou, S.; Li, C.; Gao, G.; Fan, H.; Hu, X. Bio-Based Resins with Tannin and Hydroxymethylfurfural Derived High-Yield Carbon for Zn-Ion Hybrid Supercapacitors. *J. Clean. Prod.* **2023**, *389*, 136067. [CrossRef]

139. Zhang, W.; Yin, J.; Jian, W.; Wu, Y.; Chen, L.; Sun, M.; Schwingenschlögl, U.; Qiu, X.; Alshareef, H.N. Supermolecule-Mediated Defect Engineering of Porous Carbons for Zinc-Ion Hybrid Capacitors. *Nano Energy* **2022**, *103*, 107827. [CrossRef]
140. Ma, Y.; Hou, C.; Kimura, H.; Xie, X.; Jiang, H.; Sun, X.; Yang, X.; Zhang, Y.; Du, W. Recent Advances in the Application of Carbon-Based Electrode Materials for High-Performance Zinc Ion Capacitors: A Mini Review. *Adv. Compos. Hybrid Mater.* **2023**, *6*, 59. [CrossRef]
141. Li, Y.; Zhang, X.; Lu, T.; Zhang, Y.; Li, X.; Yu, D.; Zhao, G. Boosting the Capacitance of Aqueous Zinc-Ion Hybrid Capacitors by Engineering Hierarchical Porous Carbon Architecture. *Batteries* **2023**, *9*, 429. [CrossRef]
142. Tekin, B.; Topcu, Y. Novel Hemp Biomass-Derived Activated Carbon as Cathode Material for Aqueous Zinc-Ion Hybrid Supercapacitors: Synthesis, Characterization, and Electrochemical Performance. *J. Energy Storage* **2024**, *77*, 109879. [CrossRef]
143. Yao, L.; Jiang, J.; Peng, H.; Yang, H.; Liu, S.; Wen, X.; Cai, P.; Zou, Y.; Zhang, H.; Xu, F.; et al. Glutinous Rice-Derived Carbon Material for High-Performance Zinc-Ion Hybrid Supercapacitors. *J. Energy Storage* **2023**, *58*, 106378. [CrossRef]
144. Yuksel, R.; Karakehya, N. High Energy Density Biomass-Derived Activated Carbon Materials for Sustainable Energy Storage. *Carbon* **2024**, *221*, 118934. [CrossRef]
145. Jiang, G.; Chen, M.; Sun, Y.; Pan, J. Dual N-Doped Porous Carbon Derived from Pyrolytic Carbon Black and Critical PANIs Constructing High-Performance Zn Ion Hybrid Supercapacitor. *J. Energy Storage* **2023**, *63*, 106955. [CrossRef]
146. Wang, D.; Li, Z.; Guo, D.; Sun, M. Metal-Organic Framework Derived Zinc and Nitrogen Co-Doped Porous Carbon Materials for High Performance Zinc-Ion Hybrid Supercapacitors. *Electrochim. Acta* **2022**, *427*, 140854. [CrossRef]
147. Jia, D.; Shen, Z.; Zhou, W.; Li, Y.; He, J.; Jiang, L.; Wei, Y.; He, X. Ultrahigh N-Doped Carbon with Hierarchical Porous Structure Derived from Metal-Organic Framework for High-Performance Zinc Ion Hybrid Capacitors. *Chem. Eng. J.* **2024**, *485*, 149820. [CrossRef]
148. Liu, H.; Chen, W.; Peng, H.; Huang, X.; Li, S.; Jiang, L.; Zheng, M.; Xu, M.; Zhu, J. Bioinspired Design of Graphene-Based N/O Self-Doped Nanoporous Carbon from Carp Scales for Advanced Zn-Ion Hybrid Supercapacitors. *Electrochim. Acta* **2022**, *434*, 141312. [CrossRef]
149. Zhang, X.; Tian, X.; Song, Y.; Wu, J.; Yang, T.; Liu, Z. High-Performance Activated Carbon Cathodes from Green Cokes for Zn-Ion Hybrid Supercapacitors. *Fuel* **2022**, *310*, 122485. [CrossRef]
150. Wang, J.; Huang, Y.; Han, X.; Li, Z.; Zhang, S.; Zong, M. A Flexible Zinc-Ion Hybrid Supercapacitor Constructed by Porous Carbon with Controllable Structure. *Appl. Surf. Sci.* **2022**, *579*, 152247. [CrossRef]
151. Zhang, X.; Zhang, Y.; Zhang, H.; Zhang, Y.; Ma, Z.; Sun, L. Zinc-ion Hybrid Capacitor with High Energy Density Constructed by Bamboo Shavings Derived Spongy-like Porous Carbon. *ChemistrySelect* **2021**, *6*, 6937–6943. [CrossRef]
152. Liu, Y.; Tan, H.; Tan, Z.; Cheng, X. Rice Husk-Derived Carbon Materials for Aqueous Zn-Ion Hybrid Supercapacitors. *Appl. Surf. Sci.* **2023**, *608*, 155215. [CrossRef]
153. Fan, H.; Zhou, S.; Li, Q.; Gao, G.; Wang, Y.; He, F.; Hu, G.; Hu, X. Hydrogen-Bonded Frameworks Crystals-Assisted Synthesis of Flower-like Carbon Materials with Penetrable Meso/Macropores from Heavy Fraction of Bio-Oil for Zn-Ion Hybrid Supercapacitors. *J. Colloid Interface Sci.* **2021**, *600*, 681–690. [CrossRef] [PubMed]
154. Dang, Z.; Li, X.; Li, Y.; Dong, L. Heteroatom-Rich Carbon Cathodes toward High-Performance Flexible Zinc-Ion Hybrid Supercapacitors. *J. Colloid Interface Sci.* **2023**, *644*, 221–229. [CrossRef] [PubMed]
155. Song, B.; Liu, Q.; Shi, F.; Xue, T.; Yang, C.; Zang, L. Porous Carbon Derived from a By-Product of Traditional Chinese Medicine for High-Performance Aqueous Zinc-Ion Hybrid Supercapacitors. *Diam. Relat. Mater.* **2024**, *142*, 110785. [CrossRef]
156. Lou, G.; Pei, G.; Wu, Y.; Lu, Y.; Wu, Y.; Zhu, X.; Pang, Y.; Shen, Z.; Wu, Q.; Fu, S.; et al. Combustion Conversion of Wood to N, O Co-Doped 2D Carbon Nanosheets for Zinc-Ion Hybrid Supercapacitors. *Chem. Eng. J.* **2021**, *413*, 127502. [CrossRef]
157. Liu, H.; Chen, H.; Shi, K.; Zhang, F.; Xiao, S.; Huang, L.; Zhu, H. Lignin-Derived Porous Carbon for Zinc-Ion Hybrid Capacitor. *Ind. Crops Prod.* **2022**, *187*, 115519. [CrossRef]
158. Xue, B.; Liu, C.; Wang, X.; Feng, Y.; Xu, J.; Gong, F.; Xiao, R. Urea-Boosted Gas-Exfoliation Synthesis of Lignin-Derived Porous Carbon for Zinc Ion Hybrid Supercapacitors. *Chem. Eng. J.* **2024**, *480*, 147994. [CrossRef]
159. Yang, S.; Cui, Y.; Yang, G.; Zhao, S.; Wang, J.; Zhao, D.; Yang, C.; Wang, X.; Cao, B. $ZnCl_2$ Induced Hierarchical Porous Carbon for Zinc-Ion Hybrid Supercapacitors. *J. Power Sources* **2023**, *554*, 232347. [CrossRef]
160. Xue, B.; Xu, J.; Xiao, R. Ice Template-Assisting Activation Strategy to Prepare Biomass-Derived Porous Carbon Cages for High-Performance Zn-Ion Hybrid Supercapacitors. *Chem. Eng. J.* **2023**, *454*, 140192. [CrossRef]
161. Liu, H.; Li, S.; Huang, X.; Chen, W.; Xu, M.; Ren, Y.; Zhang, R.; Miao, Z.; Zhu, J. High Energy-Power Zinc-Ion Hybrid Supercapacitors Achieved by 3D Channels Enriched Biomass-Derived N/O Co-Doped 2D Arcuate Carbon Nanosheets. *Mater. Today Chem.* **2023**, *29*, 101476. [CrossRef]
162. Muruganantham, R.; Wang, F.-M.; Yuwono, R.A.; Sabugaa, M.; Liu, W.-R. Biomass Feedstock of Waste Mango-Peel-Derived Porous Hard Carbon for Sustainable High-Performance Lithium-Ion Energy Storage Devices. *Energy Fuels* **2021**, *35*, 10878–10889. [CrossRef]
163. Ding, J.; Li, Z.; Cui, K.; Boyer, S.; Karpuzov, D.; Mitlin, D. Heteroatom Enhanced Sodium Ion Capacity and Rate Capability in a Hydrogel Derived Carbon Give Record Performance in a Hybrid Ion Capacitor. *Nano Energy* **2016**, *23*, 129–137. [CrossRef]
164. Liao, K.; Wang, H.; Wang, L.; Xu, D.; Wu, M.; Wang, R.; He, B.; Gong, Y.; Hu, X. A High-Energy Sodium-Ion Capacitor Enabled by a Nitrogen/Sulfur Co-Doped Hollow Carbon Nanofiber Anode and an Activated Carbon Cathode. *Nanoscale Adv.* **2019**, *1*, 746–756. [CrossRef]

165. Hong, K.; Qie, L.; Zeng, R.; Yi, Z.; Zhang, W.; Wang, D.; Yin, W.; Wu, C.; Fan, Q.; Zhang, W.; et al. Biomass Derived Hard Carbon Used as a High Performance Anode Material for Sodium Ion Batteries. *J. Mater. Chem. A* **2014**, *2*, 12733–12738. [CrossRef]
166. Liu, L.; Sun, X.; Dong, Y.; Wang, D.; Wang, Z.; Jiang, Z.; Li, A.; Chen, X.; Song, H. N-Doped Hierarchical Porous Hollow Carbon Spheres with Multi-Cavities for High Performance Na-Ion Storage. *J. Power Sources* **2021**, *506*, 230170. [CrossRef]
167. Han, P.; Han, X.; Yao, J.; Zhang, L.; Cao, X.; Huang, C.; Cui, G. High Energy Density Sodium-Ion Capacitors through Co-Intercalation Mechanism in Diglyme-Based Electrolyte System. *J. Power Sources* **2015**, *297*, 457–463. [CrossRef]
168. Ghanem, A.S.; Ba-Shammakh, M.; Usman, M.; Khan, M.F.; Dafallah, H.; Habib, M.A.M.; Al-Maythalony, B.A. High Gas Permselectivity in ZIF-302/Polyimide Self-consistent Mixed-matrix Membrane. *J. Appl. Polym. Sci.* **2020**, *137*, 48513. [CrossRef]
169. Xiang, A.; Shi, D.; Chen, P.; Li, Z.; Tu, Q.; Liu, D.; Zhang, X.; Lu, J.; Jiang, Y.; Yang, Z.; et al. $Na_4Fe_3(PO_4)_2(P_2O_7)$@C/$Ti_3C_2T_x$ Hybrid Cathode Materials with Enhanced Performances for Sodium-Ion Batteries. *Batteries* **2024**, *10*, 121. [CrossRef]
170. Thirumal, V.; Sreekanth, T.V.M.; Yoo, K.; Kim, J. Biomass-Derived Hard Carbon and Nitrogen-Sulfur Co-Doped Graphene for High-Performance Symmetric Sodium Ion Capacitor Devices. *Energies* **2023**, *16*, 802. [CrossRef]
171. Zhu, J.; Roscow, J.; Chandrasekaran, S.; Deng, L.; Zhang, P.; He, T.; Wang, K.; Huang, L. Biomass-Derived Carbons for Sodium-Ion Batteries and Sodium-Ion Capacitors. *ChemSusChem* **2020**, *13*, 1275–1295. [CrossRef]
172. Thangavel, R.; Kannan, A.G.; Ponraj, R.; Yoon, G.; Aravindan, V.; Kim, D.-W.; Kang, K.; Yoon, W.-S.; Lee, Y.-S. Surface Enriched Graphene Hollow Spheres towards Building Ultra-High Power Sodium-Ion Capacitor with Long Durability. *Energy Storage Mater.* **2020**, *25*, 702–713. [CrossRef]
173. Panda, M.R.; Kathribail, A.R.; Modak, B.; Sau, S.; Dutta, D.P.; Mitra, S. Electrochemical Properties of Biomass-Derived Carbon and Its Composite along with $Na_2Ti_3O_7$ as Potential High-Performance Anodes for Na-Ion and Li-Ion Batteries. *Electrochim. Acta* **2021**, *392*, 139026. [CrossRef]
174. Wang, H.; Yu, W.; Shi, J.; Mao, N.; Chen, S.; Liu, W. Biomass Derived Hierarchical Porous Carbons as High-Performance Anodes for Sodium-Ion Batteries. *Electrochim. Acta* **2016**, *188*, 103–110. [CrossRef]
175. Dong, G.; Wang, H.; Liu, W.; Shi, J.; Sun, S.; Li, D.; Zhang, H.; Yang, Y.; Cui, Y. Nitrate Salt Assisted Fabrication of Highly N-Doped Carbons for High-Performance Sodium Ion Capacitors. *ACS Appl. Energy Mater.* **2018**, *1*, 5636–5645. [CrossRef]
176. Nagmani; Satpathy, B.K.; Singh, A.K.; Pradhan, D.; Puravankara, S. Utilization of Single Biomass-Derived Micro-Mesoporous Carbon for Dual-Carbon Symmetric and Hybrid Sodium-Ion Capacitors. *New J. Chem.* **2023**, *47*, 12658–12669. [CrossRef]
177. Casal, M.D.; Díez, N.; Payá, S.; Sevilla, M. Cork-Derived Carbon Sheets for High-Performance Na-Ion Capacitors. *ACS Appl. Energy Mater.* **2023**, *6*, 8120–8131. [CrossRef]
178. Liu, H.; Jia, M.; Sun, N.; Cao, B.; Chen, R.; Zhu, Q.; Wu, F.; Qiao, N.; Xu, B. Nitrogen-Rich Mesoporous Carbon as Anode Material for High-Performance Sodium-Ion Batteries. *ACS Appl. Mater. Interfaces* **2015**, *7*, 27124–27130. [CrossRef]
179. Ding, J.; Wang, H.; Li, Z.; Cui, K.; Karpuzov, D.; Tan, X.; Kohandehghan, A.; Mitlin, D. Peanut Shell Hybrid Sodium Ion Capacitor with Extreme Energy–Power Rivals Lithium Ion Capacitors. *Energy Environ. Sci.* **2015**, *8*, 941–955. [CrossRef]
180. Shi, X.; Yu, J.; Liu, Q.; Shao, L.; Cai, J.; Sun, Z. Metal-Organic-Framework-Derived Multi-Heteroatom Doped Cu1.8Se/C Composites for High-Performance Na-Ion Hybrid Capacitor. *Sustain. Mater. Technol.* **2021**, *28*, e00275. [CrossRef]
181. Yan, R.; Leus, K.; Hofmann, J.P.; Antonietti, M.; Oschatz, M. Porous Nitrogen-Doped Carbon/Carbon Nanocomposite Electrodes Enable Sodium Ion Capacitors with High Capacity and Rate Capability. *Nano Energy* **2020**, *67*, 104240. [CrossRef]
182. Zhao, G.; Yu, D.; Zhang, H.; Sun, F.; Li, J.; Zhu, L.; Sun, L.; Yu, M.; Besenbacher, F.; Sun, Y. Sulphur-Doped Carbon Nanosheets Derived from Biomass as High-Performance Anode Materials for Sodium-Ion Batteries. *Nano Energy* **2020**, *67*, 104219. [CrossRef]
183. Li, L.; Sun, M.; Xu, Z.; Wang, Z.; Liu, K.; Chen, Y.; Wang, Z.; Chen, H.; Yang, H. Hierarchical Porous Hard Carbon Derived from Rice Husks for High-Performance Sodium Ion Storage. *Colloids Surf. A Physicochem. Eng. Asp.* **2023**, *661*, 130927. [CrossRef]
184. Zhang, L.; Wang, Y.; Yang, S.; Zhao, G.; Han, L.; Li, Y.; Zhu, G. Biomass-Derived S, P, Cl Tri-Doped Porous Carbon for High-Performance Supercapacitor. *Diam. Relat. Mater.* **2022**, *126*, 109061. [CrossRef]
185. Ajuria, J.; Redondo, E.; Arnaiz, M.; Mysyk, R.; Rojo, T.; Goikolea, E. Lithium and Sodium Ion Capacitors with High Energy and Power Densities Based on Carbons from Recycled Olive Pits. *J. Power Sources* **2017**, *359*, 17–26. [CrossRef]
186. Dos Reis, G.S.; Petnikota, S.; Subramaniyam, C.M.; de Oliveira, H.P.; Larsson, S.; Thyrel, M.; Lassi, U.; García Alvarado, F. Sustainable Biomass-Derived Carbon Electrodes for Potassium and Aluminum Batteries: Conceptualizing the Key Parameters for Improved Performance. *Nanomaterials* **2023**, *13*, 765. [CrossRef]
187. Li, B.; Dai, F.; Xiao, Q.; Yang, L.; Shen, J.; Zhang, C.; Cai, M. Activated Carbon from Biomass Transfer for High-Energy Density Lithium-Ion Supercapacitors. *Adv. Energy Mater.* **2016**, *6*, 1600802. [CrossRef]
188. Xie, K.; Zhang, W.; Ren, K.; Zhu, E.; Lu, J.; Chen, J.; Yin, P.; Yang, L.; Guan, X.; Wang, G. Electrochemical Performance of Corn Waste Derived Carbon Electrodes Based on the Intrinsic Biomass Properties. *Materials* **2023**, *16*, 5022. [CrossRef]
189. Nayem, S.M.A.; Islam, S.; Shah, S.S.; Sultana, N.; Mahfoz, W.; Ahammad, A.J.S.; Aziz, A. Biomass Based S-doped Carbon for Supercapacitor Application. In *Biomass-Based Supercapacitors*; Wiley: Hoboken, NJ, USA, 2023; pp. 315–327.
190. Wei, X.; Qiu, B.; Xu, L.; Qin, Q.; Zhang, W.; Liu, Z.; Wei, F.; Lv, Y. High Performance Hierarchical Porous Carbon Derived from Waste Shrimp Shell for Supercapacitor Electrodes. *J. Energy Storage* **2023**, *62*, 106900. [CrossRef]
191. Zeng, Y.; Zhao, W.; Li, X.; Chen, X.; Song, J.; Wu, X.; Huang, Z. Biomass Waste Derived Heteroatom Doping Porous Carbon Sheets for High Performance Supercapacitor. *Mater. Today Commun.* **2023**, *36*, 106623. [CrossRef]
192. Wahed, F.; Shah, S.S.; Hayat, K.; Shah, S.K.; Aziz, M.A. Conduction Mechanisms and Thermoelectric Applications of La1-XSrxCoO_3 Nanofibers. *J. Mater. Sci.* **2022**, *57*, 8828–8844. [CrossRef]

193. Hou, J.; Cao, C.; Idrees, F.; Ma, X. Hierarchical Porous Nitrogen-Doped Carbon Nanosheets Derived from Silk for Ultrahigh-Capacity Battery Anodes and Supercapacitors. *ACS Nano* **2015**, *9*, 2556–2564. [CrossRef]
194. Shen, Y.; Zhu, Y. One-Pot Synthesis of Biomass-Derived Porous Carbons for Multipurpose Energy Applications. *J. Mater. Chem. A* **2024**, *12*, 6211–6242. [CrossRef]
195. Li, Z.; Guo, D.; Liu, Y.; Wang, H.; Wang, L. Recent Advances and Challenges in Biomass-Derived Porous Carbon Nanomaterials for Supercapacitors. *Chem. Eng. J.* **2020**, *397*, 125418. [CrossRef]
196. Qian, Y.; Wu, B.; Li, Y.; Pan, Z.; Tian, J.; Lin, N.; Qian, Y. Pressure-Dependent Self-Template Pyrolysis Modulates the Porosity and Surface Chemical Configuration of Carbon for Potassium Ion Hybrid Capacitors. *Chem. Eng. J.* **2023**, *451*, 138579. [CrossRef]
197. Zhao, L.; Sun, S.; Lin, J.; Zhong, L.; Chen, L.; Guo, J.; Yin, J.; Alshareef, H.N.; Qiu, X.; Zhang, W. Defect Engineering of Disordered Carbon Anodes with Ultra-High Heteroatom Doping Through a Supermolecule-Mediated Strategy for Potassium-Ion Hybrid Capacitors. *Nano-Micro Lett.* **2023**, *15*, 41. [CrossRef]
198. Pan, Z.; Qian, Y.; Li, Y.; Xie, X.; Lin, N.; Qian, Y. Novel Bilayer-Shelled N, O-Doped Hollow Porous Carbon Microspheres as High Performance Anode for Potassium-Ion Hybrid Capacitors. *Nano-Micro Lett.* **2023**, *15*, 151. [CrossRef]
199. Gao, Q.; Li, T.; Liu, C.; Sun, J.; Liu, Y.; Hou, L.; Yuan, C. Hierarchically Porous N-Doped Carbon Framework with Enlarged Interlayer Spacing as Dual-Carbon Electrodes for Potassium Ion Hybrid Capacitors. *Carbon Neutrality* **2023**, *2*, 18. [CrossRef]
200. Yvenat, M.-E.; Chavillon, B.; Mayousse, E.; Perdu, F.; Azaïs, P. Development of an Adequate Formation Protocol for a Non-Aqueous Potassium-Ion Hybrid Supercapacitor (KIC) through the Study of the Cell Swelling Phenomenon. *Batteries* **2022**, *8*, 135. [CrossRef]
201. Zhu, L.; Zhang, Z.; Luo, J.; Zhang, H.; Qu, Y.; Yang, Z. Self-Templated Synthesis of Hollow Hierarchical Porous Olive-like Carbon toward Universal High-Performance Alkali (Li, Na, K)-Ion Storage. *Carbon* **2021**, *174*, 317–324. [CrossRef]
202. Yang, M.; Dai, J.; He, M.; Duan, T.; Yao, W. Biomass-Derived Carbon from Ganoderma Lucidum Spore as a Promising Anode Material for Rapid Potassium-Ion Storage. *J. Colloid Interface Sci.* **2020**, *567*, 256–263. [CrossRef] [PubMed]
203. Cao, B.; Gao, S.; Ma, Y.; Zhang, D.; Guo, Z.; Du, M.; Xin, Z.; Zhou, C.; Liu, H. Biomass-Derived Carbon-Sulfur Hybrids Boosting Electrochemical Kinetics to Achieve High Potassium Storage Performance. *J. Colloid Interface Sci.* **2024**, *661*, 598–605. [CrossRef] [PubMed]
204. Lu, S.; Xiao, Q.; Yang, W.; Wang, X.; Guo, T.; Xie, Q.; Ruan, Y. Multi-Heteroatom-Doped Porous Carbon with High Surface Adsorption Energy of Potassium Derived from Biomass Waste for High-Performance Supercapacitors. *Int. J. Biol. Macromol.* **2024**, *258*, 128794. [CrossRef] [PubMed]
205. Zhong, L.; Qiu, X.; Yang, S.; Sun, S.; Chen, L.; Zhang, W. Supermolecule-Regulated Synthesis Strategy of General Biomass-Derived Highly Nitrogen-Doped Carbons toward Potassium-Ion Hybrid Capacitors with Enhanced Performances. *Energy Storage Mater.* **2023**, *61*, 102887. [CrossRef]
206. Zhu, L.; Zhang, Z.; Zhang, H.; Wang, Y.; Luo, J.; Yu, J.; Qu, Y.; Yang, Z. Tunable 2D Tremella-Derived Carbon Nanosheets with Enhanced Pseudocapacitance Behavior for Ultrafast Potassium-Ion Storage. *Sci. China Technol. Sci.* **2021**, *64*, 2047–2056. [CrossRef]
207. Hu, W.; Xiang, R.; Lin, J.; Cheng, Y.; Lu, C. Lignocellulosic Biomass-Derived Carbon Electrodes for Flexible Supercapacitors: An Overview. *Materials* **2021**, *14*, 4571. [CrossRef]
208. Xie, K.; Qin, X.; Wang, X.; Wang, Y.; Tao, H.; Wu, Q.; Yang, L.; Hu, Z. Carbon Nanocages as Supercapacitor Electrode Materials. *Adv. Mater.* **2012**, *24*, 347–352. [CrossRef]
209. Adam, A.A.; Ojur Dennis, J.; Al-Hadeethi, Y.; Mkawi, E.M.; Abubakar Abdulkadir, B.; Usman, F.; Mudassir Hassan, Y.; Wadi, I.A.; Sani, M. State of the Art and New Directions on Electrospun Lignin/Cellulose Nanofibers for Supercapacitor Application: A Systematic Literature Review. *Polymers* **2020**, *12*, 2884. [CrossRef]
210. Yang, X.; Lv, T.; Qiu, J. High Mass-Loading Biomass-Based Porous Carbon Electrodes for Supercapacitors: Review and Perspectives. *Small* **2023**, *19*, 2300336. [CrossRef]
211. Zhang, M.; Peng, L. Research Progress of Biomass-Derived Carbon for the Supercapacitors. *Mater. Res. Express* **2024**, *11*, 012004. [CrossRef]
212. Lu, Q.; Zhou, S.; Zhang, Y.; Chen, M.; Li, B.; Wei, H.; Zhang, D.; Zhang, J.; Liu, Q. Nanoporous Carbon Derived from Green Material by an Ordered Activation Method and Its High Capacitance for Energy Storage. *Nanomaterials* **2020**, *10*, 1058. [CrossRef] [PubMed]
213. Momodu, D.; Bello, A.; Oyedotun, K.; Ochai-Ejeh, F.; Dangbegnon, J.; Madito, M.; Manyala, N. Enhanced Electrochemical Response of Activated Carbon Nanostructures from Tree-Bark Biomass Waste in Polymer-Gel Active Electrolytes. *RSC Adv.* **2017**, *7*, 37286–37295. [CrossRef]
214. Zhu, Y.; Murali, S.; Stoller, M.D.; Ganesh, K.J.; Cai, W.; Ferreira, P.J.; Pirkle, A.; Wallace, R.M.; Cychosz, K.A.; Thommes, M.; et al. Carbon-Based Supercapacitors Produced by Activation of Graphene. *Science* **2011**, *332*, 1537–1541. [CrossRef] [PubMed]
215. Li, B.; Zheng, J.; Zhang, H.; Jin, L.; Yang, D.; Lv, H.; Shen, C.; Shellikeri, A.; Zheng, Y.; Gong, R.; et al. Electrode Materials, Electrolytes, and Challenges in Nonaqueous Lithium-Ion Capacitors. *Adv. Mater.* **2018**, *30*, 1705670. [CrossRef]
216. Biswal, M.; Banerjee, A.; Deo, M.; Ogale, S. From Dead Leaves to High Energy Density Supercapacitors. *Energy Environ. Sci.* **2013**, *6*, 1249–1259. [CrossRef]
217. Zhang, Q.; Yuan, M.; Liu, L.; Li, S.; Chen, X.; Liu, J.; Pang, X.; Wang, X. Study of Zinc Diffusion Based on S, N-Codoped Honeycomb Carbon Cathodes for High-Performance Zinc-Ion Capacitors. *Langmuir* **2024**, *40*, 5326–5337. [CrossRef] [PubMed]

218. Eskusson, J.; Thomberg, T.; Lust, E.; Jänes, A. Electrochemical Characteristics of Zn-Ion Hybrid Supercapacitors Based on Aqueous Solution of Different Electrolytes. *J. Electrochem. Soc.* **2022**, *169*, 020512. [CrossRef]
219. Shao, Y.; El-Kady, M.F.; Sun, J.; Li, Y.; Zhang, Q.; Zhu, M.; Wang, H.; Dunn, B.; Kaner, R.B. Design and Mechanisms of Asymmetric Supercapacitors. *Chem. Rev.* **2018**, *118*, 9233–9280. [CrossRef]
220. Simon, P.; Gogotsi, Y. Capacitive Energy Storage in Nanostructured Carbon-Electrolyte Systems. *Acc. Chem. Res.* **2013**, *46*, 1094–1103. [CrossRef]
221. Maurya, D.K.; Murugadoss, V.; Guo, Z.; Angaiah, S. Designing $Na_2Zn_2TeO_6$-Embedded 3D-Nanofibrous Poly(Vinylidenefluoride)-Co-Hexafluoropropylene-Based Nanohybrid Electrolyte via Electrospinning for Durable Sodium-Ion Capacitors. *ACS Appl. Energy Mater.* **2021**, *4*, 8475–8487. [CrossRef]
222. Dubal, D.P.; Ayyad, O.; Ruiz, V.; Gómez-Romero, P. Hybrid Energy Storage: The Merging of Battery and Supercapacitor Chemistries. *Chem. Soc. Rev.* **2015**, *44*, 1777–1790. [CrossRef] [PubMed]
223. Tomboc, G.M.; Tesfaye Gadisa, B.; Jun, M.; Chaudhari, N.K.; Kim, H.; Lee, K. Carbon Transition-metal Oxide Electrodes: Understanding the Role of Surface Engineering for High Energy Density Supercapacitors. *Chem.—Asian J.* **2020**, *15*, 1628–1647. [CrossRef] [PubMed]
224. Kim, T.; Jung, G.; Yoo, S.; Suh, K.S.; Ruoff, R.S. Activated Graphene-Based Carbons as Supercapacitor Electrodes with Macro- and Mesopores. *ACS Nano* **2013**, *7*, 6899–6905. [CrossRef] [PubMed]
225. Shaker, M.; Ng, S.; Sadeghi Ghazvini, A.A.; Javanmardi, S.; Gaho, M.A.; Jin, Z.; Ge, Q. Carbon/Graphene Quantum Dots as Electrolyte Additives for Batteries and Supercapacitors: A Review. *J. Energy Storage* **2024**, *85*, 111040. [CrossRef]
226. Liu, X.; Sun, Y.; Tong, Y.; Wang, X.; Zheng, J.; Wu, Y.; Li, H.; Niu, L.; Hou, Y. Exploration in Materials, Electrolytes and Performance towards Metal Ion (Li, Na, K, Zn and Mg)-Based Hybrid Capacitors: A Review. *Nano Energy* **2021**, *86*, 106070. [CrossRef]
227. Iqbal, M.Z.; Zakar, S.; Haider, S.S. Role of Aqueous Electrolytes on the Performance of Electrochemical Energy Storage Device. *J. Electroanal. Chem.* **2020**, *858*, 113793. [CrossRef]
228. Kumar, K.S.; Pandey, D.; Thomas, J. High Voltage Asymmetric Supercapacitors Developed by Engineering Electrode Work Functions. *ACS Energy Lett.* **2021**, *6*, 3590–3599. [CrossRef]
229. Guo, Q.; Bai, C.; Gao, C.; Chen, N.; Qu, L. Two Dimensional Silicene Nanosheets: A New Choice of Electrode Material for High-Performance Supercapacitor. *ACS Appl. Mater. Interfaces* **2022**, *14*, 39014–39021. [CrossRef] [PubMed]
230. Wan, F.; Wang, X.; Tang, C.; Jiang, C.; Wang, W.; Li, B.; Zhang, Y.; Zhu, X. Metallic 1T-MoS_2 Coupled with MXene towards Ultra-High Rate-Capabilities for Supercapacitors. *J. Mater. Chem. A* **2022**, *10*, 12258–12268. [CrossRef]
231. Hu, Z.; Song, Z.; Huang, Z.; Tao, S.; Song, B.; Cao, Z.; Hu, X.; Wu, J.; Li, F.; Deng, W.; et al. Reconstructing Hydrogen Bond Network Enables High Voltage Aqueous Zinc-Ion Supercapacitors. *Angew. Chem.* **2023**, *135*, e202309601. [CrossRef]
232. Komaba, S.; Hasegawa, T.; Dahbi, M.; Kubota, K. Potassium Intercalation into Graphite to Realize High-Voltage/High-Power Potassium-Ion Batteries and Potassium-Ion Capacitors. *Electrochem. Commun.* **2015**, *60*, 172–175. [CrossRef]
233. Zhang, F.; Dong, H.; Li, Y.; Fu, D.; Yang, L.; Shang, Y.; Li, Q.; Shao, Y.; Gang, W.; Ding, T.; et al. In Situ Metal-Oxygen-Hydrogen Modified B-Tio_2@Co_2 P-X S-Scheme Heterojunction Effectively Enhanced Charge Separation for Photo-assisted Uranium Reduction. *Adv. Sci.* **2024**, *11*, 2305439. [CrossRef] [PubMed]
234. Huang, Z.; Wang, T.; Song, H.; Li, X.; Liang, G.; Wang, D.; Yang, Q.; Chen, Z.; Ma, L.; Liu, Z.; et al. Effects of Anion Carriers on Capacitance and Self-Discharge Behaviors of Zinc Ion Capacitors. *Angew. Chem.* **2021**, *133*, 1024–1034. [CrossRef]
235. Li, X.; Chen, F.; Zhao, B.; Zhang, S.; Zheng, X.; Wang, Y.; Jin, X.; Dai, C.; Wang, J.; Xie, J.; et al. Ultrafast Synthesis of Metal-Layered Hydroxides in a Dozen Seconds for High-Performance Aqueous Zn (Micro-) Battery. *Nano-Micro Lett.* **2023**, *15*, 32. [CrossRef] [PubMed]
236. Sen, A.; Groß, A. Promising Sensitizers for Dye Sensitized Solar Cells: A Comparison of Ru(II) with Other Earth's Scarce and Abundant Metal Polypyridine Complexes. *Int. J. Quantum Chem.* **2019**, *119*, e25963. [CrossRef]
237. dos Santos Junior, G.A.; Fortunato, V.D.S.; Bastos, G.A.A.; Silva, G.G.; Ortega, P.F.R.; Lavall, R.L. High-Performance Lithium-Ion Hybrid Supercapacitors Based on Lithium Salt/Imidazolium Ionic Liquid Electrolytes and Ni-Doped $LiMn_2O_4$ Cathode Materials. *ACS Appl. Energy Mater.* **2020**, *3*, 9028–9039. [CrossRef]
238. Shah, S.S.; Aziz, M.A.; Cevik, E.; Ali, M.; Gunday, S.T.; Bozkurt, A.; Yamani, Z.H. Sulfur Nano-Confinement in Hierarchically Porous Jute Derived Activated Carbon towards High-Performance Supercapacitor: Experimental and Theoretical Insights. *J. Energy Storage* **2022**, *56*, 105944. [CrossRef]
239. Yu, J.; Jia, X.; Peng, J.; Meng, B.; Wei, Y.; Hou, X.; Zhao, J.; Yang, N.; Xie, K.; Chu, D.; et al. Synergistic Effect of Nitrogen–Sulfur Codoping on Honeycomb-like Carbon-Based High-Energy-Density Zinc-Ion Hybrid Supercapacitors. *ACS Appl. Energy Mater.* **2023**, *6*, 2728–2738. [CrossRef]
240. Qin, J.; Sari, H.M.K.; Wang, X.; Yang, H.; Zhang, J.; Li, X. Controlled Design of Metal Oxide-Based (Mn^{2+}/Nb^{5+}) Anodes for Superior Sodium-Ion Hybrid Supercapacitors: Synergistic Mechanisms of Hybrid Ion Storage. *Nano Energy* **2020**, *71*, 104594. [CrossRef]
241. Li, J.; El-Demellawi, J.K.; Sheng, G.; Björk, J.; Zeng, F.; Zhou, J.; Liao, X.; Wu, J.; Rosen, J.; Liu, X.; et al. Pseudocapacitive Heteroatom-Doped Carbon Cathode for Aluminum-Ion Batteries with Ultrahigh Reversible Stability. *Energy Environ. Mater.* **2024**, e12733. [CrossRef]
242. Chen, J.; Liu, B.; Cai, H.; Liu, S.; Yamauchi, Y.; Jun, S.C. Covalently Interlayer-Confined Organic–Inorganic Heterostructures for Aqueous Potassium Ion Supercapacitors. *Small* **2023**, *19*, 2204275. [CrossRef] [PubMed]

243. Huang, J.; Sumpter, B.G.; Meunier, V. A Universal Model for Nanoporous Carbon Supercapacitors Applicable to Diverse Pore Regimes, Carbon Materials, and Electrolytes. *Chem.—Eur. J.* **2008**, *14*, 6614–6626. [CrossRef] [PubMed]
244. Han, C.; Wang, X.; Peng, J.; Xia, Q.; Chou, S.; Cheng, G.; Huang, Z.; Li, W. Recent Progress on Two-Dimensional Carbon Materials for Emerging Post-Lithium (Na^+, K^+, Zn^{2+}) Hybrid Supercapacitors. *Polymers* **2021**, *13*, 2137. [CrossRef] [PubMed]
245. Hao, P.; Zhao, Z.; Tian, J.; Li, H.; Sang, Y.; Yu, G.; Cai, H.; Liu, H.; Wong, C.P.; Umar, A. Hierarchical Porous Carbon Aerogel Derived from Bagasse for High Performance Supercapacitor Electrode. *Nanoscale* **2014**, *6*, 12120–12129. [CrossRef] [PubMed]
246. Xu, X.; Tang, J.; Qian, H.; Hou, S.; Bando, Y.; Hossain, M.S.A.; Pan, L.; Yamauchi, Y. Three-Dimensional Networked Metal–Organic Frameworks with Conductive Polypyrrole Tubes for Flexible Supercapacitors. *ACS Appl. Mater. Interfaces* **2017**, *9*, 38737–38744. [CrossRef]
247. Khan, A.; Sadiq, S.; Khan, I.; Humayun, M.; Jiyuan, G.; Usman, M.; Khan, A.; Khan, S.; Alanazi, A.F.; Bououdina, M. Preparation of Visible-Light Active MOFs-Perovskites (ZIF-67/$LaFeO_3$) Nanocatalysts for Exceptional CO_2 Conversion, Organic Pollutants and Antibiotics Degradation. *Heliyon* **2024**, *10*, e27378. [CrossRef] [PubMed]
248. Sadiq, S.; Khan, S.; Khan, I.; Khan, A.; Humayun, M.; Wu, P.; Usman, M.; Khan, A.; Alanazi, A.F.; Bououdina, M. A Critical Review on Metal-Organic Frameworks (MOFs) Based Nanomaterials for Biomedical Applications: Designing, Recent Trends, Challenges, and Prospects. *Heliyon* **2024**, *10*, e25521. [CrossRef] [PubMed]
249. Yuksel, R.; Yarar Kaplan, B.; Bicer, E.; Yurum, A.; Alkan Gursel, S.; Unalan, H.E. All-Carbon Hybrids for High Performance Supercapacitors. *Int. J. Energy Res.* **2018**, *42*, 3575–3587. [CrossRef]
250. Deng, Y.; Wang, X.; Wang, Z.; Wang, X.; Li, Z.; Wang, L.; Zhou, C.; Chen, D.; Luo, Y. Yolk–Shell Structured Nickel Cobalt Sulfide and Carbon Nanotube Composite for High-Performance Hybrid Supercapacitors. *Energy Fuels* **2021**, *35*, 5342–5351. [CrossRef]

batteries

MDPI

Article

Anion Intercalation/De-Intercalation Mechanism Enabling High Energy and Power Densities of Lithium-Ion Capacitors

Yang Zhang, Junquan Lao and Ping Xiao *

Department of Materials, Henry Royce Institute, University of Manchester, Manchester M13 9PL, UK; yang.zhang-26@postgrad.manchester.ac.uk (Y.Z.); junquan.lao@manchester.ac.uk (J.L.)
* Correspondence: p.xiao@manchester.ac.uk

Abstract: The growing demands for electrochemical energy storage systems is driving the exploration of novel devices, with lithium-ion capacitors (LICs) emerging as a promising strategy to achieve both high energy density and fast charge capability. However, the low capacitance of commercial activated carbon (AC) cathode based on anion absorption/desorption limits LIC applications. Herein, commercial graphite is proposed as the cathode to construct an innovative AC (−)//graphite (+) system. The graphite cathode functions as anion hosting, allowing reversible intercalation/de-intercalation of anions into/from its interlayers. The as-designed AC (−)//graphite (+) full cell achieves stable cycling with 90.6% capacity retention after 200 cycles at 0.1 A g^{-1} and a prolonged lifespan with 87.5% capacity retention after 5000 cycles at 0.5 A g^{-1} with the upper cut-off voltage of 5.0 V, yielding a high average Coulombic efficiency (CE) of 99.3%. Moreover, the full cell exhibits a high energy density (>200 Wh kg^{-1}) and power density of 7.7 kW kg^{-1} (calculated based on active mass in both electrodes). These performances exceed most LICs based on anions absorption/desorption on the surface of AC cathodes. This work explores an effective electrode revolution with the assistance of anion intercalation/de-intercalation chemistry for developing novel LICs with high energy and power densities.

Keywords: lithium-ion capacitor; hybrid dual-ion system; graphite cathode; anion intercalation/de-intercalation; energy density

1. Introduction

Addressing the increasing global energy demands requires advancements in electrochemical devices, given their wide-ranging applications from portable electronics to electric transportation to the integration of renewable energy sources for powering the electrical grid [1,2]. Lithium-ion batteries (LIBs) have emerged as the predominant energy storage and conversion device owing to its high energy density (>300 Wh kg^{-1}), yet the lower power density remains a notable challenge [3,4]. In comparison to extensively researched LIBs, supercapacitors (electrical double-layer capacitors) typically consist of a symmetrical structure composed of high surface area carbon materials such as activated carbon (AC). This configuration facilitates rapid energy storage through ion adsorption/desorption on the surface of the AC, consequently achieving high power density and a long cycling life, but their practical application is limited by lower energy density (<10 Wh kg^{-1}) [5]. Therefore, there is a strong appeal for innovative electrochemical energy storage and conversion devices that integrate the merits of both batteries (high energy density) and supercapacitors (high power density).

Lithium-ion capacitors (LICs), utilising a battery-type anode (graphite or $Li_4Ti_5O_{12}$) driven by Faraday reactions (mainly intercalation/deintercalation behaviour) and a capacitor-type cathode operating on non-Faraday reactions (absorption/desorption behaviour), have been developed to achieve enhanced energy density (20–146 Wh kg^{-1}) [6]. However, the energy density of commercial AC cathode-based LICs via anion absorption/desorption behaviour during charging/discharging processes is constrained by the limitation of low capacitance and low stable working voltage (<4.6 V versus Li/Li^+) [7–9]. Building upon this foundation, considerable efforts have been made, encompassing the exploration of high-capacitance electrode materials, surpassing commercial AC to construct high-voltage devices [10,11].

Graphite, consisting of stacked layers of graphene held together by weak van der Waals force, has the capability to electrochemically accommodate charge carriers. The graphite can form acceptor-type graphite interphase compounds (GICs) by stabilising additional holes in its bonding π-band, allowing for the intercalation/de-intercalation of anions. Typically, the charge-transfer reaction with the intercalation/de-intercalation of anions is exhibited as Equation (1) [12].

$$C + xA^{-} \leftrightarrow A_xC + xe^{-} \quad (1)$$

The graphite cathode demonstrates favourable operation at a high upper cut-off voltage of 5.3 V, yielding a medium voltage of around 4.6 V, which is significant compared to most cathode materials. The high working voltage ensures a potentially high energy density. Typically, graphite cathode is coupled with graphite anode to construct dual-graphite batteries. However, the dual-graphite batteries face challenges, including a limited rate capability and, particularly, a short cycle life. These issues are primarily due to the insufficient reaction kinetics of the graphite anode, which experiences inevitable capacity decay with increasing the current density up to 300 mA g^{-1} [13]. In contrast, the graphite cathode is expected to exhibit superior rate performance, capable of sustaining a stable capacity close to its theoretical value of approximately 100 mAh g^{-1} even at a high current density up to 2 A g^{-1} [14]. Remarkably, the high operating voltage and excellent rate performance of the graphite cathode contribute to maximising the advantages of LICs incorporating an AC anode, especially in terms of rapid and stable charge storage. This suggests promising potential for developing innovative high-energy and high-power LICs based on anion intercalation chemistry. However, this novel construction is still in the early stages of research, with limited studies primarily focusing on the low voltage range (0–4.5 V), demonstrating low energy density (<100 Wh kg^{-1}) [15–17]. Therefore, optimising and matching the graphite cathode and electrolyte to unlock performance at higher voltages and achieve a prolonged cyclic life is of great significance. Furthermore, there is substantial room for improvement in the fundamental and technical principles of this system, as well as in the exploration of its working mechanisms.

Herein, we propose the construction of novel LICs featuring an AC anode and graphite cathode in a standard concentration electrolyte (1M $LiPF_6$ EMC), employing the combination of cation absorption/desorption and anion intercalation/de-intercalation chemistry. The as-prepared AC (−)//graphite (+) full cell is operated at a high voltage range of 2.0–5.0 V, with a medium voltage above 4.2 V at the current density of 0.1 A g^{-1}. As expected, excellent cyclic stability (90.6% capacity retention after 200 cycles at 0.1 A g^{-1}) and long-term cycle life (87.5% capacity retention after 5000 cycles at 0.5 A g^{-1} with an average CE of 99.3%) can be achieved in the AC (−)//graphite (+) full cell. Notably, a high energy density of 204 Wh kg^{-1} (calculated on the active mass of both electrodes) is achieved. This exploration sheds light on the construction of novel LICs based on graphite cathode and demonstrates the potential to achieve electrochemical systems with high-performance energy storage capability through the assistance of anion intercalation chemistry.

2. Materials and Methods

2.1. Materials

Electrolyte: 1M $LiPF_6$ EC/EMC (1:1, vol.) and 1M $LiPF_6$ EMC were produced from Merck Life Science UK Limited. SLP50, SP1, and SFG6L graphite and Super P carbon black were sourced from IMERYS Graphite & Carbon (Switzerland). Lithium metal foil with a diameter of 16 mm and a thickness of 0.025 cm, along with polypropylene (PP) separator film, were obtained from Xiamen TOB New Energy Technology Co., Ltd (China). Sodium alginate (NaAl) was a product of MP Biomedicals. Commercial Ketjen Black (KB) was purchased from Shandong Gelon Lib Co., Limited (China). The glass fibre membrane separator (MG 160) was acquired from Fisher Scientific UK Ltd. Commercial TF-B520 activated carbon was supplied by MTI Corporation (UK). All chemicals were used directly without undergoing any additional purification processes.

2.2. *Experimental Procedures*

2.2.1. Graphite and Activated Carbon Electrode Preparation

For preparation of the graphite and activated carbon electrode, SLP50 graphite, Ketjen Black, and sodium alginate (NaAl) binder were first mixed by deionised water using a speed mixer (SpeedMixer DAC 600.2, Artisan technology group, USA) in a mass ratio of 8:1:1. The homogeneous slurry was then cast onto carbon-coated Al foil as the current collector, and the electrode was dried in a vacuum oven at 80 °C overnight. The electrode discs of 12 mm in diameter with a mass loading of 1.8–2.2 mg were cut. The TF-B520 activated carbon electrode was prepared using CMC binder and carbon black (Super P) in a mass ratio of 7:1.5:1.5. The CMC binder was first dissolved by deionised water, and then, activated carbon and carbon black were added to form a homogeneous slurry. The paste was coated on the Cu foil, and the desired electrode (diameter of 15 mm) was obtained after drying at 80 °C overnight. The mass ratio between graphite and the activated carbon electrode was 2:1 in this work.

2.2.2. Cell Assembly and Electrochemical Testing of the Li//AC, Li//Graphite Half Cell, and AC (−)//Graphite (+) Cells

Cell assembly: The CR2032-type coin cells were used to assemble cells, while AC or graphite with the electrode and lithium metal (diameter 16 mm) was served as the counter/reference electrode, and the polypropylene separator film or glass fibre (diameter 18 mm) was used as the separator. The electrolytes for Li//AC and Li//graphite half cells were 1M $LiPF_6$ EC/EMC (1:1, vol.) and 1M $LiPF_6$ EMC (~120 μL), respectively. For the assembly of AC (−)//graphite (+) cells, the AC anode was first discharged/charged for 5 cycles in Li//AC half cells within 0.01–3.0 V at 100 mA g^{-1}, then discharged to 0.01 V to the pre-lithiate. In the meantime, the graphite cathode was pre-cycled for 5 charge/discharge processes in Li//graphite half cells within 3.0–5.0 V at 100 mA g^{-1}, then stopped at the upper cut-off voltage of 5.0 V. The coin cell configuration is displayed in Figure S1 (Supplementary Materials). The pre-prepared AC anode and graphite cathode were then assembled in a coin cell with a glass fibre separator, and the amount of electrolyte was ~150 μL. The assembly of coin cells was conducted in the glovebox with $O_2 < 0.5$ ppm and $H_2O < 0.5$ ppm.

Electrochemical performance measurements: The galvanostatic charge/discharge test was performed with the NEWARE test system (BTS9000, Shenzhen, China), and the voltage ranges for the Li//AC, Li//graphite, and AC (−)//graphite (+) cells were 0.01–3.0 V, 3.0–5.0 V, and 2.0–5.0 V, respectively. Cyclic voltammetry (CV) was measured with the electrochemical workstation (VersaSTAT 4 Potentiostat Galvanostat, Ametek scientific instruments, USA).

2.2.3. Materials Characterisations

The morphologies of pristine graphite and activated carbon were examined by scanning electron microscope (SEM; TESCAN MIRA3 SC + OI EDS). The transmission electron microscope (TEM) images were acquired by Talos™ F200X TEM. X-ray diffraction (XRD) was conducted with an X-ray diffractometer (Powder X-ray Diffraction Phillips X'Pert) using Cu-Kα radiation at a wavelength of 0.15418 nm (V = 30 kV, I = 25 mA). Raman measurement was carried out by a Renishaw inVia Reflex Raman microscope. The surface area was analysed by acquiring nitrogen absorption/desorption isotherms using the Brunauer–Emmett–Teller (BET) method on a Surface Area Analyzer—Quadrasorb Evo.

3. Results and Discussion

3.1. Half Cell Investigations

3.1.1. Li//Graphite Half Cell via Anion Intercalation/De-Intercalation

Figure 1a shows the XRD pattern of SLP50 graphite, in which a strong peak located at 26.35° is observed. This reflection can be associated with the crystallographic (002) plane of graphite. The Raman spectrum of SLP50 graphite is exhibited in Figure 1b, appearing in three peaks positioned at 1335, 1580, and 2686 cm^{-1}, which correspond to the D band, G band, and 2D band, respectively. The G band is the specific reflection for sp^2 carbon networks, while the D band is the defect-induced Raman feature. The ratio of intensity I_D/I_G for the D and G bands is normally used for predicting the defect quantity and degree of graphitisation. The I_D/I_G for SLP50 graphite was calculated to be 0.256, suggesting few defects in the well-ordered graphitic structures. Figure 1c is the N_2 adsorption/desorption isotherms, revealing the BET surface area for SLP50 graphite to be 2.6 m^2/g. This relatively small surface suggests limited porosity or specific surface features. The pore size distribution is exhibited in Figure S2 (Supplementary Materials), indicating the presence of a small quantity of pores. The SEM technique was further employed to gain a more comprehensive understanding of the surface feature and morphology of SLP50 graphite. The SEM images are depicted in Figure 1d–f, and flake-shaped graphitic particles with a thickness of ~2 μm are observed, while the particle size distribution is mainly between the range of ~20 to 50 μm. Furthermore, the high-resolution TEM images are listed in Figure 1g,h, revealing detailed insights into the crystalline structure. The lattice distance for the (002) plane of SLP50 graphite was analysed to be ~0.338–0.346 nm, offers a suitable interlayer spacing that can effectively accommodate the intercalation of anions.

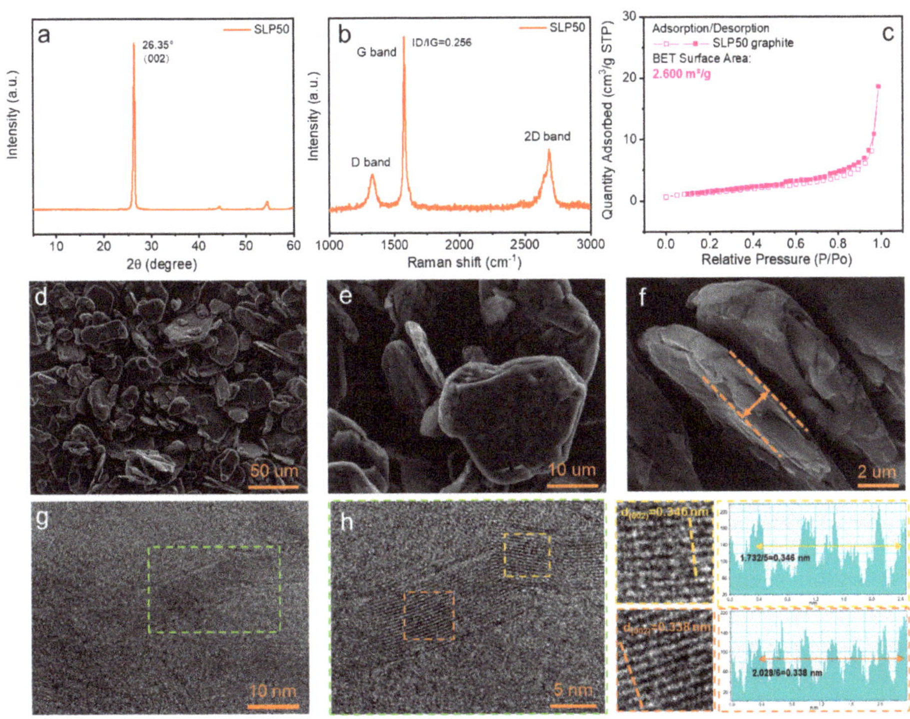

Figure 1. Characterisation of SLP50 graphite: (**a**) XRD; (**b**) Raman; (**c**) N_2 adsorption/desorption isotherm; (**d–f**) SEM images; (**g**,**h**) high-resolution TEM images.

Figure 2 displays the testing results of electrochemical PF_6^- anion intercalation into SLP50 graphite in the Li//graphite half cells based on 1M $LiPF_6$ EMC electrolyte. The PF_6^- anion intercalation/de-intercalation process was first analysed at a constant current density of 100 mA g^{-1} in the voltage range of 3.0–5.0 V. The galvanostatic charge/discharge curves for the initial three cycles are exhibited in Figure 2a. The charge platforms detected in the curves are associated with the "staging" process, which are identified as four potential regions of 4.44–4.51 V (stage I), 4.51–4.74 V (stage II), 4.74–4.92 V (stage III), and 4.92–5.0 V (stage IV), reflecting that the PF_6^- anions intercalate into graphite interlayers to form graphite intercalation compounds (GICs) [18]. Meanwhile, the anion de-intercalation platforms are observed in the discharge curves, revealing a reversible PF_6^- anion intercalation/de-intercalation process. Figure 2b shows the relevant capacitance differential (dQ/dV) curves, while these specific peaks correspond to the charge/discharge plateaus in galvanostatic charge/discharge curves. The onset voltage for anion intake is around 4.44 V, and subsequent peaks confirm the occurrence of intercalation/de-intercalation processes at certain potentials. Figure 2c lists the evaluation of the specific capacity and Coulombic efficiency (CE) within 100 cycles. Importantly, a high discharge capacity of ~120 mAh g^{-1} via this anion intercalation/de-intercalation process was achieved and was extremely stable after 100 repeated charge/discharge processes. The initial CE was 40% and increased from cycle to cycle until it was stable at ~90%, and the small CE was due to the electrolyte degradation and self-discharge reactions at the high upper cut-off voltage. Figure 2d exhibits the specific capacity of Li//graphite half cells at a high current density of 500 mA g^{-1} in the voltage range of 3.0–5.0 V. The discharge capacity decreases slightly to ~100 mAh g^{-1}, and the capacity retention after 2000 cycles is 82%, indicating the great long-term cyclic stability. This excellent cyclic capability of SLP50 graphite in a standard concentration electrolyte (1M $LiPF_6$ EMC) is comparable to other reported graphite cathodes, even those using high-concentration electrolytes or ionic liquid electrolytes [19,20]. Figure 2e concludes the medium discharge voltage of the Li//graphite half cells, and the inset displays charge/discharge curves from the 100th to 2000th cycles. Notably, the medium discharge voltage was stable at ~4.5 V after 2000 cycles, indicating excellent cyclic reversibility and stability. A slight decrease in the charge/discharge plateau voltage during the cycles further demonstrates less polarisation in the cell. Figure 2f presents the charge/discharge capacities of the Li//graphite half cells at different current densities. At the current densities of 0.1, 0.2, 0.4, 0.8, and 1.0 A g^{-1}, the discharge capacities of the cell were around 100 mAh g^{-1}, respectively, indicating great applicability with the current increasing. Figure S3 (Supplementary Materials) shows the cyclic voltammetry curves of the Li//graphite half cells at different scan rates, where the primary oxidation/reduction peaks are observed and retained well when the scan rate changes, revealing good reversibility of the electrode reaction. In addition, the capacity decayed to 75 mAh g^{-1} and then fell to 70 mAh g^{-1} when the current density increased to 1.5 and 2.0 A g^{-1}. This decrease in capacity under high current conditions is a consequence of larger concentration polarisation, hindering the electrode reactions and the effective capacity. Figure 2g shows the charge/discharge curves of the cell at different current densities. Correspondingly, the voltage hypothesis phenomenon was observed in the charge/discharge curves as the current densities increased, suggesting there was a noticeable deviation in the voltage behaviour. Notably, the discharge capacity was recovered to 100 mAh g^{-1} when the current density was reversed to 1.0 A g^{-1}, revealing a good reversibility of the Li//graphite cell. Moreover, the medium voltage and onset voltage for anion intercalation into graphite interlayers at different current densities are compared in Figure 2h. It is observed that the medium voltage was reversibly resumed to 4.51 V when the current density returned from 2 A g^{-1} to 0.1 A g^{-1}, suggesting good reversibility of the cell. The onset voltage for anion intercalation kept increasing when the current density was increased, which most likely contributed to the higher activation barrier resulting from the sluggish kinetic movement of PF_6^- at the large current condition [21]. These results demonstrate the effective intercalation and de-intercalation of PF_6^- anions in SLP50 graphite, suggesting the potential for the construction of a high-performance full-cell system.

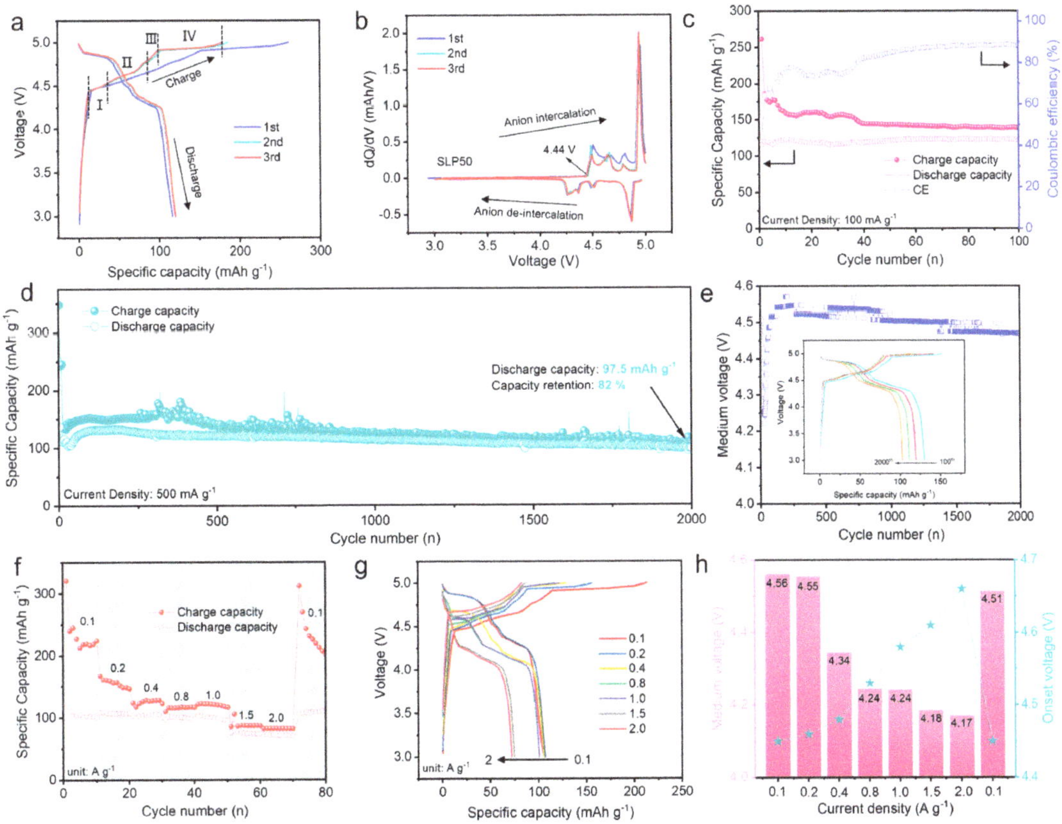

Figure 2. Electrochemical performances of Li//graphite half cells in 1M $LiPF_6$ EMC electrolytes. (**a**) Galvanostatic charge/discharge curves for the initial three cycles and (**b**) relevant initial three capacitance differential (dQ/dV) curves; (**c**) charge/discharge capacities and Coulombic efficiency at the current density of 100 mA g^{-1} within 100 cycles; (**d**) charge/discharge capacities at the current density of 500 mA g^{-1} within 2000 cycles; (**e**) medium voltage at different cycles, and the inset is the charge/discharge curves at the 100th–2000th cycles; (**f**) charge discharge capacities of the cells under various current densities from 0.1 to 2.0 A g^{-1}; (**g**) relevant charge/discharge curves; (**h**) medium and onset voltage comparisons under different current densities.

3.1.2. Li//AC Half Cells via Li^+ Adsorption/Desorption

The crystallinity of TF-B520 activated carbon (AC) was conducted using XRD, and the pattern is exhibited in Figure 3a. The absence of a diffraction peak for the reflection of the graphite (002) plane is attributed to the lack of a well-defined long-range order in the arrangement of the carbon atom [22]. The broad and weak peak around 43° is observed in the spectrum, corresponding to the diffraction of the (100) plane, while the low intensity of the diffraction peak demonstrates the amorphous structures [23]. Figure 3b displays the Raman spectra of AC, while the D band (1345.7 cm^{-1}) and G band (1598.5 cm^{-1}) are clearly observed. The ID/IG was calculated as 0.93, indicating lower fractions of the graphitic region and the disordered nature of the carbon matrix. According to the N_2 adsorption/desorption curves in Figure 3c, the BET surface area value for AC was measured to be 911.75 m^2/g, which is much larger than that of SLP50 graphite. The pore size distribution, shown in Figure S4 (Supplementary Materials), highlights a prominent microporous structure with a pore size ranging from 0.5 to 0.8 nm. A larger surface area implies more adsorption sites, thus conferring a higher adsorption capacity. This facilitates rapid adsorption and desorption of ions on the AC electrode surface. Figure 3d,e show the SEM images of the AC, and the size is mainly distributed in the range of 5–10 µm. Figure 3f illustrates the TEM image of the AC, revealing disordered microstructures characteristic of amorphous carbon.

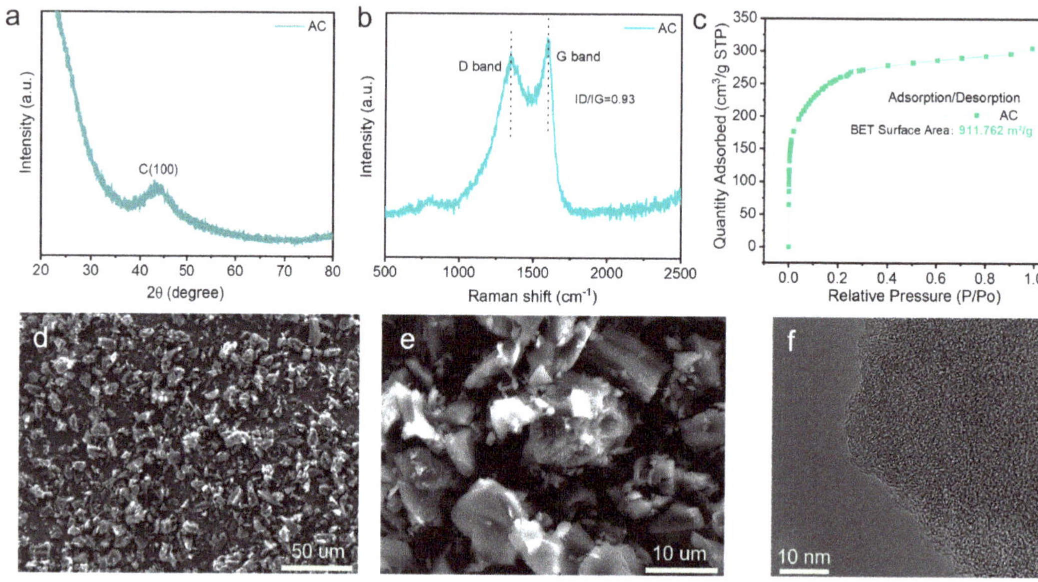

Figure 3. Characterisation of TF-B520 activated carbon: (**a**) XRD; (**b**) Raman; (**c**) N_2 adsorption/desorption isotherm; (**d**,**e**) SEM images; (**f**) high-resolution TEM image.

The electrochemical performances of the AC electrode were investigated in Li//AC half cells with the electrolyte of 1M $LiPF_6$ EC/EMC (vol. 1:1). Figure 4a displays the initial five galvanostatic charge/discharge curves for the AC electrode at the current density of 100 mA g^{-1} between the voltage range of 0.01–3.0 V. The first discharge capacity was found to be 1200 mAh g^{-1}, which reduced to ~300 mAh g^{-1} in the second cycle and was stable for five cycles. The long plateau in the first cycle is attributed to the formation of a solid electrolyte interphase (SEI) film on the AC electrode surface. Figure 4b illustrates the specific charge/discharge capacities and CE of the cell at 100 mA g^{-1} in different cycles. The discharge capacity was kept at ~260 mAh g^{-1} after 200 repeated charge/discharge processes, indicating an excellent cyclic stability. Figure 4c shows the charge/discharge capacities and CE of the cell at a large current density of 500 mA g^{-1} after the initial 10 formation cycles at 100 mA g^{-1}. The capacity was maintained above 200 mAh g^{-1} after 1000 cycles, yielding a CE near 99.7%. Notably, a good cyclic capability was confirmed by the overlapped charge/discharge curves of the cells at different cycles from the 100th to 1000th (Figure 4d). CV analysis of the Li//AC cell was first conducted at the scan rate of 0.1 mV s^{-1}, and the initial four cycle CV curves are exhibited in Figure 4e. It is obvious that the typical broad peak between 0.5 and 1.0 V is observed for each cycle, probably due to a formation of the SEI layer on the AC electrode surface. The strong current drop below the voltage of 0.5 V is associated with the reaction between Li^+ and the AC electrode. Furthermore, the CV analysis at various scan rates (0.2, 0.5, 0.8, 1.0, and 2.0 mV s^{-1}) was carried out to investigate the electrochemical dynamic behaviour of the electrode, and the curves are depicted in Figure 4f. As the scan rate increases, the peak intensity of the electrode reaction rises, resulting from an augmented electrochemical reaction rate. Notably, according to previous reports, the electrochemical storage behaviour can be specified by the equations as follows [24–26]:

$$i = av^b \tag{2}$$

$$\log (i) = b\log(v) + \log a \tag{3}$$

$$i = k_1 v + k_2 v^{1/2} \tag{4}$$

$$i/v^{1/2} = k_1 v^{1/2} + k_2 \tag{5}$$

where i is the current, v is the scan rate, and a and b are the variable parameters. The b-value, determined from the slop of log(i) versus log(v) plot, approaches 0.5 when diffusion-controlled behaviours are predominant. Conversely, a b-value close to 1.0 signifies the dominance of pseudocapacitive contributions. Figure 4g shows the fitted relationship between log(i) and log(v) of the AC electrode at different voltages. The corresponding b-values when discharged to 1.0 and 0.8 V and charged to 0.8 and 1.0 V are calculated to be 0.92, 0.86, 0.96, and 1.0, respectively. The b-value approaching 1 indicates the surface-induced capacitive behaviour of the AC electrode, revealing efficient electrode reaction kinetics. Based on Equations (4) and (5), the current (i) is expressed as the sum of k_1v and $k_2v^{1/2}$ at specific voltages corresponding to the capacitive behaviour and diffusion-controlled behaviour, respectively. The value of k_1 can be determined by plotting the curve with $i/v^{1/2}$ versus $v^{1/2}$, and the capacitive contribution is the value of k_1v. Figure 4h quantitatively analyses the surface-dominated contribution, revealing that 65.45% of the whole capacity belongs to the non-diffusion controlled capacity at the scan rate of 2 mV s^{-1}. Clearly, it is found that the diffusion-controlled behaviours mainly contribute to the whole discharge capacity in the reduction range between 0.75 and 0.01 V, suggesting a Li^+ intake process occurring on the electrode at low voltage. This behaviour is similar to that of previously reported electrode materials [27,28]. Moreover, as shown in Figure 4i, the percentage for capacitive contributions of the AC electrode exhibits an increasing trend when the scan rate rises, disclosing the more dominated capacitive behaviour at high rates.

3.2. AC (−)//Graphite (+) Full-Cell Investigations

3.2.1. Cell Configuration

The full-cell configuration of AC (−)//graphite (+) is achieved by employing an AC anode and graphite cathode. Scheme 1 illustrates a comparison of the three capacitors mentioned above (supercapacitor, conventional LICs with an AC cathode, and LICs with a graphite cathode). In these novel LICs with a graphite cathode, PF_6^- anions are forced to intercalate into graphite interlayers during the charging process, while Li^+ ions are absorbed onto the AC surface. Conversely, during the discharging process, PF_6^- anions and Li^+ ions are released and return to the electrolyte. Here, a pre-lithiation process is necessary for the AC anode to enhance the availability of Li^+ ions for charge storage and facilitate the effective formation of SEI film. The charge/discharge curves for the pre-lithiation process of the AC anode are displayed in Figure S5a (Supplementary Materials). Additionally, a pre-cycle treatment is applied to the graphite cathode to activate the anion intercalation pathways, with the charge/discharge curves of the Li//graphite half cells shown in Figure S5b (Supplementary Materials). The AC and graphite electrodes are fully discharged and fully charged to 0.01 V and 5 V, respectively, allowing for Li^+ absorption and PF_6^- anions intercalation, before being used to assemble the full cell.

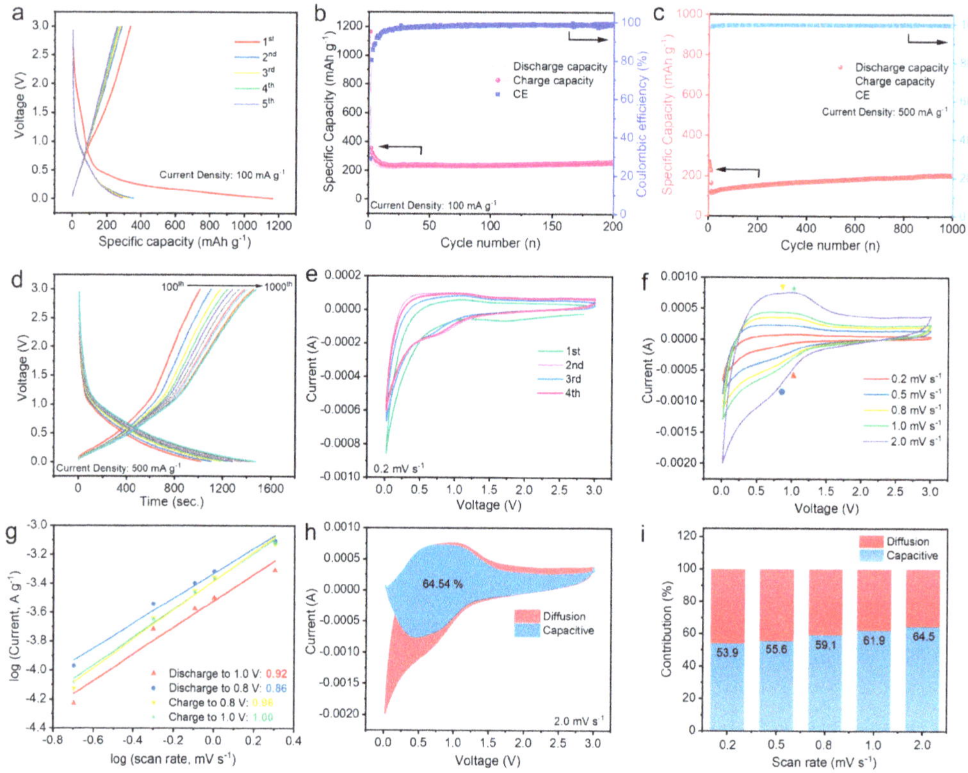

Figure 4. Electrochemical performances of Li//AC half cells in 1M $LiPF_6$ EC/EMC (vol. 1:1) electrolytes. (**a**) Galvanostatic charge/discharge curves for the initial five cycles, and (**b**) charge/discharge capacities and Coulombic efficiency at the current density of 100 mA g^{-1} within 200 cycles; (**c**) charge/discharge capacities at the current density of 500 mA g^{-1} within 1000 cycles, and (**d**) relevant charge/discharge curves at different cycles; (**e**) initial four CV curves at the scan rate of 0.1 mV s^{-1}; (**f**) CV curves at various scan rates of 0.2, 0.5, 0.8, 1.0, and 2.0 mV s^{-1}; (**g**) the linear relation between log(i) and log(v); (**h**) pseudocapacitive contributions (blue) and diffusion-controlled contributions (pink) at the scan rate of 2.0 mV s^{-1}; (**i**) ratios of capacitive and diffusion charge storage to the total charge storage at different scan rates.

3.2.2. Electrochemical Performances

The electrochemical performance of AC (−)//graphite (+) full cells was first investigated by the galvanostatic charge/discharge test. Figure 5a displays the initial five charge/discharge curves of the cell at the current density of 100 mA g^{-1} with the voltage range between 2.0 and 5.0 V. It is clear that the voltage plateaus corresponding to the anion intercalation behaviour are observed. The charge capacity of the first cycle reaches 90 mAh g^{-1}, but the first discharge capacity is lower than 50 mAh g^{-1}, yielding an initial CE of 50%. The lower CE in the first cycle is probably associated with the parasitic reactions, such as the electrolyte decomposition and self-discharge behaviour. Notably, the subsequent curves are overlapped well, which demonstrates the excellent cyclic reversibility and stability. Correspondingly, the initial five dQ/dV curves are exhibited in Figure 5b, while the peaks are attributed to the plateaus in the charge/discharge curves. The onset voltage for anion intercalation in the first cycle is 3.95 V, and the voltage decreases to 3.81 V in the second cycle and 3.76 V for subsequent cycles. The shift in the onset voltage indicates that the first uptake of anions into graphite interlayers needs to overcome the activation barrier to open the interlayer gaps [21,29,30]. This onset shift is also observed in CV curves of the cell at the scan rate of 0.2 mV s^{-1} (Figure S6a, Supplementary Materials). Meanwhile, the specific peaks observed in the CV curves are associated with the anion intercalation/de-intercalation behaviours. Moreover, Figure S6b (Supplementary Materials) exhibits CV curves at various scan rates from 0.2 to 2.0 mV s^{-1}, and good reversibility of the cell is demonstrated. Figure 5c exhibits the charge/discharge capacities and relevant CE of the AC (−)//graphite (+) cell at the current density of 100 mA g^{-1} between the voltage range of 2.0–5.0 V within the 200 cycles. The discharge capacity of 45 mAh g^{-1} is delivered after 200 repeated charge/discharge processes, yielding a CE of 91.4%. The capacity retention of the cell is higher than 90%, demonstrating an excellent cyclic stability. The cyclic stability of the AC (−)//graphite (+) cell was further explored through the galvanostatic charge/discharge cycle at a high current density of 500 mA g^{-1}. The charge/discharge capacities and CE for 5000 cycles are exhibited in Figure 5d. Obviously, the capacity was still stable after 5000 cycles, yielding a capacity retention of 87.5% with an average CE of 99.3%. Furthermore, as shown in Figure 5e, the reversible discharge capacities of 48.7, 46.1, 41.0, 33.5, 31.4, 27.4, and 26.3 mAh g^{-1} are maintained at various current densities of 0.1, 0.2, 0.5, 0.8, 1.0, 1.5, and 2.0 A g^{-1}, respectively. Particularly, the capacity of the AC (−)//graphite (+) cell recovered to 47.6 mAh g^{-1} when the current density was switched back to the initial 0.1 A g^{-1}, exhibiting excellent rate capability and cyclic reversibility. Figure S7a (Supplementary Materials) displays the charge/discharge curves of the cell cycled at various current densities. Obviously, the discharge time reduces with the increasing current density. Figure S7b (Supplementary Materials) illustrates the medium voltage of the cell at various current densities, while the medium voltage drops from 4.2 V to 3.8 V when the current density increases from 0.1 to 2.0 A g^{-1}. This result demonstrates a high operating voltage in the AC (−)//graphite (+) full-cell system.

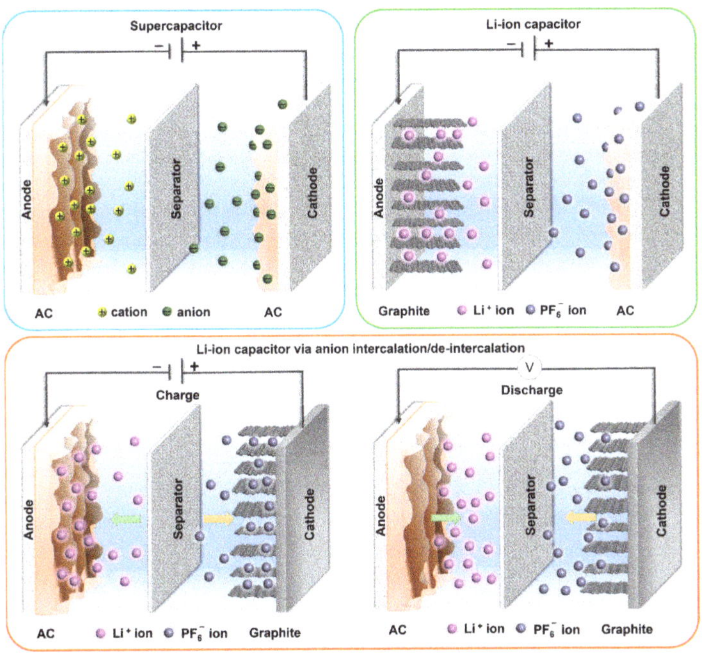

Scheme 1. The comparison among the conventional supercapacitor (AC//AC symmetric configuration), Li-ion capacitor (graphite (−)//AC (+) asymmetric configuration), and Li-ion capacitor via anion intercalation (AC (−)//graphite (+) configuration).

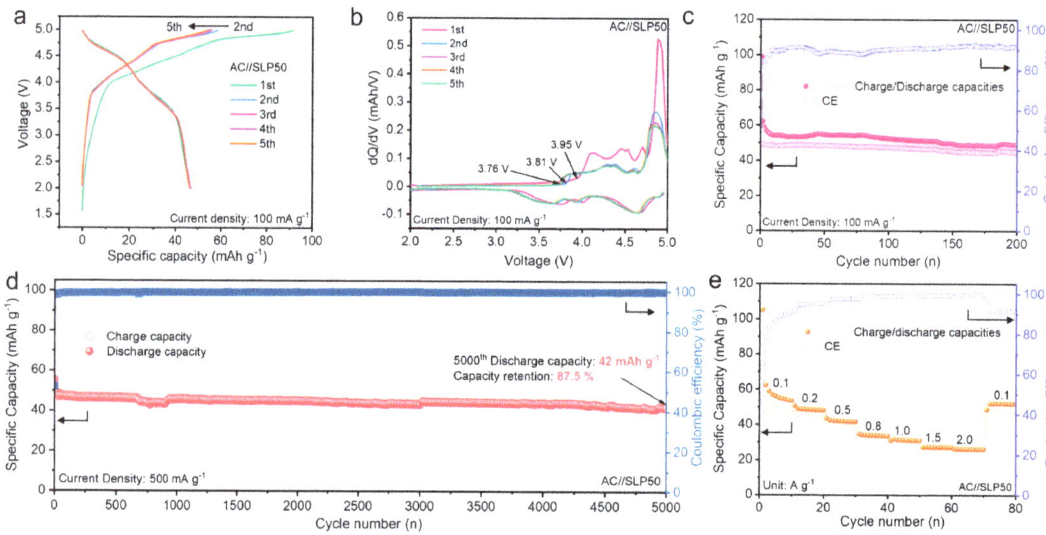

Figure 5. Electrochemical performances of the AC (−)//graphite (+) full cell in 1M $LiPF_6$ EMC electrolytes. (**a**) Galvanostatic charge/discharge curves for the initial five cycles and (**b**) relevant dQ/dV curves; (**c**) charge/discharge capacities and CE at the current density of 100 mA g^{-1} within 200 cycles; (**d**) charge/discharge capacities and CE at current density of 500 mA g^{-1} within 5000 cycles; (**e**) rate performances at different current densities of 0.1, 0.2, 0.5, 0.8, 1.0, 1.5, and 2.0 A g^{-1}.

Accordingly, the energy and power density of the AC (−)//graphite (+) full-cell system were calculated using the equations provided in the Supplementary Materials, with detailed data displayed in Table S1 (Supplementary Materials). The comparisons, shown in the Ragone plot in Figure 6a and Table S2 (Supplementary Materials), are displayed with the previously fabricated LICs. A high energy density of 204 Wh kg^{-1} (calculated based on the active mass of both electrodes) is achieved at a power density of 418 W kg^{-1}. At a high power density of 7.7 kW kg^{-1}, the AC (−)//SLP50 graphite (+) full cell delivers an energy density of 100 Wh kg^{-1}, exceeding most LICs based on anion absorption/desorption on the surface of the AC cathode [9,31–40]. Additionally, the electrochemical performance of this AC (−)//SLP50 graphite (+) full-cell system is compared to reported dual-graphite battery systems [41–50]. The comparison of the cycle life, capacity retention, and average Coulombic efficiency is presented in Figure 6b, with the detailed data summarised in Table S3. Notably, the novel AC (−)//SLP50 graphite (+) full-cell system in this work demonstrates a prolonged cycle life, along with cyclic efficiency and capacity retention comparable to those of dual-graphite batteries.

Moreover, different types of graphite, including SP1 and SFG6L, were applied as cathodes in the AC (−)//graphite (+) full-cell system. The detailed physical properties of these two graphite types are displayed in Figure S8 (Supplementary Materials). Figure S9 (Supplementary Materials) exhibits the electrochemical performance of the full cell, demonstrating a reversible and stable cyclic capability for both AC (−)//SP1 graphite (+) and AC (−)//SFG6L graphite (+) configurations. In summary, it can be concluded that various graphite cathode materials are suitable for this novel AC (−)//graphite (+) system in the electrolyte of 1 M $LiPF_6$ EMC operating at a high voltage range. This demonstrates the compatibility of graphite cathodes with anion intercalation behaviour for the improvement in the overall performances of AC-based energy storage devices. Notably, the electrochemical performance varies with different types of graphite. The potential influence of graphite properties on the overall performance of this AC (−)//graphite (+) full-cell system will be explored in our future work.

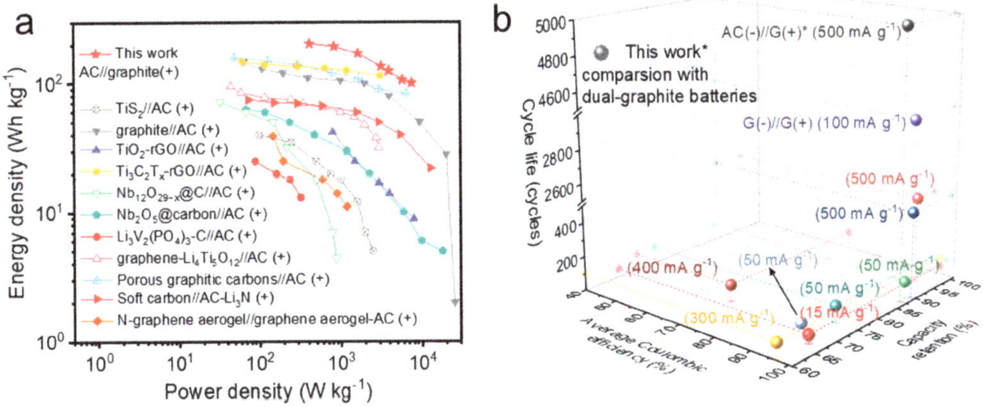

Figure 6. Comparison of performances of the AC (−)//graphite (+) full cell in this work with reported works. (**a**) Ragone plot of the AC (−)//graphite (+) full cell in this work, and comparison with previously fabricated LICs in the literature. (**b**) Cyclic performances of the AC (−)//graphite (+) full cell in this work compared with reported dual-graphite batteries.

4. Conclusions

In summary, we demonstrated graphite as the cathode material for a novel AC (−)//graphite (+) device with the assistance of anion intercalation/de-intercalation chemistry. It has been found that the graphite cathode storing an anion charge via intercalation behaviour can deliver a higher capacitance than the commercial AC based on surface adsorption. The as-designed AC (−)//graphite (+) device showed a high energy density (>200 Wh kg^{-1}) with a well-maintained power density (>7 kW kg^{-1}), surpassing most reported supercapacitors and LICs. The AC (−)//graphite (+) device was also evaluated by cyclic stability and rate capability, which achieved a high capacity retention of 87.5% after 5000 cycles at the current density of 0.5 A g^{-1}, yielding an average CE of 99.3%. Moreover, various commercial graphite with different properties were successfully applied in this AC (−)//graphite (+) system, demonstrating good compatibility. Considering the abundance, wide availability, and low cost of the commercial graphite, this work is expected to inspire the development of novel electrochemical devices with a prolonged cyclic life, high energy density, and fast energy conversion rates.

Supplementary Materials: The following supporting information can be downloaded at https://www.mdpi.com/article/10.3390/batteries10090296/s1: Figure S1. The coin cell construction for (a) Li//AC or Li//graphite half cells and (b) AC (-)//graphite (+) full cell. Figure S2. Pore size distribution of SLP50 graphite. Figure S3. The cyclic voltammetry curves of the Li//graphite half cells at different scan rates. Figure S4. Pore size distribution of TF-B520 activated carbon. Figure S5. (a) The charge/discharge curves for (a) the pre-lithiation process of Li//AC half cells, and (b) the pre-cycle treatment of Li//graphite half cells. Figure S6. (a) Initial three CV curves of the AC (-)//SLP50 (+) full cell at the scan rate of 0.1 mV s^{-1}, and (b) CV curves of the cell at different scan rates. Figure S7. (a) Charge/discharge curves of the AC (-)//SLP50 (+) full cell cycled at various current densities. (b) Medium voltage of the cell at various current densities. Table S1. Energy density and power density of the AC (-)//SLP50 graphite (+) full cell. Table S2. Comparison of the energy and power densities of the AC (-)//SLP50 graphite (+) full cell with reported AC cathode-based LICs. Table S3. Comparison of the electrochemical performance of the AC (-)//SLP50 graphite (+) full cell with reported dual-graphite batteries (G(-)//G(+)). Figure S8. (a,b) SEM images and (c) Raman spectrum of SP1 graphite, and (d,e) SEM images and (f) Raman spectrum of SFG6L graphite. Figure S9. (a) Charge/discharge capacities and CE at current density of 100 mA g^{-1} within 500 cycles, (b) rate performance at various current densities, and (c) charge/discharge capacities and CE at current density of 500 mA g^{-1} within 3000 cycles of AC (-)//SP1 (+) and AC (-)//SFG6L (+) full cells.

Author Contributions: Y.Z.: Conceptualisation, Methodology, Validation, Formal analysis, Investigation, Data curation, and Writing—original draft. J.L.: Formal analysis and Investigation. P.X.: Resources, Writing—review and editing, Supervision, and Project administration. All authors have read and agreed to the published version of the manuscript.

Funding: This work was supported by the China Scholarship Council/University of Manchester award. Ping Xiao acknowledges financial support from the Royal Academy of Engineering and Rolls-Royce for appointment of the Rolls-Royce/Royal Academy of Engineering Research Chair in Advanced Coating Technology.

Data Availability Statement: Data will be made available on request.

Conflicts of Interest: The authors declare that they have no known competing financial interests or personal relationships that could have appeared to influence the work reported in this paper.

References

1. Zou, K.; Deng, W.; Silvester, D.S.; Zou, G.; Hou, H.; Banks, C.E.; Li, L.; Hu, J.; Ji, X. Carbonyl Chemistry for Advanced Electrochemical Energy Storage Systems. *ACS Nano* **2024**, *18*, 19950–20000. [CrossRef]
2. Liu, J.; Wang, J.; Xu, C.; Jiang, H.; Li, C.; Zhang, L.; Lin, J.; Shen, Z.X. Advanced Energy Storage Devices: Basic Principles, Analytical Methods, and Rational Materials Design. *Adv. Sci.* **2018**, *5*, 1700322. [CrossRef] [PubMed]
3. Zou, K.; Jiang, M.; Ning, T.; Tan, L.; Zheng, J.; Wang, J.; Ji, X.; Li, L. Thermodynamics-directed bulk/grain-boundary engineering for superior electrochemical durability of Ni-rich cathode. *J. Energy Chem.* **2024**, *97*, 321–331. [CrossRef]
4. Khan, F.M.N.U.; Rasul, M.G.; Sayem, A.S.M.; Mandal, N. Maximizing energy density of lithium-ion batteries for electric vehicles: A critical review. *Energy Rep.* **2023**, *9*, 11–21. [CrossRef]
5. Divya, M.L.; Natarajan, S.; Lee, Y.S.; Aravindan, V. Biomass-Derived Carbon: A Value-Added Journey Towards Constructing High-Energy Supercapacitors in an Asymmetric Fashion. *ChemSusChem* **2019**, *12*, 4353–4382. [CrossRef]
6. Wang, G.; Oswald, S.; Löffler, M.; Müllen, K.; Feng, X. Beyond Activated Carbon: Graphite-Cathode-Derived Li-Ion Pseudocapacitors with High Energy and High Power Densities. *Adv. Mater.* **2019**, *31*, 1807712. [CrossRef]
7. Zhang, S.; Li, C.; Zhang, X.; Sun, X.; Wang, K.; Ma, Y. High Performance Lithium-Ion Hybrid Capacitors Employing Fe_3O_4-Graphene Composite Anode and Activated Carbon Cathode. *ACS Appl. Mater. Interfaces* **2017**, *9*, 17136–17144. [CrossRef]
8. Eleri, O.E.; Lou, F.; Yu, Z. Lithium-Ion Capacitors: A Review of Strategies toward Enhancing the Performance of the Activated Carbon Cathode. *Batteries* **2023**, *9*, 533. [CrossRef]
9. Khomenko, V.; Raymundo-Piñero, E.; Béguin, F. High-energy density graphite/AC capacitor in organic electrolyte. *J. Power Sources* **2008**, *177*, 643–651. [CrossRef]
10. Sun, H.; Mei, L.; Liang, J.; Zhao, Z.; Lee, C.; Fei, H.; Ding, M.; Lau, J.; Li, M.; Wang, C.; et al. Three-dimensional holey-graphene/niobia composite architectures for ultrahigh-rate energy storage. *Science* **2017**, *356*, 599–604. [CrossRef]
11. Kim, H.S.; Cook, J.B.; Lin, H.; Ko, J.S.; Tolbert, S.H.; Ozolins, V.; Dunn, B. Oxygen vacancies enhance pseudocapacitive charge storage properties of MoO_{3-x}. *Nat. Mater.* **2017**, *16*, 454–462. [CrossRef] [PubMed]
12. Zhang, L.; Wang, H.; Zhang, X.; Tang, Y. A Review of Emerging Dual-Ion Batteries: Fundamentals and Recent Advances. *Adv. Funct. Mater.* **2021**, *31*, 2010958. [CrossRef]
13. Xiang, W.; Chen, M.; Zhou, X.; Chen, J.; Huang, H.; Sun, Z.; Lu, Y.; Zhang, G.; Wen, X.; Li, W. Highly Enforced Rate Capability of a Graphite Anode via Interphase Chemistry Tailoring Based on an Electrolyte Additive. *J. Phys. Chem. Lett.* **2022**, *13*, 5151–5159. [CrossRef]
14. Wang, G.; Yu, M.; Wang, J.; Li, D.; Tan, D.; Löffler, M.; Zhuang, X.; Müllen, K.; Feng, X. Self-Activating, Capacitive Anion Intercalation Enables High-Power Graphite Cathodes. *Adv. Mater.* **2018**, *30*, 1800533. [CrossRef]
15. Gao, J.; Tian, S.; Qi, L.; Yoshio, M.; Wang, H. Hexafluorophosphate intercalation into graphite electrode from gamma-butyrolactone solutions in activated carbon/graphite capacitors. *J. Power Sources* **2015**, *297*, 121–126. [CrossRef]
16. Yoshio, M.; Nakamura, H.; Wang, H. Novel megalo-capacitance capacitor based on graphitic carbon cathode. *Electrochem. Solid-State Lett.* **2006**, *9*, 561–563. [CrossRef]
17. Rajalekshmi, A.R.; Divya, M.L.; Lee, Y.S.; Aravindan, V. High-performance Li-ion capacitor via anion-intercalation process. *Batter. Energy* **2022**, *1*, 20210005. [CrossRef]
18. Ji, B.; Zhang, F.; Song, X.; Tang, Y. A Novel Potassium-Ion-Based Dual-Ion Battery. *Adv. Mater.* **2017**, *29*, 1700519. [CrossRef]
19. Beltrop, K.; Qi, X.; Hering, T.; Röser, S.; Winter, M.; Placke, T. Enabling bis(fluorosulfonyl)imide-based ionic liquid electrolytes for application in dual-ion batteries. *J. Power Sources* **2018**, *373*, 193–202. [CrossRef]
20. Wu, L.N.; Peng, J.; Sun, Y.K.; Han, F.M.; Wen, Y.F.; Shi, C.G.; Fan, J.J.; Huang, L.; Li, J.T.; Sun, S.G. High-Energy Density Li metal Dual-Ion Battery with a Lithium Nitrate-Modified Carbonate-Based Electrolyte. *ACS Appl. Mater. Interfaces* **2019**, *11*, 18504–18510. [CrossRef]

21. Placke, T.; Fromm, O.; Lux, S.F.; Bieker, P.; Rothermel, S.; Meyer, H.-W.; Passerini, S.; Winter, M. Reversible Intercalation of Bis(trifluoromethanesulfonyl)imide Anions from an Ionic Liquid Electrolyte into Graphite for High Performance Dual-Ion Cells. *J. Electrochem. Soc.* **2012**, *159*, A1755–A1765. [CrossRef]
22. Luo, X.F.; Yang, C.H.; Peng, Y.Y.; Pu, N.W.; Der Ger, M.; Hsieh, C.T.; Chang, J.K. Graphene nanosheets, carbon nanotubes, graphite, and activated carbon as anode materials for sodium-ion batteries. *J. Mater. Chem. A* **2015**, *3*, 10320–10326. [CrossRef]
23. Zięzio, M.; Charmas, B.; Jedynak, K.; Hawryluk, M.; Kucio, K. Preparation and characterization of activated carbons obtained from the waste materials impregnated with phosphoric acid(V). *Appl. Nanosci.* **2020**, *10*, 4703–4716. [CrossRef]
24. Zou, K.; Cai, P.; Liu, C.; Li, J.; Gao, X.; Xu, L.; Zou, G.; Hou, H.; Liu, Z.; Ji, X. A kinetically well-matched full-carbon sodium-ion capacitor. *J. Mater. Chem. A* **2019**, *7*, 13540–13549. [CrossRef]
25. Ge, P.; Li, S.; Xu, L.; Zou, K.; Gao, X.; Cao, X.; Zou, G.; Hou, H.; Ji, X. Hierarchical Hollow-Microsphere Metal–Selenide@Carbon Composites with Rational Surface Engineering for Advanced Sodium Storage. *Adv. Energy Mater.* **2019**, *9*, 1803035. [CrossRef]
26. Tan, L.; Huang, X.; Yin, T.; Guo, Y.; Ning, T.; Mei, Y.; Zou, K.; Li, L.; Ji, X.; Zou, G. A 5 V ultrahigh energy density lithium metal capacitor enabled by the fluorinated electrolyte. *Energy Storage Mater.* **2024**, *71*, 103692. [CrossRef]
27. Tan, X.; Zhang, F.; Chen, D.; Wang, P.; Liu, Y.; Meng, C.; Zhang, Y. Modulating NH4+ in vanadium oxide framework for high-efficient aqueous NH_4^+ storage. *Chem. Eng. J.* **2024**, *489*, 151119. [CrossRef]
28. Sun, L.; Zhang, Y.; Si, H.; Zhang, Y.; Liu, J.; Liu, J.; Zhang, Y. TiO_2-modified red phosphorus nanosheets entangled in carbon nanotubes for high performance lithium ion batteries. *Electrochim. Acta* **2019**, *297*, 319–327. [CrossRef]
29. Balabajew, M.; Reinhardt, H.; Bock, N.; Duchardt, M.; Kachel, S.; Hampp, N.; Roling, B. In-Situ Raman Study of the Intercalation of Bis(trifluoromethylsulfonyl)imid Ions into Graphite inside a Dual-Ion Cell. *Electrochim. Acta* **2016**, *211*, 679–688. [CrossRef]
30. Lu, Y.; Zhang, D.; Wang, L.; Xu, M.; Song, J.; Goodenough, J.B. Electrochemical Behavior of a Graphite Electrode Prepared by Anodic Electrophoretic Deposition. *J. Electrochem. Soc.* **2012**, *159*, A321–A324. [CrossRef]
31. Satish, R.; Aravindan, V.; Ling, W.C.; Madhavi, S. Carbon-coated $Li_3V_2(PO_4)_3$ as insertion type electrode for lithium-ion hybrid electrochemical capacitors: An evaluation of anode and cathodic performance. *J. Power Sources* **2015**, *281*, 310–317. [CrossRef]
32. Kim, H.; Cho, M.Y.; Kim, M.H.; Park, K.Y.; Gwon, H.; Lee, Y.; Roh, K.C.; Kang, K. A novel high-energy hybrid supercapacitor with an anatase TiO_2-reduced graphene oxide anode and an activated carbon cathode. *Adv. Energy Mater.* **2013**, *3*, 1500–1506. [CrossRef]
33. Leng, K.; Zhang, F.; Zhang, L.; Zhang, T.; Wu, Y.; Lu, Y.; Huang, Y.; Chen, Y. Graphene-based Li-ion hybrid supercapacitors with ultrahigh performance. *Nano Res.* **2013**, *6*, 581–592. [CrossRef]
34. Li, G.; Yin, Z.; Guo, H.; Wang, Z.; Yan, G.; Yang, Z.; Liu, Y.; Ji, X.; Wang, J. Metalorganic Quantum Dots and Their Graphene-Like Derivative Porous Graphitic Carbon for Advanced Lithium-Ion Hybrid Supercapacitor. *Adv. Energy Mater.* **2019**, *9*, 1802878. [CrossRef]
35. Wang, X.; Wang, Z.; Zhang, X.; Peng, H.; Xin, G.; Lu, C.; Zhong, Y.; Wang, G.; Zhang, Y. Nitrogen-Doped Defective Graphene Aerogel as Anode for all Graphene-Based Lithium Ion Capacitor. *ChemistrySelect* **2017**, *2*, 8436–8445. [CrossRef]
36. Lim, E.; Jo, C.; Kim, H.; Kim, M.H.; Mun, Y.; Chun, J.; Ye, Y.; Hwang, J.; Ha, K.S.; Roh, K.C.; et al. Facile Synthesis of Nb_2O_5@Carbon Core-Shell Nanocrystals with Controlled Crystalline Structure for High-Power Anodes in Hybrid Supercapacitors. *ACS Nano* **2015**, *9*, 7497–7505. [CrossRef] [PubMed]
37. Chaturvedi, A.; Hu, P.; Aravindan, V.; Kloc, C.; Madhavi, S. Unveiling two-dimensional TiS_2 as an insertion host for the construction of high energy Li-ion capacitors. *J. Mater. Chem. A* **2017**, *5*, 9177–9181. [CrossRef]
38. Sun, C.; Zhang, X.; Li, C.; Wang, K.; Sun, X.; Ma, Y. High-efficiency sacrificial prelithiation of lithium-ion capacitors with superior energy-storage performance. *Energy Storage Mater.* **2020**, *24*, 160–166. [CrossRef]
39. Yi, S.; Wang, L.; Zhang, X.; Li, C.; Liu, W.; Wang, K.; Sun, X.; Xu, Y.; Yang, Z.; Cao, Y.; et al. Cationic intermediates assisted self-assembly two-dimensional $Ti_3C_2T_x$/rGO hybrid nanoflakes for advanced lithium-ion capacitors. *Sci. Bull.* **2021**, *66*, 914–924. [CrossRef]
40. Liu, C.; Wang, B.; Xu, L.; Zou, K.; Deng, W.; Hou, H.; Zou, G.; Ji, X. Novel Nonstoichiometric Niobium Oxide Anode Material with Rich Oxygen Vacancies for Advanced Lithium-Ion Capacitors. *ACS Appl. Mater. Interfaces* **2023**, *15*, 5387–5398. [CrossRef]
41. Wang, G.; Wang, F.; Zhang, P.; Zhang, J.; Zhang, T.; Müllen, K.; Feng, X. Polarity-Switchable Symmetric Graphite Batteries with High Energy and High Power Densities. *Adv. Mater.* **2018**, *30*, 1802949. [CrossRef] [PubMed]
42. Wrogemann, J.M.; Haneke, L.; Ramireddy, T.; Frerichs, J.E.; Sultana, I.; Chen, Y.I.; Brink, F.; Hansen, M.R.; Winter, M.; Glushenkov, A.M.; et al. Advanced Dual-Ion Batteries with High-Capacity Negative Electrodes Incorporating Black Phosphorus. *Adv. Sci.* **2022**, *9*, 2201116. [CrossRef]
43. Rothermel, S.; Meister, P.; Schmuelling, G.; Fromm, O.; Meyer, H.-W.W.; Nowak, S.; Winter, M.; Placke, T. Dual-graphite cells based on the reversible intercalation of bis(trifluoromethanesulfonyl)imide anions from an ionic liquid electrolyte. *Energy Environ. Sci.* **2014**, *7*, 3412–3423. [CrossRef]
44. Read, J.A.; Cresce, A.V.; Ervin, M.H.; Xu, K. Dual-graphite chemistry enabled by a high voltage electrolyte. *Energy Environ. Sci.* **2014**, *7*, 617–620. [CrossRef]
45. Qiao, Y.; Jiang, K.; Li, X.; Deng, H.; He, Y.; Chang, Z.; Wu, S.; Guo, S.; Zhou, H. A Hybrid Electrolytes Design for Capacity-Equivalent Dual-Graphite Battery with Superior Long-Term Cycle Life. *Adv. Energy Mater.* **2018**, *8*, 1801120. [CrossRef]
46. Heckmann, A.; Thienenkamp, J.; Beltrop, K.; Winter, M.; Brunklaus, G.; Placke, T. Towards high-performance dual-graphite batteries using highly concentrated organic electrolytes. *Electrochim. Acta* **2018**, *260*, 514–525. [CrossRef]

47. Wang, A.; Yuan, W.; Fan, J.; Li, L. A Dual-graphite Battery with Pure 1-Butyl-1-methylpyrrolidinium bis(trifluoromethylsulfonyl) Imide as the Electrolyte. *Energy Technol.* **2018**, *6*, 2172–2178. [CrossRef]
48. Huang, Y.; Xiao, R.; Ma, Z.; Zhu, W. Developing Dual-Graphite Batteries with Pure 1-Ethyl-3-methylimidazolium Trifluoromethanesulfonate Ionic Liquid as the Electrolyte. *ChemElectroChem* **2019**, *6*, 4681–4688. [CrossRef]
49. Ghosh, S.; Sarma, D.; Mahata, A.; Martha, S.K. Differences between cation and anion storage electrochemistry of graphite and its impact on dual graphite battery. *J. Power Sources* **2024**, *589*, 233721. [CrossRef]
50. Zhang, L.; Wang, H. Dual-Graphite Batteries with Flame-Retardant Electrolyte Solutions. *ChemElectroChem* **2019**, *6*, 4637–4644. [CrossRef]

batteries

MDPI

Article

One-Step Hydrothermally Synthesized $Ni_{11}(HPO_3)_8(OH)_6/Co_3(HPO_4)_2(OH)_2$ Heterostructure with Enhanced Rate Performance for Hybrid Supercapacitor Applications

Mingjun Jing [1,*], Kaige Long [1], Rui Liu [1], Xingyu Wang [1], Tianjing Wu [1], Yirong Zhu [2,*], Lijie Liu [3], Sheng Zhang [4], Yang Zhang [5] and Cheng Liu [6]

1 The College of Chemistry, Xiangtan University, Xiangtan 411105, China
2 College of Materials and Advanced Manufacturing, Hunan University of Technology, Zhuzhou 412007, China
3 College of Science, Henan Agricultural University, 63 Agricultural Road, Zhengzhou 450002, China
4 Institute of Nanoscience and Engineering, Henan University, Kaifeng 475004, China
5 Department of Materials, Henry Royce Institute, University of Manchester, Manchester M13 9PL, UK; yang.zhang-26@postgrad.manchester.ac.uk
6 Institut de Ciència de Materials de Barcelona (ICMAB), CSIC, Campus UAB, 08193 Bellaterra, Spain
* Correspondence: jingmingjun@xtu.edu.cn (M.J.); zhuyirong@hut.edu.cn (Y.Z.)

Abstract: Transition metal phosphate is the prospective electrode material for supercapacitors (SCs). It has an open frame construction with spacious cavities and wide aisles, resulting in excellent electric storage capacity. However, the inferior rate behavior and cycling stability of transition metal phosphate materials in alkaline environments pose significant barriers to their application in SCs. Herein, $Ni_{11}(HPO_3)_8(OH)_6/Co_3(HPO_4)_2(OH)_2$ heterostructured materials synthesized through a one-step hydrothermal process exhibiting remarkable rate capability coupled with exceptional cycling endurance. $Ni_{11}(HPO_3)_8(OH)_6/Co_3(HPO_4)_2(OH)_2$ samples exhibit a micron-scale structure composed of sheet-like compositions and unique pore structure. The multistage pore structure is favorable for promoting the diffusion of protons and ions, enhancing the sample's electrochemical storage capacity. Upon conducting electrochemical tests, it was observed that $Ni_{11}(HPO_3)_8(OH)_6/Co_3(HPO_4)_2(OH)_2$ composite electrode surpassed both the standalone $Ni_{11}(HPO_3)_8(OH)_6$ and $Co_3(HPO_4)_2(OH)_2$ electrode, achieving a remarkable specific capacity of 163 mAh g^{-1} with exceptional stability and efficiency at 1 A g^{-1}. Notably, this electrode also exhibits superior rate performance, maintaining 82.5% and 71% of its original full capacity even at 50 A g^{-1} and 100 A g^{-1}, respectively. Furthermore, it demonstrates superior stability in cycling, retaining a capacity of 92.7% at 10 A g^{-1} after 5000 cycles. Moreover, $Ni_{11}(HPO_3)_8(OH)_6/Co_3(HPO_4)_2(OH)_2$ and porous carbon (PC) were assembled into a hybrid supercapacitor (HSC). Electrochemical tests reveal an impressive power density of up to 36 kW kg^{-1} and an exceptional energy density of up to 47.4 Wh kg^{-1} for the HSC. Moreover, $Ni_{11}(HPO_3)_8(OH)_6/Co_3(HPO_4)_2(OH)_2$//PC HSC exhibits robust capacity retention stability of 92.9% after enduring 10,000 cycles at 3 A g^{-1}, demonstrating its remarkable durability. This work imparts viewpoints into the design of transition metal phosphate heterostructured materials.

Keywords: high-rate performance; hybrid supercapacitor; nickel cobalt phosphate; heterostructure

1. Introduction

As electrochemical energy storage technologies evolve, supercapacitors (SCs) have distinguished themselves among various storage devices because of their environmental friendliness, fast discharging and charging rates, lengthy lifespan, and superior safety features [1–3]. Nevertheless, despite these numerous advantages, there remains a substantial gap between the energy density of SCs and the requirements of commercial applications. Researchers have proposed a feasible approach in the form of a hybrid supercapacitor (HSC) [4,5]. HSC exhibits an exceptional combination of high energy density alongside robust power density, offering a unique balance in performance [6,7]. The performance of HSC is contingent upon the electrode material utilized. However, conventional cathode materials with a battery type exhibit slow ionic kinetics, while capacitive cathode materials display fast kinetic behavior, rendering it challenging to achieve a suitable match between the two kinetic behaviors [8,9]. In response to the issue, further investigations have revealed that oxides, hydroxides, sulfides, phosphides, and phosphates are all suitable cathode materials for HSC [10–16]. Therefore, developing cathode materials that possess enhanced high kinetics and durability can have a positive effect on overcoming current challenges.

Among various materials, transition metal phosphates have recently attracted much attention, mainly because of their unique properties as both transition metal oxides and transition metal phosphides. This mixture of properties gives them excellent electrochemical performance and high chemical stability [17–19]. Besides, it also has an open structural framework and rich redox behavior, which makes it exhibit excellent energy storage capacity [20,21]. For example, the specific capacitance for $Ni_3P_2O_8$ at 0.5 A g^{-1} is 1464 F g^{-1} [22]. This measured specific capacitance for $Co_{11}(HPO_3)_8(OH)_6$ at 0.5 A g^{-1} is 1200 F g^{-1} [23]. Therefore, transition metal phosphates are more attractive when choosing materials for electrodes. Compared with other metal phosphates that have been extensively studied, nickel-based phosphates have a high specific capacitance [24,25]. However, their poor rate behaviors and cyclic stability in alkaline electrolytes limit their large-scale application.

Researchers have implemented various strategies to address the issue, including refining the morphology of materials, constructing a suitable pore structure, designing composite materials and so on [17,26]. For example, an ultra-thin nanosheet-based nickel phosphate lamellar structure attained at 2.0 A g^{-1} current conditions the remarkable specific capacity of 131.6 mAh g^{-1} [27]. $Ni_3(PO_4)_2$/GF composite nanostructures were prepared, demonstrating a capacitance of 48 mAh g^{-1} at 0.5 A g^{-1} current conditions [28]. While these materials display favorable electrochemical characteristics, there is a continued necessity to identify straightforward and efficacious methodologies for improving nickel-based phosphate rate properties and cycling stability. However, constructing a heterostructure has attracted much attention as a novel approach to creating materials with unique interfaces, elastic structures, and synergistic effects that enhance SCs energy/power density [29–31]. Meanwhile, embedding an internal electric field across the interfaces of heterogeneous structures can potentially expedite ion diffusion, enhancing overall ion transport efficiency [32].

Moreover, the reallocation of charge within the heterostructure components fosters the creation of additional energetic storage sites, enhancing the electrode's reversible capacity and optimizing its energy storage performance [33]. Hence, rationally designed combinations of two materials in heterostructures may yield unprecedented performance due to the complementary effects of different materials. Although previous studies have achieved enhanced electrochemical performance by constructing heterostructures with different components, the synthesis process is complicated. For instance, the amorphous/crystalline nickel manganese phosphate octahydrate heterostructured samples were obtained solely through a hydrothermal process with subsequent annealing in argon gas, demonstrating a specific capacitance of 2351.6 F g^{-1} at 1 A g^{-1} [34]. This specific capacity for $Fe_{0.4}Co_{0.6}Se_2$@NiCo-P heterostructured materials was prepared using the two-step electrodeposition method at 1 A g^{-1} with a current condition of 202.3 mAh g^{-1} [35]. Reports on the simple and rapid synthesis of nickel-based metallic phosphate heterostructured materials are scarce. Therefore, it is worthwhile to explore a simple approach to construct a heterostructure of transition metal phosphates.

Herein, the $Ni_{11}(HPO_3)_8(OH)_6/Co_3(HPO_4)_2(OH)_2$ heterojunction composites were synthesized through a straightforward one-step hydrothermal process, utilizing red phosphorus (RP) as the phosphorous source. This heterostructure effectively has optimized its electronic structure. The redistribution of charges in the heterostructure will induce a greater number of active sites to participate in energy storage, improving the capacity performance of the electrodes. A further examination and analysis of the composition, properties, and structure of the materials is carried out through a series of physical characterization and electrochemical tests. Electrochemical assessments indicate that the $Ni_{11}(HPO_3)_8(OH)_6/Co_3(HPO_4)_2(OH)_2$ composite electrode boasts enhanced specific capacitance, exceptional rate performance (the capacity retention attains 82.5% and 71%, respectively, when operated at 50 A g^{-1} and 100 A g^{-1} current densities), and impressive recycling capacity. Furthermore, the HSC configured with $Ni_{11}(HPO_3)_8(OH)_6/Co_3(HPO_4)_2(OH)_2$ and porous carbon (PC) demonstrates outstanding energy/power density and robust cycling endurance. The work gives a practical and effective generalized approach for the preparation of metal-phosphate heterostructured materials.

2. Experimental Section

2.1. Chemicals and Materials

Red phosphorus (RP, AR), nickel nitrate hexahydrate ($Ni(NO_3)_2{\cdot}6H_2O$, AR), potassium hydroxide (KOH, AR), cobalt nitrate hexahydrate ($Co(NO_3)_2{\cdot}6H_2O$, AR), ammonia ($NH_3{\cdot}H_2O$, AR), and anhydrous ethanol (C_2H_5OH, AR) were sourced from Sinopharm Chemical Reagent Co. Ltd. (Shanghai, China) Acetylene black (C, AR) was procured from Shenzhen Kejing Zhida Technology Co., Shenzhen, China, porous carbon (PC, 2 nm) was obtained from Xianfeng Nanomaterials Technology Co., Ltd., Nanjing, China, polytetrafluoroethylene (PTFE, 60%) was acquired from Daikin, Japan, N-methylpyrrolidone (NMP, battery grade) was purchased from Shenzhen Kejing Zhi Technology Co., Ltd., Shenzhen, China. And Nickel foam was obtained from Changde Liyuan New Material Co., Changde, China.

2.2. Synthesis of $Ni_{11}(HPO_3)_8(OH)_6$ Cathode Material

After dissolving 0.67 mmol of $Ni(NO_3)_2{\cdot}6H_2O$ into 50 mL distilled water, 103.75 mg of RP was added and magnetically mixed at 30 min for the solution of deep red color. Then, $NH_3{\cdot}H_2O$ was dropped into this solution with the pH adjusted to 8–9. The final solution was moved to the hydrothermal kettle heated to 180 °C and held for 12 h. This reacted solution was naturally cooled to ambient temperature, and the resulting precipitated samples were fully centrifuged alternately into distilled water and subsequently into ethanol to separate any unreacted residue and dried at the temperature of 50 °C, lasting for a period of 12 h to obtain the $Ni_{11}(HPO_3)_8(OH)_6$ product.

2.3. Synthesis of $Ni_{11}(HPO_3)_8(OH)_6/Co_3(HPO_4)_2(OH)_2$ Heterojunction Composites

After dissolving 0.67 mmol $Ni(NO_3)_2{\cdot}6H_2O$ and 0.67 mmol $Co(NO_3)_2{\cdot}6H_2O$ into 50 mL of distilled water, 103.75 mg of RP was added and magnetically mixed at 30 min for the solution of deep red color. Then, $NH_3{\cdot}H_2O$ was dropped into this solution with pH adjusted to 8–9. The final solution was moved to the hydrothermal kettle heated to 180 °C and held for 12 h. This reacted solution was naturally cooled to ambient temperature, and resulting precipitated samples were fully centrifuged alternately with distilled water and ethanol to separate any unreacted residue and dried at the temperature of 50 °C, lasting for a period of 12 h to afford the $Ni_{11}(HPO_3)_8(OH)_6/Co_3(HPO_4)_2(OH)_2$ product.

2.4. Material Characteristics

The physical properties of the studied materials were analyzed using a Rigaku D/max 2550 VB$^+$ X-ray diffractometer. The physical and structural properties of materials, such as elemental composition, valence states, etc., were studied in depth using an ESCALab 250 X-ray photoelectron spectrometer with Thermo K-Alpha as a light source. The prepared samples were analyzed in detail for morphology, composition, and microstructure using FESEM with Signa 500. The morphological characteristics and the intricate microstructural features of the samples were thoroughly investigated using the advanced Titan G2 60–300 TEM. Pore sizing and the surface characteristics of materials were analyzed using a specific surface analyzer of Micrometrics ASAP 2020.

2.5. Electrochemical Tests

These working electrodes were produced by combining the $Ni_{11}(HPO_3)_8(OH)_6/Co_3(HPO_4)_2(OH)_2$ sample with acetylene black, serving as a conductive agent, and polytetrafluoroethylene (PTFE), acting as a binder, in a 7:2:1 ratio within an n-methyl-2-pyrrolidinone solution, which served as the dispersant. After mixing, preparation of the paste was uniformly applied to a cleaning nickel foam (collector) and dried in a vacuum oven. Finally, nickel foam was squeezed into working electrodes using a powder press at 10 MPa. The loading of $Ni_{11}(HPO_3)_8(OH)_6/Co_3(HPO_4)_2(OH)_2$ is ~1 mg cm^{-2}. The CHI 760E electrochemical workstation was used for all electrochemical tests. Therein, the Hg/HgO electrode was used as the reference electrode and the platinum sheet electrode was used as the counter electrode. The cyclic voltammetry (CV) test, alternating current impedance test (EIS), and constant current charge/discharge test (GCD) were used to analyze the electrochemical properties of electrodes. CV curves were acquired over a voltage range of 0~0.6 V, EIS was performed over a frequency range of 100 kHz to 0.01 Hz, and the GCD tests were performed over a potential range of 0~0.55 V.

In addition, HSCs were assembled and used to test their potential for practical applications. The HSC was assembled utilizing $Ni_{11}(HPO_3)_8(OH)_6/Co_3(HPO_4)_2(OH)_2$ as a cathode, PC as an anode, 2 mol L^{-1} KOH as an electrolyte, and NKK as the separator in a 2016-type coin cell configuration. To maintain charge equilibrium between the positive and negative components in the HSC, the optimal mass ratio of these two electrodes can be determined using the formula: $m^+/m^- = (C^- \times \Delta V^-)/(C^+ \times \Delta V^+)$, where + and − symbols represent the cathode and anode, respectively, while C and ΔV signify capacity and voltage range, respectively.

3. Results and Discussion

3.1. The Characterizations of Morphology, Composition, and Structure

Figure 1 presents the schematic for the preparation of $Ni_{11}(HPO_3)_8(OH)_6$ and $Ni_{11}(HPO_3)_8(OH)_6/Co_3(HPO_4)_2(OH)_2$ composites using a one-step hydrothermal process. First, $Ni_{11}(HPO_3)_8(OH)_6$ material is prepared using red phosphorus as a phosphorus source through the hydrothermal method. To elevate both the specific capacity and cycling stability, the $Ni_{11}(HPO_3)_8(OH)_6$ material has been further improved by constructing a heterostructure. Cobalt metal salt, nickel metal salt, RP, and $NH_3{\cdot}H_2O$ are thoroughly mixed, and the $Ni_{11}(HPO_3)_8(OH)_6/Co_3(HPO_4)_2(OH)_2$ product with micrometer structure of flake composition is obtained by a simple hydrothermal method (hydrothermal reaction at 180 °C in a hydrothermal autoclave for 12 h) and centrifugal drying. It is noteworthy that the $Ni_{11}(HPO_3)_8(OH)_6/Co_3(HPO_4)_2(OH)_2$ composites show a significant change in morphology compared with $Ni_{11}(HPO_3)_8(OH)_6$, and their regular lamellar composition structure enhances the energy storage capacity because it enlarges the specific surface area, providing lots of active sites. The $Ni_{11}(HPO_3)_8(OH)_6/Co_3(HPO_4)_2(OH)_2$ sample is synthesized as follows:

$$Co^{2+} + Ni^{2+} + 4OH^- \rightarrow Co(OH)_2 + Ni(OH)_2 \quad (1)$$

$$\begin{gathered} 33Ni(OH)_2 + 9Co(OH)_2 + 64P + 36H_2O \\ \rightarrow 3Ni_{11}(HPO_3)_8(OH)_6/Co_3(HPO_4)_2(OH)_2 + 34PH_3 \end{gathered} \quad (2)$$

As depicted in Figure 2a, the crystalline nature and structural configuration of the $Ni_{11}(HPO_3)_8(OH)_6/Co_3(HPO_4)_2(OH)_2$ specimen were characterized using XRD. The XRD diffraction peaks of $Ni_{11}(HPO_3)_8(OH)_6/Co_3(HPO_4)_2(OH)_2$ sample correspond to the standard cards of $Ni_{11}(HPO_3)_8(OH)_6$ (JCPDS No. 81-1065) and $Co_3(HPO_4)_2(OH)_2$ (JCPDS No. 80-1996). This suggests that the obtained material is a $Ni_{11}(HPO_3)_8(OH)_6/Co_3(HPO_4)_2(OH)_2$ heterogeneous structure. The XPS conducted on $Ni_{11}(HPO_3)_8(OH)_6/Co_3(HPO_4)_2(OH)_2$ sample offered detailed insights into its valence states and surface chemical composition. As illustrated in Figure 2b, the broad XPS spectrum validates the presence of Co, Ni, P,

C, and O, confirming the purity of the specimen without any discernible traces of unintended elements beyond these constituents. Ni 2p spectroscopy with high resolution (Figure 2c) displays two peaks situated at 856.67 eV and 874.52 eV, corresponding to Ni $2p_{3/2}$ and Ni $2p_{1/2}$, respectively. Each of these main signals is accompanied by a satellite peak [36,37]. This matches well with the electronic state of Ni^{2+}. Co 2p spectroscopy with high resolution depicted in Figure 2d reveals two distinct peaks at 781.79 eV and 797.83 eV, which are indicative of Co $2p_{3/2}$ and Co $2p_{1/2}$ orbitals. This spectral signature confirms the existence of Co^{2+} ions within the system. The high-resolution P 2p spectra (Figure 2e) displays this peak at 133.4 eV attributable to P 2p, indicating the presence of $(PO_4)^{3-}$ on the $Ni_{11}(HPO_3)_8(OH)_6/Co_3(HPO_4)_2(OH)_2$ surface [38]. The $(PO_4)^{3-}$ form improves the surface activity on $Ni_{11}(HPO_3)_8(OH)_6/Co_3(HPO_4)_2(OH)_2$. This is consistent with the composition in $Ni_{11}(HPO_3)_8(OH)_6/Co_3(HPO_4)_2(OH)_2$ according to the above tests.

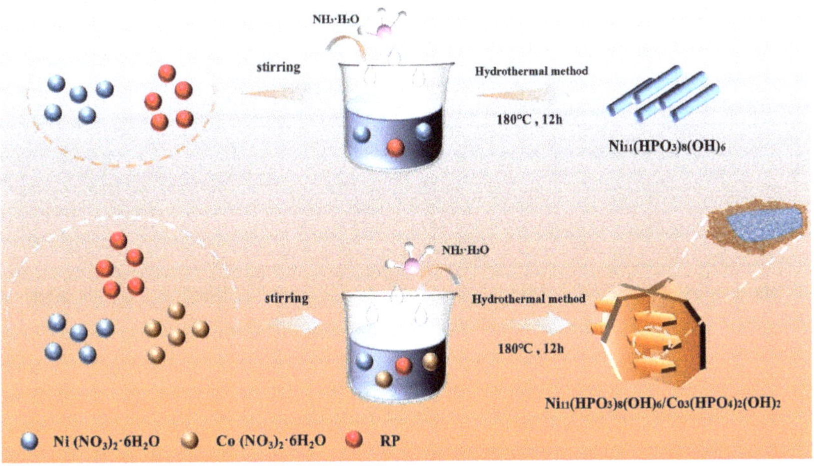

Figure 1. The schematic illustration for hydrothermal preparation of the $Ni_{11}(HPO_3)_8(OH)_6$ and $Ni_{11}(HPO_3)_8(OH)_6/Co_3(HPO_4)_2(OH)_2$ samples.

The N_2 adsorption/desorption test of the $Ni_{11}(HPO_3)_8(OH)_6/Co_3(HPO_4)_2(OH)_2$ heterogeneous structural material is shown in Figure 2f. The $Ni_{11}(HPO_3)_8(OH)_6/Co_3(HPO_4)_2$ $(OH)_2$ material displays a clear hysteresis ring in the adsorption and desorption isotherms and a type IV isotherm, indicating the existence of their mesoporous structure in $Ni_{11}(HPO_3)_8$ $(OH)_6/Co_3(HPO_4)_2(OH)_2$. At higher pressures ($P/P_0 > 0.9$), N_2 adsorption increased rapidly, indicating the existence of microporous structures. The specific surface area of BET for the $Ni_{11}(HPO_3)_8(OH)_6/Co_3(HPO_4)_2(OH)_2$ material is measured to be 43.2 $m^2\ g^{-1}$. For the $Ni_{11}(HPO_3)_8(OH)_6/Co_3(HPO_4)_2(OH)_2$ samples, the pore structure is analyzed by BJH. The $Ni_{11}(HPO_3)_8(OH)_6/Co_3(HPO_4)_2(OH)_2$ sample exhibits a distinctive hierarchical pore structure, further illuminated by its pore size distribution. This primarily comprises 4 nm mesopores, accompanied by a minimal presence of macropores. The $Ni_{11}(HPO_3)_8(OH)_6/Co_3(HPO_4)_2(OH)_2$ material's unique pore structure, featuring a dominance of mesopores and a small fraction of macropores, could boost electrochemical energy storage efficiency by facilitating proton and ion diffusion.

Morphology and microstructure are observed for $Ni_{11}(HPO_3)_8(OH)_6$ and $Ni_{11}(HPO_3)_8$ $(OH)_6/Co_3(HPO_4)_2(OH)_2$ samples using FESEM. As depicted in Figure 3a–d, the findings demonstrate that the synthesized samples all display a uniform and regular morphology. The $Ni_{11}(HPO_3)_8(OH)_6$ sample displays a rod shape with a diameter of around 300 nm. In contrast, the $Ni_{11}(HPO_3)_8(OH)_6/Co_3(HPO_4)_2(OH)_2$ samples exhibit notable alterations. In particular, the $Ni_{11}(HPO_3)_8(OH)_6/Co_3(HPO_4)_2(OH)_2$ samples display a micrometer-scale structure comprising flake-like units with an overall size of approximately 9 μm. The EDS and elemental distribution of $Ni_{11}(HPO_3)_8(OH)_6$ are signaled in Figure 3e,f. EDS pattern (Figure 3g) reveals that the $Ni_{11}(HPO_3)_8(OH)_6/Co_3(HPO_4)_2(OH)_2$ material is consisted of Ni, Co, and P. In the atomic ratio of Ni, Co, and P, the approximate ratio is 3:11:10. Figure 3h provides further evidence that the elements Ni, Co and P have a single-form distribution over the structure. According to the above data, the prepared $Ni_{11}(HPO_3)_8(OH)_6/Co_3(HPO_4)_2(OH)_2$ material has the outstanding advantages of micrometer structure with lamellar composition, high specific surface area and abundant number of mesopores/macropores. Its rich active sites and unique hierarchical pore structure offer the possibility to realize electrochemical properties for the $Ni_{11}(HPO_3)_8(OH)_6/Co_3(HPO_4)_2(OH)_2$ electrode materials, such as high kinetics and high stability.

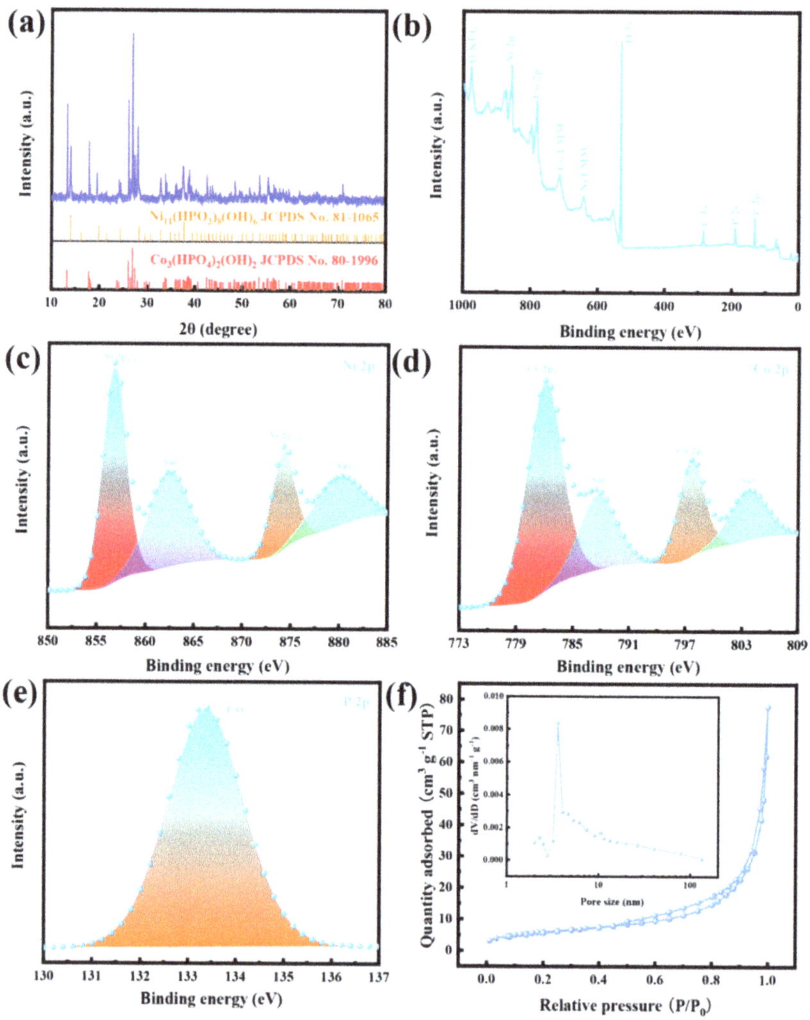

Figure 2. Characterization of $Ni_{11}(HPO_3)_8(OH)_6/Co_3(HPO_4)_2(OH)_2$ samples: (**a**) XRD plots; (**b–e**) XPS spectrum; (**f**) pore distribution (insert) and sorption–desorption isotherms curves of N_2.

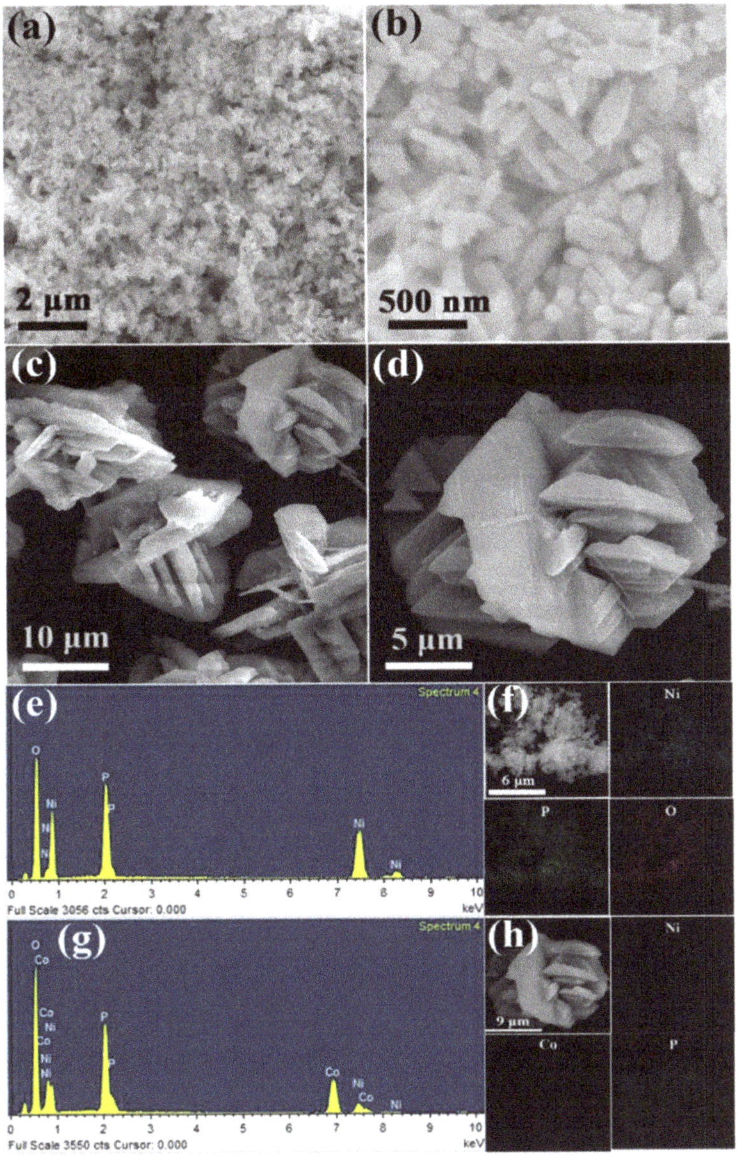

Figure 3. (**a**,**b**) FESEM images for $Ni_{11}(HPO_3)_8(OH)_6$ at different magnifications. (**c**,**d**) FESEM images for $Ni_{11}(HPO_3)_8(OH)_6/Co_3(HPO_4)_2(OH)_2$ samples with different magnifications. (**e**,**f**) EDS and elemental mappings for $Ni_{11}(HPO_3)_8(OH)_6$. (**g**,**h**) EDS and elemental mappings for the $Ni_{11}(HPO_3)_8(OH)_6/Co_3(HPO_4)_2(OH)_2$ samples.

3.2. The Electrochemical Performances of Samples

The CV curves for the $Ni_{11}(HPO_3)_8(OH)_6$ and $Ni_{11}(HPO_3)_8(OH)_6/Co_3(HPO_4)_2(OH)_2$ electrodes at 2 mV s^{-1} are presented in Figure 4a. These CV curves display clear redox peaks, demonstrating typical Faraday characteristics. The equations for the redox reactions could be as follows:

$$Ni_{11}(HPO_3)_8(OH)_6/Co_3(HPO_4)_2(OH)_2 + 14OH^- \leftrightarrow Ni_{11}(HPO_3)_8(OH)_{17} + Co_3(HPO_4)_2(OH)_5 + 14e^- \quad (3)$$

$$Ni_{11}(HPO_3)_8(OH)_{17} + Co_3(HPO_4)(OH)_5 + 14OH^- \leftrightarrow Co_3(HPO_4)(O)_4 + Ni_{11}(HPO_3)_8(O)_{14} + 18H_2O + 14e^- \quad (4)$$

Meanwhile, the $Ni_{11}(HPO_3)_8(OH)_6/Co_3(HPO_4)_2(OH)_2$ electrode displays a greater area of integration compared to the CV curve for the $Ni_{11}(HPO_3)_8(OH)_6$ electrode, indicating the $Ni_{11}(HPO_3)_8(OH)_6/Co_3(HPO_4)_2(OH)_2$ electrode possesses larger specific capacitance. Comparison is made between two electrodes with GCD curves between 0 and 0.55 V at 1 A g^{-1} is shown in Figure 4b. Compared to the $Ni_{11}(HPO_3)_8(OH)_6$ electrode, the $Ni_{11}(HPO_3)_8(OH)_6/Co_3(HPO_4)_2(OH)_2$ electrodes demonstrated superior symmetry in their charge-discharge profiles, longer discharge durations, and a broader voltage plateau. These phenomena indicate that the $Ni_{11}(HPO_3)_8(OH)_6/Co_3(HPO_4)_2(OH)_2$ electrode possesses greater capacity scalability and higher charge conversion efficiency. This specific capacity for the $Ni_{11}(HPO_3)_8(OH)_6/Co_3(HPO_4)_2(OH)_2$ electrode is determined to be 163.2 mAh g^{-1}, representing a significant increase compared to 79.4 mAh g^{-1} exhibited by the $Ni_{11}(HPO_3)_8(OH)_6$ electrode. The GCD curves (Figure 4c,d) for the $Ni_{11}(HPO_3)_8(OH)_6$ and $Ni_{11}(HPO_3)_8(OH)_6/Co_3(HPO_4)_2(OH)_2$ electrodes show nonlinear charge/discharge profiles featuring distinct plateaus, attributable to reversible faraday redox reactions occurring both on the surface and within the electrodes. This further demonstrates their excellent Faraday cell performance. The specific capacities for the $Ni_{11}(HPO_3)_8(OH)_6/Co_3(HPO_4)_2(OH)_2$ electrodes were 163.2, 161.9, 161.4, 160, 158.3, 150, 134.6, and 115.9 mAh g^{-1} at 1, 2, 3, 5, 10, 20, 50, and 100 A g^{-1}, respectively, and those of $Ni_{11}(HPO_3)_8(OH)_6$ electrode are 79.4, 78.2, 77.6, 76.9, 76.2, 73.3, 65.3 and 54.8 mAh g^{-1}. The capacity retention for the $Ni_{11}(HPO_3)_8(OH)_6/Co_3(HPO_4)_2(OH)_2$ electrodes is about 71% at 100 A g^{-1}, which exhibits a notably elevated percentage in comparison to 68.9% for $Ni_{11}(HPO_3)_8(OH)_6$ electrodes, are exhibited within Figure 4e. The $Ni_{11}(HPO_3)_8(OH)_6/Co_3(HPO_4)_2(OH)_2$ electrode exhibits high specific capacity at ultra-high current densities, establishing its excellence in practical applications as a high-rate cell-typical electrode material. As shown in Figure 4f, Cyclic stability tests of the $Ni_{11}(HPO_3)_8(OH)_6/Co_3(HPO_4)_2(OH)_2$ electrodes were conducted through cyclic discharging and charging at 10 A g^{-1}. This capacity retention of the $Ni_{11}(HPO_3)_8(OH)_6/Co_3(HPO_4)_2(OH)_2$ electrode initially increases to 119% over the first 2000 cycles due to the electrode undergoing a gradual activation process. Subsequently, the capacity retention rate declines to 110.4% and remains stable for approximately 3700 cycles. After 5000 cycles, the $Ni_{11}(HPO_3)_8(OH)_6/Co_3(HPO_4)_2(OH)_2$ electrode demonstrated remarkable cycling performance, exhibiting an initial capacity retention of 110.4% and a maximum capacity retention of 92.7%. Additionally, the $Ni_{11}(HPO_3)_8(OH)_6/Co_3(HPO_4)_2(OH)_2$ electrode exhibits a Coulombic efficiency approaching 100%. The $Ni_{11}(HPO_3)_8(OH)_6/Co_3(HPO_4)_2(OH)_2$ electrode's excellent cycling stabilization may be due to the construction of heterostructures and unique multistage pore structure, along with the -OH groups incorporated within [39].

As illustrated in Table 1, the galvanic properties for the $Ni_{11}(HPO_3)_8(OH)_6/Co_3(HPO_4)_2(OH)_2$ electrodes are in comparison with other reported metal phosphide and phosphate electrodes. From the results of the comparison, it is found that the $Ni_{11}(HPO_3)_8(OH)_6/Co_3(HPO_4)_2(OH)_2$ electrode exhibits superior rate performance compared to other already reported electrodes. Furthermore, this method of constructing heterostructures to enhance material properties electrochemically is still rarely reported.

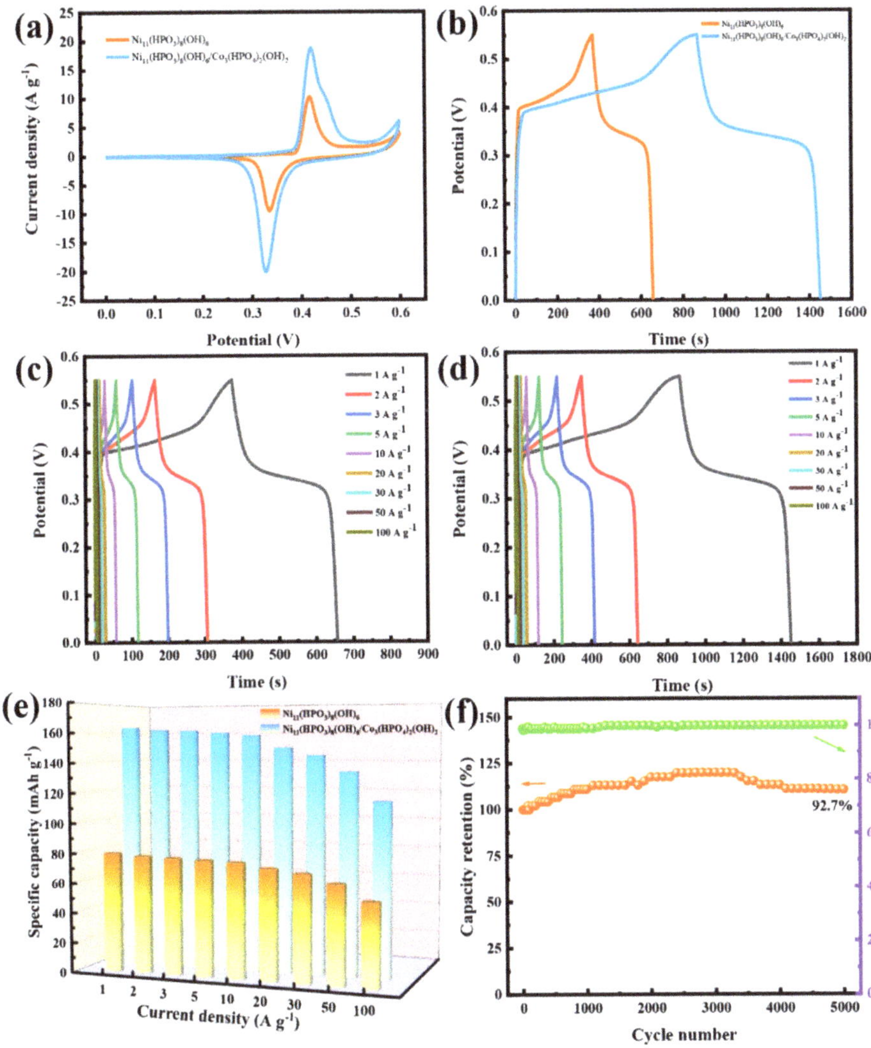

Figure 4. Electrochemical performance of the $Ni_{11}(HPO_3)_8(OH)_6$ and $Ni_{11}(HPO_3)_8(OH)_6/Co_3(HPO_4)_2(OH)_2$ electrodes in the three-electrode system: (**a**) CV curves at 2 mV s^{-1}, (**b**) GCD curves at 1 A g^{-1}. (**c**) GCD curves for the $Ni_{11}(HPO_3)_8(OH)_6$ electrode from 1 to 100 A g^{-1}. (**d**) GCD curves for the $Ni_{11}(HPO_3)_8(OH)_6/Co_3(HPO_4)_2(OH)_2$ electrode from1 to 100 A g^{-1}. (**e**) Rate performance and (**f**) cyclic stability and Coulombic efficiency for the $Ni_{11}(HPO_3)_8(OH)_6/Co_3(HPO_4)_2(OH)_2$ electrode.

Figure 5a,b presents display the CV curves for the $Ni_{11}(HPO_3)_8(OH)_6$ and $Ni_{11}(HPO_3)_8(OH)_6/Co_3(HPO_4)_2(OH)_2$ electrodes, respectively, within a range from 5 to 200mV s^{-1}. The redox peaks are visible on the CV curves, implying classical Faraday cell-type characteristics. Furthermore, the peak current distortion for electrodes is extremely small at 200 mV s^{-1}, displaying that the $Ni_{11}(HPO_3)_8(OH)_6/Co_3(HPO_4)_2(OH)_2$ electrode possesses rapid charging and discharging characteristics. Figure 5c displays the I_p-$v^{1/2}$ slopes of the oxidation peaks for the $Ni_{11}(HPO_3)_8(OH)_6$ and $Ni_{11}(HPO_3)_8(OH)_6/Co_3(HPO_4)_2(OH)_2$ electrodes, which are 17.8 and 32.1, respectively. Additionally, the I_p-$v^{1/2}$ slopes of the reduction peaks for these electrodes are −17.3 and −30.6, respectively. The two electrodes undergo a semi-quantitative analysis according to the charge storage mechanism to facilitate a deeper understanding of their electrochemical behavior, where this peak current I_p on scanning rate v is shown in the below equation: $I_p = a\,v^b$, $\log I_p = b \log v + \log a$, where the peak current denoted by the symbol I_p. The scan velocity is represented by v. Meanwhile, a and b serve as constants. Figure 5d displays the log I_p-log v fit line. The b-values for electrodes $Ni_{11}(HPO_3)_8(OH)_6$ and $Ni_{11}(HPO_3)_8(OH)_6/Co_3(HPO_4)_2(OH)_2$ are 0.7 and 0.75, respectively. Their values, ranging between 0.5 and 1, suggest a combined influence of diffusion and capacitive-controlled storage mechanisms, with capacitive control predominating. The above conclusions are subject to further verification.

Table 1. A summary of metal phosphides and phosphates used in SCs together with their corresponding electrochemical performance.

Material	Specific Capacity ($mAh\ g^{-1}$)	Rate Retention	Cyclic Property	Ref.
NiCoP/NiCo-OH	152.8 at 1 A g^{-1}	60% (1–10 A g^{-1})	88% after 1000 cycles	[40]
NiCoP	160.8 at 1 A g^{-1}	81% (1–16 A g^{-1})	/	[41]
Porous NiCoP	158.6 at 1 A g^{-1}	72.8% (1–20 A g^{-1})	72% after 3000 cycles	[42]
Ni-Co phosphate	125.8 at 1 A g^{-1}	63.4% (1–10 A g^{-1})	93% after 8000 cycles	[43]
$NH_4Co_{0.33}Ni_{0.67}PO_4{\cdot}H_2O$	158 at 1.5 A g^{-1}	66% (1.5–30 A g^{-1})	57% after 1000 cycles	[44]
$Ni_xCo_{3-x}(PO_4)_2$	94.2 at 1 A g^{-1}	81.4% (1–10 A g^{-1})	85% after 1000 cycles	[45]
$Co_{0.5}Ni_{0.5}$ pyrophosphate	161 at 1.5 A g^{-1}	/	/	[46]
Cobalt-doped Ni phosphite	83.6 at 0.5 A g^{-1}	85% (0.5–5 A g^{-1})	93% after 8000 cycles	[47]
$(Ni,Co)_3(PO_4)_2{\cdot}8H_2O/(NH_4)$ $(Ni, Co)PO_4{\cdot}0.67H_2O$	141 at 0.5 A g^{-1}	88% (0.5–24 A g^{-1})	/	[48]
Ni-Co phosphate	147 at 0.2 A g^{-1}	85% (0.2–10 A g^{-1})	/	[49]
Co(P, S)/CC	101.6 at 1 A g^{-1}	56% (1–20 A g^{-1})	/	[50]
$Ni_{11}(HPO_3)_8(OH)_6/Co_3(HPO_4)_2(OH)_2$	163.2 at 1 A g^{-1}	96.9% (1–10 A g^{-1}) 91.9% (1–20 A g^{-1}) 71% (1–100 A g^{-1})	92.7% after 5000 cycles	This work

Furthermore, the diffusion-controlled/surface capacitive charge contributions can be quantified by the below equations: $I_p\,(v) = k_1 v + k_2 v^{1/2}$. Here, $k_1 v$ is for surface capacitance, and $k_2 v^{1/2}$ is for diffusion control. Figure 5e,f depicts comparative contributions of capacitive and diffusive processes in the $Ni_{11}(HPO_3)_8(OH)_6$ and $Ni_{11}(HPO_3)_8(OH)_6/Co_3(HPO_4)_2(OH)_2$ electrodes at 10 mV s^{-1}, respectively. This capacitance-controlled contribution for $Ni_{11}(HPO_3)_8(OH)_6$ is 72.1%, while that of $Ni_{11}(HPO_3)_8(OH)_6/Co_3(HPO_4)_2(OH)_2$ is 80.6%. It is reiterated that the capacitive characteristic predominates over the overall capacity of the electrodes. Figure 5g displays capacitive influences from both electrodes at varying scan velocities. These distinct capacitive behaviors of the two electrodes demonstrated an increase as the scanning velocity escalated. Moreover, the capacitive contributions of the $Ni_{11}(HPO_3)_8(OH)_6/Co_3(HPO_4)_2(OH)_2$ electrode are observed to be greater than those of the $Ni_{11}(HPO_3)_8(OH)_6$ electrode at varying scan rates. Consistent with the rate performance trend, the $Ni_{11}(HPO_3)_8(OH)_6/Co_3(HPO_4)_2(OH)_2$ electrode exhibits a dominant surface capacitance of 74.6% at 2 mV s^{-1}, indicative of its typical cell behavior under low scanning rates. The surface capacitance contribution at 100 mV s^{-1}, at 99.8%, is concordant with CV analysis outcomes. EIS tests were conducted to investigate the electrode kinetics further. Figure 5h showcases Nyquist plots for the $Ni_{11}(HPO_3)_8(OH)_6$ and $Ni_{11}(HPO_3)_8(OH)_6/Co_3(HPO_4)_2(OH)_2$ electrodes. The radius of the semicircular arc signifies the impedance to charge migration (R_{ct}), and the x-intercept represents internal resistance (R_s) in the high-frequency region. Furthermore, the $Ni_{11}(HPO_3)_8(OH)_6$ and $Ni_{11}(HPO_3)_8(OH)_6/Co_3(HPO_4)_2(OH)_2$ electrodes demonstrated a markedly low R_{ct}, indicating the process of charge translocation transpires across the electrode-electrolyte juncture. The line slope of the $Ni_{11}(HPO_3)_8(OH)_6/Co_3(HPO_4)_2(OH)_2$ electrode is closer to 90° across the lower frequency spectrum, which suggests that the $Ni_{11}(HPO_3)_8(OH)_6/Co_3(HPO_4)_2(OH)_2$ electrode exhibits greater capacitive properties. Moreover, $Ni_{11}(HPO_3)_8(OH)_6/Co_3(HPO_4)_2(OH)_2$ exhibits a more pronounced linear gradient than that observed in $Ni_{11}(HPO_3)_8(OH)_6$. The electrode's enhanced ion diffusion rate stems from the unique hierarchical porous architecture for $Ni_{11}(HPO_3)_8(OH)_6/Co_3(HPO_4)_2(OH)_2$, coupled with its superior conductivity. The EIS analysis results serve to further corroborate the assertion that the $Ni_{11}(HPO_3)_8(OH)_6/Co_3(HPO_4)_2(OH)_2$ electrode exhibits superior internal resistance and commendable reaction kinetics in comparison to the $Ni_{11}(HPO_3)_8(OH)_6$ electrode.

For evaluating the potential application for $Ni_{11}(HPO_3)_8(OH)_6/Co_3(HPO_4)_2(OH)_2$ in energy storage, the HSC was constructed in 2 mol L^{-1} KOH utilizing PC as anode and this material as a cathode. Figure 6a illustrates the total voltage window of the HSC, derived from analyzing the CV profiles for the PC and $Ni_{11}(HPO_3)_8(OH)_6/Co_3(HPO_4)_2(OH)_2$ electrodes independently at 10 mV s^{-1} in 2 mol L^{-1} KOH electrolytes. The CV curves of the PCs exhibit a rectangular shape and demonstrate excellent electric double layer (EDL) behavior. The potential ranges for the PC and $Ni_{11}(HPO_3)_8(OH)_6/Co_3(HPO_4)_2(OH)_2$ electrodes are confined to -1 to 0 V as well as 0 to 0.6 V. Figure 6b displays rectangular shape for the CV curves for $Ni_{11}(HPO_3)_8(OH)_6/Co_3(HPO_4)_2(OH)_2$//PC HSC within 2~100 mV s^{-1}, with significant redox peak characteristics within the 0~1.5 V. Meanwhile, CV curves exhibited minimal variation across varying scan rates, signifying the device's outstanding rate performance. The capacitance characteristics for the $Ni_{11}(HPO_3)_8(OH)_6/Co_3(HPO_4)_2(OH)_2$//PC HSC underwent additional examination through performing GCD tests between 2~100 A g^{-1}. Figure 6c makes known that the symmetric GCD curve of the $Ni_{11}(HPO_3)_8(OH)_6/Co_3(HPO_4)_2(OH)_2$//PC HSC shows good electrochemical reversibility. Figure 6d reveals that the $Ni_{11}(HPO_3)_8(OH)_6/Co_3(HPO_4)_2(OH)_2$//PC HSC boasts a specific capacitance value of 151.6 F g^{-1} when tested at 2 A g^{-1}, and retains a capacitance of 58.8 F g^{-1} at 100 A g^{-1}. This underscores the device's high specific capacitance, attributable to the optimal chemical composition, distinctive layered porous architecture, and exceptional electronic conductivity for the $Ni_{11}(HPO_3)_8(OH)_6/Co_3(HPO_4)_2(OH)_2$ materials. Figure 6e illustrates the Ragone diagram of the $Ni_{11}(HPO_3)_8(OH)_6/Co_3(HPO_4)_2(OH)_2$//PC HSC. The energy density for $Ni_{11}(HPO_3)_8(OH)_6/Co_3(HPO_4)_2(OH)_2$//PC HSC reaches 47.4 Wh kg^{-1} at 660 Wh kg^{-1}, and maintains 18.4 Wh kg^{-1} even at a super-high 33 kW kg^{-1}. This surpasses the energy density reported for previous cobalt-based phosphide and nickel-based phosphate HSCs. For example, $Co_3(PO_4)_2$//AC (26.66 Wh kg^{-1}/750 W kg^{-1}) [51], $Co_3(PO_4)_2{\cdot}4H_2O$/GF//C-FP (24 Wh kg^{-1}/468 W kg^{-1}) [52], $NH_4CoPO_4{\cdot}H_2O$//AC (10.4 Wh kg^{-1}/141.3 W kg^{-1}) [53], NiCoP@NF//AC (27 Wh kg^{-1}/647 W kg^{-1}) [54], Ni-Co-P/PO_x//RGO(36.84 Wh kg^{-1}/727.8 W kg^{-1}) [55], $NiCoO_2$/NiCoP//AC (40.32Wh kg^{-1}/800.18 W kg^{-1}) [56], $Co_3(PO_4)_2{\cdot}8H_2O$//AC (29.29 Wh kg^{-1}/468.75 W kg^{-1}) [57]. The cycling properties for $Ni_{11}(HPO_3)_8(OH)_6/Co_3(HPO_4)_2(OH)_2$//PC HSC were examined through the implementation of cycling tests (Figure 6f). After 10,000 cycles, $Ni_{11}(HPO_3)_8(OH)_6/Co_3(HPO_4)_2(OH)_2$//PC HSC demonstrated remarkable cycling performance, exhibiting capacity retention of up to 92.9% at 3 A g^{-1}. Their feasibility in practical applications is confirmed by these test data.

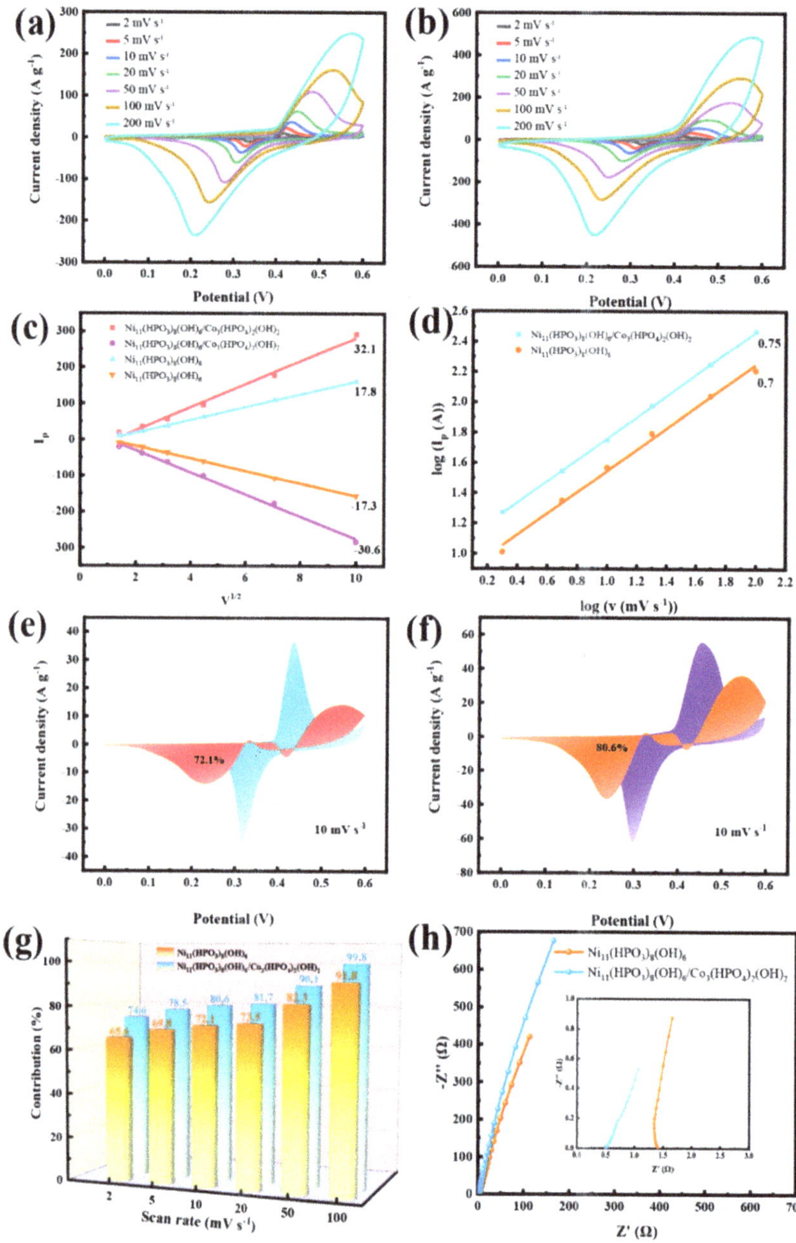

Figure 5. Electrochemical performance of the $Ni_{11}(HPO_3)_8(OH)_6$ and $Ni_{11}(HPO_3)_8(OH)_6/Co_3(HPO_4)_2(OH)_2$ electrodes: (**a**) $Ni_{11}(HPO_3)_8(OH)_6$ CV curves between 2 and 200 mV s^{-1}, (**b**) $Ni_{11}(HPO_3)_8(OH)_6/Co_3(HPO_4)_2(OH)_2$ CV curves between 2 and 200 mV s^{-1}, (**c**) linear relation of I_p-$v^{1/2}$, (**d**) log *Ip*-log *v* linear relationship, (**e**,**f**) diffusion and capacitive contributions at 10 mV s^{-1}, (**g**) capacitive contribution comparison from 2 to 100 mV s^{-1}, and (**h**) the EIS plots.

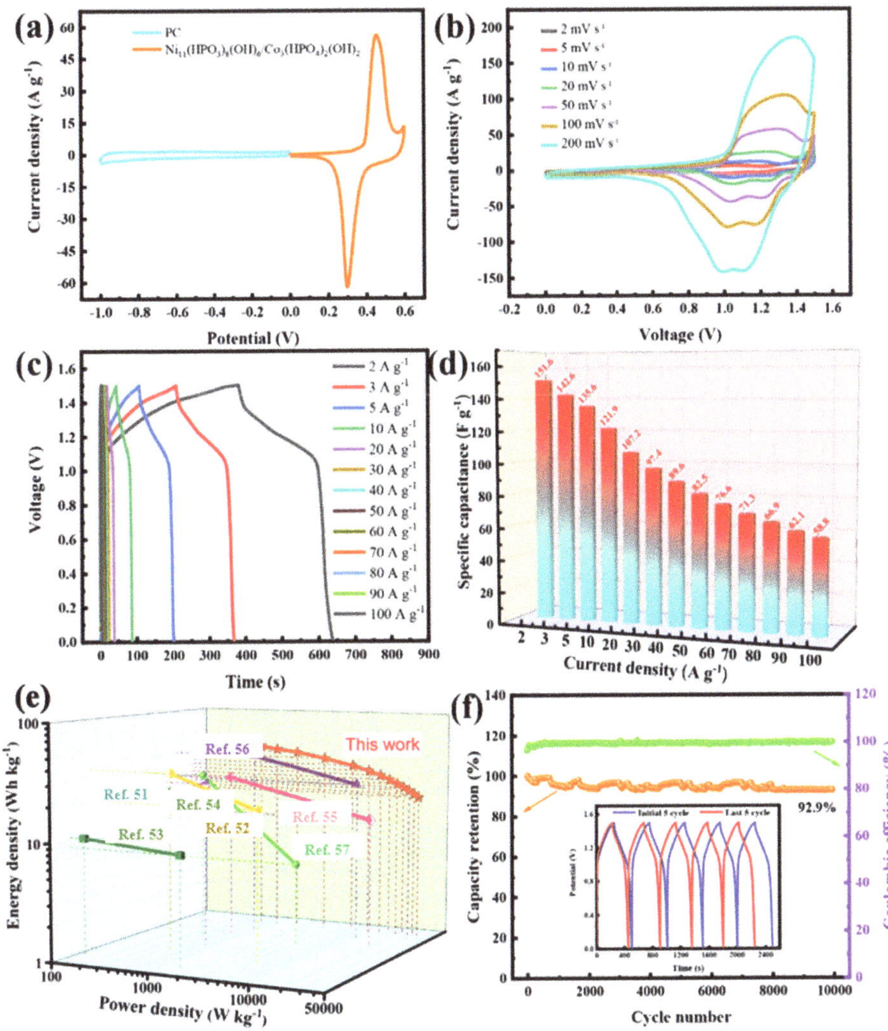

Figure 6. Electrochemical performance of $Ni_{11}(HPO_3)_8(OH)_6/Co_3(HPO_4)_2(OH)_2$//PC HSC: (**a**) CV profiles of PC and $Ni_{11}(HPO_3)_8(OH)_6/Co_3(HPO_4)_2(OH)_2$ electrodes at 10 mV s^{-1}. (**b**) CV profiles, (**c**) GCD profiles, (**d**) rate performance diagram, (**e**) Ragone plot, and (**f**) circulation characteristics and the Coulombic efficiency at 3 A g^{-1} after 10,000 cycles, inset picture shows the pre/post five GCD curves.

4. Conclusions

In summary, the $Ni_{11}(HPO_3)_8(OH)_6/Co_3(HPO_4)_2(OH)_2$ heterostructured material consisting of flakes has been synthesized through one-step hydrothermal process. Benefiting from its intricate multi-scale porosity architecture and the synergistic effect, the prepared $Ni_{11}(HPO_3)_8(OH)_6/Co_3(HPO_4)_2(OH)_2$ composite electrode demonstrates excellent cycling endurance and rate capability in alkaline conditions. Specifically, the $Ni_{11}(HPO_3)_8(OH)_6/Co_3(HPO_4)_2(OH)_2$ composite electrode exhibits superior capacity retention capabilities, with 82.5% retention at 50 A g^{-1} and 71% even at the elevated currently densities of 100 A g^{-1}. Moreover, it sustains an impressive 92.7% capacity retention after enduring 5000 cycles at 10 A g^{-1}. In addition, the $Ni_{11}(HPO_3)_8(OH)_6/Co_3(HPO_4)_2(OH)_2$//PC

HSC device showcases exceptional performance, featuring a heightened energy density of 47.4 Wh kg^{-1} alongside a superlative power density of 36 kW kg^{-1}. Notably, after enduring 10,000 cycles, it retains a good 92.9% capacity at 3 A g^{-1}. Obtained from the above data, the $Ni_{11}(HPO_3)_8(OH)_6/Co_3(HPO_4)_2(OH)_2$ heterostructured materials are likely candidates for electrode materials for high-performance HSC. Additionally, this hydrothermal synthesis method, with its low cost and simple process, provides a feasible idea for constructing phosphate heterostructured materials with excellent properties that may be applied to future storage systems.

Author Contributions: Writing—original draft preparation, K.L.; writing—review and editing, M.J., Y.Z. (Yirong Zhu) and C.L.; project administration, M.J. and Y.Z. (Yirong Zhu); formal analysis T.W. and Y.Z. (Yang Zhang); investigation, R.L. and X.W.; resources, L.L. and S.Z.; data curation, K.L. and Y.Z (Yang Zhang). All authors have read and agreed to the published version of the manuscript.

Funding: This research was funded by the Distinguished Young Scholar Fund Project of Hunan Province Natural Science Foundation (No. 2023JJ10041), the Hunan Provincial Education Office Foundation of China (No. 22A0114), the Natural Science Foundation of Henan Province (232300421228), the Topnotch Talents Program of Henan Agricultural University (30501032), and the National Natural Science Foundation of China (22305070).

Data Availability Statement: The dataset is available on request from the authors.

Conflicts of Interest: The authors declare no conflicts of interest.

References

1. Pomerantseva, E.; Bonaccorso, F.; Feng, X.; Cui, Y.; Gogotsi, Y. Energy storage: The future enabled by nanomaterials. *Science* **2019**, *366*, eaan8285. [CrossRef] [PubMed]
2. Bi, S.; Banda, H.; Chen, M.; Niu, L.; Chen, M.; Wu, T.; Wang, J.; Wang, R.; Feng, J.; Chen, T.; et al. Molecular understanding of charge storage and charging dynamics in supercapacitors with MOF electrodes and ionic liquid electrolytes. *Nat. Mater.* **2020**, *19*, 552–558. [CrossRef] [PubMed]
3. Zhang, Q.-Z.; Zhang, D.; Miao, Z.-C.; Zhang, X.-L.; Chou, S.-L. Research Progress in MnO_2–Carbon Based Supercapacitor Electrode Materials. *Small* **2018**, *14*, 1702883. [CrossRef] [PubMed]
4. Lee, J.-H.; Yang, G.; Kim, C.-H.; Mahajan, R.L.; Lee, S.-Y.; Park, S.-J. Flexible solid-state hybrid supercapacitors for the internet of everything (IoE). *Energy Environ. Sci.* **2022**, *15*, 2233–2258. [CrossRef]
5. Zhao, Z.; Ye, Y.; Zhu, W.; Xiao, L.; Deng, B.; Liu, J. Bismuth oxide nanoflake@carbon film: A free-standing battery-type electrode for aqueous sodium ion hybrid supercapacitors. *Chin. Chem. Lett.* **2018**, *29*, 629–632. [CrossRef]
6. Amiri, A.; Bruno, A.; Polycarpou, A.A. Configuration-dependent stretchable all-solid-state supercapacitors and hybrid supercapacitors. *Carbon Energy* **2023**, *5*, e320. [CrossRef]
7. Sadavar, S.; Wang, K.J.; Kang, T.; Hwang, M.; Saeed, G.; Yu, X.; Park, H.S. Anion storage for hybrid supercapacitor. *Mater. Today Energy* **2023**, *37*, 101388. [CrossRef]
8. Wang, X.; Yan, C.; Yan, J.; Sumboja, A.; Lee, P.S. Orthorhombic niobium oxide nanowires for next generation hybrid supercapacitor device. *Nano Energy* **2015**, *11*, 765–772. [CrossRef]
9. Sun, L.; Liu, Y.; Yan, M.; Yang, Q.; Liu, X.; Shi, W. Lewis acid etched $Ni_xCo_{1-x}Se_2$ derived from ZIF-L on CoO nanowires for hybrid-supercapacitors. *Chem. Eng. J.* **2022**, *431*, 133472. [CrossRef]
10. Roy, A.; Inta, H.R.; Ghosh, S.; Koppisetti, H.V.S.R.M.; Mondal, A.; Verma, B.R.; Bag, S.; Mahalingam, V. Electrochemical surface reconstruction of nickel cobalt pyrophosphate to Ni/Co-hydroxide-(oxy)hydroxide: An efficient and highly durable battery-type supercapacitor electrode material. *J. Mater. Chem. A* **2024**, *12*, 4086–4098. [CrossRef]
11. Salunkhe, R.R.; Kaneti, Y.V.; Yamauchi, Y. Metal–Organic Framework-Derived Nanoporous Metal Oxides toward Supercapacitor Applications: Progress and Prospects. *ACS Nano* **2017**, *11*, 5293–5308. [CrossRef] [PubMed]
12. Yu, X.Y.; Lou, X.W. Mixed Metal Sulfides for Electrochemical Energy Storage and Conversion. *Adv. Funct. Mater.* **2018**, *8*, 1701592. [CrossRef]
13. Zou, K.; Deng, W.; Silvester, D.S.; Zou, G.; Hou, H.; Banks, C.E.; Li, L.; Hu, J.; Ji, X. Carbonyl Chemistry for Advanced Electrochemical Energy Storage Systems. *ACS Nano* **2024**, *18*, 19950–20000. [CrossRef] [PubMed]
14. Zan, G.; Li, S.; Chen, P.; Dong, K.; Wu, Q.; Wu, T. Mesoporous Cubic Nanocages Assembled by Coupled Monolayers With 100% Theoretical Capacity and Robust Cycling. *ACS Cent. Sci.* **2024**, *10*, 1283–1294. [CrossRef]
15. Manikandan, R.; Savariraj, A.D.; Nagaraju, G.; Kale, A.M.; Puigdollers, J.; Park, H.; Kim, H.-S.; Oh, J.-M.; Raj, C.J.; Kim, B.C. Mixed-phase composites derived from cobalt terephthalate as efficient battery-type electrodes for high-performance supercapattery. *J. Mater. Sci. Technol.* **2023**, *157*, 220–233. [CrossRef]

16. Manikandan, R.; Raj, C.J.; Goli, N.; Oh, J.-M.; Kim, B.C.; Periyasamy, S.; Lee, J. Interconnected Vanadyl Pyrophosphate Nanonetworks as a Flexible Electrode for High-Voltage and Long-Life Li-Ion Supercapacitors. *Appl. Mater. Interfaces* **2023**, *15*, 25452–25461. [CrossRef]
17. Zhao, H.; Yuan, Z.-Y. Design Strategies of Transition-Metal Phosphate and Phosphonate Electrocatalysts for Energy-Related Reactions. *Chem. Eng. Sci* **2021**, *14*, 130–149. [CrossRef]
18. Cheng, Q.; Zhao, X.; Yang, G.; Mao, L.; Liao, F.; Chen, L.; He, P.; Pan, D.; Chen, S. Recent advances of metal phosphates-based electrodes for high-performance metal ion batteries. *Energy Storage Mater.* **2021**, *41*, 842–882. [CrossRef]
19. Wang, D.; Wang, Y.; Fu, Z.; Xu, Y.; Yang, L.-X.; Wang, F.; Guo, X.; Sun, W.; Yang, Z.-L. Cobalt–Nickel Phosphate Composites for the All-Phosphate Asymmetric Supercapacitor and Oxygen Evolution Reaction. *Appl. Mater. Interfaces* **2021**, *13*, 34507–34517. [CrossRef]
20. Xie, L.; Zong, Q.; Zhang, Q.; Sun, J.; Zhou, Z.; He, B.; Zhu, Z.; E, S.; Yao, Y. Hierarchical NiCoP nanosheet arrays with enhanced electrochemical properties for high-performance wearable hybrid capacitors. *J. Alloys Compd.* **2019**, *781*, 783–789. [CrossRef]
21. Fan, J.; Wu, K.; Chen, A.; Wang, M.; Xie, X.; Fan, J. Fingerprint-like NiCoP electrode material with rapid charging/discharging performance under large current density. *J. Alloys Compd.* **2022**, *902*, 163866. [CrossRef]
22. Liu, M.; Li, J.; Han, W.; Kang, L. Simple synthesis of novel phosphate electrode materials with unique microstructure and enhanced supercapacitive properties. *J. Energy Chem.* **2016**, *25*, 601–608. [CrossRef]
23. Zhang, Y.; Zheng, M.; Qu, M.; Sun, M.; Pang, H. Core–shell $Co_{11}(HPO_3)_8(OH)_6$–Co_3O_4 hybrids for high-performance flexible all-solid-state asymmetric supercapacitors. *J. Alloys Compd.* **2015**, *651*, 214–221. [CrossRef]
24. Li, X.; Elshahawy, A.M.; Guan, C.; Wang, J. Metal Phosphides and Phosphates-based Electrodes for Electrochemical Supercapacitors. *Small* **2017**, *13*, 1701530. [CrossRef] [PubMed]
25. Wang, X.; Zhai, X.; Lv, P.; Jiao, Y.; Wang, S.; Chi, J. Amorphous nickel phosphate as a high performance electrode material for supercapacitor. *Synth. Met.* **2023**, *292*, 117217. [CrossRef]
26. Zhang, Y.-N.; Su, C.-Y.; Chen, J.-L.; Huang, W.-H.; Lou, R. Recent progress of transition metal-based biomass-derived carbon composites for supercapacitor. *Rare Met.* **2023**, *42*, 769–796. [CrossRef]
27. Zhang, G.C.; Feng, M.; Li, Q.; Wang, Z.; Fang, Z.; Niu, Z.; Qu, N.; Fan, X.; Li, S.; Gu, J.; et al. High Energy Density in Combination with High Cycling Stability in Hybrid Supercapacitors. *Appl. Mater. Interfaces* **2022**, *14*, 2674–2682. [CrossRef]
28. Mirghni, A.A.; Madito, M.J.; Oyedotun, K.O.; Masikhwa, T.M.; Ndiaye, N.M.; Ray, S.J.; Manyala, N. A high energy density asymmetric supercapacitor utilizing a nickel phosphate/graphene foam composite as the cathode and carbonized iron cations adsorbed onto polyaniline as the anode. *RSC Adv.* **2018**, *8*, 11608–11621. [CrossRef]
29. Chen, G.; Yan, L.; Luo, H.; Guo, S. Nanoscale Engineering of Heterostructured Anode Materials for Boosting Lithium-Ion Storage. *Adv. Mater.* **2016**, *28*, 7580–7602. [CrossRef]
30. Lin, Y.; Sun, K.; Liu, S.; Chen, X.; Cheng, Y.; Cheong, W.-C.; Chen, Z.; Zheng, L.; Zhang, J.; Li, X.; et al. Construction of CoP/NiCoP Nanotadpoles Heterojunction Interface for Wide pH Hydrogen Evolution Electrocatalysis and Supercapacitor. *Adv. Funct. Mater.* **2019**, *9*, 1901213. [CrossRef]
31. Liu, X.; Zhu, Y.; Lu, Z.; Xiao, J.; Zou, G.; Hou, H.; Ji, X. Heterostructured flower-like NiO/Co_3O_4 microspheres modified by bifunctional carbon quantum dots as a battery-type cathode for high energy and power density hybrid supercapacitors. *Carbon Neutralization* **2023**, *2*, 721–737. [CrossRef]
32. Li, Y.; Zhang, J.; Chen, Q.; Xia, X.; Chen, M. Emerging of Heterostructure Materials in Energy Storage: A Review. *Adv. Mater.* **2021**, *33*, 2100855. [CrossRef] [PubMed]
33. Yun, X.; Lu, T.; Zhou, R.; Lu, Z.; Li, J.; Zhu, Y. Heterostructured $NiSe_2/CoSe_2$ hollow microspheres as battery-type cathode for hybrid supercapacitors: Electrochemical kinetics and energy storage mechanism. *Chem. Eng. J.* **2021**, *426*, 131328. [CrossRef]
34. Li, M.; Li, M.; Zhao, M.; Wu, B.; Mourdikoudis, S.; Wei, S.; Oliveira, F.M.; He, J.; Děkanovský, L.; Luxa, J.; et al. Rational design of crystalline/amorphous nickel manganese phosphate octahydrate heterostructure for high-performance aqueous and all-solid-state asymmetric supercapacitors. *Chem. Eng. J.* **2024**, *482*, 148895. [CrossRef]
35. Jiang, T.; Zhang, Y.; Du, C.; Xiao, T.; Wan, L. Two-step electrodeposition synthesis of iron cobalt selenide and nickel cobalt phosphate heterostructure for hybrid supercapacitors. *J. Colloid Interface Sci.* **2023**, *629*, 1049–1060. [CrossRef]
36. Davidson, A.; Tempere, J.F.; Che, M.; Roulet, H.; Dufour, G. Spectroscopic Studies of Nickel(II) and Nickel(III) Species Generated upon Thermal Treatments of Nickel/Ceria-Supported Materials. *J. Phys. Chem.* **1996**, *100*, 4919–4929. [CrossRef]
37. Casella, I.G.; Gatta, M. Electrochemical and XPS Characterization of Composite Modified Electrodes Obtained by Nickel Deposition on Noble Metals. *Anal. Chem.* **2000**, *72*, 2969–2975. [CrossRef]
38. Huang, J.; Xiong, Y.; Peng, Z.; Chen, L.; Wang, L.; Xu, Y.; Tan, L.; Yuan, K.; Chen, Y. A General Electrodeposition Strategy for Fabricating Ultrathin Nickel Cobalt Phosphate Nanosheets with Ultrahigh Capacity and Rate Performance. *ACS Nano* **2020**, *14*, 14201–14211. [CrossRef]
39. Dan, H.; Tao, K.; Zhou, Q.; Gong, Y.; Lin, J. Ni-Doped Cobalt Phosphite, $Co_{11}(HPO_3)_8(OH)_6$, with Different Morphologies Grown on Ni Foam Hydro(solvo)thermally for High-Performance Supercapacitor. *Appl. Mater. Interfaces* **2018**, *10*, 31340–31354. [CrossRef]
40. Li, X.; Wu, H.; Elshahawy, A.M.; Wang, L.; Pennycook, S.J.; Guan, C.; Wang, J. Cactus-Like NiCoP/NiCo-OH 3D Architecture with Tunable Composition for High-Performance Electrochemical Capacitors. *Adv. Funct. Mater.* **2018**, *28*, 1800036. [CrossRef]

41. Ding, R.; Li, X.; Shi, W.; Xu, Q.; Liu, E. One-pot solvothermal synthesis of ternary Ni-Co-P micro/nano-structured materials for high performance aqueous asymmetric supercapacitors. *Chem. Eng. J.* **2017**, *320*, 376–388. [CrossRef]
42. Hu, Y.-M.; Liu, M.-C.; Hu, Y.-X.; Yang, Q.-Q.; Kong, L.-B.; Kang, L. One-pot hydrothermal synthesis of porous nickel cobalt phosphides with high conductivity for advanced energy conversion and storage. *Electrochim. Acta* **2016**, *215*, 114–125. [CrossRef]
43. Li, B.; Gu, P.; Feng, Y.; Zhang, G.; Huang, K.; Xue, H.; Pang, H. Ultrathin Nickel–Cobalt Phosphate 2D Nanosheets for Electrochemical Energy Storage under Aqueous/Solid-State Electrolyte. *Adv. Funct. Mater.* **2017**, *27*, 1605784. [CrossRef]
44. Liang, B.; Chen, Y.; He, J.; Chen, C.; Liu, W.; He, Y.; Liu, X.; Zhang, N.; Roy, V.A.L. Controllable Fabrication and Tuned Electrochemical Performance of Potassium Co–Ni Phosphate Microplates as Electrodes in Supercapacitors. *Appl. Mater. Interfaces* **2018**, *10*, 3506–3514. [CrossRef]
45. Zhang, J.; Yang, Y.; Zhang, Z.; Xu, X.; Wang, X. Rapid synthesis of mesoporous $Ni_xCo_{3-x}(PO_4)_2$ hollow shells showing enhanced electrocatalytic and supercapacitor performance. *J. Mater. Chem. A* **2014**, *2*, 20182–20188. [CrossRef]
46. Chen, C.; Zhang, N.; He, Y.; Liang, B.; Ma, R.; Liu, X. Controllable Fabrication of Amorphous Co—Ni Pyrophosphates for Tuning Electrochemical Performance in Supercapacitors. *Appl. Mater. Interfaces* **2016**, *8*, 23114–23121. [CrossRef]
47. Li, B.; Shi, Y.; Huang, K.; Zhao, M.; Qiu, J.; Xue, H.; Pang, H. Cobalt-Doped Nickel Phosphite for High Performance of Electrochemical Energy Storage. *Small* **2018**, *14*, 1703811. [CrossRef]
48. Zhao, Y.; Chen, Z.; Xiong, D.-B.; Qiao, Y.; Tang, Y.; Gao, F. Hybridized Phosphate with Ultrathin Nanoslices and Single Crystal Microplatelets for High Performance Supercapacitors. *Sci. Rep.* **2016**, *6*, 17613. [CrossRef]
49. Tang, Y.; Liu, Z.; Guo, W.; Chen, T.; Qiao, Y.; Mu, S.; Zhao, Y.; Gao, F. Honeycomb-like mesoporous cobalt nickel phosphate nanospheres as novel materials for high performance supercapacitor. *Electrochim. Acta* **2016**, *190*, 118–125. [CrossRef]
50. Elshahawy, A.M.; Guan, C.; Li, X.; Zhang, H.; Hu, Y.; Wu, H.; Pennycook, S.J.; Wang, J. Sulfur-doped cobalt phosphide nanotube arrays for highly stable hybrid supercapacitor. *Nano Energy* **2017**, *39*, 162–171. [CrossRef]
51. Mao, H.; Zhang, F.; Liu, X.; Qiu, J.; Li, B.; Jin, Z. Synthesis of cobalt phosphate nanoflakes for high-performance flexible symmetric supercapacitors. *J. Mater. Sci. Mater. Electron.* **2018**, *29*, 16721–16729. [CrossRef]
52. Pang, H.; Wang, S.; Shao, W.; Zhao, S.; Yan, B.; Li, X.; Li, S.; Chen, J.; Du, W. Few-layered $CoHPO_4{\cdot}3H_2O$ ultrathin nanosheets for high performance of electrode materials for supercapacitors. *Nanoscale* **2013**, *5*, 5752–5757. [CrossRef] [PubMed]
53. Pang, H.; Yan, Z.; Wang, W.; Chen, J.; Zhang, J.; Zheng, H. Facile fabrication of $NH_4CoPO_4{\cdot}H_2O$ nano/microstructures and their primarily application as electrochemical supercapacitor. *Nanoscale* **2012**, *4*, 5946–5953. [CrossRef] [PubMed]
54. Lan, Y.; Zhao, H.; Zong, Y.; Li, X.; Sun, Y.; Feng, J.; Wang, Y.; Zheng, X.; Du, Y. Phosphorization boosts the capacitance of mixed metal nanosheet arrays for high performance supercapacitor electrodes. *Nanoscale* **2018**, *10*, 11775–11781. [CrossRef] [PubMed]
55. Chen, H.C.; Jiang, S.; Xu, B.; Huang, C.; Hu, Y.; Qin, Y.; He, M.; Cao, H. Sea-urchin-like nickel–cobalt phosphide/phosphate composites as advanced battery materials for hybrid supercapacitors. *J. Mater. Chem. A* **2019**, *7*, 6241–6249. [CrossRef]
56. Ning, W.-W.; Chen, L.-B.; Wei, W.-F.; Chen, Y.-J.; Zhang, X.-Y. $NiCoO_2$/NiCoP@Ni nanowire arrays: Tunable composition and unique structure design for high-performance winding asymmetric hybrid supercapacitors. *Rare Met.* **2020**, *39*, 1034–1044. [CrossRef]
57. Li, H.; Yu, H.; Zhai, J.; Sun, L.; Yang, H.; Xie, S. Self-assembled 3D cobalt phosphate octahydrate architecture for supercapacitor electrodes. *Mater. Lett.* **2015**, *152*, 25–28. [CrossRef]

batteries

MDPI

Article

Honeycomb-like N-Doped Carbon Matrix-Encapsulated $Co_{1-x}S/Co(PO_3)_2$ Heterostructures for Advanced Lithium-Ion Capacitors

Yutao Liu [1], Xiaopeng Xie [2], Zhaojia Wu [2], Tao Wen [3], Fang Zhao [4], Hao He [3], Junfei Duan [3] and Wen Wang [5,*]

1 International Institute of Engineering, Changsha University of Science and Technology, Changsha 410114, China; 19520009053@163.com
2 Hunan Datang Xianyi Technology Co., Ltd., Changsha 410114, China; xnhqin@126.com (X.X.); win32logon@163.com (Z.W.)
3 Materials Science and Engineering, Changsha University of Science and Technology, Changsha 410114, China; 19108471202@163.com (T.W.); hehao0520@gmail.com (H.H.); junfei_duan@csust.edu.cn (J.D.)
4 Guangdong Power Grid Co., Ltd., Guangzhou Power Supply Bureau, Guangzhou 510660, China; 15603072389@163.com
5 State Key Laboratory Disaster Prevention & Reduction for Power Grid, Changsha University of Science and Technology, Changsha 410114, China
* Correspondence: wew@csust.edu.cn

Abstract: Lithium-ion capacitors (LICs) are emerging as promising hybrid energy storage devices that combine the high energy densities of lithium-ion batteries (LIBs) with high power densities of supercapacitors (SCs). Nevertheless, the development of LICs is hindered by the kinetic imbalances between battery-type anodes and capacitor-type cathodes. To address this issue, honeycomb-like N-doped carbon matrices encapsulating $Co_{1-x}S/Co(PO_3)_2$ heterostructures were prepared using a simple chemical blowing-vulcanization process followed by phosphorylation treatment ($Co_{1-x}S/Co(PO_3)_2$@NC). The $Co_{1-x}S/Co(PO_3)_2$@NC features a unique heterostructure engineered within carbon honeycomb structures, which efficiently promotes charge transfer at the interfaces, alleviates the volume expansion of Co-based materials, and accelerates reaction kinetics. The optimal $Co_{1-x}S/Co(PO_3)_2$@NC composite demonstrates a stable reversible capacity of 371.8 mAh g^{-1} after 800 cycles at 1 A g^{-1}, and exhibits an excellent rate performance of 242.9 mAh g^{-1} even at 8 A g^{-1}, alongside enhanced pseudocapacitive behavior. The assembled $Co_{1-x}S/Co(PO_3)_2$@NC//AC LIC delivers a high energy density of 90.47 Wh kg^{-1} (at 26.28 W kg^{-1}), a high power density of 504.94 W kg^{-1} (at 38.31 Wh kg^{-1}), and a remarkable cyclic stablitiy of 86.3% retention after 5000 cycles. This research is expected to provide valuable insights into the design of conversion-type electrode materials for future energy storage applications.

Keywords: metal sulfides; anode material; heterostructures; honeycomb carbons; lithium-ion capacitors

1. Introduction

Lithium-ion capacitors (LICs) are considered promising candidates in the field of electrochemical energy storage due to their hybrid energy storage capabilities, which combine the high energy characteristics of lithium-ion batteries (LIBs) with the high power densities of supercapacitors (SCs). However, the practical implementation of LICs remains challenging due to the distinct energy storage mechanisms: the high-energy anodes operate through a slower, repeated Faraday reaction of lithium ions, while the cathodes undergo rapid anion absorption and desorption [1–3].

To address this challenge, researchers have devoted significant efforts to discovering new battery-type anode materials that offer both rapid rate performance and high capacities. Among these materials, cobalt sulfides, which are typical transition metal sulfides, show promise as conversion-type anodes due to their potential application on account of their high theoretical specific capacities (590 mAh g^{-1}), weaker metal–sulfur bonds, and Co/Li_2S interfaces offering high electron transfer mobility [4,5]. However, cobalt sulfides experience significant volume expansion and contraction during repeated lithiation/delithiation cycles, leading to particles aggregation and the detrimental pulverization of electrode structures. This results in a rapid decline in rate performance, creating a mismatch with the capacitor-type cathodes in LICs. Numerous approaches have been proposed to improve reaction kinetics and structural stability, including downsizing particle size [6,7], constructing hierarchical porous structures [8,9], integrating with heteroatom-doped conductive carbonaceous materials [10–12], and fabricating heterogeneous architectures [13–15]. These approaches are expected to alleviate internal stress, increase contact with the electrolyte and shorten electronic and ionic transport pathways, enhance conductivity and maintain the integrity of the electrode materials, and regulate the electronic structure, ultimately aiming to enhance electrochemical performance, including increasing capacity, rate, and cycling stability. It is noteworthy that heterostructures composed of coupling components with large energy bandgap differences can induce a strong intrinsic electric field at the heterogeneous interfaces, thereby promoting reaction kinetics. Moreover, the as-formed heterointerfaces, characterized by abundant phase boundaries, have been shown to significantly influence rate capability and reversible capacity compared to individual components [16–18]. Recently, Dong and co-workers demonstrated that the CoS_2/Co_4S_3@NC heterogeneous composites exhibit high pseudocapacitive behavior and accelerate the diffusion rate of ions and electrons, thus improving the overall performance of the electrodes [19]. Wan and colleagues prepared a hierarchical binary metal sulfide (CoSn)S/C heterostructure, where the micro/nanostructures of (CoSn)S/C shorten the ion diffusion distance, facilitating the rapid transfer of Na^+ and enhancing the mechanical strength of the electrode [20]. Although cobalt-based sulfides play a significant role in improving the electrochemical properties of LICs, unlocking the potential of heterogeneous materials still requires a great deal of work to be explored.

Herein, honeycomb-like N-doped carbon matrices encapsulating $Co_{1-x}S/Co(PO_3)_2$ heterostructures are prepared using a simple chemical blowing-vulcanization process followed by phosphorylation treatment. The selection of $Co_{1-x}S$ and $Co(PO_3)_2$ components is based on their bandgaps, which are 0 eV for $Co_{1-x}S$ and approximately 2.6 eV for $Co(PO_3)_2$. The heterointerfaces between $Co_{1-x}S$ and $Co(PO_3)_2$ components induce an internal electric field, increasing the number of electrochemical active sites, improving electrochemical kinetics, and stabilizing the structure during the repeated charge/discharge. Additionally, the integrated honeycomb-structured N-doped carbon matrices provide physical confinement to reduce nanograins aggregation, accommodate volume expansion during repeated long cycling, and ensure the integrity of the electrode materials, resulting in a high reversible capacity. As expected, the optimal $Co_{1-x}S/Co(PO_3)_2$@NC composite demonstrates a stable reversible capacity of 371.8 mAh g^{-1} after 800 cycles at 1 A g^{-1}, and an excellent rate performance of 242.9 mAh g^{-1} even at 8 A g^{-1}, accompanied with enhanced pseudocapacitive behavior. Furthermore, the assembled $Co_{1-x}S/Co(PO_3)_2$@NC//AC LIC delivers a high energy density of 90.47 Wh kg^{-1} (at 26.28 W kg^{-1}), a high power density of 504.94 W kg^{-1} (at 38.31 Wh kg^{-1}), and a high cyclic stability of 86.3% retention after 5000 cycles.

2. Results and Discussion

The synthetic strategy for the honeycomb-like N-doped carbon matrix decorated with $Co_{1-x}S/Co(PO_3)_2$ heterostructures is schematically illustrated in Figure 1a. Initially, polyvinylpyrrolidone (PVP), serving as the carbon source, weakly coordinates with Co^{2+} to form a sol–gel coordination complex precursor [21,22]. The precursor provides physical confinement to prevent the agglomeration of $Co_{1-x}S$ nanograins during the calcination process. Subsequently, the precursor and mixed sulfur powder undergo thermal blowing-vulcanization treatment. Due to the high viscosity of molten PVP, the pyrolysis gasses (NO_2, NO, etc.) released from nitrate ions generate numerous bubbles as the temperature rises. These bubbles rapidly rupture, promoting the growth of interconnected thin carbon walls, resulting in the formation of N-doped carbon matrices decorated with cobalt sulfides that resemble a 3D honeycomb-like structure. Afterwards, $Co_{1-x}S/Co(PO_3)_2$@NC is obtained through a facile phosphorylation process of the $Co_{1-x}S$/@NC composite, yielding conformal $Co_{1-x}S/Co(PO_3)_2$ heterostructures. FESEM results from Figure 1b–d show that the as-prepared $Co(PO_3)_2$@NC, $Co_{1-x}S$@NC, and $Co_{1-x}S/Co(PO_3)_2$@NC composites exhibit a typical 3D interconnected honeycomb-like framework comprising numerous thin carbon walls with smooth surfaces. This indicates sulfides/phosphides are uniformly encapsulated within the carbon matrices. This structure facilitates electrolyte penetration and effectively mitigates the volume change in active species during repeated cycling, thereby maintaining structural stability. TEM was employed to further examine the morphology and chemical properties. As shown in Figures 1e–g and S1–S3, numerous small nanoparticles are found within the carbon matrices of all the $Co_{1-x}S/Co(PO_3)_2$@NC, $Co_{1-x}S$@NC, and $Co(PO_3)_2$@NC composites. Notably, the $Co_{1-x}S/Co(PO_3)_2$@NC contains nanoparticles with an average size of less than 5 nm, demonstrating shortened electronic and ionic transport paths and effectively relieving internal strain during the electrochemical process. The lattice spacings of 0.268 and 0.254 nm in the higher magnification image (Figure 1e,f) correspond to the (130) plane of $Co(PO_3)_2$ and the (101) plane of $Co_{1-x}S$, respectively. Similarly, the lattice spacings of 0.194 and 0.187 nm in Figure 1g correspond to the (102) plane of $Co_{1-x}S$ and the (−242) plane of $Co(PO_3)_2$, respectively, confirming the successful fabrication of heterostructures. The energy dispersive X-ray spectrometer (EDS) spectrum in Figure 1h confirmed the uniform dispersion of Co, S, P, N, and O in the carbon matrices.

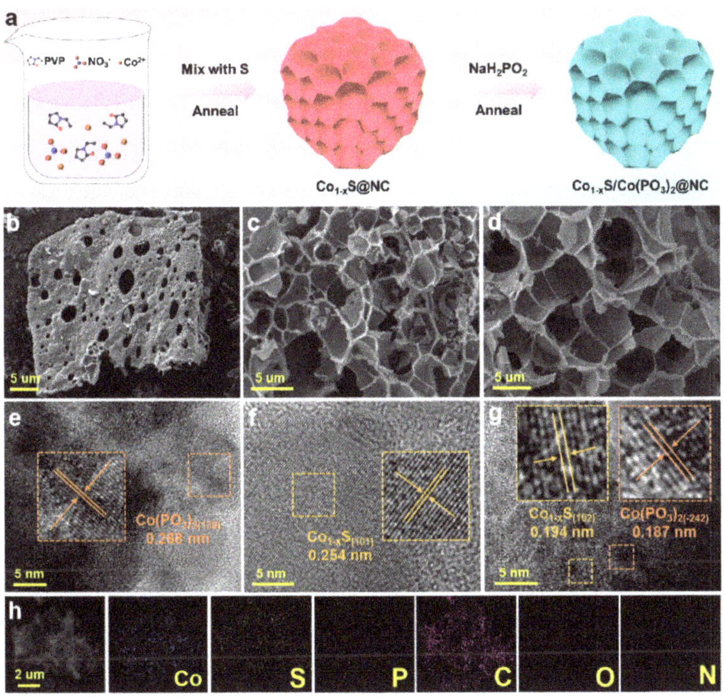

Figure 1. (**a**) Schematic illustration of synthetic process. SEM images of (**b**) $Co(PO_3)_2$@NC, (**c**) $Co_{1-x}S$@NC, and (**d**) $Co_{1-x}S/Co(PO_3)_2$@NC. HRTEM images of (**e**) $Co(PO_3)_2$@NC, (**f**) $Co_{1-x}S$@NC, and (**g**) $Co_{1-x}S/Co(PO_3)_2$@NC. (**h**) Elemental mapping of $Co_{1-x}S/Co(PO_3)_2$@NC.

The crystal structure of the synthesized materials was characterized using X-ray diffraction (XRD) patterns, as depicted in Figure 2a. The diffraction peaks of the $Co_{1-x}S/Co(PO_3)_2$@NC composites were consistently indexed to the monoclinic $Co(PO_3)_2$ phase (JCPDS No. 27-

1120) and the hexagonal $Co_{1-x}S$ phase (JCPDS No. 42-0826). For the $Co_{1-x}S$@NC and $Co(PO_3)_2$@NC composites, only single phases of $Co_{1-x}S$ and $Co(PO_3)_2$ were observed, without any impurity peaks. Moreover, the Raman spectra for three samples (Figure 2b) exhibit two prominent peaks at ~1351 and ~1371 cm^{-1}, which represent the graphitic carbon (G-band) and disordered carbon (D-band), respectively. This confirms the presence of N-doped carbon matrices. The relatively larger intensity ratio of I_D/I_G for $Co_{1-x}S/Co(PO_3)_2$@NC (0.989) implies a higher degree of structural defects compared to $Co_{1-x}S$@NC (0.973) and $Co(PO_3)_2$@NC (0.969), which is conducive for fast electron transfer [23]. Thermogravimetric analysis (TGA) was conducted to determine the contents of $Co_{1-x}S$, $Co(PO_3)_2$, and carbon in $Co_{1-x}S/Co(PO_3)_2$@NC (Figure 2c). As reported in our previous work, $Co_{1-x}S$ undergoes a multi-stage reaction involving complex oxidation processes in the air followed by carbon combustion [24,25]. The evaporation of adsorbed water accounts for the slight mass loss in the range of 50–200 °C. The oxidation of $Co_{1-x}S$ to $CoSO_4$ and Co_3O_4 is the primary cause of a notable weight loss, as well as the combustion of NC, as the temperature increases from 200 to 650 °C. Notably, the differences between the curves of $Co_{1-x}S/Co(PO_3)_2$@NC and $Co_{1-x}S$@NC are primarily attributed to the greater weight loss from NC combustions compared to the increase from oxidation of $Co_{1-x}S$. The TGA curves indicate the percentages of $Co_{1-x}S$, $Co(PO_3)_2$, and NC in $Co_{1-x}S/Co(PO_3)_2$@NC, which were 40.9 wt%, 29.4 wt%, and 29.7 wt%, respectively.

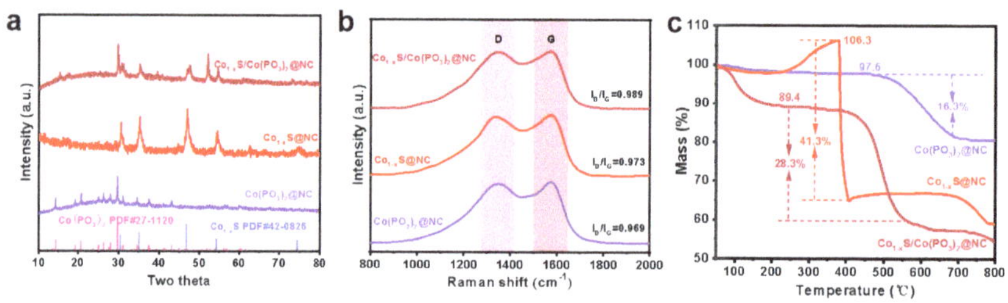

Figure 2. (**a**) XRD patterns, (**b**) Raman spectra, and (**c**) TGA curvesof $Co(PO_3)_2$@NC, $Co_{1-x}S$@NC, and $Co_{1-x}S/Co(PO_3)_2$@NC, respectively.

Furthermore, the surface chemical compositions of the three samples were analyzed using X-ray photoelectron spectroscopy (XPS) measurements. As shown in Figure 3a, elements such as cobalt, sulfur, phosphorus, oxygen, carbon, and nitrogen were clearly detected. The Co 2p high-resolution spectra, presented in Figure 3b, reveal the characteristic peaks at ~798.4 eV and ~782.6 eV, corresponding to Co $2p_{1/2}$ and Co $2p_{3/2}$ of Co^{2+}, along with their satellite peaks [26,27]. An additional set of peaks in the $Co_{1-x}S$@NC composites, located at ~793.4 eV and 778.4 eV, is associated with Co^0 and suggests that the vulcanization process of the PVP-Co^{2+} precursors is not fully completed. Figure 3c shows the S 2p spectra, where characteristic peaks at approximately 165.2 eV and 164.0 eV are associated with S $2p_{1/2}$ and S $2p_{3/2}$ of the C-S-C bond, while peaks at ~162.9 eV and ~161.8 eV correspond to S $2p_{1/2}$ and S $2p_{3/2}$ of S^{2-} [28]. Additionally, a peak at ~168.5 eV corresponds to SO_x likely due to inevitable surface oxidation. The strong chemical bonds between active species and carbon substrate are indicated by the presence of C-S-C, ensuring robust structures and excellent cycling stability during the intercalation and de-intercalation of Li^+ [29]. The characteristic peak at ~134.4 eV (Figure 3d) corresponds to P-O bonds. The C 1s peaks can be deconvoluted into three distinct peaks at ~284.8, ~285.9, and ~289.0 eV (Figure 3e), corresponding to C-C, C-N/C-O/C-S, and C=O bonds, respectively. The successful nitrogen substitution in the carbon matrix is confirmed by the presence of the C-N bond, and this is supported by the N 1s high-resolution spectra (Figure 3f). The nitrogen dopant can not only enhance interfacial wettability, create more defects and active sites, and accelerate reaction kinetics [30].

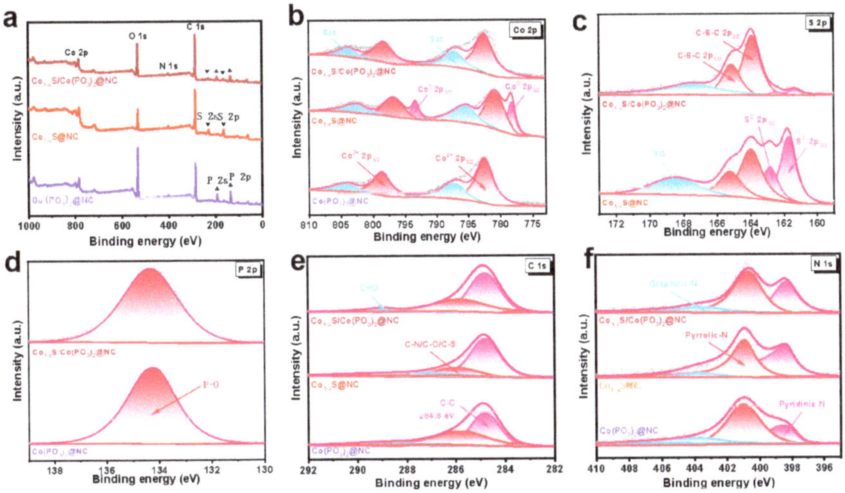

Figure 3. (**a**) The wide scan XPS of $Co(PO_3)_2$@NC, $Co_{1-x}S$@NC, and $Co_{1-x}S/Co(PO_3)_2$@NC, (**b**) Co 2p of $Co(PO_3)_2$@NC, $Co_{1-x}S$@NC, and $Co_{1-x}S/Co(PO_3)_2$@NC, (**c**) S 2p and (**d**) P 2p of $Co_{1-x}S$@NC and $Co_{1-x}S/Co(PO_3)_2$@NC, (**e**) C 1s and (**f**) N 1s of $Co_{1-x}S$@NC and $Co_{1-x}S/Co(PO_3)_2$@NC.

The Li^+ storage behavior of the $Co_{1-x}S/Co(PO_3)_2$@NC electrodes was evaluated using lithium metal as a counter electrode in an assembled 2032 coin-type half-cell. First, as shown in Figure 4a, cyclic voltammetry (CV) curves were measured at a sacn rate of 0.2 mV s^{-1}, revealing the electrochemical process over the first five cycles. In the initial cathodic scan, the sample showed significant broad peaks ~0.52 V and ~0.94 V, indicative of irreversible transformations and the formation of a solid electrolyte interface (SEI). In the subsequent scans, a set of new characteristic cathodic peaks appeared at 1.98 and 1.24 V related to the stepwise conversion reaction of $Co_{1-x}S$ into Co/Li_2S, $Co(PO_3)_2$ to CoP_2/Li_2O intermediates, and Co/Li_3P products; additionally, notable oxidation peaks were observed at 1.32, 2.05, and 2.56 V, corresponding to the oxidation reactions of Co/Li_2S to $Co_{1-x}S$ and Co/Li_3P to CoP_2 [31], which persisted in the following scans. The galvanostatic charge/discharge (GCD) profiles of the $Co_{1-x}S/Co(PO_3)_2$@NC electrode displayed initial charge/discharge capacities of 1194.0/719.1 mAh g^{-1} with an initial Coulombic efficiency of 60.2%; in subsequent cycles, the capacities remained stable over 200 cycles with Coulombic efficiency, approaching to 99.6% (Figure 4b). The cycling performance reveals that the $Co_{1-x}S/Co(PO_3)_2$@NC electrode maintains the highest reversible capacity of 643.7 mAh g^{-1} after 200 cycles at 0.2 A g^{-1}, significantly exceeding those of $Co_{1-x}S$@NC (336.5 mAh g^{-1}), $Co(PO_3)_2$@NC (274.0 mAh g^{-1}) (Figure 4c), $Co_{1-x}S/Co(PO_3)_2$@NC-1v3 (478.4 mAh g^{-1}), and $Co_{1-x}S/Co(PO_3)_2$@NC-1v8 (426.8 mAh g^{-1}) (Figure S4). The rate capability of all samples was further examined across different current densities ranging from 0.2 to 8.0 A g^{-1}. The $Co_{1-x}S/Co(PO_3)_2$@NC exhibited the best rate performance regardless of current density, with a high capacity of 642.7 mAh g^{-1} at 0.2 A g^{-1} and 242.9 mAh g^{-1} at 8.0 A g^{-1} (Figure 4d,e); the rate performance of $Co_{1-x}S/Co(PO_3)_2$@NC is superior to some previously reported cobalt sulfide-based anode materials (Figure 4f). The long-term cycling performance of $Co_{1-x}S/Co(PO_3)_2$@NC, $Co_{1-x}S$@NC, and $Co(PO_3)_2$@NC electrodes was further compared at a current density of 1 A g^{-1}. As shown in Figures 4g and S4, the $Co_{1-x}S/Co(PO_3)_2$@NC electrode delivers an obvious higher reversible capacity and improved cycling stability (371.8 mAh g^{-1} after 800 cycles), while $Co(PO_3)_2$@NC maintains a capacity of only 230.1 mAh g^{-1}, which is lower than the reversible capacity for $Co_{1-x}S/Co(PO_3)_2$@NC-1v3 and $Co_{1-x}S/Co(PO_3)_2$@NC-1v8. This difference in performance is mainly due to the synergistic effect of the heterogeneous interface and the integrated honeycomb-structured carbon, which efficiently promote charge transfer at the interfaces and accelerate the reaction kinetics, accommodate volume changes during prolonged long cycling, and ensure the structural integrity of the electrode materials.

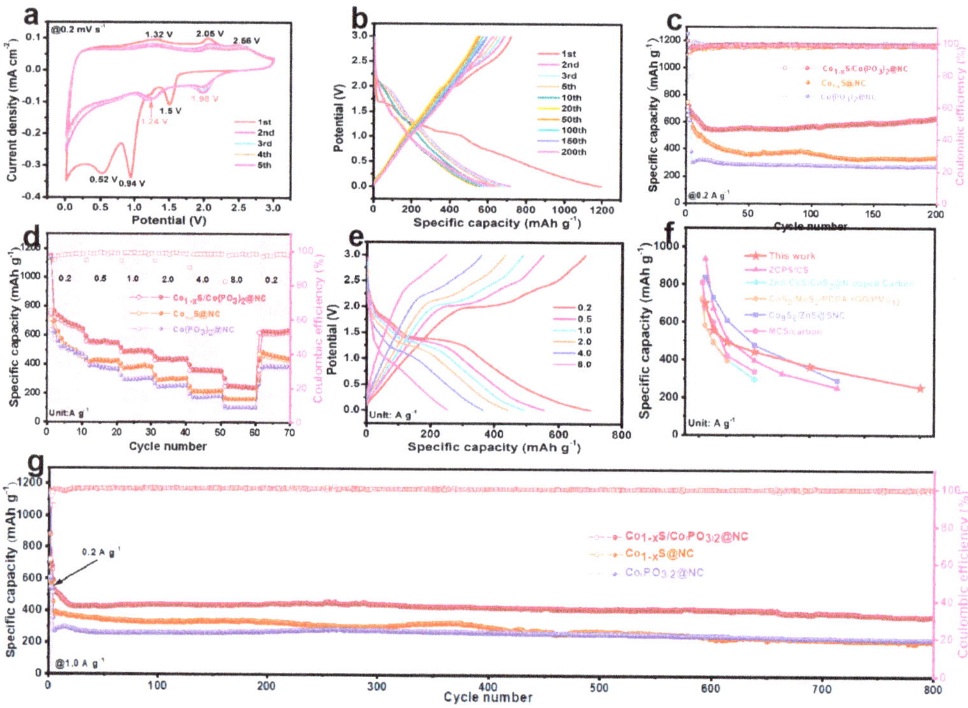

Figure 4. (**a**) The initial three CV curves of $Co_{1-x}S/Co(PO_3)_2$@NC at 0.2 mV s^{-1}, (**b**) galvanostatic charge/discharge profiles of $Co_{1-x}S/Co(PO_3)_2$@NC at 0.2 A g^{-1}, (**c**) cycling stability and coulombic efficiency at 0.2 A g^{-1} for 200 cycles, (**d**,**e**) rate capability and galvanostatic charge/discharge profiles of three samples, (**f**) corresponding comparison with reported studies, (**g**) long-term performance at 1.0 A g^{-1} for 800 cycles [32–36].

In order to assess the superior lithium storage capabilities of the $Co_{1-x}S/Co(PO_3)_2$@NC, cyclic voltammetry (CV) curves were recorded at various scanning rates, ranging from 0.1 to 5.0 mV S^{-1} (Figure 5a). As the scan rate increased, there were no discernible changes in the number or peak voltage of the observed peaks in the CV curves, indicating that $Co_{1-x}S/Co(PO_3)_2$@NC has good reversibility. It is widely accepted that the following Equation (1) could be used to explain the relationship between peak current (i) and scan rate (v):

$$i = av^{b} \tag{1}$$

where both a and b are variable parameters. The value of b can be used to diagnose whether the reaction is a diffusion-controlled reaction (for b = 0.5) or surface-controlled (for b = 1.0). As shown in Figure 5b, the b values at the $O1$ peak and $R1$ peak of the $Co_{1-x}S/Co(PO_3)_2$@NC electrodes were determined to be 0.66 and 0.67, respectively, suggesting a combination of diffusion-controlled and capacitive-controlled behaviors. Furthermore, the proportion of pseudocapacitive contribution could be quantified using the following Equation (2),

$$i = k_1 v + k_2 v^{1/2} \tag{2}$$

where $k_1 v$ and $k_2 v^{1/2}$ indicate the current response from the capacitive capacity and diffusion-controlled capacity, respectively. As shown in Figure 5c,d, the pseudocapacitive contributions of the $Co_{1-x}S/Co(PO_3)_2$@NC electrodes at the scanning rate of 2.0 mV s^{-1} were evaluated to be 70.0%. With the increase in v, the pseudocapacitive contribution gradually increases from 52.3% to 74.7%, demonstrating the important effect of the heterostructure in facilitating the electrode reaction kinetics. Figure 5e shows the Nyquist plots of three samples after three cycles at 0.1 A g^{-1}. All EIS spectra display similar characteristics, including two sections: a semicircular region at a medium- to high-frequency corresponding to R_{ct} and a linear region at low-frequency associated with Li^+ diffusion. Compared to $Co(PO_3)_2$@NC and $Co_{1-x}S$@NC, the $Co_{1-x}S/Co(PO_3)_2$@NC electrode exhibits significantly smaller R_{ct} in the medium- to high-frequency range (186 Ω compared to 304 Ω, 201 Ω), indicating that the unique heterostructures and effective particle refinement indeed enhance charge transfer and reaction kinetics. Moreover, based on the following equations related to diffusion coefficient of Li^+:

$$Z' = R_s + R_{ct} + \sigma\omega^{-1/2} \quad (3)$$

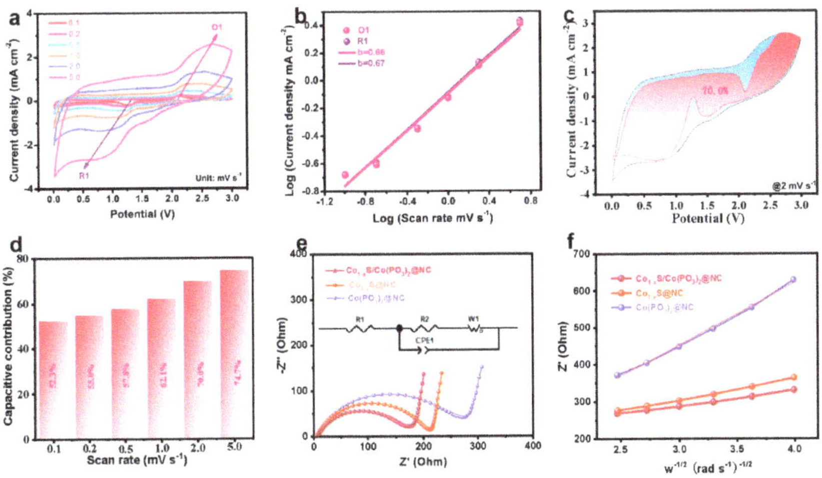

Figure 5. (**a**) CV curves at different sweep rates, (**b**) linear plot of Log v-Log i, (**c**,**d**) pseudocapacitive contribution of the $Co_{1-x}S/Co(PO_3)_2$@NC electrode at 2.0 mV s^{-1} and at different sweep rates, (**e**,**f**) Nyquist plots and the Z′ as a function of $\omega^{-1/2}$ of of $Co(PO_3)_2$@NC, $Co_{1-x}S$@NC and $Co_{1-x}S/Co(PO_3)_2$@NC.

According to Equation (3), the Warburg factor (σ) of the two electrodes was calculated based on the linear relationship between Z' and $\omega^{-1/2}$ (Figure 5f). The value of σ for the $Co_{1-x}S/Co(PO_3)_2$@NC electrode (41.0) is much lower than the $Co_{1-x}S$@NC (57.6) and $Co(PO_3)_2$@NC electrodes (168.9). As the D_{Li+} diffusion coefficient is inversely proportional to the square of the Warburg factor (σ) [3], this lower σ value indicates a faster Li^+ diffusion coefficient for the $Co_{1-x}S/Co(PO_3)_2$@NC electrode.

Considering the exceptional electrochemical performance of the $Co_{1-x}S/Co(PO_3)_2$@NC in half-cells, a prelithiated $Co_{1-x}S/Co(PO_3)_2$@NC anode and an AC cathode were combined in an optimal mass ratio of 2:1 to assemble a $Co_{1-x}S/Co(PO_3)_2$@NC@NC//AC LIC device. This device was subsequently evaluated for its electrochemical energy storage properties within a voltage window of 0–4.0 V. The working principle of the assembled LIC is illustrated in Figure 6a. During charging, Li^+ ions either intercalate into $Co_{1-x}S/Co(PO_3)_2$@NC or adsorb onto the anode surface, while PF_6^- anions are absorbed onto the AC surface. The discharge process reverses these actions. Figure 6b presents the CV curves of the LICs at scan rates of 5, 10, 20, 30, and 50 mV s^{-1}. Even at 50 mV s^{-1}, a rectangular profile is observed, indicating the capacitive behavior of the LICs [37,38]. The rate performance of the LICs was assessed under various current densities, revealing reversible capacities of 19.15 mAh g^{-1} even at 2 A g^{-1}, respectively (Figure 6c,d). The assembled optimal $Co_{1-x}S/Co(PO_3)_2$@NC//AC LIC delivers a high energy density of 90.47 Wh kg^{-1} (at 26.28 W kg^{-1}) and a high power density of 504.94 W kg^{-1} (at 38.31 Wh kg^{-1}). Furthermore, Figure 6e illustrates the long-term cycling stability of the LICs, showing a high cyclic stability of 86.3% retention after 5000 cycles at a current of 2 A g^{-1}, accompanied by a Coulombic efficiency close to 100%. The superiorities of the as-prepared $Co_{1-x}S/Co(PO_3)_2$ heterostructures were compared with the reported literature in terms of synthetic methods and electrochemical performance (Table S1). In practical applications, the LICs successfully power a light-emitting diode (LED) to illuminate a "wang" character, further demonstrating the potential of the $Co_{1-x}S/Co(PO_3)_2$ heterostructures as an electrode material for Li-ion capacitors.

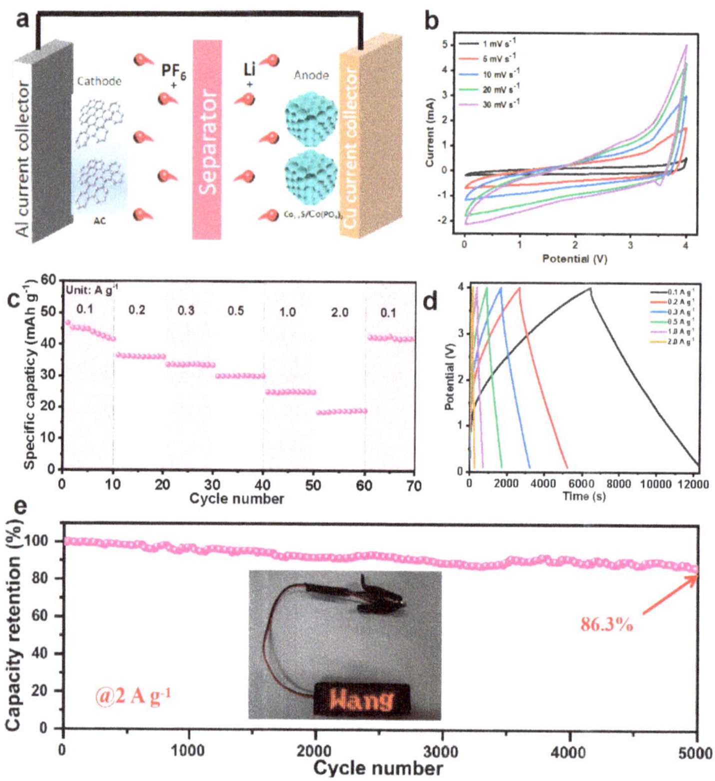

Figure 6. (**a**) Schematic illustration of the assembled $Co_{1-x}S/Co(PO_3)_2$@NC//AC LICs, (**b**) CV curves, (**c**) galvanostatic charge/discharge profiles, (**d**) rate capability, (**e**) long-term durability at 2 A/g of $Co_{1-x}S/Co(PO_3)_2$@NC//AC LICs.

3. Conclusions

In summary, we have constructed high-performance LICs using $Co_{1-x}S/Co(PO_3)_2$@NC as the anode material. This material is synthesized through the chemical blowing-vulcanization of a Co^{2+}-PVP precursor followed by phosphorylation treatment. The characteristic heterointerface and 3D honeycomb-like N-doped carbon matrix efficiently boost interfacial charge dynamics, provide spatial confinement for refined particles, and prevent structural degradation during repeated charge/discharge cycles. As a result, the optimal $Co_{1-x}S/Co(PO_3)_2$@NC composite demonstrates a stable reversible capacity of 371.8 mAh g^{-1} after 800 cycles at 1 A g^{-1}, showcasing significant pseudocapacitive contributions. The assembled $Co_{1-x}S/Co(PO_3)_2$@NC//AC LIC delivers a power density of 504.94 W kg^{-1} and an energy density of 90.47 Wh kg^{-1} within the voltage range of 0–4.0 V. Moreover, they exhibit a high cyclic stability of 86.3% retention after 5000 cycles at 2 A g^{-1}. The facile synthetic method opens new avenues for the fabrication of cost-effective, advanced energy storage devices utilizing Co-based sulfides heterostructures with balanced performance.

Supplementary Materials: The following supporting information can be downloaded at https://www.mdpi.com/article/10.3390/batteries10100346/s1: Figure S1: SEM image of $Co(PO_3)_2$@NC composites; Figure S2: SEM image of $Co_{1-x}S$@NC composites; Figure S3: SEM image of $Co_{1-x}S/Co(PO_3)_2$@NC composites; Figure S4: (a) cycling performance at 0.2 A g^{-1}, (b) rate capability, (c) long-term performance at 1.0 A g^{-1} of $Co_{1-x}S/Co(PO_3)_2$@NC, $Co_{1-x}S/Co(PO_3)_2$@NC-1v3, and $Co_{1-x}S/Co(PO_3)_2$@NC-1v8 electrodes; Table S1: The LICs performance comparison among Co1-xS/Co(PO3)2@NC and other materials [39–46].

Author Contributions: Conceptualization, J.D. and W.W.; methodology, Y.L. and X.X.; validation, T.W. and H.H.; investigation, Z.W. and F.Z.; writing—original draft preparation, Y.L., X.X. and T.W.; writing—review and editing, Y.L., Z.W. and T.W.; visualization, Y.L. and W.W.; supervision, J.D. and W.W. All authors have read and agreed to the published version of the manuscript.

Funding: The authors gratefully acknowledge the financial support from the Hunan Provincial Natural Science Foundation of China (No. 2024JJ5012), Research Foundation of Education Bureau of Hunan Province (No. 23A0251), Changsha Natural Science Foundation (No. kq2402009).

Data Availability Statement: The data that supported the findings of this study are available from the corresponding author upon reasonable request.

Conflicts of Interest: Authors Xiaopeng Xie and Zhaojia Wu were employed by the company Hunan Datang Xianyi Technology Co., Ltd. Author Fang Zhao was employed by the company Guangdong Power Grid Co., Ltd., Guangzhou Power Supply Bureau. The remaining authors declare that the research was conducted in the absence of any commercial or financial relationships that could be construed as a potential conflict of interest.

References

1. Sun, J.L.; Luo, B.C.; Li, H.X. A Review on the Conventional Capacitors, Supercapacitors, and Emerging Hybrid Ion Capacitors: Past, Present, and Future. *Adv. Energy Sustain. Res.* **2022**, *3*, 2000216–2100191. [CrossRef]
2. Liang, J.; Wang, D.W. Design rationale and device configuration of lithium-ion capacitors. *Adv. Energy Mater.* **2022**, *12*, 2200920. [CrossRef]
3. Wang, L.; Zhang, X.; Li, C.; Sun, X.Z.; Wang, K.W.; Su, F.Y.; Liu, F.Y.; Ma, Y.W. Recent advances in transition metal chalcogenides for lithium-ion capacitors. *Rare Met.* **2022**, *41*, 2971–2984. [CrossRef]
4. Yan, T.; Hu, H.; Duan, J.; Zhu, C.; Wang, Y.; Wen, J.; Li, L.; Xu, Z.; Wen, T.; Yang, P.; et al. Achieving superior sodium storage performance of sulfide heterostructures via copper-driven and electrochemical reconstruction strategy. *Chem. Eng. J.* **2024**, *499*, 155871. [CrossRef]
5. Liang, L.P.; Li, J.C.; Zhu, M.Y.; Li, Y.; Chou, S.L.; Li, W.X. Cobalt Chalcogenides/Cobalt Phosphides/Cobaltates with Hierarchical Nanostructures for Anode Materials of Lithium-Ion Batteries: Improving the Lithiation Environment. *Small* **2021**, *17*, 1903418. [CrossRef]
6. Liu, X.; Zhang, K.; Lei, K.X.; Li, F.J.; Tao, Z.L.; Chen, J. Facile synthesis and electrochemical sodium storage of CoS_2 micro/nano-structures. *Nano Res.* **2016**, *9*, 198–206. [CrossRef]
7. Wang, X.; Chen, Y.; Fang, Y.J.; Zhang, J.T.; Gao, S.Y.; Lou, X.W. Synthesis of cobalt sulfide multi-shelled nanoboxes with precisely controlled two to five shells for sodium-ion batteries. *Angew. Chem. Int. Ed.* **2019**, *58*, 2675–2679. [CrossRef]
8. Xiao, Y.; Hwang, J.Y.; Belharouak, I.; Sun, Y.K. Superior Li/Na-storage capability of a carbon-free hierarchical CoSx hollow nanostructure. *Nano Energy* **2017**, *32*, 320–328. [CrossRef]
9. Zheng, Y.Y.; He, L.; Kong, X.R.; Song, Y.; Zhao, Y. Three-dimensional porous N-doped graphite carbon with embedded CoS_2 nanoparticles as advanced anode for sodium-ion batteries. *Appl. Surf. Sci.* **2022**, *603*, 154481. [CrossRef]
10. Wang, L.; Zhang, X.; Kong, Y.Y.; Li, C.; An, Y.B.; Sun, X.Z.; Wang, K.; Ma, Y.W. Metal-organic framework-derived $CoSe_2$@ N-doped carbon nanocubes for high-performance lithium-ion capacitors. *Rare Met.* **2024**, *43*, 2150–2160. [CrossRef]
11. Li, Q.; Jiao, Q.Z.; Yan, Y.; Li, H.J.; Zhou, W.; Gu, T.T.; Shen, X.R.; Lu, C.X.; Zhao, Y.; Zhang, Y.Y.; et al. Optimized Co-S bonds energy and confinement effect of hollow MXene@CoS_2/NC for enhanced sodium storage kinetics and stability. *Chem. Eng. J.* **2022**, *450*, 137922. [CrossRef]
12. Lei, D.; Gao, Y.; Hou, Z.; Ren, L.; Jiang, M.; Cao, Y.; Zhang, Y.; Wang, J.-G. A Superior Lithium-Ion Capacitor Based on Ultrafine MnO/Dual N-Doped Carbon Anode and Porous Carbon Cathode. *Batteries* **2023**, *9*, 241. [CrossRef]

13. Lin, Y.; Qiu, Z.; Li, D.; Ullah, S.; Hai, Y.; Xin, H.; Liao, W.; Yang, B.; Fan, H.; Xu, J.; et al. NiS_2@CoS_2 nanocrystals encapsulated in N-doped carbon nanocubes for high performance lithium/sodium ion batteries. *Energy Storage Mater.* **2018**, *11*, 67–74. [CrossRef]
14. Liu, J.; Xu, Y.G.; Kong, L.B. Cleverly embedded CoS_2/NiS_2 on two-dimensional graphene nanosheets as high-performance anode material for improved sodium ion batteries and sodium ion capacitors. *J. Mater. Sci. Mater. Electron.* **2020**, *31*, 9946–9959. [CrossRef]
15. Wang, L.; Zhao, S.S.; Zhang, X.; Xu, Y.N.; An, Y.B.; Li, C.; Yi, S.; Liu, C.; Wang, K.; Sun, X.Z.; et al. In Situ Construction of Bimetallic Selenides Heterogeneous Interface on Oxidation-Stable $Ti_3C_2T_x$ MXene Toward Lithium Storage with Ultrafast Charge Transfer Kinetics. *Small* **2024**, 2403078. [CrossRef] [PubMed]
16. Xue, F.F.; Fan, F.F.; Zhu, Z.C.; Zhang, Z.G.; Gu, Y.F.; Li, Q.H. MoS_2/CoS heterostructures grown on carbon cloth as free-standing anodes for high-performance sodium-ion batteries. *Nanoscale* **2023**, *15*, 6822–6829. [CrossRef]
17. Cai, J.M.; Zhou, Y.L.; Tao, S.S.; Liu, Y.C.; Deng, W.T.; Hou, H.H.; Zou, G.Q.; Ji, X.B. Nanocrystalline Heterostructure with Low Voltage Hysteresis for Ultrahigh-Power Sodium-Ion Capacitors. *Energy Storage Mater.* **2024**, *71*, 103582. [CrossRef]
18. Xiao, Y.H.; Kong, Y.; Wang, X.Z.; Luo, H.R.; Yuan, G.Z.; Zhang, S.W.; Zhang, A.Q.; Zhou, J.; Fan, Y.Y.; Xin, L.; et al. Synergetic interface engineering and space-confined effect in $CoSe_2$@$Ti_3C_2T_x$ heterostructure for high power and long life sodium ion capacitors. *J. Colloid Interface Sci.* **2025**, *677*, 577–586. [CrossRef]
19. Dong, C.F.; Guo, L.J.; Li, H.B.; Zhang, B.; Gao, X.; Tian, F.; Qian, Y.T.; Wang, D.B.; Xu, L.Q. Rational fabrication of CoS_2/Co_4S_3@N-doped carbon microspheres as excellent cycling performance anode for half/full sodium ion batteries. *Energy Storage Mater.* **2020**, *25*, 679–686. [CrossRef]
20. Wan, S.Y.; Cheng, M.; Chen, H.Y.; Zhu, H.J.; Liu, Q.M. Nanoconfined bimetallic sulfides (CoSn)S heterostructure in carbon microsphere as a high-performance anode for half/full sodium-ion batteries. *J. Colloid Interface Sci.* **2022**, *609*, 403–413. [CrossRef]
21. Zhao, M.T.; Lu, Q.P.; Ma, Q.L.; Zhang, H. Two-Dimensional Metal-Organic Framework Nanosheets. *Small Methods* **2017**, *1*, 1600030. [CrossRef]
22. Yang, M.; Yan, Z.H.; Zhang, H.; Li, J.W.; Zhu, Z.Q.; Liu, L.; Jiao, L.F. Controllable Synthesis of Sub-10 nm ZnS Nanograins Confined in Micro-Size Carbon Skeleton for Aqueous Zn-S Batteries. *Adv. Funct. Mater.* **2024**, 2406077–2406085. [CrossRef]
23. Wen, F.; Zhu, C.; Li, L.; Zeng, G.; Duan, J.; Chen, Z. Enhanced pseudocapacitive behaviors of Sb-based anodes for lithium ion batteries via dual modification approach of Fe doping combined with double carbon coatings. *J. Alloys Compd.* **2021**, *889*, 161658. [CrossRef]
24. Wu, Y.L.; Zhua, C.; Shu, L.; Duan, J.F.; Wei, D.H.; Xu, J.X.; Zhu, Z.Y.; Li, L.J.; Peng, Z.Y.; Chen, Z.Y. Co_9S_8 confined in bifunctional N/S co-doped carbon/carbon with high electrochemical performance for lithium-ion batteries. *Appl. Surf. Sci.* **2019**, *489*, 528–537. [CrossRef]
25. Yan, T.C.; Wen, F.; Duan, J.F.; Zhu, C.; Wen, J.H.; Wang, Y.X.; Tong, J.T.; Chen, Z.Y. Fabricating tunable metal sulfides embedded with honeycomb-structured N-doped carbon matrices for high-performance lithium-ion capacitors. *Chem. Eng. J.* **2023**, *474*, 145839–145850. [CrossRef]
26. Huang, G.; Hu, M.; Xu, X.T.; Alothman, A.A.; Mushab, M.S.S.; Ma, S.J.; Shen, P.K.; Zhu, J.L.; Yamauchi, Y. Optimizing Heterointerface of Co_2P-Co_xO_y Nanoparticles within a Porous Carbon Network for Deciphering Superior Water Splitting. *Small Struct.* **2023**, *4*, 2200235–2200244. [CrossRef]
27. Tan, L.; Huang, X.; Yin, T.; Guo, Y.; Ning, T.; Mei, Y.; Zou, K.; Li, L.; Ji, X.; Zou, G. A 5 V ultrahigh energy density lithium metal capacitor enabled by the fluorinated electrolyte. *Energy Storage Mater.* **2024**, *71*, 103692. [CrossRef]
28. Cheng, Q.; Liu, X.Z.; Deng, Q.; Chen, C.D.; Zhong, W.T.; Yang, C.H. Heterostructured Ni_3S_4/Co_9S_8 Encapsulated in Nitrogen-Doped Carbon Nanocubes for Advanced Potassium Storage. *Chem. Eng. J.* **2022**, *446*, 136829–136836. [CrossRef]
29. Zhang, J.F.; Sun, C.H.; Qu, S.Q.; Qian, M.M.; Zhan, W.; Su, A.Q.; Zhang, K.; Liu, Q.; Shao, R.W.; Wang, J.; et al. Paradigm metallothermic-sulfidation-carbonization constructing ZIFs-derived TMSs@Graphene/CNx heterostructures for high-capacity and long-life energy storage. *Nano Energy* **2023**, *111*, 108401–108410. [CrossRef]
30. Shen, M.; Ma, H. Metal-organic frameworks (MOFs) and their derivative as electrode materials for lithium-ion batteries, Coordination. *Chem. Rev.* **2022**, *470*, 214715.
31. Shi, X.; Li, G.; Liu, X.; Li, J.; Zhang, X.; Guo, J. Carbon coated $Co(PO_3)_2/CoSe_2$ heterostructure as high performance sodium storage anode. *J. Alloys Compd.* **2023**, *951*, 169989. [CrossRef]
32. Zhong, H.Y.; Lu, X.; Zhong, Y.; Zhao, Y.; Liu, X.M.; Cheng, D.H.; Huang, X.Y.; Du, K.Z.; Wu, X.H. Lithium Storage Performance Boosted via Delocalizing Charge in $Zn_xCo_{1-x}PS_3/CoS_2$ of 2D/3D Heterostructure. *Small* **2022**, *18*, 2104295. [CrossRef]
33. Cheng, W.; Di, H.F.; Shi, Z.; Zhang, D. Synthesis of ZnS/CoS/CoS_2@N-doped carbon nanoparticles derived from metal-organic frameworks via spray pyrolysis as anode for lithium-ion battery. *J. Alloys Compd.* **2020**, *831*, 154607. [CrossRef]
34. Li, Q.; Xu, M.Q.; Wang, T.; Wang, H.J.; Sun, J.W.; Sha, J.Q. Nanohybridization of CoS_2/MoS_2 Heterostructure with Polyoxometalate on Functionalized Reduced Graphene Oxide for High-Performance LIBs. *Chem. A Eur. J.* **2022**, *28*, e202200207. [CrossRef] [PubMed]
35. Nong, Y.T.; Zhang, M.; Li, Q.Q.; Pan, Q.C.; Huang, Y.G.; Wang, H.Q.; Zheng, F.H.; Li, Q.Y. Carbon coated bimetallic sulfides Co_9S_8/ZnS heterostructures microrods as advanced anode materials for lithium ion batteries. *J. Taiwan Inst. Chem. Eng.* **2022**, *141*, 104601. [CrossRef]

36. Zhu, L.; Lu, H.Y.; Xiao, F.P.; Yao, T.H.; Liu, T.; Li, F.; Wang, J.K.; Han, X.G.; Cheng, Y.H.; Wang, H.K. Flower-like Mn/Co Glycerolate-Derived α-MnS/Co_9S_8/Carbon Heterostructures for High-Performance Lithium-Ion Batteries. *ACS Appl. Energy Mater.* **2020**, *3*, 10215. [CrossRef]
37. Hou, J.B.; Shao, Y.Y.; Ellis, M.W.; Moore, R.B.; Yi, B.L. Graphene-based electrochemical energy conversion and storage: Fuel cells, supercapacitors and lithium ion batteries. *Phys. Chem. Chem. Phys.* **2011**, *13*, 15384–15402. [CrossRef]
38. Tao, S.S.; Momen, R.; Luo, Z.; Zhu, Y.R.; Xiao, X.H.; Cao, Z.W.; Xiong, D.Y.; Deng, W.T.; Liu, Y.C.; Hou, H.S.; et al. Trapping Lithium Selenides with Evolving Heterogeneous Interfaces for High-Power Lithium-Ion Capacitors. *Small* **2023**, *19*, 2207975. [CrossRef]
39. Shi, Z.C.; Wei, S.; Zuo, H.; Huang, M.H.; Shi, J.; Wang, H.L. Boosting capacitance and energy density by construction $NiCoO_2/CoS_2$ nanocomposites arrays as pseudocapacitor. *J. Alloys Compd.* **2021**, *881*, 160627. [CrossRef]
40. Wang, S.Q.; Song, Y.P.; Ma, Y.; Zhu, Z.Q.; Zhao, C.H.; Zhao, C.J. Attaining a high energy density of 106 Wh kg^{-1} for aqueous supercapacitor based on VS_4/rGO/CoS_2@Co electrode. *Chem. Eng. J.* **2019**, *365*, 88. [CrossRef]
41. Sonia, Y.K.; Paliwal, M.K.; Meher, S.K. The rational design of hierarchical $CoS_2/CuCo_2S_4$ for three-dimensional all-solid-state hybrid supercapacitors with high energy density, rate efficiency, and operational stability. *Sustain. Energy Energy Fuels* **2021**, *5*, 973. [CrossRef]
42. Li, X.N.; Zhang, W.Y.; Kang, H.W.; Liu, H.L.; Yang, B.C.; Li, Z.J.; Li, Z.K. Self-assembled $CoS_2/NiCo_2S_4$/RGO nanohybrids as advanced electrode for hybrid supercapacitor with enhanced energy density and ultra-long durability. *J. Energy Storage* **2023**, *67*, 107528. [CrossRef]
43. Wang, Y.K.; Liu, M.C.; Cao, J.Y.; Zhang, H.J.; Kong, L.B.; Trudgeon, D.P.; Li, X.H.; Frank, C.W. 3D Hierarchically Structured CoS Nanosheets: Li+ Storage Mechanism and Application of the High-Performance Lithium-Ion Capacitors. *ACS Appl. Mater. Interfaces* **2020**, *12*, 3709. [CrossRef] [PubMed]
44. Wang, H.Y.; Tian, L.H.; Zhao, X.; Ali, M.; Feng, H.M.; Han, S.Y.; Xing, Z.C.; Kumar, S.; Ding, J. Synthesis of MoS_2/CoS composite electrode and its application for supercapacitors. *J. Alloys Compd.* **2023**, *960*, 170835. [CrossRef]
45. Reddy, P.A.K.; Han, H.; Kim, K.C.; Bae, S. Synthesis of ZIF-67-derived CoS_2@graphitic carbon/reduced graphene oxide for supercapacitor application. *Chem. Eng. J.* **2023**, *471*, 144608. [CrossRef]
46. Zhang, K.; Gao, X.; Yao, F.; Xie, Y.Q.; Bai, H.; Sun, Y.J.; Liu, R.R.; Yue, H.Y. Construction of hierarchical MOF-derived CoS_2 microsheet arrays@$NiMo_2S_4$ nanoflakes on Ni foam as a high-performance supercapacitor electrode. *J. Colloid Interface Sci.* **2023**, *650*, 105. [CrossRef]

batteries

MDPI

Review

Recent Advances in High-Performance Carbon-Based Electrodes for Zinc-Ion Hybrid Capacitors

Ying Liu [1,*], Lechun Song [1], Chenze Li [1], Caicheng Song [1] and Xiang Wu [2,*]

1 SINOPEC Dalian (Fushun) Research Institute of Petroleum and Petrochemicals, Dalian 116045, China; songlechun.fshy@sinopec.com (L.S.); lichenze.fshy@sinopec.com (C.L.); songcaicheng.fshy@sinopec.com (C.S.)
2 School of Materials Science and Engineering, Shenyang University of Technology, Shenyang 110870, China
* Correspondence: liuying.fshy@sinopec.com (Y.L.); wuxiang05@sut.edu.cn (X.W.)

Abstract: Aqueous zinc-ion hybrid capacitors (ZIHCs) have emerged as a promising technology, showing superior energy and power densities, as well as enhanced safety, inexpensive and eco-friendly features. Although ZIHCs possess the advantages of both batteries and supercapacitors, their energy density is still unsatisfactory. Therefore, it is extremely crucial to develop reasonably matched electrode materials. Based on this challenge, a surge of studies has been conducted on the modification of carbon-based electrode materials. Herein, we first summarize the progress of the related research and elucidate the energy storage mechanism associated with carbon-based electrodes for ZIHCs. Then, we investigate the influence of the synthesis routes and modification strategies of the electrode materials on electrochemical stability. Finally, we summarize the current research challenges facing ZIHCs and predict potential future research pathways. In addition, we suggest key scientific questions to focus on and potential directions for further exploration.

Keywords: aqueous zinc-ion hybrid capacitors; carbon-based materials; cathode; energy storage device; electrochemical stability

1. Introduction

With the rapid changes in the energy patterns of the world, the efficient utilization and conversion of fossil fuels is an inevitable trend to achieve high-quality and sustainable energy development [1]. Recently, lithium-ion batteries (LIBs) have occupied a dominant position in the electronics industry, especially in portable electronics and electric vehicles [2]. However, limited Li resources and flammable organic electrolytes cause cost and safety concerns [3]. Thus, aqueous energy storage devices (metal-ion batteries, hybrid capacitors, and supercapacitors) have risen as potential solutions to the existing dilemma due to their high safety [4–9]. Among them, sodium and potassium ions possess ionic radii of 0.95 nm and 1.33 nm, respectively. It leads to slow diffusion kinetics and serious electrode polarization. In contrast, the ion radius of the zinc ion is relatively small (0.74 nm). Also, zinc anodes are characterized by their high theoretical capacity of 820 mAh g^{-1}, economical price, easy availability, and a low redox potential of -0.762 eV relative to the standard hydrogen electrode [10–12]. Nonetheless, the practical application of zinc anodes is limited by the scarcity of matching cathode materials and low power density [13–16]. Energy storage systems with high power densities and extremely long lifetimes can be fully recharged in a short period of time. They are also less susceptible to volume changes during cycling, which meets the needs of society for energy development. Zinc-ion hybrid capacitors (ZIHCs) incorporate their many benefits over traditional zinc-ion batteries. Generally, the device is featured by integrating the properties of a capacitive material with those of a battery-type material in a mild Zn-based electrolyte [17–21]. Due to the different operating potential ranges of these two types, the voltage of the matched device is extended.

As a result, there is an increase in the energy and power density of the devices. This is facilitated by the favorable combination of batteries and supercapacitors [22,23].

Since 2016, Wang and collaborators [24] have designed an aqueous zinc-ion capacitor assembled with oxidized carbon nanotubes and zinc metal as electrodes. It can be repeatedly charged and discharged for 5000 cycles (10 mV s^{-1}) with the working voltage of 0–1.8 V. Then, Kang et al. [25] reported the study of aqueous activated carbon (AC)/Zn capacitors with ultra-long lifetimes. ZIHCs are rapidly growing as a research hotspot [26]. Further, Kang's group [27] constructed a hybrid ion capacitor with MnO_2 and AC as cathode and anode, respectively. The device with Zn^{2+} aqueous solution as electrolyte can output a maximum power density of 13 kW kg^{-1}. During the past decade, carbon-based materials have been exploited as cathode materials owing to their high conductivity and customizable pore structure, including biomass and polymer-derived carbon, metal organic compounds derived carbon, graphene, and carbon nanotubes [28–30]. For example, Alshareef's group [31] prepared hybrid capacitors with oxygen-enriched porous carbon cathodes and carbon-coated zinc anodes. Benefiting from the EDLC of the porous carbon and the pseudocapacitance provided by the oxygen functional groups, the assembled ZIHCs obtain a specific capacity of 179.8 mAh g^{-1}. The energy density is up to 104.8 Wh kg^{-1} (0.1 A g^{-1}) with a voltage window of 0–1.9 V. Fan and co-workers [32] designed honeycomb porous carbon using polyacrylonitrile (PAN) as a precursor through a pre-oxidation and activation process. This abundant pore and interconnected structure facilitate rapid charge transfer and generate numerous ion adsorption sites. The assembled capacitor shows a specific capacity of 238 mAh g^{-1} at 0.1 A g^{-1}. Furthermore, it achieves an energy density of 193.6 Wh kg^{-1}, corresponding to a power density of 76.6 W kg^{-1}. Li et al. [33] fabricated porous carbon rich in pyridine nitrogen (PrN) with Zn and Co bimetallic organic skeletons as precursors. The content of PrN can be controlled to 56% by changing the annealing temperature. This high pyridine N content provides rich reaction sites for the structure of materials. Moreover, the presence of residual cobalt nanoparticles contributes to accelerating electron transfer. Based on this, the assembled ZIHCs deliver a specific capacity of 302 mAh g^{-1} (1 A g^{-1}) and an energy density of 157.6 Wh kg^{-1}.

The energy density of these reported devices is significantly weaker than expected. It can be ascribed to the distinct energy storage mechanisms of these two electrodes. This leads to the dynamic imbalance of ZIHCs, which fails to acquire high energy and power density [34]. This asymmetric device tends to increase the overpotentials of hydrogen and oxygen evolution in aqueous solutions. Its voltage range is generally restricted to 2.0 V [35,36]. Nevertheless, the energy density of aqueous ZIHCs is still far below that of batteries. Therefore, the charge storage capacity and voltage window can be increased by designing the electrode structure, electrolyte, and electrode/electrolyte interface [37].

Herein, we focus on the recent developments of carbon-based materials as electrodes for aqueous ZIHCs. First, we distinguish the different types of ZIHCs according to the properties of electrode materials. Then, we discuss the structure and morphology of carbon-based materials. Their cycling, rate performance, and durability are demonstrated when they are assembled as positive electrodes into capacitors. Further, we highlight the modification strategies of carbon-based materials, including surface modification, morphology and size tuning, and heterogeneous atom doping. We also summarize the energy storage mechanism in conjunction with the electrode structural changes. Finally, we present a conclusion and prospect the future development of ZIHCs.

2. The Bases of Zinc-Ion Capacitors

2.1. The Growth of Energy Storage Technologies

Among the various energy storage systems, supercapacitors and batteries are two highly representative electrochemical energy storage technologies. The former is a promising device with the merits of rapid charge capabilities, longevity, and excellent power density. The latter demonstrates high energy density and voltage window. Research on energy storage devices began in the 1800s (Figure 1). H. I. Becker first discovered electrochemical capacitors with a high farad in 1954. In the following 20 years, Whittingham reported reversible electrochemical embedding reactions of layered TiS_2 with lithium in Li/TiS_2 batteries [38]. However, Li/TiS_2 batteries cannot be developed into commercially promising secondary batteries. It can be attributed to the low open-circuit voltage of the TiS_2 cathode (2.2 V). Goodenough et al. [39] proposed that the layered $LiCoO_2$ materials can be used as the cathodes for LIBs. It presents an open-circuit voltage of 4.0 V, almost twice that of the TiS_2 electrode. In the 21st century, research on various metal-ion capacitors has also become increasingly attractive with the massive growth in energy demand [40].

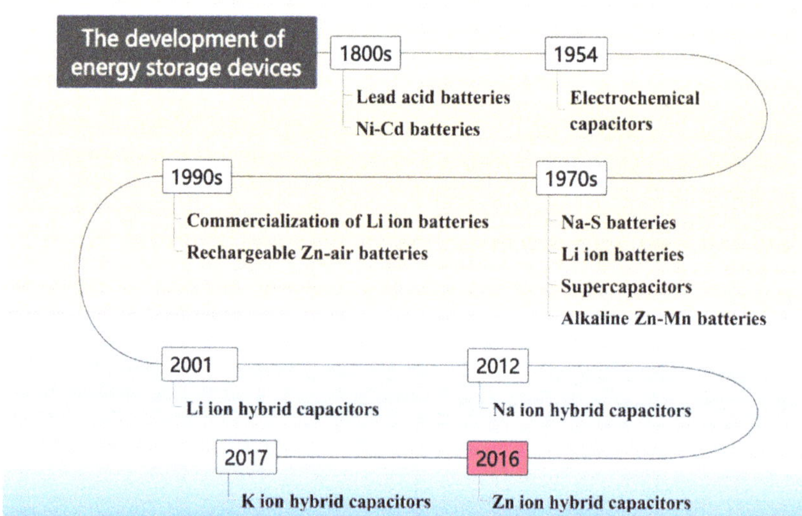

Figure 1. The growth of energy storage technologies.

Supercapacitors are mainly employed in portable electronic devices, rail public transportation, heavy trucks, and backup energy for uninterruptible power supply (UPS) [41]. Based on the principles of charge accumulation, the cathode can be categorized into two main types: double electric layer capacitors (EDLCs) and faradaic electrode materials [42,43]. The former involves the formation of a double layer by electrostatic adsorption of charges at the electrode/electrolyte interface. The latter represents a prompt reversible redox process occurring on/near the surface. The difference between the two is that EDLC is not undergoing a faradaic process [44,45]. The electrode materials for supercapacitors include carbonaceous materials, conducting polymers, and transition metal oxides and hydroxides [46–49]. Notably, carbon-based materials are characterized by abundant resources, inexpensiveness, extensive specific surface area, and superior electrical conductivity [50,51]. During the past few years, a multitude of studies have concentrated on the design of the pore structure of such electrodes to modulate the electrochemical performance of the devices [52]. Liu et al. [53] designed an activated porous carbon material with a specific surface area of 843 m^2 g^{-1} via rapid high-temperature shock carbonization and the HTS-KOH activation approach. The prepared electrodes are assembled into supercapacitors with an energy density of 25 Wh kg^{-1} and a power density of 582 W kg^{-1}. Prabakar's group [54] synthesized activated carbon nanosheets with a specific surface area up to 2943 m^2 g^{-1} by one-step activation method. The assembled symmetric supercapacitor achieves a capacitance of 403 F g^{-1} with an energy density of 32.9 Wh kg^{-1} (0.5 A g^{-1}).

2.2. The Types of ZIHCs

Benefiting from the synergistic strengths of supercapacitors and batteries, the novel ZIHCs are regarded as critical candidates for energy storage technologies. This is attributed to their ability to satisfy market requirements for high energy and power density [55]. Among them, electrodes can be divided into capacitor-type and battery-type materials. The former is mainly carbon-based electrodes (derived carbon and graphene-based materials) and pseudocapacitive electrodes (MXene and polymer–carbon materials) [56]. Generally, zinc ion adsorption/desorption reactions occur on their surfaces during charging and discharging. The latter is generally metallic zinc, Mn, and V oxides, which mainly occur in zinc ion insertion/extraction reactions. Based on the unique characteristics of carbon-based electrodes, the cathode and anode materials of ZIHCs can be classified into carbon materials/Zn foils and manganese-based or vanadium-based oxides/carbon materials.

Figure 2 illustrates the dual energy storage mechanisms of ZIHCs. From Figure 2a, carbon-based cathodes utilize the EDLC mechanism to facilitate highly reversible ion adsorption/desorption processes, thus providing excellent power densities. In fact, their total capacitance is contributed by EDLC and pseudocapacitance. Therefore, a redox reaction can occur on the surface or near the surface of a carbon electrode. On the anode side, zinc ions in the electrolyte can be deposited and stripped on the zinc foil with the charging and discharging process to obtain excellent energy density.

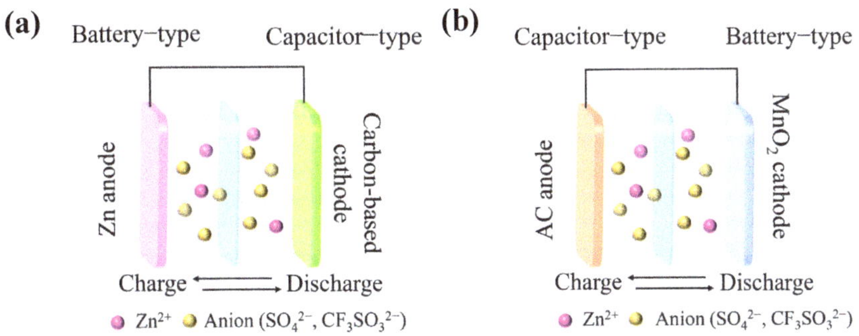

Figure 2. The energy storage mechanism of different electrodes. (**a**) Carbon-based cathode/Zn anode capacitor. (**b**) MnO_2 cathode/AC anode capacitor.

In Figure 2b, a Mn-based oxide/carbon material hybrid capacitor is taken as an example. When charging, zinc ions escape from the battery-type cathode and adhere to the surface of the capacitor-type anode. In the discharge phase, zinc ions are desorbed from the anode and embedded in the structure of the cathode. This reversible transfer of zinc ions contributes to the stable electrochemical performance of ZIHCs.

Compared with the two systems, the capacitive material as the cathode can provide rapid ion adsorption/desorption, thus achieving high power density. However, zinc anode is prone to producing zinc dendrites, which leads to potential safety hazards. The battery material as cathode enables a dendrite-free system. The voltage range of the device can be extended up to 2 V, showing high energy density. However, their electrochemical performance suffers from issues such as low rate capability and cycling stability, which are attributed to the dissolution of the cathode material.

3. Carbon-Based Electrode Materials

Carbon is an indispensable material to promote social development and progress. In the last two decades, carbon materials with various allotropic structures (graphene, diamond, carbon fiber, and carbon nanotubes) have grown rapidly in a wide range of fields. In particular, they stand out in new energy sources owing to their highly theoretically specific surface area and superior electrical conductivity [57]. From Figure 3, carbon-based cathodes mainly include commercial activated carbon (AC), biomass or polymer-derived carbon, metallic organic compound-derived carbon, nanofibers, and other graphite-like carbon materials. This section mainly introduces the structure and morphology of these carbon materials and their corresponding electrochemical performance.

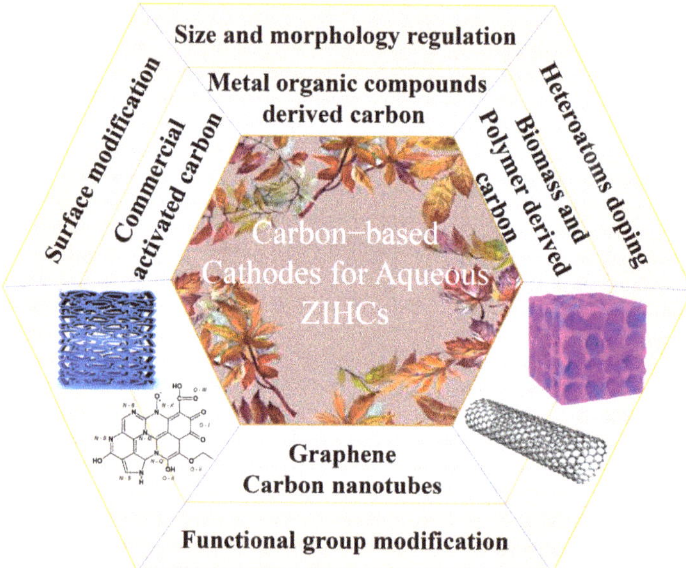

Figure 3. Classification and synthetic approaches of carbon-based cathodes for aqueous ZIHCs.

3.1. Amorphous Carbon Materials

3.1.1. Commercial Activated Carbon

In recent years, AC has attracted great interest as a promising candidate electrode material for ZIHCs because of its specific surface area of 3000 $m^2 g^{-1}$, rich pore structure, high chemical stability, and wide working potential window. They can usually be prepared from various precursors such as polymers, biomass, coal, organometallic frameworks, and organic salts. However, commercial AC has a specific capacity of only 100–200 mAh g^{-1}. The electrochemical performance of ZIHCs depends on the pore structure, conductivity, and surface properties of carbon-based electrodes. Pore structure is critical for charge storage and diffusion of hydrated Zn^{2+} ions [58]. Strategic designs of pore structures (including pore volume, size distribution, and surface functional group chemistry) are effective approaches to enhance the specific capacitance of electrodes.

Pore size distribution is an important parameter affecting zinc-ion storage. The pore structures in AC are usually categorized into micropores with a size less than 2 nm, mesopores with a size of 2–50 nm, and macropores with a size over 50 nm [59–61]. However, not all pores in the material are beneficial to charge storage. Both micropores and mesopores contribute to the EDLC. However, a part of the micropores prevents the embedding of electrolyte ions because of the distorted pore channel or small pore size. It also deprives their electrochemical activity at high loadings. The solvated $[Zn(H_2O)_6]^{2+}$ and SO_4^{2-} are the main carriers in the charging and discharging process of ZIHCs [62]. Its ion diameters are 0.86 and 0.59 nm, respectively. After solvation, the radius of Zn^{2+} is increased compared with the original one. For carbonaceous materials with pore sizes less than 1 nm, most of the solvated Zn^{2+} ions are limited by the pore structure. This leads to their inability to provide capacitive contributions.

From Figure 4, when there are various pores in the structure of carbon materials, they are prone to diffuse in the one with large pore size. It contributes to enhancing the specific capacitance of the devices. The pore size structure affects the ion diffusion resistance of the electrode. Small pore size enhances the diffusion resistance of electrolyte ions. The presence of mesopores and macropores can shorten the ion migration path and contribute to the rate capacity [63]. Zhang and colleagues [64] systematically compared various samples with similar specific surface areas but distinct pore structures. It is found that the materials with abundant micropore structure (0.6 nm) are generally lower in capacitance and rate performance than those with mesoporous channels. This phenomenon is associated with the partial failure of the microporous structure, which hinders the accommodation and storage of hydrated zinc ions. Hence, it is the key to stabilizing the cycle performance to match the appropriate pore channels.

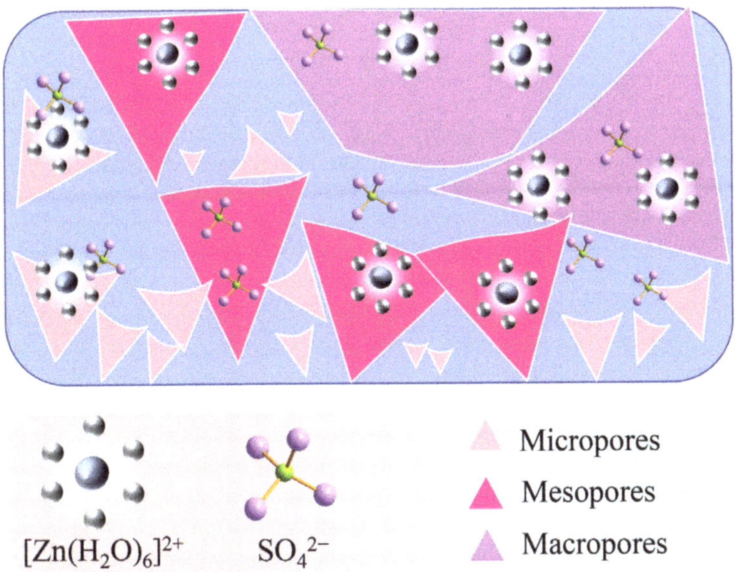

Figure 4. Schematic of charge carriers embedded in various pore structures.

To address the issue of the low specific capacitance of commercial AC, researchers have made many efforts. Dong et al. [25] designed the AC/Zn devices with a capacity of 121 mAh g^{-1} and an energy density of 84 Wh kg^{-1}. In fact, there is still space for improvement in cycle performance. Surface-functionalized AC with reasonable pore size enhances chemical binding with carriers, which contributes to an increase in specific capacity [65]. For instance, An's group [66] prepared the AC material with a mesoporous structure by polymerization PVP dehydrogenation process. This increases the number of oxygen groups on the surface of AC. The modified materials, such as cathodes, show an abundance of electrochemical active sites and rapid ion diffusion and transmission ability. Therefore, the assembled ZIHCs can still deliver a specific capacity of 72 mAh g^{-1} even at a current density of 10 A g^{-1}. In addition, the cycle life can reach 40,000 cycles with a capacity retention rate of 78% (Figure 5a).

The specific capacity of AC electrodes is enhanced based on surface modulation techniques, but there are limited improvements in their conductivity, wettability of the electrode and electrolyte surfaces, and specific surface area. Therefore, researchers have developed some effective modification strategies and superior electrode materials. Heterogeneous atom doping can introduce O-/N-containing functional groups. This offers fast redox reactions for the electrodes and increases the pseudocapacitive contribution to charge storage [67,68]. The construction of novel nanostructures or hierarchically porous carbon materials can effectively increase the adsorption sites. This accelerates ion/electron transport and favors the rate capacity [69]. Further detailed discussions are presented in the subsequent sections.

3.1.2. Metal–Organic Compound Derived Carbon

Metal–organic compounds are converted into nanostructured carbon materials through high-temperature calcination or controlled chemical reactions. This conversion process is conducive to the formation of porous architecture. This hierarchical porous structure facilitates the optimization of electrochemical performance. Zhang and collaborators [70] fabricated ultra-thin layered porous carbon nanoplates (Ca-900 samples) using calcium gluconate precursor by in situ template method. The tuning of calcination temperature can optimize the specific surface area and porosity of carbon materials. The assembled device shows an energy density of 75.22 Wh kg^{-1} and a power density of 879.12 W kg^{-1}. After 4000 times of charge and discharge, the capacitance retention rate is 90.9%.

Metal–organic frameworks (MOFs) constitute a class of crystalline compounds synthesized by coordination of metal ions with organic linkers. MOF-derived carbon materials can possess a wide range of specific surface areas and rich porous structures by high-temperature calcination and etching processes [71]. Li et al. [72] designed a rod-like porous carbon fabricated by pyrolysis of MOF (MIL-88B). Based on the structural features of MOF and the synergistic effect of alkali activation, the carbon rod material possesses a specific surface area (SSA) of 1272.4 m^2 g^{-1}. The assembled device can still show a specific capacity of 73.5 mAh g^{-1} at 50 A g^{-1}. The capacity retention of 96.5% (10 A g^{-1}) is achieved after 10,000 cycles. Our group [20] prepared V metal–organic framework (MOF)-derived porous carbon anodes. The assembled devices present specific capacities of 168 and 140 mAh g^{-1} with a capacity loss of 16.7% at 0.1 and 2 A g^{-1}, respectively. These kinds of materials are often treated by strong acid/alkali etching, which increases the technical difficulty and cost. Therefore, there is an urgent need to design eco-friendly, high-performance carbonaceous materials.

3.1.3. Polymer-Derived Porous Carbon

Polymers are complex structures composed of many macromolecular chains. They possess nanoscale self-assembly ability. Compared with small molecular hydrocarbons, polymers as carbon sources show controllability in structure and doping amount. A variety of porous carbonaceous structures, including spherical particles, fibrous materials, films, or membrane-type materials, can be prepared by matching suitable polymer precursors, such as polyacrylonitrile, polyphenylamine, and conjugated polymers. Due to the unique atomic composition of polymer precursors, non-metallic atoms can be doped by pyrolysis, including N, B, P, and S. These types of atoms are combined into the carbon framework individually or jointly to optimize the energy storage capability of carbon materials [73]. Yang et al. [74] developed a hierarchical porous carbon fiber derived from polyvinylpyrrolidone/polyacrylonitrile (PVP/PAN) by electrospinning. The device assembled with PVP-/PAN-derived porous carbon fibers shows a specific capacity of 208 mAh g^{-1} and a rate retention of 49.5% (0.5~5 A g^{-1}). After 10,000 stable cycles, the capacity maintains 72.25% of the initial capacity. This multi-channel structure is beneficial to the specific capacity and rate performance of ZIHCs. However, the specific surface area obtained by this direct pyrolysis is only 449.05 m^2 g^{-1}. Therefore, its cyclic stability is limited. Liu group's [75] studied the pyrolysis of chitosan aerogels to produce porous carbon with 3D honeycomb lattice. The combination of nitrogen and oxygen realizes the co-doping of carbon materials. However, the specific surface area of carbonized chitosan is only 1000 m^2 g^{-1}, and it lacks essential micropores and mesopores. Combined with the KOH activation strategy, its specific surface area can be increased to 2267 m^2 g^{-1}. Therefore, the capacity retention of the assembled device can remain at 86.3% after 10,000 cycles.

Carbon materials derived from asphalt have garnered extensive research attention for their applications in supercapacitors and LIBs [76–78]. However, the activation of asphalt with complex compositions at high temperatures usually leads to violent volatilizing of light components and polymerization of polycyclic aromatic hydrocarbons [79]. To regulate the pore size during fabrication, Liu and co-workers [80] proposed two steps of carbonization and alkali activation to enhance its structural stability. In Figure 5b, compared with commercial AC, the specific surface area of this material can be increased to 3525 m^2 g^{-1}. The porous structure is composed of micropores and abundant mesopores, which effectively increase the ion occupation sites. The fabricated device achieves a specific capacity of 172.7 mAh g^{-1} at 1 A g^{-1}, and it retains a capacity of 119.4 mAh g^{-1} at 20 A g^{-1} (Figure 5c). Lu and collaborators [81] designed an N-doped asphalt-derived carbon material (NPC) as a cathode. In Figure 5d, incorporating N into the carbon electrode material enhances its electrical conductivity, and the addition of pyridine N effectively increases the electrochemical kinetics. They further proposed a nitrogen redox mechanism. From Figure 5e, the activation energy (E_a2) of the NPC electrode is much superior to that of PC one. This indicates that the addition of N triggers multiple redox processes in favor of increasing the pseudocapacitance. The assembled device possesses a capacity of 136.2 mAh g^{-1}. When the current density increases from 0.3 A g^{-1} to 15 A g^{-1}, the capacity retention rate is 50.8% (Figure 5f). In addition, it only experienced 4.6% self-discharge capacity loss within 24 h.

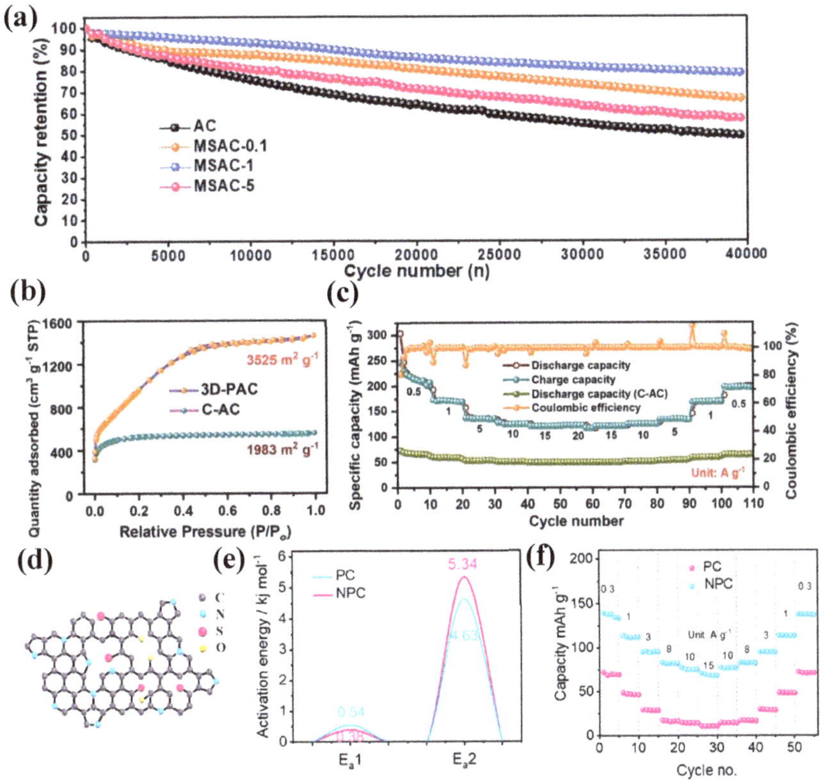

Figure 5. Effect of specific surface area on electrochemical performance of materials (**a**) The capacity retention (10.0 A g^{-1}) for the various electrodes. Reproduced with permission [66], Copyright 2020, Elsevier B.V. (**b**) Nitrogen adsorption–desorption isotherms of the two materials. (**c**) Rate capability of the assembled device. Reproduced with permission [80], Copyright 2020, Wiley-VCH GmbH. (**d**) Schematic structure of NPC. (**e**) The activation energy comparison of PC and NPC. (**f**) Rate performance. Reproduced with permission [81], Copyright 2021, American Chemical Society.

3.1.4. Biomass-Derived Porous Carbon

Biomass-derived porous carbon materials are frequently applied as cathodes for capacitors due to their cost-effectiveness, structural diversity, and preparation convenience [82]. Li and co-workers [83] synthesized pencil shaving-derived carbon materials and obtained optimal PSC-A600 electrodes by adjusting the pyrolysis temperature. This porous structure enables a fast ion transport rate and provides sufficient charge storage space. Therefore, the Zn/PSC-A600 device presents a specific capacity of 183.7 mAh g^{-1}. From Figure 6a, it can perform 10,000 cycles at 10 A g^{-1} and maintain a retention rate of 92.2%. Qiu's group [84] prepared porous carbon positive electrodes (LHPCs) by the self-templating method employing lignin as a carbon precursor. The electrode material demonstrates fast electrochemical kinetic processes. This is due to the rich pore structure and the introduction of carboxyl functional groups. The assembled Zn/LHPCs capacitors obtain an energy density of 135 Wh kg^{-1} at a power density of 101 W kg^{-1}.

However, the impurities and functional groups of biomass carbon-based raw materials accelerate the self-discharge speed of the assembled hybrid capacitor. This is closely related to the pore structure of the carbon cathode. This pore effect exacerbates charge redistribution and carbon oxidation [85]. Li et al. [86] synthesized oxygen-enriched multistage porous carbon by olive pyrolysis through chemical activation. Oxygen-enriched functional groups provide additional capacitance and boost the wettability and conductivity of porous carbon. The as-built devices obtain an energy density of 136.3 Wh kg^{-1} and a self-discharge rate of 2.3 mV h^{-1} (Figure 6b). They can achieve 20,000 charge and discharge cycles at 10 A g^{-1} (Figure 6c).

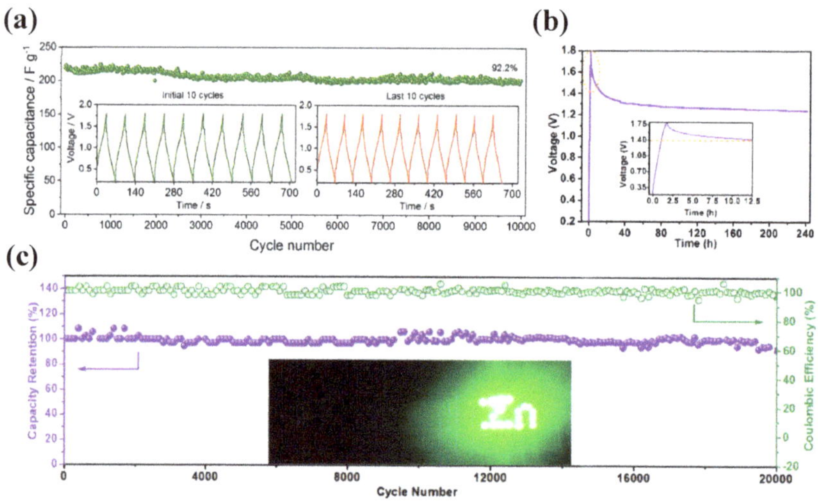

Figure 6. The cycle capability of biomass-derived porous carbon. (**a**) Cycle performance. Reproduced with permission [83], Copyright 2020, Elsevier B.V. (**b**) Self-discharge curves. (**c**) Cycling durability at 10 A g^{-1}. Reproduced with permission [86], Copyright 2023, Wiley-VCH GmbH.

3.2. Carbon Nanofibers

Flexible wearable devices have been a hot research topic. Therefore, there is a demand for the exploitation of flexibility electrodes. Typically, the fabrication of electrodes depends on the support of current collectors and additives in addition to active materials, which can reduce the total energy density. One-dimensional (1D) carbon nanofibers are highly promising for the construction of freestanding cathodes, which is attributed to their favorable flexibility and abundant pore structure. However, their hydrophobicity hinders charge transport at the electrolyte–electrode interface, demonstrating a large interfacial resistance.

Surface modification improves the hydrophilicity of carbon-based materials. For instance, this is a viable approach to enhance the chemisorption of metal ions on electrode materials. Zhang and collaborators [87] synthesized flexible oxygen-enriched carbon fiber films by electrostatic spinning and acid-assisted oxidation. ZIHCs with this carbon fiber cathode show an energy density of 97.7 Wh kg^{-1} and a power density of 9.9 kW kg^{-1}. Also, they maintain long-term cycling stability (81%) of 50,000 cycles. Figure 7a demonstrates the near-zero superhydrophilicity angle. It reduces the interfacial resistance between electrolyte and electrode. The adsorption of zinc ions on the electrode surface is greatly enhanced. Density functional theory (DFT) calculations show the chemisorption relationship between zinc ions and C=O functional groups (Figure 7b). Among them, zinc ions are more easily adsorbed between two carbonyl groups. The adsorption process is shown below:

$$2C{=}O + Zn^{2+} + 2e^{-} \rightarrow C\text{-}O\text{-}Zn\text{-}O\text{-}C, \tag{1}$$

The existence of oxygen-enriched functional groups significantly improves the wettability of carbon fiber and promotes the chemisorption of zinc ions. In addition, these cathodes could operate stably with a loading of 20 mg cm^{-2} and 180° bending. Their specific capacity decreases by 19% (5 A g^{-1}) after 50,000 cycles (Figure 7c).

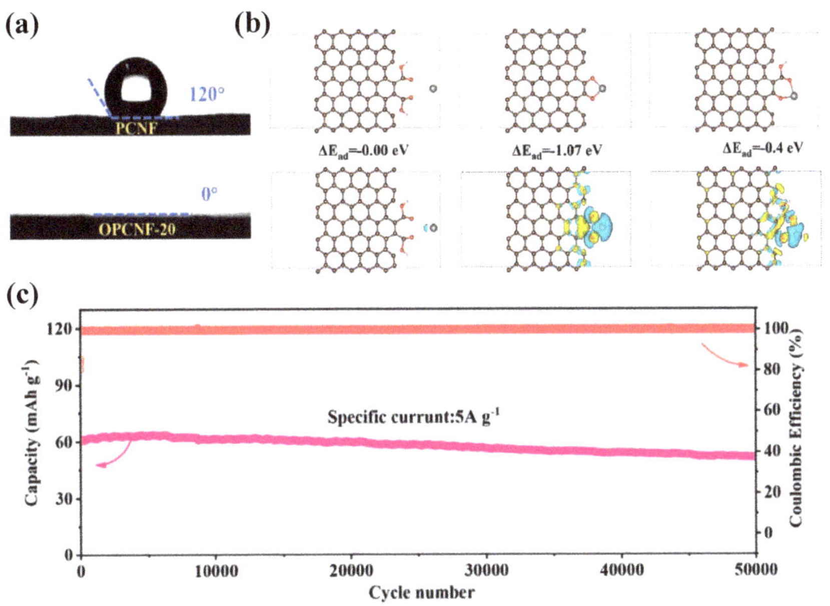

Figure 7. Wetting behavior and cycling stability. (**a**) Contact angles. (**b**) DFT calculations for zinc storage behavior. Red, brown, white and gray balls represent oxygen, carbon, hydrogen and zinc atoms, respectively. The regions indicate the increase (yellow) and decrease (blue) of electron density. (**c**) Long-term cycle at 5 A g^{-1}. The pink and orange curves represent capacity and coulombic efficiency, respectively. Reproduced with permission [87], Copyright 2021, Elsevier B.V.

3.3. Graphite-Type Carbon Materials

This refers to a high-performance material with a graphite-like microcrystalline structure. Representative members include graphene, carbon nanotubes, and carbon fibers [24,88]. These materials are characterized by a layered accumulation of carbon atoms interconnected by van der Waals interactions. This unique layered structure endows graphite-like carbon materials with favorable electrical and thermal properties, as well as excellent chemical stability. Therefore, they show wide application potential in various fields, including energy storage, electronic devices, composite materials, and catalytic technology [89].

3.3.1. Carbon Nanotubes

One-dimensional carbon nanotube (CNT) materials offer unique fast carrier transfer pathways, but the limited specific surface area (less than 500 m^2 g^{-1}) hinders their further development. Previous studies have demonstrated that the introduction of pseudocapacitive active sites on carbon nanotubes can increase the specific capacitance of carbon nanotubes [24,90,91]. Wang's group [24] used oxidized CNT (oCNT) as cathode and revealed the role of oxygen groups in ion storage by electrochemical characterization. As shown in Figure 8a, Zn^{2+} on the oxygen-containing groups of the oCNT electrode undergoes electrochemical adsorption/desorption. The surface-endowed ZICs are confirmed to provide better performance than the pure carbon nanotube cathode.

3.3.2. Graphene

Graphene, featuring a single atom thickness, superior electrical conductivity (10^6 S cm^{-1}), mechanical properties, and a specific surface area of 2630 m^2 g^{-1}, is a suitable positive material for ZIHCs. However, two-dimensional (2D) graphene tends to aggregate. Moreover, there are morphological defects such as holes, edges, cracks, and heteroatoms on the surface of graphene [92]. This greatly limits its properties. Three-dimensional (3D) graphene is a novel class of carbon nanomaterial composed of 2D graphene on a macroscopic scale. When graphene nanosheets are interconnected as building blocks, they can be assembled into 3D structures from the bottom up. Han et al. [93] fabricated a ZIHC based on 3D nanostructures graphene and polyaniline composites as cathodes (Figure 8b). It shows an energy density of 205 Wh kg^{-1} (0.1 A g^{-1}). From Figure 8c, it can still maintain 80.5% of the initial capacity after 6000 cycles of repeated charging and discharging. However, these high-porosity materials always present low bulk density and high weight capacitance. It reduces the volume capacitance and hinders the practical application of the device [88]. Therefore, it is very necessary to develop highly dense structures. Zheng's group [94] prepared a high-density nanoparticle 3D porous graphene (3D-PG-1) through capillary evaporation. The pores formed through the evaporation process provide extensive pathways for the rapid diffusion of Zn^{2+}. It also weakens the strong electrostatic force of Zn^{2+} embedded in the structure. This ensures that the electrode achieves 30,000 stable cycles (Figure 8d).

Recently, graphene-derived 3D network/aerogel structures have been identified to enhance the number of active sites in electrode materials. Compared with conventional graphene films, graphene aerogels feature stable 3D networks and high pore volumes [95]. Okhay et al. [96] synthesized reduced graphene oxide aerogels with optimal properties by tuning the freeze-drying conditions. The graphene oxide composite electrode demonstrates favorable uniformity and surface wettability with a specific capacitance of 129 F g^{-1} at 0.1 A g^{-1}.

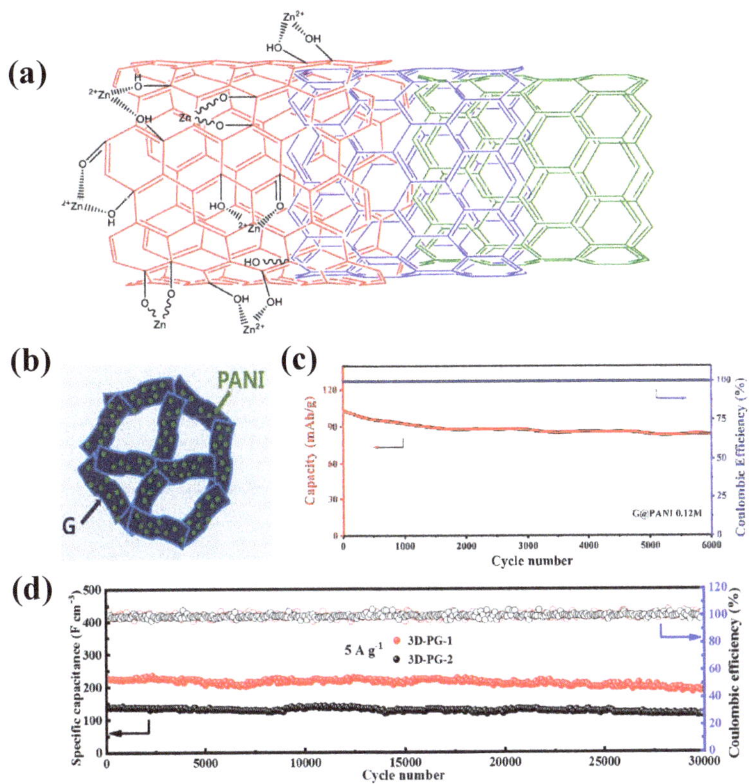

Figure 8. The electrochemical performance of the other carbon electrodes. (**a**) Schematic of zinc ion adsorption on the surface of oCNTs electrode, Reproduced with permission [24], Copyright 2016, Frontiers. (**b**) 3D cross-linked structure of graphene@PANI materials (**c**) The cycle stability of graphene@PANI electrode at 5 A g^{-1}. Reproduced with permission [93], Copyright 2018, Royal Society of Chemistry. (**d**) Long cycle stability at 5 A g^{-1}. Reproduced with permission [94], Copyright 2021, Elsevier Ltd.

4. Modification Strategies

These carbon materials suffer from many issues as electrode materials, including structural instability, poor wettability, and slow electrochemical kinetic. This hinders efficient carrier transport. Therefore, it is essential to modify the carbon-based materials to mitigate these adverse effects. Researchers have proposed several effective strategies, including surface modification, morphology and size tuning, and heterogeneous atom doping.

4.1. Surface Modification

To further improve the cycle stability, it is a feasible method to construct novel carbon-based composites. Xin et al. [97] prepared a poly(4,40-thiodiphenol, TDP)-modified nanoporous AC cathode material. The introduction of this polymer increases the capacitance contribution and broadens the voltage window to 1.9 V through the faraday process. The areal energy density of the device is 1.03 mWh cm^{-2}, and the corresponding power density can reach 0.9 mW cm^{-2}. The increase in voltage is related to their charge storage mechanism. During the charging, Zn^{2+} first extracts from the cathode due to strong electrostatic interaction. Then, H^+ is released from the carbonyl group of poly(4,40-TDP). This continuous extraction leads to a wide voltage window of the device. The interfacial redox reaction is also a viable strategy for increasing the energy density of the device.

Ramanujam's group [98] synthesized a porous AC from chitosan and N, N′-bis(glycinyl) naphthalene diimide (H_2BNDI) composite electrode material. The porous structure can reduce the transmission path of ions to facilitate the rapid transfer of ions. The presence of oxygen functional groups and N-enriched sites on the electrode surface can provide additional pseudocapacitance. Therefore, the fabricated devices deliver a specific capacitance of 500 F g^{-1} with an energy density of 250 Wh kg^{-1} at 0.1 A g^{-1}.

Graphene and CNT materials produce severe agglomeration and re-stacking. This significantly reduces their actual specific surface area and causes a reduction in the specific capacity. To address these drawbacks, it is necessary to fabricate composites with optimal structures. It is a common approach to prevent agglomeration by functionalizing graphene oxide with redox-active organic molecules. For example, Chen's group [99] introduced the p-phenylenediamine (PPD) into the interlayer of the graphene film. This enlarges the interlayer distance and exposes more ion-accessible regions. In addition, the partial reduction of the amino group improves the conductivity of the RGO film. Benefiting from the redox activity of PPD, the RGO@PPD film electrode shows high pseudocapacitance and EDLC for charge storage. Yang et al. [100] prepared a composite electrode of electrochemical graphene oxide (EGO) and polypyrrole (PPy) by one-step electrochemical co-deposition. The Zn-PPy/EGO hybrid capacitor obtains an energy density of 72.1 Wh kg^{-1} at 12.4 kW kg^{-1} and maintains the long-term cycle stability of 81% after 5000 cycles. Compared with chemical graphene oxide (CGO), EGO material possesses strong π-π interaction with PPy due to less structural destruction of graphene [101]. It is beneficial to enhance the conductivity of the composite material.

The lamination issue between 2D nanosheets can be effectively resolved by introducing additional dimensional materials, such as 1D carbon nanotubes [102,103]. Zhang and co-workers [104] developed kapok-derived N- and P-rich 2D thin-walled microporous carbon tiles (CTs). Leveraging the superior conductivity of single-walled carbon nanotubes (SWNTs), they constructed CT/SWNT porous paper cathodes with considerable specific surface area. The uniformly doped thin walled (~700 nm) structure can effectively shorten the ion diffusion path. The submicron layer spacing between the CTs with suitable curvature and carbon nanotubes builds a fast ion/charge transport channel. Based on this synergistic effect, this carbon cathode delivers a specific capacity of 127.6 mAh g^{-1} at 0.1 A g^{-1}. Even at 20 A g^{-1}, its specific capacity can still be kept at 72.4 mAh g^{-1}.

4.2. Morphology Regulation

The microstructure of carbon materials plays a pivotal role in regulating the electrochemical performance of energy storage devices. Carbon materials can be generally categorized into 0D carbon spheres, 1D carbon nanofibers, 2D carbon nanosheets, and 3D porous structures [30]. According to various preparation approaches, carbon spheres can be adjusted into hollow, yolk–shell, and core–shell structures. It possesses favorable conductivity and controllable particle size distribution. The hollow architecture facilitates the rapid transport of carriers and relieves the volume expansion in the long-term cycle [105].

The template method is commonly utilized to prepare hollow structures. Yan's group [106] synthesized mesoporous hollow carbon spheres by the SiO_2 template method. They are used as cathodes and MCHSs-coated Zn foils as anodes to assemble hybrid capacitors in 2M $ZnSO_4$ electrolyte. The nanostructures of MCHSs electrodes provide abundant ion diffusion channels. In addition, the cladding layer inhibits the growth of dendrites/protrusions on the Zn anode surface. Based on the above advantages, the devices show a high energy density of 129.3 Wh kg^{-1} at 266.4 W kg^{-1} and a cycle retention of 96% after 10,000 cycles at 1 A g^{-1}. Further, they found that the capacity of the devices with different mass loadings decayed as the mass of the active substance increased. This is attributed to the sluggish ion and charge transport of thick electrodes. Nonetheless, the device can still achieve a specific capacity of 110 mAh g^{-1} at a high mass loading (20 mg cm^{-2}). Chen et al. [107] reported a copolymer-derived (polyaniline-co-polypyrrole) hollow carbon sphere using surfactant as a soft template. This cavity sphere possesses a high surface area, which facilitates the increase of the active site of the electrode. The hollow structure expands the contact area between electrode and electrolyte. Therefore, the assembled capacitor shows an energy density of 59.7 Wh kg^{-1} and a power density of 447.8 W kg^{-1}. In addition, it can be stably cycled 15,000 times at 1 A g^{-1} with less capacity loss (2%).

4.3. Heteroatom Doping

Carbon materials doped with N, S, P, B, and other heteroatoms are increasingly gaining attention due to their high electron mobility and enhanced energy storage capacity [67,108,109]. The interaction between the heteroatoms and the host material improves the adsorption of electrolyte ions as well as optimizing the surface wettability and conductivity. This promotes rapid charge transfer, which effectively accelerates the point electrochemical kinetics. We mentioned above that the AC cathode suffers from a low practical specific capacity. In addition, it shows weak wettability at the interface with the electrolyte. This constrains the longevity of the device at high current densities [110]. Due to the formation of polar functional groups, the combination of N and O in carbon materials enhances the interaction with electrolytes, thus improving wettability. Wang et al. [111] reported that doping onion-like carbon with N and P elements significantly increased the surface area available for ion adsorption. The assembled N, P-OLC-based devices provide 63% capacity retention at 20 A g^{-1}. Their energy and power densities are 149.5 Wh kg^{-1} and 26.7 kW kg^{-1}. Among them, the doping of N significantly increases the hydrophilicity of the electrode interface, thus optimizing the contact resistance between the electrode and the electrolyte. Also, a nitric acid treatment can provide oxygen-containing functional groups on the surface of porous carbon, which contributes to the improvement of wettability [112]. Therefore, the device can obtain superior rate performance at high current density [113].

Dong's group [114] constructed a fiber-carbon cathode HPCF with a layered porous surface and O/N functional groups. The surface of the electrode contains a large number of micropores to provide high capacity, and the mesopores and macropores can ensure superior rate capability (Figure 9a). The CV curve further reveals that HPCF cathodes store more Zn^{2+} cations and $SO_4{}^{2-}$ anions at high scanning rates. This indicates that most of the stored charge of the cathode comes from $SO_4{}^{2-}$ anion storage. Figure 9b shows the capacitance contribution of the electrode gradually increases with the scanning rate. The reason could be that the rich N/O heteroatoms in HPCF cathode can enhance the interaction between pseudocapacitance behavior and electrolyte ions. The mechanism of charge storage is explored through combining various characterization methods. When the electrode is in the discharge state, $Zn_4SO_4(OH)_6{\cdot}5H_2O$ flakes (BZS) appear. At charging, most of the flakes disappear. This may be due to the following reactions:

$$4Zn^{2+} + 6OH^- + SO_4{}^{2-} + 5H_2O \rightarrow Zn_4SO_4(OH)_6{\cdot}5H_2O\downarrow, \quad (2)$$

In addition, the H+ provided by the acidic electrolyte adsorbs on the electrode surface, which results in an increase in the pH of the electrolyte. This suppresses the hydrogen evolution potential of the electrolyte. Only BZS products are detected in the structure of the HPCF cathode. This further confirms the energy storage mechanism of ion adsorption/desorption occurring at the surface of the electrode. Zheng et al. [69] designed an O-doped 3D porous carbon material. The combination of 3D porous structure and oxygen-containing functional group can endow the electrode with fast ion/electron transport and strong pseudocapacitance. The assembled flexible device can be recharged and discharged 10,000 times at 1.0 A g^{-1} with a capacity retention rate of 87.6%. Zhang's group [115] utilized potassium chloroacetate ($ClCH_2COOK$) as a carbon source and template to form oxygen-enriched porous carbon (ORCs) with an average pore size of 1.95–2.19 nm. This strategy can construct ORCs with high oxygen doping (4.38 atomic%). As a cathode material, it shows high EDLC and pseudocapacitance.

However, excessive oxygen functional groups can diminish the interaction between electrolyte ions and the carbon surface, which induces the aggravation of the self-discharge behavior. This directly inhibits the dynamic process of the device [85]. Yuan et al. [116] modulated the temperature range during their hydrogen thermal reduction process to prepare superior starch-based porous carbon materials. It reveals the structural evolution of the reactive deoxygenation process. The results indicate that the pyrolysis and hydrogenation of oxygen functional groups may occur simultaneously. Among them, the HAC-700 electrode can be maintained for 10 h at an open-circuit voltage of 2.29 V, demonstrating strong resistance to self-discharge (Figure 9c). This may be associated with the suitable pore size in the carbon skeleton and the consumption of instable COOH and R-COOR groups. They can effectively improve the wettability of the electrode and prevent the electrolyte from irreversibly reacting with the unstable oxygen functional groups on the carbon electrode surface.

The addition of N can reduce the energy barrier of C-O interaction with Zn ions [117]. Lu's group [37] constructed capacitors with a lifetime of 100,000 cycles based on nitrogen-doped layered porous carbon (HNPC) as the cathode (Figure 9d). The existence of the N element optimizes the conductivity, electrochemical active surface area (ECSA) of the electrode, and surface wettability. It also improves the efficient transfer of ions/electrons and ensures sufficient storage space for charges. DFT calculation reveals the whole process of zinc ion adsorption and desorption. The adsorption of zinc ions causes the O-H bond to form a C-O-Zn bond, resulting in the release of H^+ ions. In Figure 9e, the reaction of an N-doped HNPC electrode only needs to overcome the energy barrier of 1.25 eV (2.13 eV for an undoped electrode). It effectively promotes the chemical adsorption of Zn ions on the electrode surface.

Biomass-derived porous carbon possesses inherent functional groups, or heteroatoms. It can offer sufficient active sites and surface wettability for the reaction. Therefore, the addition of chemical activators can further increase the specific surface area and optimize the pore structure, which is beneficial to the improvement of electrochemical performance. Wang et al. [118] proposed a novel potassium thioacetate activation technique. The pine needles are pyrolyzed as a carbon source to produce sulfur-doped 3D porous carbon (S-3DPCs). The prepared S-3DPCs material shows a specific surface area of 2336.9 $m^2\ g^{-1}$ and a certain amount of sulfur (0.88–3.60 at%). Therefore, the capacitor assembled with the S-3DPC-800 positive electrode can provide a specific capacity of 203.3 mAh g^{-1} (0.2 A g^{-1}) and 81 mAh g^{-1} (20 A g^{-1}). In addition, it can be recycled 18,000 times at 10 A g^{-1}, and the specific capacity retention reaches 96.8%.

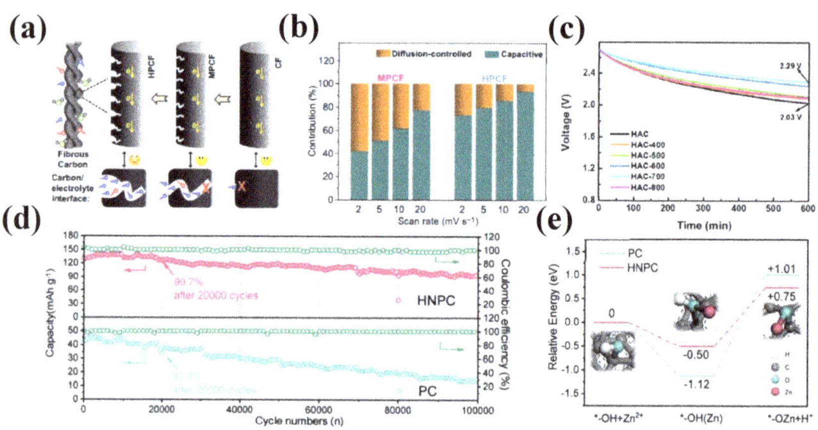

Figure 9. Performance of doped carbon-based materials. (**a**) Schematic of energy storage on cathodes with various surface conditions. (**b**) Contribution ratio of charge storage behavior. Reproduced with permission [114], Copyright 2021, Springer Nature. (**c**) Leakage current curves. Reproduced with permission [116], Copyright 2020, Elsevier B.V. and Science Press. (**d**) Cycling stability. (**e**) The energy profiles.* represents C atom. Reproduced with permission [37], Copyright 2019, Wiley-VCH.

Lu and his collaborators [119] designed the activation mechanism of microporous carbon (DSPCs) derived from sodium alginate through double templates. Highly active micropores (about 0.7–2.0 nm) endow the material with reasonable specific surface area and rich reaction sites. The assembled device shows an energy density of 99.22 Wh kg^{-1} (200 W kg^{-1}). The reversible specific capacity of the electrode is 87.5 mAh g^{-1} at 0.2 A g^{-1}. The capacity of 18.1 mAh g^{-1} can be maintained when the current density rises to 40 A g^{-1}.

In Table 1, we summarize the electrochemical performance of the various carbon-based electrodes. All of these electrode materials can achieve a long cycle life up to more than 10,000 cycles. Among them, the specific capacity of the hierarchical porous material is higher than that of the AC material. Oxygen-rich carbon electrodes can show a cycle life of more than 20,000 cycles. The atom-doped electrode materials demonstrate a long cycle life of up to 60,000 cycles with a capacity retention of 98.9% and a maximum power density of 48.8 kW kg^{-1}.

Table 1. The electrochemical performance of the various electrodes for ZIHCs.

Cathode	Electrolyte	Voltage (V)	Capacity/ Current Density ($mAh\ g^{-1}/A\ g^{-1}$)	Capacity Retention/ Current Density/ The Long-Term Cycle ($\%/A\ g^{-1}$/cycles)	Energy Density ($Wh\ kg^{-1}$)	Power Density ($kW\ kg^{-1}$)	Refs.
AC	2M $ZnSO_4$	0.2–1.8	121/0.1	91/10/10,000	30	14.9	[25]
O-rich porous carbon	3 M $Zn(ClO_4)_2$	0–1.9	179.8/0.1	99.2/20/30,000	40.4	48.8	[31]
Mesoporous AC	2M $ZnSO_4$	0.3–1.8	176/0.5	78/10/40,000	77	10.7	[66]
Porous AC	3M $Zn(CF_3SO_3)_2$	0.01–1.8	231/0.5	70/10/18,000	77.5	11.4	[80]
N doped pitch-derived carbon	1 M $ZnSO_4$	0–1.8	136.2/0.3	98.9/10/60,000	81.1	12.8	[81]
Porous carbon	1M $Zn(CF_3SO_3)_2$	0.2–1.8	183.7/0.2	92.2/10/10,000	65.4	15.7	[83]
O-rich porous carbon	2M $ZnSO_4$	0.2–1.8	317.8/0.5	91/10/20,000	136.3	20	[86]
O-enriched carbon fiber	gelatin/$ZnSO_4$ gel	0.2–1.8	136.4/0.1	81/5/50,000	97.7	9.9	[89]
3D graphene@PANI composite	2M $ZnSO_4$	0.3–1.6	154/0.1	80.5/5/6000	205	45.8	[93]
3D graphene	3M $Zn(CF_3SO_3)_2$	0.1–1.8	299 F cm^{-3}/0.1	85/5/30,000	61 Wh/L	23.2 kW/L	[94]
H_2BNDI@Ch-C	2 M $ZnSO_4$	0.1–1.9	500/0.1	80/5/10,000	55	9.5	[98]
PPy/EGO	1 M $ZnCl_2$	0.5–1.5	444.2/0.35	81/5 mA cm^{-2}/5000	72.1	12.4	[100]
Hollow carbon spheres	2M $ZnSO_4$	0.2–1.8	174.7/0.1	96/1/10,000	36.8	13.7	[106]
S-doped 3D porous carbon	2M $ZnSO_4$	0.2–1.8	203.3/0.2	96.8/10/18,000	64.8	16	[118]

5. Summary and Outlook

In summary, carbon-based materials are of interest owing to their structural diversity, superior electrical conductivity, and large specific surface area. These features render them crucial electrode materials in ZIHCs. We concentrate on the research advances of carbon cathode and discuss their energy storage mechanism, electrochemical performance, and electrode material structure optimization strategies. Some progress has been achieved, but there are still challenges in applications, especially in improving the energy power density and specific capacitance. At present, ZIHCs are in the infancy stage of growth, and more efforts are needed to propose new innovations to overcome these problems (Figure 10).

1. It is still difficult to identify the energy storage mechanism of ZIHCs. In the current study, carbon materials are often compounded with some materials for structural optimization in order to increase the specific capacity of electrodes. The pseudocapacitance during charge storage provides a partial contribution rather than simple ion adsorption/desorption. Furthermore, there is no complete clarification about the contribution ratio of the charge storage behavior. The combination of first-principles calculations, electrochemical kinetic analysis, and in situ structural characterization can also illustrate the structural changes of the electrode during charging and discharging and reveal the energy storage mechanism. The effect of by-product generation on electrode performance is still controversial.
2. Carbon-based electrode materials inherently show low specific capacitance and poor energy density. Suitable structural modifications, including heteroatom doping, microstructure modulation, and surface modification, can effectively improve the charge storage capacity of carbon-based materials. However, the constitutive relationship and intrinsic connection between structural improvement and electrochemical performance are still not clearly explained.

3. Commercially available zinc foils are usually the counter electrodes of the devices. Although they present high theoretical capacity, issues such as dendrite growth, corrosion, and HER reactions limit the high performance of ZIHCs. Therefore, it is essential to develop the structural design of anode materials to facilitate their advancement. For example, 3D nanostructured zinc electrodes possess rich active sites and a large electrolyte/electrode contact area. The uniformly distributed electric field can retard the growth of dendrites and promote the diffusion of ions.
4. Fully charged hybrid capacitors often suffer from spontaneous charge loss and voltage drop, resulting in self-discharge. The existence of the self-discharge process limits its market application. For carbon-based electrodes, ash and oxygen-containing functional groups can weaken the interaction between electrolyte ions and the electrode surface and enhance the self-discharge rate. In addition, the pore effect of porous structures leads to the redistribution of charges in carbon materials. We should also focus on the self-discharge behavior of devices in diverse working conditions. The corresponding approach is established to suppress the self-discharge of the device.

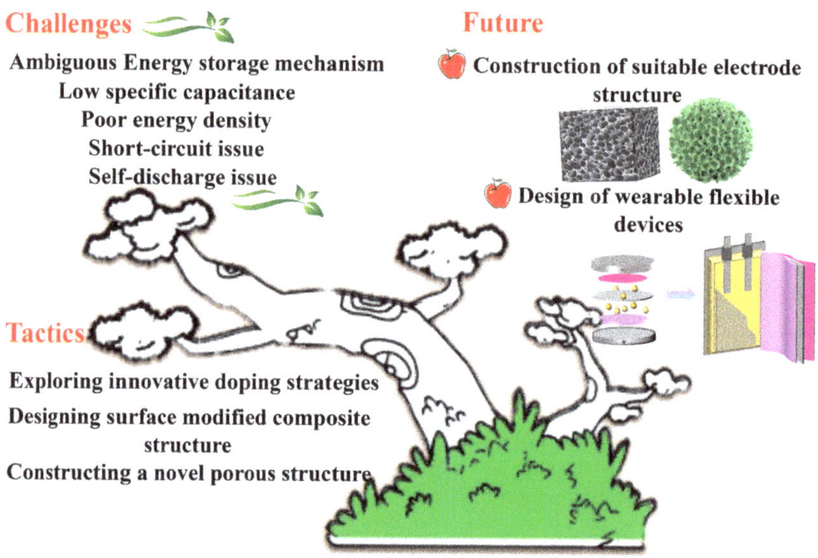

Figure 10. Challenges, tactics, and future goals for carbon-based electrodes of ZIHCs.

In the future, we consider that ZIHCs should concern the following aspects:

1. It is necessary to explore novel electrode material synthesis tactics to obtain porous carbon with favorable pore size and electrolyte ion compatibility. Various carbon-based precursor materials, such as biomass, asphalt, and polymers, possess diverse physical and chemical properties. It is helpful to design reasonable approaches according to their properties to give full play to the characteristics of raw materials and avoid their shortcomings.
2. The development of gel electrolyte is in line with the future development trend. Ordinary diaphragm is easy to be pierced by zinc branches in long-term circulation, resulting in a short circuit. Hydrogel electrolyte shows certain flexibility and mechanical strength, which can effectively slow down the growth of zinc dendrites. In addition, with the development of wearable flexible devices, positive and negative electrodes as well as the matching of the diaphragm are a key concern. The electrodes of flexible capacitors need to meet the characteristics of superior mechanical flexibility, light weight, and high-pressure resistance. For instance, Wu's group [120] prepared a super-folded conductive carbon material based on the bionic design idea of silkworm

spinning–cocooning–reeling, which can withstand 1 million or even unlimited folds without any damage. These structural materials possess superior mechanical properties. Their integration with ZIHCs can greatly improve the durability and life of devices. This synergy can promote the technological innovation of wearable devices.

3. Combining advanced in situ characterization techniques, it allows an intensive investigation into the structure of the electrodes and the dynamic changes in the energy storage process. It is essential for revealing the electrode energy storage process. Also, the structural design of electrode materials combined with theoretical calculation can provide theoretical guidance for structural stability, reaction-free energy, ion adsorption, diffusion, and reaction processes.

Author Contributions: Conceptualization, investigation, and resources, Y.L., L.S. and C.L.; writing—original draft preparation, Y.L. and C.S.; writing—review and editing, Y.L., C.S., L.S., C.L. and X.W.; visualization, Y.L. and C.S.; supervision, Y.L. and X.W. All authors have read and agreed to the published version of the manuscript.

Funding: This research was funded by the Technology Development Contract of Sinopec, grant number: 122013.

Data Availability Statement: All data used for this work are included within the manuscript.

Conflicts of Interest: Authors Ying Liu, Lechun Song, Chenze Li and Caicheng Song were employed by the company SINOPEC Dalian (Fushun) Research Institute of Petroleum and Petrochemicals. The remaining author declare that the research was conducted in the absence of any commercial or financial relationships that could be construed as a potential conflict of interest.

References

1. Huang, Y.; Zhu, M.; Huang, Y.; Pei, Z.; Li, H.; Wang, Z.; Xue, Q.; Zhi, C. Multifunctional Energy Storage and Conversion Devices. *Adv. Mater.* **2016**, *28*, 8344–8364. [CrossRef] [PubMed]
2. Zheng, M.; Tang, H.; Hu, Q.; Zheng, S.; Li, L.; Xu, J.; Pang, H. Tungsten-Based Materials for Lithium-Ion Batteries. *Adv. Funct. Mater.* **2018**, *28*, 1707500. [CrossRef]
3. Ue, M.; Sakaushi, K.; Uosaki, K. Basic knowledge in battery research bridging the gap between academia and industry. *Mater. Horiz.* **2020**, *7*, 1937–1954. [CrossRef]
4. Wang, H.; Zhu, C.; Chao, D.; Yan, Q.; Fan, H.J. Nonaqueous Hybrid Lithium-Ion and Sodium-Ion Capacitors. *Adv. Mater.* **2017**, *29*, 1702093. [CrossRef] [PubMed]
5. Dong, L.; Yang, W.; Yang, W.; Li, Y.; Wu, W.; Wang, G. Multivalent metal ion hybrid capacitors: A review with a focus on zinc-ion hybrid capacitors. *J. Mater. Chem. A* **2019**, *7*, 13810–13832. [CrossRef]
6. Liu, H.; Zhao, D.; Hu, P.; Wu, X. Ternary core-shell structured transition metal chalcogenide for hybrid electrochemical capacitor. *Chin. Chem. Lett.* **2018**, *29*, 1799–1803. [CrossRef]
7. Yang, Y.; Yang, S.; Xue, X.; Zhang, X.; Li, Q.; Yao, Y.; Rui, X.; Pan, H.; Yu, Y. Inorganic All-Solid-State Sodium Batteries: Electrolyte Designing and Interface Engineering. *Adv. Mater.* **2024**, *36*, 2308332. [CrossRef]
8. Sun, J.; Zhang, Q.; Liu, C.; Zhang, A.; Hou, L.; Yuan, C. Conductive Zinc-Based Metal–Organic Framework Nanorods as Cathodes for High-Performance Zn-Ion Capacitors. *Batteries* **2024**, *10*, 222. [CrossRef]
9. Zhou, J.; Hu, H.-Y.; Li, H.-Q.; Chen, Z.-P.; Yuan, C.-Z.; He, X.-J. Advanced carbon-based materials for Na, K, and Zn ion hybrid capacitors. *Rare Met.* **2023**, *42*, 719–739. [CrossRef]
10. Liu, Y.; Liu, Y.; Wu, X.; Cho, Y.-R. Enhanced Electrochemical Performance of Zn/VO_x Batteries by a Carbon-Encapsulation Strategy. *ACS Appl. Mater. Interfaces* **2022**, *14*, 11654–11662. [CrossRef]
11. Zhao, D.; Wang, X.; Zhang, W.; Zhang, Y.; Lei, Y.; Huang, X.; Zhu, Q.; Liu, J. Unlocking the Capacity of Vanadium Oxide by Atomically Thin Graphene-Analogous $V_2O_5{\cdot}nH_2O$ in Aqueous Zinc-Ion Batteries. *Adv. Funct. Mater.* **2023**, *33*, 2211412. [CrossRef]
12. Liu, Y.; Liu, Y.; Wu, X. Rational design of bi-phase CaV_2O_6/NaV_6O_{15} cathode materials for long-life aqueous zinc batteries. *EcoMat* **2023**, *5*, e12409. [CrossRef]
13. Blanc, L.E.; Kundu, D.; Nazar, L.F. Scientific Challenges for the Implementation of Zn-Ion Batteries. *Joule* **2020**, *4*, 771–799. [CrossRef]
14. Liu, Y.; Wu, X. Review of vanadium-based electrode materials for rechargeable aqueous zinc ion batteries. *J. Energy Chem.* **2021**, *56*, 223–237. [CrossRef]
15. Tang, B.; Shan, L.; Liang, S.; Zhou, J. Issues and opportunities facing aqueous zinc-ion batteries. *Energy Environ. Sci.* **2019**, *12*, 3288–3304. [CrossRef]

16. Liu, Y.; Wu, X. High durable aqueous zinc ion batteries by synergistic effect of V_6O_{13}/VO_2 electrode materials. *J. Energy Chem.* **2023**, *87*, 334–341. [CrossRef]
17. Sun, Y.; Liu, B.; Liu, L.; Lang, J.; Qiu, J. A Low-Concentration and High Ionic Conductivity Aqueous Electrolyte toward Ultralow-Temperature Zinc-Ion Hybrid Capacitors. *Small Struct.* **2023**, *4*, 2200345. [CrossRef]
18. Liu, Y.; Zhao, D.; Liu, H.; Umar, A.; Wu, X. High performance hybrid supercapacitor based on hierarchical MoS_2/Ni_3S_2 metal chalcogenide. *Chin. Chem. Lett.* **2019**, *30*, 1105–1110. [CrossRef]
19. Muzaffar, A.; Ahamed, M.B.; Deshmukh, K.; Thirumalai, J. A review on recent advances in hybrid supercapacitors: Design, fabrication and applications. *Renew. Sust. Energy Rev.* **2019**, *101*, 123–145. [CrossRef]
20. Liu, Y.; Umar, A.; Wu, X. Metal-organic framework derived porous cathode materials for hybrid zinc ion capacitor. *Rare Met.* **2022**, *41*, 2985–2991. [CrossRef]
21. Yu, L.; Li, J.; Ahmad, N.; He, X.; Wan, G.; Liu, R.; Ma, X.; Liang, J.; Jiang, Z.; Zhang, G. Recent progress on carbon materials for emerging zinc-ion hybrid capacitors. *J. Mater. Chem. A* **2024**, *12*, 9400–9420. [CrossRef]
22. Miao, L.; Lv, Y.; Zhu, D.; Li, L.; Gan, L.; Liu, M. Recent advances in zinc-ion hybrid energy storage: Coloring high-power capacitors with battery-level energy. *Chin. Chem. Lett.* **2023**, *34*, 107784. [CrossRef]
23. Lal, M.S.; Arjunan, A.; Balasubramanian, V.; Sundara, R. Redox-active polymer hydrogel electrolyte in biowaste-derived microporous carbon-based high capacitance and energy density ultracapacitors. *J. Electroanal. Chem.* **2020**, *870*, 114236. [CrossRef]
24. Tian, Y.; Amal, R.; Wang, D.-W. An Aqueous Metal-Ion Capacitor with Oxidized Carbon Nanotubes and Metallic Zinc Electrodes. *Front. Energy Res.* **2016**, *4*, 34. [CrossRef]
25. Dong, L.; Ma, X.; Li, Y.; Zhao, L.; Liu, W.; Cheng, J.; Xu, C.; Li, B.; Yang, Q.H.; Kang, F. Extremely safe, high-rate and ultralong-life zinc-ion hybrid supercapacitors. *Energy Storage Mater.* **2018**, *13*, 96–102. [CrossRef]
26. Li, Z.; An, Y.; Dong, S.; Chen, C.; Wu, L.; Sun, Y.; Zhang, X. Progress on zinc ion hybrid supercapacitors: Insights and challenges. *Energy Storage Mater.* **2020**, *31*, 252–266. [CrossRef]
27. Ma, X.; Cheng, J.; Dong, L.; Liu, W.; Mou, J.; Zhao, L.; Wang, J.; Ren, D.; Wu, J.; Xu, C.; et al. Multivalent ion storage towards high-performance aqueous zinc-ion hybrid supercapacitors. *Energy Storage Mater.* **2019**, *20*, 335–342. [CrossRef]
28. Sun, Y.; Shi, X.L.; Yang, Y.L.; Suo, G.Q.; Zhang, L.; Lu, S.Y.; Chen, Z.G. Biomass-Derived Carbon for High-Performance Batteries: From Structure to Properties. *Adv. Funct. Mater.* **2022**, *32*, 2201584. [CrossRef]
29. Li, X.; Wei, B.Q. Supercapacitors based on nanostructured carbon. *Nano Energy* **2013**, *2*, 159–173. [CrossRef]
30. Liu, Y.; Wu, L. Recent advances of cathode materials for zinc-ion hybrid capacitors. *Nano Energy* **2023**, *109*, 108290. [CrossRef]
31. Yin, J.; Zhang, W.; Wang, W.; Alhebshi, N.A.; Salah, N.; Alshareef, H.N. Electrochemical Zinc Ion Capacitors Enhanced by Redox Reactions of Porous Carbon Cathodes. *Adv. Energy Mater.* **2020**, *10*, 2001705. [CrossRef]
32. Fan, X.; Liu, P.; Ouyang, B.; Cai, R.; Chen, X.; Liu, X.; Liu, W.; Wang, J.; Liu, K. Polyacrylonitrile Derived Porous Carbon for Zinc-Ion Hybrid Capacitors with High Energy Density. *ChemElectroChem* **2021**, *8*, 3572–3578. [CrossRef]
33. Li, Y.; Lu, P.; Shang, P.; Wu, L.; Wang, X.; Dong, Y.; He, R.; Wu, Z.-S. Pyridinic nitrogen enriched porous carbon derived from bimetal organic frameworks for high capacity zinc ion hybrid capacitors with remarkable rate capability. *J. Energy Chem.* **2021**, *56*, 404–411. [CrossRef]
34. Huang, Z.; Wang, T.; Song, H.; Li, X.; Liang, G.; Wang, D.; Yang, Q.; Chen, Z.; Ma, L.; Liu, Z.; et al. Effects of Anion Carriers on Capacitance and Self-Discharge Behaviors of Zinc Ion Capacitors. *Angew. Chem. Int. Ed.* **2020**, *60*, 1011–1021. [CrossRef]
35. Kühnel, R.-S.; Reber, D.; Battaglia, C. A High-Voltage Aqueous Electrolyte for Sodium-Ion Batteries. *ACS Energy Lett.* **2017**, *2*, 2005–2006. [CrossRef]
36. Tan, J.; Liu, J. Electrolyte Engineering Toward High-Voltage Aqueous Energy Storage Devices. *Energy Environ. Mater.* **2020**, *4*, 302–306. [CrossRef]
37. Zhang, H.; Liu, Q.; Fang, Y.; Teng, C.; Liu, X.; Fang, P.; Tong, Y.; Lu, X. Boosting Zn-Ion Energy Storage Capability of Hierarchically Porous Carbon by Promoting Chemical Adsorption. *Adv. Mater.* **2019**, *31*, 1904948. [CrossRef]
38. Whittingham, M.S.; Siu, C.; Ding, J. Can Multielectron Intercalation Reactions Be the Basis of Next Generation Batteries? *Acc. Chem. Res.* **2018**, *51*, 258–264. [CrossRef]
39. Thomas, M.G.S.R.; Bruce, P.G.; Goodenough, J.B. Lithium Mobility in the Layered Oxide $Li_{1-x}COO_2$. *Solid State Ion.* **1985**, *17*, 13–19. [CrossRef]
40. Amatucci, G.G.; Badway, F.; Pasquier, A.D.; Zheng, T. An Asymmetric Hybrid Nonaqueous Energy Storage Cell. *J. Electrochem. Soc.* **2001**, *148*, A930. [CrossRef]
41. Raza, W.; Ali, F.; Raza, N.; Luo, Y.; Kim, K.-H.; Yang, J.; Kumar, S.; Mehmood, A.; Kwon, E.E. Recent advancements in supercapacitor technology. *Nano Energy* **2018**, *52*, 441–473. [CrossRef]
42. Simon, P.; Gogotsi, Y.; Dunn, B. Where Do Batteries End and Supercapacitors Begin? *Science* **2014**, *343*, 1210–1211. [CrossRef] [PubMed]
43. Zan, G.; Li, S.; Chen, P.; Dong, K.; Wu, Q.; Wu, T. Mesoporous Cubic Nanocages Assembled by Coupled Monolayers With 100% Theoretical Capacity and Robust Cycling. *ACS Cent. Sci.* **2024**, *10*, 1283–1294. [CrossRef] [PubMed]
44. Gui, Q.; Ba, D.; Li, L.; Liu, W.; Li, Y.; Liu, J. Recent advances in materials and device technologies for aqueous hybrid supercapacitors. *Sci. China Mater.* **2021**, *65*, 10–31. [CrossRef]
45. Zuo, W.; Li, R.; Zhou, C.; Li, Y.; Xia, J.; Liu, J. Battery-Supercapacitor Hybrid Devices: Recent Progress and Future Prospects. *Adv Sci* **2017**, *4*, 1600539. [CrossRef]

46. Wu, X.; Yao, S. Flexible electrode materials based on WO_3 nanotube bundles for high performance energy storage devices. *Nano Energy* **2017**, *42*, 143–150. [CrossRef]
47. Yan, J.; Wang, Q.; Wei, T.; Fan, Z. Recent Advances in Design and Fabrication of Electrochemical Supercapacitors with High Energy Densities. *Adv. Energy Mater.* **2014**, *4*, 1300816. [CrossRef]
48. Liu, C.; Wu, X.; Wang, B. Performance modulation of energy storage devices: A case of Ni-Co-S electrode materials. *Chem. Eng. J.* **2020**, *392*, 123651. [CrossRef]
49. Liu, H.; Dai, M.; Zhao, D.; Wu, X.; Wang, B. Realizing Superior Electrochemical Performance of Asymmetric Capacitors through Tailoring Electrode Architectures. *ACS Appl. Energy Mater.* **2020**, *3*, 7004–7010. [CrossRef]
50. Lv, Y.; Zhang, L.; Wei, X.; Qiu, B.; Zhang, W.; Qin, Q.; Jia, D.; He, X.; Liu, Z.; Wei, F. The emerging of zinc-ion hybrid supercapacitors: Advances, challenges, and future perspectives. *Sustain. Mater. Technol.* **2023**, *35*, e00536. [CrossRef]
51. Wang, Y.; Sun, S.; Wu, X.; Liang, H.; Zhang, W. Status and Opportunities of Zinc Ion Hybrid Capacitors: Focus on Carbon Materials, Current Collectors, and Separators. *Nano Micro Lett.* **2023**, *15*, 78. [CrossRef] [PubMed]
52. Sui, D.; Wu, M.; Shi, K.; Li, C.; Lang, J.; Yang, Y.; Zhang, X.; Yan, X.; Chen, Y. Recent progress of cathode materials for aqueous zinc-ion capacitors: Carbon-based materials and beyond. *Carbon* **2021**, *185*, 126–151. [CrossRef]
53. Liu, Z.; Duan, C.; Dou, S.; Yuan, Q.; Xu, J.; Liu, W.D.; Chen, Y. Ultrafast Porous Carbon Activation Promises High-Energy Density Supercapacitors. *Small* **2022**, *18*, 2200954. [CrossRef] [PubMed]
54. Selvaraj, A.R.; Muthusamy, A.; Inho, C.; Kim, H.-J.; Senthil, K.; Prabakar, K. Ultrahigh surface area biomass derived 3D hierarchical porous carbon nanosheet electrodes for high energy density supercapacitors. *Carbon* **2021**, *174*, 463–474. [CrossRef]
55. Liu, Y.; Liu, Y.; Wu, X.; Cho, Y.-R. General Carbon Modification Avenue to Construct Highly Stable V_2O_5 Electrodes for Aqueous Zinc-Ion Batteries. *ACS Sustain. Chem. Eng.* **2023**, *11*, 13298–13305. [CrossRef]
56. Jin, J.; Geng, X.; Chen, Q.; Ren, T.-L. A Better Zn-Ion Storage Device: Recent Progress for Zn-Ion Hybrid Supercapacitors. *Nano Micro Lett.* **2022**, *14*, 64. [CrossRef]
57. Pomerantseva, E.; Bonaccorso, F.; Feng, X.; Cui, Y.; Gogotsi, Y. Energy storage: The future enabled by nanomaterials. *Science* **2019**, *366*, eaan8285. [CrossRef]
58. Borchardt, L.; Leistenschneider, D.; Haase, J.; Dvoyashkin, M. Revising the Concept of Pore Hierarchy for Ionic Transport in Carbon Materials for Supercapacitors. *Adv. Energy Mater.* **2018**, *8*, 1800892. [CrossRef]
59. Lee, J.; Kim, J.; Hyeon, T. Recent Progress in the Synthesis of Porous Carbon Materials. *Adv. Mater.* **2006**, *18*, 2073–2094. [CrossRef]
60. Presser, V.; Heon, M.; Gogotsi, Y. Carbide-Derived Carbons-From Porous Networks to Nanotubes and Graphene. *Adv. Funct. Mater.* **2011**, *21*, 810–833. [CrossRef]
61. Borchardt, L.; Oschatz, M.; Kaskel, S. Tailoring porosity in carbon materials for supercapacitor applications. *Mater. Horiz.* **2014**, *1*, 157–168. [CrossRef]
62. Wang, C.; Pei, Z.; Meng, Q.; Zhang, C.; Sui, X.; Yuan, Z.; Wang, S.; Chen, Y. Toward Flexible Zinc-Ion Hybrid Capacitors with Superhigh Energy Density and Ultralong Cycling Life: The Pivotal Role of $ZnCl_2$ Salt-Based Electrolytes. *Angew. Chem. Int. Ed.* **2020**, *60*, 990–997. [CrossRef] [PubMed]
63. Lv, K.; Zhang, J.; Zhao, X.; Kong, N.; Tao, J.; Zhou, J. Understanding the Effect of Pore Size on Electrochemical Capacitive Performance of MXene Foams. *Small* **2022**, *18*, 2202203. [CrossRef] [PubMed]
64. Jian, W.; Zhang, W.; Wei, X.; Wu, B.; Liang, W.; Wu, Y.; Yin, J.; Lu, K.; Chen, Y.; Alshareef, H.N.; et al. Engineering Pore Nanostructure of Carbon Cathodes for Zinc Ion Hybrid Supercapacitors. *Adv. Funct. Mater.* **2022**, *32*, 2209914. [CrossRef]
65. Hulicova Jurcakova, D.; Seredych, M.; Lu, G.Q.; Bandosz, T.J. Combined Effect of Nitrogen- and Oxygen-Containing Functional Groups of Microporous Activated Carbon on its Electrochemical Performance in Supercapacitors. *Adv. Funct. Mater.* **2009**, *19*, 438–447. [CrossRef]
66. An, G.H. Ultrafast long-life zinc-ion hybrid supercapacitors constructed from mesoporous structured activated carbon. *Appl. Surf. Sci.* **2020**, *530*, 147220. [CrossRef]
67. Liu, Y.; Liu, Y.; Wu, X. Defect engineering of vanadium-based electrode materials for zinc ion battery. *Chin. Chem. Lett.* **2023**, *34*, 107839. [CrossRef]
68. Song, C.; Liu, K.; Wang, T.; Zhao, P.-Y.; Huang, H.; Liu, Y.; Lu, R.; Zhang, S. Spontaneous template approach towards nitrogenous multi-shelled hollow carbon spheres with unique onion-like architecture. *Mater. Today Chem.* **2024**, *40*, 102196. [CrossRef]
69. Zheng, Y.; Zhao, W.; Jia, D.; Liu, Y.; Cui, L.; Wei, D.; Zheng, R.; Liu, J. Porous carbon prepared via combustion and acid treatment as flexible zinc-ion capacitor electrode material. *Chem. Eng. J.* **2020**, *387*, 124161. [CrossRef]
70. Zhang, Y.; Wang, Z.; Li, D.; Sun, Q.; Lai, K.; Li, K.; Yuan, Q.; Liu, X.; Ci, L. Ultrathin carbon nanosheets for highly efficient capacitive K-ion and Zn-ion storage. *J. Mater. Chem. A* **2020**, *8*, 22874–22885. [CrossRef]
71. Deng, S.; Yuan, Z.; Tie, Z.; Wang, C.; Song, L.; Niu, Z. Electrochemically Induced MOF-Derived Amorphous V_2O_5 for Superior Rate Aqueous Zn-Ion Batteries. *Angew Chem. Int. Ed.* **2020**, *59*, 22002. [CrossRef] [PubMed]
72. Li, H.; Liao, Q.; Liu, Y.; Li, Y.; Niu, X.; Zhang, D.; Wang, K. Hierarchically Porous Carbon Rods Derived from Metal-Organic Frameworks for Aqueous Zinc-Ion Hybrid Capacitors. *Small* **2023**, *20*, 2307184. [CrossRef] [PubMed]
73. Wang, H.; Shao, Y.; Mei, S.; Lu, Y.; Zhang, M.; Sun, J.-k.; Matyjaszewski, K.; Antonietti, M.; Yuan, J. Polymer-Derived Heteroatom-Doped Porous Carbon Materials. *Chem. Rev.* **2020**, *120*, 9363–9419. [CrossRef] [PubMed]
74. Yang, S.-H.; Fu, W.-Q.; Cui, Y.-W.; Cao, B.-Q. PVP/PAN-derived porous carbon fiber for zinc-ion hybrid supercapacitors. *Rare Met.* **2024**, *43*, 3066–3073. [CrossRef]

75. Zhang, Y.; Wang, L.; Jia, D.; Yue, L.; Zhang, H.; Liu, J. Hierarchical Porous Doped Carbon Plates Derived from Chitosan Aerogel as Cathode for High Performance Zn-ion Hybrid Capacitor. *ChemElectroChem* **2023**, *10*, e202200972. [CrossRef]
76. Pan, L.; Wang, Y.; Hu, H.; Li, X.; Liu, J.; Guan, L.; Tian, W.; Wang, X.; Li, Y.; Wu, M. 3D self-assembly synthesis of hierarchical porous carbon from petroleum asphalt for supercapacitors. *Carbon* **2018**, *134*, 345–353. [CrossRef]
77. Zhang, M.; Sun, Z.; Zhang, T.; Qin, B.; Sui, D.; Xie, Y.; Ma, Y.; Chen, Y. Porous asphalt/graphene composite for supercapacitors with high energy density at superior power density without added conducting materials. *J. Mater. Chem. A* **2017**, *5*, 21757–21764. [CrossRef]
78. Zhang, X.; Tian, X.; Song, Y.; Wu, J.; Yang, T.; Liu, Z. Boosting Zn-ion storage capacity of pitch coke-based activated carbon via pre-oxidation assisted KOH activation strategy. *Micropor. Mesopor. Mater.* **2022**, *333*, 111721. [CrossRef]
79. Yang, Y.; Li, Z.; Zhao, J.; Qu, S. Structural engineering of pitch-based porous carbon and its application in supercapacitors: A review. *J. Energy Storage* **2023**, *74*, 109334. [CrossRef]
80. Zhou, Z.; Zhou, X.; Zhang, M.; Mu, S.; Liu, Q.; Tang, Y. In Situ Two-Step Activation Strategy Boosting Hierarchical Porous Carbon Cathode for an Aqueous Zn-Based Hybrid Energy Storage Device with High Capacity and Ultra-Long Cycling Life. *Small* **2020**, *16*, 2003174. [CrossRef]
81. Shi, X.; Zhang, H.; Zeng, S.; Wang, J.; Cao, X.; Liu, X.; Lu, X. Pyrrolic-Dominated Nitrogen Redox Enhances Reaction Kinetics of Pitch-Derived Carbon Materials in Aqueous Zinc Ion Hybrid Supercapacitors. *ACS Mater. Lett.* **2021**, *3*, 1291–1299. [CrossRef]
82. Saini, S.; Chand, P.; Joshi, A. Biomass derived carbon for supercapacitor applications: Review. *J. Energy Storage* **2021**, *39*, 102646. [CrossRef]
83. Li, Z.; Chen, D.; An, Y.; Chen, C.; Wu, L.; Chen, Z.; Sun, Y.; Zhang, X. Flexible and anti-freezing quasi-solid-state zinc ion hybrid supercapacitors based on pencil shavings derived porous carbon. *Energy Storage Mater.* **2020**, *28*, 307–314. [CrossRef]
84. Zhao, L.; Jian, W.; Zhang, X.; Wen, F.; Zhu, J.; Huang, S.; Yin, J.; Lu, K.; Zhou, M.; Zhang, W.; et al. Multi-scale self-templating synthesis strategy of lignin-derived hierarchical porous carbons toward high-performance zinc ion hybrid supercapacitors. *J. Energy Storage* **2022**, *53*, 105095. [CrossRef]
85. Liu, K.; Yu, C.; Guo, W.; Ni, L.; Yu, J.; Xie, Y.; Wang, Z.; Ren, Y.; Qiu, J. Recent research advances of self-discharge in supercapacitors: Mechanisms and suppressing strategies. *J. Energy Chem.* **2021**, *58*, 94–109. [CrossRef]
86. Li, H.; Su, P.; Liao, Q.; Liu, Y.; Li, Y.; Niu, X.; Liu, X.; Wang, K. Olive Leaves-Derived Hierarchical Porous Carbon as Cathode Material for Anti-Self-Discharge Zinc-Ion Hybrid Capacitor. *Small* **2023**, *19*, 2304172. [CrossRef]
87. He, H.; Lian, J.; Chen, C.; Xiong, Q.; Zhang, M. Super hydrophilic carbon fiber film for freestanding and flexible cathodes of zinc-ion hybrid supercapacitors. *Chem. Eng. J.* **2021**, *421*, 129786. [CrossRef]
88. Li, Z.; Gadipelli, S.; Li, H.; Howard, C.A.; Brett, D.J.L.; Shearing, P.R.; Guo, Z.; Parkin, I.P.; Li, F. Tuning the interlayer spacing of graphene laminate films for efficient pore utilization towards compact capacitive energy storage. *Nat. Energy* **2020**, *5*, 160–168. [CrossRef]
89. Hu, X.; Bao, X.; Zhang, M.; Fang, S.; Liu, K.; Wang, J.; Liu, R.; Kim, S.H.; Baughman, R.H.; Ding, J. Recent Advances in Carbon Nanotube-Based Energy Harvesting Technologies. *Adv. Mater.* **2023**, *35*, 2303035. [CrossRef]
90. Wei, F.; Wei, Y.; Wang, J.; Han, M.; Lv, Y. N, P dual doped foamy-like carbons with abundant defect sites for zinc ion hybrid capacitors. *Chem. Eng. J.* **2022**, *450*, 137919. [CrossRef]
91. Sun, Y.; Zheng, J.; Tong, Y.; Wu, Y.; Liu, X.; Niu, L.; Li, H. Construction of three-dimensional nitrogen doped porous carbon flake electrodes for advanced potassium-ion hybrid capacitors. *J. Colloid Interface Sci.* **2022**, *606*, 1940–1949. [CrossRef] [PubMed]
92. Fang, X.Y.; Yu, X.X.; Zheng, H.M.; Jin, H.B.; Wang, L.; Cao, M.S. Temperature- and thickness-dependent electrical conductivity of few-layer graphene and graphene nanosheets. *Phys. Lett. A* **2015**, *379*, 2245–2251. [CrossRef]
93. Han, J.; Wang, K.; Liu, W.; Li, C.; Sun, X.; Zhang, X.; An, Y.; Yi, S.; Ma, Y. Rational design of nano-architecture composite hydrogel electrode towards high performance Zn-ion hybrid cell. *Nanoscale* **2018**, *10*, 13083–13091. [CrossRef] [PubMed]
94. Xu, X.; Zhao, X.; Yang, Z.; Lin, Q.; Jian, B.; Li, N.; Zheng, C.; Lv, W. High-density three-dimensional graphene cathode with a tailored pore structure for high volumetric capacity zinc-ion storage. *Carbon* **2022**, *186*, 624–631. [CrossRef]
95. Mao, J.; Iocozzia, J.; Huang, J.; Meng, K.; Lai, Y.; Lin, Z. Graphene aerogels for efficient energy storage and conversion. *Energy Environ. Sci.* **2018**, *11*, 772–799. [CrossRef]
96. Okhay, O.; Tkach, A.; Gallo, M.J.H.; Otero-Irurueta, G.; Mikhalev, S.; Staiti, P.; Lufrano, F. Energy storage of supercapacitor electrodes on carbon cloth enhanced by graphene oxide aerogel reducing conditions. *J. Energy Storage* **2020**, *32*, 101839. [CrossRef]
97. Xin, T.; Wang, Y.; Wang, N.; Zhao, Y.; Li, H.; Zhang, Z.; Liu, J. A high-capacity aqueous Zn-ion hybrid energy storage device using poly(4,4′-thiodiphenol)-modified activated carbon as a cathode material. *J. Mater. Chem. A* **2019**, *7*, 23076–23083. [CrossRef]
98. Potham, S.; Ramanujam, K. A novel hierarchical porous activated carbon-organic composite cathode material for high performance aqueous zinc-ion hybrid supercapacitors. *J. Power Sources* **2023**, *557*, 232551. [CrossRef]
99. Xu, Y.; Chen, X.; Huang, C.; Zhou, Y.; Fan, B.; Li, Y.; Hu, A.; Tang, Q.; Huang, K. Redox-active p-phenylenediamine functionalized reduced graphene oxide film through covalently grafting for ultrahigh areal capacitance Zn-ion hybrid supercapacitor. *J. Power Sources* **2021**, *488*, 229426. [CrossRef]
100. Yang, J.; Cao, J.; Peng, Y.; Bissett, M.; Kinloch, I.A.; Dryfe, R.A.W. Unlocking the energy storage potential of polypyrrole via electrochemical graphene oxide for high performance zinc-ion hybrid supercapacitors. *J. Power Sources* **2021**, *516*, 230663. [CrossRef]

101. Cao, J.; Wang, Y.; Chen, J.; Li, X.; Walsh, F.C.; Ouyang, J.H.; Jia, D.; Zhou, Y. Three-dimensional graphene oxide/polypyrrole composite electrodes fabricated by one-step electrodeposition for high performance supercapacitors. *J. Mater. Chem. A* **2015**, *3*, 14445–14457. [CrossRef]
102. Xu, T.; Yang, D.; Fan, Z.; Li, X.; Liu, Y.; Guo, C.; Zhang, M.; Yu, Z.-Z. Reduced graphene oxide/carbon nanotube hybrid fibers with narrowly distributed mesopores for flexible supercapacitors with high volumetric capacitances and satisfactory durability. *Carbon* **2019**, *152*, 134–143. [CrossRef]
103. Ni, T.; Wang, S.; Shi, J.; Du, X.; Cheng, Q.; Dong, Z.; Ruan, L.; Zeng, W.; Guo, X.; Ren, X.; et al. Highly Flexible and Self-Healable Zinc-Ion Hybrid Supercapacitors Based on MWCNTs-RGO Fibers. *Adv. Mater. Technol.* **2020**, *5*, 2000268. [CrossRef]
104. Cao, Y.; Tang, X.; Liu, M.; Zhang, Y.; Yang, T.; Yang, Z.; Yu, Y.; Li, Y.; Di, J.; Li, Q. Thin-walled porous carbon tile-packed paper for high-rate Zn-ion capacitor cathode. *Chem. Eng. J.* **2022**, *431*, 133241. [CrossRef]
105. Li, J.; Zhang, J.; Yu, L.; Gao, J.; He, X.; Liu, H.; Guo, Y.; Zhang, G. Dual-doped carbon hollow nanospheres achieve boosted pseudocapacitive energy storage for aqueous zinc ion hybrid capacitors. *Energy Storage Mater.* **2021**, *42*, 705–714. [CrossRef]
106. Liu, P.; Liu, W.; Huang, Y.; Li, P.; Yan, J.; Liu, K. Mesoporous hollow carbon spheres boosted, integrated high performance aqueous Zn-Ion energy storage. *Energy Storage Mater.* **2020**, *25*, 858–865. [CrossRef]
107. Chen, S.; Ma, L.; Zhang, K.; Kamruzzaman, M.; Zhi, C.; Zapien, J.A. A flexible solid-state zinc ion hybrid supercapacitors based on copolymer derived hollow carbon spheres. *J. Mater. Chem. A* **2019**, *7*, 7784–7790. [CrossRef]
108. Li, W.; Zhou, M.; Li, H.; Wang, K.; Cheng, S.; Jiang, K. A high performance sulfur-doped disordered carbon anode for sodium ion batteries. *Energy Environ. Sci.* **2015**, *8*, 2916–2921. [CrossRef]
109. Lee, Y.-G.; An, G.-H. Synergistic Effects of Phosphorus and Boron Co-Incorporated Activated Carbon for Ultrafast Zinc-Ion Hybrid Supercapacitors. *ACS Appl. Mater. Interfaces* **2020**, *12*, 41342–41349. [CrossRef]
110. Zhou, H.; Liu, C.; Wu, J.-C.; Liu, M.; Zhang, D.; Song, H.; Zhang, X.; Gao, H.; Yang, J.; Chen, D. Boosting the electrochemical performance through proton transfer for the Zn-ion hybrid supercapacitor with both ionic liquid and organic electrolytes. *J. Mater. Chem. A* **2019**, *7*, 9708–9715. [CrossRef]
111. Wang, H.; Chen, Q.; Xiao, P.; Cao, L. Unlocking Zinc-Ion Energy Storage Performance of Onion-Like Carbon by Promoting Heteroatom Doping Strategy. *ACS Appl. Mater. Interfaces* **2022**, *14*, 9013–9023. [CrossRef] [PubMed]
112. Nian, Y.-R.; Teng, H. Nitric Acid Modification of Activated Carbon Electrodes for Improvement of Electrochemical Capacitance. *J. Electrochem. Soc.* **2002**, *149*, A1008–A1014. [CrossRef]
113. Li, X.; Jiang, Y.; Wang, P.; Mo, Y.; Lai, W.; Li, Z.; Yu, R.; Du, Y.; Zhang, X.; Chen, Y. Effect of the oxygen functional groups of activated carbon on its electrochemical performance for supercapacitors. *New Carbon Mater.* **2020**, *35*, 232–243. [CrossRef]
114. Li, Y.; Yang, W.; Yang, W.; Wang, Z.; Rong, J.; Wang, G.; Xu, C.; Kang, F.; Dong, L. Towards High-Energy and Anti-Self-Discharge Zn-Ion Hybrid Supercapacitors with New Understanding of the Electrochemistry. *Nano Micro Lett.* **2021**, *13*, 95. [CrossRef] [PubMed]
115. Zhao, L.; Jian, W.; Zhu, J.; Zhang, X.; Wen, F.; Fei, X.; Chen, L.; Huang, S.; Yin, J.; Chodankar, N.R.; et al. Molten Salt Self-Template Synthesis Strategy of Oxygen-Rich Porous Carbon Cathodes for Zinc Ion Hybrid Capacitors. *ACS Appl. Mater. Interfaces* **2022**, *14*, 43431–43441. [CrossRef]
116. Yuan, S.; Huang, X.; Wang, H.; Xie, L.; Cheng, J.; Kong, Q.; Sun, G.; Chen, C.-M. Structure evolution of oxygen removal from porous carbon for optimizing supercapacitor performance. *J. Energy Chem.* **2020**, *51*, 396–404. [CrossRef]
117. Lou, G.; Pei, G.; Wu, Y.; Lu, Y.; Wu, Y.; Zhu, X.; Pang, Y.; Shen, Z.; Wu, Q.; Fu, S.; et al. Combustion conversion of wood to N, O co-doped 2D carbon nanosheets for zinc-ion hybrid supercapacitors. *Chem. Eng. J.* **2021**, *413*, 127502. [CrossRef]
118. Wang, D.; Wang, S.; Lu, Z. S-doped 3D porous carbons derived from potassium thioacetate activation strategy for zinc-ion hybrid supercapacitor applications. *Int. J. Energy Res.* **2020**, *45*, 2498–2510. [CrossRef]
119. Lu, Y.; Liu, L.; Zhang, R.; Jiang, Z.; Li, Y.; Sun, Z.; Chen, X.; Song, H. Sodium alginate-derived micropore dominated carbon 3D architectures through dual template engineering for high-performance Zn-ion hybrid capacitors. *Appl. Surf. Sci.* **2022**, *604*, 154631. [CrossRef]
120. Zan, G.; Wu, T.; Zhu, F.; He, P.; Cheng, Y.; Chai, S.; Wang, Y.; Huang, X.; Zhang, W.; Wan, Y.; et al. A biomimetic conductive super-foldable material. *Matter* **2021**, *4*, 3232–3247. [CrossRef]

batteries

MDPI

Review

Heteroatom Doping Strategy of Advanced Carbon for Alkali Metal-Ion Capacitors

Ti Yin, Yaqin Guo, Xing Huang *, Xinya Yang, Leixin Qin, Tianxiang Ning, Lei Tan, Lingjun Li and Kangyu Zou *

School of Materials Science and Engineering, Changsha University of Science and Technology, Changsha 410114, China; 202228020409@stu.csust.edu.cn (T.Y.); 202228020413@stu.csust.edu.cn (Y.G.); 202328020426@stu.csust.edu.cn (X.Y.); 202328020438@stu.csust.edu.cn (L.Q.); ningtianxiang@csust.edu.cn (T.N.); tanlei@csust.edu.cn (L.T.); lingjun.li@csust.edu.cn (L.L.)

* Correspondence: 22206031250@stu.csust.edu.cn (X.H.); ky-zou@csust.edu.cn (K.Z.)

Abstract: Alkali metal-ion capacitors (AMICs) combine the advantages of the high specific energy of alkali metal-ion batteries (AMIBs) and the high power output of supercapacitors (SCs), which are considered highly promising and efficient energy storage devices. It is found that carbon has been the most widely used electrode material of AMICs due to its advantages of low cost, a large specific surface area, and excellent electrical conductivity. However, the application of carbon is limited by its low specific capacity, finite kinetic performance, and few active sites. Doping heteroatoms in carbon materials is an effective strategy to adjust their microstructures and improve their electrochemical storage performance, which effectively helps to increase the pseudo-capacitance, enhance the wettability, and increase the ionic migration rate. Moreover, an appropriate heteroatom doping strategy can purposefully guide the design of advanced AMICs. Herein, a systematic review of advanced heteroatom (N, S, P, and B)-doped carbon, which has acted as a positrode and negatrode in AMICs (M = Li, Na, and K) in recent years, has been summarized. Moreover, emphasis is placed on the mechanism of single-element doping versus two-element doping for the enhancement in the performance of carbon positrodes and negatrodes, and an introduction to the use of doped carbon in dual-carbon alkali metal-ion capacitors (DC-AMICs) is discussed. Finally, an outlook is given to solve the problems arising when using doped carbon materials in practical applications and future development directions are presented.

Keywords: alkali metal-ion capacitor; carbon material; heteroatom doping; electrochemical performance

1. Introduction

The evolution of energy storage technology has led to increased demands for advanced energy storage devices that offer higher energy and power densities, along with extended lifecycles. Constructing such advanced energy storage devices is crucial for the efficient utilization and storage of sustainable energy sources. Currently, lithium-ion batteries (LIBs) and supercapacitors (SCs) dominate the energy storage market [1]. However, LIBs, characterized by their low specific power, and SCs, characterized by their low specific energy, are increasingly unable to meet the escalating energy demands of the market for energy storage technologies [2–4]. In 2001, Amatucci et al. [5] reported lithium-ion capacitors (LICs) for the first time, which are constructed by an activated carbon (AC) positrode and a $Li_4Ti_5O_{12}$ (LTO) negatrode. Surprisingly, LICs exhibit a specific energy three times greater than that of conventional SCs (20 Wh kg^{-1}), heralding a new era in energy storage systems. LICs combine SC-type positrodes and LIB-type negatrodes [6], achieving high

 https://doi.org/10.3390/batteries11020069

specific energy/power and a long lifecycle. Compared to lithium, which is present in low concentrations (0.002wt.%), sodium (2.36wt.%) and potassium (2.09wt.%) are more abundant on Earth and exhibit analogous chemical properties to lithium. They demonstrate significant potential for large-scale storage and cost-effectiveness. Therefore, sodium-ion capacitors (SICs) and potassium-ion capacitors (KICs) exhibit similar mechanisms to LICs, which have become one of the most promising electrochemical energy storage devices. The relationship between the specific energy and specific power of common electrochemical energy storage devices is depicted in Figure 1a [7]. Electrode materials play a crucial role in the electrochemical performance of alkali metal-ion capacitors (AMICs) [8]. Carbon materials, known for their low cost, large specific surface area, and excellent electrical conductivity, are used as electrode materials in AMICs to store energy by absorbing electrolyte ions on the electrode surface, forming an electric double-layer structure. Nevertheless, the application of carbon materials is still constrained due to their low specific capacity, limited kinetic performance, and scarcity of active sites, which result in the inferior specific energy of the assembled devices [9,10].

Heteroatom doping is a rational design method, used in the realm of advanced material development, involving the substitution of certain carbon atoms with other elements [11–14]. This process can alter the electronic structure of the carbon matrix, thereby promoting charge transfer [15]. Additionally, it can increase the number of redox-active sites and defect locations, as well as expand the interlayer distance within the carbon structure. Doping with various heteroatoms can further enhance specific physical and chemical properties of carbon materials, such as electrical conductivity and surface wettability [16,17]. Superior electrical conductivity ensures the swift passage of electrical currents through the material, while excellent surface wettability fosters stronger interactions between the material and its surrounding environment. Figure 1b,c [18] illustrate the comparative positive and negative impacts of different doping atoms in batteries and SCs, respectively. In the context of AMICs, carbon negatrodes and carbon positrodes operate through distinct energy storage mechanisms. For carbon negatrodes, specific heteroatom doping may be essential to bolster stability and energy storage capacity, whereas for carbon positrodes, a different type of heteroatom doping might be necessary to enhance conductivity and reactivity. Consequently, their doping modification strategies and performance enhancement pathways can differ significantly.

Selecting the right dopant atoms is a critical research area for researchers, as it can significantly enhance the electrochemical performance of battery-type carbon negatrodes and capacitor-type carbon positrodes in a targeted manner. This review offers a systematic examination of the recent advancements in the use of advanced heteroatom (N, S, P, and B)-doped carbons as both positrodes and negatrodes in AMICs (M = Li, Na, and K). It concentrates on the application of single-element-doped and double-doped carbons in AMICs, delivering a thorough analysis of the distinct performance benefits offered by various elemental doping strategies and elucidating the mechanisms behind performance enhancement. Furthermore, this discussion tackles the challenges that doped carbon materials face in practical applications and proposes potential avenues for future research and development. The aim is to improve the performance of LICs, SICs, and KICs, and by extension, other MICs. By identifying the unique advantages and mechanisms of different doping elements, researchers can better tailor the design of carbon materials to meet the specific demands of high-performance energy storage devices.

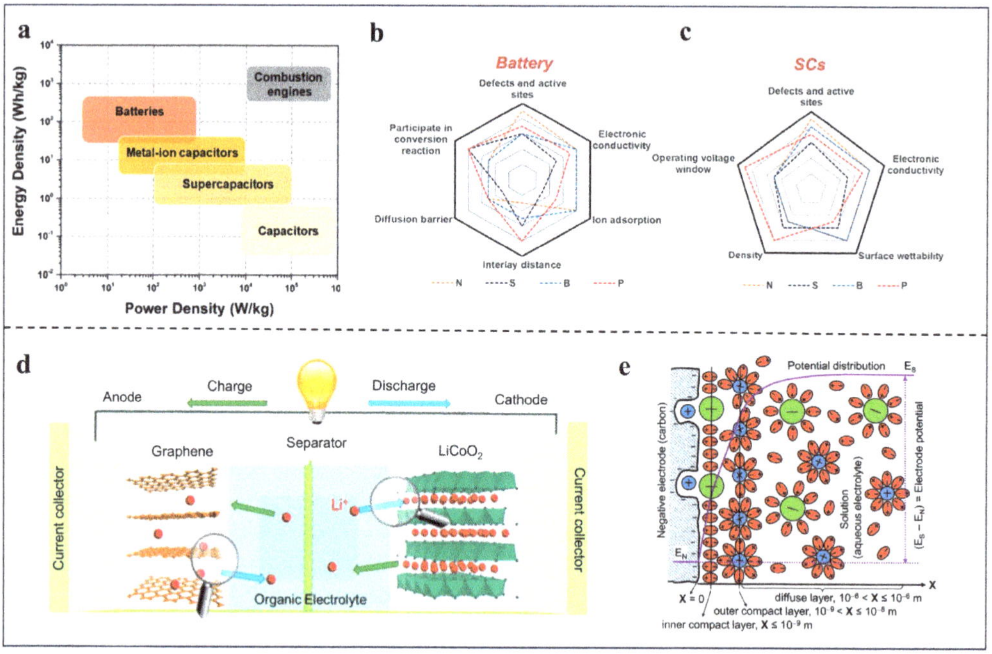

Figure 1. (**a**) A Ragone diagram of the energy storage devices as indicated. Reproduced with permission from Ref. [7]. Copyright 2022, The Royal Society of Chemistry. A comparison of the positive and negative effects of the doping of different heteroatoms in (**b**) batteries and (**c**) SCs. Reproduced with permission from Ref. [18]. Copyright 2021, The Royal Society of Chemistry. (**d**) A schematic of the charging and discharging process in typical MIBs. Reproduced with permission from Ref. [19]. Copyright 2019, Wiley. (**e**) Schematic representations of the EDL structure (cross-section) of the interface between a porous-carbon negative electrode and an aqueous electrolyte. Reproduced with permission from Ref. [20]. Copyright 2017, Taylor & Francis.

2. Energy Storage Mechanisms in Electrochemical Storage Devices

AMICs are hybrid energy storage devices that integrate the energy storage mechanisms of both AMIBs and SCs. Typically, an AMIC comprises a battery-type negatrode and a capacitor-type positrode. The battery-type negatrode supplies sustained power to the device, while the capacitive positrode is designed for the swift absorption and release of energy. MICs are characterized by their high energy and power densities, as well as their extended operational lifetimes. However, there is a kinetic rate mismatch between the Li^+ intercalation/de-intercalation of the battery-type negatrode and the ion adsorption/desorption behavior of the capacitor-type negatrode. Thus, it is necessary to understand the two different energy storage mechanisms to ensure the rational engineering of carbon materials for advanced AMICs.

2.1. Embedding–Removal Mechanism of AMIBs

Rechargeable AMIBs have indeed become prominent contenders in the realm of new energy technologies, offering portability, energy efficiency, and environmental sustainability. The energy storage in AMIBs is predicated according to the reversible intercalation/de-intercalation of alkali metal ions (M^+) within the electrode materials, a process often referred to as the "rocking chair" mechanism. This mechanism is aptly likened to a magical rocking chair, where M^+ oscillates between the negatrode and positrode materials during the charging and discharging cycles.

Let us take lithium-ion batteries (LIBs) as an example. Figure 1d [19] shows the charge–discharge process of a conventional lithium-ion battery. During charging, electrons follow the external circuit from the positive electrode to the negative electrode, while Li^+ located in the lithium layer undergoes removal and enters the electrolyte. Firstly, Li^+ is dissolved by the solvent molecules in the electrolyte, and then migrates across the separator to the negatrode under the action of the electric field to undergo de-solvation behavior and is intercalated in the graphite layer. The discharge process of a lithium-ion battery is opposite to the charging process; electrons flow from the negative electrode to the positive electrode, and Li^+ is removed from the graphite and finally intercalated in the positive electrode.

There exists a significant correlation between the reversible intercalation/de-intercalation properties of AMIBs and their specific energy. If the electrode material possesses a limited number of sites that restrict the quantity of M^+ that can be stored, this limitation diminishes the amount of energy that can be released by the AMIB, resulting in reduced specific energy. Conversely, unfavorable external conditions, such as temperature, pressure, and electrolyte environment, can also impact these reversible intercalation/de-intercalation properties, thereby affecting the specific energy of the battery.

2.2. Electric Double-Layer Capacitors

Electric double-layer capacitors (EDLCs), a subset of supercapacitors, play a critical role in modern energy storage solutions. Based on the electrical double-layer model shown in Figure 1e [20], EDLCs store electrical energy through electrostatic interactions, which facilitate charge separation within the Helmholtz double layer at the interface between the electrode surface and the electrolyte [21–23]. When subjected to an electric field, the positive and negative ions in the electrolyte migrate toward the respective oppositely charged electrodes, creating an electric double layer on the electrode surfaces that enables charge storage [24,25]. During the charging phase, anions migrate to the negatrode, while cations move toward the positrode. Conversely, during the discharging phase, these ions move in the opposite direction, releasing the stored charge.

Table 1 presents a comparison of the primary performance metrics of commercialized AMIBs, SCs, and AMICs. It is evident that EDLCs, a type of SC, exhibit superior specific power and cycling stability in comparison to AMIBs [26,27]. EDLCs store energy through a non-Faradaic process that does not involve electron transfer or redox reactions. This distinct energy storage mechanism endows EDLCs with higher efficiency and stability during both the energy storage and release processes. Consequently, double-layer capacitors can serve as an alternative or supplement to secondary rechargeable batteries in applications where high power transfer or rapid energy harvesting is necessary.

Table 1. Comparison of main properties of commercialized AMIBs, SCs, and AMICs [26,27].

Properties	AMIB	SC	AMIC
Discharging time	0.3–3 h	~30 s	min to h
Charging time	1–5 h	s to min	min to h
Specific energy (Wh kg^{-1})	150–250	20–30	20–200
Specific power (W kg^{-1})	<1000	500–10,000	100–20,000
Cycling efficiency (%)	70–85	85–98	70–95
Lifecycle (Cycles)	<4000	>500,000	>25,000

3. Element Doping for Carbon Negatrode

The carbon negatrode, known for its abundant availability, high conductivity, and stability, is a key component in many energy storage systems. Graphite, the most common negatrode material for lithium-ion batteries, has a theoretical specific capacity of

372 mAh g^{-1}. Since Na^+ has a larger ionic radius than Li^+, it cannot be intercalated or de-intercalated between graphite layers. Consequently, amorphous carbon with significant disorder is employed for its negative electrode [28], among which hard carbon is widely utilized due to its high theoretical specific capacity (>300 mAh g^{-1}) [29]. In contrast, potassium can form intercalation compounds with graphite, allowing graphite to serve as a negatrode material for potassium-ion batteries that exhibit a high theoretical capacity of 279 mAh g^{-1} and demonstrate considerable potassium-embedding capabilities. However, its theoretical capacity, which stands at 372 mAh g^{-1}, and its suboptimal rate performance are limitations that prevent it from meeting the specific energy and specific power demands expected from AMICs. During rapid charging and discharging cycles, carbon negatrodes struggle to respond swiftly to fluctuations in current, further highlighting their inability to fulfill the specific energy and specific power requirements of AMICs. Within the electrochemical energy storage processes of AMICs, there is a significant mismatch between the slow ion diffusion at the negatrode and the rapid electrostatic accumulation at the positrode [30–33], leading to an imbalance in negatrode/positrode kinetics. Therefore, it is necessary to strategically develop carbon negatrodes with high multiplicity performance [34–36], which will increase the specific energy and specific power of AMICs at the same time.

3.1. N Doping

Nitrogen, as the most widely studied doping element, is easier to chemically bond with carbon compared to other elemental doping. The atomic radius of nitrogen is slightly smaller than that of carbon, which means that doping with nitrogen results in minimal disruption to the backbone structure of the carbon material, thereby preserving its stability [37]. Nitrogen atoms tend to form new micropores in the carbon skeleton and increase the specific surface area of the material [38]. In addition, the electronegativity of the nitrogen atoms is high (N is 3.04 and C is 2.55), generating very small lattice distortions [39,40], which gives doped carbon materials excellent electrical conductivity [41].

There are five types of doped N (Figure 2a), namely pyrrolic N, nitroso group, amino group, pyridinic N, and graphitic N [42,43]. Positively charged pyridinic N oxides and graphite N elevate electrical conductivity and facilitate electron transport within the carbon electrode. Negatively charged pyridinic N and pyrrolic N generate pseudo-capacitance and enhance the specific capacity and rate performance of the negatrode carbon material [44]. Despite these benefits, nitrogen doping is not without its drawbacks; for instance, graphitic N can diminish the cycling stability and initial coulombic efficiency (ICE) of doped carbon materials. Thus, the choice of nitrogen doping type and quantity is of paramount importance. Tang et al. [45] developed a method to synthesize high-nitrogen-doped, low-graphitic-N carbons (LGNCs) that exhibit improved ICE by eliminating graphitic nitrogen from graphitic carbon nitride (g-C_3N_4) through a denitrification process. As shown in Figure 2b, the LGNC (C:N = 4:1) has higher pyridinic N and pyrrolic N contents and a lower graphitic N content compared to the other two LGNCs (2:1 and 8:1). Therefore, it shows excellent lithium storage capacity and good cycle stability. By calculating the contribution of LGNC diffusion (intercalation) and capacitance (surface-controlled) at different scanning rates (Figure 2c,d), it is evident that the capacitive contribution of LGNC is directly proportional to the scanning rate.

Incorporating nitrogen into the graphite structure of carbon nanomaterials, such as graphene, carbon nanotubes, and activated carbon, can significantly enhance the atomic and electronic structure of these substrates. This modification not only boosts their electrical conductivity but also introduces a conjugated carbon system with sp2 hybridization. These changes can markedly alter the properties of the material, including its reactivity, mechanical strength, and electrocatalytic performance. Zhao et al. [46] utilized N_2 gas as a dopant to achieve in situ nitrogen doping of carbon substrates. The Raman spectroscopy results indicate a progressive decrease in the intensity ratio (I_D/I_G) of the D-band to the G-band as gaseous N_2 is introduced (Figure 2e). Additionally, a two-dimensional band near 2669 cm^{-1} emerged in the spectra of C-N-30-8 h (introducing N_2 with a flow rate of 30 $mL \cdot min^{-1}$ for 8 h), demonstrating high crystallinity and intensity within the range of 1000–3000 cm^{-1} [47,48]. This suggests that the doping of N atoms enhances the graphitization of the carbon material, which in turn improves the sodium storage capabilities of C-N-30-8h at high rates.

Furthermore, nitrogen doping in carbon negatrodes enhances the anion-exchange capacity [49], creates additional defects and active sites, and bolsters the catalytic activity of redox reactions [50–52], which in turn improves the specific capacitance [53–57]. For instance, the introduction of substantial nitrogen content significantly boosts the pseudocapacitance contribution of reticulated nitrogen-doped carbon nanosheets' negatrode, which stands at 58.5%, compared to the 38.3% contribution from carbon nanosheets' negatrode without nitrogen doping [58]. The researchers [59] induced graphite nanodomains and micro mesopores into N-doped carbon to prepare tubular N-doped porous carbons (NPCs), called As_8Mg. As shown in Figure 2f, As_8Mg exhibits a capacity retention rate of up to 98.7% with close to 100% CE at a specific current of 0.2 A g^{-1}. The cycling performance of As_8Mg is exceptionally outstanding among a variety of NPC negatrodes. Based on As_8Mg, SICs also achieve high specific energy (51 Wh kg^{-1}) and specific power (10,410 W kg^{-1}). This indicates that N-doped As_8Mg has high capacitive energy storage properties and is one of the ideal negatrodes for sodium-ion capacitors.

Nitrogen doping exerts a substantial influence on the storage of metal ions, with a particular emphasis on three forms of doped nitrogen: pyrrolic N, pyridinic N, and graphitic N. Taking Li^+ as an example, Chen et al. [60] prepared nitrogen-doped interconnected carbon networks (INCs) which revealed high reversible capacity and excellent electrochemical kinetics. The prepared INC//AC LIB has high specific energy (172 Wh kg^{-1} at 260 W kg^{-1}) and large specific power (26 kW kg^{-1} at 108 Wh kg^{-1}). The XPS results showed that the N element of the INCs prepared at 800 °C is mainly divided into three components: pyridinic N (N-6, 398.4 eV), pyrrolic N (N-5, 400.2 eV), and graphitic N (N-Q, 401.2 eV) (Figure 2g). Among the N-doped structures, the study concludes that the electronic defects of pyrrolic N (N-5) and pyridinic N (N-6) enhance the adsorption of Li atoms [61,62], while graphitic N doping (N-Q) weakens the adsorption of Li atoms.

3.2. S Doping, P Doping, and B Doping

Owing to the larger atomic radius of sulfur, sulfur doping may occupy a larger space in the pore volume. This causes the expansion of layer spacing and a significant decrease in the specific surface area of the carbon material [63]. Sulfur-doped carbon typically results in larger defects, distortions, and increased interlayer distances within the carbon structure [64–67]. S atoms are often found at the edges or defects of carbon planes, which can offer a greater number of active sites. Using the synthetic strategy of extracting sulfur-doped carbon flakes from softwood (Figure 3a), Casal and co-workers produced S-doped carbon flakes used as negatrodes for SICs [68]. The Galvanostatic charge–discharge (GCD) potential curves in Figure 3b indicate that for S-doped carbon sheets at discharge potentials of 1.2–1.5 V, the curve lengths are proportional to the sulfur content. Meanwhile, the GCD curve of the undoped sheets (Figure 3c) has a potential plateau below 0.1 V. The addition of sulfur doubles the CS-700m (prepared at 700 °C, in the absence of sulfur and subjected as well to a ball milling process) capacity, providing an excellent performance at 420 mAh g^{-1} (403 mAh cm^{-3}, ICE = 69%). At a high rate of 10 A g^{-1} (Figure 3d), the S-doped carbon sheets (55–40 mAh g^{-1}) possess a significantly larger reversible capacity (15 mAh g^{-1}) than the undoped carbon sheet, indicating that the redox reaction brought about by sulfur can still play a significant role at high rates. The above results suggest that sulfur doping provides abundant vacant defect locations for the carbon sheets. Due to the defects, the interlayer spacing of carbon is increased [69], and the resistance to Na^+ diffusion is greatly reduced, accelerating its embedding/removal time.

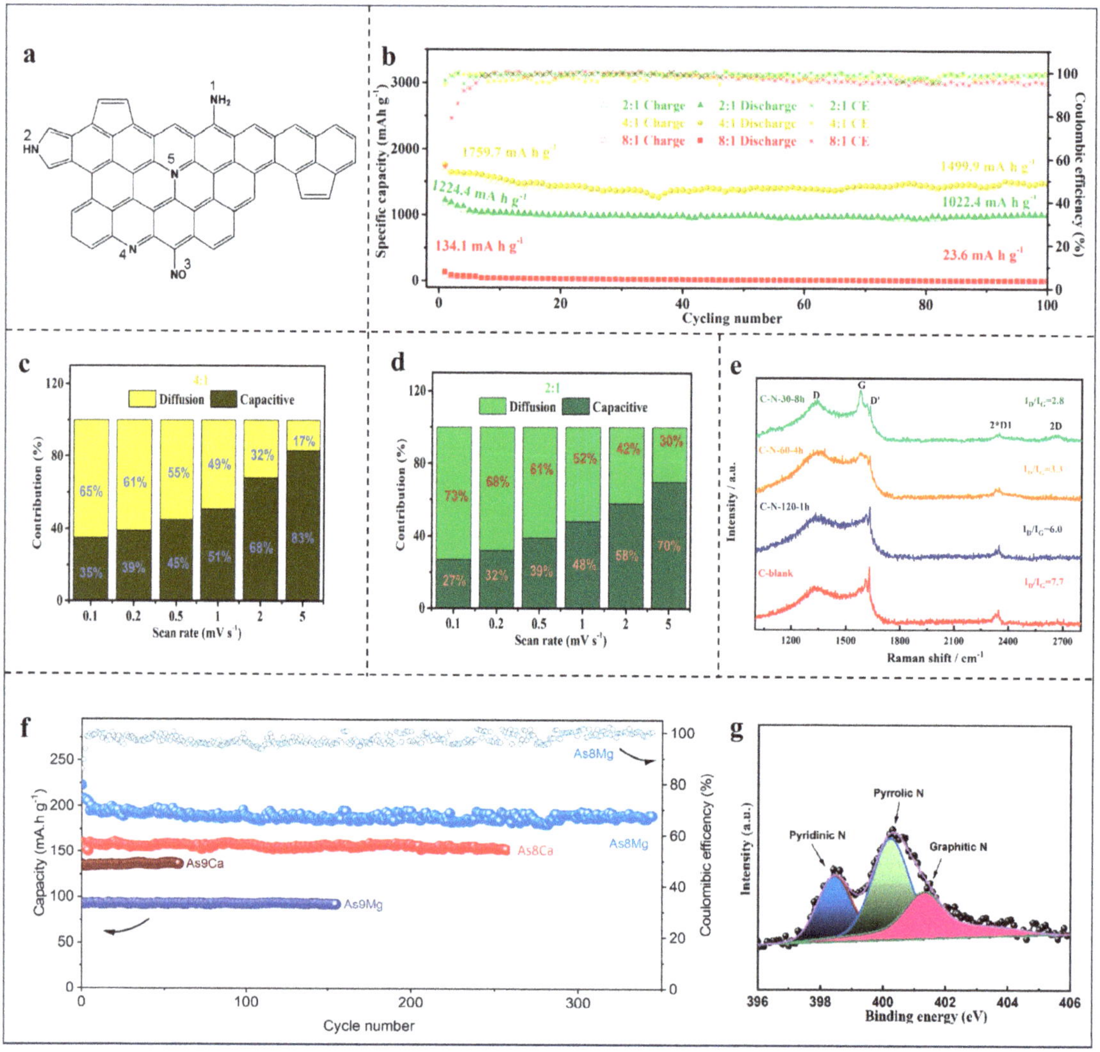

Figure 2. (**a**) Schematic overview of possible N sites of carbon backbone. Reproduced with permission from Ref. [43]. Copyright 2010, The Royal Society of Chemistry. (**b**) Galvanostatic charge/discharge cycling performances. Contribution of capacitive and diffusion-controlled capacity for LGNC at (**c**) 4:1 and (**d**) 2:1. LGNC, low-graphitic-N carbon. Reproduced with permission from Ref. [45]. Copyright 2022, Wiley. (**e**) Raman spectra of C-blank, C-N-120-1h, C-N-60-4h, and C-N-30-8h. Reproduced with permission from Ref. [46]. Copyright 2023, Elsevier. (**f**) Cycling performance of samples with specific current of 0.2 A·g^{-1}. Reproduced with permission from Ref. [59]. Copyright 2024, Wiley. (**g**) High-resolution XPS for N1s of INC prepared at 800 °C. Reproduced with permission from Ref. [60]. Copyright 2023, Elsevier.

Phosphorus (P) atoms are regarded as one of the most promising doping elements. Compared to nitrogen, which shares the same number of valence electrons, P atoms have a larger radius (~0.110 nm). Therefore, phosphorus atoms will cause pore enlargement and increase pore spacing during the doping process [70]. This characteristic endows phosphorus-doped carbon with significant structural distortions and a robust adsorption capacity for M^+ [71–74]. Yang et al. [75] successfully synthesized phosphorus-doped hard carbon (PHC-4), which contains a substantial amount of phosphorus (3.44%), through esterification and carbonation reactions between starch and phytic acid. PHC-4 is based on the phytic acid volume of 4 mL. PHC-4 has three main components. Li_xPC_y with P-C bonds is involved in the electrochemical redox reaction and provides high-capacitance performance [76]; meanwhile, P-O and P=O bonds enhance the adsorption of M^+ by the carbon material [77]. The binding of phosphorus atoms and active sites of carbon increases the chemical stability. Moreover, it is conducive to the insertion/extraction of Li^+, and the presence of P-doped atoms inhibits the oxidation of the carbon materials [78].

Boron (B) doping can significantly alter the characteristics of carbon-based materials in several ways: It disrupts the electroneutrality of these materials, leading to the generation of numerous active sites [79,80], and it also manages the charge distribution, which in turn expedites electron transfer [81,82]. Moreover, B doping can substitute for carbon atoms within the structure of the material, altering its electronic configuration and introducing more micropores into the carbon skeleton [83,84]. These modifications result in improved electrochemical performance for B-doped carbon in supercapacitors. Researchers [85] prepared boron-doped graphene nanosheets (BGNSs) by incorporating B into graphene nanosheets. B doping indispensably boosted the unique lithium storage mechanism and the adsorption properties of microporous capacitive carbon in BGNS. As shown in Figure 3e of the pore size distributions (PSDs), the addition of B prevented the agglomeration of graphene sheets. Moreover, it induced the formation of p-type conductive networks [86,87]. The mesoporous structure of BGNS observed in TEM images (Figure 3f) also supported the above conclusion. Meanwhile, the lithiation/de-lithiation ability of BGNS was greatly enhanced compared to undoped TGNS, as shown in Figure 3g. This enhancement is likely a result of boron doping increasing the disorder within the matrix, which in turn promotes porosity and the creation of Li^+ storage sites.

3.3. N/S Co-Doping

N/S co-doped carbon has seen rapid development in recent years, capitalizing on the synergistic effects between nitrogen and sulfur. The co-doping of N and S atoms enhances the conductivity of carbon materials, increases interlayer distances, and introduces more defect sites within the carbon structure. These changes lead to an improved reversible storage capacity and rate capability, enhancing the overall performance of M^+ ion storage [88–91]. Researchers [92] synthesized an N/S double-doped multistage porous carbon (NSC) with excellent reversibility. Based on this premise, the highly stable LIC (Figure 3h) constructed achieves both high specific energy and high specific power simultaneously. As shown in the XPS spectrum (Figure 3i), the N1 peaks of NSC1–3 include pyridinic N, pyrrolic N, graphitic N, and nitrogen-containing oxides [93,94]. Among these, pyridinic N and pyrrolic N demonstrate strong adsorption properties for M^+, such as Li^+ and Na^+, due to their electron-deficient characteristics [95]. Additionally, S doping enhances the surface wettability of the carbon material and the electrolyte, facilitating rapid charge transfer [96,97]. Moreover, the incorporation of both N and S doping in porous carbon significantly reduces interfacial resistance, thereby improving electron conduction [98,99]. The presence of nitrogen and sulfur heteroatoms contributes to superior sodium-ion storage capabilities, allowing carbon negatrodes doped with N/S to achieve a high reversible capacity and exhibit remarkable cycle stability. As a result, N/S-doped carbon has been widely utilized as a negatrode in SICs [100–106].

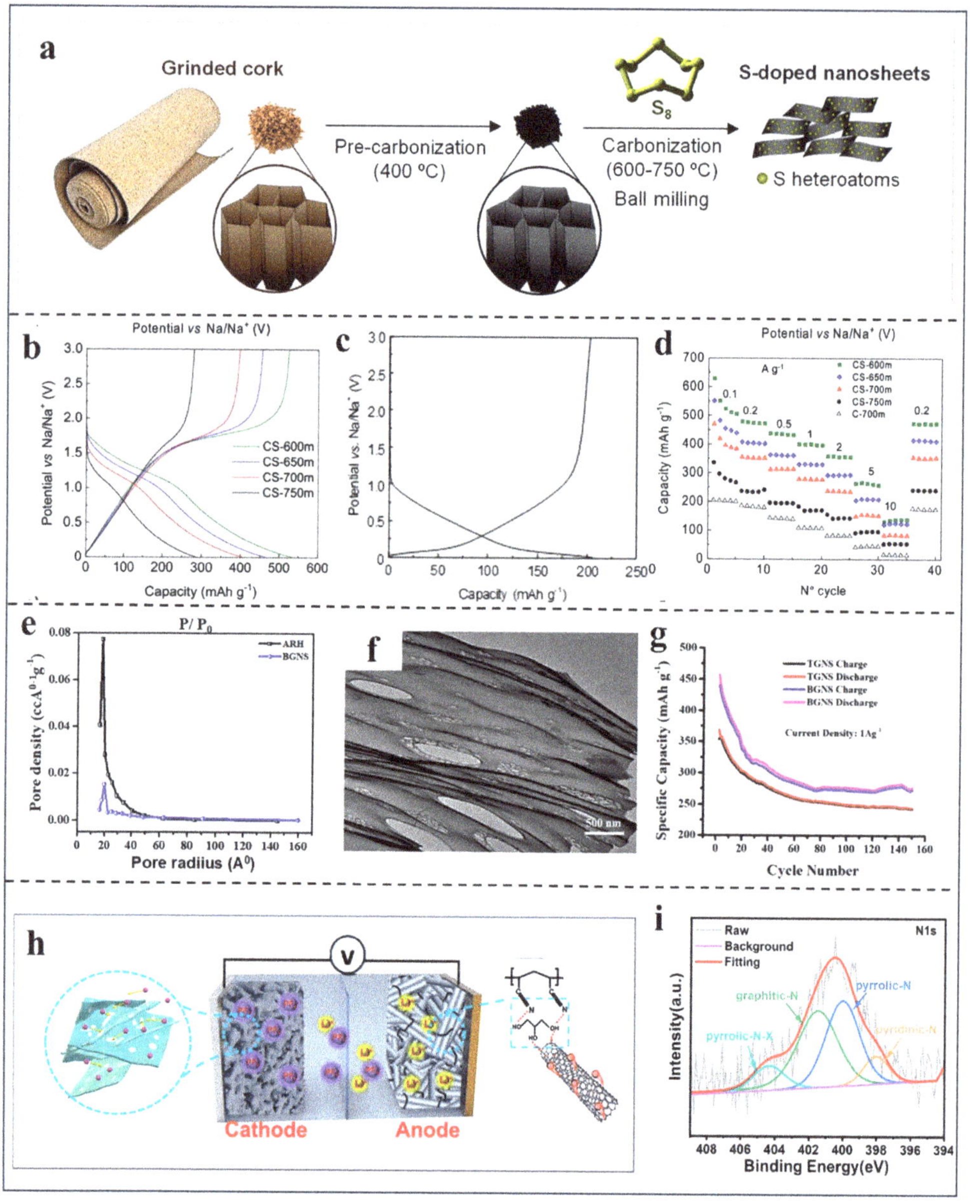

Figure 3. (**a**) An illustration of the synthesis strategy of S-doped carbon sheets from cork. (**b**,**c**) GCD profiles in the third cycle for the different S-doped sheets and un-doped carbon sheets C-700m. (**d**) The rate capability of the different S-doped sheets. Reproduced with permission from Ref. [68]. Copyright 2023, American Chemical Society. (**e**) Pore size distribution plot of BGNSs and ARH. (**f**) TEM images of BGNSs. (**g**) Cycling of BGNSs and TGNSs at a specific current of 1 A g^{-1}. Reproduced with permission from Ref. [85]. Copyright 2022, Elsevier. (**h**) A schematic illustration of LIC (NSC//GPN@CNTs). (**i**) N1s Entire XPS spectra of NSC1–3 carbonized at 800 °C. Reproduced with permission from Ref. [92]. Copyright 2023, Elsevier.

4. Element Doping for Carbon Positrode

The capacitance of metal-ion capacitors is often limited by the properties of the positrode material, making the selection of a suitable positrode material a critical decision. Carbon, being the most prevalent positrode material, is known for its significant capacitance contribution to fast ion storage and its excellent chemical stability, which allows it to maintain its structure and performance across a broad spectrum of electrolyte environments. However, the low specific capacity of carbon can lead to reduced specific energy in devices that incorporate it. Consequently, researchers have dedicated considerable effort to designing positrode carbon materials with enhanced specific capacity. There is an ongoing quest to discover new electrode material structures and preparation techniques aimed at surmounting the challenges related to the low specific capacity of carbon materials. This continuous exploration is essential for advancing the performance and specific energy of AMICs.

4.1. N Doping

To achieve a balance between electrochemical kinetics and specific capacity in the negatrode and positrode of AMICs, Xu and colleagues proposed a preparation method, combining the advantages of elemental doping and nanostructure, and also successfully prepared nitrogen-doped (3.65%) multistage porous carbon sphere (NHPCS) positrodes [107]. Nitrogen doping endows the NHPCS electrode with superior wettability compared to other electrodes, as shown in Figure 4a. This improved wettability is anticipated to reduce interfacial impedance, thereby enhancing the rate performance. Furthermore, Figure 4b illustrates that the synergistic effect of N/O doping markedly enhances the electrochemical performance of the NHPCS. The unique electronic configuration of nitrogen allows N-doped carbon materials to more readily attract ions from the electrolyte [108], which in turn boosts their capacitive characteristics.

4.2. N/S Co-Doping and N/B Co-Doping

The introduction of different dopants into materials can significantly alter their electrochemical properties, leading to substantial changes in both physical and chemical characteristics [109]. Among these dopants, N is the most extensively used and is often co-doped with other elements to achieve synergistic effects, such as in the case of N/S co-doping. Recently, Shi et al. [110] prepared S/N co-doped porous carbon (SNC) by using a simple method. The addition of S and N enhanced the specific capacity and wettability of the SNC samples. The ratio of S to N can be fine-tuned to achieve the optimal synergistic effect, including on the pore structure of the resulting materials. Table 2 confirms the small pore sizes of SNC-3 and SNC-4 micropores. Simultaneously, as can be seen in the CV curves of Figure 4c, all SNC electrodes, particularly SNC-3, exhibit excellent capacitive behavior [111,112]. This is due to the ability of S to significantly enhance the reactivity between the electrode and the electrolyte. However, with an increase in S content and a corresponding decrease in carbon content, the specific capacity of SNC-4 diminishes, preventing the achievement of optimal performance. Therefore, the proportion of non-metallic elements in the doped material can be easily adjusted by modifying the ligand ratio to optimize performance. As depicted in Figure 4d, the LIC using SNC-3 as a positrode and pre-lithiation HC as a negatrode still outperforms others, and it can achieve powerful performance with maximum specific power (6072 W kg^{-1}) and maximum specific energy (52 Wh kg^{-1}).

The electronegativity of carbon lies between that of B and N, which makes it a versatile element for doping. By substituting some carbon atoms with N and B, the surface polarity and electronic structure of the carbon substrate can be synergistically modulated.

Consequently, co-doped carbon materials exhibit greater catalytic and electrochemical activity compared to their singly doped counterparts [113–118]. Zhang et al. [119] utilized N/B co-doped swollen reduced oxidized graphite (NBEG) as a positrode for SICs; the preparation process is shown in Figure 4e. N-containing functional groups are beneficial for increasing the electronic conductivity of carbon-based materials, inducing the creation of unbalanced charged regions in the carbon materials. This action not only enhances electronic conductivity but also provides additional active sites [120], thereby improving the overall electrochemical performance.

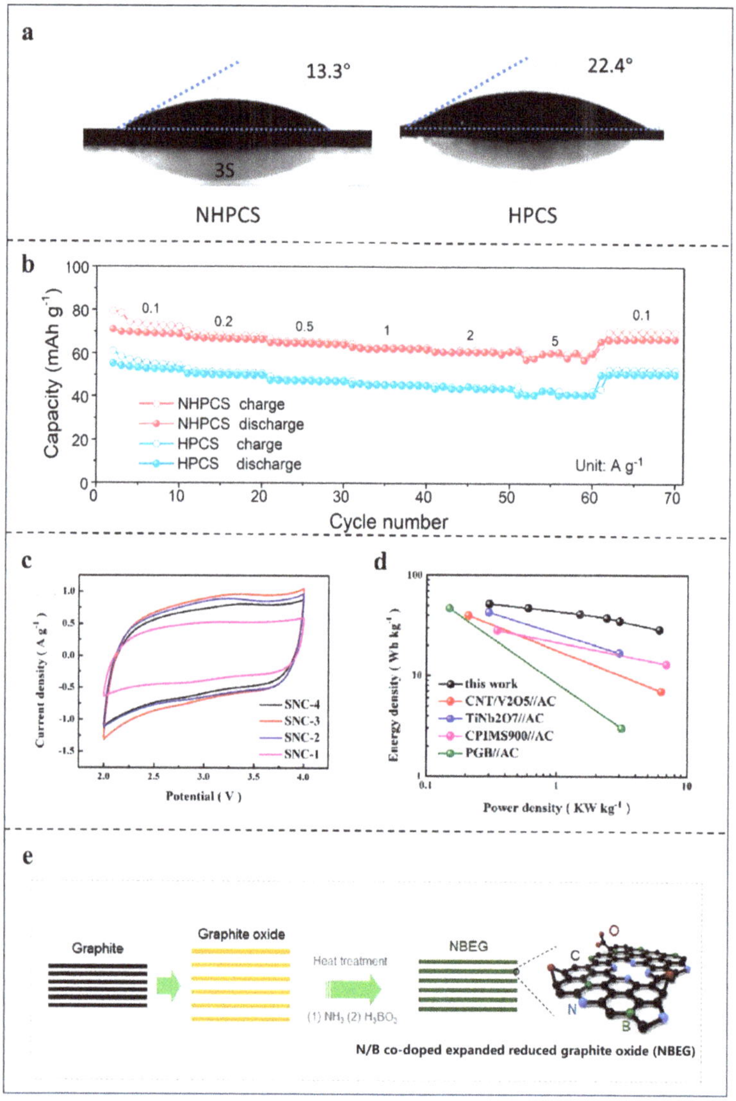

Figure 4. (**a**) The contact angles and (**b**) rate capabilities of the NHPCS and HPCS. Reproduced with permission from Ref. [107]. Copyright 2020, Springer Nature. (**c**) The CV curve of SNC1–4 samples at a scan rate of 10 mV s^{-1}. (**d**) A performance comparison of this LIC with other works. Reproduced with permission from Ref. [110]. Copyright 2023, Elsevier. (**e**) N/B co-doped expanded reduced graphite oxide (NBEG). Reproduced with permission from Ref. [119]. Copyright 2020, Elsevier.

Table 2. Estimated parameters from BET and EDS analysis.

Material	S_{BET} (m^2 g^{-1})	V_{tot} (cm^3 g^{-1})	V_{mic} (cm^3 g^{-1})	D_{DA} (nm)	Elemental Analysis (at%)			
					C1s	N1s	O1s	S2p
SNC-1	547	1.32	0.11	10.91	72.59	12.92	12.7	1.81
SNC-2	619	1.32	0.03	9.68	71.03	11.91	10.52	6.55
SNC-3	616	1.18	0.013	9.72	67.94	7.98	16.67	7.42
SNC-4	514	1.43	0.017	11.58	56.15	5.7	24.62	13.54

S_{BET}: BET surface area. V_{tot}: total volume in pores. V_{mic}: micropore area. D_{DA}: desorption average pore diameter.

5. Element Doping for Dual-Carbon Electrodes

Carbon materials have broad prospects in the field of energy storage due to their large capacity and environmental friendliness [121]. The double-carbon alkali metal-ion capacitor (DC-AMIC), which utilizes carbon materials for both the positrode and negatrode, boasts the benefits of low cost, excellent stability, and adjustable surface compositions and structural morphology. For instance, dual-carbon sodium-ion capacitors (DC-SICs) select similar material systems for their negatrodes and positrodes, addressing the issues of capacity imbalance and kinetic mismatch between the two electrodes. The double-carbon system manifests in two primary forms: one where two distinct carbon materials with different properties serve as the negatrode and positrode, and another where the same carbon material is used for both. The former can fully utilize the advantages of different carbon materials, resulting in a significant improvement in both the capacity and power performance of DC-AMICs. The latter has a lower production cost, which is conducive to mass production and application, while the synergistic effect between the positrode and negatrode in the charging and discharging process will be enhanced.

5.1. Dual Electrodes Made of the Same Carbon Material

5.1.1. LIC

Research has demonstrated that atomic doping can significantly enhance the charge storage capacity of carbon materials. Li et al. [122] prepared a two-dimensional (2D) nitrogen-doped carbon nanosheet (NCN). The high-resolution transmission electron microscopy (HRTEM) image depicted in Figure 5a reveals the lattice stripe extension of the NCN (0.36 nm), providing evidence that the large interlayer distance of the carbon layer contributes to the high electrochemical conductivity of the NCN. This observation is closely associated with the abundant and uniformly distributed nitrogen within the NCN (Figure 5b). Furthermore, the porous structure of the carbon substrate (CNF) endows the NCN positrode with a high specific capacity, enabling excellent pseudo-capacitance storage dynamics even at high current rates. Using an NCN as the negatrode and positrode of the LIC not only reduces the cost but also effectively alleviates the dynamic imbalance, improving the electrochemical performance of the capacitor. The dual-carbon NCN//NCN LIC system can provide a high specific energy of 167 Wh kg^{-1} at a specific power of 240 W kg^{-1}. Moreover, it exhibits good cycling stability (capacity retention of 85.6% for 9000 cycles at 1 A g^{-1}).

Jiang and colleagues have developed coal-based nitrogen-doped porous carbon materials (CNPCs) that feature a folded carbon nanosheet structure [123]. The use of g-C3N4 as both a templating agent and a nitrogen source plays a pivotal role in determining the pore size and nitrogen content of CNPCs, which in turn significantly impacts the electrochemical performance of these materials. In Figure 5c,d, the optimized CNPCs (CTK-1.0) display good pore regulation, high conductivity, and a large specific surface area. Scanning transmission electron microscopy (STEM) and energy-dispersive X-ray spectroscopy (EDS) analyses of CTK-1.0 confirm the uniform distribution of carbon and nitrogen elements within the CNPCs, signifying the successful doping with heteroatomic nitrogen (Figure 5e–g). The nitrogen-doped structures, specifically pyridinic N and pyrrolic N, offer additional active sites for pseudo-capacitance and ion adsorption, thereby enhancing the fast-charging capabilities of the materials. Moreover, ammonium N facilitates electron transfer within the carbon lattice [124], which enhances the capacity of dual-carbon lithium-ion capacitors (DC-LICs). The constructed CTK-1.0//CTK-1.0 LIC achieves a balanced capacity and kinetic balance, with a maximum specific energy of 137.6 Wh kg^{-1} and a long lifetime of 3000 cycles.

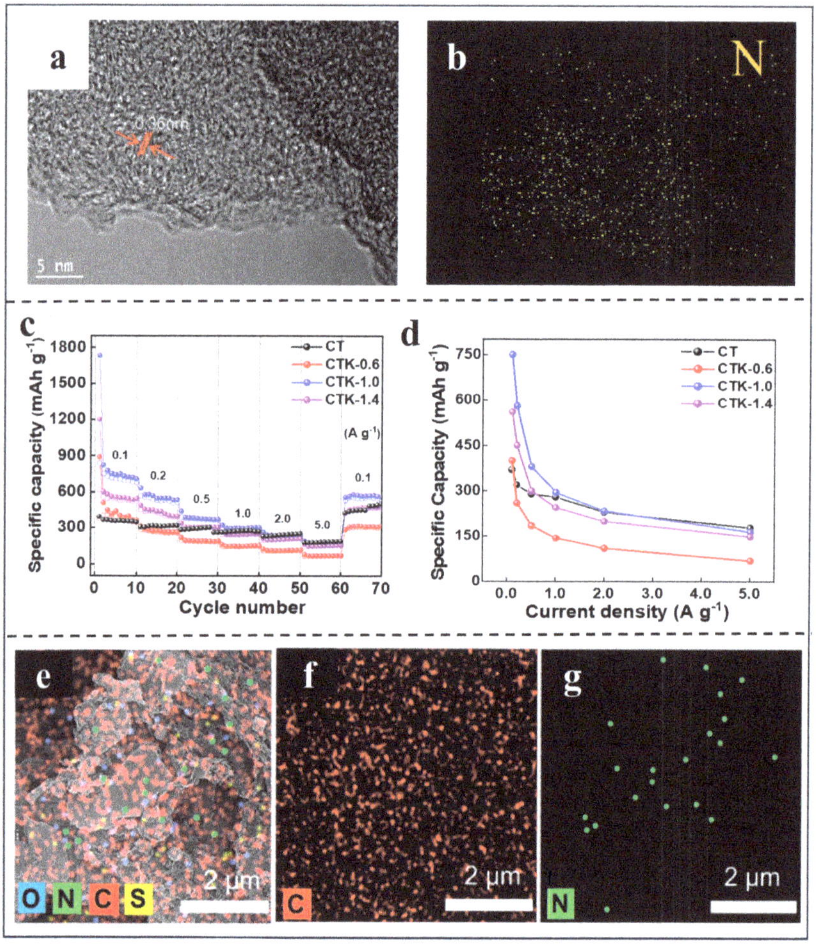

Figure 5. (**a**) HRTEM pictures of the NCN. TEM images (**b**) of the corresponding EDS elemental analysis of the NCN. Reproduced with permission from Ref. [122]. Copyright 2022, Elsevier. (**c**) The rate performance and (**d**) specific capacities at different current densities of CNPCs with different g-C additions. (**e**) STEM images of CTK-1.0 with the corresponding (**f**) C and (**g**) N elemental mappings. Reproduced with permission from Ref. [123]. Copyright 2023, Multidisciplinary Digital Publishing Institute.

Enhancing the low-temperature performance of LICs is a significant challenge in the energy storage industry. At low temperatures, the electrochemical reaction kinetics of capacitors slow down, adversely affecting ion transport and charge transfer, which in turn impedes the normal functioning of LICs. To address this problem, Xiao et al. [125]

prepared S-doped graphene nanocapsules (SGCs) as the negatrode and positrode of LICs. The storage mechanism of the prepared SGCs//SGCs LICs is shown in Figure 6a. As illustrated in Figure 6b, SGCs//SGCs LICs embody a capacity retention rate of more than 83% at −30 °C (10,000 cycles at 5 A g^{-1} and 10 A g^{-1}), revealing excellent ultra-long cryogenic cycle performance. The superior performance of the SGC electrodes is attributed to their stable nanocapsule morphology, abundant pore volume, and S-doped structure. The presence of S contributes to a gradual decrease in the Li^+ migration rate under low-temperature test conditions. As shown in Figure 6c, the Li^+ diffusion coefficients of the SGC negatrode are significantly higher than those of the GC negatrode at −20 and −30 °C. Furthermore, by forming Li-S chemical bonds, the S-doped structure can also facilitate the storage of more Li^+ ions and generate larger pseudo-capacitance, enhancing the overall performance of the LICs under low-temperature conditions.

5.1.2. SIC

Thangavel et al. [126] used graphene hollow nanospheres (GHNSs) doped with double heteroatoms (N and S) for the first time in the fabrication of symmetric DC-SICs. Initially, the NS-GHNSs exhibited a relatively low initial coulombic efficiency of only 36%, which was attributed to the formation of the solid electrolyte interphase (SEI) layer and irreversible reactions between carbon functional groups and sodium ions. However, the NS-GHNSs demonstrated stable performance in subsequent cycles (Figure 6d), achieving a coulombic efficiency of greater than 99% and maintaining a discharge capacity of 264 mAh g^{-1}. This improvement is credited to the presence of nitrogen and sulfur dopants, which enhance the electrical conductivity of the carbon substrate and help mitigate irreversible capacity loss during the initial cycling period. Additionally, there are a large number of heteroatom functional groups on the NS-GHNSs (Figure 6e,f), which have an irreplaceable advantage, for example, providing excellent wettability conducive to the transfer storage of Na^+ [127,128].

5.1.3. KIC

Nitrogen-doped porous carbon nanosheets (NPCNs), serving a dual role, were crafted through a combination of carbonization and self-assembly techniques [129]. In Figure 6g, NPCN-900 shows a type IV curve, and the specific surface area is 1586 m^2 g^{-1}. This suggests that NPCN-900, with its ample pores and expansive surface area, offers swift diffusion channels and adsorption sites for potassium ions (K^+), thereby bolstering its capacitive performance. Similarly, the graphitic N in NPCN-900 can promote the formation of ordered structures, which gives NPCN-900 an ordered graphitic carbon structure [130]. Figure 6h illustrates the remarkable rate capability of NPCN-900, which substantiates its pre-eminent position among negatrode materials [124]. The superior performance can be attributed to three primary factors. The addition of nitrogen not only enriches the active sites on the carbon material, providing more adsorption sites for K^+, but also increases the spacing between the carbon layers, which reduces resistance and deformation during K^+ intercalation/decalcification. Likewise, the abundant pores of the porous carbon nanosheets improve the capacitance performance. The NPCN//NPCN KIC exhibits 128 Wh kg^{-1} with high specific energy and excellent cycling stability (90.8% of the discharge capacity after 9000 cycles at 1.0 A g^{-1}). Notably, the capacity of the KIC shows a trend of decreasing and then increasing. It is shown that a large amount of hidden nitrogen is distributed inside the NPCN electrode, which is gradually activated during the long cycling process, thus releasing more active sites [131,132].

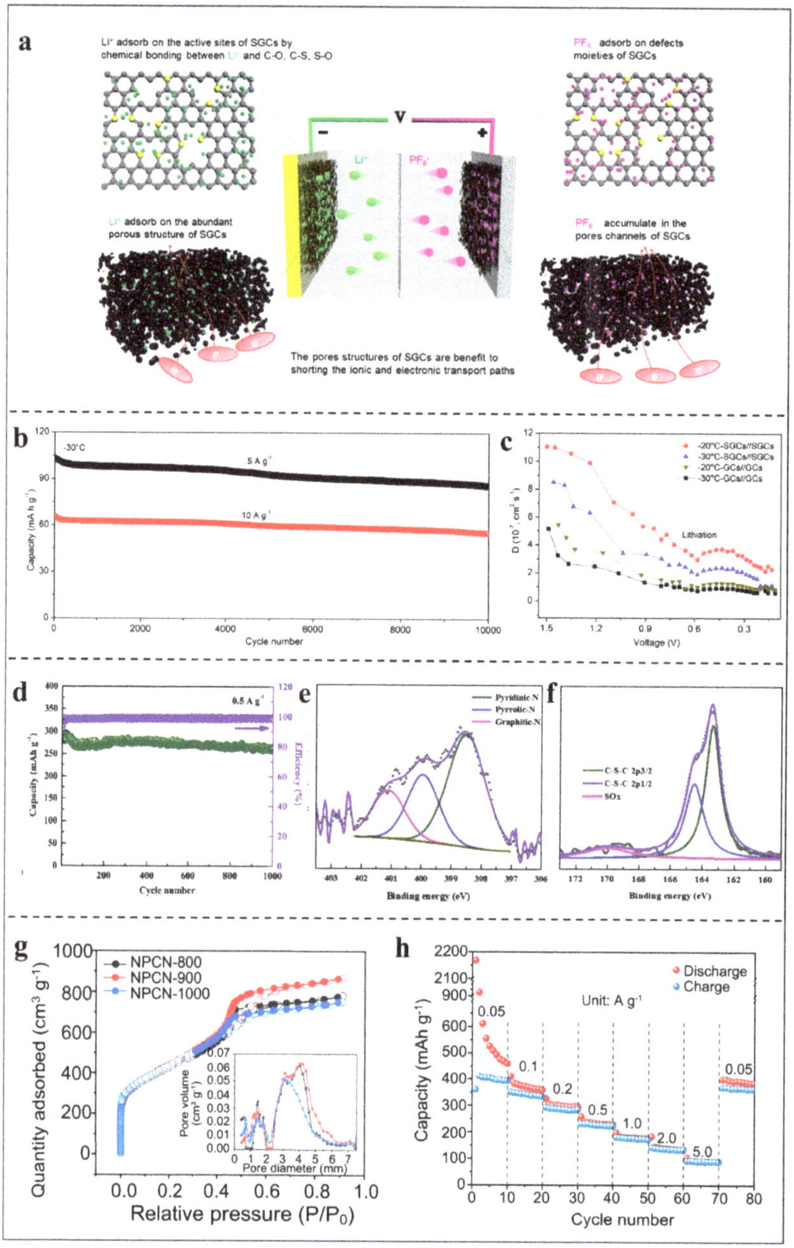

Figure 6. (**a**) A schematic illustration of the entire charge–discharge storage mechanisms of the as-prepared all-graphene symmetric SGCs//SGCs LIC. (**b**) The cycling performance of SGCs//SGCs after 10,000 cycles at −30 °C. (**c**) Li^+ diffusion coefficients of the SGC negatrode compared with GCs at low temperature. Reproduced with permission from Ref. [125]. Copyright 2022, Elsevier. (**d**) The anodic performance of NS-GHNSs, with cycling stability at 0.5 A g^{-1}. XPS survey spectra of NS-GHNSs, and high-resolution deconvoluted XPS spectra of (**e**) N1s, and (**f**) S2p. Reproduced with permission from Ref. [126]. Copyright 2019, Elsevier. (**g**) N_2 adsorption and desorption isotherms of NPCN-T (T = 800, 900, and 1000). (**h**) The rate capability of NPCN-900 as a negatrode for potassium-ion capacitors. Reproduced with permission from Ref. [129]. Copyright 2023, Elsevier.

5.2. Different Carbon Materials as Electrodes

5.2.1. LIC

In AMICs, the selection of negatrode materials with ample layer spacing and rapid kinetics is essential to facilitate the swift embedding/removal of M^+ [133,134]. For the positrode materials, in order to have an outstanding capacitive capacity, abundant active sites and a large specific surface area are necessary [135–137].

Gao et al. [138] synthesized phosphorus-doped activated carbon (PAC-6) by carbonation. Figure 7a shows the manufacturing process of PAC, and the schematic diagram of PHC-4//PAC-6 LIC is shown in Figure 7b. The elemental mapping is shown in Figure 7c, which indicates that P derived from phytic acid (PA) is uniformly doped throughout PAC-6. A higher P content (3.57%) creates active sites for a redox reaction with PAC-6, thus enhancing the capacity of the sample. As a result, the specific capacitance of PAC-6 reaches 147.1 F g^{-1} at 0.05 A g^{-1} with excellent rate capability (90% capacitance retention after 5000 cycles at 1.6 A g^{-1}). Due to the appropriate PA content, the high specific surface area (SSA) and ideal power spectral density (PSD) characteristics of PAC-6 contribute to its superior areal capacitance. The assembled LIC shows excellent energy power characteristics (maximum 122 Wh k g^{-1}, 9775 W k g^{-1}) and operating life.

The electrode kinetics of the negatrode can be significantly constrained by the low conductivity of the electrode material. Fortunately, researchers [139–141] found that S-doped carbon nanotubes (SCNTs) have high conductivity which can improve positrode and negatrode performance. This improvement is attributed to the ability of SCNTs to form a conductive network within the electrodes, thereby accelerating the ion transport rate. Zhao et al. [142] developed an integrated process suitable for sulfur doping in carbon materials, which led to the preparation of sulfur-doped porous carbon (SPC) and sulfur-doped carbon nanotubes (SCNTs). Due to sulfur doping, SPC exhibits an increased energy storage capacity and accelerated electrode kinetics. Then, Li^+ enters the SPC structure by adsorbing on S-containing active sites and inserts into interlayers/pores. The unique hydrophilicity and conductivity of SCNTs enable the construction of a uniformly distributed three-dimensional conductive network within the electrode, significantly enhancing the charge transfer rate and migration speed of Li^+ and PF_6^- during the charging/discharging processes. The constructed AC-SCNTs//SPC-SCNT LIC possesses excellent electrochemical performance (Figure 7d), displaying a specific energy of 169 Wh kg^{-1}, excellent specific power of 1022 W kg^{-1} at 136 Wh kg^{-1}.

5.2.2. SIC

For DC-SIC, doping with heteroatom layers and constructing porous structures are two effective strategies that can significantly enhance the kinetic performance of carbon materials [91,143].

An NHPC-800//NHPAC SIC device was constructed by Zou et al. [144] using nitrogen-doped hierarchical porous carbon (NHPC) as the negatrode and nitrogen-doped hierarchical porous activated carbon (NHPAC) as the positrode. By virtue of the appropriate layer spacing (maximum at 0.42 nm) brought about by N doping and the hierarchical pores. This SIC has excellent specific energy (115 Wh kg^{-1} at 200 W kg^{-1}) and a long lifecycle. Figure 7e demonstrates that the NHPC-800 electrode has the initial charge/discharge capacities of 392.4 mAh g^{-1}/ 536.3 mAh g^{-1}, respectively, with an ICE of 73.2%. Upon activation of NHPC to form NHPAC, both the porosity and specific surface area are enhanced, which is beneficial for the swift adsorption and desorption of anions (ClO_4^-) on the surface. Figure 7f shows the rate performance of the NHPAC electrode at various current densities. The specific capacity of the NHPAC electrode at 0.1 A g^{-1} is 110 mAh g^{-1}, which is much higher than that of commercial activated carbon [145].

While the porous structure of electrode materials is indeed critical for Na^+ storage, pore sizes that are not appropriately tailored can adversely affect the performance of the electrodes.

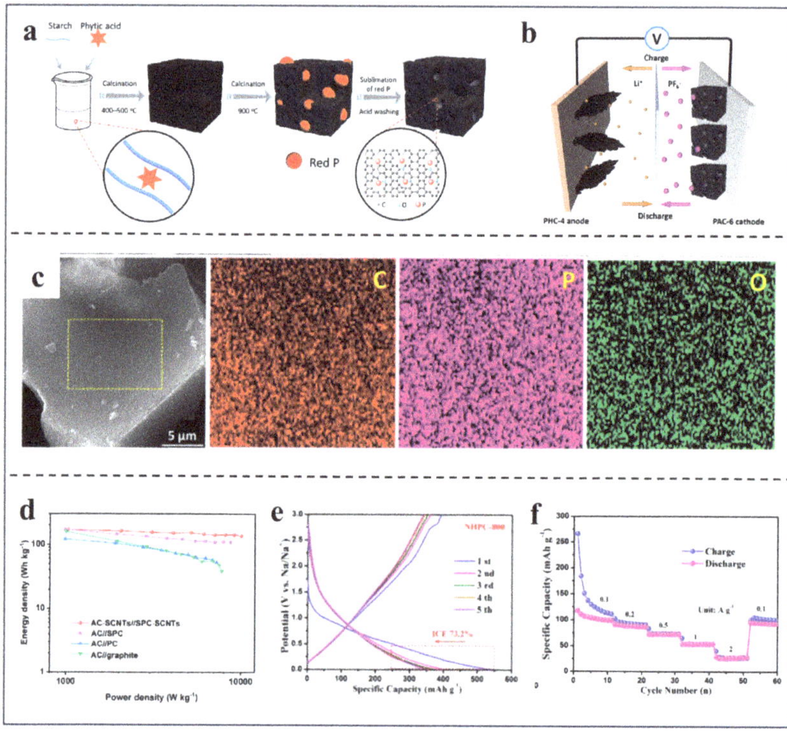

Figure 7. (**a**) Fabrication process for PACs. (**b**) Schematic of fabricated PHC-4//PAC-6 LIC. (**c**) SEM image of PAC-6 and corresponding elemental mapping images. Reproduced with permission from Ref. [138]. Copyright 2021, Elsevier. (**d**) Ragone diagram of assembled two-carbon AC-SCNT//SPC-SCNT LIC. Reproduced with permission from Ref. [142]. Copyright 2022, Elsevier. (**e**) GCD profiles of NHPC-800. (**f**) Rate capability of NHPAC positrode. Reproduced with permission from Ref. [144]. Copyright 2019, The Royal Society of Chemistry.

Wang et al. [146] doped hard carbon (HC) with N/S to enhance SIC performance and specific capacity. The structure of HC bent graphite was tuned by adding thiophene S, which has a characteristic heterocyclic structure, reducing the degree of graphitization and improving the wettability. Introducing more defects and edges during the carbonization process activates the electrochemical potential of the carbon, thereby boosting its electrical conductivity. Furthermore, the incorporation of nitrogen in the form of pyridinic N increases pseudo-capacitance and the capacity of the material. NS-pHC-1.348 (N and S co-doped porous hard carbon with 1.348 g $MgCl_2$ precursor) was prepared to provide reversible capacity, increased rate performance, and excellent cycling stability, which balances the electrochemical kinetics of the positrode and negatrode within the SIC. The NS-pHC-1.348//NPC SIC shows excellent specific power, reaching 92.03 kW kg^{-1} at 5.98 Wh kg^{-1}.

5.2.3. KIC

For KIC, it is important to enhance the K^+ storage capacity to construct a porous carbonaceous negatrode with reasonable-sized defects and abundant active sites.

Liu and colleagues [147] expanded microporous defects while increasing the number of chemisorbed sites, and finally synthesized phosphorus–nitrogen double-doped foamy porous carbon (PN-FPC). The formation process of the PN-FPC material is shown in Figure 8a. P/N double doping constructs a defect-rich structure in PN-FPC, facilitating the reversible adsorption/desorption behaviors of K^+. The rich mesoporous structure and fine size of

PN-FPC forms an inter-connection structure between the carbon bubbles (Figure 8b,c), which allows K^+ to move fast. As the scanning rate increases, the capacitance contribution of the PN-FPC electrode gradually increases, demonstrating that the synergy between P/N co-doping and the carbon matrix results in excellent enhancement in reaction kinetics. Meanwhile, by observing the P-C (133.5 eV) and P-O (132.5 eV) peaks in the P 2p spectra (Figure 8g), it is demonstrated that the P-O functional group could improve the wettability of PN-FPC, shortening the charge transfer process of K^+ [148]. Figure 8d illustrates that PN-FAPC, used as the positrode, maintains a reversible capacity of 110.2 mAh g^{-1} at a high rate of 5 A g^{-1}, showcasing its excellent structural stability. It exhibits a long-term lifecycle of 86.3% capacity retention over 1000 charge/discharge cycles (Figure 8e). The PN-FPC//PN-FAPC KIC shows ultra-high specific energy (155.6 Wh kg^{-1}), specific power (17,000 W kg^{-1}), and an ultra-long lifetime (90.9% capacity retention after 10,000 cycles), as shown in Figure 8f.

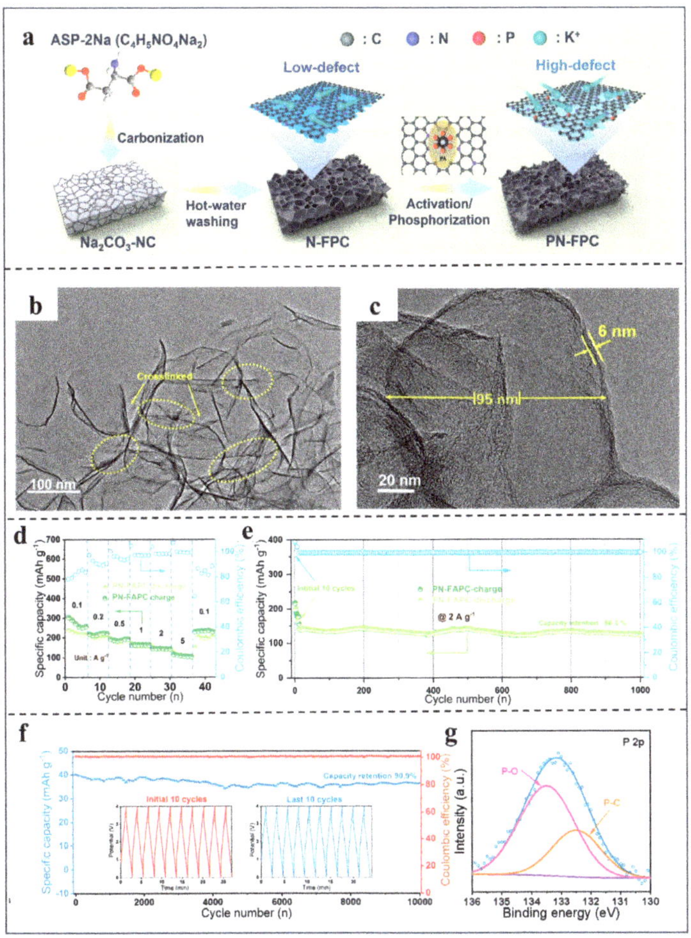

Figure 8. (**a**) A schematic illustration of the formation process of PN-FPC. (**b**,**c**) TEM images of PN-FPC. Half-cell performance of the PN-FAPC electrode tested between 1.5 and 4.2 V vs. K. (**d**) The rate performance and (**e**) lifecycle performance of the PN-FPC positrode (initial 10 cycles worked at 0.1 A g^{-1}). (**f**) The long-term cycling performance of the PN-FPC//PN-FAPC KIC at 2 A g^{-1} in full-cell configuration. (The inset shows GCD curves for the first 10 cycles and the last 10 cycles.) (**g**) The high-resolution XPS spectrum of PN-FPC of P 2p and its deconvolution. Reproduced with permission from Ref. [147]. Copyright 2024, Elsevier.

6. Discussion and Conclusions

This review has systematically summarized the recent advancements in doped carbon materials for alkali metal-ion capacitors (AMICs) with M = Li, Na, and K, as well as the mechanisms by which doping atoms such as nitrogen (N), sulfur (S), phosphorus (P), and boron (B) influence the properties of AMICs. Table 3 reports a summary of the synthesis methods and electrochemical performance of heteroatom co-doped carbon-based materials. By employing diverse synthesis strategies and combining unique precursors with dopants, carbon substrates have been crafted into a variety of electrode materials with superior characteristics. These include single-element-doped carbons (N-, S-, P-, and B-doped) and materials with double-element doping. While there has been a surge in research on heteroatom doping in carbon materials in recent years, leading to significant progress, numerous challenges remain in the practical application of these doped carbon materials as electrode materials for AMICs.

Table 3. A summary of the synthesis methods and electrochemical performance of heteroatom co-doped carbon-based materials.

Materials	Synthesis Method	Heteroatom Content%	Electrochemical System	Working Window	Cycling Data [a]	Rate Capacity [b]	Ref.
LGNC (4:1)	The denitrification and carbonization	N: 23.1at%	LIB	0.01–3.0 V	1499/100	1556/1	[45]
C-N-30-8h	molten salt method	N: 1–2at%	SIB	0.01–3.0 V	179/500	179/1	[46]
As_8Mg	Template-assisted synthesis	N: 17wt.%	SIC	0.01–2.0 V	192/600	159/1	[59]
INC	calcination	N: 5.63wt.%	LIC	0.01–3.0 V	280/3000	316/1	[60]
CS-700m	Ball milling process	S: 19.8wt.%	SIC	0.01–3.0 V	370/200	280/1	[68]
PHC-4	Esterification and carbonation reactions	P: 3.44at%	LIC	0.01–3.0 V	84.7%/5000	770.4/0.1	[75]
BGNS	Thermal reduction	B: 3.01wt.%	LIC	1.9–4.0 V	404/1000	800/0.1	[85]
NSC	Molten salt method and carbonization	N: 5.6wt.% S: 2.3wt.%	LIC	2.0–4.5 V	91.5%/10,000	81/1	[92]
NHPCS	Template-assisted methods	N: 3.65at%		2.0–4.3 V	88.7%/10,000	74/0.1	[107]
NBEG	Calcination (purify by heating)		SIC	1.5–4.2 V		328/0.5	[119]
SNC-3	Template-assisted methods	S: 7.42at% N: 7.98at%	LIC	2.0–4.0 V	90%/5000	78/0.1	[110]
NCN	Template-assisted methods	N: 7.98at%	LIC	Negatrode: 0.01–3.0 V Positrode: 0.5–4.3 V	Negatrode: 620/1000 Positrode: 110/1000	Negatrode: 1129/0.1 Positrode: 110/1.0	[122]
CTK-1.0	Carbonization in argon atmosphere	N: 3.03at%	LIC	0.01–3.0 V	97%/800	750/0.1	[123]
SGCs	Chemical Vapor Deposition (CVD) methods	S: 3.3at%	LIC	Negatrode: 0–1.5V Positrode: 1.5–4.5V	Negatrode: 98.8%/10,000 Positrode: 89.1%/10,000	Negatrode: 336/0.1 Positrode: 1668/0.1	[125]
NS-GHNS		N: 3.34wt.% S: 1.82wt.%	SIC	Negatrode: 0.01–3.0 V Positrode: 2.5–4.2 V	Negatrode: 86%/10,000 Positrode: 92%/10,000	Negatrode: 630/0.1 Positrode: 52/0.2	[126]
NPCNs	Carbonation Self-Assembly method	N: 6.60wt.%	PIC	0.01–3.0 V 0.01–4.0 V		176/1.0 82.8/0.1	[129]
PAC-6	Esterification and carbonation reactions	P: 3.57wt.%	LIC	0.01–3.0 V	90%/10,000	147.1/0.05	[138]
SPC-SCNTs	Template-assisted methods	S: 17.3wt.%	LIC	0.01–1.5 V		348/0.5	[142]
NHPC-800	Pyrolysis	N: 10.4wt.%	SIC	0.01–3.0 V		392.4/0.1	[144]
AC-SCNTs	Activation method	N: 2.1wt.%	SIC	2.0–4.5 V		110/0.1	[144]
NS-pHC-1.348	Carbonization and pickling	S: 7.77wt.%	SIC	0.01–2.5 V		383.9/0.05	[146]
PN-FPC	One-pot activation and phosphatization	N: 7.2wt.% P: 1.9wt.%	PIC	0.01–3.0 V	88.9%/1000	201.1/5	[147]
PN-FAPC	Carbonation	N: 5.1wt.%	PIC	1.5–4.2 V	86.3%/1000	110.2/5	[147]

[a] The cycling data are summarized as capacity (mAh g^{-1})/corresponding cycle numbers. [b] The rate capacity is summarized as capacity (mAh g^{-1}) at a specific current (A g^{-1}).

(1) The incorporation of various doping elements can enhance distinct chemical properties of carbon materials. For instance, the concentration of B doping is pivotal to

altering the electronic mobility within B-containing carbon and the mobility of metal ions. Similarly, different forms of nitrogen doping, such as pyrrolic N and pyridinic N, exert specific influences on the chemical properties of the carbon material. In the case of double-element-doped carbon, selecting the appropriate elements and the optimal doping ratio is crucial for achieving a better balance between the negatrode and positrode. It is important to note that while doping elements are effective in enhancing the properties of carbon materials, suitable doping levels and precise doping locations require extensive and in-depth investigation.

(2) Carbon materials from different groups possess distinct structural features, and these variations can markedly influence how doping elements affect their properties. For instance, the two-dimensional planar structure of graphene endows it with a high specific surface area and superior electrical conductivity. N doping can further enhance its electron transport capacity, thereby improving the overall electrical conductivity of the material. This makes graphene particularly suitable for applications that necessitate swift charge transfer. On the other hand, activated carbon, known for its rich pore structure, experiences an increase in surface active sites and improved ion adsorption capacity when doped with elemental phosphorus. This enhancement bolsters its capacitive performance, making it an effective electrode material. At present, due to the limitations in our research understanding, further exploration and investigation are needed to uncover how the structure of carbon materials can enhance specific chemical properties. Advanced characterization techniques and theoretical calculations can provide a deeper understanding of the interaction mechanisms between doping elements and the structure of carbon substrates.

(3) Doped carbon materials with a high specific surface area indeed offer a greater number of active sites, which enhances the adsorption and storage capabilities for M^+. However, an excessively large specific surface area can lead to excessive consumption of the electrolyte in the initial cycle. The numerous active sites can interact strongly with the electrolyte, leading to the formation of a thick solid electrolyte interphase (SEI) film on the electrode surface. This SEI film formation consumes a significant amount of Li^+, causing substantial energy loss for the electrochemical device. Additionally, the overconsumption of the electrolyte results in a large amount of Li^+ being irreversibly intercalated into the electrode material during the first charge/discharge process, thereby reducing the initial coulombic efficiency (ICE). Therefore, designing doped carbon with an appropriate specific surface area is essential to strike a balance between achieving high capacity and maintaining high ICE.

In summary, the key to the design of doped carbon materials with excellent performance lies in mastering the type of doping atoms, the choice of concentration, and the design of a suitable specific surface area. At present, the practical application of doped carbon materials in AMICs has shown some potential but is still immature. Future research should continue to explore optimization strategies of doped carbon materials to overcome the current challenges and to promote the overall performance of AMICs for wider and more efficient applications in the field of energy storage.

Author Contributions: Conceptualization, K.Z.; methodology, T.Y. and X.H.; software, X.Y.; validation, X.H.; formal analysis, Y.G. and T.N.; investigation, Y.G. and L.Q.; resources, K.Z. and L.T.; data curation, T.Y.; writing—original draft preparation, T.Y. and Y.G.; writing—review and editing, X.H.; visualization, T.Y.; supervision, K.Z. and X.H.; project administration, L.L.; funding acquisition, K.Z. and L.L. All authors have read and agreed to the published version of the manuscript.

Funding: This work was supported by the National Natural Science Foundation of China (52374299, 52304320, and 52204306), the Outstanding Youth Foundation of Hunan Province (2023JJ10044), the

Scientific Research Fund of Hunan Provincial Department of Education (24B0295), the Natural Science Foundation of Hunan Province (2023JJ40014), and the Innovation and Entrepreneurship Training Program of Hunan Province (S202310536164).

Conflicts of Interest: The author declares no conflicts of interest.

References

1. Liu, Z.; Wang, J.; Lu, B. Plum pudding model inspired $KVPO_4F$@3DC as high-voltage and hyperstable cathode for potassium ion batteries. *Sci. Bull.* **2020**, *65*, 1242–1251. [CrossRef]
2. Zou, K.; Jiang, M.; Ning, T.; Tan, L.; Zheng, J.; Wang, J.; Ji, X.; Li, L. Thermodynamics-directed bulk/grain-boundary engineering for superior electrochemical durability of Ni-rich cathode. *J. Energy Chem.* **2024**, *97*, 321–331. [CrossRef]
3. Zhang, X.; Wang, L.; Liu, W.; Li, C.; Wang, K.; Ma, Y. Recent advances in MXenes for lithium-ion capacitors. *ACS Omega* **2020**, *5*, 75–82. [CrossRef] [PubMed]
4. Dai, M.; Zhao, D.; Wu, X. Research progress on transition metal oxide based electrode materials for asymmetric hybrid capacitors. *Chin. Chem. Lett.* **2020**, *31*, 2177–2188. [CrossRef]
5. Amatucci, G.G.; Badway, F.; Pasquier, D.A.; Zheng, T. An asymmetric hybrid nonaqueous energy storage cell. *J. Electrochem. Soc.* **2001**, *148*, A930–A939. [CrossRef]
6. Soltani, M.; Beheshti, S.H. A comprehensive review of lithium ion capacitor: Development, modelling, thermal management and applications. *J. Energy Storage* **2021**, *34*, 102019. [CrossRef]
7. Nagamuthu, S.; Zhang, Y.; Xu, Y.; Sun, J.; Zhang, Y.; Zaman, F.; Denis, D.; Hou, L.; Yuan, C. Non-lithium-based metal ion capacitors: Recent advances and perspectives. *J. Mater. Chem. A* **2022**, *10*, 357–378. [CrossRef]
8. Liu, W.; Zhang, X.; Li, C.; Wang, K.; Sun, X.; Ma, Y. Carbon-coated Li_3VO_4 with optimized structure as high capacity anode material for lithium-ion capacitors. *Chin. Chem. Lett.* **2020**, *31*, 2225–2229. [CrossRef]
9. Cao, L.; Li, H.; Liu, X.; Liu, S.; Zhang, L.; Xu, W.; Yang, H.; Hou, H.; He, S.; Zhao, Y.; et al. Nitrogen, sulfur co-doped hierarchical carbon encapsulated in graphene with "sphere-in-layer" interconnection for high-performance supercapacitor. *J. Colloid Interface Sci.* **2021**, *599*, 443–452. [CrossRef]
10. Ma, X.; Gao, D. High capacitive storage performance of sulfur and nitrogen codoped mesoporous graphene. *ChemSusChem* **2018**, *11*, 1048–1055. [CrossRef] [PubMed]
11. Jiang, M.; Wang, P.; Chen, Q.; Zhang, Y.; Wu, Q.; Tan, L.; Ning, T.; Li, L.; Zou, K. Enabling the Nb/Ti co-doping strategy for improving structure stability and rate capability of Ni-rich cathode. *Chin. Chem. Lett.* **2024**, *35*, 110040. [CrossRef]
12. Wang, B.; Gao, X.; Xu, L.; Zou, K.; Cai, P.; Deng, X.; Yang, L.; Hou, H.; Zou, G.; Ji, X. Advanced carbon materials for sodium-ion capacitors. *Batter. Supercaps* **2021**, *4*, 538–553. [CrossRef]
13. Zhou, J.; Hu, H.; Li, H.; Chen, Z.; Yuan, C.; He, X. Advanced carbon-based materials for Na, K, and Zn ion hybrid capacitors. *Rare Met.* **2023**, *42*, 719–739. [CrossRef]
14. Li, L.; Chen, Q.; Jiang, M.; Ning, T.; Tan, L.; Zhang, X.; Zheng, J.; Wang, J.; Wu, Q.; Ji, X.; et al. Uncovering mechanism behind tungsten bulk/grain-boundary modification of Ni-rich cathode. *Energy Storage Mater.* **2025**, *75*, 104016. [CrossRef]
15. Peng, Y.; Bai, Y.; Liu, C.; Cao, S.; Kong, Q.; Pang, H. Applications of metal-organic framework-derived N, P, S doped materials in electrochemical energy conversion and storage. *Coord. Chem. Rev.* **2022**, *466*, 214602. [CrossRef]
16. Zhang, J.; Fang, J.; Han, J.; Yan, T.; Shi, L.; Zhang, D. N, P, S co-doped hollow carbon polyhedra derived from MOF-based core-shell nanocomposites for capacitive deionization. *J. Mater. Chem. A* **2018**, *6*, 15245–15252. [CrossRef]
17. Zou, K.; Xie, S.; Jiang, M.; Wang, P.; Ning, T.; Tan, L.; Li, H.; Zhou, Y.; Wang, W.; Li, L. Insights into the precursor specific surface area for engineering Co-free Ni-rich cathodes with tailorable properties. *Chem. Eng. J.* **2024**, *483*, 149189. [CrossRef]
18. Feng, X.; Bai, Y.; Liu, M.; Li, Y.; Yang, H.; Wang, X.; Wu, C. Untangling the respective effects of heteroatom-doped carbon materials in batteries, supercapacitors and the ORR to design high performance materials. *Energy Environ. Sci.* **2021**, *14*, 2036–2089. [CrossRef]
19. Shao, T.; Liu, C.; Deng, W.; Li, C.; Wang, X.; Xue, M.; Li, R. Recent research on strategies to improve ion conduction in alkali metal-ion batteries. *Batter. Supercaps* **2019**, *2*, 403–427. [CrossRef]
20. Chen, Z. Supercapacitor and supercapattery as emerging electrochemical energy stores. *Int. Mater. Rev.* **2017**, *62*, 173–202. [CrossRef]
21. Shao, H.; Wu, Y.; Lin, Z.; Taberna, P.; Simon, P. Nanoporous carbon for electrochemical capacitive energy storage. *Chem. Soc. Rev.* **2020**, *49*, 3005–3039. [CrossRef] [PubMed]
22. Mirzaeian, M.; Abbas, Q.; Gibson, D.; Mazur, M. Effect of nitrogen doping on the electrochemical performance of resorcinol-formaldehyde based carbon aerogels as electrode material for supercapacitor applications. *Energy* **2019**, *173*, 809–819. [CrossRef]
23. Rudra, S.; Seo, H.; Sarker, S.; Kim, D. Supercapatteries as hybrid electrochemical energy storage devices: Current status and future prospect. *Molecules* **2024**, *29*, 243. [CrossRef] [PubMed]

24. Sun, Y.; Wang, H.; Wei, W.; Zheng, Y.; Tao, L.; Wang, Y.; Huang, M.; Shi, J.; Shi, Z.; Mitlin, D. Sulfur-rich graphene nanoboxes with ultra-high potassiation capacity at fast charge: Storage mechanisms and device performance. *ACS Nano* **2021**, *15*, 1652–1665. [CrossRef] [PubMed]
25. Zou, K.; Jiang, M.; Zhao, Z.; Xie, S.; Ning, T.; Tan, L.; Li, H.; Zhou, Y.; Wang, W.; Wu, X.; et al. Mechanistic insights into suppressing microcracks by regulating grain size of precursor for high-performance Ni-rich cathodes. *Chem. Eng. J.* **2023**, *476*, 146793. [CrossRef]
26. González, A.; Goikolea, E.; Barrena, J.A.; Mysyk, R. Review on supercapacitors: Technologies and materials. *Renew. Sustain. Energy Rev.* **2016**, *58*, 1189–1206. [CrossRef]
27. Wu, M., Hu, X.; Zheng, W.; Chen, L.; Zhang, Q. Recent advances in porous carbon nanosheets for high-performance metal-ion capacitors. *Chem. Eng. J.* **2023**, *466*, 143077. [CrossRef]
28. Cao, B.; Li, X. Recent progress on carbon-based anode materials for Na-ion batteries. *Acta Phys.-Chim. Sin.* **2020**, *36*, 1905003. [CrossRef]
29. Zhao, C.; Wei, Y.; Pan, Y.; Chen, C. A novel strategy for capacity judgement of hard carbon in sodium-ion batteries: Ensuring the consistency of the available anode capacity between half-cell and full-cell. *Solid State Ion.* **2024**, *412*, 116586. [CrossRef]
30. Han, P.; Xu, G.; Han, X.; Zhao, J.; Zhou, X.; Cui, G. Lithium ion capacitors in organic electrolyte system: Scientific problems, material development, and key technologies. *Adv. Energy Mater.* **2018**, *8*, 1801243. [CrossRef]
31. Li, B., Zheng, J.; Zhang, H.; Jin, L.; Yang, D.; Lv, H.; Shen, C.; Shellikeri, A.; Zheng, Y.; Gong, R.; et al. Electrode materials, electrolytes, and challenges in nonaqueous lithium-ion capacitors. *Adv. Mater.* **2018**, *30*, 1705670. [CrossRef]
32. Wei, Q.; Liu, J.; Feng, W.; Sheng, J.; Tian, X.; He, L.; An, Q.; Mai, L. Hydrated vanadium pentoxide with superior sodium storage capacity. *J. Mater. Chem. A* **2015**, *3*, 8070–8075. [CrossRef]
33. Wang, H.; Zhu, C.; Chao, D.; Yan, Q.; Fan, H. Nonaqueous hybrid lithium-ion and sodium-ion capacitors. *Adv. Mater.* **2017**, *29*, 1702093. [CrossRef] [PubMed]
34. Shi, R.; Han, C.; Li, H.; Xu, L.; Zhang, T.; Li, J.; Lin, Z.; Wong, C.; Kang, F.; Li, B. NaCl-templated synthesis of hierarchical porous carbon with extremely large specific surface area and improved graphitization degree for high energy density lithium ion capacitors. *J. Mater. Chem. A* **2018**, *6*, 17057–17066. [CrossRef]
35. Li, N.; Du, X.; Shi, J.; Zhang, X.; Fan, W.; Wang, J.; Zhao, S.; Liu, Y.; Xu, W.; Li, M.; et al. Graphene@hierarchical meso-/microporous carbon for ultrahigh energy density lithium-ion capacitors. *Electrochim. Acta* **2018**, *281*, 459–465. [CrossRef]
36. Ma, H.; Geng, H.; Yao, B.; Wu, M.; Li, C.; Zhang, M.; Chi, F.; Qu, L. Highly ordered graphene solid: An efficient platform for capacitive sodium-ion storage with ultrahigh volumetric capacity and superior rate capability. *ACS Nano* **2019**, *13*, 9161–9170. [CrossRef]
37. Cao, Y.; Mao, S.; Li, M.; Chen, Y.; Wang, Y. Metal/porous carbon composites for heterogeneous catalysis: Old catalysts with improved performance promoted by N-doping. *ACS Catal.* **2017**, *7*, 8090–8112. [CrossRef]
38. Inagaki, M.; Toyoda, M.; Soneda, Y.; Morishita, T. Nitrogen-doped carbon materials. *Carbon* **2018**, *132*, 104–140. [CrossRef]
39. Du, J.; Liu, L.; Yu, Y.; Zhang, Y.; Lv, H.; Chen, A. N-doped ordered mesoporous carbon spheres derived by confined pyrolysis for high supercapacitor performance. *J. Mater. Sci. Technol.* **2019**, *35*, 2178–2186. [CrossRef]
40. Lv, Q.; Si, W.; He, J.; Sun, L.; Zhang, C.; Wang, N.; Yang, Z.; Li, X.; Wang, X.; Deng, W.; et al. Selectively nitrogen-doped carbon materials as superior metal-free catalysts for oxygen reduction. *Nat. Commun.* **2018**, *9*, 3376. [CrossRef]
41. Tan, H.; Tang, J.; Kim, J.; Kaneti, Y.; Kang, Y.; Sugahara, Y.; Yamauchi, Y. Rational design and construction of nanoporous iron- and nitrogen-doped carbon electrocatalysts for oxygen reduction reaction. *J. Mater. Chem. A* **2019**, *7*, 1380–1393. [CrossRef]
42. Wang, X.; Li, X.; Zhang, L.; Yoon, Y.; Weber, P.; Wang, H.; Guo, J.; Dai, H. N-doping of graphene through electrothermal reactions with ammonia. *Science* **2009**, *324*, 768–771. [CrossRef]
43. Paraknowitsch, P.; Thomas, A.; Antonietti, M. A detailed view on the polycondensation of ionic liquid monomers towards nitrogen doped carbon materials. *J. Mater. Chem.* **2010**, *20*, 6746–6758. [CrossRef]
44. Seredych, M.; Hulicova-Jurcakova, D.; Lu, G.; Bandosz, T. Surface functional groups of carbons and the effects of their chemical character, density and accessibility to ions on electrochemical performance. *Carbon* **2008**, *46*, 1475–1488. [CrossRef]
45. Tang, Y.; Chen, J.; Mao, Z.; Roth, C.; Wang, D. Highly N-doped carbon with low graphitic-N content as anode material for enhanced initial coulombic efficiency of lithium-ion batteries. *Carbon Energy* **2022**, *5*, e257. [CrossRef]
46. Zhao, J.; Song, Q.; Zhao, H.; Xie, H.; Ning, Z.; Yu, K. Molten salt splitting of CO_2 with CaC_2 and in-situ nitrogen doping for carbon anode of lithium/sodium-ion batteries. *J. Electroanal. Chem.* **2023**, *950*, 117845. [CrossRef]
47. Zhang, C.; Zhong, X.; Chen, P.; Sun, S.; Jiang, Y.; Yan, X. Facile synthesis of porous graphite by calcium carbide and nitrogen gas for lithium-ion batteries. *J. Energy Storage* **2023**, *66*, 107386. [CrossRef]
48. Sadezky, A.; Muckenhuber, H.; Grothe, H.; Niessner, R.; Pöschl, U. Raman microspectroscopy of soot and related carbonaceous materials: Spectral analysis and structural information. *Carbon* **2005**, *43*, 1731–1742. [CrossRef]
49. Rivera–Utrilla, J.; Ferro–García, M.A. Effect of carbon-oxygen and carbon-nitrogen surface complexes on the adsorption of cations by activated carbons. *Adsorpt. Sci. Technol.* **1986**, *3*, 293–302. [CrossRef]

50. Keegel, F.; Suruda, A.; Schwob, C. The catalytic properties of charcoal. III. A comparison of the oxidative properties of various charcoals. The effect of charcoal hydrosols on hydrogen peroxide and oxygen. *J. Am. Chem. Soc.* **1938**, *60*, 2483–2486. [CrossRef]
51. Stöhr, B.; Boehm, H.; Schlögl, R. Enhancement of the catalytic activity of activated carbons in oxidation reactions by thermal treatment with ammonia or hydrogen cyanide and observation of a superoxide species as a possible intermediate. *Carbon* **1991**, *29*, 707–720. [CrossRef]
52. Boehm, H. Catalytic properties of nitrogen-containing carbons. *Carbon Mater. Catal.* **2008**, *7*, 219–265. [CrossRef]
53. Fu, W.; Zhang, K.; Chen, M.; Zhang, M.; Shen, Z. One-pot synthesis of N-doped hierarchical porous carbon for high-performance aqueous capacitors in a wide pH range. *J. Power Sources* **2021**, *491*, 229587. [CrossRef]
54. Luo, X.; Yang, Q.; Dong, Y.; Huang, X.; Kong, D.; Wang, B.; Liu, H.; Xiao, Z.; Zhi, L. Maximizing pore and heteroatom utilization within N, P-co-doped polypyrrole-derived carbon nanotubes for high-performance supercapacitors. *J. Mater. Chem. A* **2020**, *8*, 17558–17567. [CrossRef]
55. Xu, Y.; Zhang, C.; Zhou, M.; Fu, Q.; Zhao, C.; Wu, M.; Lei, Y. Highly nitrogen doped carbon nanofibers with superior rate capability and cyclability for potassium ion batterie. *Nat. Commun.* **2018**, *9*, 1720. [CrossRef]
56. Chen, Z.; Zhao, S.; Zhou, Y.; Yu, C.; Zhong, W.; Yang, W. Nacre-like laminate nitrogen-doped porous carbon/carbon nanotubes/graphene composite for excellent comprehensive performance supercapacitors. *Nanoscale* **2018**, *10*, 15229–15237. [CrossRef]
57. Sanchez-Sanchez, A.; Izquierdo, M.; Mathieu, S.; González-Álvarez, J.; Celzard, A.; Fierro, V. Outstanding electrochemical performance of highly N- and O-doped carbons derived from pine tannin. *Green Chem.* **2017**, *19*, 2653–2665. [CrossRef]
58. Li, Z.; Cao, L.; Chen, W.; Huang, Z.; Liu, H. Mesh-like carbon nanosheets with high-level nitrogen doping for high-energy dual-carbon lithium-ion capacitors. *Small* **2019**, *15*, e1805173. [CrossRef] [PubMed]
59. Li, C.; Song, Z.; Liu, M.; Lepre, E.; Antonietti, M.; Zhu, J.; Liu, J.; Fu, Y.; López-Salas, N. Template-induced graphitic nanodomains in nitrogen-doped carbons enable high-performance sodium-ion capacitors. *Energy Environ. Mater.* **2024**, *7*, e12695. [CrossRef]
60. Chen, J.; Zhu, Y.; Luo, D.; Ma, G.; Zhao, Y.; Liu, B.; Yang, B.; Li, Y. Interconnected N-doped carbon network as high-performance anode material for superior lithium ion hybrid capacitor. *Mater. Today Commun.* **2023**, *34*, 105142. [CrossRef]
61. Ma, C.; Shao, X.; Cao, D. Nitrogen-doped graphene nanosheets as anode materials for lithium ion batteries: A first-principles study. *J. Mater. Chem.* **2012**, *22*, 8911–8915. [CrossRef]
62. Lu, Z.; Gao, D.; Yi, D.; Yang, Y.; Wang, X.; Yao, J. sp^2/sp^3 hybridized carbon as an anode with extra Li-ion storage capacity: Construction and origin. *ACS Cent. Sci.* **2020**, *6*, 1451–1459. [CrossRef]
63. Yan, J.; Li, W.; Feng, P.; Wang, R.; Jiang, M.; Han, J.; Cao, S.; Wang, K.; Jiang, K. Enhanced Na^+ pseudocapacitance in a P, S co-doped carbon anode arising from the surface modification by sulfur and phosphorus with C-S-P coupling. *J. Mater. Chem. A* **2020**, *8*, 422–432. [CrossRef]
64. Gopalakrishnan, A.; Raju, D.; Badhulika, S. Green synthesis of nitrogen, sulfur-co-doped worm-like hierarchical porous carbon derived from ginger for outstanding supercapacitor performance. *Carbon* **2020**, *168*, 209–219. [CrossRef]
65. Li, G.; Mao, K.; Liu, M.; Yan, M.; Zhao, J.; Zeng, Y.; Yang, L.; Wu, Q.; Wang, X.; Hu, Z. Achieving ultrahigh volumetric energy storage by compressing nitrogen and sulfur dual-doped carbon nanocages via capillarity. *Adv. Mater.* **2020**, *32*, 2004632. [CrossRef] [PubMed]
66. Zhao, G.; Yu, D.; Zhang, H.; Sun, F.; Li, J.; Zhu, L.; Sun, L.; Yu, M.; Besenbacher, F.; Sun, Y. Sulphur-doped carbon nanosheets derived from biomass as high-performance anode materials for sodium-ion batteries. *Nano Energy* **2020**, *67*, 104219. [CrossRef]
67. Li, D.; Chang, G.; Zong, L.; Xue, P.; Wang, Y.; Xia, Y.; Lai, C.; Yang, D. From double-helix structured seaweed to S-doped carbon aerogel with ultra-high surface area for energy storage. *Energy Storage Mater.* **2018**, *17*, 22–30. [CrossRef]
68. Casal, M.; Díez, N.; Payá, S.; Sevilla, M. Cork-derived carbon sheets for high-performance Na-ion capacitors. *ACS Appl. Energy Mater.* **2023**, *6*, 8120–8131. [CrossRef] [PubMed]
69. Hong, Z.; Zhen, Y.; Ruan, Y.; Kang, M.; Zhou, K.; Zhang, J.; Huang, Z.; Wei, M. Rational design and general synthesis of S-Doped hard carbon with tunable doping sites toward excellent Na-ion storage performance. *Adv. Mater.* **2018**, *30*, 1802035. [CrossRef]
70. Liu, F.; Niu, J.; Chuan, X.; Zhao, Y. Nitrogen and phosphorus co-doped porous carbon: Dopant, synthesis, performance enhancement mechanism and versatile applications. *J. Power Sources* **2024**, *601*, 234308. [CrossRef]
71. Wang, H.; Chen, H.; Chen, C.; Li, M.; Xie, Y.; Zhang, X.; Wu, X.; Zhang, Q.; Lu, C. Tea-derived carbon materials as anode for high-performance sodium ion batteries. *Chin. Chem. Lett.* **2023**, *34*, 107465. [CrossRef]
72. Liu, H.; Wang, M.; Zhai, D.; Chen, X.; Zhang, Z. Design and theoretical study of carbon-based supercapacitors especially exhibiting superior rate capability by the synergistic effect of nitrogen and phosphor dopants. *Carbon* **2019**, *155*, 223–232. [CrossRef]
73. Wang, X.; Hou, M.; Shi, Z.; Liu, X.; Mizota, I.; Lou, H.; Wang, B.; Hou, X. Regulate phosphorus configuration in high P-doped hard carbon as a superanode for sodium storage. *ACS Appl. Mater. Interfaces* **2021**, *13*, 12059–12068. [CrossRef]
74. Guo, J.; Xu, X.; Hill, J.; Wang, L.; Dang, J.; Kang, Y.; Li, Y.; Guan, W.; Yamauchi, Y. Graphene-carbon 2D heterostructures with hierarchically-porous P, N-doped layered architecture for capacitive deionizatio. *Chem. Sci.* **2021**, *12*, 10334–10340. [CrossRef]
75. Yang, Z.; Gao, Y.; Zhao, Z.; Wang, Y.; Wu, Y.; Wang, X. Phytic acid assisted formation of P-doped hard carbon anode with enhanced capacity and rate capability for lithium ion capacitors. *J. Power Sources* **2020**, *474*, 228500. [CrossRef]

76. Qian, Y.; Jiang, S.; Li, Y.; Yi, Z.; Zhou, J.; Li, T.; Han, Y.; Wang, Y.; Tian, J.; Lin, N.; et al. In situ revealing the electroactivity of P-O and P-C bonds in hard carbon for high-capacity and long-life Li/K-ion batteries. *Adv. Energy Mater.* **2019**, *9*, 1901676. [CrossRef]
77. Li, Y.; Yuan, Y.; Bai, Y.; Liu, Y.; Wang, Z.; Li, L.; Wu, F.; Amine, K.; Wu, C.; Lu, J. Insights into the Na^+ storage mechanism of phosphorus-functionalized hard carbon as ultrahigh capacity anodes. *Adv. Energy Mater.* **2018**, *8*, 1702781. [CrossRef]
78. McKee, D. Oxidation protection of carbon materials. In *Chemistry & Physics of Carbon*; CRC Press: Boca Raton, FL, USA, 2021; pp. 173–232. [CrossRef]
79. Yao, J.; Li, F.; Zhou, R.; Guo, C.; Liu, X.; Zhu, Y. Phosphorous-doped carbon nanotube/reduced graphene oxide aerogel cathode enabled by pseudocapacitance for high energy and power zinc-ion hybrid capacitors. *Chin. Chem. Lett.* **2024**, *35*, 108354. [CrossRef]
80. Zhao, Y.; Yang, L.; Chen, S.; Wang, X.; Ma, Y.; Wu, Q.; Jiang, Y.; Qian, W.; Hu, Z. Can boron and nitrogen co-doping improve oxygen reduction reaction activity of carbon nanotubes? *J. Am. Chem. Soc.* **2013**, *135*, 1201–1204. [CrossRef]
81. Kim, Y.; Fujisawa, K.; Muramatsu, H.; Hayashi, T.; Endo, M.; Fujimori, T.; Kaneko, K.; Terrones, M.; Behrends, J.; Eckmann, A.; et al. Raman spectroscopy of boron-doped single-layer graphene. *ACS Nano* **2012**, *6*, 6293–6300. [CrossRef]
82. Zhang, Y.; Zuo, L.; Zhang, L.; Yan, J.; Lu, H.; Fan, W.; Liu, T. Immobilization of NiS nanoparticles on N-doped carbon fiber aerogels as advanced electrode materials for supercapacitors. *Nano Res.* **2016**, *9*, 2747–2759. [CrossRef]
83. Han, J.; Zhang, L.; Lee, S.; Oh, J.; Lee, K.; Potts, J.; Ji, J.; Zhao, X.; Ruoff, R.; Park, S. Generation of B-doped graphene nanoplatelets using a solution process and their supercapacitor application. *ACS Nano* **2013**, *7*, 19–26. [CrossRef]
84. Panchakarla, S.; Subrahmanyam, S.; Saha, K.; Govindaraj, A.; Krishnamurthy, R.; Waghmare, V.; Rao, R. Synthesis, structure, and properties of boron- and nitrogen-doped graphene. *Adv. Mater.* **2009**, *21*, 4726–4730. [CrossRef]
85. Bhattacharjee, U.; Bhowmik, S.; Ghosh, S.; Vangapally, N.; Martha, S. Boron-doped graphene anode coupled with microporous activated carbon cathode for lithium-ion ultracapacitors. *Chem. Eng. J.* **2022**, *430*, 132835. [CrossRef]
86. Wang, X.; Sun, G.; Routh, P.; Kim, D.; Huang, W.; Chen, P. Heteroatom-doped graphene materials: Syntheses, properties and applications. *Chem. Soc. Rev.* **2014**, *43*, 7067–7098. [CrossRef] [PubMed]
87. Naresh, V.; Bhattacharjee, U.; Martha, K. Boron doped graphene nanosheets as negative electrode additive for high-performance lead-acid batteries and ultracapacitors. *J. Alloys Compd.* **2019**, *797*, 595–605. [CrossRef]
88. Miao, X.; Sun, D.; Zhou, X.; Lei, Z. Designed formation of nitrogen and sulfur dual-doped hierarchically porous carbon for long-life lithium and sodium ion batteries. *Chem. Eng. J.* **2019**, *364*, 208–216. [CrossRef]
89. Ruan, J.; Zhao, Y.; Luo, S.; Yuan, T.; Yang, J.; Sun, D.; Zheng, S. Fast and stable potassium-ion storage achieved by in situ molecular self-assembling N/O dual-doped carbon network. *Energy Storage Mater.* **2019**, *23*, 46–54. [CrossRef]
90. Zou, G.; Hou, H.; Foster, C.; Banks, C.; Guo, T.; Jiang, Y.; Zhang, Y.; Ji, X. Advanced hierarchical vesicular carbon co-doped with S, P, N for high-rate sodium storage. *Adv. Sci.* **2018**, *5*, 1800241. [CrossRef]
91. Huang, R.; Cao, Y.; Qin, S.; Ren, Y.; Lan, R.; Zhang, L.; Yu, Z.; Yang, H. Ultra-high N-doped open hollow carbon nano-cage with excellent Na^+ and K^+ storage performances. *Mater. Today Nano* **2022**, *18*, 100217. [CrossRef]
92. Hu, S.; Ge, Z.; Tan, J.; Lai, H.; Feng, T.; Zhang, S.; Xu, Z.; Zhou, H.; Cao, X.; Zhu, G.; et al. Dual-functionally modified N/S doped hierarchical porous carbon and glycerol-engineered polyacrylonitrile carbon nanofibers combine for high-performance lithium-ion capacitors. *J. Power Sources* **2023**, *558*, 232624. [CrossRef]
93. Guan, L.; Pan, L.; Peng, T.; Gao, C.; Zhao, W.; Yang, Z.; Hu, H.; Wu, M. Synthesis of biomass-derived nitrogen-doped porous carbon nanosheests for high-performance supercapacitors. *ACS Sustain. Chem. Eng.* **2019**, *7*, 8405–8412. [CrossRef]
94. Pei, Z.; Li, H.; Huang, Y.; Xue, Q.; Huang, Y.; Zhu, M.; Wang, Z.; Zhi, C. Texturing in situ: N, S-enriched hierarchically porous carbon as a highly active reversible oxygen electrocatalyst. *Energy Environ. Sci.* **2017**, *10*, 742–749. [CrossRef]
95. Ghosh, S.; Barg, S.; Jeong, S.; Ostrikov, K. Heteroatom-doped and oxygen-functionalized nanocarbons for high-performance supercapacitors. *Adv. Energy Mater.* **2020**, *10*, 2001239. [CrossRef]
96. Chen, M.; Le, T.; Zhou, Y.; Kang, F.; Yang, Y. Thiourea-induced N/S dual-doped hierarchical porous carbon nanofibers for high-performance lithium-ion capacitors. *ACS Appl. Energy Mater.* **2020**, *3*, 1653–1664. [CrossRef]
97. Hu, L.; Hou, J.; Ma, Y.; Li, H.; Zhai, T. Multi-heteroatom self-doped porous carbon derived from swim bladders for large capacitance supercapacitors. *J. Mater. Chem. A* **2016**, *4*, 15006–15014. [CrossRef]
98. Wang, J.; Liu, H.; Zhang, X.; Shao, M.; Wei, B. Elaborate construction of N/S-co-doped carbon nanobowls for ultrahigh-power supercapacitors. *J. Mater. Chem. A* **2018**, *6*, 17653–17661. [CrossRef]
99. Xu, F.; Tang, Z.; Huang, S.; Chen, L.; Liang, Y.; Mai, W.; Zhong, H.; Fu, R.; Wu, D. Facile synthesis of ultrahigh-surface-area hollow carbon nanospheres for enhanced adsorption and energy storage. *Nat. Commun.* **2015**, *6*, 7221. [CrossRef] [PubMed]
100. Xu, D.; Chen, C.; Xie, J.; Zhang, B.; Miao, L.; Cai, J.; Huang, Y.; Zhang, L. A hierarchical N/S-codoped carbon anode fabricated facilely from cellulose/polyaniline microspheres for high-performance sodium-ion batterie. *Adv. Energy Mater.* **2016**, *6*, 1501929. [CrossRef]
101. Ye, J.; Zang, J.; Tian, Z.; Zheng, M.; Dong, Q. Sulfur and nitrogen co-doped hollow carbon spheres for sodium-ion batteries with superior cyclic and rate performance. *J. Mater. Chem. A* **2016**, *4*, 13223–13227. [CrossRef]

102. Chen, C.; Li, G.; Zhu, J.; Lu, Y.; Jiang, M.; Hu, Y.; Shen, Z.; Zhang, X. In-situ formation of tin-antimony sulfide in nitrogen-sulfur co-doped carbon nanofibers as high performance anode materials for sodium-ion batteries. *Carbon* **2017**, *120*, 380–391. [CrossRef]
103. Sheng, J.; Yang, L.; Zhu, Y.; Li, F.; Zhang, Y.; Zhou, Z. Oriented SnS nanoflakes bound on S-doped N-rich carbon nanosheets with a rapid pseudocapacitive response as high-rate anodes for sodium-ion batteries. *J. Mater. Chem. A* **2017**, *5*, 19745–19751. [CrossRef]
104. Zou, L.; Lai, Y.; Hu, H.; Wang, M.; Zhang, K.; Zhang, P.; Fang, J.; Li, J. N/S Co-doped 3D porous carbon nanosheet networks enhancing anode performance of sodium-ion batteries. *Chem. A Eur. J.* **2017**, *23*, 14261–14266. [CrossRef]
105. Ruan, J.; Yuan, T.; Pang, Y.; Luo, S.; Peng, C.; Yang, J.; Zheng, S. Nitrogen and sulfur dual-doped carbon films as flexible free-standing anodes for Li-ion and Na-ion batteries. *Carbon* **2018**, *126*, 9–16. [CrossRef]
106. Yang, J.; Zhou, X.; Wu, D.; Zhao, X.; Zhou, Z. S-doped N-rich carbon nanosheets with expanded interlayer distance as anode materials for sodium-ion batteries. *Adv. Mater.* **2017**, *29*, 1604108. [CrossRef]
107. Xu, Y.; Jiang, J.; Li, Z.; Yang, Z.; Zhang, Y.; An, Y.; Zhu, Q.; Dou, H.; Zhang, X. Aerosol-assisted preparation of N-doped hierarchical porous carbon spheres cathodes toward high-stable lithium-ion capacitors. *J. Mater. Sci.* **2020**, *55*, 13127–13140. [CrossRef]
108. Sun, L.; Zhou, H.; Yao, Y.; Qu, H.; Zhang, C.; Liu, S.; Zhou, Y. Double soft-template synthesis of nitrogen/sulfur-codoped hierarchically porous carbon materials derived from protic ionic liquid for supercapacitor. *ACS Appl. Mater. Interfaces* **2017**, *9*, 26088–26095. [CrossRef] [PubMed]
109. Deng, X.; Zou, K.; Momen, R.; Cai, P.; Chen, J.; Hou, H.; Zou, G.; Ji, X. High content anion (S/Se/P) doping assisted by defect engineering with fast charge transfer kinetics for high-performance sodium ion capacitors. *Sci. Bull.* **2021**, *66*, 1858–1868. [CrossRef] [PubMed]
110. Shi, L.; Zeng, Q.; Liu, J.; Yang, Y. S, N co-doped porous carbon derived from metal-organic frameworks as cathode for lithium-ion capacitor. *J. Electroanal. Chem.* **2023**, *944*, 117675. [CrossRef]
111. Bhagwan, J.; Nagaraju, G.; Ramulu, B.; Sekhar, S.; Yu, J. Rapid synthesis of hexagonal $NiCo_2O_4$ nanostructures for high-performance asymmetric supercapacitors. *Electrochim. Acta* **2019**, *299*, 509–517. [CrossRef]
112. Gu, J.; Sun, L.; Zhang, Y.; Zhang, Q.; Li, X.; Si, H.; Shi, Y.; Sun, C.; Gong, Y.; Zhang, Y. MOF-derived Ni-doped CoP@C grown on CNTs for high-performance supercapacitors. *Chem. Eng. J.* **2020**, *385*, 123454. [CrossRef]
113. Zou, K.; Deng, W.; Silvester, D.S.; Zou, G.; Hou, H.; Banks, C.E.; Li, L.; Hu, J.; Ji, X. Carbonyl chemistry for advanced electrochemical energy storage systems. *ACS Nano* **2024**, *18*, 19950–20000. [CrossRef]
114. Hu, Y.; Tang, C.; Li, H.; Du, A.; Luo, W.; Wu, M.; Zhang, H. B-incorporated, N-doped hierarchically porous carbon nanosheets as anodes for boosted potassium storage capability. *Chin. Chem. Lett.* **2022**, *33*, 480–485. [CrossRef]
115. Choi, C.H.; Park, S.H.; Woo, S.I. Binary and ternary doping of nitrogen, boron, and phosphorus into carbon for enhancing electrochemical oxygen reduction activity. *ACS Nano* **2012**, *6*, 7084–7091. [CrossRef]
116. Wang, S.; Iyyamperumal, E.; Roy, A.; Xue, Y.; Yu, D.; Dai, L. Vertically aligned BCN nanotubes as efficient metal-free electrocatalysts for the oxygen reduction reaction: A synergetic effect by co-doping with boron and nitrogen. *Angew. Chem. Int. Ed.* **2011**, *50*, 11756–11760. [CrossRef]
117. Zheng, Y.; Jiao, Y.; Ge, L.; Jaroniec, M.; Qiao, S. Two-step boron and nitrogen doping in graphene for enhanced synergistic catalysis. *Angew. Chem. Int. Ed.* **2013**, *52*, 3110–3116. [CrossRef] [PubMed]
118. Wu, Z.; Winter, A.; Chen, L.; Sun, Y.; Turchanin, A.; Feng, X.; Müllen, K. Three-dimensional nitrogen and boron co-doped graphene for high-performance all-solid-state supercapacitors. *Adv. Mater.* **2012**, *24*, 5130–5135. [CrossRef]
119. Zhang, H.; Hu, M.; Huang, Z.; Kang, F.; Lv, R. Sodium-ion capacitors with superior energy-power performance by using carbon-based materials in both electrodes. *Prog. Nat. Sci. Mater. Int.* **2020**, *30*, 13–19. [CrossRef]
120. Putri, K.; Ong, W.; Chang, S.; Chai, S. Heteroatom doped graphene in photocatalysis: A review. *Appl. Surf. Sci.* **2015**, *358*, 2–14. [CrossRef]
121. Subramanyan, K.; Divya, M.; Aravindan, V. Dual-carbon Na-ion capacitors: Progress and future prospects. *J. Mater. Chem. A* **2021**, *9*, 9431–9450. [CrossRef]
122. Li, S.; Liu, P.; Zheng, X.; Wu, M. High-performance dual carbon lithium-ion capacitors based on nitrogen-doped 2D carbon nanosheets as both anode and cathode. *Electrochim. Acta* **2022**, *428*, 140921. [CrossRef]
123. Jiang, J.; Shen, Q.; Chen, Z.; Wang, S. Nitrogen-doped porous carbon derived from coal for high-performance dual-carbon lithium-ion capacitors. *Nanomaterials* **2023**, *13*, 2525. [CrossRef] [PubMed]
124. Jiang, J.; Zhang, Y.; An, Y.; Wu, L.; Zhu, Q.; Dou, H.; Zhang, X. Engineering ultrathin MoS_2 nanosheets anchored on N-doped carbon microspheres with pseudocapacitive properties for high-performance lithium-ion capacitors. *Small Methods* **2019**, *3*, 1900081. [CrossRef]
125. Xiao, Z.; Yu, Z.; Gao, Z.; Li, B.; Zhang, M.; Xu, C. S-doped graphene nano-capsules toward excellent low-temperature performance in Li-ion capacitors. *J. Power Sources* **2022**, *535*, 231404. [CrossRef]
126. Thangavel, R.; Kannan, G.; Ponraj, R.; Yoon, G.; Aravindan, V.; Kim, D.; Kang, K.; Yoon, W.; Lee, Y. Surface enriched graphene hollow spheres towards building ultra-high power sodium-ion capacitor with long durability. *Energy Storage Mater.* **2019**, *25*, 702–713. [CrossRef]

127. Zou, G.; Hou, H.; Zhao, G.; Huang, Z.; Ge, P.; Ji, X. Preparation of S/N-codoped carbon nanosheets with tunable interlayer distance for high-rate sodium-ion batteries. *Green Chem.* **2017**, *19*, 4622–4632. [CrossRef]
128. Dong, S.; Xu, Y.; Wu, L.; Dou, H.; Zhang, X. Surface-functionalized graphene-based quasi-solid-state Na-ion hybrid capacitors with excellent performance. *Energy Storage Mater.* **2018**, *11*, 8–15. [CrossRef]
129. Du, Q.; Zhao, Y.; Chen, Y.; Liu, J.; Li, H.; Bai, G.; Zhou, K.; Wang, J. Nitrogen-doped porous carbon nanosheets as both anode and cathode for advanced potassium-ion hybrid capacitors. *Green Energy Environ.* **2023**, *8*, 579–588. [CrossRef]
130. Zhang, W.; Yin, J.; Sun, M.; Wang, W.; Chen, C.; Altunkaya, M.; Emwas, A.; Han, Y.; Schwingenschlögl, U.; Alshareef, H. Direct pyrolysis of supermolecules: An ultrahigh edge-nitrogen doping strategy of carbon anodes for potassium-ion batteries. *Adv. Mater.* **2020**, *32*, 2000732. [CrossRef] [PubMed]
131. Pham, H.; Mahale, K.; Hoang, T.; Mundree, S.; Gomez-Romero, P.; Dubal, D. Dual carbon potassium-ion capacitors: Biomass-derived graphene-like carbon nanosheet cathodes. *ACS Appl. Mater. Interfaces* **2020**, *12*, 48518–48525. [CrossRef]
132. Hu, X.; Liu, Y.; Chen, J.; Yi, L.; Zhan, H.; Wen, Z. Fast redox kinetics in Bi-heteroatom doped 3D porous carbon nanosheets for high-performance hybrid potassium-ion battery capacitors. *Adv. Energy Mater.* **2019**, *9*, 1901533. [CrossRef]
133. Chang, X.; Zhou, X.; Ou, X.; Lee, C.; Zhou, J.; Tang, Y. Ultrahigh nitrogen doping of carbon nanosheets for high capacity and long cycling potassium ion storage. *Adv. Energy Mater.* **2019**, *9*, 1902672. [CrossRef]
134. Tai, Z.; Liu, Y.; Zhang, Q.; Zhou, T.; Guo, Z.; Liu, H.K.; Dou, S.X. Ultra-light and flexible pencil-trace anode for high performance potassium ion and lithium-ion batteries. *Green Energy Environ.* **2017**, *2*, 278–284. [CrossRef]
135. Tan, L.; Huang, X.; Yin, T.; Guo, Y.; Ning, T.; Mei, Y.; Zou, K.; Li, L.; Ji, X.; Zou, G. A 5 V ultrahigh energy density lithium metal capacitor enabled by the fluorinated electrolyte. *Energy Storage Mater.* **2024**, *71*, 103692. [CrossRef]
136. Cao, J.; Xu, H.; Zhong, J.; Li, X.; Li, S.; Wang, Y.; Zhang, M.; Deng, H.; Wang, Y.; Cui, C.; et al. Dual-carbon electrode-based high-energy-density potassium-ion hybrid capacitor. *ACS Appl. Mater. Interfaces* **2021**, *13*, 8497–8506. [CrossRef] [PubMed]
137. Cui, Y.; Liu, W.; Wang, X.; Li, J.; Zhang, Y.; Du, Y.; Liu, S.; Wang, H.; Feng, W.; Chen, M. Bioinspired mineralization under freezing conditions: An approach to fabricate porous carbons with complicated architecture and superior K^+ storage performance. *ACS Nano* **2019**, *13*, 11582–11592. [CrossRef] [PubMed]
138. Gao, Y.; Yang, Z.; Wang, Y.; Wang, X. Boosting capacitive storage of cathode for lithium-ion capacitors: Combining pore structure with P-doping. *Electrochim. Acta* **2021**, *368*, 137646. [CrossRef]
139. Huynh, N.; Tran, D.; Nguyen, A.; Nguyen, T.; Tran, M.; Grag, A.; Le, P. Carbon-coated $LiFePO_4$-carbon nanotube electrodes for high-rate Li-ion battery. *J. Solid State Electrochem.* **2018**, *22*, 2247–2254. [CrossRef]
140. Shah, A.; Zahid, A.; Subhan, H.; Munir, A.; Iftikhar, F.; Akbar, M. Heteroatom-doped carbonaceous electrode materials for high performance energy storage devices. *Sustain. Energy Fuels* **2018**, *2*, 1398–1429. [CrossRef]
141. Qi, C.; Ma, X.; Ning, G.; Song, X.; Chen, B.; Lan, X.; Li, Y.; Zhang, X.; Gao, J. Aqueous slurry of S-doped carbon nanotubes as conductive additive for lithium ion batteries. *Carbon* **2015**, *92*, 245–253. [CrossRef]
142. Zhao, L.; Sun, D.; Cao, Q.; Xiao, Z.; Yu, Z.; Qi, C.; Li, X.; Ning, G.; Ma, X.; Peng, C.; et al. Green and universal sulfur doping technique coupled with construction of conductive network for enhanced kinetics of Li-ion capacitors. *Chem. Eng. Sci.* **2022**, *258*, 117749. [CrossRef]
143. Yang, C.; Xiong, J.; Ou, X.; Wu, C.; Xiong, X.; Wang, J.; Huang, K.; Liu, M. A renewable natural cotton derived and nitrogen/sulfur co-doped carbon as a high-performance sodium ion battery anode. *Mater. Today Energy* **2018**, *8*, 37–44. [CrossRef]
144. Zou, K.; Cai, P.; Liu, C.; Li, J.; Gao, X.; Xu, L.; Zou, G.; Hou, H.; Liu, Z.; Ji, X. A kinetically well-matched full-carbon sodium-ion capacitor. *J. Mater. Chem. A* **2019**, *7*, 13540–13549. [CrossRef]
145. Que, L.; Yu, F.; He, K.; Wang, Z.; Gu, D. Robust and conductive $Na_2Ti_2O_{5-x}$ nanowire arrays for high-performance flexible sodium-ion capacitor. *Chem. Mater.* **2017**, *29*, 9133–9141. [CrossRef]
146. Wang, C.; Zhao, N.; Li, B.; Yu, Q.; Shen, W.; Kang, F.; Lv, R.; Huang, Z. Pseudocapacitive porous hard carbon anode with controllable pyridinic nitrogen and thiophene sulfur co-doping for high-power dual-carbon sodium ion hybrid capacitors. *J. Mater. Chem. A* **2021**, *9*, 20483–20492. [CrossRef]
147. Liu, Z.; Peng, H.; Xie, X.; Wang, X.; Hou, W.; Miao, W.; Tao, B.; Ma, G.; Lei, Z. Coupling of reasonable micro-defect structure and multiple chemisorption sites for boosting the K^+ storage capacity in dual-carbon potassium ion hybrid capacitors. *Energy Storage Mater.* **2024**, *65*, 103100. [CrossRef]
148. Peng, Z.; Bannov, G.; Li, S.; Huang, Y.; Tang, L.; Tan, L.; Chen, Y. Coupling uniform pore size and multi-chemisorption sites: Hierarchically ordered porous carbon for ultra-fast and large zinc ion storage. *Adv. Funct. Mater.* **2023**, *33*, 2303205. [CrossRef]

batteries

MDPI

Article

Construction of $CoNi_2S_4@Ni(OH)_2$ Nanosheet Structures for Asymmetric Supercapacitors with Excellent Performance

Yongli Tong [1,*], Baoqian Chi [1], Yu Jiang [1] and Xiang Wu [2,*]

[1] School of Science, Shenyang Ligong University, Shenyang 110159, China; baoqianchi@sylu.edu.cn (B.C.); jiangyu@sylu.edu.cn (Y.J.)
[2] School of Materials Science and Engineering, Shenyang University of Technology, Shenyang 110870, China
* Correspondence: tongyongli@sylu.edu.cn (Y.T.); wuxiang05@sut.edu.cn (X.W.)

Abstract: It is crucial for energy storage devices to construct electrode materials with excellent performance. However, enhancing energy density and cycling stability for supercapacitors is a significant challenge. We successfully synthesized $CoNi_2S_4@Ni(OH)_2$ nanosheets on the surface of Ni foam substrate by a two-step hydrothermal approach. The obtained products exhibit a remarkable areal capacitance of 1534 F g^{-1} at a current density of 1 A g^{-1}. Moreover, even after 10,000 cycles, the specific capacitance remains 90% of its initial value, highlighting the exceptional long-term stability and durability. Furthermore, an asymmetric supercapacitor (ASC) device incorporating the $CoNi_2S_4@Ni(OH)_2$ material shows remarkable electrochemical performance. It delivers an energy density of 58.5 mW h g^{-1} at a power density of 2700 W kg^{-1}. The outstanding performance mainly arises from the selection of materials, the design of the structure, and the synergistic interaction between the materials. The result suggests that this material holds great potential as an energy storage material.

Keywords: supercapacitors; $CoNi_2S_4@Ni(OH)_2$; energy density; composite structure; specific capacity

1. Introduction

The world is currently facing significant environmental and energy challenges [1–3]. These issues pose a serious threat to both human life and wildlife. To mitigate these problems, the development of green renewable energy sources has become essential [4–6]. Investing in green energy can help to reduce carbon footprints, create jobs, and promote energy independence, while preserving natural resources for future generations. Supercapacitors have garnered considerable interest due to their outstanding advantages over traditional batteries. Unlike conventional batteries, supercapacitors offer extremely fast charge/discharge rates, which make them an excellent choice for applications that demand rapid bursts of energy. Additionally, supercapacitors have an extended cycle life, which can endure hundreds of thousands to millions of cycles without significant degradation. Supercapacitors feature high power density, enabling them to deliver large amounts of power in a short time [7–10]. Moreover, supercapacitors are more environmentally friendly and have fewer safety concerns. Based on the charge storage mechanism, asymmetric supercapacitors are classified into two types. One is a double-layer capacitor, which stores energy by electrostatically adsorbing charges between the electrolyte and the interface of the electrode. During this process, ions gather on the surface of electrode, creating a double electric layer that facilitates energy storage. The other is a pseudocapacitor, where energy is stored through fast surface redox reactions, involving electron transfer between the electrolyte and the electrode material. The material type is essential in determining the charge storage capacity. The advancement of these supercapacitors is crucial for meeting the increasing need for energy storage solutions. They are not only environmentally sustainable, but also efficient [11–14].

Transition metal sulfides, like Co_3S_4 and $CoNi_2S_4$, have already been extensively studied as electrode materials, due to their outstanding electrochemical properties [15–17]. Among these, $CoNi_2S_4$ has been extensively researched. This material exhibits a high specific capacitance, primarily due to its extensive surface area, which allows for greater charge storage. Additionally, it can undergo reversible redox reactions when the charge–discharge reaction is repeated. This contributes to its high capacitance. These reactions, which involve the transfer of electrons between the electrolyte and the electrode, provide an additional mechanism for charge storage beyond just physical adsorption. Furthermore, the material's relatively high electrical conductivity is essential for boosting its performance. This enhanced conductivity facilitates the efficient movement of charge. Good conductivity improves the flow of charge, leading to a better rate capability. This means that the material can charge and discharge more quickly without significant loss of efficiency, which makes it ideal for high-power applications [18]. Li et al. [19] report the creation of r-$CoNi_2S_4$ nanosheets in transition metal dichalcogenides though a moderate-reduction method. The material demonstrated a remarkable improvement in conductivity and an increased number of active sites, which were mainly attributed to the abundant sulfur vacancies that were deliberately introduced. These vacancies are key to improving the electrochemical properties of the material, leading to better charge storage and faster electron transfer. The r-$CoNi_2S_4$ nanosheets, in particular, demonstrated a considerably higher specific capacitance, reaching 1117 C g^{-1} at 2 A g^{-1}. In addition, the device utilizing these nanosheets achieved an outstanding energy density of 55.4 Wh kg^{-1} at a power density of 8 kW kg^{-1}, along with 80% capacitance retention after 10,000 cycles. However, like many transition metal sulfides, $CoNi_2S_4$ experiences volume expansion during cycling, which can lead to mechanical stress and degradation of the material, reducing its long-term performance.

To improve this drawback, a composite structure has been designed. When transition metal sulfides are combined with other components, the composite structure can provide better mechanical stability, reducing the effects of volume changes and improving cycling performance. Additionally, by combining these materials with conductive materials, the resulting composite improves the overall electrical conductivity, facilitating better electron transport and faster charge/discharge rates [20,21]. The combination of different materials can lead to a synergistic effect, where the properties of the individual components complement each other [22–24]. Thus, composites can be specifically engineered to improve multiple aspects of supercapacitor performance, making them a highly versatile and effective solution. In this work, a hybrid $CoNi_2S_4$@$Ni(OH)_2$ electrode is fabricated on the surface of nickel foam through a two-step hydrothermal approach. The composite material exhibits outstanding electrochemical performance. The assembled device also possesses excellent power density and energy density.

2. Experimental Section

2.1. Preparation of Materials

Cobalt nitrate, nickel nitrate, ammonium fluoride, urea, and nickel foam were sourced from Aladdin. All chemical reagents were used without any modifications. Before the experiment, a piece of Ni foam (4 × 4 × 0.3 cm) was immersed in a 1 M hydrochloric acid solution for half an hour to remove surface impurities, then rinsed three times with ethanol and deionized (DI) water, respectively. Finally, it was placed in an oven to dry at 60 °C overnight.

Total volumes of 60 mL DI water, 3 mmol NH_4F, 2 mmol $Co(NO_3)_2{\cdot}6H_2O$, 7 mmol urea, and 1 mmol $Ni(NO_3)_2{\cdot}6H_2O$ were placed in a beaker, followed by magnetic stirring. When the solution became clear at room temperature, stirring was stopped. The prepared mixture was subsequently introduced into an autoclave of 100 mL, to which prepared Ni foam was added. The autoclave was heated at 120 °C for 6 h in an oven, to facilitate the reaction. After the heating process was complete, the product was left to cool to room temperature. The product was then washed thoroughly with ethanol and DI water multiple times to eliminate any remaining impurities, resulting in the formation of the cobalt–nickel precursor.

After that, 0.40 g Na_2S and the as-prepared sample above, along with 40 mL DI water, were transferred to an autoclave, where they were heated at 120 °C for 4 h. The sample was dried according to the steps mentioned above, to obtain the $CoNi_2S_4$ product. Finally, 0.12 g of urea, 0.29 g of $Ni(NO_3)_2{\cdot}6H_2O$, and 60 mL of DI water were mixed thoroughly to obtain clear solution. The prepared $CoNi_2S_4$ sample was then immersed in the mixture and heated at 100 °C for 8 h. $CoNi_2S_4$@$Ni(OH)_2$ samples were successfully obtained. The average mass loading on the nickel foam was approximately 2.32 mg cm^{-2}.

2.2. Morphology and Structure Characterization

The structural features of the products were analyzed through scanning electron microscopy (SEM, Sigma500 field emission, ZEISS, Oberkochen, Germany). X-ray diffraction (XRD) was performed with a BRUKER D8 X-ray diffractometer(BRUKER, Karlsruhe, Germany), utilizing Cu Kα radiation (λ = 1.5406 Å). The elemental composition and structural details were examined through X-ray photoelectron spectroscopy (XPS, ESCALAB250, Thermo Fisher Scientific, Waltham, MA, USA).

2.3. Fabrication of ASC Device

A mixture of activated carbon, acetylene black, and polytetrafluoroethylene (PTFE), in a mass ratio of 7:2:1, was applied to the surface of the treated nickel foam with a diameter of 1 cm, followed by heating at 80 °C for 24 h. The carbon electrode obtained as described above was used as the cathode of the ASC, and the $CoNi_2S_4$@$Ni(OH)_2$ product served as the cathode. The electrolyte was bended by first heating 50 mL of deionized (DI) water to 95 °C. Then, 5 mg of PVA and 5 g of KOH were added gradually. The solution was stirred until it became clear. A piece of NKK paper was used to separate the anode and cathode to form a sandwich structure. The mass ratio of the anode and cathode can be determined using the following formula, based on charge balance [25]:

$$m^+/m^- = C_- \times V_-/C_+ \times V_+ \quad (1)$$

2.4. Electrochemical Characterization

A range of electrochemical tests were carried out on an electrochemical workstation (Shanghai Chenhua CHI660e, Shanghai Chenhua Instrument Co., Ltd., Shanghai, China) to characterize the material's performance. They included Galvanostatic Charge–Discharge (GCD) tests, Electrochemical Impedance Spectroscopy (EIS) and Cyclic Voltammetry (CV). Specific capacitance (C_s) was obtained using the following formula [26,27]:

$$C_s = I\Delta t/mV \quad (2)$$

The energy density (E) and power density (P) of the ASC were constructed by the following formula:

$$E = 1/2C_s(\Delta V)^2 \quad (3)$$

$$P = 3600E/t \quad (4)$$

in which I serves as current density, and m stands for the mass of the electrode.

3. Results and Discussion

The phase component of the $CoNi_2S_4@Ni(OH)_2$ sample was analyzed by XRD; the results are shown in Figure 1a. The diffraction peaks at 44.5°, 51.8°, and 76.4° correspond to those characteristic of nickel foam (JCPDS no. 04-0850). These peaks indicate that the structure of the material is consistent with the well-established crystallographic patterns of nickel foam. The peaks at 38.2 °, 47.2 °, 50.3 °, 55.1 °, and 68.9 ° correspond well to the (400), (422), (511), (440), and (444) planes of $CoNi_2S_4$ (JCPDS no. 86–2247). The peaks observed at 33.2°, 38.6°, 52.1°, 59.3°, and 73.0° are attributed to the specific crystallographic planes of $Ni(OH)_2$, namely the (100), (011), (012), (110), and (201) planes, respectively. These peaks serve as clear indicators of the presence of $Ni(OH)_2$ within the material, confirming its formation and crystallinity. Each of these diffraction angles corresponds to a distinct plane in the crystal structure of $Ni(OH)_2$, providing valuable information about its phase and orientation.

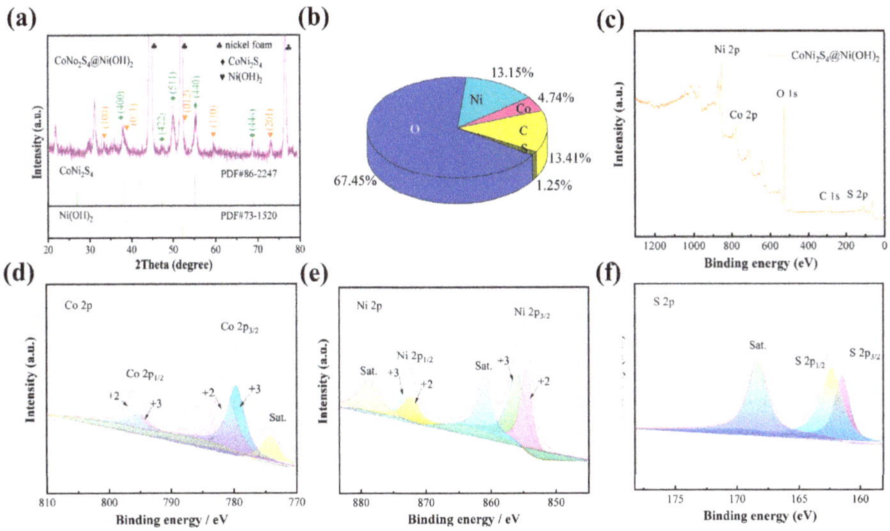

Figure 1. (**a**) XRD patterns of $CoNi_2S_4@Ni(OH)_2$. (**b**) Element content pie chart. (**c**) XPS survey. (**d**) Co 2p. (**e**) Ni 2p. (**f**) S 2p.

The surface chemical states of the $CoNi_2S_4@Ni(OH)_2$ sample were determined by XPS. Figure 1b shows the proportions of the five elements, as follows: O—67.45%, Ni—13.15%, Co—4.74%, C—13.41%, and S—1.25%. The survey spectrum in Figure 1c indeed confirms the presence of these elements. As shown in Figure 1d, the spectra exhibit two characteristic spin–orbit doublet peaks, which correspond to Co $2p_{3/2}$ and Co $2p_{1/2}$, with binding energies of 796.1 eV and 780.2 eV, respectively. These peaks suggest the presence of cobalt in its characteristic oxidation states, with the measured binding energies closely matching expectations for Co 2p electron states in cobalt-containing compounds. The spin–orbit splitting value is approximately 15.9 eV, which is attributed to the presence of Co^{3+} and Co^{2+} ionic states [28,29]. The energies at 778.2 and 792.6 eV are attributed to Co^{3+}, while the energies at 796.4 and 780.9 eV are associated with Co^{2+}.

The Ni 2p spectra (Figure 1e) exhibit four peaks, at 855.4, 873.4, 857.1, and 875.6 eV, corresponding to Ni^{2+} and Ni^{3+} [30,31]. Additionally, two satellite peaks at 862.1 and 880.4 eV can be attributed to the higher oxidation state. Among the four peaks shown on the Ni spectra, the two energies of 872.8 eV and 854.9 eV are the orbital peaks, while the energies at 862.1 and 880.4 eV are satellite peaks. The peaks at 855.4 and 873.4 eV are attributed to Ni^{2+}, while the energies at 857.3 and 875.7 eV correspond to Ni^{3+}. The S 2p spectra display two orbital peaks, at 156.6 eV and 157.2 eV, corresponding to S $2p_{3/2}$ and S $2p_{1/2}$, respectively.

The morphology of the prepared samples is illustrated in Figure 2. It can be seen that the nanosheets evenly cover the whole surface of the Ni foam (Figure 2a). From the high-magnification SEM image in Figure 2b, it can be seen that the thickness of the $CoNi_2S_4$ product is approximately 20 nm, and nanowires grow uniformly at the edges of the nanosheets. This means that the sample has a larger specific surface area compared to pure sheets. Figure 2c,d show the uniform morphology of $Ni(OH)_2$ directly grown on Ni foam. The average thickness of its nanosheets is around 10 nm. Figure 2e,f show how the $Ni(OH)_2$ nanosheets are uniformly attached to the surface of the $CoNi_2S_4$ material. The $CoNi_2S_4$@ $Ni(OH)_2$ nanosheets reinforce the electrode structure and exhibit a large specific surface area, suggesting excellent electrochemical performance.

The electrochemical data of the prepared products were evaluated though a traditional three-electrode system, with a 3 M potassium hydroxide solution serving as the electrolyte. Figure 3a shows the CV curves of the three samples, $Ni(OH)_2$, $CoNi_2S_4$, and $CoNi_2S_4$@$Ni(OH)_2$, obtained at 50 mV s^{-1}. The clearly defined redox peaks suggest that the materials exhibit battery-type behavior. The $CoNi_2S_4$@$Ni(OH)_2$ samples show the largest area among the three materials, which suggests a significant improvement in capacitance following the formation of the composite. This improvement can be attributed to the shortening of the electron and ion transport pathways, as well as the increase in specific surface area following the formation of the composite material. The large surface area provides abundant active sites, allowing for more efficient electrochemical reactions during charge and discharge cycles. With more available sites, the material can accommodate a greater number of ions, enhancing its overall capacitance. Additionally, the shorter transport pathways facilitate faster electron and ion movement, which allow for more efficient electron and ion movement. This combination enhances the material's overall electrochemical performance, leading to better charge storage capacity, faster charge–discharge rates, and improved energy efficiency.

Figure 3b exhibits the GCD curves of the $CoNi_2S_4$@$Ni(OH)_2$, $CoNi_2S_4$, and $Ni(OH)_2$ materials at 1 A g^{-1}. The curves show distinct plateaus, which correspond to the charge–discharge redox processes of the electrode, consistently with the CV curves. The $CoNi_2S_4$@$Ni(OH)_2$ material possesses the longest discharge time, suggesting that it has the highest capacitance. Therefore, it shows significant improvement compared to the $CoNi_2S_4$ and $Ni(OH)_2$ materials. This suggests that the composite material is more efficient at storing and releasing charge.

Nyquist plots of three electrodes were created to further assess the kinetic performance, as shown in Figure 3c. The linear segment indicates the rate at which ions are transported between the electrolyte and the electrode surface. A greater slope of the line signifies a more rapid ion transport rate. The figure shows that the $CoNi_2S_4$@$Ni(OH)_2$ material possesses a faster rate compared to the other two materials. The equivalent resistance of the $CoNi_2S_4$@$Ni(OH)_2$ sample at the intersection with the x-axis is approximately 0.65 Ω, which is smaller than that of the other two materials.

CV curves were created to further assess the performance of the $CoNi_2S_4$@$Ni(OH)_2$ sample at various scan rates, as shown in Figure 3d. As the scan rate increases, the general shape of the curve stays largely unchanged, suggesting that the material maintains a stable electrochemical behavior and can function effectively across a range of operating conditions. Moreover, with an increase in scan rate, the redox peaks display a slight shift in both directions. This shift is often attributed to limitations in ion diffusion at higher scan rates, where the material may experience slight delays in the movement of ions to the active sites.

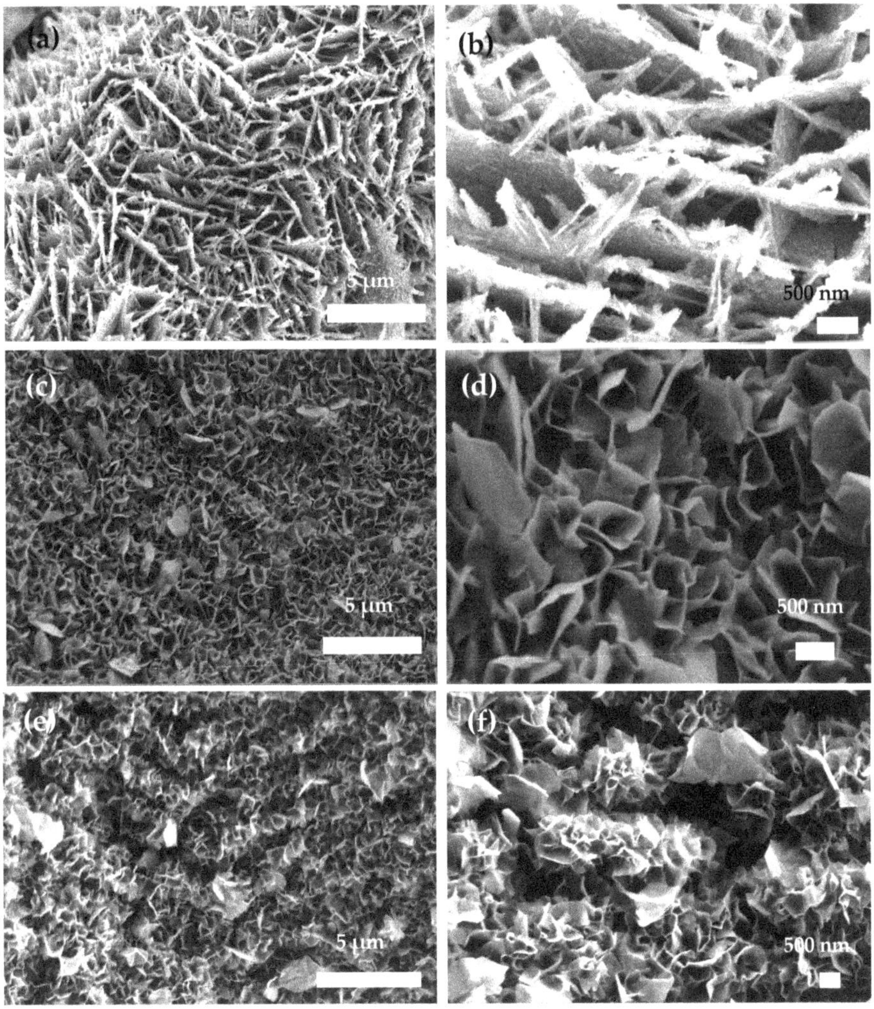

Figure 2. SEM images of samples: (**a**,**b**) $CoNi_2S_4$ material; (**c**,**d**) $Ni(OH)_2$ sample; (**e**,**f**) $CoNi_2S_4@Ni(OH)_2$ sample.

From these CV curves, the charge–discharge mechanism is further revealed in Figure 3e. CV curves serve as a powerful method for analyzing the reaction kinetics of the as-prepared samples. The equation for this is presented below:

$$i = av^b \tag{5}$$

In this equation, i serves as the current, v stands for scan rate, and b indicates the reaction kinetics speed.

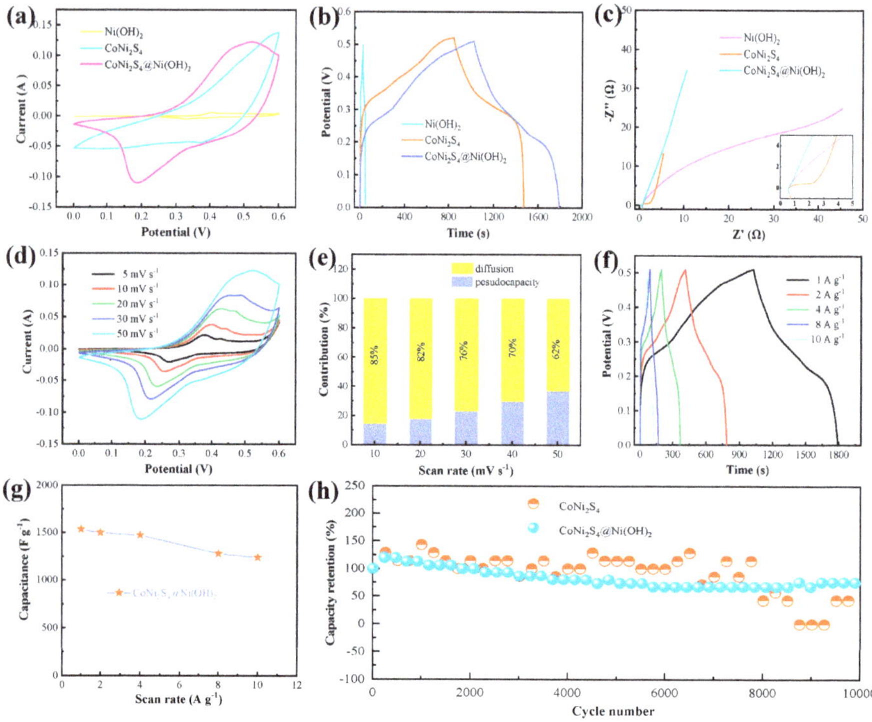

Figure 3. Electrochemical performance: (**a**) CV curves of prepared samples at scan rate of 50 mV/s. (**b**) GCD curves of prepared samples at 1 A g^{-1}. (**c**) Nyquist plots of prepared samples. (**d**) CV curves at different scan rates. (**e**) Contribution ratio. (**f**) GCD curves at different current densities. (**g**) Capacitance at different scan rates. (**h**) Cycling performance.

To explore the proportional contribution of the two charge storage mechanisms, surface capacitance and diffusion capacitance, the following formula is used:

$$i = k_1 v + k_2 v^{1/2} \tag{6}$$

where the values of k_1 and k_2 can be determined from the CV curves. The charge storage caused by diffusion reaches 85% at 10 mV s^{-1}, and decreases continuously when the scan rate increases. The charge storage caused by pseudocapacitance increases continuously with an increase in scan rate.

The charging and discharging portions of the GCD curves are basically symmetrical, indicating that the material has high recoverability. According to Equation (2), the specific capacitance is calculated to be 1534 F g^{-1}, based on the discharge time at 1 A g^{-1} (Figure 3f). The material maintains a capacitance retention rate of up to 80%, even at a high current density of 10 A g^{-1}, which indicates that the phenomenon of polarization of the material is not significant. As observed from Figure 3g, the curve drops gently. This indicates that the specific capacitance gradually decreases with an increase in current density. This demonstrates the material's excellent ability to withstand rapid charge and discharge cycles.

The stability and durability of the $CoNi_2S_4@Ni(OH)_2$ electrode material were explored through long-cycle testing, as shown in Figure 3h. As the number of cycles increases, the curve shows an upward trend, reaching a maximum value at approximately 200 cycles. This indicates the activation process of the material. The electrode material continuously absorbs the electrolyte, leading to an increase in active sites, more particles participating in the reaction, and a rise in specific capacitance. Subsequently, the curve shows a downward trend, which is due to the repeated adsorption/desorption of the material, causing volume expansion or collapse, and even convergence in some cases. This leads to a gradual reduction in reaction sites. After 10,000 cycles, the specific capacitance retains 90% of its original value, demonstrating impressive cycling stability. This excellent retention indicates that the material undergoes minimal degradation during the repeated redox process, which is a key factor for long-term performance. After the composite formation, it is evident that the material's stability is significantly improved. This is because sulfides have poor stability. By coating them with a layer of $Ni(OH)_2$ material, this can alleviate issues, such as volume expansion and material collapse, caused by the repeated reversible redox process.

To explore the applicability of the $CoNi_2S_4@Ni(OH)_2$ electrode material, a device was assembled. As shown in Figure 4a, the shape of the curves remains largely unchanged when the operating voltage varies from 1.2 to 1.6 V. This indicates that the device functions normally. Meanwhile, the shape of the curve remains largely unchanged when the scan rate varies between 10 and 100 mV s^{-1} (Figure 4b). This suggests that the material possesses a low polarization rate and excellent electrochemical performance [32,33].

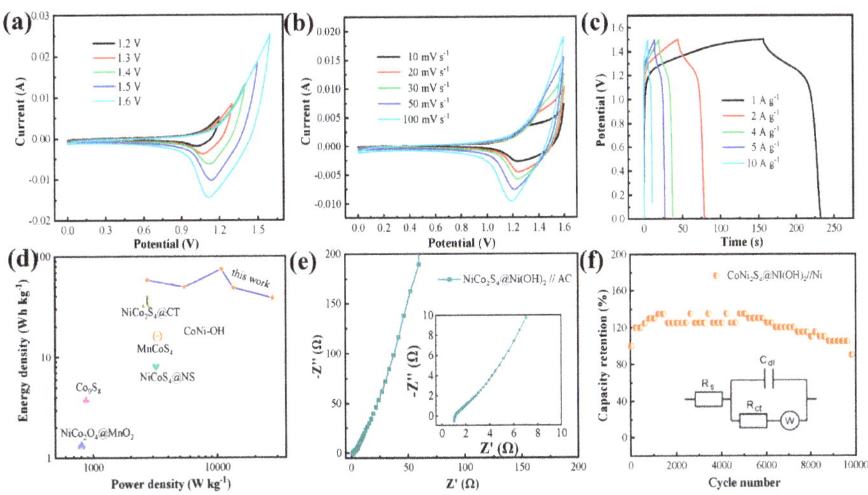

Figure 4. Electrochemical performance of device: (**a**) CV curves at different scan rates. (**b**) CV curves at 50 mV/s. (**c**) GCD curves at different current densities. (**d**) Nyquist plots. (**e**) Ragone plot. (**f**) Cycling performance.

It can be observed that the GCD curve shows some deviation in the charge and discharge times, which means lower Coulombic efficiency. This may be caused by the following factors: 1. Charge leakage and electrode polarization; 2. Inconsistent electrochemical kinetics between the anode and cathode. The kinetic rate of the $CoNi_2S_4@Ni(OH)_2$ material is slower than that of the carbon material used for the cathode; 3. The use of a separator and binder during the device assembly process increases the resistance of the electrode and hinders the transfer of electrons and charge; 4. At high potentials, the ion transport rate in the electrolyte slows down, requiring a longer time for charge accumulation. The specific capacitance of the device is determined to be 52.4 F g^{-1} at 1 A g^{-1}, according to Figure 4c. To further assess the device's performance, its energy density and power density were calculated by Equations (2) and (3). The energy density of the device is calculated to be 58.5 Wh kg^{-1}, while the power density reaches 2700 W kg^{-1}. These calculations highlight the material's ability in terms of storing and delivering energy efficiently, making it a match in terms of both high power output and substantial energy storage. As shown in Figure 4d, the values for these are higher in this work than in some previously reported data [34–39]. Table 1 shows a data comparison between this sample and other materials. Figure 4e shows that the equivalent resistance of the device is about 1 Ω. The results of EIS and inset in Figure 4e show that the device possesses a small equivalent resistance and a fast transmission speed. The cyclic characteristic of the device is also an important indicator. The device can still maintain a specific capacitance that equals 90% of the initial capacitance after 10,000 cycles, which reveals its outstanding electrochemical properties.

Table 1. A comparison of the current device with previously reported electrode materials.

Materials	Energy Density ($Wh{\cdot}kg^{-1}$)	Power Density ($W{\cdot}kg^{-1}$)	Capacitance ($F\ g^{-1}$)	Stability	Refs.
$NiCo_2S_4$@CT	36.1	798	59.2	89.2% (3000 cycles)	[34]
$NiCo_2S_4$@NS	25.5	334	71.6	85.6% (4000 cycles)	[35]
$MnCo_2S_4$	31.3	800	41.2	89% (5000 cycles)	[36]
Co_9S_8	13.3	1110	48.4	73% (4000 cycles)	[37]
$NiCo_2O_4$@MnO_2	5.8	175	31.3	93% (3000 cycles)	[38]
CoNi-OH	34.2	800	62.5	70% (6000 cycles)	[39]
$CoNi_2S_4$@$Ni(OH)_2$	58.5	2700	52.4	90% (10,000 cycles)	This work

4. Conclusions

In summary, we prepared $CoNi_2S_4$@$Ni(OH)_2$ samples using a straightforward hydrothermal synthesis method. $Ni(OH)_2$ nanosheets are interspersedly grown on the surface of $CoNi_2S_4$ nanosheets, significantly enhancing the material's electrochemical performance. This approach not only simplifies the production process, but also yields a material with exceptional electrochemical performance, exhibiting a specific capacitance of 1534 F g^{-1}. The assembled device still retains 90% of the initial specific capacitance after 10,000 cycles, and it can achieve an energy density of 58.5 Wh kg^{-1} at 2700 W kg^{-1}. Through this innovative design, this method not only effectively enhances the material's specific surface area and shortens the reaction path, but also strengthens the surface strength of the sulfide, greatly improving cycling performance.

Author Contributions: Conceptualization, Y.T., B.C., Y.J. and X.W.; Software, B.C. and Y.J.; Validation, B.C.; Formal analysis, Y.T. and X.W.; Investigation, B.C.; Resources, X.W.; Data curation, Y.T. and Y.J.; Writing—original draft, Y.T.; Writing—review & editing, X.W.; Project administration, Y.T. and X.W.; Funding acquisition, Y.T. All authors have read and agreed to the published version of the manuscript.

Funding: The work is supported by the Special Fund of Basic scientific Research Business expenses of undergraduate universities in Liaoning Province (LJ212410144004) from Liaoning Provincial Department of Education, and the College student innovation and entrepreneurship project of Shenyang Ligong University (X202410144013) from Shenyang Ligong University.

Data Availability Statement: Data are contained within the article.

Conflicts of Interest: The authors declare no conflicts of interest.

References

1. Zhang, Y.; Chen, Y.; Liang, K.; Zhang, Y.; Wang, D.; Wang, W.; Wang, J.; Mitsuzaki, N.; Chen, Z. Synthesis of nickel cobalt sulfide electrode materials for high-performance asymmetric supercapacitors. *J. Energy Storage* **2025**, *109*, 115176. [CrossRef]
2. Iqbal, M.; Saykar, N.G.; Alegaonkar, P.S.; Mahapatra, S.K. Synergistically modified WS_2@PANI binary nanocomposite-based all-solid-state symmetric supercapacitor with high energy density. *New J. Chem.* **2022**, *46*, 7043–7054. [CrossRef]
3. Yao, S.Y.; Jiao, Y.; Lv, C.D.; Kong, Y.; Ramakrishna, S.; Chen, G. Lattice-strain engineering of CoOOH induced by NiMn-MOF for high-efficiency supercapacitor and water oxidation electrocatalysis. *J. Colloid Interf. Sci.* **2022**, *623*, 1111–1121. [CrossRef]
4. De Adhikari, A.; Shauloff, N.; Turkulets, Y.; Shalish, I.; Jelinek, R. Tungsten-Disulfide/Polyaniline High Frequency Supercapacitors. *Adv. Electron. Mater.* **2021**, *7*, 2100025. [CrossRef]
5. Crisci, M.; Boll, F.; Merola, L.; Pflug, J.J.; Liu, Z.M.; Gallego, J.; Lamberti, F.; Gatti, T. Nanostructured 2D WS_2@PANI nanohybrids for electrochemical energy storage. *Front. Chem.* **2022**, *10*, 1000910. [CrossRef]
6. Tong, Y.; Liu, X.; Qi, D.; Chi, B. Constructing high performance supercapacitor electrode through PPy decorated Co_3O_4@$CoMoO_4$ sandwich structure. *J. Energy Storage* **2024**, *97*, 112798. [CrossRef]
7. Shi, C.; Wang, Y.; Kulaots, I.; Zhu, H.; Sheldon, B. Water-in-Salt Battery Electrolyte for High-Voltage Supercapacitors: A Fundamental Study on Biomass and Carbon Fiber Electrodes. *J. Electrochem. Soc.* **2024**, *171*, 110526. [CrossRef]
8. Yao, S.Y.; Jiao, Y.; Sun, S.F.; Wang, L.X.; Li, P.Y.; Chen, G. Vertically Co-oriented Mn-Metal–Organic Framework grown on 2D cation-intercalated manganese oxide via a self-sacrificing template process for a high performance asymmetric supercapacitor. *ACS Sustain. Chem. Eng.* **2020**, *8*, 3191–3199. [CrossRef]
9. Li, R.Y.; Xu, S.L.; Ai, Z.Q.; Qi, J.G.; Wu, F.F.; Zhao, R.D.; Zhao, D.P. Interface Engineering Accelerated Surface Reconstruction for Electrocatalytic Water Splitting and Energy Storage Device through Hybrid Structured $ZnCo_2O_4$@NiCo-LDH Nanocomposite. *Int. J. Hydrog. Energy* **2024**, *91*, 867–876. [CrossRef]
10. Li, R.Y.; Shen, X.Y.; Li, J.; Zhao, D.P.; Zhao, R.D.; Wu, F.F. Enhanced Electrochemical Performance of NiCo-Layered Double Hydroxides: Optimal Synthesis Conditions and Supercapacitor Applications. *Adv. Sustain. Syst.* **2024**, 2400753.
11. Yao, S.Y.; Ramakrishna, S.; Chen, G. Recent advances in metal–organic frameworks based on electrospinning for energy storage. *Adv. Fiber Mater.* **2023**, *5*, 1592–1617. [CrossRef]
12. Dhandapani, E.; Thangarasu, S.; Ramesh, S.; Ramesh, K.; Vasudevan, R.; Duraisamy, N. Recent development and prospective of carbonaceous material, conducting polymer and their composite electrode materials for supercapacitor—A review. *J. Energy Storage* **2022**, *52*, 104937. [CrossRef]
13. Zhag, M.; Du, H.; Wei, Z.; Zhang, X.; Wang, R. Ultrafast microwave synthesis of nickel-cobalt sulfide/graphene hybrid electrodes for high-performance asymmetrical supercapacitors. *ACS Appl. Energy Mater.* **2021**, *4*, 8262–8274.
14. Zhu, J.; Wu, Q.; Li, J. Review and prospect of Mn_3O_4-based composite materials for supercapacitor electrodes. *ChemistrySelect* **2020**, *5*, 10407–10423. [CrossRef]
15. Chen, L.; Wan, J.f.; Fan, L.; Wei, Y.; Zou, J.l. Construction of $CoNi_2S_4$ hollow cube structures for excellent performance asymmetric supercapacitors. *Appl. Surf. Sci.* **2021**, *570*, 151174. [CrossRef]
16. Aisha, S.; Nagaraju, D.H.; Mahesh, P. High-energy-density asymmetric supercapacitor based on layered-double-hydroxide-derived $CoNi_2S_4$ and eco-friendly biomass-derived activated carbon. *Energy Fuels* **2022**, *36*, 13286–13295.
17. Zhu, J.; Han, C.; Song, X. Facile synthesis of novel $CoNi_2S_4$/carbon nanofibers composite for high-performance supercapacitor. *Mater. Chem. Phys.* **2022**, *283*, 126038. [CrossRef]
18. Patil, S.J.; Kim, J.H.; Lee, D.W. Self-assembled Ni_3S_2//$CoNi_2S_4$ nanoarrays for ultra high-performance supercapacitor. *Chem. Eng. J.* **2017**, *322*, 498–509. [CrossRef]
19. Li, Z.; Zhao, D.; Xu, C.; Ning, J.; Zhong, Y.; Zhang, Z.; Wang, Y.; Hu, Y. Reduced $CoNi_2S_4$ nanosheets with enhanced conductivity for high-performance supercapacitors. *Electrochim. Acta* **2018**, *278*, 33–41. [CrossRef]
20. Visakh, C.; Mohan, V.; Rakhi, R.B. WS_2/Conducting polymer nanocomposite-based flexible and binder-free electrodes for high-performance supercapacitors. *Electrochim. Acta* **2024**, *498*, 144657.
21. Wang, J.; Zheng, F.; Yu, Y.; Hu, P.; Li, M.; Wang, J.; Fu, J.; Zhen, Q.; Bashir, S.; Liu, J.L. Symmetric supercapacitors composed of ternary metal oxides ($NiO/V_2O_5/MnO_2$) nanoribbon electrodes with high energy storage performance. *Chem. Eng. J.* **2021**, *426*, 131804. [CrossRef]
22. Chevulamaddi, H.; Kalagadda, V.R. The improved electrochemical performance of MnO_2 nanorods decorated with polyaniline as efficient electrode towards high performance supercapacitors. *J. Mater. Sci. Mater. Electron.* **2024**, *35*, 2273. [CrossRef]
23. Yue, T.; Shen, B.; Gao, P. Carbon material/MnO_2 as conductive skeleton for supercapacitor electrode material: A review. *Renew. Sustain. Energy Rev.* **2022**, *158*, 112131. [CrossRef]
24. Song, M.; Zhou, Y.; Ren, X.; Wan, J.; Du, Y.; Wu, G.; Ma, F. Biowaste-based porous carbon for supercapacitor: The influence of preparation processes on structure and performance. *J. Colloid Interface Sci.* **2019**, *535*, 276–286. [CrossRef] [PubMed]
25. Agyapong, A.D.; Cooley, K.A.; Mohney, S.E. Reactivity of contact metals on monolayer WS_2. *J. Appl. Phys.* **2020**, *128*, 055306. [CrossRef]

26. Zubairi, H.; Lu, Z.; Zhu, Y.; Reaney, I.M.; Wang, G. Current development, optimisation strategies and future perspectives for lead-free dielectric ceramics in high field and high energy density capacitors. *Chem. Soc. Rev.* **2024**, *53*, 10761–10790. [CrossRef]
27. Yang, W.; Wang, P.; Tu, Z.; Hou, L.; Yan, L.; Jiang, B.; Zhang, C.; Huang, G.; Yang, F.; Li, Y. Heteroatoms-doped hierarchical porous carbon with multi-scale structure derived from petroleum asphalt for high-performance supercapacitors. *Carbon* **2022**, *187*, 338–348. [CrossRef]
28. Zhang, M.; Du, H.; Wei, Z.; Zhang, X.; Wang, R. Facile electrodeposition of Mn-CoP nanosheets on Ni foam as high-rate and ultrastable electrodes for supercapacitors. *ACS Appl. Energy Mater.* **2022**, *5*, 186–195. [CrossRef]
29. Luo, L.; Lan, Y.; Zhang, Q.; Deng, J.; Luo, L.; Zeng, Q.; Gao, H.; Zhao, W. A review on biomass-derived activated carbon as electrode materials for energy storage supercapacitors. *J. Energy Storage* **2022**, *55*, 105839. [CrossRef]
30. Shah, S.S. Biomass-Derived Carbon Materials for Advanced Metal-Ion Hybrid Supercapacitors: A Step Towards More Sustainable Energy. *Batteries* **2024**, *10*, 168. [CrossRef]
31. Sun, L.; Zhuo, K.; Chen, Y.; Du, Q.; Zhang, S.; Wang, J. Ionic Liquid-Based Redox Active Electrolytes for Supercapacitors. *Adv. Funct. Mater.* **2022**, *32*, 2203611. [CrossRef]
32. Liu, R.; Wang, J.-X.; Yang, W.-D. Hierarchical Porous Heteroatoms—Co-Doped Activated Carbon Synthesized from Coconut Shell and Its Application for Supercapacitors. *Nanomaterials* **2022**, *12*, 3504. [CrossRef] [PubMed]
33. Li, Y.; Zhang, D.; Zhang, Y.; He, J.; Wang, Y.; Wang, K.; Xu, Y.; Li, H.; Wang, Y. Biomass-derived microporous carbon with large micropore size for high-performance supercapacitors. *J. Power Sources* **2020**, *448*, 227396. [CrossRef]
34. Hao, L.; Shen, L.; Wang, J.; Xu, Y.; Zhang, X. Hollow $NiCo_2S_4$ nanotube arrays grown on carbon textile as a self-supported electrode for asymmetric supercapacitors. *RSC Adv.* **2016**, *6*, 9950–9957. [CrossRef]
35. Wu, Z.; Pu, X.; Ji, X.; Zhu, Y.; Jing, M.; Chen, Q.; Jiao, F. High energy density asymmetric supercapacitors from mesoporous $NiCo_2S4$ nanosheets. *Electrochim. Acta* **2015**, *174*, 238–245. [CrossRef]
36. Liu, S.; Jun, S. Hierarchical manganese cobalt sulfide core–shell nanostructures for high-performance asymmetric supercapacitors. *J. Power Sources* **2017**, *342*, 629–637. [CrossRef]
37. Mao, X.; Wang, Z.; Kong, W.; Wang, W. Nickel foam supported hierarchical Co_9S_8 nanostructures for asymmetric supercapacitors. *New J. Chem.* **2017**, *41*, 1142–1148. [CrossRef]
38. Kuang, M.; Wen, Z.; Guo, X.; Zhang, S.; Zhang, Y. Engineering firecracker-like beta-manganese dioxides@spinel nickel cobaltates nanostructures for high-performance supercapacitors. *J. Power Sources* **2014**, *270*, 426–433. [CrossRef]
39. Tang, Y.; Liu, Y.; Yu, S.; Guo, W.; Mu, S.; Wang, H.; Zhao, Y.; Hou, L.; Fan, Y.; Gao, F. Template-free hydrothermal synthesis of nickel cobalt hydroxide nanoflowers with high performance for asymmetric supercapacitor. *Electrochim. Acta* **2015**, *161*, 279–289. [CrossRef]

batteries

MDPI

Review

One Stone, Three Birds: Innovations and Challenges of Layered Double Hydroxides in Batteries, Supercapacitors, and Hydrogen Production

Syed Shaheen Shah *, Manisha Das and Takaya Ogawa *

Socio-Environmental Energy Science Department, Graduate School of Energy Science, Kyoto University, Yoshida-Honmachi, Sakyo-Ku, Kyoto 606-8501, Japan; nisha.ma.25s@st.kyoto-u.ac.jp
* Correspondence: shah.syedshaheen.3d@kyoto-u.ac.jp (S.S.S.); ogawa.takaya.8s@kyoto-u.ac.jp (T.O.)

Abstract: Layered double hydroxides (LDHs), notable for their unique two-dimensional layered structures, have attracted significant research attention due to their exceptional versatility and promising performance in energy storage and conversion applications. This comprehensive review systematically addresses the fundamentals and diverse synthesis strategies for LDHs, including co-precipitation, hydrothermal synthesis, electrochemical deposition, sol-gel processes, ultrasonication, and exfoliation techniques. The synthesis methods profoundly influence the physicochemical properties, morphology, and electrochemical performance of LDHs, necessitating a detailed understanding to optimize their applications. In this paper, the role of LDHs in batteries, supercapacitors, and hydrogen production is critically evaluated. We discuss their incorporation in various battery systems, such as lithium-ion, lithium–sulfur, sodium-ion, chloride-ion, zinc-ion, and zinc–air batteries, highlighting their structural and electrochemical advantages. Additionally, the superior pseudocapacitive behavior and high energy densities offered by LDHs in supercapacitors are elucidated. The effectiveness of LDHs in hydrogen production, particularly through electrocatalytic water splitting, underscores their significance in renewable energy systems. This review paper uniquely integrates these three pivotal energy technologies, outlining current innovations and challenges, thus fulfilling a critical need for the scientific community by providing consolidated insights and guiding future research directions.

Keywords: layered double hydroxides; energy storage and conversion; batteries and supercapacitors; hydrogen evolution reaction; sustainable energy materials

1. Introduction

Layered double hydroxides (LDHs), known as hydrotalcite-like compounds or anionic clays, represent a versatile class of two-dimensional nanostructured materials characterized by their unique layered architecture and compositional tunability [1–4]. Structurally, LDHs are composed of positively charged metal hydroxide layers balanced by interlayer anions and water molecules, forming a stacked, sandwich-like configuration. The general chemical formula of LDHs is expressed as $[M(II)_{1-x}M(III)_x(OH)_2]^{x+}(A^{n-})_{x/n}{\cdot}mH_2O$, where M(II) and M(III) denote divalent and trivalent metal cations, respectively, and A^{n-} represents exchangeable interlayer anions [5]. This intrinsic flexibility in structure and composition has made LDHs attractive candidates for various scientific and industrial applications, especially in energy storage, conversion, and environmental remediation.

The urgent global demand for efficient, sustainable energy technologies has intensified research into materials that enhance performance in batteries, supercapacitors, and hydrogen production systems [4,6]. LDHs have emerged as promising materials due to their advantageous properties, including high specific surface area, tunable interlayer chemistry, rich redox-active sites, and robust electrochemical stability [7]. These features improve ion mobility, energy storage capacity, and catalytic efficiency. In battery technologies, especially those supporting electric vehicles and large-scale energy storage, materials with high energy density, long cycle life, and stable charge/discharge behavior are essential [8]. LDHs have demonstrated substantial potential to meet these requirements. Their layered structure facilitates efficient ion transport, while their ability to accommodate multiple metal cations and interlayer species enables tailored electrochemical properties. Moreover, the hybridization of LDHs with carbon-based materials such as graphene and carbon nanotubes (CNTs) further enhances electrical conductivity and structural integrity, thereby extending their practical utility in lithium-ion batteries (LIBs) and other emerging battery systems [9,10]. In addition to LIBs, LDHs have been applied in different battery chemistries, including lithium–sulfur batteries (LSBs), sodium-ion batteries (NIBs), chloride-ion batteries (CIBs), zinc-ion batteries (ZIBs), and zinc–air batteries (ZABs) [11–16]. In LSBs, LDHs help mitigate the polysulfide shuttle effect by effectively adsorbing and immobilizing dissolved polysulfides. Their high surface area and interlayer spacing also enhance sodium-ion diffusion in NIBs, which is critical for their energy storage performance. Furthermore, their incorporation into aqueous batteries, such as ZIB and ZAB systems, has improved rate capability and cyclic stability. LDHs also offer compelling advantages in supercapacitor applications, particularly due to their pseudocapacitive behavior [17]. Their electrochemical performance stems from fast and reversible surface redox reactions facilitated by the layered architecture that supports rapid ion transport. Consequently, LDHs exhibit high capacitance, fast charge/discharge rates, and long cycle life, making them suitable for high-power applications. Developing hybrid LDH materials, especially those combined with conductive polymers or carbon nanomaterials, has led to further energy and power density enhancements, expanding their practical viability in real-world devices [18,19]. Parallel to their electrochemical applications, LDHs have gained increasing recognition in hydrogen production, particularly through electrocatalytic water splitting [20]. Efficient and sustainable hydrogen generation depends on active, stable, and cost-effective catalysts for the hydrogen evolution reaction (HER) and oxygen evolution reaction (OER). LDHs, owing to their tunable composition and catalytic activity, have proven to be excellent candidates. Their performance can be further optimized by compositional modifications or integration with nanoparticles, semiconductors, or metal–organic frameworks (MOFs) to enhance electronic conductivity and surface reactivity [21–26].

The synthesis route plays a vital role in tailoring the physicochemical and electrochemical properties of LDHs. Conventional methods such as co-precipitation and hydrothermal synthesis have been widely adopted for their scalability and simplicity. Co-precipitation enables precise control over metal cation ratios and particle morphology, while hydrothermal methods offer improved crystallinity and uniform particle distribution [27]. Other advanced techniques, including electrochemical deposition, sol-gel processing, ultrasonication-assisted synthesis, and exfoliation, have shown promise in producing LDHs with higher surface areas, controlled nanoscale morphologies, and enhanced electrochemical behavior [28]. Furthermore, successfully doping various metal cations into LDH frameworks allows for the rational design of materials that address specific challenges in energy applications, such as conductivity limitations and ion diffusion barriers [29,30]. This adaptability highlights the strategic importance of synthesis optimization in maximizing LDH performance for targeted functionalities.

Beyond energy applications, LDHs also exhibit strong potential in environmental remediation due to their anion exchange capabilities and high chemical reactivity [31]. Recent advancements demonstrate that LDHs are pivotal in advancing circular economy principles and sustainable material practices. Researchers have significantly reduced environmental burdens while creating high-value functional materials by utilizing industrial by-products such as red mud, slag, and fly ash for LDH synthesis [32,33]. Waste-derived LDHs, particularly Ni-based variants, have shown remarkable efficiency in electrocatalysis for oxygen evolution reactions, simultaneously addressing waste valorization and promoting green hydrogen production [34]. In carbon capture, LDH-based photocatalysts facilitate CO_2 reduction to value-added chemicals under solar irradiation, supporting carbon circularity strategies [35]. Moreover, LDHs have been successfully applied in agriculture for controlled-release fertilizers, enhancing nutrient efficiency and minimizing environmental runoff [36]. These examples underscore the growing importance of LDHs in enabling resource recovery, pollution mitigation, and sustainable energy solutions, making them crucial materials in the transition toward greener technologies and closed-loop material systems. These characteristics make them effective in adsorbing pollutants, such as heavy metals and organic contaminants, from aqueous solutions. The dual functionality of LDHs as both energy materials and environmental agents underscores their relevance in supporting sustainability goals and circular economy models. In recent advancements, LDH-derived composite materials, especially those integrated with MOFs, have shown exceptional improvements in surface area, electron/ion transport pathways, and structural robustness [37]. These composites combine the best properties of both components, opening up new directions for multifunctional material systems capable of simultaneously addressing energy and environmental challenges.

LDHs have witnessed a significant surge in global research interest over the past decade due to their structural versatility and broad application potential. As of 2023, the global LDH market was valued at approximately USD 30.57 billion and is projected to reach USD 47.91 billion by 2032, growing at a compound annual growth rate (CAGR) of 5.12%. Research trends show that LDHs are extensively explored in biomedical fields (e.g., drug delivery, cancer therapy, and biosensors), environmental remediation (such as selenium and heavy metal removal from water), energy storage (as precursors for battery and supercapacitor electrodes), and catalytic applications. In the biomedical sector alone, there has been an average annual increase of ~8% in LDH-related publications between 2018 and 2024 [38]. Environmental applications are also expanding, with LDH-based materials achieving over 90% removal efficiencies for specific contaminants like selenium [39]. Research publications in LDH-based energy applications increased by over 250% between 2017 and 2023 (Scopus database). NiFe-LDHs are leading candidates for water splitting in electrocatalysis and hydrogen production with doping strategies (e.g., Co, Mn, Cr) and interlayer anion modifications significantly boosting activity and stability [40]. LDHs are also increasingly explored in supercapacitors, where their pseudocapacitive behavior and fast ion diffusion are enhanced through hybridization with graphene and carbon nanotubes; this field has seen a CAGR of approximately 10% in publications over the past five years. In photocatalysis, LDH-based heterostructures such as LDH/TiO_2 and LDH/g-C_3N_4 composites are under intense study for solar-driven water splitting and CO_2 reduction, with key focus areas including bandgap tuning and charge carrier separation [41]. Furthermore, due to their high surface area and redox-active sites, calcined LDH-derived mixed metal oxides are increasingly used for Li-ion and Zn–air batteries [42]. Novel research also advances LDHs as hosts for anion intercalation batteries and fuel cell electrocatalysts for the ORR [43]. LDHs are at the center of next-generation energy technologies, driven by their structural flexibility, high activity, and earth-abundant composition. Future research directions focus on designing magnetic or structurally modified LDHs for easier recovery and enhanced performance, aiming to expand their role in advanced catalysis, CO_2 capture, and sustainable technologies.

Despite the vast promise of LDHs, several challenges hinder their widespread application. These include their relatively low intrinsic electronic conductivity, structural degradation under prolonged electrochemical cycling, and difficulties scaling up industrial synthesis techniques. To overcome these limitations, continued efforts are required to understand the fundamental relationships between LDH structure, composition, and electrochemical performance. Addressing these gaps will be crucial to unlocking the full potential of LDHs across diverse energy technologies. This review provides a comprehensive overview of the recent progress and emerging trends in applying LDHs in energy-related fields. By systematically discussing the role of LDHs in batteries, supercapacitors, and hydrogen production, alongside synthesis strategies and hybrid material developments, this work aims to consolidate fragmented research findings and provide an integrated perspective. Although several recent reviews have discussed the role of LDHs in electrochemical applications [44–49], most have focused on individual systems or narrowly defined topics. This review uniquely integrates the applications of LDHs across batteries, supercapacitors, and hydrogen production within a single framework, offering a comprehensive and comparative analysis that highlights interconnections, emerging challenges, and future research opportunities. The focus on batteries, supercapacitors, and hydrogen production as application areas for LDHs stems from their central role in advancing clean energy technologies. These three domains, although distinct, share critical commonalities in their underlying electrochemical processes, including charge storage, ion transport, and surface redox reactions, processes where LDHs exhibit remarkable properties such as tunable composition, high surface area, and excellent ion exchange capabilities. The synergy among them is evident via the advancements in LDH-based electrode materials for batteries, which often directly inform the design strategies for supercapacitors, given the need for high conductivity and cycling stability. Similarly, the catalytic properties of LDHs that enhance HER are closely tied to the electron and ion mobility mechanisms essential for energy storage devices. Thus, studying these three applications together allows a more integrated understanding of how LDH structural tuning can optimize performance across multiple energy technologies, paving the way for multifunctional materials that can simultaneously address storage and generation needs. The review also identifies existing challenges and future research directions, offering valuable insights for advancing LDH-based technologies toward scalable and sustainable energy solutions.

2. Fundamentals of LDHs

LDHs, often called anionic clays or hydrotalcite-like materials, represent a unique class of compounds characterized by their layered structure and variable ionic composition. These materials have attracted considerable attention across various fields, including catalysis, energy storage, and environmental remediation, owing to their distinctive properties and potential applications [23,50–52]. LDHs adopt a general structural formula of $[M_{1-x}{}^{2+}M_x{}^{3+}(OH)_2]^{x+}(A^{n-})_{x/n}\cdot mH_2O$, where M^{2+} and M^{3+} are metal cations (i.e., divalent (e.g., Fe^{2+}, Mg^{2+}, Ni^{2+}) and trivalent (e.g., Al^{3+}, Fe^{3+}, Cr^{3+}), while A^{n-} denotes interlayer anions (such as $CO_3{}^{2-}$, $NO_3{}^{-}$, or Cl^{-}) [5,40,50,53,54]. The schematic representation for LDH basic structures is shown in Figure 1a,b. The octahedral sheets formed by metal hydroxides create anionic layers that house exchangeable anions and water molecules within their interlayer spaces [51,52]. This unique structural arrangement provides a large surface area for adsorption and allows for easy intercalation of various anions, making LDHs versatile for numerous applications. LDHs can form various morphologies, the most prominent being flower-like structures [55], as illustrated in Figure 1c. Different types of LDHs have been synthesized, including those based on magnesium–aluminum (MgAl), nickel–aluminum (NiAl), nickel–cobalt (NiCo), and zinc–aluminum (ZnAl) combinations. The versatility of LDHs is primarily attributed to the variety of cations that can exist in the mixed metal sheets, leading to a diverse range of properties and functionalities. For instance, the intercalation of anionic species can significantly affect the catalytic activity, thermal stability, and solubility of LDHs, tailoring their usability in specific applications [56–58]. One notable property of LDHs is their hydrophilic nature, which enhances their interaction with water and biological systems, making them suitable for drug delivery and biocompatibility applications [58,59]. Furthermore, LDHs exhibit excellent acid–base characteristics, which can be optimized for processes such as CO_2 capture and adsorption of pollutants from aqueous solutions [60,61]. For example, studies have shown that LDHs can effectively remove azo dyes and other organic contaminants from water through adsorption mechanisms [57,62,63]. In energy storage applications, LDHs have emerged as promising candidates for electrode materials in supercapacitors and batteries. Their layered structure allows for easy ion diffusion, essential for charge/discharge processes [64,65]. Research has demonstrated that LDH-based composites incorporating materials such as graphene can significantly enhance the performance and efficiency of energy storage devices [56,66]. This dual functionality as ion storage matrices and conductive frameworks bolsters their application relevance in renewable energy systems. Another significant aspect of LDHs is their utility in hydrogen production, particularly in catalysis. LDHs, when modified with transition metals such as nickel or cobalt, can serve as highly efficient electrocatalysts for water splitting, an essential reaction to produce hydrogen [67,68]. Incorporating these metals into the LDH framework improves catalytic activity due to increased surface area and enhanced electronic properties. Studies have revealed that such modified LDHs can maintain stability and activity over prolonged periods, making them suitable candidates for industrial applications [66,69]. Several examples highlight the utility and properties of specific LDHs. The MgAl-LDH, often synthesized through coprecipitation methods, has shown remarkable adsorption capabilities for various pollutants due to its high surface area and effective ion exchange properties [70,71]. Similarly, ZnAl-LDH composites have been used for effective dye removal owing to their favorable ion exchange capacities and adsorptive kinetics, demonstrating high removal rates in wastewater treatments [72,73]. Additionally, NiFe-LDHs have been utilized as electrocatalysts for HER, achieving commendable efficiencies and stability in alkaline electrolytes [67,68]. LDHs are a vital material with extensive applications in various fields, such as water purification, energy storage, and catalysis. Their unique structural and compositional characteristics allow for adaptable performance based on specific application requirements. This adaptability, coupled with ongoing research to optimize their properties, ensures continued advancements in the functionality and applicability of LDHs in enhancing environmental sustainability and energy efficiency [74–76].

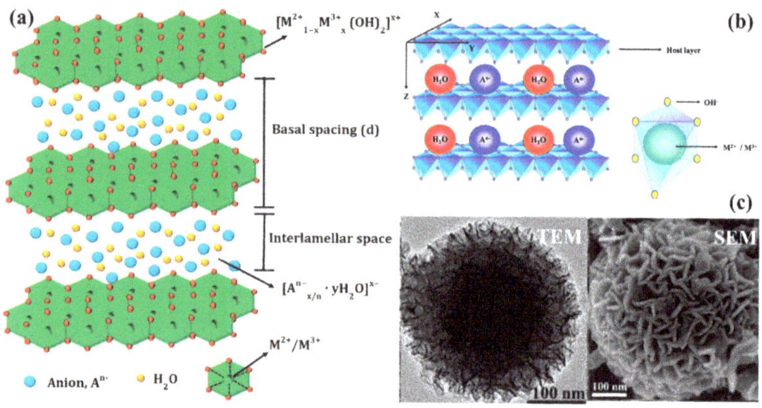

Figure 1. (**a**,**b**) Schematic representation for LDH structures. Reproduced with permission [77,78]. (**c**) TEM and FESEM images of LDH. Reproduced with permission [55].

The deployment of LDHs in practical applications faces several significant challenges, particularly when scaling up synthesis methods, ensuring stability during long-term operation, and addressing low intrinsic conductivity. These issues must be thoroughly examined and mitigated through ongoing research to realize the full potential of LDH-based materials in large-scale systems. One of the critical hurdles in deploying LDHs at scale is the synthesis methods employed. Traditional synthesis techniques often lack the efficiency required for large production. However, recent advancements, such as the "on-the-fly" synthesis method demonstrated by Mattera et al., utilize controlled microfluidic reaction/diffusion conditions to rapidly produce self-supported LDH hollow structures [79]. This technique not only enhances the scalability of production but also maintains the unique structural characteristics of LDHs, potentially overcoming current limitations in synthesis yields and uniformity. Furthermore, the separate nucleation and aging steps method has been explored, providing additional strategies for scalable synthesis while maintaining the structural integrity and performance of the resulting LDHs [80]. Stability under operational conditions represents another significant concern for the practical application of LDHs. The structural integrity of LDH materials can degrade over time in various environments, particularly in aqueous solutions, where traditional LDH configurations may disintegrate [81]. Research focusing on hybrid materials that combine LDHs with conductive carbon matrices has shown promising results. Integrating LDHs with conductive materials significantly enhances their electrochemical performance and stability over extended periods [82]. Moreover, innovative designs, such as incorporating porous structures into LDH composites, exhibit improved performance metrics, addressing the inherent limitations of LDHs regarding durability and efficiency [83]. The intrinsic low conductivity of LDHs is a well-documented limitation that impedes their performance in many electrochemical applications. Recent studies indicate that hybrid LDH composites, such as those combined with conductive polymers or carbon-based materials, can ameliorate these issues [84,85]. The enhancement of electric conductivity through these composites and the careful design of the pore structure allow for better electrolyte penetration and lower charge transfer resistance, significantly boosting the electrochemical activity of LDH-based systems [86]. Hybrid LDH–polymer membranes have also demonstrated substantial improvements in permeability and selectivity for CO_2 separation, showcasing their potential for real-world applications while addressing the conductivity concerns inherent to traditional LDH materials [87]. Significant strides have been made in addressing the challenges of deploying LDH-based materials. Innovations in synthesis techniques, material design, and stability enhancements are crucial to overcoming these barriers. Continued research will be essential to refine these approaches further and ensure that LDHs can fulfil their promise in large-scale applications.

3. Synthesis Methods of LDHs

LDHs can be synthesized through various methods, each significantly influencing their structural, morphological, and functional properties. Common synthesis techniques include the co-precipitation method [88], valued for its simplicity and scalability, typically producing LDHs with uniform morphology and high purity; hydrothermal synthesis [89], favored for precise control over crystallinity and particle size, resulting in materials with enhanced thermal stability; and electrochemical deposition [90], advantageous for fabricating thin films with controllable thickness and adherence. Advanced techniques such as sol-gel, ultrasonication, and exfoliation offer further control over LDH structures; the sol-gel approach produces highly homogeneous and porous LDHs, ultrasonication facilitates smaller particle sizes and improved dispersion, while exfoliation allows isolation of single-layer nanosheets with exceptional surface activity [23]. Figure 2 depicts the structural framework of LDHs, highlighting common synthesis methods and their wide-ranging applications. Overall, the choice of synthesis method profoundly affects LDHs' properties, including crystallinity, surface area, porosity, interlayer spacing, and electrochemical performance, thus determining their suitability for diverse applications like batteries, supercapacitors, and catalysis.

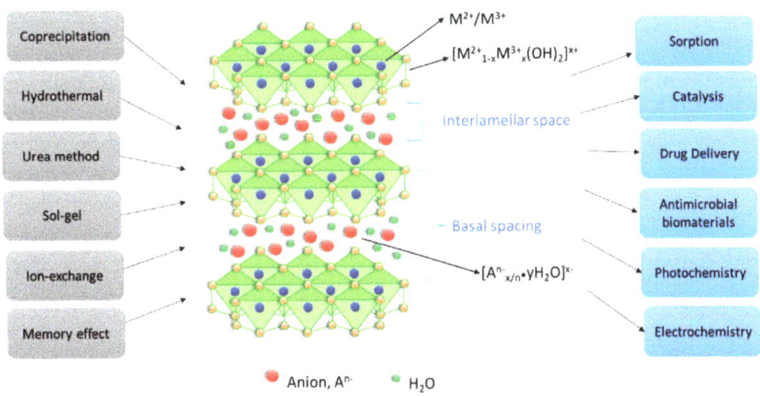

Figure 2. Schematic illustration of the synthesis methods, layered structure, and diverse applications of LDHs. Reproduced with permission [91].

3.1. Co-Precipitation

The co-precipitation method is a widely utilized synthesis technique for LDHs. This method involves the simultaneous precipitation of two or more metal salts from an aqueous solution in a specific pH-controlled environment. The goal is to create a solid precipitate comprising cations with differing oxidation states that can form a layered structure upon aging and further processing. The co-precipitation approach offers advantages such as simplicity, cost-efficiency, and the capability of producing homogeneous materials [92,93]. The fundamental mechanism underlying the co-precipitation of LDHs includes the generation of hydroxides by reacting an alkaline solution (e.g., sodium hydroxide) with metal salt solutions. When mixed, metal ions, typically divalent metals (M^{2+}) and trivalent metals (M^{3+}), condense to form brucite-like sheets, a precursor structure of LDHs. The excess positive charge of the brucite-like layers is balanced by the intercalation of anions within the interlayer spacing, such as carbonate, nitrate, or organic molecules [94,95]. Thus, the co-precipitation method is fundamental for synthesizing LDHs with diverse properties suitable for various applications, including catalysis, ion exchange, and drug delivery systems [96,97].

Recent advancements in solid-state reactions have revealed numerous strategies for improving the structural and morphological qualities of LDHs synthesized via the co-precipitation method. For instance, a meticulous study by Wijitwongwan et al. [98] demonstrated that varying layer charge densities effectively alter the microstructure and performance of NiFe-LDH. Similarly, Zhang et al. [99] expounded on the preparation of chromium-modified ultrathin CoFe-LDH, noting that enhanced electrocatalytic performance can be attributed to its nanosheet structure, achieved efficiently through the co-precipitation route. Additionally, other research has indicated the potential for co-precipitation in the fabrication of ZnAl-LDH, particularly in enhancing the corrosion protection of steel-reinforced concrete. This study determined that the layering structure includes intercalated organic anions, which augment the protective qualities of the resulting material [88,98]. Their findings emphasize the importance of co-precipitation in producing multifunctional layered structures designed for specific engineering applications. In the domain of catalysis, the co-precipitation method allows for creating various catalytic layered materials with particular properties tailored for high activity. For example, NiAl-LDHs synthesized via the co-precipitation method exhibited significant electrochemical activity and stability, making them suitable candidates for application in batteries and supercapacitors [96,99]. The correlation between the morphology of LDHs and their catalytic behaviors has been systematically reviewed, indicating that factors such as particle size, crystallinity, and layering impact performance significantly [92,100]. Furthermore, approaches involving the hybridization of various LDH materials have emerged, expanding the range of applications and performance metrics of these materials. For instance, research has explored Co–Ni–Fe oxyhydroxide composites synthesized via co-precipitation, modifying their catalytic capabilities for various reactions [101]. These hybrid systems demonstrate that co-precipitation can yield single-component LDHs and complex multi-component systems with advanced functionality [97]. Ongoing studies focus on optimizing synthesis parameters to enhance performance metrics. Key considerations include the concentration of metal salts, pH level, temperature, and stirring conditions during synthesis. Attempts to generate various forms of zinc–Al LDHs have shown that altering these conditions can lead to significant variations in structural integrity and interlayer spacing, characterized through advanced spectroscopy techniques [93,102]. The versatility of the co-precipitation method is evident in its application; in a recent study exploring hybrid nanoparticles, the uniformity and dispersion of mesoporous ZnAl-LDHs were ensured by carefully controlling synthesis parameters [93,103]. Innovative adaptations of the co-precipitation method, including using surfactants and polymers as structure-directing agents, have progressively unlocked pathways for producing new LDH varieties with tailored layer spacing and enhanced chemical reactivity [96,104].

Thorough examinations have also been directed at using co-precipitation to produce nanostructured materials like mesoporous catalysts. For example, Zhang et al. [99] elucidated the synthesis of cobalt-LDH nanostructures displaying significant efficacy in HER. The nanosheet morphology achieved through co-precipitation allows for higher surface area and better catalytic performance than bulk materials [99,105]. The emergence of LDHs as effective adsorbents for environmental remediation has garnered considerable attention, with co-precipitation playing a critical role. Various authors have demonstrated the removal of toxic anionic contaminants such as fluoride and heavy metals using LDHs synthesized through this method [104,106]. The selective adsorption capabilities are primarily attributed to the structural characteristics of LDHs, wherein the intercalated anions can be exchanged with environmental pollutants, applying this functionality in wastewater treatment [96,98]. The co-precipitation method is not limited to traditional LDH compositions; recent works have shown its viability in synthesizing more complex layered materials that integrate organic and inorganic components. Such multifunctional materials can effectively combine properties, enabling application in fields ranging from nanotechnology to biochemistry and medicine [107,108]. The co-precipitation method remains a powerful and versatile technique for synthesizing LDHs, allowing significant control over their structure and properties. The evolution and innovation in this methodology reflect ongoing academic and industrial pursuits toward developing sophisticated materials capable of addressing contemporary challenges in catalysis, energy storage, and environmental sustainability.

Flower-like NiCo-LDH nanoparticles were synthesized via the co-precipitation method using triethanolamine as the alkali source and hydrogen peroxide (H_2O_2) as the size-controlling reagent [109]. A subsequent heat treatment step (at 200 °C) was implemented to create a nanocrystal@amorphousness core/shell structure. This precise temperature-controlled process effectively induced amorphization at the surface of LDH crystals, significantly refining the grain size. As illustrated in Figure 3a, the LDH structure transforms with increasing temperature in an inert (Ar) atmosphere, leading to increased amorphous content and improved Li^+ diffusion channels due to the formation of disordered regions. Figure 3b presents the nitrogen adsorption/desorption isotherms of LDH samples treated at different temperatures. It can be observed that surface area increases significantly with temperature, reaching up to 178.9 m^2/g for LDH. In Figure 3c, the pore size distribution reveals that higher temperatures increase pore volume and broader pore distribution, which benefits ion transport and electrochemical performance. FESEM and TEM analyses revealed flower-like nanoparticles (~100 nm diameter) composed of small nanosheets (~20 nm × 5 nm), forming a unique nanocrystal@amorphous core/shell structure. The BET surface area measurement indicated a mesoporous structure with a specific surface area of approximately 76.6 m^2/g, conducive to efficient electrolyte interaction and lithium-ion diffusion. The LDH amorphous-crystalline core/shell structure and enhanced surface area promote rapid lithium-ion transport, making it highly suitable for high-performance LIBs anodes.

The MgAl-LDH was synthesized via a straightforward co-precipitation method and incorporated with CNTs to form a hybrid composite [110]. This LDH@CNT hybrid was subsequently employed to create a sulfur composite cathode. The resulting material exhibited a well-dispersed morphology with nanoscale features and layered structure, a large surface area, and strong polysulfide adsorption characteristics. These features make the LDH highly suitable for LSB applications by effectively mediating polysulfide conversion and suppressing the shuttle effect. In another study, NiFe-NO_3 LDH was synthesized via a one-pot co-precipitation method by titrating a mixed nitrate solution of Ni and Fe with NaOH under an inert atmosphere, then aging at 70 °C for three days [111]. The obtained product was filtered, washed, dried, and ground to obtain the final LDH powder. The material revealed a typical hexagonal LDH structure with low crystallinity, turbostratic stacking, and a nanostructured surface. These structural features, combined with the high theoretical capacity (477 mAh/g) and dual redox/active metal centers (Ni^{2+}/Fe^{3+}), make the LDH highly suitable as an anode for NIB applications. The $Ni_3Mn_{0.7}Fe_{0.3}$-LDH was synthesized via a co-precipitation method and subsequently activated electrochemically to introduce hydrogen vacancies, resulting in the Hv-$Ni_3Mn_{0.7}Fe_{0.3}$-LDH [16]. The activation process involved CV in 1 M KOH to generate hydrogen-deficient lamellae that enhance zinc-ion accommodation. The reported LDH structure is highly suitable for ZIB applications due to its exposed terminal-O active sites and reduced Zn^{2+} diffusion barriers. NiFe

LDH nanosheets with varying interlayer spacings were synthesized through a simple co-precipitation method by intercalating different anionic species (Cl^-, $C_7H_5O_3^-$, and $C_{17}H_{25}O_3^-$) into the interlayer space [112]. The interlayer spacing was tuned from 7.695 Å to 24.114 Å by substituting the interlayer anions. The materials exhibited layered sheet-like morphology, preserved crystallinity, distinct interlayer anion fingerprints, uniform elemental distribution, consistent oxidation states, and enhanced interlayer water retention, all indicative of structural integrity and ion transport capacity. The expanded interlayer spacing and water facilitate rapid chloride ion diffusion, making these LDH materials highly suitable as cathodes in CIBs.

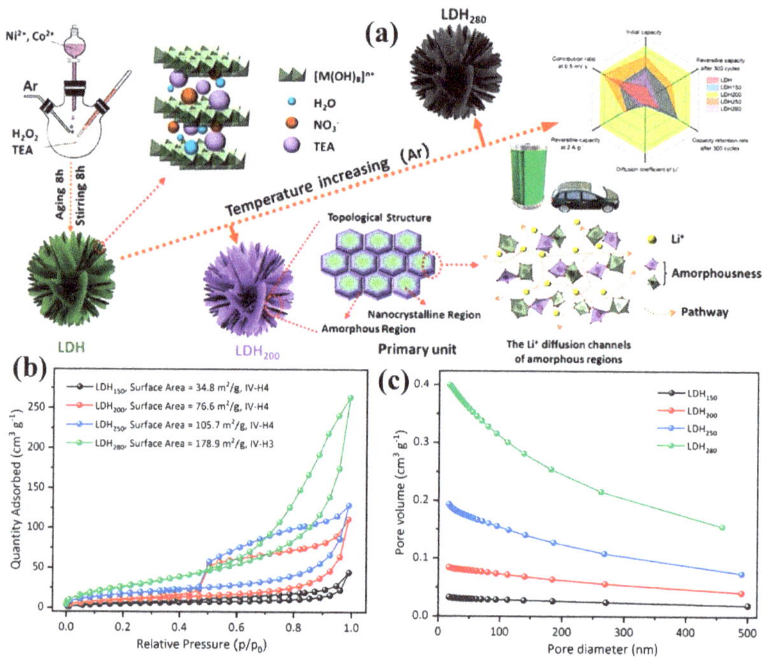

Figure 3. (**a**) Illustration of the synthesis pathway of temperature-regulated LDH and its transformation into an amorphous structure through thermal treatment. (**b**) Nitrogen adsorption/desorption curves and (**c**) pore size distribution profiles corresponding to the prepared samples. Reproduced with permission [109].

The NiMn-Cl LDH was synthesized via a co-precipitation method involving grafting NiMn-LDH nanosheets with chloride ion intercalation onto a CNT framework [113]. This process yielded a cross-linked network with a core/shell structure, enhancing ion transport and mechanical stability. These features make the material highly suitable for reversible chloride-ion storage in CIB applications. Similarly, Ni_5Ti-Cl LDH was synthesized via a simple co-precipitation method using a high nickel-to-titanium molar ratio (5:1), followed by an anion exchange process in HCl/NaCl solution to replace interlayer carbonate with chloride ions [114]. This resulted in successful Cl^- intercalation, evidenced by XRD peak shifts and IR spectra indicating carbonate removal. Figure 4 highlights the structural and chemical characteristics of Ni_5Ti-Cl LDH, designed to improve stability during cycling and enhance chloride ion adsorption. As seen in Figure 4a, the XRD analysis confirms the well-ordered layered structure. The FTIR results in Figure 4b show carbonate replacement with chloride, indicating successful ion exchange. Figure 4c presents a model structure of the LDH, showing expanded interlayer spacing favorable for Cl^- ion interaction. The TEM image and elemental maps in Figure 4d reveal a uniform distribution of key elements and confirm nanosheet morphology. Figure 4e displays XPS spectra, verifying the oxidation states of Ni and Ti and supporting the material's chemical composition and structure. The high redox activity (Ni^{2+}/Ni^{3+} and Ti^{3+}/Ti^{4+}) and excellent structural integrity with minimal volume change (~1.006%) make this LDH suitable for reversible chloride-ion intercalation in CIBs. The LDH was synthesized via a two-step method where ZIF-67 was first formed and then partially converted into Co-Fe-LDH by ion exchange and structural transformation under hydrothermal treatment at 80 °C for 48 h [115]. This Co-Fe-LDH was further hybridized with Fe-MOF and subsequently pyrolyzed at 900 °C under an argon atmosphere to yield the final CoFe/Fe_3C@CN heterojunction catalyst. The material possessed a nanosphere morphology with well-dispersed metal nanoparticles encapsulated by graphitic carbon layers, a high BET surface area of 247 m^2/g, abundant mesoporosity, and a heterojunction structure. The resulting structure exhibits high surface area, abundant oxygen vacancies, and fast electron transfer, making it highly suitable as a bifunctional cathode catalyst in ZABs. Fe/Co LDH was synthesized via a simple in situ co-precipitation reaction over a cobalt rich nitrogen-doped carbon framework (Co/N–C) derived by annealing ZIF-67 [116]. The resulting composite (CoL 2:1) was formed by varying mass ratios of LDH and Co/N–C precursors to optimize bifunctional catalytic performance. The material reveals a flaky LDH morphology on a porous carbon matrix with a high specific surface area of 230 m^2/g and a meso/microporous structure. This structure enhances the availability of catalytic sites and is highly suitable for ZAB applications due to its excellent bifunctionality and electrochemical stability.

3.2. Hydrothermal Synthesis

Hydrothermal synthesis is a versatile method involving aqueous reactions at elevated temperatures and pressures, promoting controlled nucleation, dissolution/recrystallization, and growth of crystalline LDHs. This approach enables precise size, morphology, composition, and crystallinity tuning, making it highly effective for fabricating LDHs with tailored properties suitable for various applications, including catalysis and environmental remediation. Hydrothermal synthesis of LDHs typically involves an initial co-precipitation of metal salts with a base, followed by treatment under controlled temperature and pressure within an autoclave to enhance crystallinity and phase purity [117]. During hydrothermal processing, diffusion-controlled nucleation and growth mechanisms facilitate the oriented assembly of metal hydroxide layers and the systematic intercalation of anions, resulting in highly ordered and defect-minimized structures [118,119]. The precise regulation of synthesis parameters such as temperature, pressure, reaction time, and precursor concentration enables fine-tuning of LDH composition, morphology, and functional properties, optimizing them for catalytic, ion exchange, and environmental remediation applications [120,121]. The hydrothermal synthesis of LDHs involves an intricate balance between nucleation and growth, beginning with rapidly forming a gelatinous, disordered metal hydroxide precipitate. Under hydrothermal conditions, this initial phase undergoes a dissolution/reprecipitation cycle driven by thermodynamic stability, resulting in a well-ordered, defect-free lamellar structure through controlled alignment and stacking of hydroxide layers [122]. Furthermore, hydrothermal synthesis uniquely facilitates the incorporation of multiple metal cations into LDH frameworks, enabling the development of multi-cationic or high-entropy LDHs with distinct electronic and structural attributes [77]. These high-entropy LDHs exhibit superior electrocatalytic properties due to synergistic interactions among diverse cationic species, making them highly promising for advanced applications such as CO_2 methanation and other catalytic processes [123,124].

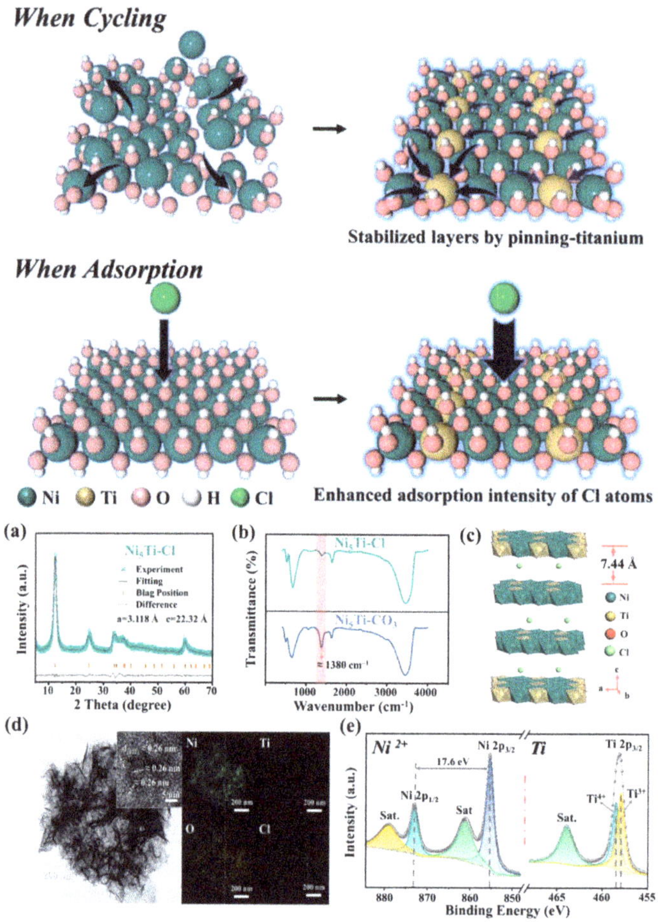

Figure 4. Demonstration of the structural stabilization and improved Cl adsorption in Ni_5Ti-Cl LDH. (**a**) XRD pattern refined using the Rietveld method for Ni_5Ti-Cl LDH. (**b**) FTIR spectra compare the Ni_5Ti-Cl LDH and its carbonate-based precursor (Ni_5Ti-CO_3 LDH). (**c**) Structural model illustrating the layer arrangement in Ni_5Ti-Cl LDH. (**d**) TEM image with SAED inset and elemental mapping showing distributions of Ni, Ti, O, and Cl. (**e**) Ni and Ti XPS profiles, confirming the oxidation states present in Ni_5Ti-Cl LDH. Reproduced with permission [114].

Hydrothermal synthesis provides precise control over morphology, particle size, and structural ordering of LDHs, enabling the tailored formation of uniform nanosheets, nanorods, hierarchical architectures, and ultrathin nanosheets with large surface areas beneficial for catalytic, adsorption, and energy-related applications [125]. This method facilitates controlled intercalation of anions, defect healing through ion migration, incorporation of multiple cations, and utilization of structure-directing additives or surfactants, enhancing LDH crystallinity, phase purity, ion exchange capacities, and electrochemical performance [126]. Additionally, the adaptability and scalability of hydrothermal synthesis make it environmentally attractive and economically viable for industrial-scale production of multifunctional LDHs optimized for diverse applications, including catalysis, environmental remediation, sensors, and energy storage systems [127]. Recent advancements in hydrothermal synthesis have enabled the fabrication of compositionally complex and highly crystalline LDHs with tunable properties for advanced applications [128].

By precisely controlling parameters such as temperature, pressure, pH, and precursor composition, this method facilitates the formation of defect-minimized, high-surface-area materials with tailored morphology, ion distribution, and interlayer characteristics [129]. High-entropy LDHs, multi-metallic composites, and nitrogen-doped structures synthesized hydrothermally demonstrate enhanced catalytic, adsorption, and electrochemical performance, particularly in environmental remediation and energy storage. The technique's scalability, energy efficiency, and use of water as a benign solvent further highlight its potential as a sustainable and industrially viable approach for producing multifunctional LDHs with minimal environmental impact [130]. Hydrothermal synthesis offers a highly versatile and controllable route for fabricating LDHs with tailored morphology, interlayer spacing, crystallinity, and multi-metal compositions [131]. Fine-tuning reaction parameters such as temperature, pressure, pH, and precursor concentration produces defect-minimized, thermally and chemically stable LDHs with enhanced ion exchange, catalytic, and electrochemical properties. The process supports in situ functionalization, hybridization with organic or carbonaceous materials, and integrating bulky anions or multiple cations [132]. It is ideal for diverse applications in catalysis, energy storage, environmental remediation, and advanced composites. Its scalability, reproducibility, and use of water as a benign solvent further establish hydrothermal synthesis as a sustainable and industrially viable strategy for developing high-performance, multifunctional LDH-based materials [133].

ZIF-67@NiCo-LDH heterostructures were synthesized using a straightforward one-pot hydrothermal method involving the reaction of cobalt and nickel nitrate precursors with 2-methylimidazole in methanol at 90 °C for 11 h [134]. This synthesis resulted in the formation of unique flower-shaped structures composed of ZIF-67 MOF cores uniformly decorated with NiCo-LDH nanosheets. The controlled reaction time allowed morphology tuning and ensured good dispersion and stability of the composite materials. Characterizations confirmed the heterostructures' flower-like morphology, high crystallinity, and uniform elemental distribution of Ni, Co, O, C, and N. The resulting porous structure exhibited enhanced electrolyte accessibility and improved electrical conductivity compared to the pristine LDH. The flower-shaped morphology and porous structure of ZIF-67@NiCo-LDH enhance ionic and electronic conductivity, making it suitable for LIB anode applications. $NiFe_2O_4$@NiCo-LDH was synthesized using a two-step hydrothermal process [135]. Initially, NiFe-MOF precursors were hydrothermally treated and annealed to obtain hollow $NiFe_2O_4$ nanocubes. Subsequently, NiCo-LDH nanosheets were hydrothermally grown onto the surface of these nanocubes, creating a hierarchical $NiFe_2O_4$@NiCo-LDH structure. The $NiFe_2O_4$@NiCo-LDH composite featured hollow hierarchical nanocubes with uniformly distributed NiCo-LDH nanosheets grown on $NiFe_2O_4$ surfaces. XRD and BET analyses revealed a crystalline structure with enhanced surface area (58.4 m^2/g) and mesoporous characteristics (pore size 2–3 nm), advantageous for electrochemical reactions. Due to its hierarchical hollow structure, high surface area, and mesoporosity, the $NiFe_2O_4$@NiCo-LDH is highly suitable for batteries, facilitating rapid electron and ion transport and alleviating volumetric expansion during cycling. NiZn-LDH intercalated with dodecyl sulfate anions was synthesized using a facile hydrothermal method [136]. Nickel nitrate and zinc nitrate precursors were dissolved under an inert atmosphere, followed by a controlled addition of sodium dodecyl sulfate and sodium hydroxide solution. The mixture was then subjected to hydrothermal treatment at 100 °C for 4 h, centrifuged, washed, and freeze-dried. The synthesized LDH exhibited ultra-thin nanosheet morphology with significantly expanded interlayer spacing due to the intercalation of dodecyl sulfate anions. BET analysis revealed a mesoporous structure with a specific surface area of around 53.4 m^2/g, indicating good accessibility and interaction sites for lithium ions. Due to its expanded interlayer spacing, ultra-thin nanosheet morphology, and favorable surface area, the synthesized NiZn-LDH provides effective diffusion pathways for lithium ions, thus making it suitable for battery applications.

LDH was synthesized initially through a solvothermal process, creating NiFe glyceric acid precursors using $Ni(NO_3)_2{\cdot}6H_2O$ and $Fe(NO_3)_3{\cdot}9H_2O$ in glycerol/isopropanol solution [137]. These precursors were then coated with polydopamine and subjected to high-temperature carbonization and vulcanization, forming NiS_2/FeS_2@NC structures. Finally, NiFe LDH/FeO(OH) nanosheets were grown on these structures via a hydrothermal reaction, resulting in hollow yolk/shell NiS_2/FeS_2@NC@NiFe LDH/FeO(OH) microspheres. The synthesized LDH exhibited a hollow yolk/shell structure with nanoflower morphology. The LDH's porous hollow yolk/shell structure and high specific surface area enable efficient ion transport and structural stability, making it highly suitable for battery applications. NiCo-LDH nanowires were synthesized using a simple one-step hydrothermal method [138]. Initially, MXene nanosheets were prepared by etching and delaminating Ti_3AlC_2. Subsequently, NiCo-LDH nanowires were uniformly grown in situ onto MXene nanosheets, forming a hierarchical 1D/2D heterostructure [139]. The composite exhibited a 1D nanowire-on-2D nanosheet morphology with increased specific surface area (14.50 m^2/g) and enhanced structural stability. The NiCo-LDH/MXene composite, with its hierarchical structure and enhanced conductivity, is highly suitable for battery applications as it ensures rapid lithium-ion diffusion and stable electrochemical cycling. MgAl LDH (Mg_2Al_1–CO_3-LDH) was synthesized using a simple hydrothermal method [140]. Carbonate ions (CO_3^{2-}) were intercalated within Mg and Al layers as interlayer anions. The synthesized LDH exhibited a typical hexagonal flake-like morphology with uniform elemental distribution. The Mg_2Al_1–CO_3-LDH reveals a well-defined hexagonal layered structure, a flake-like morphology, homogeneous elemental dispersion, and an interlamellar spacing of 7.45 Å. These analyses confirmed its composition as $[Mg_{0.68}Al_{0.32}(OH)_2](CO_3^{2-})_{0.16}{\cdot}0.66H_2O$ with a high structural integrity. The Mg_2Al_1–CO_3-LDH demonstrates suitable properties such as high specific surface area, structural stability, and active interlayer spaces, making it a promising candidate as an anode material for LIBs. The NiCo-LDH nanoflowers were synthesized via a hydrothermal method using nickel nitrate, cobalt nitrate, and CTAB, with in situ growth on either nickel foam (NF) or jute-derived AC (JAC-2) as substrates [89]. The optimal composite, NiCoLDH-1@JAC-2, was prepared by hydrothermally treating a mixture of Ni/Co in a 1:1 molar ratio with JAC-2 at 180 °C for 12 h. Figure 5a,b presents the synthetic routes for preparing JAC and the subsequent in situ formation of NiCo-LDH@JAC nanoflowers. The resulting NiCoLDH nanostructures exhibit a hierarchical nanoflower morphology composed of ultrathin nanosheets (~10 nm) with high crystallinity and a well-defined layered structure confirmed by XRD and HRTEM. JAC-2 support contributed significantly with its nanosheet structure, high surface area (~2600 m^2/g), and rich micro/mesoporous texture, enhancing ion transport. These structural and morphological features make the composite highly suitable for supercapacitor applications by maximizing active surface area and minimizing ion diffusion resistance.

The CoLa-LDH (CL) was synthesized via a hydrothermal method followed by a second hydrothermal step to grow NiCoOOH (NC) onto the CL, forming a hierarchical NiCoOOH@CoLa-LDH (NC@CL) heterostructure [141]. The NC was tightly coated over the CL nanosheets, yielding a flower-like 3D nanosheet architecture. Morphologically, the NC@CL exhibited interconnected ultrathin nanosheets with a high BET surface area of 316.4 m^2/g and mesoporous features (mainly 2–10 nm pores). Structural characterizations confirmed successful heterojunction formation and lattice deformation beneficial for ion/electron transport. The NC@CL structure provides abundant active sites and excellent electron/ion kinetics, making it highly suitable for ZIB applications. The NiCoMo LDH was initially synthesized via a hydrothermal method, followed by an anion exchange reaction to replace interlayer carbonate ions with chloride ions [142]. Subsequent temperature-differential phosphorus doping using NaH_2PO_2 vapor at different controlled temperatures introduced P heteroatoms into the LDH structure. Morphologically, the LDH retained a sea urchin-like microsphere architecture composed of nanowires, while XRD, TEM, and XPS confirmed structural integrity and successful P-doping, with increased oxygen vacancies and enhanced electronic properties. These features collectively support its application in CIBs by improving conductivity and diffusion kinetics.

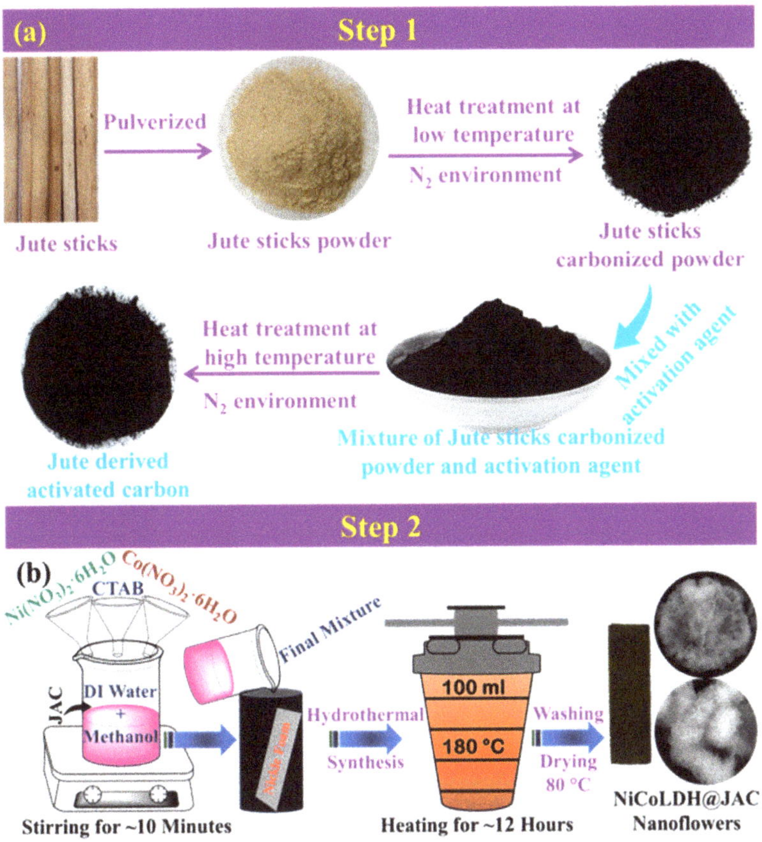

Figure 5. (**a**) Illustration of the JAC fabrication process involving a two-step pyrolysis method, and (**b**) detailed schematic of the in situ synthesis approach for growing NiCoLDH@JAC nanoflower structures. Reproduced with permission [89].

The NiAl-LDH was synthesized by first depositing Al_2O_3 on graphene using atomic layer deposition (ALD), followed by a hydrothermal transformation into NiAl-LDH@G [143]. This two-step process enabled the vertical growth of ultrathin LDH nanosheets on graphene, forming a nanoarray architecture. Characterization revealed a well-defined mesoporous structure with vertical LDH nanosheets (~3 nm thick), high surface area (230.44 m^2/g), and a uniform elemental distribution. This architecture enhances electron/ion transport and structural stability, making it ideal for CIB cathode applications.

The NiCo-based LDH was synthesized via a hydrothermal method, and chloride ions were introduced through a post-synthesis ion exchange process by treating the carbonate-intercalated LDH with an NaCl/HCl solution [144]. This treatment replaced CO_3^{2-} ions with Cl^- in the interlayer, enabling its function as a chloride host. The Mo-doped NiCo LDH exhibited a well-defined sea urchin-like morphology composed of ultrafine nanowires, with increased BET surface area (74.6 m^2/g), preserved LDH structure confirmed by XRD, and homogenous elemental distribution. This structural configuration provides enhanced active surface area and diffusion pathways, making the material highly suitable as a CIB cathode. The CoFe–Cl-LDH/CNT composite was synthesized via a modified co-precipitation method followed by hydrothermal treatment under N_2 protection [145]. The process involved the preparation of a Co/Fe/CNT precursor solution, pH adjustment, and hydrothermal aging at 130 °C, followed by Cl^- intercalation in NaCl/HCl solution. Characterizations revealed nanoflake-like LDH morphology with high crystallinity, uniform Cl^- intercalation, and improved conductivity due to CNTs. The material's layered structure, reversible Cl^- storage, and enhanced electrochemical performance make it suitable as an anode for CIBs. NiFe-LDH was synthesized on electrospun carbon nanofibers (CNFs) via hydrothermal treatment, followed by the in situ growth of conductive Ni-CAT nanorods using nickel acetate and HHTP at 85 °C for 12 h [146]. This hierarchical assembly yielded a 3D Ni-CAT/NiFe-LDH/CNFs composite. The morphology revealed interconnected nanosheets and nanorods forming a porous architecture, while XRD and XPS verified structural and chemical composition. The BET surface area of Ni-CAT/NiFe-LDH/CNFs was significantly high at 258.84 m^2/g. This structure ensures high conductivity, active site exposure, and porosity, making it highly suitable for ZAB applications.

Using a two-step hydrothermal process, the NiMn-LDH was synthesized over a $ZnCo_2O_4$ (ZCO) nanowire array pre-grown on nickel foam [147]. The first step involved forming ZCO nanowires, followed by hydrothermal deposition of NiMn-LDH nanosheets with varied Ni/Mn precursor ratios, optimized at 2:1. Characterization revealed that the LDH formed a uniform core/shell structure with high crystallinity, a flower-like morphology, and a large surface area (BET: 57.91 m^2/g), along with favorable pore distribution, layered structure, and high electrochemical surface area. This tailored architecture with synergistic interactions and abundant electroactive sites renders the LDH ideal for use as an efficient air cathode catalyst in ZABs. The NiCo-LDH/NCM@NF catalyst was synthesized through a two-step process [148]. First, nitrogen-doped carbon nanosheets (NCMs) were formed via self-assembly and calcination of perylene tetracarboxylic dianhydride and dicyandiamide under hydrothermal conditions; subsequently, NiCo-LDH nanoplates were grown in situ on these NCM-modified nickel foam substrates using hydrothermal treatment. The incorporation of NCMs during LDH growth modulated the morphology into rough-edged, ultrathin nanoflakes and enhanced nucleation and dispersion. Characterization revealed a 2D flake morphology, enlarged interlayer spacing, a high BET surface area (235.6 m^2/g for NiCo-LDH/NCM vs. 140.3 m^2/g for pristine NiCo-LDH), uniform elemental distribution, and rich oxygen vacancies. This LDH architecture ensures enhanced electron transport and more exposed catalytic sites, making it highly suitable for ZAB applications.

First, Co_9S_8 nanospheres synthesized NiFe-LDH@Co_9S_8 via hydrothermal treatment, were grown, followed by in situ growth of NiFe-LDH nanosheets on their surface [149]. Pt nanoparticles were then anchored onto the NiFe-LDH nanoflakes through a spontaneous redox reaction, exploiting cation vacancies in the LDH. NiFe-LDH nanosheet flowers were synthesized via a hydrothermal route [150], and two modified versions, CoNiFe-LDH (via one-step hydrothermal doping) and Co@NiFe-LDH (via two-step heterostructure formation with $Co(OH)_2$ nanoarrays) were developed. The morphology was characterized by FESEM and TEM, revealing spherical structures with smooth or flocculated surfaces; XRD, XPS, and SAED provided structural confirmation, while BET analysis showed Co@NiFe-LDH had the highest surface area (34.14 m^2/g) and hierarchical porosity. These structural and surface properties make the LDH, particularly Co@NiFe-LDH, well-suited for ZAB air cathode applications due to enhanced active site availability and electronic conductivity.

NiCo-LDH was synthesized via a one-step hydrothermal method by optimizing reaction temperature, time, and Ni/Co molar ratio, as shown in Figure 6a [119]. The resulting LDH showed a uniform sheet-like morphology with high porosity, well-defined crystalline structure, and the presence of Ni^{2+}/Ni^{3+} and Co^{2+}/Co^{3+} redox states, while FESEM revealed high surface area features favoring electrolyte diffusion and charge transport. Due to enhanced electron conductivity and electrochemical activity, these structural attributes make it an ideal candidate for supercapacitor applications. NiCo-LDH nanosheets were synthesized on nickel foam using a one-pot hydrothermal method involving nickel and cobalt nitrate precursors in the presence of HMTA at 110 °C [151]. As shown in Figure 6b, the basic structure of an LDH comprises stacked layers with positively charged metal hydroxides, separated by interlayer anions and water molecules. This layered structure is essential for the material's chemical and physical properties. Meanwhile, Figure 6c presents the hydrothermal synthesis approach for preparing NiCo-LDH nanosheets. It highlights how the morphology and structure of the nanosheets evolve depending on the duration of the hydrothermal treatment. Different reaction times can change sheet thickness, uniformity, and surface characteristics, significantly affecting the material's performance in various applications. The resulting LDH exhibited a binder-less, ultrathin, porous nanosheet morphology with high crystallinity and a large specific surface area of 46.8 m^2/g (for the optimized sample). These structural and morphological features facilitated rapid ion transport and abundant electroactive sites, rendering the material highly suitable for supercapacitor applications. The CoLa-LDH (CL) was synthesized via a hydrothermal method, followed by a secondary hydrothermal deposition of NiCoOOH (NC) to form the hierarchical NC@CL heterostructure [141]. As seen in Figure 6d, the process involves two hydrothermal steps. First, the CL nanosheets are synthesized at 120 °C for 5 h. Then, a second hydrothermal treatment at the same temperature for 3 h enables the formation of NC@CL, where a second layer is formed on top of the initial CL structure. Figure 6e–g presents FESEM images of the as-prepared CL nanosheets at different magnifications. These images reveal a highly porous, interconnected sheet-like structure that facilitates ion and electron transport. In comparison, Figure 6h–j displays the FESEM images of the NC@CL nanosheet arrays. The surface becomes more complex and flower-like, indicating the successful growth of the secondary NC layer. This hierarchical structure can enhance surface area and active sites, making it beneficial for energy storage or catalysis applications. The resulting NC@CL nanosheets displayed a 3D hierarchical flower-like morphology with a high BET surface area (316.4 m^2/g), mesoporous structure, and uniform elemental distribution; XRD and XPS analyses confirmed the formation of a mixed-phase structure with strong electronic interactions and lattice distortions beneficial for charge transport. These characteristics render the NC@CL heterostructure particularly advantageous for supercapacitor applications due to enhanced ion/electron transport and abundant electroactive sites.

The Ni_2Co-LDH was synthesized via a simple co-precipitation method, where nickel and cobalt chloride precursors reacted with hexamethylenetetramine under reflux conditions at 80 °C for 8 h [152]. Expanded graphite (EG) was separately synthesized by rapid microwave irradiation of expandable graphite powder. Subsequently, Ni_2Co-LDH/EG composites were prepared through an in situ growth technique, anchoring Ni_2Co-LDH nanosheets onto the surface of EG. Structural and morphological characterizations revealed layered, crystalline nanosheets of Ni_2Co-LDH anchored uniformly onto EG surfaces, with lattice spacings consistent with Ni_2Co-LDH. BET analysis indicated enhanced surface area from 26.1 m^2/g for pure LDH to 30.3 m^2/g for the composite, confirming effective integration with EG. The enhanced surface area and stable layered morphology confirmed by characterizations facilitate improved ionic and electronic transport, making Ni_2Co-LDH/EG composites suitable for battery electrode applications. NiAl LDH was synthesized via a one-pot hydrothermal reaction using nickel nitrate, aluminum nitrate, and urea dissolved in deionized water [153]. The solution was transferred to a Teflon-lined autoclave and heated at 100 °C for 24 h. The high surface area, porous nano-lamellar morphology, and structural stability of the synthesized NiAl LDH make it a promising candidate for LIB anodes. The CoNi-LDH/C composite was synthesized by first preparing ZIF-67 polyhedrons, which were then subjected to laser-induced carbonization to obtain porous Co/C structures [154]. These carbonized materials were reacted with $Ni(NO_3)_2{\cdot}6H_2O$ in ethanol under hydrothermal conditions at 90 °C to form the final CoNi-LDH/C composite. This process was optimized to retain polyhedral morphology while introducing amorphous carbon and catalytic metal sites. The resulting material exhibited a mesoporous structure with a high specific surface area (24.5 m^2/g), tubular morphology, and crystalline CoNi-LDH domains enclosed by conductive carbon layers. Due to increased carbon content, structural characterizations confirmed the uniform distribution of Co/Ni, porous architecture, and enhanced conductivity. This hierarchical and conductive CoNi-LDH/C framework provides abundant active sites for LiPS adsorption and conversion, making it highly suitable for LSB cathodes. Li-Al LDH was synthesized via hydrothermal $Al(OH)_3$ and LiCl treatment at 90 °C for 24 h [155]. The LDH was exfoliated in formamide and reassembled with varying amounts of polystyrene sulfonate (PSS) to form LDH@PSS. The optimal composition (LDH/PSS = 1:1.5) was obtained through self-assembly under N_2 atmosphere and centrifugation. The notable structural and electrochemical properties render LDH@PSS highly suitable for stabilizing Li metal anodes in LSBs by suppressing polysulfide shuttling and dendrite formation.

The FeNi-LDH was synthesized hydrothermally by mixing $NiCl_2{\cdot}6H_2O$, $FeCl_2{\cdot}6H_2O$, and L-cysteine in water/ethanol, followed by ammonia addition, then heating in a Teflon-lined autoclave at 180 °C for 12 h [156]. The resulting cysteine-functionalized FeNi-LDH (Cys/FeNi-LDH) was further reacted with $NaVO_3$ in methanol under reflux to yield the ternary V_2O_5/Cys/FeNi-LDH (VCFN) composite. This process ensured the uniform decoration of V_2O_5 nanoparticles on the LDH disks through strong coordination with cysteine functional groups. This LDH-based composite, with its synergistic combination of polysulfide adsorption, catalytic conversion capability, and mesoporous framework, is highly suitable for suppressing the shuttle effect and enhancing sulfur redox kinetics in LSBs. The NiFe-LDH, NiAl-LDH, and ZnAl-LDH were synthesized via a hydrothermal method by co-dissolving respective metal nitrates and urea in deionized water, followed by heating under controlled temperatures for different durations [157]. Sulfur was incorporated into the LDHs using the classical melt/impregnation method at 155 °C, achieving ~75% sulfur loading. Characterization revealed that the LDHs had nanoflake morphology with a 3D network-like architecture, uniform sulfur distribution, and surface areas ranging from 52.6 to 54.8 m^2/g. This structure with high surface area and interlayer spacing enabled strong interaction with polysulfides, making the LDHs suitable sulfur hosts for LSBs. NiCoAl-LDH nanosheets were synthesized on carbon cloth (CC) via a hydrothermal method [158]. Subsequently, vertical graphene was grown on the LDH/CC surface using plasma-enhanced chemical vapor deposition (PECVD), forming a 3D interconnected structure. The morphology showed uniformly distributed nanosheets and vertically aligned graphene, with lattice spacings confirming NiCo alloy and graphite planes. This nanocomposite offers strong polysulfide adsorption and fast redox kinetics, making it a suitable host for LSBs. The NiCr-Cl LDH/rGO composite was synthesized via a one-step hydrothermal method to produce NiCr-CO_3 LDH/rGO, followed by ion exchange in a NaCl/HCl solution to replace CO_3^{2-} with Cl^- ions [159]. This approach involved urea-assisted hydrothermal reaction at 150 °C for 24 h, forming a well-dispersed LDH on GO sheets. Morphological and structural characterizations revealed flower-like nanosheet morphology, increased interlayer spacing after Cl^- exchange, and intense interaction between LDH and rGO; the BET surface area of the NiCr-CO_3 LDH/rGO composite was 170.44 m^2/g, supporting good electrochemical performance. The resulting LDH composite demonstrated enhanced suitability for NIB applications due to its improved conductivity, mesoporous structure, and structural stability.

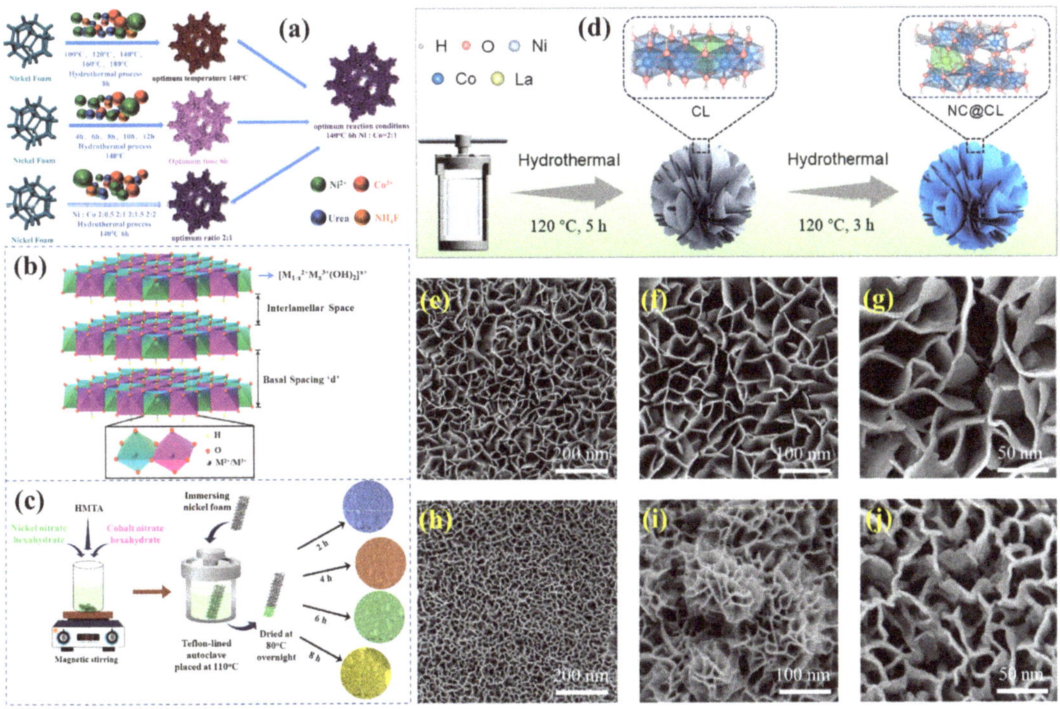

Figure 6. (**a**) Schematic representation for the NiCo-LDH synthesis. Reproduced with permission [119]. (**b**) Illustration showing the structural layout of an LDH material, and (**c**) diagram depicting the hydrothermal synthesis process of Ni-Co LDH nanosheets at various time intervals. Reproduced with permission [151]. (**d**) Conceptual diagram outlining the stepwise hydrothermal synthesis of NC@CL nanosheet arrays, beginning with CL formation and followed by growth of the NC layer. (**e–g**) FESEM images showing the surface morphology of the CL nanosheets at increasing magnifications. (**h–j**) FESEM images displaying the detailed nanostructure of the NC@CL hierarchical flower-like nanosheet arrays. Reproduced with permission [141].

The NiCo-LDH was synthesized via a hydrothermal method by immersing pre-treated sulfide-hollow ZIF-67 structures into a solution of $Ni(NO_3)_2$, $Co(NO_3)_2$, and hexamethylenetetramine at 90 °C for 24 h [160]. The resulting LDH featured a nanoflower-like morphology, layered structure, and a BET surface area of 266.76 m^2/g with a mesoporous architecture, enhancing electron/ion transport and active surface sites. Due to its superior redox activity and efficient charge storage capability, these characteristics render the NiCo-LDH highly suitable for high-performance supercapacitor applications. NiFe-LDH/$MnCO_3$/MXene (NFMM) ternary composite was synthesized via a hydrothermal method by reacting Ni, Fe, and Mn nitrates with urea in the presence of MXene [161]. The resulting NFMM exhibited a layered and granular sandwich-like structure with uniformly distributed $MnCO_3$ and NiFe-LDH on MXene, showing a high BET surface area of 32.62 m^2/g and a hierarchical mesoporous network with improved electronic conductivity. These properties make NFMM an ideal candidate for supercapacitor electrodes due to its enhanced ion/electron transport and abundant active sites. The FeCoNi-LDH nanosheets were synthesized via a secondary hydrothermal reaction onto pre-formed $MnCo_2O_4$ nanowires anchored on CC [162]. The resulting $MnCo_2O_4$@FeCoNi-LDH exhibited a nanoflower-like, hierarchical morphology with high crystallinity, uniform nanosheet distribution, and a mesoporous structure, showing a specific surface area of 33.9 m^2/g and distinct XRD and XPS signatures confirming multi-metal synergy. Figure 7 displays the synthesis steps and morphology changes of MnCo before and after FCN coating on CC, highlighting structural differences observed under SEM. This architecture enhances ion diffusion and redox activity, making it highly suitable for high-performance supercapacitor applications.

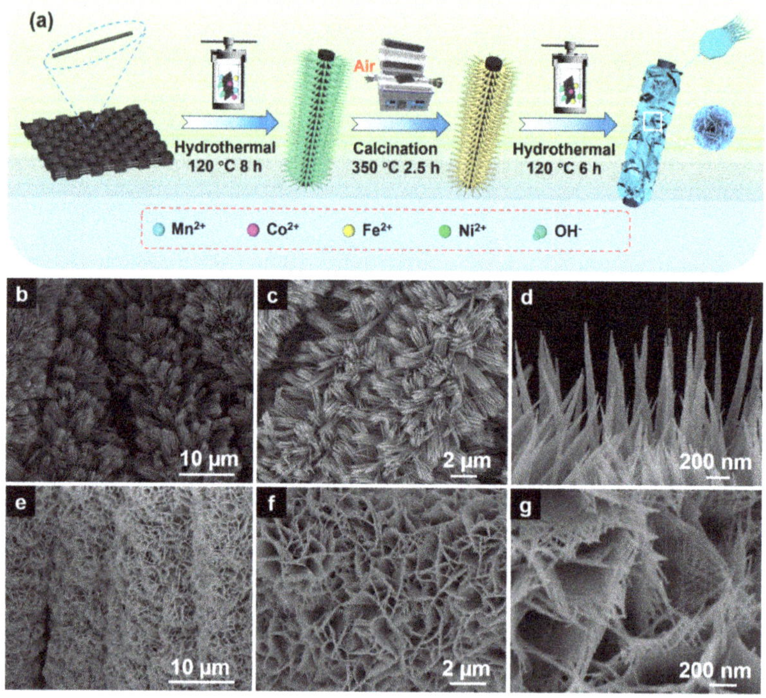

Figure 7. (**a**) Illustration of the step-by-step process to make MnCo@FCN on CC via hydrothermal synthesis. (**b–d**) FESEM images showing the surface of MnCo on CC at various zoom levels. (**e–g**) FESEM images of MnCo@FCN coated on CC under different magnifications. Reproduced with permission [162].

CoMo-LDH was synthesized via hydrothermal treatment of cobalt nitrate and sodium molybdate in the presence of urea, followed by coating with polydopamine (PDA) through a polymerization process and subsequent carbonization at 800 °C to form CoMoLDH@C-PDA composites [163]. The duration of the hydrothermal and polymerization steps was optimized to achieve controlled morphology and appropriate carbon layer thickness. The carbon-coated CoMoLDH is considered suitable as an anode material for NIBs due to enhanced conductivity, structural stability, and sodium-ion diffusion. Co-Co LDH was synthesized via a hydrothermal method using ZIF-67 as the precursor and grown onto nitrogen-doped hollow carbon nanospheres (NHCNS) [164]. The resultant Co-Co LDH@NHCNS was then selenized and coated with a carbon shell to obtain the final NHCNS@$CoSe_2$@C composite. The composite exhibited a nanoflower-like morphology with uniformly distributed $CoSe_2$ nanoparticles on the NHCNS surface. XRD validated the successful transformation to $CoSe_2$; XPS and Raman revealed the chemical states and defect structures, and TGA estimated a carbon content of 43.4%. This unique sandwich structure offers enhanced conductivity, stability, and sodium storage capability, making it highly suitable for NIB applications. The CoNi-LDH was synthesized via a two-step hydrothermal method, where cobalt phosphate (CP) was first grown on nickel foam, followed by in situ growth of molybdate-intercalated CoNi-LDH (MCN-LDH) on the CP surface [165]. Incorporating molybdate ions was achieved using urea and ammonium fluoride as structure-directing agents. The prepared LDH exhibited interconnected nanosheets on microsheets; XRD confirmed phase structure; BET analysis showed a surface area of 212.68 m^2/g; Raman and XPS confirmed successful MoO_4^{2-} intercalation. These structural and textural improvements make the MCN-LDH highly suitable for high-performance ZIB cathode applications. ZnCo-LDH was synthesized via a hydrothermal process by dissolving $CoCl_2 \cdot 6H_2O$, $ZnCl_2$, urea, and PVP in deionized water, then heating at 120 °C for 12 h [166]. The resulting product was collected, washed, and dried to obtain ZnCo-LDH microspheres. BET measurements showed a surface area of 37.1 m^2/g with mesopores averaging 28.2 nm in diameter. These structural features, including high surface area and mesoporosity, make the material highly suitable for electrochemical applications in ZIBs.

NiAl LDH was synthesized using a one-step hydrothermal method [167]. The resulting LDH exhibited a porous nanosheet morphology with oxygen vacancies, high crystallinity, hierarchical nanostructures, and a BET surface area of up to 300 m^2/g, promoting efficient ion transport and electron transfer. These features make it a suitable candidate for supercapacitor applications due to its enhanced redox activity, mechanical integrity, and electrochemical performance. MoB@NiCo-LDH was synthesized via a hydrothermal process by growing NiCo-LDH nanosheets on layered MoB templates derived from NaOH-etched MoAlB; subsequent phosphorization and $NaBH_4$ treatment introduced phosphorus vacancies, resulting in MP2 material [168]. The synthesized LDH exhibits an amorphous structure with a nanosheet-to-nanoparticle morphological transition, a layered architecture, uniform elemental distribution, enhanced electrical conductivity, and abundant surface-active sites due to phosphorus vacancies. These features contribute to fast charge transport, high redox activity, and structural stability, making the LDH ideal for supercapacitor applications.

The Te-doped NiMn LDH was synthesized via hydrothermal growth on CTAB-modified MXene-coated nickel foam, followed by tellurium doping through immersion in $TeCl_4$ ethanol solution [169]. The step-by-step design strategy and structural characterization of the Te-doped NiMn LDH system are illustrated in Figure 8. The CTAB-modified MXene coating on nickel foam (C-MXene/NF) is shown in Figure 8, which offers a rough, layered texture that supports better growth of the active material. Figure 8a,b shows the FESEM image of NiMn LDH directly grown on bare nickel foam, revealing a relatively smoother and flatter structure. With the addition of the MXene layer, Figure 8c shows that the NiMn LDH forms a more clustered, 3D nanostructure on C-MXene/NF, offering increased surface area and potential active sites. Figure 8d displays the final morphology after tellurium doping, which results in a highly porous, flower-like structure. This modification enhances electrochemical activity by introducing more redox-active centers and improving conductivity. XRD confirms the structural phases of the different samples in Figure 8e, where distinct peaks are observed for NiMn LDH/NF, NiMn LDH/C-MXene/NF, and Te-NiMn LDH/C-MXene/NF. Reference patterns for $Ni(OH)_2$ and $Mn(OH)_2$ are also compared, confirming the successful formation of the LDH phases and the influence of doping and substrate modification. The prepared LDH displayed vertically aligned, nanosheet-like morphology with enhanced crystallinity, a significantly expanded interlayer spacing, increased surface roughness, and higher electrochemical surface area; BET analysis confirmed elevated surface area due to CTAB intercalation. Such morphological and structural characteristics, combined with improved conductivity from Te doping and MXene support, make this LDH ideal for high-performance supercapacitor applications.

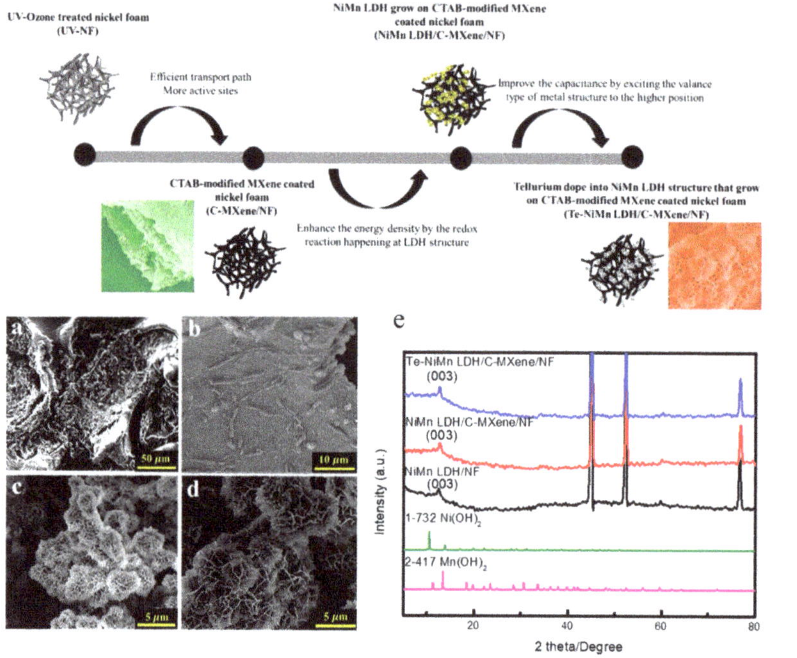

Figure 8. Overview of the synthesis process for Te-doped NiMn LDH on CTAB-modified MXene-coated nickel foam. FESEM images showing surface morphologies at various stages: (**a**) C-MXene/NF, (**b**) NiMn LDH on bare NF, (**c**) NiMn LDH on C-MXene/NF, and (**d**) Te-doped NiMn LDH on C-MXene/NF. (**e**) XRD patterns comparing NiMn LDH/NF, NiMn LDH/C-MXene/NF, and Te-NiMn LDH/C-MXene/NF, with standard reference peaks of $Ni(OH)_2$ and $Mn(OH)_2$ included for comparison. Reproduced with permission [169].

3.3. Electrochemical Deposition

Electrochemical deposition reduces metal ions from a solution with the potential to generate thin lamellar films on substrates for LDHs. Binders and post-treatment are unnecessary with this method [170–172]. Faradaic interactions at the electrode/electrolyte interface generate localized hydroxide ions, which combine with metal cations in solution to precipitate LDHs' layered structure [92,173]. This approach directly synthesizes LDHs with controlled composition and form by changing electrochemical parameters, making it a strong functional material creation tool. Electrochemical environments improve deposition kinetics, enabling fast nucleation for nanostructured coatings with high surface areas. Environmental friendliness and operational simplicity approach a viable alternative to existing chemical precipitation procedures. Electrochemical deposition for LDH synthesis reduces oxygen-containing species at the cathode, quickly raising local pH and forming metal hydroxide layers. Metal hydroxide layers form a lamellar structure held together by intercalated anions like carbonates or nitrates. The water reduction rate and hydroxide ion supply for LDH production rely on the applied voltage. Current density and deposition duration can often change the LDHs' deposit shape and crystallinity. Researchers can tune LDH film performance for catalysis and energy storage.

Water electrolysis raises local hydroxide concentrations, which react with divalent and trivalent metal cations to synthesize LDHs by electrochemical deposition [174]. Metal hydroxide layers form in situ and stack into an organized array. The interaction between nucleation and growth controls deposition, ensuring uniform deposition across the electrode surface with controlled current or potential. The mechanism is complicated since deposition kinetics and ion transport compete to determine film structure. Understanding and regulating these parameters is necessary for making high-quality LDH films with desirable qualities. LDH film shape and crystallinity depend on electrodeposition parameters such as applied voltage, current density, and deposition time. Researchers can optimize these settings to create LDH deposits with consistent thickness and well-defined platelets [174]. Changing the electrolyte composition and pH controls LDH nanofilm deposition kinetics and structural evolution. These optimized conditions give synthesized films excellent electrochemical performance, which is helpful for catalytic applications and energy storage devices. Extensive parameter optimization is necessary to synthesize materials with specific functional properties [175]. Electrochemical deposition of LDHs commonly has hexagonal platelet-like formations after alkaline treatment, as documented in the literature [176]. Reorientation and crystallization of amorphous deposits under alkaline circumstances create an ordered lamellar pattern. Ion concentration and water decrease support platelet formation. Morphological aspects directly affect functional properties like ion intercalation and film catalytic activity. Optimizing LDH-based electrode performance requires precise control over deposition and post-treatment. Alternating metal hydroxide layers and intercalated anions or water molecules give LDHs their lamellar structure. Recent research confirms this tendency [93]. The systematic stacking of LDHs is caused by metal cations reacting with hydroxide ions during electrochemical deposition. Interlayer anions balance the positive charge of metal hydroxide layers and improve structural stability and functionality. The deposition environment can also be modified to these intercalated species, further controlling material characteristics. The deposition method naturally forms highly structured, anion-rich structures, enabling rapid ion exchange and high specific surface areas.

When comparing electrodeposition with traditional synthesis methods, such as co-precipitation or hydrothermal techniques, it becomes evident that electrochemical deposition offers significant advantages for the fabrication of LDHs [177]. This technique enables the direct deposition of LDHs onto conductive substrates, thereby eliminating the need for binder agents and enhancing electron transfer efficiency. In addition, the in situ formation of hydroxide films during electrodeposition leads to rapid synthesis and improved uniformity relative to conventional methods. Such direct synthesis is crucial in applications where interface quality between the active material and the substrate determines overall device performance. Consequently, electrochemical deposition has become a preferred method for preparing LDH films in many high-performance electrochemical systems. Electrochemical deposition presents several notable advantages, including precise control over film thickness, composition, and morphology, which are critical factors for optimizing the properties of LDHs [174,177]. This approach may change reactant concentration and deposition rate by carefully adjusting the applied voltage and current. This control guarantees uniform nucleation density across the substrate, decreasing flaws and improving electrode performance. Electrodeposition also makes LDH integration on metal foams and carbon-based electrodes easy. These tunable characteristics allow researchers to customize LDHs' physical and chemical features for energy storage and electrocatalysis applications. Recent studies have exemplified the successful application of electrodeposition in synthesizing NiCo-LDH films directly onto nickel foam substrates, a configuration that significantly enhances electrocatalytic performance [174,178]. Direct development of LDHs on structured substrates provides excellent electrical connection and a high electrochemically active surface area for effective catalytic reactions. Interconnected nanosheets in the hierarchical microstructure enable quick charge transfer and mechanical stability under operational conditions. The substrate/active LDH material synergy boosts device performance, making them attractive for water splitting and supercapacitor electrodes. Electrodeposition helps synthesize improved composite electrodes for energy conversion and storage. Pulse electrodeposition, a variant of the traditional electrochemical deposition technique, involves applying cyclic pulses of current or potential, which are sometimes interspersed with reverse pulses [179]. Incorporating pauses between deposition pulses alleviates the local depletion of metal ions, thereby allowing a more uniform ion concentration to be maintained across the deposition zone. This technique effectively enhances the morphological quality of the deposited LDH films, leading to more homogeneous structures with reduced defects. Additionally, reverse pulses can help remove undesirable surface charges that may otherwise impede further deposition or lead to non-uniform film growth. As a result, pulse electrodeposition is a valuable method for fine-tuning the properties of LDH films in various electrochemical applications.

Recent research using in situ techniques, such as atomic force microscopy, has revealed the dynamic morphological evolution of Co-based LDH nanosheets during electrodeposition [180]. These studies have shown that the initial nucleation step is critical in establishing the eventual sheet-like structure and that subsequent growth is susceptible to the applied deposition parameters. The evolution process is characterized by the formation of nanoscale features that promote efficient charge transport and ion diffusion. Moreover, the real-time observation of the growth process has provided critical insights into the nucleation and growth mechanisms, enabling further optimization of deposition protocols. This understanding is vital for designing LDHs with enhanced electrocatalytic or supercapacitive properties, as the nanosheet structure is key to their high surface area and functionality. The impact of deposition time and current density is well documented, with studies demonstrating that prolonged deposition or increased current can significantly enhance the crystallinity and overall quality of LDH films [181]. Longer deposition times allow for gradually reorganising initially amorphous deposits into highly ordered, crystalline platelets suitable for various electrochemical applications. Similarly, higher current densities can accelerate hydroxide ion generation, boosting nucleation rates and leading to finer nanostructures. However, excessive current may also lead to inhomogeneities; thus, optimizing these parameters is crucial to maintaining the lattice structure. Therefore, systematic studies that vary these factors are essential for mapping the optimal conditions that yield LDHs with superior structural and electrochemical characteristics.

A significant body of work has centered on the electrochemical deposition of NiFe and NiCo LDHs, which have been shown to provide high performance for water oxidation and other electrocatalytic reactions [182,183]. In these systems, the deposition process is optimized to ensure the formation of ultrathin nanosheets, whose abundant active sites contribute to enhanced catalytic activity. The simultaneous incorporation of two types of metal ions in the LDH structure is particularly beneficial, as it facilitates synergistic effects that lower overpotentials and improve reaction kinetics. Moreover, such bimetallic systems benefit from the tunability offered by electrodeposition, allowing researchers to balance the composition and morphology for maximum efficiency. These studies provide compelling evidence that electrodeposited LDHs can meet the rigorous demands of modern electrocatalytic applications.

In another compelling example, the synthesis of CoNi-LDH films via electrodeposition onto copper substrates has been reported for high-performance supercapacitor applications [184]. Direct deposition promotes strong adhesion between the active material and substrate, improving device stability and charge/discharge cycling. Hierarchical nanostructures created by electrodeposition improve accessible surface area, boosting pseudocapacitive performance, according to researchers. Active sites are evenly distributed over the electrode surface, enabling fast ion transport and energy storage. According to such studies, electrodeposition methods are widely applicable and work well in LDH-based energy storage devices. The pulsed deposition methods, as discussed earlier, also play a critical role in refining LDH film morphology by allowing for periodic replenishment of metal ions and the removal of excessive surface charge [179]. This controlled deposition method produces uniform, defect-free coatings with outstanding adhesion and electrochemical characteristics. The intermittent deposition approach slows reactant depletion near the cathode, assuring film growth. Such methods benefit complex multimetal systems that require homogeneous distribution for desirable electrical properties. Thus, pulse electrodeposition provides a versatile and practical approach for synthesizing improved LDH materials.

Electrochemical deposition has also been extended to synthesizing CoFe and CoNi LDH catalysts, which show superior OER performance owing to their enhanced conductivity and abundant active sites [177,185]. In these studies, precise control over the deposition conditions results in ultrathin films that maximize the exposure of catalytically active sites while maintaining excellent adherence to the substrate. The voltage and current parameters are optimized to balance the nucleation and growth processes, ensuring the resulting film is uniform and highly porous. Such optimization is key to reducing the overpotential required for OER and enhancing overall catalytic efficiency. The resulting electrode architectures, therefore, present a promising route for developing cost-effective and efficient water splitting systems. Consistent with the critical role of processing parameters, numerous experimental studies have underscored that the nucleation rate and the growth kinetics during electrodeposition are pivotal in determining the overall quality of LDH films [174]. Optimal equilibrium between these elements ensures continuous, defect-free, nanostructured deposited layers. Furthermore, systematic modification of deposition time and applied voltage has greatly affected film shape and crystallinity. Finely tailored deposition techniques increase electrochemical performance and reproducibility across substrate materials. Optimizing the electrodeposition process is necessary to produce high-quality LDH films for advanced energy and catalytic applications.

The versatility of electrodeposition has allowed for the synthesis of LDH films on various substrates, including metal alloys and conductive carbon materials, thereby broadening the scope of potential applications [186,187]. Growth on substrates like titanium and aluminum alloys has been demonstrated to improve corrosion resistance and enhance the material's durability in aggressive environments. Similarly, deposition on carbon-based substrates offers benefits such as improved conductivity and flexibility, which are crucial for fabricating next-generation electrodes. The inherent adaptability of the electrochemical deposition process accommodates modifications to suit various substrate types while retaining high film quality and consistency. Thus, the method's applicability to various substrates significantly expands potential applications for LDH-based materials. The electrodeposition method inherently facilitates the formation of binder-free LDH films, thereby eliminating potential interference from insulating binders and enhancing electrical connectivity between the active material and the substrate [188,189]. This direct deposition onto conductive substrates results in improved electron transfer and overall device performance, which is particularly beneficial in electrochemical sensing and energy storage devices. Without additional processing steps, the simplicity and efficiency of the process are markedly enhanced, leading to lower production costs and higher material uniformity. Moreover, the direct interface formed by binder-free deposition ensures intimate contact between the catalyst and the electrode, a crucial factor for high-performance devices. Therefore, electrodeposition is an efficient route to achieving high-quality, binder-free LDH films with optimized properties.

The fundamental mechanism driving the electrochemical deposition of LDHs pivots on the local generation of hydroxide ions from water reduction, which quickly interact with dissolved metal cations to create precipitated hydroxide layers [175]. This process is initiated at the cathode and progresses by the continuous supply of hydroxide ions, resulting in sustained growth of the layered structure as long as the applied potential is maintained. The reaction environment is finely tuned so that the concentration gradients created during deposition facilitate the uniform nucleation of the LDH phase across the electrode surface. Additionally, the ease of ion migration and fluid dynamics within the electrolyte further contributes to the modernization of the deposition front. Thus, the coupling between electrochemical reactions and diffusive transport is essential in establishing the layered architecture characteristic of LDHs. Under optimized electrochemical deposition conditions, it is possible to generate LDH films exhibiting diverse morphologies, including nanosheets, nanoframes, and porous networks [180,183]. Deposition parameters, including current density, pulse duration, and electrolyte composition, can be adjusted to control these morphological differences to match the application. Rapid nucleation and anisotropic growth in the substrate plane produce ultrathin nanosheets with excellent lateral connectivity. Slower deposition rates and higher ion diffusion can create a more porous network, which is helpful for quick ion transport applications. Thus, controlling the deposition environment allows researchers to construct LDH films with unique morphologies for specific functional tasks.

The deposition process is susceptible to external parameters like pH and temperature, which influence the reaction kinetics and the structural outcome of the synthesized LDH films [190]. High pH levels enhance nucleation by increasing hydroxide ion concentration, forming hydroxide layers that assemble into lamellar structures quickly. Temperature also helps constituent ions move, enabling rearrangement to create a crystal lattice. This sensitivity to deposition circumstances requires rigorous experimental control because even small variations might impact material characteristics. The interaction between pH, temperature, and applied voltage is crucial for regulated deposition and high-quality LDH films. The overall synthesis of LDHs by electrodeposition has been widely employed in developing electrocatalysis and energy storage device catalysts [181,182]. The layered structure of LDHs, alternating metal hydroxide sheets intercalated with anions, provides many active sites and improved ion transport routes in various applications. These functional applications require accurate film thickness and composition control, which the deposition process offers. Electrodeposited LDHs adhere well to substrates, making them stable under high-current or long-term operation. In addition to being a straightforward synthetic pathway, the technology allows the integration of high-performance LDH materials in innovative technological applications [191]. Structural and electrochemical analyses have indicated that the LDH films produced through electrodeposition feature a unique lamellar morphology and highly porous structure, essential for rapid ion transport and high surface reactivity [177,192]. These materials have large interlayer spacing because the deposition process creates loosely packed layers intercalated by anions or solvent molecules. Electrocatalysis and supercapacitors require rapid ion diffusion and high charge storage capacities, making porousness favorable. The films' wide pore channels minimize ion transport resistance, improving electrochemical performance. The structure of electrodeposited LDHs allows for adequate charge storage and robust catalytic processes.

Recent examples in the literature have demonstrated that strategic modulation of the pulse deposition sequence, by incorporating rest periods and reverse pulses, can significantly enhance the quality of LDH films [174,179]. Incorporating pauses between active deposition pulses allows for the equilibrium re-establishment of metal ion concentration near the electrode surface, thereby avoiding depletion effects and promoting uniform growth. Furthermore, reverse pulses help remove excess charge accumulation on the surface, which can otherwise lead to undesirable side reactions or film defects. The precise control over these deposition parameters enables the creation of LDH films with highly uniform thicknesses and minimized defect densities, both of which are crucial for ensuring high electrochemical performance. Therefore, advanced pulse modulation strategies have emerged as indispensable tools in the electrodeposition synthesis of high-quality LDH materials.

Certain studies have highlighted the deposition of MnCo-based films, where incorporating cobalt into the electrodeposited film significantly affected the electrocatalytic performance of OERs [175,193]. This research demonstrated that by varying the deposition parameters, the crystallinity and morphology of the LDH films could be tailored to enhance their catalytic activity. In these cases, the presence of cobalt was shown to improve the films' electronic conductivity and redox capabilities, thereby reducing the overpotential required for catalysis. Additionally, the synergistic effect of the mixed metal composition resulted in enhanced nucleation rates and more uniform film formation during deposition. These findings underscore the importance of compositional tuning in the electrodeposition process to optimize the functional properties of LDHs [175,193].

The transformation of precursor materials such as MOFs into LDHs via electrochemical deposition is another emerging avenue that broadens the synthesis strategies available for these materials [190,194]. In this approach, the MOF serves as a sacrificial template, which, upon exposure to an alkaline environment and an applied potential, is converted into an LDH. This method offers the advantage of controlling the porosity and morphology of the final product by precisely tuning the conversion process. The reconstruction of MOFs into LDH structures simplifies the synthetic steps and results in materials with enhanced catalytic and electrochemical properties. As a result, this technique has garnered attention as a promising strategy for producing advanced LDHs with tailored features for diverse applications. In a related innovation, the reconstruction of Prussian blue analogs into thin LDHs has been meticulously demonstrated, offering a pathway toward synthesizing efficient electrocatalysts [194,195]. One of the foremost benefits of electrodeposition is the direct integration of LDH films onto conductive substrates, eliminating the need for non-conductive binders that might hinder electron transport [177,196]. The resulting electrodes exhibit excellent electrical connectivity and reduced interfacial resistance by depositing LDHs directly onto substrates such as titanium carbide or nickel foam. This direct synthesis approach enhances electrochemical performance, particularly in applications requiring high-rate charge transport, such as supercapacitors and electrolysis cells. Moreover, the robust interface formed through electrodeposition contributes significantly to the devices' mechanical stability and operational longevity. Consequently, the method provides a streamlined route to high-performance electrodes by ensuring that the active material is directly and intimately connected with the current collector.

The inherent precision of the electrodeposition method allows for exceptional tunability in the thickness, composition, and morphology of the resulting LDH films, which is essential for correlating structure with performance [174,177]. Researchers have effectively exploited this precision to fabricate films optimized for specific applications, whether enhanced catalytic activity or improved storage capacity. By adjusting parameters such as deposition potential and electrolyte concentration, the process can be customized to yield films with distinct microstructures ranging from smooth continuous layers to highly porous networks. This level of control is particularly beneficial in energy-related applications, where the surface area and active site accessibility largely dictate the device's performance. Therefore, electrodeposition remains a desirable method due to its flexibility and effectiveness in producing high-quality LDH films tailored to meet specific performance criteria. In addition to conventional electrodeposition, recent advancements have illustrated how integrated deposition strategies can further enhance the functional properties of LDH films [197,198]. During LDH electrodeposition, nanostructured carbon materials can increase the composite electrode's electrical conductivity and surface area. This synergistic combination creates an electrode with better charge transport and effective capacitance, which is ideal for energy storage. Dual nanocarbon supports ensure structural integrity and excellent catalytic reactivity of deposited LDHs. Such hybrid structures emphasize the need for multifarious deposition techniques to generate integrated materials with improved performance.

Advanced methodologies also include the in situ formation of NiAl-LDHs within confined microreactor environments, where tunable interlayer spacing can be achieved by adjusting the impinging jet parameters during deposition [186]. Microreactor-assisted electrodeposition enables precise nanoscale control of LDH structures, enhancing their electrochemical performance for energy applications. The unique design of deposition setups, including pauses between deposition pulses, further ensures the uniform regeneration of metal ion concentrations near the electrode surface [173,179]. Pulse-controlled electrodeposition enhances LDH film uniformity and structural order, yielding high-performance electrodes with reduced defects and improved interlayer connectivity.

Electrodeposition techniques have also been highly effective in synthesizing uniformly mixed LDH films incorporating multimetal systems, such as NiCo-LDH, which are critical for high-performance electrocatalysis [174,196]. Electrodeposition enables uniform multimetal integration in LDH films, enhancing conductivity, redox activity, and catalytic performance for advanced energy applications. Detailed interfacial analyses have shown that the layer-by-layer growth achieved during the electrodeposition of LDHs results in exceptional interfaces crucial for efficient electron transport and ion accessibility [183,192]. The well-organized lamellar structure in these systems ensures that charge carriers can traverse the electrode with minimal impedance due to the abundant and organized active sites. The formation of these homogeneous interfaces is critical in reducing the electrode's internal resistance and enhancing the overall electrochemical performance. Furthermore, the intimate contact facilitated by the direct deposition method minimizes potential interfacial delamination issues, thereby contributing to the longevity and robustness of the device. Consequently, the layer-by-layer growth mechanism achieved through electrodeposition is a key feature that reinforces the practical viability of LDHs in complex electrochemical applications.

The electrodeposition synthesis strategy is strongly influenced by the competition between nucleation and growth processes, which are both sensitive to the deposition parameters set during the process [193]. A higher applied potential can increase the nucleation rate, leading to a greater density of small nuclei, which eventually merge to form a continuous film. However, if the growth process is not carefully managed, rapid nucleation can lead to the formation of overly rough and non-uniform surfaces. Therefore, selecting the appropriate deposition conditions, such as current density and pulse duration, is essential to achieve the desired balance between nucleation and growth.

This delicate balance is fundamental for synthesizing high-quality LDH films with optimal microstructural characteristics [181,193].

Among the innovative examples in recent literature are ternary LDH composites that have been synthesized via electrodeposition, demonstrating significantly improved OER performance [199,200]. The introduction of a third metal component not only enhances the electronic structure of the LDH but also introduces a new level of active sites that facilitate more efficient catalytic processes. Detailed electrochemical studies have revealed that such ternary systems lower the overpotential required for OER and improve reaction kinetics due to the synergistic interactions among the incorporated metal species. Moreover, these composite materials were obtained by adjusting the deposition parameters to ensure uniformity and high dispersion of the active species across the electrode. Therefore, the evolution of electrodeposition techniques to include multimetal systems represents a significant advancement toward developing next-generation catalysts for clean energy applications.

An intriguing dimension of electrodeposition is its adaptability in fabricating flexible energy storage devices. LDH films are deposited onto flexible substrates to yield electrodes with outstanding mechanical integrity and high capacitance [189,201]. Such flexible electrodes not only sustain large mechanical deformations but also maintain excellent electrochemical performance during long-term cycling. The direct growth of LDHs onto flexible substrates ensures that the interface is robust and the ion transport channels remain unimpeded despite mechanical bending or stretching. In addition, the combination of electrodeposited LDHs with flexible conductive materials opens up new possibilities for wearable energy storage devices and portable electronics. This innovative approach underscores the versatility of electrochemical deposition in producing advanced electrodes that combine mechanical flexibility with exceptional electrochemical properties.

The careful selection of deposition parameters, such as electrolyte composition and deposition time, is essential to unlocking the theoretical structural advantages of LDHs, including high surface areas, accessible interlayer spaces, and controlled redox properties [196,202]. For instance, tuning the electrolyte composition can promote the selective intercalation of specific anions, thereby optimizing the interlayer spacing and significantly enhancing the electrochemical performance. Similarly, adjusting the deposition time allows for precise control over film thickness, directly impacting the electron and ion transport kinetics within the LDH film. These controlled modifications enable the synthesis of LDHs that exhibit tailored properties for applications ranging from water splitting to supercapacitors. In this context, the electrodeposition technique stands out for its flexibility and effectiveness in producing materials with finely tuned structural and functional characteristics.

Numerous studies have demonstrated the advantages of pulsed electrodeposition, wherein reverse pulses are utilized to mitigate the accumulation of surface charges and to ensure a steady state of deposition ion concentration [174]. By carefully calibrating the pulse duration and the intervals between pulses, the deposition process can be optimized to yield smoother, more uniform LDH films with reduced defect densities. This method also minimizes undesirable side reactions, thereby enhancing the deposited materials' chemical integrity and electrochemical stability. The resulting improvement in uniformity is directly related to improved device performance, as fewer defects translate to fewer pathways for unwanted charge recombination or degradation. As such, pulsed electrodeposition represents a sophisticated synthesis strategy well-suited for preparing high-quality LDHs for advanced electrochemical applications.

The high degree of tunability in electrodeposition also permits the fabrication of LDH films with diverse morphologies, ranging from densely packed continuous layers to highly porous, exfoliated nanosheets [180,183]. By systematically adjusting deposition conditions, researchers have induced a controlled exfoliation of the layered structure, thereby exposing more active sites and increasing the effective surface area. These morphological features are particularly beneficial for supercapacitors and electrocatalysts, where rapid ion diffusion and extensive charge storage are critical. Moreover, synthesizing distinct nanostructures within a single processing technique allows for integrating materials with complementary properties, further enhancing device performance. Consequently, the tunability of the electrodeposition process serves as a key advantage in the strategic design and synthesis of advanced LDH-based materials.

Complementary to the direct electrodeposition approach, in situ reconstruction methodologies have emerged as powerful strategies to convert existing compounds into LDH structures with desired morphologies and enhanced functionalities [92,194]. Such methods involve transforming precursor materials under the influence of applied potential and alkaline conditions, leading to the formation of LDH structures from an initially non-layered phase. This reconstruction process streamlines the overall synthesis and results in LDH materials with improved crystallinity and increased active surface areas. Integrating reconstruction techniques with electrodeposition offers additional control over the deposition kinetics and final structure. As a result, these coupled strategies enable the fabrication of high-performance LDH-based electrodes suitable for an array of advanced applications [92].

These advanced synthesis strategies have been further refined by incorporating design templates that direct the formation of hierarchical LDH structures with enhanced catalytic and energy storage properties [178,197]. Whether sacrificial or permanent, templates allow the growth of LDH films with controlled porosity, uniform layer thickness, and precisely engineered interlayer distances. Such templated electrodeposition enhances the material's electrochemical performance by increasing the effective surface area, improving ion transport, and imparting mechanical robustness to the composite electrode. Moreover, the templating approach facilitates the formation of multidimensional architectures tailored to specific functional requirements. Consequently, these design strategies underscore the critical role of electrodeposition in synthesizing advanced LDH materials with tailored structure/property relationships for various applications. The electrochemical deposition and synthesis mechanism for LDHs embodies a multifaceted process that leverages controlled faradaic reactions to generate well-defined, nanostructured films with tunable compositions and morphologies [173,181,193]. By meticulously adjusting parameters such as deposition potential, current density, ionic concentration, and pulse duration, researchers can achieve high-quality LDH films that are directly integrated with conductive substrates, making them readily applicable in electrocatalysis, energy storage, and corrosion protection [177,184,196]. Recent advances have further demonstrated that innovative approaches, ranging from pulsed electrodeposition to reconstruction of precursor materials, significantly enhance the performance attributes of LDH films by optimizing the exposure of active sites and ensuring efficient ion transport. The versatility and scalability of the electrochemical deposition technique promise to further advance the field of nanostructured material synthesis and are expected to play a pivotal role in developing next-generation devices. Thus, this synthesis process's robust and adaptable nature ultimately underscores its importance and potential for future applications in various scientific and technological domains.

NiCo LDH was synthesized via an electrodeposition method, enhanced by incorporating $Ti_3C_2T_x$ MXene on the nickel foam (NF) substrate and in the plating solution [203]. This dual role of MXene improved interfacial contact and conductivity, facilitating the deposition of ultrathin LDH nanosheets with high areal loading (~65 mg/cm^2). Characterization revealed a porous nanosheet morphology, polycrystalline hexagonal structure via HRTEM, uniform Ni and Co distribution, and MXene confirmed through XRD and XPS analyses. These features collectively make the MXene-enhanced LDH electrode highly suitable for high-performance ZIB cathodes. The CoNi-LDH was synthesized on flexible CC via in situ electrochemical deposition from a Co^{2+}/Ni^{2+} precursor, followed by electrochemical activation (ECA) in aqueous KOH to induce H-vacancy formation, resulting in CoNi-LDH(v) [204]. Figure 9a outlines the two-step process used to create the CoNi-LDH electrode. First, CoNi-LDH is deposited onto CC via electrochemical deposition, then it undergoes electrochemical activation to introduce hydrogen vacancies, forming CoNi-LDH(v)@CC. As shown in Figure 9b, the deposition occurs in a solution containing Co^{2+}, Ni^{2+}, and OH^- ions. The corresponding CV curve shows redox behavior, and the resulting material has a layered crystal structure with nitrate ions. Figure 9c illustrates the activation step, where CoNi-LDH is treated under an applied voltage to create hydrogen vacancies. The CV plot shows enhanced electrochemical activity after activation, and the structure model confirms the formation of H vacancies that improve the material's electrochemical properties. The morphology revealed vertically aligned nanosheets forming a porous 3D network; structural analysis via XRD and TEM confirmed partial amorphization and reduced interlayer spacing due to NO_3^- extravasation. BET surface area was not directly stated, but the induced H vacancies and 2D open channels significantly increased the active surface and conductivity. These features enhanced Zn^{2+} diffusion and intercalation, making the LDH highly suitable for Zn-ion battery applications.

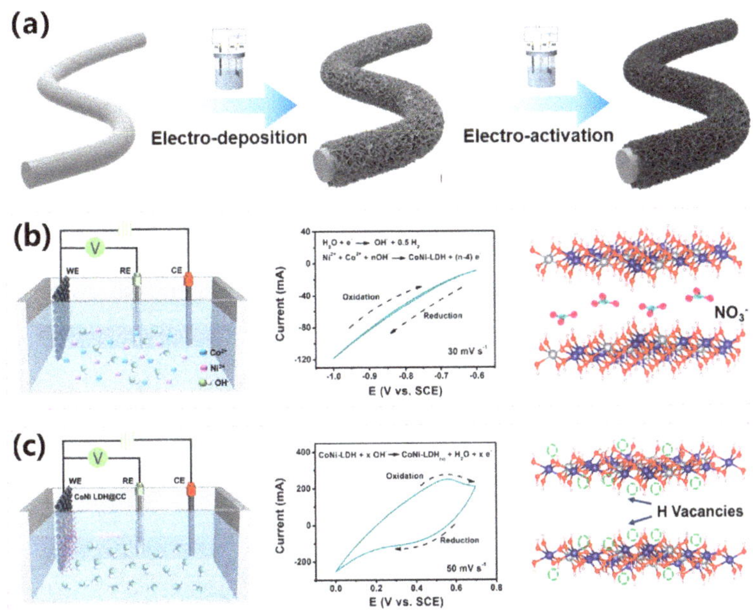

Figure 9. (**a**) Illustration of the fabrication steps involving electrochemical deposition of CoNi-LDH on CC (CoNi-LDH@CC), followed by electrochemical activation to generate hydrogen-deficient CoNi-LDH(v)@CC. (**b**) Diagram of the deposition setup (left), corresponding CV curve (middle), and the crystal structure of the as-prepared CoNi-LDH (right). (**c**) Electrochemical activation process (left), CV curve under optimized activation conditions (middle), and structural model showing hydrogen vacancies in CoNi-LDH(v) (right). Reproduced with permission [204].

A novel synthesis pathway involves the reconstruction of MOFs or Prussian blue analogs into LDHs, which can also be achieved through an electrochemical deposition process [190,194]. This approach uses the structural change of a precursor molecule in an alkaline environment under an applied voltage to create an LDH. The reconstruction method simplifies synthesis and increases product surface area and reactivity. Using MOFs as sacrificial templates also allows fine control over the LDH structure's porosity and shape. This comprehensive technique shows how electrochemical deposition can synthesize customized LDHs with enhanced functional characteristics. Adjusting the deposition parameters enables the production of ultrathin LDH films with exceptional surface area and catalytic activity, which are crucial for their application in sensors and energy devices. By fine-tuning factors such as ionic strength, deposition time, and pulse frequency, researchers can achieve controlled growth that yields films with homogeneous thickness and low defect densities [173]. The resulting LDHs exhibit enhanced electron mobility and rapid ion diffusion, vital for high-performance catalytic and supercapacitor applications. Furthermore, the deposition technique provides a direct route to engineering the film architecture, which supports efficient charge storage and transfer. As a result, the method stands out for its ability to customize nanostructured films for a wide range of electrochemical applications.

The formation of hexagonal platelets, a hallmark of well-crystallized LDHs, is significantly influenced by the alkaline treatment following electrodeposition, during which an initial disordered film is reorganized into an ordered lamellar structure [93]. This transition is attributed to the enhanced mobility of constituent ions within the deposit, which facilitates the rearrangement into a thermodynamically favorable configuration. The development of hexagonal platelets indicates high crystallinity and correlates with improved functional properties, such as ion exchange capacity and catalytic performance. Therefore, the post-deposition treatment must be carefully controlled to maximize these desirable features. Consequently, combining electrodeposition with subsequent alkaline treatment forms a robust strategy for synthesizing ordered and functional LDH films. Control over interlayer composition during the synthesis of LDHs via electrodeposition can be effectively achieved by modulating the electrolyte composition and deposition parameters [190]. Selecting electrolyte anions affects LDH charge balance, interlayer spacing, and structural stability. This process is useful when improved conductivity or selective adsorption is needed. The deposition potential also affects intercalating species production and integration, fine-tuning the final layered structure [186]. Thus, deliberate control over the ionic environment during deposition is essential for crafting LDHs with tailored interlayer characteristics and optimized performance. Co-deposition techniques have been effectively employed in more complex systems to synthesize hierarchical LDH structures by concurrently depositing multiple metal ions, which leads to synergistic enhancements in electrocatalytic performance [182,205]. The simultaneous incorporation of different metal species into the LDH framework generates a material with a multifaceted electronic structure that supports rapid redox kinetics. This hierarchical organization improves not only the catalytic activity but also the mechanical stability of the electrode under rigorous operational conditions. The co-deposition strategy thus enables fine control over the relative concentrations and spatial distribution of the metal ions, ensuring that each contributes efficiently to the overall performance. Such advanced synthesis techniques highlight the power of electrodeposition in crafting multifunctional LDH catalysts tailored for high-demand electrochemical processes.

The ZnAl-LDH coating layer was synthesized on commercial Zn foil using a galvanostatic electrochemical deposition method in a nitrate aqueous solution [172]. Optimized parameters included a current density of 3 mA/cm^2 for 2 min and a Zn^{2+}/Al^{3+} molar ratio of 2:1. The LDH layer exhibited a smooth and uniform morphology with a hydrotalcite-like structure. The coating had a thickness of ~1.5 μm and displayed excellent wettability (contact angle ~5.96°) and strong zincophilicity, indicating its suitability as an artificial interfacial layer. This tailored ZnAl-LDH effectively stabilizes Zn plating/stripping and inhibits dendrite growth, making it highly suitable for ZIB anode protection. Figure 10a–c illustrates the critical balance between zincophilicity and zincophobicity in stabilizing Zn anodes [206], a concept highly relevant to LDH-based strategies where their tunable surface chemistry and layered structures can be engineered to regulate Zn^{2+} deposition and electrolyte interactions in both ZIBs and ZABs. The CoNi LDH was synthesized via in situ electrochemical deposition on an exfoliated graphite (EG) substrate, followed by CV-based electrochemical activation in KOH to introduce hydrogen vacancies [207]. During activation, interlayer anions (NO_3^-) were removed, and O–H bonds in the hydroxide lattice were cleaved to form a defect-rich CoNi LDH(v) structure. The resulting material exhibited ultrathin mesoporous nanosheets (~5 nm thick), a hierarchically porous architecture, increased surface area (from 28.3 to 40.7 m^2/g), and good elemental distribution and crystallinity. These structural and surface properties make this LDH well-suited for Zn-ion battery cathode applications due to enhanced ion accessibility, conductivity, and defect-driven cation adsorption.

The synthesis of NiCo-LDH involved a two-step process [208]. Initially, NiCoS nanotubes were formed on nickel foam using hydrothermal sulfidation, followed by constant potential electrodeposition of ultrathin NiCo-LDH nanosheets to construct a core/shell heterostructure. Characterization through FESEM and TEM confirmed a 3D nanotube@nanosheet architecture, XRD and Raman verified phase composition, and XPS analyzed valence states; the morphology provided abundant active sites and fast ion/electron pathways. This architecture is particularly suitable for ZIB cathodes due to its enhanced conductivity and electrochemical kinetics. The LDH-based composite was synthesized by first hydrothermally growing Ni_3S_2/NiS hetero-nanowire arrays on nickel foam, followed by electrodeposition of NiCo-LDH nanosheets to form a hierarchical core/shell structure [209]. This two-step process enables intimate interfacial contact between the conductive sulfide core and the active LDH shell. The resulting composite exhibited a 3D nanoarray morphology and chemical states characterized by XPS. The uniform architecture and rich surface functionalities render this LDH ideal for rapid ion/electron transport and structural stability, making it highly suitable for ZIB applications.

NiFe LDH was synthesized via a multi-step electrochemical approach where $Cu(OH)_2$ nanowires were first electroplated onto Cu foam, converted to CuO by annealing, reduced to Cu nanowires, and, finally, used as a scaffold for the electrochemical deposition of NiFe LDH nanosheets [210]. This process formed a 3D hierarchical core/shell structure of Cu@Cu NWs@LDH. The morphology presents uniform vertical nanosheet growth; XRD verified the crystalline structure, and XPS confirmed the elemental composition and valence states. The high surface area, strong bonding, and excellent electron conductivity make this LDH suitable for ZAB applications. CuCo LDH was synthesized by potentiostatic electrodeposition on nickel foam, followed by CV-based electrodeposition of Ni_3S_2 to form the CuCo LDH@Ni_3S_2 composite [211]. The resulting composite featured a tremella-like interconnected nanosheet morphology, exhibited a layered LDH structure confirmed by XRD, weak crystallinity with abundant active sites, and had a high electrochemical double-layer capacitance (Cdl = 2.74 mF/cm^2), indicating excellent charge storage potential. This unique architecture provided enhanced conductivity, lower diffusion resistance, and increased redox sites, making it highly suitable for high-performance supercapacitor applications.

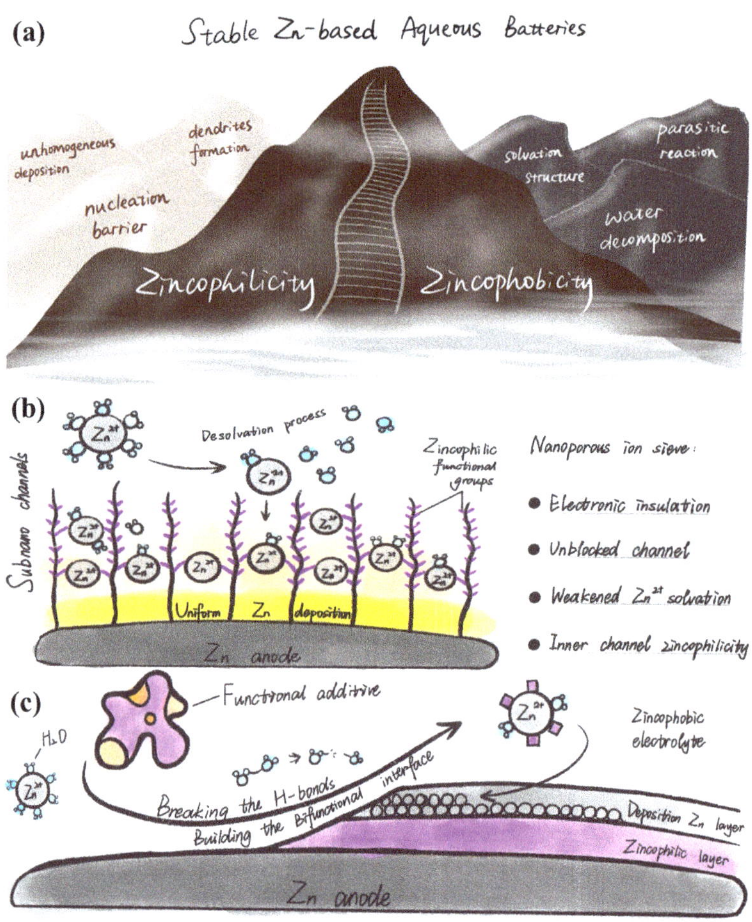

Figure 10. (**a**) Schematic illustration of the trade-off between zincophilicity and zincophobicity in achieving stable Zn-based aqueous batteries. (**b**) Design of a nanoporous ion sieve with zincophilic functional groups to enable uniform Zn deposition through controlled Zn^{2+} desolvation and selective ion transport. (**c**) Formation of a bifunctional interface using functional additives to simultaneously break water hydrogen bonds and construct zincophilic/zincophobic regions on the Zn anode. Reproduced with permission [206].

NiCo-LDH was synthesized via a simple electrochemical deposition process onto CC, followed by CV activation at various scan rates to induce hydrogen vacancies [212]. The as-prepared LDH retained its nanosheet morphology with a polycrystalline layered structure, showed reduced interlayer spacing (from 0.71 to 0.60 nm), preserved elemental homogeneity (Ni, Co, O), and exhibited enhanced M–O bonding indicative of H vacancy formation. These structural features, particularly the vacancy-rich domains, significantly improved the redox kinetics and ion diffusion, making this LDH highly suitable for supercapacitor applications. Khalafallah et al. [213] presented a low-cost, binder-free Zn-doped cobalt/lanthanum LDH (Zn/CoLa LDH) nanoarray-on-nanoarray electrode fabricated via a two-step electrodeposition method (Figure 11a), demonstrating superior electrochemical performance and durability for high-energy, high-power supercapacitor applications. NiAl-LDH was synthesized by a single-step electrochemical deposition on AC cloth using a chronoamperometry method at −1.0 V for 15 min [214]. Figure 11b depicts the stepwise electrochemical strategy to grow NiAl-LDH nanosheets on a CC substrate. The resulting NiAl-LDH@CC composite exhibited a porous, vertically aligned nanosheet array structure with a high BET surface area of 437.5 m^2/g, crystalline hydrotalcite-like layered structure, well-distributed elements (Ni, Al, O), and uniform coating thickness of ~1.11 μm. This combination of high surface area, structural integrity, and improved electron/ion transport characteristics renders the LDH highly suitable for supercapacitor applications. LDH was synthesized via in situ electrochemical deposition of nickel and cobalt nitrates onto carbon felt at −1 V for 10 min using a chronoamperometry technique [215]. The prepared NiCo-LDH formed a coral reef-inspired ultrathin mesoporous nanosheet architecture vertically aligned on the carbon felt, displaying a high BET surface area of 14.56 m^2/g and a narrow average pore diameter of 5.52 nm, with enhanced structural features confirmed by SEM, FTIR, and XRD analyses. This morphology, porosity, and strong electronic interaction between Ni and Co make the LDH ideal for efficient ion diffusion and redox kinetics, which is suitable for high-performance supercapacitor applications.

3.4. Other Advanced Synthesis Techniques (e.g., Sol-Gel, Ultrasonication, Exfoliation)

Various other methods have been reported for synthesizing LDHs, which are anionic materials characterized by a unique layered structure of hydroxides intercalated with anions. Among these methods, sol-gel synthesis, ultrasonication, and exfoliation stand out due to their distinct advantages in producing LDHs with controlled morphology and composition. The sol-gel method is a widely employed technique for synthesizing LDHs because of its ability to control the chemical composition, morphology, and crystallinity of the resultant materials. This method involves the hydrolysis and polycondensation of metal alkoxides or salts, forming a gel that can then be converted into an oxide or hydroxide material upon thermal treatment. For instance, Klydziute et al. [216] highlighted the capacity of the sol-gel approach to yield MgAl-LDHs with tunable properties, emphasizing its versatility across different metal compositions and operating conditions. Additionally, Valente et al. [217] reported that the sol-gel method allows for precise control over parameters such as pH and temperature during synthesis, which facilitates particle size and distribution adjustments, making it suitable for various applications in catalysis and environmental remediation. Ultrasonication is another promising synthesis technique that has gained popularity for preparing LDHs due to its ability to enhance reaction rates and promote homogeneous nucleation. This method employs high-frequency sound waves to create cavitation bubbles, which, upon collapsing, generate localized high temperatures and pressures conducive to the chemical reactions needed to form LDHs. Mallakpour et al. [218] demonstrated that using ultrasonic waves shortened reaction times and improved the characteristics of the synthesized LDHs. Further, Sokol et al. [219] showcased that ultrasonication could accelerate the formation of Mg–Al–phosphate LDHs, signifying the method's effectiveness in synthesizing complex layered structures. Exfoliation refers to methods to produce nanosheets or dispersed layers from bulk LDH materials, enhancing their surface area and accessibility for various applications. This technique often involves delining layered structures in suitable solvents or via chemical treatments. According to Ibrahimova [92], several methods, including exfoliation, can alter the interlayer spacing of LDHs, facilitating the intercalation of functional anions. Additionally, Butenko et al. [220] provided evidence for the potential of exfoliated LDHs as effective adsorbents for environmental contaminants like phenols, thereby highlighting the practical implications of such synthesized materials. Specific examples of LDH synthesis using these advanced techniques are abundant in the literature. For instance, research by Mahjoubi et al. [221] employed sol-gel synthesis to produce ZnAl-LDHs intercalated with various anionic species, affirming the method's capability in tailoring materials for specific functionalities such as dye removal. In the context of ultrasonication, the study by Mallakpour [218] emphasized the synthesis of bioactive intercalated LDHs with accelerated reaction rates and enhanced functional properties, which can significantly improve material performance in biomedical applications. Furthermore, the application of exfoliation in generating high-surface-area LDHs has been validated in environmental science, where these materials demonstrated promising performances in adsorbing toxic pollutants [220]. Sol-gel synthesis, ultrasonication, and exfoliation represent significant advancements in producing LDHs, allowing for tailored properties and functionalities. Each method contributes unique benefits that enhance the applicability of LDHs in diverse fields such as catalysis, environmental remediation, and materials science, leading to growing interest and research in this area.

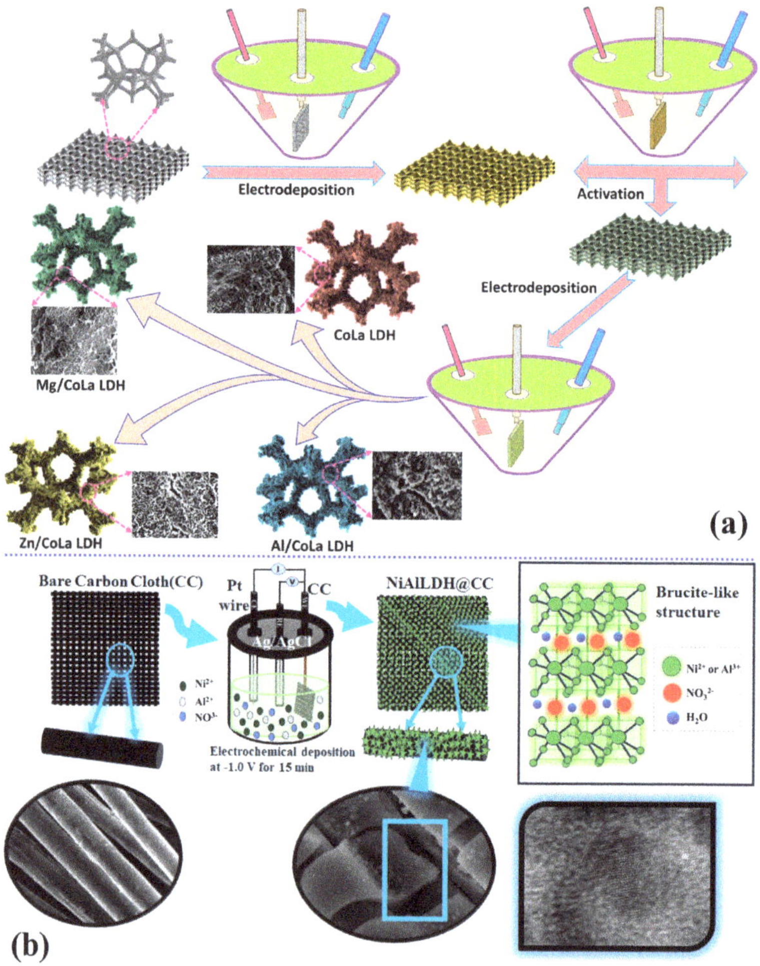

Figure 11. (**a**) Schematic illustration of the electrodeposition-based synthesis of binder-free CoLa LDH nanoarray-on-nanoarray networks doped with Zn^{2+}, Mg^{2+}, and Al^{3+} ions. Reproduced with permission [213]. (**b**) Illustration of the electrochemical synthesis procedure for fabricating NiAl-LDH nanosheets directly on CC (NiAl-LDH@CC). Reproduced with permission [214].

The nickel microwire aerogel (NMWAs) was synthesized via a sol-gel method under a magnetic field, followed by vacuum drying to obtain a porous and flexible aerogel structure [222]. Subsequently, ultrathin NiCo-LDH nanosheets were grown in situ on the NMWAs surface via a solvothermal route, resulting in a core/shell NMWAs@NiCo-LDH architecture. The sulfur was then infused into the composite to form NMWAs@NiCo-LDH/S via a melt/diffusion method. Characterization revealed a 3D porous network of aligned nickel microwires coated with ultrathin LDH nanosheets (~6.8 nm thick). BET analysis showed mesopores (4–10 nm) and high surface area, while XRD, Raman, and XPS confirmed the phase purity and the presence of active functional groups like hydroxyls. This core/shell structure with high porosity, hydrophilicity, and strong chemical affinity for polysulfides makes it highly suitable for LSB applications by enhancing electrolyte access, trapping LiPSs, and promoting redox kinetics.

NiCo-LDHs were synthesized via a self-growth aging method in the presence of Ti_3C_2-MXene [156]. Initially, few-layered MXene nanosheets were prepared through LiF/HCl etching of Ti_3AlC_2, followed by ultrasonic exfoliation. These MXene sheets then served as substrates for the in situ growth of NiCo-LDHs through the reaction of Ni^{2+} and Co^{2+} with $NH_3{\cdot}H_2O$ under an inert atmosphere, resulting in Vo-LDHs-MXenes with abundant oxygen vacancies. Morphologically, TEM and FESEM revealed a homogeneous 2D nanosheet structure, while BET analysis showed a high surface area (35.22 m^2/g) with mesoporosity, ideal for electrolyte access and sulfur redox reaction sites. XRD, XPS, and EPR further confirmed structural integration, oxygen vacancies, and electronic synergy between LDHs and MXenes. These features collectively enhance the immobilization of polysulfides and facilitate faster redox kinetics, making Vo-LDHs-MXenes highly suitable for LSB applications.

The NiCo-LDH was first synthesized by refluxing a mixed aqueous solution of $Ni(NO_3)_2{\cdot}6H_2O$, $Co(NO_3)_2{\cdot}6H_2O$, hexamethylenetetramine, and oxalic acid at 100 °C for 6 h [223]. This precursor was then converted into NiCo–MOF/LDH nanorods through a solvothermal reaction with 2,5-dihydroxyterephthalic acid (DHTA) in a DMF/ethanol/water mixture at 140 °C for 24 h. The synthesized NiCo–MOF/LDH nanorods exhibit a rod-like morphology with homogeneous elemental distribution, a high surface area (512.3 m^2/g), meso/microporosity, and coexisting crystalline (NiCo–MOF) and amorphous (NiCo–LDH) phases. These features make the material ideal for LSB applications by combining strong polysulfide adsorption and electrocatalytic conversion properties.

NiCo-LDH was synthesized by etching ZIF-67 nanocrystals using $Ni(NO_3)_2{\cdot}6H_2O$ in ethanol at elevated temperature, which led to the formation of hollow NiCo-LDH nanostructures [224]. Sulfur was subsequently incorporated into the LDH via a melt/diffusion process. The composite exhibited a hollow polyhedral morphology, confirmed by SEM/TEM, with a high BET surface area of 76.089 m^2/g, mesoporosity (2–8 nm), and phase purity validated by XRD and XPS. This structural design, featuring abundant active sites and high sulfur content (~71.9%), is highly beneficial for Li–S batteries due to enhanced polysulfide confinement and catalytic redox behavior.

Ni-Co LDH was synthesized by ion-exchanging Co from ZIF-67 nanocubes with Ni^{2+} ions in ethanol, forming ZIF-67@LDH nanoboxes [225]. These were subsequently carbonized and sulfided at 400 °C to form $NiCo_2S_4$-NiS_2 nano-heterostructures embedded in hollow carbon boxes. Characterization using SEM, TEM, and XRD confirmed well-maintained nanobox morphology, distinct lattice spacings for $NiCo_2S_4$ and NiS_2, and uniform elemental distribution; BET analysis showed a high surface area of 111.5 m^2/g. The resulting LDH-derived nano-heterostructure offers abundant catalytic sites and strong lithium polysulfide (LiPS) adsorption capabilities, making it ideal for LSB applications. The CuCo-LDH was synthesized via a chemical etching strategy, where ZIF-67 (a cobalt-based MOF) acted as a sacrificial template and was treated with $Cu(NO_3)_2{\cdot}3H_2O$ in ethanol to form a hollow and hierarchical structure composed of vertically aligned LDH nanosheets [226]. Figure 12 shows the synthesis process and morphological evolution of CuCo-LDH materials used in lithium–sulfur batteries. The structure is derived from ZIF-67 by introducing copper nitrate, forming a CuCo-LDH composite that enhances sulfur redox kinetics and polysulfide adsorption. FESEM images in Figure 12A–G reveal how particle morphology changes with different Co/Cu ratios. As the Co content decreases, the structure becomes less defined and more fragmented, indicating that the metal ratio significantly influences the material's architecture, which, in turn, may affect its electrochemical performance. The optimal Co/Cu ratio (0.7) facilitated complete etching and formation of well-defined CuCo-LDH polyhedrons. Characterization results revealed that the synthesized LDH has a hollow dodecahedral morphology, layered nanosheet microstructure, mesoporous texture (BET surface area: 104.1 m^2/g), and a mixed Cu^{2+}/Co^{3+} oxidation state. CuCo-LDH is highly suitable for LSBs due to its strong polysulfide anchoring capability and catalytic activity.

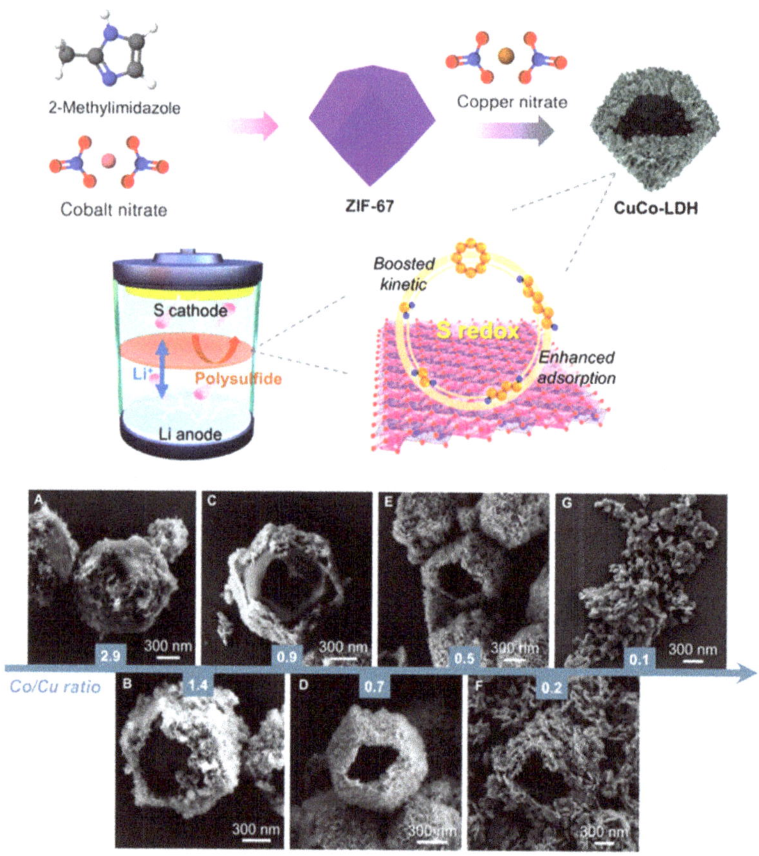

Figure 12. Schematic representation of the synthesis route for CuCo-LDH derived from ZIF-67 and its role in LSBs, highlighting its ability to accelerate redox reactions and trap polysulfides. FESEM images show the morphology of CuCo-LDH synthesized at various Co/Cu molar ratios: (**A**) 2.9, (**B**) 1.4, (**C**) 0.9, (**D**) 0.7, (**E**) 0.5, (**F**) 0.2, and (**G**) 0.1. Reproduced with permission [226].

The synthesis of PPy@LDH-S involved a multi-step process beginning with the formation of ZIF-67 nanocrystals, which were then transformed into hollow NiCo-LDH through a nickel ion exchange process [227]. Subsequently, sulfur was infiltrated via the solid-state melting method, and finally, a thin polypyrrole (PPy) film was polymerized onto the LDH-S composite. BET analysis showed a progressive reduction in surface area from 135.4 m^2/g (LDH) to 3.3 m^2/g (PPy@LDH-S) due to sulfur loading. The structural and chemical features—such as strong polysulfide adsorption, mesoporosity, and catalytic activity—make PPy@LDH-S a highly effective sulfur host for LSBs. The $NixCa_{2-x}$Al-Cl LDH and $CoyCa_{2-\gamma}$Al-Cl LDH were synthesized through a simple one-step ion exchange method, using Ca_2Al-Cl LDH derived from chlorine-containing wastewater as a precursor [228]. Ni^{2+} and Co^{2+} ions replaced Ca^{2+} ions under ambient conditions with stirring and sonication, followed by filtration and drying. The resulting LDHs exhibited a thin, porous, tunnel-like morphology with abundant surface folds; XRD confirmed the layered structure, BET indicated high surface area ($\sim$4$\times$ compared to the precursor), and XPS verified successful Ni^{2+}/Co^{2+} substitution. Due to improved conductivity, active sites, and ion diffusion kinetics, this structural and electronic modification makes the LDHs highly suitable as anode materials for NIBs. CoSn-LDH@MXene nanocomposites were synthesized via a one-step solvothermal process where $Co(NO_3)_2{\cdot}6H_2O$ and $SnCl_2{\cdot}2H_2O$ were reacted in a mixture of isopropanol and glycerol at 180 °C, followed by ultrasonic treatment with few-layer Ti_3C_2 MXene in ethanol [229]. The CoSn-LDH microspheres self-assembled and anchored onto the MXene nanosheets through hydrogen bonding and electrostatic interactions, forming a uniform hybrid structure. BET analysis showed a high surface area (44.15 m^2/g) and mesoporous structure, and XPS confirmed successful chemical integration. This architecture facilitates electron/ion transport and suppresses agglomeration, making it well-suited for Na-ion battery applications. Ni_3Ti-CO_3 LDH was synthesized via a urea precipitation method, whereas Ni_3Cr-CO_3 LDH was prepared using a conventional hydrothermal method [230]. Both were further converted into Cl^--intercalated LDHs through an anion exchange process using HCl/NaCl solution. The developed LDHs exhibit high Cl^- insertion capacity and structural integrity, making them well-suited for CIB applications.

The CoMn-LDH/NPGA was synthesized via a combined cross-linking gelation, hydrothermal self-assembly, and freeze-drying process [231]. Vertically grown CoMn-LDH nanosheets were anchored onto a 3D hierarchical porous N,P co-doped graphene aerogel framework (NPGA). Characterizations revealed a lamellar morphology, well-defined crystalline structure, mesoporous nature with a BET surface area of 105.8 m^2/g, and a high degree of N,P doping. This configuration provides high surface area, exposed active sites, and effective charge/mass transport, making it well-suited for ZAB applications. CoNi LDH nanoflowers (NFs) and nanosheets (NSs) were synthesized via a simple solvothermal method using a mixed solvent of deionized water and ethylene glycol in varying ratios [232]. The nanoflowers exhibited a hierarchical flower-like morphology with high crystallinity, a mesoporous structure, and a large BET surface area of 101.28 m^2/g (vs. 38.54 m^2/g for NSs), with XRD confirming α-phase $Co(OH)_2$ and $Ni(OH)_2$, and XPS revealing mixed oxidation states of Co^{2+}/Co^{3+} and Ni^{2+}. These structural and surface characteristics endow the CoNi LDH, especially the NFs, with abundant active sites, good charge transport properties, and superior suitability for high-performance supercapacitor applications. NiCo-MOF@LDH hybrid nanosheets were synthesized via a one-step solvothermal method using 1,4-terephthalic acid and urea [233]. The resulting material displayed ultrathin, interconnected nanosheet morphology with high porosity, alongside the presence of Ni, Co, O, and C through EDS and XPS; the XRD patterns confirmed a hybrid structure of NiCo-MOF and LDH, and the sample exhibited high surface area and abundant active sites. These structural features enhance ion diffusion and electron transfer, making the material highly suitable for supercapacitor applications.

NiCo-LDH nanosheets were synthesized by transforming Co-based MOFs on AC cloth (ACC) using a simple room-temperature ion exchange method in a mixed water and ethylene glycol solution [234]. The resulting material exhibited vertically aligned ultrathin amorphous nanosheets interconnected in a 3D mesoporous network, with a surface area of 25.8 m^2/g and a pore diameter of 3.9 nm. This unique structure and amorphous nature facilitate high electrochemical activity and ion transport, making it highly suitable for high-performance supercapacitor applications. Co-Ni LDH was synthesized through a facile cation exchange reaction between $Mg(OH)_2$ and Co^{2+}/Ni^{2+} ions, followed by a controlled anion exchange sulfurization using Na_2S [235]. The resulting CoNi LDH displayed a nanosheet morphology, layered crystal structure, and a BET surface area of 58 m^2/g, progressively decreasing with sulfurization. These structural and chemical features, coupled with improved conductivity due to sulfur incorporation, make the material well-suited for high-performance supercapacitor applications. $NiCo_2S_4$@CoAl-LDH (NCS@CA) heterojunction nanosheets were synthesized on nickel foam using a two-step solvothermal process [236]. The resulting material exhibited a unique flower-like core/shell nanosheet morphology with high crystallinity and a well-defined layered structure; it also showed enhanced light absorption (bandgap ~2.23 eV) and high redox-active surface area ideal for electrochemical applications. This rationally designed nanostructure facilitates fast ion diffusion and superior redox activity, making it highly suitable for supercapacitor applications. The NiCo-LDH was synthesized via a one-step microwave hydrothermal co-assembly method using tannin-derived carbon microspheres as a substrate [237]. The resulting LDH exhibited a well-defined sea urchin-like morphology with a core/shell nanostructure composed of uniformly distributed NiCo-LDH nanoneedles anchored on smooth carbon microspheres, possessing mesoporous features and a BET surface area of 56 m^2/g. This structured design provides a high active surface area and improved charge transport, making it highly suitable for supercapacitor applications.

3.5. Impact of Synthesis Methods on LDH Properties

The synthesis methods employed significantly influence the structural, morphological, and functional characteristics of LDHs, directly impacting their performance in batteries, supercapacitors, and hydrogen production applications. Co-precipitation, a widely adopted synthesis method, excels due to its simplicity, scalability, and cost-efficiency. It typically results in LDHs with uniform morphology, high purity, and controllable interlayer spacing. These attributes are particularly beneficial in battery applications, where homogeneity and controlled particle size ensure efficient ion diffusion and high electrochemical stability. Recent advancements, such as incorporating surfactants, polymers, and organic anions during co-precipitation, have enabled further refinement of LDHs, enhancing their corrosion protection capabilities in reinforced concrete, boosting electrochemical activity, and improving stability in NIB and LIB systems. Furthermore, co-precipitated LDHs demonstrate remarkable adsorption characteristics, making them suitable for environmental remediation, a critical supplementary property in battery and supercapacitor electrodes where impurity management is crucial.

Hydrothermal synthesis, characterized by elevated temperatures and pressures, provides precise control over the crystallinity, particle size, and hierarchical architectures of LDHs, yielding materials with enhanced thermal stability and defect-free structures. This technique is particularly advantageous in synthesizing high-entropy LDHs and multimetallic systems, where synergistic interactions among multiple cationic species significantly improve electrocatalytic performance. Hydrothermally synthesized LDHs display superior catalytic activities for hydrogen production applications due to their high surface area, controlled porosity, and finely tuned morphology, collectively facilitating efficient HER. Hybridization approaches combining LDHs with MOFs or carbonaceous substrates via hydrothermal methods further enhance their suitability for energy storage and catalytic processes by improving conductivity, structural integrity, and electrolyte accessibility.

Electrochemical deposition offers direct, binder-free growth of LDH films onto conductive substrates, thereby improving electron transport, adherence, and structural stability crucial for high-performance supercapacitors and catalytic electrodes. The tunability of electrochemical parameters enables precise control over film thickness, crystallinity, and porosity, directly correlating with enhanced electrochemical and catalytic performance. Techniques such as pulsed electrodeposition provide superior morphological control, reducing defects and optimizing the exposure of catalytic active sites, which is significantly beneficial for hydrogen production and supercapacitor applications.

Advanced synthesis methods such as sol-gel, ultrasonication, and exfoliation further expand the potential of LDHs. Sol-gel synthesis enables highly homogeneous and porous structures with tailored compositions, significantly enhancing electrode performance in supercapacitors and batteries due to improved ion transport pathways and increased active sites. Ultrasonication accelerates LDH formation, yielding finer particle sizes and improved dispersion, which critically benefits applications requiring rapid charge/discharge capabilities and high surface reactivity. Exfoliation methods generate ultrathin LDH nanosheets with exceptional surface area, which is crucial for enhancing catalytic activities in hydrogen production by maximizing exposed active sites and facilitating efficient electron transfer. Collectively, the meticulous selection and optimization of LDH synthesis methodologies are essential for tailoring materials with targeted functional properties, ensuring optimal performance across energy storage and catalytic applications.

LDHs are a class of materials characterized by a unique double-layered structure formed by two distinct metallic cations and an interlayer of anions, typically hydroxides or organic moieties. They exhibit properties that make them suitable for energy storage and hydrogen production applications. This response categorizes LDHs based on structural composition, defect engineering, and their composites, examining their properties relevant to energy applications. LDHs can be broadly classified based on their cationic composition. Common cations include transition metals such as nickel, cobalt, magnesium, aluminum, and iron, which can form various configurations [44,238]. Synthesis methods, such as hydrothermal and co-precipitation techniques, can induce defects like vacancies and irregular stacking, significantly influencing their electrochemical properties [239,240]. For instance, NiFe-LDHs exhibit enhanced OER activity when defects are strategically engineered, allowing for improved electronic conductivity and catalytic performance [241,242]. Structural modifications enable tailoring LDHs for specific application requirements. For example, engineering heterostructures, combining LDHs with other materials (such as carbon or metal oxides), can enhance catalytic activity and stability for water-splitting applications [242,243]. Using pulsed laser ablation or ion intercalation can produce nanosheets that provide a larger surface area for ion exchange, which is crucial for efficient energy storage systems like supercapacitors [244,245].

LDHs can also be integrated into composite matrices, enhancing their performance in energy applications. For instance, composites of LDHs with conductive polymers or carbonaceous materials show improved capacitance and energy density relative to their standalone counterparts [243,246]. The synergistic effects observed in such composites lead to enhanced electrochemical performance, making them suitable candidates for hybrid supercapacitors and batteries [247].

LDHs stand out due to their impressive electrochemical properties relevant to energy storage and hydrogen production. LDHs demonstrate significant catalytic activity toward the OER and HER due to their tunable electronic structures and high specific surface areas [248–250]. The presence of transition metals in their structure allows them to participate actively in redox reactions, enhancing their functionality as electrocatalysts [241,251]. The layered structure of LDHs contributes to their stability under operational conditions. They can maintain structural integrity during cycling in energy devices, making them reliable choices for applications in supercapacitors and batteries [240,252]. Furthermore, proper intercalation and defect engineering can improve their ionic conductivity and structural resilience, leading to a prolonged lifespan and consistent performance [251,253]. The unique interlayer space in LDHs permits the exchange of various anions to enhance energy storage capabilities. For instance, LDHs can be intercalated with anions suited for specific electrochemical applications, improving overall electrochemical properties [40,254]. The classification of LDHs into structural categories, the strategic use of defects and engineering approaches, and the development of composite materials are pivotal in leveraging their properties for enhanced energy storage and hydrogen production. Ongoing research continues to explore this fascinating class of materials, further unlocking their potential across various applications.

Among various synthesis strategies for LDHs, hydrothermal synthesis has emerged as the most widely employed method, particularly for energy-related applications. The hydrothermal route allows for precise control over crystallinity, morphology, and composition, enabling the fabrication of highly ordered LDH nanostructures with tailored porosity and surface area, properties that are critical for high-performance battery electrodes, supercapacitors, and electrocatalysts for water splitting and CO_2 reduction. Co-precipitation methods remain popular for producing LDHs at large scales, especially for supercapacitor electrodes and lithium-ion batteries, due to their simplicity and compositional flexibility. Electrodeposition techniques are increasingly favored for directly growing LDH thin films on conductive substrates, particularly in water-splitting electrocatalysis and flexible energy devices, where strong substrate adhesion and tunable thickness are desirable. Emerging methods such as spray pyrolysis, microemulsion synthesis, and mechanochemical activation are being explored for large-scale production of LDH-derived oxides for HER and carbon dioxide reduction reactions (CO_2RR). Ultimately, the choice of synthesis method critically influences the structural features, ion transport properties, and overall electrochemical performance of LDHs, thereby determining their suitability for specific energy storage and conversion applications. Figure 13 highlights the key advantages of various synthesis methods for LDHs, emphasizing how each technique offers unique benefits such as compositional flexibility, controlled morphology, high purity, and structural tailoring.

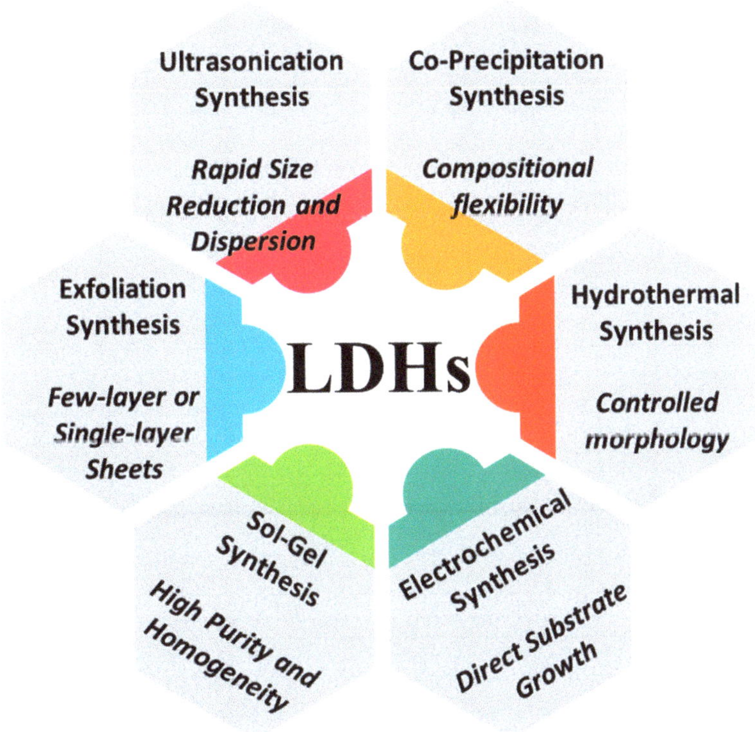

Figure 13. Schematic illustration of the advantages associated with different synthesis methods for LDHs.

4. LDHs in Batteries

LDHs have garnered significant interest in recent years due to their unique structural properties and wide-ranging applications, particularly in batteries. Composed of positively charged metal hydroxide layers interleaved with charge-compensating anions, LDHs can host a variety of cations and anions, which endow them with tunable electrochemical properties essential for battery technologies. Optimizing LDH materials for battery applications involves enhancing their structural integrity, electrochemical performance, and conductivity. Several strategies have emerged in recent studies. Incorporating high-valence metal ions, such as molybdenum, into LDH structures can stimulate lattice oxygen and improve stability and performance in applications such as CIBs [144]. Adjusting the metal/ion ratio within LDHs can optimize charge storage capabilities and reduce phase instability, enhancing electrochemical properties [255]. Research has also shown that hybridizing LDHs with conductive materials, such as graphene or carbon nanotubes, further enhances electrical conductivity and facilitates electron transport, providing superior performance in high-capacity batteries [256]. Furthermore, computational models for high-throughput screening of LDHs enable the identification of optimal stoichiometric ratios and structures, streamlining the development process [230]. Collectively, these strategies represent a multifaceted approach to optimizing LDH materials for next-generation battery technologies.

4.1. Lithium-Ion Batteries

In LIBs, LDHs exhibit promising behavior as cathode materials. For instance, nickel–cobalt LDHs have been shown to provide higher capacity and stability than conventional materials. Research by Bai et al. [257] highlights that these LDHs achieve about 160 mAh/g capacities at high current densities, showcasing excellent rate capability and cycling stability, which are crucial for practical applications. Furthermore, the study emphasizes how the layered structure and the ability to intercalate lithium ions enhance ionic conductivity and electron transport, improving overall battery performance. ZIF-67@NiCo-LDH was utilized as an anode material in LIBs, displaying high electrochemical performance [134]. Initially, the composite delivered a high discharge capacity of 1997.1 mAh/g at 100 mA/g and maintained a stable reversible capacity of 807.9 mAh/g after 100 cycles. It demonstrated excellent rate capability, achieving capacities of approximately 1303.9, 931.5, 601.3, and 445.0 mAh/g at current densities of 100, 200, 500, and 1000 mA/g, respectively. The excellent cyclic stability, high specific capacity, and superior rate performance were attributed to the synergistic effect between ZIF-67 MOF and NiCo-LDH nanosheets and the unique heterostructure morphology's enhanced conductivity and structural integrity. $NiFe_2O_4$@NiCo-LDH was utilized as an anode in LIBs [135], exhibiting superior electrochemical performance characterized by a high specific capacity of 636.9 mAh/g after 100 cycles at 0.3 A/g, along with high coulombic efficiency (~98%). The composite anode demonstrated good rate performance, maintaining stable capacities at various current densities and temperatures (−10 °C, 25 °C, 50 °C), thus confirming its robustness. Electrochemical impedance and galvanostatic intermittent titration techniques indicated that the $NiFe_2O_4$@NiCo-LDH possessed significantly lower internal resistance and higher conductivity, resulting in efficient ion/electron transfer and improved cyclic stability compared to pristine $NiFe_2O_4$. The synthesized NiZn-LDH, intercalated with DS− anions, was employed as an anode material for LIBs [136]. Specifically, the optimized LDH (NZDS-20) demonstrated superior electrochemical performance, delivering a high specific capacity of 850 mAh/g at a current density of 0.5 A/g, which remained stable after 400 cycles. Furthermore, the material showed excellent rate capability, maintaining a capacity of approximately 473.5 mAh/g at a higher current density of 3 A/g. EIS and galvanostatic intermittent titration technique (GITT) measurements further confirmed improved lithium-ion diffusivity and reduced reaction resistance, highlighting the beneficial effects of DS− intercalation. Moreover, a significant pseudocapacitance contribution was observed, underscoring rapid reaction kinetics and enhanced electrochemical stability. The hollow yolk/shell NiS_2/FeS_2@NC@NiFe LDH/FeO(OH) microspheres were utilized as an anode in LIBs, demonstrating excellent electrochemical performance [137]. At 0.2 A/g, the LDH electrode delivered a specific 709.9 mAh/g capacity after 200 cycles, highlighting robust cycling stability. Furthermore, at a higher current density (1.0 A/g), it maintained a capacity of 403.3 mAh/g even after 500 cycles, showcasing superior cyclic and rate stability compared to individual NiS_2/FeS_2 or NiFe LDH/FeO(OH) electrodes. The improved performance is attributed to the hollow yolk/shell structure, N-doped carbon layer, and enhanced conductivity and stability. The NiCo-LDH/MXene composite was used as an anode material for LIBs, showing excellent electrochemical performance [138]. It achieved a high specific capacity of 1081 mAh/g after 100 cycles at 100 mA/g and demonstrated outstanding rate capability, retaining 439 mAh/g even at a high current density of 1000 mA/g. Additionally, the material exhibited superior cycling stability, with minimal capacity fading, demonstrating the effectiveness of the 1D/2D structure in enhancing battery performance. Mg_2Al_1–CO_3-LDH was utilized as an anode in LIBs [140], exhibiting an impressive initial discharge-specific capacity of 814 mAh/g at a current density of 200 mA/g. Despite initial capacity fading, the material recovered significantly, stabilizing at 203.8 mAh/g after 300 cycles. Additionally, it demonstrated reasonable rate capabilities (up to 61.3 mAh/g at 2000 mA/g), attributing electrochemical performance primarily to reversible Li^+ storage through valence change of Mg and conversion reactions involving LiOH and LiH/Li_2O.

An LDH material (LDH200) was employed as an anode material for LIBs, demonstrating exceptional electrochemical properties [109]. Specifically, the LDH anode delivered a high specific capacity of 1821.3 mAh/g at a current density of 0.1 A/g. Impressively, the material maintained a robust cyclic stability, retaining a reversible capacity of about 687.7 mAh/g after 500 cycles at 0.5 A/g, corresponding to a very low capacity attrition rate of 0.092%. Figure 14A shows the CV curves of LDH200, where the redox peaks remain stable after the first few cycles, indicating good electrochemical reversibility. In Figure 14B, the charge/discharge profiles of LDH200 over various cycles at 0.5 A/g show consistent voltage plateaus, demonstrating stable cycling behavior. Figure 14C compares the long-term cycling performance of LDH200 with other samples. LDH200 shows the highest capacity retention and coulombic efficiency over 500 cycles. Figure 14D presents rate performance data, where LDH200 retains superior capacity across different current densities compared to the other LDH samples. Figure 14E provides a radar plot comparing all electrode types across six performance parameters. LDH200 shows balanced and enhanced electrochemical behavior, confirming it as the most promising candidate among the studied materials. Additionally, this material exhibited a significantly improved lithium-ion diffusion coefficient of approximately 4.08×10^{-11} cm^2/s, indicative of its excellent rate capability and suitability for practical high-performance battery applications.

The synthesized Ni_2Co-LDH/EG composite served as an anode material for LIBs [152], demonstrating superior electrochemical performance compared to pristine Ni_2Co-LDH. It exhibited a high specific capacity of 1880 mAh/g at 0.05 A/g and maintained an excellent specific capacity of 919 mAh/g at a high current density of 1 A/g. Additionally, it showed remarkable cyclic stability with minimal capacity fading, retaining a specific capacity of approximately 973 mAh/g after 100 cycles at 1 A/g. The integration of expanded graphite significantly enhanced electronic conductivity and effectively suppressed volume expansion during the lithiation/delithiation process, supported by reduced charge transfer resistance measured through EIS. NiAl-LDH was used as a negative electrode (anode) material for LIBs and tested with two types of binders: sodium alginate (SA) and PVDF [153]. The SA-based electrode exhibited a remarkable initial discharge capacity of 2586 mAh/g at 0.05 A/g with a reversible 1578 mAh/g capacity. After 400 cycles at 0.5 A/g, the capacity retained was 697 mAh/g (42.2%), and at 1.0 A/g, it remained 388 mAh/g after 1400 cycles (27.6% retention). In contrast, the PVDF-based electrode showed significantly lower retention, indicating the superior performance of the SA binder. The conversion-type mechanism involving Ni^{2+} to Ni^0 and forming a stable SEI layer (rich in Li_2O and LiOH) contributed to the enhanced cycling stability and rate capability.

4.2. Lithium–Sulfur Batteries

In LSBs, utilizing LDHs helps mitigate the polysulfide dissolution issue, which is crucial for improving cycling stability [258]. Bai et al. [257] demonstrated that a NiCo-LDH-based cathode effectively retains capacity over 500 cycles, outperforming conventional sulfur cathodes. The LDH structure helps configure polysulfide species, thus facilitating a more stable electrochemical reaction environment during charge/discharge cycles, addressing a significant challenge in LSBs [202]. LDH@PSS was applied as an artificial solid electrolyte interphase (ASEI) on lithium metal anodes in LSBs [155]. The composite SEI was fabricated via spin-coating, providing mechanical robustness and ionic selectivity. Li||Li symmetric cells with LDH@PSS-Li electrodes demonstrated an ultra-long cycle life of over 3100 h at 1 mA/cm^2 with significantly reduced overpotential. When paired with sulfur cathodes, the full Li–S cells showed excellent performance: an initial specific capacity of 1247.2 mAh/g and a reversible capacity of 1032.6 mAh/g after 200 cycles at 0.2 C, with a minimal capacity decay of 0.086% per cycle. At higher rates (1 C), the cell retained 65.2% of its capacity after 1000 cycles. Additionally, the system exhibited suppressed shuttle currents, lower self-discharge, enhanced Li^+ transference numbers ($tLi^+ = 0.81$), and improved charge transfer kinetics, highlighting its potential for stable, high-performance LSBs.

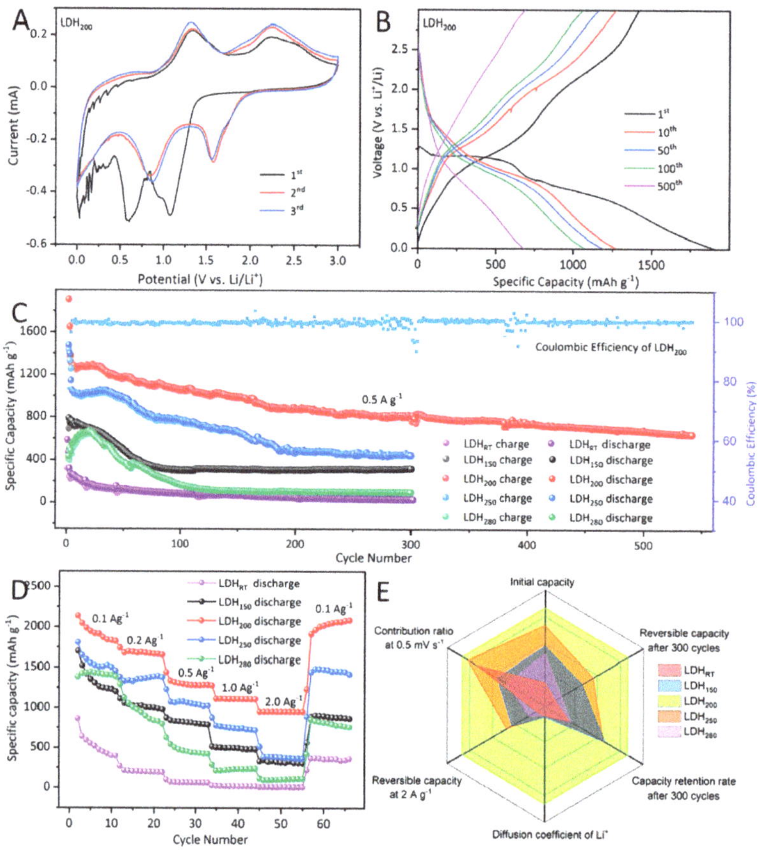

Figure 14. (**A**) CV curves of the LDH sample recorded at 0.1 mV/s within a voltage range of 0.01–3.0 V vs. Li/Li^+. (**B**) Charge/discharge voltage profiles of LDH at a current density of 0.5 A/g over multiple cycles. (**C**) Long-term cycling performance and coulombic efficiency of LDH200 compared to other LDH-based electrodes at 0.5 A/g. (**D**) Rate capability comparison for different samples under various current densities. (**E**) The radar chart displays normalized electrochemical metrics, including initial capacity, rate performance, and capacity retention, for all LDH variants. Reproduced with permission [109].

NiCo-LDH-based material was used as a freestanding cathode (NMWAs@NiCo-LDH/S) in LSBs [222]. It demonstrated exceptional electrochemical performance, with an initial specific capacity of 1238.4 mAh/g at 0.1 C, and 805.8 mAh/g at 5.0 C, reflecting excellent rate capability. Over 700 cycles at 5.0 C, it retained 647.1 mAh/g with a minimal capacity decay rate of 0.018% per cycle. The unique structure facilitated effective sulfur confinement, rapid Li^+ diffusion, and robust redox kinetics while alleviating volume expansion and suppressing the polysulfide shuttle effect. This makes NMWAs@NiCo-LDH/S a competitive candidate for high-rate and long-life LSB cathodes. The Vo-LDHs-MXenes were employed as a functional separator coating in LSBs [156]. The heterostructure significantly improved the electrochemical performance through enhanced adsorption and catalytic conversion of lithium polysulfides. The assembled LSBs exhibited an initial specific capacity of 1549 mAh/g at 0.1 C and retained 701 mAh/g after 300 cycles at 1.0 C, with a minimal capacity decay rate of 0.084% per cycle. Under high sulfur loading (6.7 mg/cm^2) and a low electrolyte/sulfur ratio (6 μL/mg), the areal capacity reached 6.09 mAh/cm^2. The cell demonstrated superior rate capability, ranging from 1530 to 342.9 mAh/g across 0.1 C to 3.0 C. These results underscore the effectiveness of the Vo-LDHs-MXenes in suppressing shuttle effects and enhancing redox kinetics in LSBs.

A VCFN composite was employed not as a cathode or anode material but as a separator modification layer (VCFN/SP/PVDF) on a glass fiber (GF) separator in LSBs [156]. The modified separator enabled superior polysulfide confinement and redox kinetics due to the synergistic effects of V_2O_5 catalysis, LDH's high surface area, and cysteine's functional linkage. The battery exhibited a high initial specific capacity of 1035.2 mAh/g at 0.5 C and retained 920.1 mAh/g after 300 cycles, demonstrating excellent cycling stability with a minimal capacity decay of 0.039% per cycle. Even at elevated sulfur loadings (up to 4.5 mg/cm^2), the system maintained stable capacities above 900 mAh/g, highlighting its scalability. Moreover, the cell exhibited improved Li-ion diffusion, reduced charge/transfer resistance (15.3 Ω), high ionic conductivity (4.1×10^{-2} S/cm), and minimized polarization, underlining the role of the VCFN-modified separator in promoting efficient electrochemical performance.

In LSBs, NiCo–MOF/LDH was applied as a functional interlayer between the sulfur cathode and separator [223]. The cell delivered remarkable electrochemical performance when used with a sulfur cathode composed of sulfur-encapsulated nitrogen-doped carbon nanocages (S@hNCNC). Specifically, it exhibited a high specific capacity of 950 mAh/g after 200 cycles at 1 C with a minimal capacity decay of 0.033% per cycle. It also maintained 982 mAh/g at 0.5 C after 200 cycles and 633 mAh/g over 1000 cycles at 1 C. The rate capability was also excellent, achieving 640 mAh/g at 5 C and returning to 1103 mAh/g when the current returned to 0.1 C. The interlayer significantly reduced polarization and enhanced polysulfide conversion, which was further validated by CV, Tafel plots, and EIS. These enhancements are attributed to the synergistic effect of NiCo-LDH's strong chemisorption and NiCo–MOF's catalytic activity.

In LSBs, the synthesized NiCo-LDH was used as a cathode host after sulfur loading [224]. The S/NiCo-LDH cathode delivered a high initial specific capacity of 1540 mAh/g at 0.1 C and retained 485 mAh/g even at 5 C, showing excellent rate performance. At 0.2 C, the electrode retained a stable 650 mAh/g capacity over 100 cycles with nearly 100% coulombic efficiency. Most notably, after 500 cycles at 1 C, the capacity remained at 475 mAh/g with 78% retention, indicating robust cycling stability. The synergistic effect of strong chemical polysulfide adsorption and catalytic activity of NiCo-LDH significantly improved the redox kinetics and suppressed shuttle effects, validating its potential as a high-performance cathode host in LSB systems.

The synthesized L-CoNi-LDH/C was employed as a sulfur host in the cathode (S@L-CoNi-LDH/C) of LiSB [154]. The composite enabled high sulfur loading (~66.4 wt.%) and delivered a high initial discharge capacity of 1574 mAh/g, retaining 1097 mAh/g after 100 cycles at 0.1 C with nearly 100% coulombic efficiency. At a high rate of 2 C, the electrode exhibited a stable capacity of 503 mAh/g after 200 cycles. Additionally, it showed excellent rate capability, sustaining 486 mAh/g at 5 C and recovering to 730 mAh/g when returned to 0.2 C. Figure 15 highlights the electrochemical performance and mechanism of the S@L-CoNi-LDH/C cathode for LSBs. As shown in Figure 15a, the charge/discharge profiles at 0.1 C indicate that the S@L-CoNi-LDH/C electrode delivers a higher capacity than the

S@CoNi-LDH/C, confirming enhanced redox activity and sulfur utilization. Figure 15b demonstrates the superior cycling stability of S@L-CoNi-LDH/C at 0.1 C, while Figure 15c shows its high capacity retention even at 2 C, along with excellent coulombic efficiency. The performance at various current densities is displayed in Figure 15d, and the rate capability in Figure 15e confirms the electrode's ability to recover its capacity when the current is returned to a lower value. Figure 15f presents EIS measurements at different voltages, illustrating the charge transfer behavior and confirming the effective catalytic role of the CoNi-LDH/C framework in promoting polysulfide conversion. This performance is attributed to the enhanced catalytic activity, conductivity, and strong LiPS confinement offered by the L-CoNi-LDH/C composite. In applying LSBs, the $NiCo_2S_4$-NiS_2 nano-heterostructure was used as a sulfur host in the cathode [225]. The designed cathode exhibited an outstanding initial specific capacity of 1207 mAh/g at 0.2 C and excellent rate capability with 766 mAh/g at 2 C. Under high sulfur loading conditions (5.0 mg/cm^2) and lean electrolyte usage (E/S ratio of 8.0 mL/g), the system achieved a high areal capacity of 6.09 mAh/cm^2. Long-term cycling at 1 C showed a capacity retention of 60.23% after 450 cycles (fading rate of 0.089% per cycle), and further tests confirmed suppression of LiPS shuttle and lithium dendrite formation, validating its superior electrochemical stability and catalytic effect. CuCo-LDH was utilized as a separator modifier rather than a direct cathode or anode material [226]. Its hollow, mesoporous architecture provided a strong barrier against polysulfide shuttle and enhanced sulfur redox kinetics via metal–sulfur interactions. As a result, the CuCo-LDH-modified separator enabled high reversible specific capacities (~1262.8 mAh/g initially at 0.2 C and ~697.0 mAh/g after 500 cycles at 1 C), excellent rate capability up to 5 C, and an extremely low capacity decay of 0.049% per cycle over 500 cycles. Under practical conditions with high sulfur loading (4.2 mg/cm^2) and lean electrolyte (E/S = 6 mL/g), a remarkable areal capacity of 4.39 mAh/cm^2 was achieved and retained 3.42 mAh/cm^2 after 200 cycles. These results highlight its dual-functionality as a physical barrier and a chemical catalyst in improving LSB performance. MgAl-LDH@CNT was utilized as a cathode component in Li–S batteries, combined with sulfur to form a composite cathode and paired with a dual-functional graphene/PP/Al_2O_3 (DF-GPA) separator [110]. This system significantly enhanced the electrocatalytic conversion of polysulfides and stabilized the redox kinetics. As a result, the Li–S cell delivered a high specific capacity and showed excellent cycling performance with minimal capacity decay over 200 cycles at 1 C. With a high sulfur loading of 4.0 mg/cm^2, the battery exhibited stable operation and high coulombic efficiency, making it a promising design for high-energy-density sulfur cathodes in practical applications.

LDHs were used as sulfur host materials in the cathode [157]. Among the variants, NiFe-LDH@S demonstrated superior electrochemical performance due to the stronger coordination between Ni/Fe sites and lithium polysulfides, facilitating better adsorption and catalytic conversion. It delivered an initial 676 mAh/g capacity at 0.2 C and maintained 503 mAh/g after 100 cycles. At a higher rate of 2 C, it still retained 386 mAh/g after 500 cycles. In contrast, NiAl-LDH@S and ZnAl-LDH@S showed poorer capacity retention and higher polarization, highlighting the critical role of transition metal composition in tuning LDH performance. Moreover, NiFe-LDH@S exhibited the lowest charge transfer resistance (4.9 Ω) and the highest polysulfide adsorption capacity, reinforcing its effectiveness as a high-performance sulfur host.

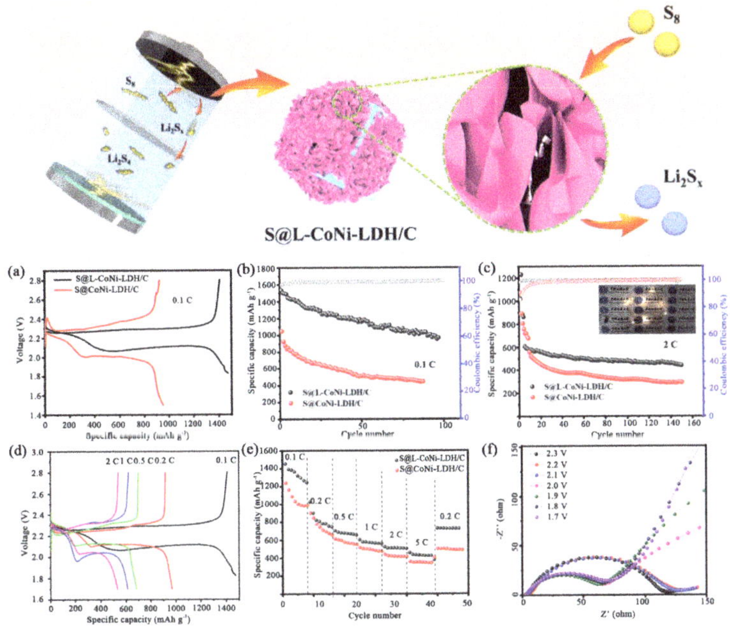

Figure 15. Illustration showing the role of S@CoNi-LDH/C in enhancing polysulfide adsorption and catalytic conversion. (**a**) Charge/discharge voltage profiles of S@L-CoNi-LDH/C and S@CoNi-LDH/C electrodes tested at 0.1 C. (**b**) Cycling stability at 0.1 C and (**c**) high-rate cycling at 2 C for both cathodes, including coulombic efficiency. (**d**) GCD curves at various current rates, and (**e**) corresponding rate capability of S@L-CoNi-LDH/C. (**f**) Nyquist plots of S@L-CoNi-LDH/C cathode collected at different voltage stages. Reproduced with permission [154].

The PPy@LDH-S was employed as a cathode in LSBs [227], demonstrating superior performance due to its unique multi-functional architecture. The composite cathode delivered a high initial discharge capacity of 907.2 mAh/g at 1 C. It retained 633.4 mAh/g after 500 cycles with a minimal fading rate of 0.06% per cycle, maintaining ~99% coulombic efficiency. At various rates from 0.2 C to 5 C, the electrode exhibited excellent rate capability and reversible capacity up to 983.2 mAh/g when returned to 0.2 C. It also achieved a reversible areal capacity of 535.7 mAh/g (3.5 mg/cm^2 sulfur loading) and 416.9 mAh/g (5.2 mg/cm^2) after 500 cycles. The flexible pouch cell configuration further validated its mechanical robustness and suitability for wearable electronics, maintaining functionality under bending and twisting conditions. In the LSB configuration, the VG@LDH/CC composite was used as a sulfur host for the cathode [158]. This structure provides both physical confinement and chemical adsorption of polysulfides, minimizing shuttle effects and improving reaction kinetics. The battery delivered a high specific capacity of 1182.4 mAh/g at 0.1 C, and retained 441.3 mAh/g after 750 cycles at 0.5 C, with a very low capacity decay rate of 0.0755% per cycle. The rate performance was also excellent, sustaining 780.8 mAh/g at 3 C, and maintaining capacity recovery upon returning to lower rates. The design supports high-rate and high-capacity operation, even under high sulfur loading conditions (e.g., 547.7 mAh/g after 300 cycles at 3.5 mg/cm^2 loading at 1 C).

4.3. Sodium-Ion Batteries

For NIBs, LDHs have shown notable electrochemical performance. Research indicates that ZnAl-LDHs can be effective anode materials, exhibiting around 200 mAh/g capacities with good cycling stability over extended periods [102]. The ability to intercalate sodium ions while maintaining structural integrity enhances energy storage capabilities. The intrinsic layered structure also allows for efficient ion diffusion and electron transport, leading to faster charge/discharge rates than traditional anode materials. The synthesized NiCr-Cl LDH/rGO was utilized as an anode material in NIBs [159]. Electrochemical tests revealed a high initial discharge capacity of 632 mAh/g, with a reversible capacity of 218 mAh/g after 200 cycles at 100 mA/g, indicating good cyclic stability. Moreover, it delivered 219 mAh/g at a high rate of 2 A/g, highlighting its excellent rate capability. The improved electrochemical performance is attributed to the expanded interlayer spacing, enhancing Na^+ ion diffusion, and the rGO matrix, which boosts conductivity and buffers volume changes. EIS confirmed the lowest charge transfer resistance among the compared samples, supporting fast kinetics. Capacitive contribution analysis showed dominant pseudocapacitive behavior, with 55.1% capacitive contribution at 0.2 mV/s, confirming its favorable kinetic profile for high-rate applications.

The CoMoLDH@C-PDA material was evaluated as an anode in NIBs [163]. The composite synthesized with 6-h PDA polymerization (CoMoLDH-24@C-PDA-6) exhibited superior electrochemical performance among various samples. This anode delivered an initial specific capacity of 779.9 mAh/g at 0.05 A/g and maintained 310.9 mAh/g after 100 cycles at 0.1 A/g, with a 70% capacity retention. High rate capability and long-term stability were attributed to the formation of α-$CoMoO_4$ and conductive carbon coating, which improved Na^+ diffusion and minimized charge transfer resistance. Figure 16a presents the fabrication steps for CoMoLDH-based materials with and without carbon and PDA modifications. The composite CoMoLDH-24@C-PDA-6 is prepared by combining polymerized dopamine with carbonized CoMoLDH. The electrochemical behavior of these electrodes is shown in Figure 16b,c. The CoMoLDH-24@C-PDA-6 electrode exhibits more stable and extended charge/discharge profiles, indicating enhanced capacity and reaction reversibility. In Figure 16d, cycling performance over 100 cycles at 0.1 A/g shows that CoMoLDH-24@C-PDA-6 retains 70% of its capacity, significantly outperforming the unmodified CoMoLDH-24 (43.7%). Nyquist plots in Figure 16e,f demonstrate reduced charge transfer resistance for the modified electrode, both before and after cycling, confirming improved conductivity and electrochemical stability. Further, EIS and GITT analysis confirmed a lower R_{ct} and higher Na^+ diffusion coefficient for the coated material, reinforcing its viability for high-performance NIB applications.

In NIB applications, both $Ni_{0.8}Ca_{1.2}Al$-Cl LDH and $Co_{0.6}Ca_{1.4}Al$-Cl LDH were evaluated as anode materials in CR2032 coin-type cells [228]. These materials demonstrated promising electrochemical properties, with initial discharge capacities of 819.2 and 746.6 mAh/g and coulombic efficiencies of 65.2% and 56.8%, respectively. $Ni_{0.8}Ca_{1.2}Al$-Cl LDH delivered a stable specific capacity of 256.9 mAh/g, and $Co_{0.6}Ca_{1.4}Al$-Cl LDH achieved 292.8 mAh/g at 0.2 A/g after 200 cycles. Notably, they maintained 102 and 118.1 mAh/g after 600 cycles at 2 A/g, demonstrating excellent cycling stability. Both materials exhibited enhanced rate capability and reduced impedance, attributable to the porous structure and bimetallic synergies, and their performance was comparable to that of conventional hard carbon. The sodium storage mechanism involved both diffusion-controlled intercalation and surface capacitive effects, supported by ex situ XRD and XPS analyses. NHCNS@$CoSe_2$@C was employed as an anode material. It demonstrated an initial discharge capacity of 465.6 mAh/g at 0.1 A/g, and retained 373.8 mAh/g after 100 cycles, with a coulombic efficiency of 59.7% in the first cycle [164]. Even at a high current density of 5 A/g, the electrode retained a specific capacity of 314.2 mAh/g, reflecting outstanding rate capability. Additionally, long-term cycling at 0.5 A/g for 1000 cycles showed stabilized performance at 325 mAh/g, despite some capacity fade due to volume expansion. The hybrid's superior performance is attributed to its high conductivity, abundant active sites, and structural robustness against sodium insertion/extraction.

NiAl-LDH@G was employed as a cathode material in CIBs. Compared to pure NiAl-LDH, the NiAl-LDH@G cathode demonstrated significantly enhanced electrochemical performance [143]. It delivered a high specific capacity of 223.3 mAh/g during the first discharge and a stable reversible capacity of 107 mAh/g over 500 cycles at 100 mA/g with a coulombic efficiency of ~96%. It also retained 72 mAh/g after 120 days, indicating low self-discharge and excellent long-term stability. The composite exhibited good rate capability with capacities of 148.5, 67.8, 30.9, and 15.9 mAh/g at increasing current densities, recovering to 120.8 mAh/g when returned to 100 mA/g. EIS showed reduced charge transfer resistance and higher chloride ion diffusion coefficients (up to 3.22×10^{-12} cm^2/s), confirming improved conductivity and ion mobility. Electrochemical properties of NiAl-LDH@G are shown in Figure 17j to Figure 17n, with CV and impedance tests confirming improved charge storage and ion diffusion compared to NiAl-LDH. Long-term cycling behavior is presented in Figure 17o, and rate capability at different current densities is plotted in Figure 17p, showing the superior performance of the NiAl-LDH@G composite. These results underscore the superiority of the nanoarray structure in facilitating fast redox kinetics and durable performance in CIBs.

4.5. Zinc-Ion Batteries

ZIBs also benefit from LDH incorporation. Studies have shown that LDH compounds like Zn–Al serve as structural promoters and improve charge storage capabilities, achieving specific capacities upward of 300 mAh/g [259]. Furthermore, their layered structure contributes to the selective intercalation of zinc ions, which further optimizes battery performance and longevity [260]. This characteristic is particularly advantageous in developing versatile and efficient energy storage solutions. Hv-$Ni_3Mn_{0.7}Fe_{0.3}$-LDH was employed as a cathode material in aqueous ZIBs (AZIBs) [16]. The modified LDH displayed exceptional performance, delivering a high specific capacity of 328 mAh/g at 50 mA/g and maintaining 122 mAh/g over 500 cycles at 1 A/g with 85% capacity retention. The enhanced electrochemical behavior stems from introducing hydrogen vacancies that expose terminal oxygen atoms, promoting Zn^{2+} intercalation and diffusion. The material also exhibited robust rate capability and reversible Zn^{2+}/H^+ intercalation mechanism. The synergy among Ni, Mn, and Fe contributed to conductivity and redox activity and effectively suppressed Jahn–Teller distortion, ensuring structural stability during long-term cycling. The Ragone plot revealed competitive energy and power densities, positioning this LDH among the top-performing ZIB cathodes.

The synthesized NiCo LDH, supported with MXene and electrodeposited on nickel foam, served as a high-performance cathode in alkaline Zn||NiCo batteries [203]. The battery exhibited a remarkable areal specific capacity of 20.2 mAh/cm^2 (equivalent to 311.7 mAh/g), with exceptional rate capabilities maintaining 9.6 mAh/cm^2 at 120 mA/cm^2 (148 mAh/g). It also delivered ultrahigh areal and gravimetric energy densities of 31.2 mWh/cm^2 and 465 Wh/kg, respectively, alongside a peak power density of 80 mW/cm^2 (2752 W/kg). The electrode retained 88.6% of its capacity after 10,000 cycles, demonstrating outstanding long-term cycling stability, fast kinetics, and high ion diffusion coefficients due to the unique MXene-assisted structure. MCN-LDH@CP was used as the cathode in an aqueous CoNi//Zn battery [165]. The hybrid electrode exhibited a high areal specific capacity of 1.74 mAh/cm^2, an areal energy density of 2.89 mWh/cm^2, and a peak power density of 111.22 mW/cm^2. It demonstrated excellent cycling stability, maintaining 97.8% capacity retention after 7000 cycles at 40 mA/cm^2. These superior performances are attributed to the molybdate-induced increase in surface area, interlayer spacing, and improved diffusion kinetics. The battery also showed outstanding mechanical resilience and safety in quasi-solid-state form, maintaining functionality under physical damage.

The ZnCo-LDH was employed as a cathode material in Zn-ZnCo batteries, using Ni foam as the current collector and a 6 M KOH + 0.2 M $Zn(CH_3COO)_2$ electrolyte [166]. The batteries exhibited a linear increase in specific capacity with increased ZnCo-LDH loading, reaching up to 2.5 mAh/cm^2 at a 60 mg/cm^2 loading. Remarkable rate performance was maintained across various current densities, with over 75% capacity retention at high rates and full recovery upon returning to lower current densities. The ZnCo-LDH cathode demonstrated exceptional cycling stability, retaining 86% of its capacity after 5000 cycles at 30 mg/cm^2 loading. Additionally, hybrid Zn batteries incorporating an air-accessible Pt/C-coated gas diffusion layer alongside ZnCo-LDH delivered dual-stage voltage plateaus, enabling a combination of Zn–ZnCo and Zn–air functionalities. The hybrid cell exhibited stable cycling over 270 cycles and enhanced Zn–ZnCo segment capacity due to improved electrode wettability.

The H-vacancy-rich CoNi-LDH(v) was used as a cathode in aqueous and flexible solid-state Zn-ion batteries (ZIBs), paired with Zn@CC as the anode [204]. The aqueous CoNi-LDH(v)@CC//Zn@CC battery delivered a high specific capacity of 225 mAh/g at 0.5 A/g, with energy and power densities of 353.62 Wh/kg and 3056.51 W/kg, respectively. It retained 53.9% capacity after 900 cycles, significantly outperforming the non-activated CoNi-LDH. Figure 18a shows the layered structure of the flexible CoNi-LDH(v)@CC//Zn@CC solid-state battery. Electrochemical behavior is displayed in Figure 18b, where the GCD curves demonstrate good performance at different current rates. The corresponding specific capacity and coulombic efficiency over cycling are shown in Figure 18c. Figure 18d presents the cycling stability, with the battery maintaining 47.22% of its capacity after 1000 cycles. Mechanical flexibility is evaluated in Figure 18e, where the battery retains over 85% capacity even when bent at 150°. Figure 18f shows it remains functional after 100 bending cycles with 74.82% capacity retention. Figure 18g–i visually demonstrates the battery's robustness: it powers a digital device while bent, continues operating even after being cut, and maintains output for over 80 h in an open environment. Moreover, a flexible, solid-state version of the battery demonstrated excellent deformability and stability under mechanical stress and sustained operation over 80 h without encapsulation, underlining its promise for wearable electronics.

ZnAl-LDH was applied as an artificial interface layer on Zn anodes (SZ/LDH@Zn) to enhance the performance of aqueous ZIBs [172]. The SZ/LDH@Zn||SZ/LDH@Zn symmetric cell delivered an ultra-long cycling lifespan of 5500 h at 1 mA/cm^2 with low voltage hysteresis (16.4 mV). In asymmetric Zn||Cu cells, it achieved a high average coulombic efficiency of 99.85% over 3800 cycles, outperforming bare Zn and ZnAl-LDH@Zn. Furthermore, in an entire cell with $NaV_3O_8{\cdot}xH_2O$ cathode and a low N/P ratio of 3.62, the SZ/LDH@Zn||NVOH cell exhibited a specific capacity of 251.45 mAh/g, an areal capacity of 5.36 mAh/cm^2, and stable cycling over 1500 cycles at 1 A/g. These results confirm the superior reversibility, capacity retention, and anti-corrosion performance of SZ/LDH@Zn in practical ZIB systems.

H-vacancy-enriched CoNi LDH (denoted as CoNi LDH(v)) was employed as a cathode material for aqueous ZIBs in a mild $ZnSO_4$ electrolyte (pH $\approx$ 4) [207]. The electrode demonstrated a high specific capacity of 185 mAh/g at 1.2 A/g, with an average discharge voltage of 1.6 V (vs. Zn^{2+}/Zn), leading to an excellent energy density of 296.2 Wh/kg at a power density of 1894 W/kg. Even at a high current density of 12 A/g, it maintained a 103 mAh/g capacity, indicating good rate performance. Furthermore, the electrode showed impressive cyclic stability with 80% capacity retention over 1000 cycles at 6 A/g. The superior performance was attributed to the Zn^{2+}/H^+ co-intercalation mechanism enabled by the introduced hydrogen vacancies, which improved ion diffusion and electronic conductivity.

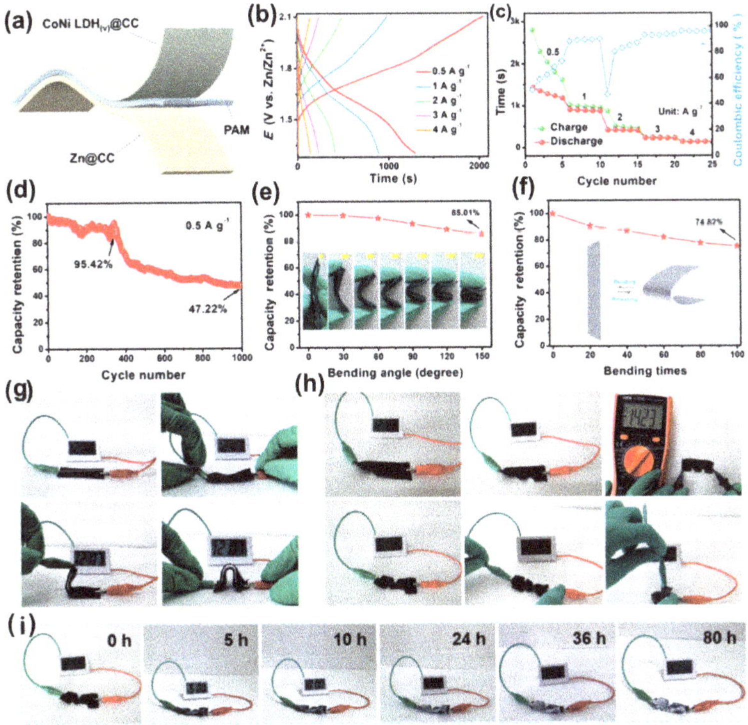

Figure 18. Design and performance of a flexible solid-state battery based on a CoNi-LDH(v)@CC//Zn@CC layered configuration. (**a**) Schematic of the battery's structural composition. (**b**) Charge/discharge profiles at various current densities. (**c**) Specific capacity and coulombic efficiency during charge/discharge cycling. (**d**) Long-term cycling performance at 0.5 A/g. (**e**) Capacity retention under different bending angles. (**f**) Capacity retention after multiple bending cycles. (**g**) Photographs showing operation under various bending and twisting states. (**h**) Device functionality after being cut. (**i**) Duration of continuous power output to a digital watch in open-air conditions. Reproduced with permission [204].

In the alkaline ZIB configuration, NC@CL was employed as the cathode against a Zn anode in an aqueous electrolyte (1 M KOH + 5 mM ZnO) [141]. The NC@CL//Zn battery demonstrated outstanding electrochemical performance with a high specific capacity of 381.1 mAh/g at 0.5 A/g, maintaining 243.4 mAh/g even at 10 A/g, confirming excellent rate capability. The device showed a robust energy density of 66.56 Wh/kg and power density of 15,375 W/kg, with superior cyclic stability—retaining 98% capacity over 2000 cycles at 5 A/g. These performances are attributed to the hierarchical nanosheet morphology, heterostructure-induced interfacial charge redistribution, and enhanced OH^- adsorption facilitated by La-doping and NC–CL synergism. The synthesized NiCoS@NiCo-LDH was utilized as the cathode in aqueous alkaline zinc-based batteries (AAZBs), assembled in a full cell with zinc foil as the anode [208]. The device delivered an impressive specific capacity of 312 mAh/g at 2 mA/cm^2. Moreover, it achieved a high energy density of 435.3 Wh/kg at a power density of 4.1 kW/kg, maintaining 95.9% capacity retention over 3000 cycles at 20 mA/cm^2. These values highlight excellent rate capability and stability and surpass the performance of many previously reported nickel-based cathode materials, showcasing its potential for high-performance, long-life ZIBs.

The synthesized NF\Ni_3S_2/NiS@NiCo-LDH was employed as the cathode in an aqueous alkaline Ni–Zn battery, paired with a zinc anode and a mixed KOH/$Zn(CH_3COO)_2$

electrolyte [209]. The battery delivered a high specific capacity of 434.5 mAh/g at 3 mA/cm^2, showcasing excellent rate performance with capacities ranging from 317.9 to 100.3 mAh/g across 2–15 A/g. The device achieved a maximum energy density of 556.3 Wh/kg and a peak power density of 26.3 kW/kg. The cycling stability was outstanding, retaining 116.7% of its original capacity after 5000 cycles at 60 mA/cm^2, highlighting its activation behavior and robust long-term performance.

4.6. Zinc–Air Batteries

In the context of ZABs, LDHs demonstrate exceptional catalytic activity for OER. Yan et al. [261] highlighted the efficiency of Ni-Fe LDHs as bifunctional electrocatalysts, achieving low overpotentials that significantly enhance the performance of ZABs. Their superior activity can be attributed to the highly active sites the layered structure provides, which facilitate oxygen reduction and evolution reactions. Consequently, integrating LDHs in air battery systems can lead to enhanced energy densities and improved operational lifetimes. In the ZAB application, Co@NiFe-LDH was employed as the air cathode catalyst, while a zinc plate was used as the anode [150]. The battery demonstrated outstanding performance metrics, surpassing even Pt/C + RuO_2-based systems. Specifically, Co@NiFe-LDH achieved a specific capacity of 652 mAh/g, power density of 165 mW/cm^2, and an open circuit voltage of 1.44 V. It exhibited excellent cyclic stability over 90 h of continuous GCD cycling with a minor voltage gap increase (from 0.78 V to 0.80 V), maintaining about 59.5% round-trip efficiency. These results confirm Co@NiFe-LDH's promise as a cost-effective and efficient catalyst for rechargeable ZABs.

The Pt-anchored NiFe-LDH@Co_9S_8 was used as a bifunctional air cathode catalyst in both liquid-state and flexible quasi-solid ZABs [149]. The catalyst demonstrated superior electrocatalytic performance with a high specific capacity of 796.6 mAh/g and a 164.3 mW/cm^2 peak power density in liquid-state ZABs. It also exhibited an energy density of 76.4 Wh/kg and excellent cyclic stability, maintaining performance over 240 cycles (80 h) in flexible ZABs. Figure 19a presents the structural design of the flexible ZAB, which includes an air cathode and a NiFe-LDH@Co_9S_8 electrocatalyst that supports both OER and ORR processes. Figure 19b compares the impedance of three different polymer electrolytes, showing that the PANa-CMC-IL hydrogel offers the lowest resistance. Figure 19c confirms its superior water retention over time, supporting stable operation. The practical application of the battery is shown in Figure 19d, where it successfully powers an LED display. The discharge and power density performance are detailed in Figure 19e, showing a peak power density of 78.4 mW/cm^2. Figure 19f illustrates the battery's ability to maintain high capacities across different current densities, and Figure 19g confirms cycling durability through stable voltage over repeated charge/discharge cycles. Figure 19h demonstrates excellent mechanical flexibility, as the device retains stable output after multiple bending cycles. The catalyst outperformed commercial Pt/C and RuO_2 references, attributed to its synergistic core/shell architecture, high electrochemical surface area (28.2 mF/cm^2), and enhanced ORR/OER activity. The NiCo-LDH/NCM@NF was employed as a bifunctional air cathode in ZABs [148]. It demonstrated a high open-circuit voltage (1.374 V), a peak power density of 158 mW/cm^2, and a specific capacity of 685 mAh/g at 10 mA/cm^2, outperforming commercial Pt/C + IrO_2 catalysts. The catalyst also showed excellent cycling stability, maintaining consistent charge/discharge voltage profiles over 500 cycles and sustaining high round-trip efficiency (~62%) with minimal voltage gap increase. The four-electron ORR pathway, high ECSA (194 cm^2), and low OER overpotential (0.352 V at 50 mA/cm^2) contribute to its robust and durable ZAB performance, attributed to the synergistic integration of NCM and LDH structures enriched with oxygen vacancies.

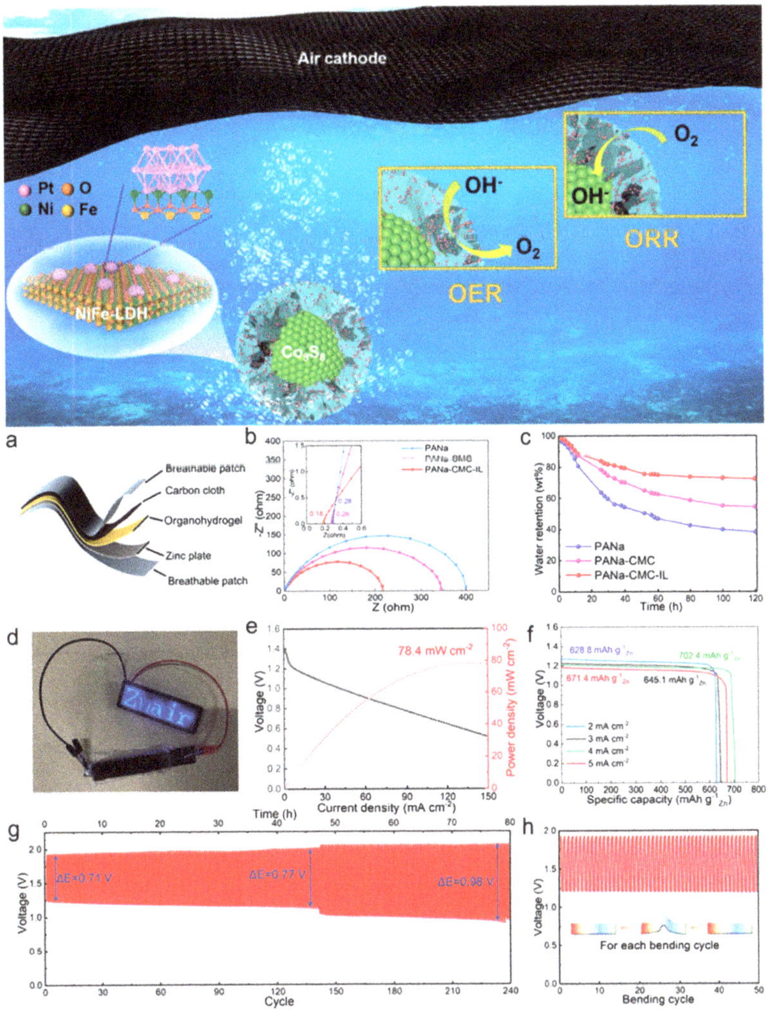

Figure 19. Design and performance of a flexible quasi-solid ZAB using NiFe-LDH@Co_9S_8 nanoflowers with Pt nanocrystals as bifunctional electrocatalysts. (**a**) Schematic of the overall FZAB configuration and its air cathode mechanism involving OER and ORR. (**b**) Nyquist plots comparing PANa, PANa-CMC, and PANa-CMC-IL hydrogels. (**c**) Water retention characteristics over time for each hydrogel type. (**d**) Photograph showing the FZAB powering an LED panel. (**e**) Discharge polarization curve and power density output of the battery. (**f**) Discharge profiles of the device at varying current densities. (**g**) Charge/discharge cycling stability at 2 mA/cm^2. (**h**) Charge/discharge behavior after repeated bending cycles, highlighting mechanical flexibility. Reproduced with permission [149].

In the ZAB application, the synthesized $ZnCo_2O_4$@NiMn-LDH (2:1) was employed as the air cathode [147]. It demonstrated a significantly lower overpotential of 390 mV at a current density of 50 mA/cm^2, outperforming pristine $ZnCo_2O_4$. The battery exhibited a peak power density of 43 mW/cm^2 and delivered a high specific capacity of 639 mAh/g. Notably, it maintained excellent stability, operating continuously for up to 200 h under repeated charge/discharge cycling. These results highlight the material's suitability for long-term and efficient OER catalysis in ZABs. The optimized Fe/CoLDH-based composite (CoL 2:1) was employed as a bifunctional cathode catalyst in a rechargeable ZAB [116].

The assembled battery exhibited a high open-circuit voltage (OCV) of 1.503 V and delivered a peak power density of 161 mW/cm^2. It achieved a specific discharge capacity of 804 mAh/gZn at 5 mA/cm^2 and a calculated energy density of 965 Wh/kg. Notably, the device maintained stable operation over 200 h with a minimal voltage gap of 0.76 V at 10 mA/cm^2, corresponding to a high voltage efficiency of 96%. Even under high current densities (up to 50 mA/cm^2), the system exhibited remarkable rate capability and durability, highlighting the catalyst's robustness and practical viability in real-world ZAB systems.

In application, the CoFe/Fe_3C@CN-900 served as the air cathode catalyst in both aqueous and all-solid-state ZABs [115]. The aqueous ZAB delivered a specific 795 mAh/g capacity, an energy density of 982 Wh/kg, and a maximum power density of 190 mW/cm^2, outperforming the benchmark Pt/C + RuO_2 catalyst. Additionally, it retained 96.5% current after 10 h in stability tests and displayed excellent charge/discharge cycling performance with minimal voltage hysteresis at 5 and 10 mA/cm^2. Solid-state ZABs also demonstrated competitive performance, with a specific capacity of 750 mAh/g and energy density of 929 Wh/kg, along with robust cycling stability, confirming their strong potential for practical applications. The Ni-CAT/NiFe-LDH/CNFs composite was used as a bifunctional catalyst at the air cathode of both liquid and solid-state ZABs [146]. Liquid ZABs demonstrated a high peak power density of 292.1 mW/cm^2, with an open-circuit voltage of 1.407 V, and superior cyclic stability up to 66 h with a low voltage gap. In solid-state flexible ZABs, it retained structural integrity under bending. It showed even better performance, reaching a power density of 299.48 mW/cm^2 and a cycle life of 30 h at a current density of 1 mA/cm^2. These results highlight its promising bifunctional catalytic activity and robust mechanical flexibility for advanced energy storage systems.

In applying ZABs, the Cu@Cu NWs@LDH composite was employed as the air cathode [210]. The assembled battery exhibited a high open-circuit potential of 1.336 V and achieved a peak power density of 179.3 mW/cm^2. It showed superior charge/discharge performance with a low potential gap of 0.852 V at 50 mA/cm^2. The battery maintained stable cycling performance for over 250 h at 10 mA/cm^2 with negligible voltage decay. The NiFe LDH coating on Cu nanowires offered high catalytic activity for OER and ORR, attributed to its enhanced surface area, conductivity, and corrosion resistance. The synthesized CoMn-LDH/NPGA was used as the air cathode in a rechargeable liquid ZAB, where it demonstrated outstanding bifunctional catalytic activity for both ORR and OER [231]. The ZAB showed an open-circuit voltage of 1.514 V and delivered a maximum power density of 206 mW/cm^2 at 226 mA/cm^2, surpassing the performance of Pt/C + IrO_2. It also achieved a high energy density and excellent rate capability, maintaining a discharge voltage of ~1.37 V across varied current densities. Impressively, it exhibited exceptional cycling stability over 72 h (432 cycles) with minimal voltage decay, attributed to the structural robustness of the 3D NPGA framework and uniform CoMn-LDH dispersion.

LDHs have emerged as highly versatile and tunable materials across a wide spectrum of battery chemistries, including LIBs, LSBs, NIBs, CIBs, ZIBs, and ZABs. Their unique 2D lamellar structures, high surface area, compositional flexibility, and anion exchange capacity enable exceptional electrochemical performance through enhanced ion diffusion, redox activity, and structural integrity. In LIBs and NIBs, LDH-based electrodes exhibit high specific capacities, excellent rate performance, and long-term cycling stability due to their ability to accommodate lithium or sodium ions via intercalation and conversion mechanisms. In LSBs, LDHs serve as cathode hosts, interlayers, and separator coatings, effectively suppressing polysulfide shuttling and catalyzing redox reactions, leading to outstanding capacity retention and coulombic efficiency. Their role in CIBs is expanding, with interlayer-engineered LDHs showing promising Cl^- transport properties and structural robustness. Similarly, in ZIBs, LDHs contribute to superior zinc ion intercalation kinetics, energy density, and mechanical resilience, particularly in flexible and high-loading configurations. Moreover, in ZABs, LDHs, often combined with conductive or catalytic materials, act as efficient bifunctional electrocatalysts for oxygen evolution and reduction, offering low overpotentials, high power densities, and extended cycle life. Collectively, the multifunctional role of LDHs across these systems underscores their transformative potential in advancing next-generation, high-performance, and sustainable energy storage technologies. A comprehensive comparison of LDH-based materials in diverse battery systems, including their synthesis methods and electrochemical performance, is presented in Table 1.

Table 1. Summary of various LDH-based electrode materials, their synthesis methods, and electrochemical performance parameters for battery applications.

LDH Name	Synthesis Procedure	Battery Type	Electrolyte	Specific Capacity (mAh/g)	Stability	Ref.
Hv-$Ni_3Mn_{0.7}Fe_{0.3}$-LDH	Co-precipitation + Electrochemical Activation	ZIB	0.2 M $ZnSO_4$	328	85% capacity retention over 500 cycles at 1 A/g	[16]
NiCo-LDH nanocrystal@amorphousness core/shell structure (LDH200)	Co-precipitation followed by heat treatment at 200 °C	LIB	1.0 M $LiPF_6$	1821.3 (0.1 A/g)	~687.7 mAh/g at 0.5 A/g after 500 cycles (attrition rate: 0.092%)	[109]
MgAl-LDH@CNT	Co-precipitation	LSB	Li_6PS_5Cl	…	Stable over 200 cycles at 1 C with 4.0 mg/cm^2 S loading	[110]
NiFe-NO_3 LDH	Co-precipitation	NIB	1 M $NaClO_4$	972 (initial); 388 (60 cycles at 50 mA/g); 122 (500 cycles at 500 mA/g)	388 mAh/g after 60 cycles; 122 mAh/g after 500 cycles	[111]
NiFe-Cl LDH	Co-precipitation	CIB	0.5 M $Bpy_{14}Cl$/PC	27.1	64.2mAh/g after 200 cycles	[112]
Ni_5Ti-Cl LDH	Co-precipitation + anion exchange	CIB	0.5 M $Bpy_{14}Cl$	257.8	127.9 mAh/g after 1000 cycles @1 A/g	[114]
CoFe/Fe_3C@CN-900	MOF-derived + Co-precipitation	ZAB	0.1 M KOH	795	96.5% current retention after 10 h discharge	[115]
Fe/Co LDH (CoL 2:1)	Co-precipitation	ZAB	0.1 M KOH	804	200 h (307 cycles)	[116]
ZIF-67@NiCo-LDH	One-pot hydrothermal synthesis	LIB	1 M $LiPF_6$	1997.1 (initial), 807.9 (after 100 cycles)	Stable up to 100 cycles, excellent rate capability	[134]
NiFe2O4@NiCo-LDH	Two-step hydrothermal method (MOF-derived)	LIB	1 M $LiPF_6$	636.9	Stable after 100 cycles (636.9 mAh/g at 0.3 A/g), coulombic efficiency ~98%	[135]
Ni-Zn LDH intercalated with DS^-	Hydrothermal synthesis	LIB	1 M $LiPF_6$	850 at 0.5 A/g	850 mAh/g over 400 cycles	[136]
NiS_2/FeS_2@NC@NiFe LDH/FeO(OH)	Solvothermal followed by hydrothermal synthesis	LIB	1 M $LiPF_6$	709.9	709.9 mAh/g at 0.2 A/g after 200 cycles; 403.3 mAh/g at 1.0 A/g after 500 cycles	[137]
NiCo-LDH/MXene	Hydrothermal	LIB	1 M $LiPF_6$	1081 (at 100 mA/g)	100 cycles at 0.1 A/g with excellent capacity retention	[138]
Mg_2Al_1–CO_3-LDH	Hydrothermal synthesis	LIB	1 M $LiPF_6$	814	203.8 mAh/g after 300 cycles	[140]
NiCoOOH@CoLa-LDH (NC@CL)	Hydrothermal synthesis (2-step)	ZIB	1 M KOH + 5 mM ZnO	381.1	98% capacity retention after 2000 cycles	[141]

Table 1. *Cont.*

LDH Name	Synthesis Procedure	Battery Type	Electrolyte	Specific Capacity (mAh/g)	Stability	Ref.
NiCoMo-P150 LDH	Hydrothermal + Anion exchange + Temperature-differential phosphorus doping	CIB	0.5 M Bpy14Cl	363.4 (initial), 150.2 (stable)	800 cycles at 300 mA/g	[142]
NiAl-LDH@G	Atomic layer deposition + Hydrothermal	CIB	0.5 M Bpy14Cl	223.3	107 mAh/g after 500 cycles; 72 mAh/g after 120 days	[143]
$Mo_{0.3}NiCo_2$-Cl LDH	Hydrothermal synthesis + Ion exchange	CIB	0.5 M Bpy14Cl	352.5	159.7 mAh/g (after 300 cycles)	[144]
CoFe–Cl-LDH/CNT	Co-precipitation + Hydrothermal synthesis	CIB	1.0 M NaCl	190	125 mAh/g after 200 cycles	[145]
Ni-CAT/NiFe-LDH/CNFs	Hydrothermal synthesis + in situ MOF growth	ZAB	6 M KOH	...	66 h (liquid)/30 h (solid)	[146]
$ZnCo_2O_4$@NiMn-LDH (2:1)	Two-step hydrothermal	ZAB	6 M KOH	639	200 h cycling	[147]
NiCo-LDH/NCM@NF	Hydrothermal synthesis	ZAB	1 M KOH	685	Stable over 500 cycles	[148]
Pt-NiFe-LDH@Co_9S_8	Hydrothermal + Spontaneous-redox strategy	ZAB	6.0 M KOH/ 0.2 M $Zn(OAc)_2$	796.6	240 cycles (80 h, flex)	[149]
Co@NiFe-LDH	Two-step hydrothermal	ZAB	0.2 M $Zn(OAC)_2$ + 6 M KOH	652	90 h cycling with minor degradation	[150]
Ni_2Co-LDH/EG composites	Co-precipitation and Microwave irradiation	LIB	1 M $LiPF_6$	1880 (0.05 A/g); 919 (1 A/g)	~973 mAh/g after 100 cycles at 1 A/g	[152]
NiAl LDH	Hydrothermal	LIB	1 M $LiPF_6$	2586 (initial), 697 (400th cycle @ 0.5 A/g)	27.6% after 1400 cycles @ 1.0 A/g	[153]
L-CoNi-LDH/C	Laser-induced + Hydrothermal synthesis	LSB	1.0 M LiTFSI	1574 (initial), 1097 (100th)	503 mAh/g after 200 cycles @ 2 C	[154]
LDH@PSS (1:1.5)	Hydrothermal + Exfoliation + Self-assembly	LSB	1 M LiTFSI	1247.2 (initial), 1032.6 (after 200 cycles @ 0.2 C)	0.086% capacity decay per cycle over 200 cycles (0.2 C); 65.2% retention after 1000 cycles (1 C)	[155]
Vo-LDHs-MXenes	Self-growth aging	LSB	1.0 M LiTFSI	1549 (initial), 701 (after 300 cycles at 1 C)	300 cycles at 1.0 C, 0.084% decay/cycle	[156]
V_2O_5/Cys/FeNi-LDH	Hydrothermal + Reflux	LSB	1.0 M LiTFSI	1035.2 (initial), 920.1 (after 300 cycles)	88.9% retention after 300 cycles; 0.039% decay per cycle	[156]
NiFe-LDH@S	Hydrothermal	LSB	1 M LiTFSI	676 (initial), 386 (after 500 cycles at 2 C)	386 mAh/g after 500 cycles @2 C	[157]
NiAl-LDH@S	Hydrothermal	LSB	1 M LiTFSI	432 (initial), 238 (after 500 cycles at 2 C)	238 mAh/g after 500 cycles @2 C	[157]
ZnAl-LDH@S	Hydrothermal	LSB	1 M LiTFSI	357 (initial), 198 (after 500 cycles at 2 C)	198 mAh/g after 500 cycles @2 C	[157]

Table 1. *Cont.*

LDH Name	Synthesis Procedure	Battery Type	Electrolyte	Specific Capacity (mAh/g)	Stability	Ref.
NiCoAl-LDH	Hydrothermal + PECVD (for VG growth)	LSB	1.0 M LiTFSI with 1 wt.% $LiNO_3$	1182.4 (initial), 441.3 (after 750 cycles at 0.5 C)	0.0755% capacity decay per cycle (750 cycles at 0.5 C)	[158]
NiCr-Cl LDH/rGO	Hydrothermal + Ion Exchange	NIB	1 M $NaCF_3SO_3$	218 (200 cycles @ 100 mA/g)	34.5% capacity retention after 200 cycles	[159]
CoMoLDH@C-PDA-6	Hydrothermal + Polymerization + Carbonization	NIB	1 M $NaClO_4$	779.9 (initial); 310.9 (after 100 cycles)	70% retention after 100 cycles at 0.1 A/g	[163]
Co-Co LDH-derived $CoSe_2$@NHCNS@C	Hydrothermal synthesis + selenization + carbon coating	NIB	1 M $NaClO_4$	465.6 (initial), 373.8 (after 100 cycles)	373.8 mAh/g after 100 cycles; 325 mAh/g after 1000 cycles at 0.5 A/g	[164]
MCN-LDH@CP	Two-step hydrothermal	ZIB	3.0 M KOH	1.74 mAh/cm^2	97.8% @ 7000 cycles	[165]
ZnCo-LDH	Hydrothermal	Zn-ZnCo/Hybrid Zn	6 M KOH + 0.2 M $Zn(AC)_2$	2.5 mAh/cm^2	86% after 5000 cycles	[166]
NiCo LDH/MXene@NF	Electrodeposition	ZIB	1 M KOH	311.7	capacity retention of 88.6% after 10,000 cycles at 2 A/g	[203]
CoNi-LDH(v)	Electrochemical deposition + Electrochemical activation (ECA)	Aqueous and Solid-state ZIB	3 M $ZnSO_4$	225	53.9% after 900 cycles (aqueous); 47.22% after 1000 cycles (solid-state)	[204]
ZnAl-LDH@Zn	Electrochemical deposition	ZIB	3 M $ZnSO_4$	251.45	Stable after 1500 cycles	[172]
CoNi LDH(v)	Electrochemical deposition + CV activation	ZIB	1 M $ZnSO_4$	185	80% retention after 1000 cycles	[207]
NiCoS@NiCo-LDH	Hydrothermal + Electrodeposition	ZIB	6 M KOH	312	95.9% after 3000 cycles @ 20 mA/cm^2	[208]
Ni_3S_2/NiS@NiCo-LDH	Hydrothermal + Electrochemical deposition	ZIB	6 M KOH	434.5	116.7% after 5000 cycles	[209]
NiFe LDH on Cu NWs	Electrochemical deposition	ZAB	6 M KOH and 0.2 M zinc acetate	...	>250 h @ 10 mA/cm^2	[210]
NiCo-LDH	Sol–gel (Ni aerogel) + Solvothermal	LSB	1.0 M LITFSI	1238.4 (0.1 C); 805.8 (5.0 C)	647.1 mAh/g after 700 cycles at 5.0 C (0.018% decay/cycle)	[222]
NiCo–MOF/LDH	Solvothermal synthesis	LSB	1 M LITFSI	950 (after 200 cycles @ 1 C)	0.033% capacity decay per cycle; 633 mAh/g after 1000 cycles @ 1 C	[223]
NiCo-LDH	ZIF-67 template etching + melt-diffusion with sulfur	LSB	1.0 M LiTFSI	1540 @ 0.1 C, 485 @ 5 C	475 mAh/g after 500 cycles @ 1 C (78%)	[224]
$NiCo_2S_4$–NiS_2 NH@C	Ion exchange + in situ transformation	LSB	5 mM Li_2S_6	1207 at 0.2 C; 766 at 2 C	60.23% retention after 450 cycles at 1 C; 6.09 mAh/cm^2 at 5 mg/cm^2 sulfur loading	[225]
CuCo-LDH	Chemical etching (ZIF-67 template)	LSB	1 M LiTFSI	1262.8 (initial), 697.0 (after 500 cycles at 1 C)	500 cycles, 0.049% capacity decay per cycle	[226]
PPy@LDH-S	Ion exchange + solid-state melting + polymerization	LSB	1 M LiTFSI with 1% $LiNO_3$	907.2 (initial, 1 C); 633.4 after 500 cycles	0.06% capacity fading per cycle over 500 cycles	[227]

Table 1. *Cont.*

LDH Name	Synthesis Procedure	Battery Type	Electrolyte	Specific Capacity (mAh/g)	Stability	Ref.
$Ni_{0.8}Ca_{1.2}$Al-Cl LDH	Ion exchange	NIB	1.0 M $NaPF_6$	256.9 (200 cycles at 0.2 A/g)	102 mAh/g after 600 cycles at 2 A/g	[228]
$Co_{0.6}Ca_{1.4}$Al-Cl LDH	Ion exchange	NIB	1.0 M $NaPF_6$	292.8 (200 cycles at 0.2 A/g)	118.1 mAh/g after 600 cycles at 2 A/g	[228]
CoSn-LDH@MXene	Solvothermal + Ultrasonic	NIB	1 M $NaClO_4$	976.1 (0.1 A/g)	87.6% retention after 1000 cycles	[229]
Ni_3Ti-Cl LDH	Urea precipitation + anion exchange	CIB	0.5 M $PP_{14}Cl$	346.4	131.8 mAh/g over 200 cycles	[230]
CoMn-LDH/NPGA	Cross-linking gelation + Hydrothermal self-assembly + Freeze drying	ZAB	1 M KOH	…	72 h (~432 cycles) stable operation	[231]

5. LDHs in Supercapacitors

LDHs have attracted significant attention in electrochemical energy storage, particularly as electrode materials in supercapacitors, owing to their tunable chemical composition, two-dimensional layered structure, and abundant redox-active sites [18,182]. The distinctive crystal structure of LDHs features divalent and trivalent metal cations coordinated by hydroxyl groups with intercalated anions and water molecules. This architecture offers multiple pathways for charge storage through both surface redox reactions (pseudocapacitance) and ion intercalation mechanisms, as well as enabling the formation of the electric double layer (EDLC) at the electrode/electrolyte interface [262]. In particular, the exchangeable interlayer anions and the expansive interlayer spacing facilitate rapid diffusion of electrolyte ions, which is critical for achieving high-rate capabilities in supercapacitor applications. Structural modifications and composite strategies have been extensively exploited to enhance the electrochemical performance of LDH-based electrodes. Chemical modifications, such as doping with different metal ions (e.g., Ni, Co, Al, Mn) and anion intercalation, allow precise tuning of the redox properties and electrical conductivity [18,263]. For example, constructing NiCo-LDH microspheres derived from MOFs (Ni-MOFs) has significantly enhanced ion transport kinetics, reduced the charge transfer resistance, and increased the accessible electroactive surface area. Moreover, the integration of LDHs with conductive carbonaceous materials (e.g., graphene, CNTs) or conductive polymers has emerged as a promising strategy to overcome the intrinsic low electronic conductivity of LDHs, thereby improving the overall charge storage efficiency and cycling durability [264,265]. These composite strategies involve synergistic effects whereby the carbon matrix provides an effective conductive network while the LDH component contributes high pseudocapacitance from its Faradaic reactions [18,266].

Electrochemical performance metrics such as capacitance, cycling stability, and rate capability have been a primary focus in assessing the viability of LDH-based supercapacitors. High specific capacitances have been achieved by designing nanostructured LDH architectures that maximize ion diffusion and electron transfer; for example, ultrathin LDH nanosheets demonstrate significantly enhanced pseudocapacitive behavior due to the exposure to a high density of active sites [267]. The cycling stability of these materials is also improved by engineering composite structures that mitigate the typically observed volume changes and agglomeration during charge/discharge cycles [268,269]. Additionally, the rate capability, crucial for high-power applications, is enhanced when LDHs are integrated with conductive scaffolds, resulting in faster charge/transfer kinetics and high capacitance retention at increased current densities [263,266].

Hybrid LDH-based supercapacitors represent a sophisticated approach to delivering high energy and power density by synergistically combining materials that exhibit EDLC and pseudocapacitive behaviors. In hybrid devices, incorporating LDHs as pseudocapacitive components in conjunction with high-conductivity materials (such as MXenes, CNTs, or metal tellurides) has demonstrated promising improvements in energy and power performance [270,271]. For instance, the fabrication of hybrid architectures involving CuCo LDHs coated on conductive substrates, as well as the assembly of molecular-level heterostructures between titanium carbide MXene and Ni-Co-Al-LDH nanosheets, has led to devices capable of delivering rapid charge/discharge cycles while maintaining high specific capacitance and excellent cyclic stability [271,272]. These hybrid systems effectively leverage the ultrafast redox reactions of LDHs and the robust electrical conductivity of the secondary phases, making them attractive for applications in portable electronics and electric vehicles where both high energy and power demands are critical [270,271].

The application of LDHs in supercapacitors is underpinned by their unique layered structures that facilitate efficient charge storage via EDLC and pseudocapacitive mechanisms. The evolution of structural modifications and composite strategies, including the integration with conductive carbonaceous materials and the formation of hybrid heterostructures, has advanced the electrochemical performance of LDH-based devices in terms of high capacitance, improved cycling stability, and superior rate capability. These innovations pave the way for the next generation of high-energy and high-power supercapacitors. Shah et al. [89] reported that NiCoLDH-1@JAC-2 served as the positive electrode, while JAC-2 was employed as the negative electrode in an all-solid-state asymmetric HSC, using a PVA/KOH gel as the solid electrolyte. Figure 20a,b outlines the design and assembly of both symmetric and ASC devices based on JAC and NiCoLDH@JAC materials. The device demonstrated a specific capacitance of 750 F/g, with a specific capacity of 209 mAh/g at a current density of 0.5 A/g. Figure 20c–f demonstrates the electrochemical behavior of the NiCoLDH-1@JAC-2//JAC-2 HSC under different scan rates, potential windows, and current densities. It achieved a maximum energy density of 100 Wh/kg at a power density of 250 W/kg, significantly outperforming many conventional systems. The supercapacitor maintained excellent cyclic stability, with negligible performance degradation over repeated cycles. The synergistic interaction between the high surface area carbon scaffold and the electroactive LDH nanoflowers contributed to the remarkable electrochemical behavior.

The synthesized NiCo-LDH was employed as the positive electrode in an ASC with AC as the negative electrode, using PVA-KOH gel as the electrolyte [119]. Under optimal conditions (140 °C, 6 h, Ni/Co = 2:1), the device demonstrated a high specific capacitance of 400.2 C/g at 1 A/g. The ASC exhibited a maximum energy density of 51.59 $\mu Wh/cm^2$ at a power density of 1.125 mW/cm^2. GCD profiles were highly symmetric, and impedance analysis revealed low internal resistance (0.861–0.915 Ω), indicating efficient charge transport. Furthermore, the device maintained 70% of its initial capacitance after 10,000 cycles, highlighting excellent cycling stability and long-term operational reliability. The CoNi LDH materials were employed as positive electrodes in HSC devices, with AC as the negative electrode and 2 M KOH as the electrolyte [232]. The CoNi LDH nanoflowers (NFs) demonstrated superior performance, delivering a specific capacity of 768.3 C/g at 1 A/g and maintaining 643.1 C/g at 10 A/g, compared to 669.3 C/g for nanosheets (NSs). The HSC assembled with CoNi LDH NFs//AC exhibited a high energy density of 37.1 Wh/kg at a power density of 748.0 W/kg while retaining 21.6 Wh/kg even at 7.12 kW/kg. Both HSCs showed excellent long-term cyclic stability, maintaining over 93% capacity retention after 4000 cycles at 10 A/g, demonstrating their strong potential as advanced battery-type materials in energy storage devices.

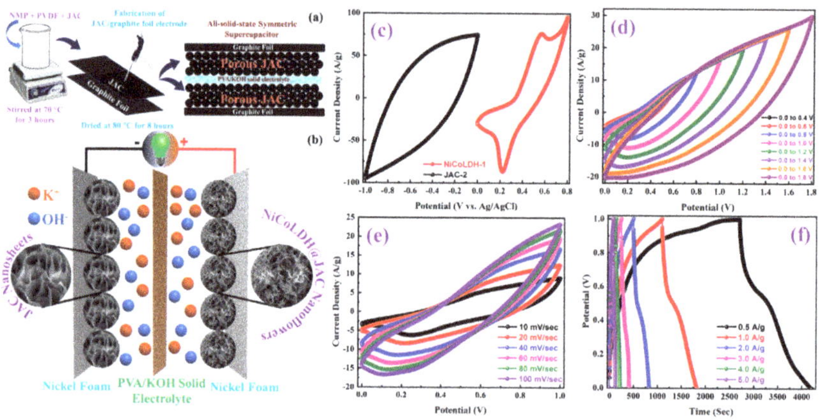

Figure 20. (a) Schematic overview of the fabrication process for JAC-based electrodes and assembly of the corresponding all-solid-state symmetric supercapacitor. (b) Diagram illustrating the configuration of the asymmetric HSC constructed using NiCoLDH@JAC and JAC electrodes. (c) CV curves of JAC-2 and NiCoLDH-1 electrodes measured in a three-electrode setup at a scan rate of 50 mV/s, (d) CV responses of the NiCoLDH-1@JAC-2//JAC-2 HSC at different operating potential windows (OPWs) at 50 mV/s, (e) CV curves of the same HSC recorded at various scan rates, and (f) GCD profiles of the HSC at different current densities. Reproduced with permission [89].

The electrochemical performance of the synthesized NiCo-LDH/NF (NCL3 for the optimized 6 h synthesis) was assessed in three-electrode and asymmetric two-electrode configurations [151]. In the three-electrode system with 1 M KOH, the NiCo-LDH served as the positive electrode, demonstrating a high specific capacitance of 1577 F/g at 1 mA/cm^2, with an energy density of 107.34 Wh/kg and a power density of 159.09 W/kg. Excellent cyclic stability was observed, with 96.17% retention after 5000 cycles. For the asymmetric device (NiCo-LDH/NF as positive and AC as negative electrode), the assembled AHSC device delivered an energy density of 38 Wh/kg and a power density of 1702 W/kg, retaining 76% capacitance after 2000 cycles, highlighting its practical viability for high-performance supercapacitor applications.

The synthesized NiCo-MOF@LDH-2 was employed as the positive electrode in an ASC device, with AC serving as the negative electrode and 2 M KOH as the electrolyte [233]. Electrochemical tests revealed excellent pseudocapacitive behavior with a high specific capacitance of 1873.9 F/g at 0.5 A/g and a maintained capacitance of 694.5 F/g at 20 A/g, indicating good rate capability. The ASC device (NiCo-MOF@LDH-2//AC) delivered an energy density of 49.8 Wh/kg at a power density of 422.4 W/kg and retained 21 Wh/kg even at 9318 W/kg, demonstrating superior energy/power performance. Furthermore, the device exhibited impressive cyclic stability with 83% capacitance retention after 10,000 cycles at 5 A/g, and coulombic efficiency of 99.3%, confirming its durability and practical viability. The synthesized NiCo-LDH/ACC composite was employed as a bifunctional electrode (cathode and anode) in a symmetric aqueous supercapacitor configuration with 1 M KOH as the electrolyte [234]. As a cathode, it achieved a high areal capacitance of 3770 mF/cm^2 at 1 mA/cm^2 and retained 79.5% of its capacity at 50 mA/cm^2 with 88% capacitance retention after 5000 cycles. As an anode, the electrode showed 1480 mF/cm^2 at 1 mA/cm^2, retaining 80% at 50 mA/cm^2 and demonstrating 100% stability after 10,000 cycles. The symmetric device delivered a high areal energy density of 0.352 mWh/cm^2 and a volumetric energy density of 5.61 mWh/cm^3, with a maximum power density of 559.5 mW cm^{-3}, maintaining performance over 15,000 cycles, indicating exceptional durability and fast ion/electron transport due to the amorphous, ultrathin architecture.

The synthesized NiCo-LDH-based heterostructure (S-NCCO) was employed as the positive electrode in an ASC, paired with AC as the negative electrode and PVA/KOH gel as the solid-state electrolyte [160]. The ASC device, configured as S-NCCO//AC, demonstrated exceptional performance, achieving a high specific capacitance of 3744.4 F/g at 1 A/g. It delivered an energy density of 92.3 Wh/kg and a power density of 750 W/kg, significantly outperforming comparable materials. Moreover, the device retained 87% of its capacitance after 5000 cycles at 10 A/g, indicating excellent long-term cycling stability and practical viability in energy storage applications. In supercapacitor applications, the CuCo LDH@Ni_3S_2 composite (specifically the CCNS-15 sample) was used as the positive electrode, while AC served as the negative electrode in an ASC configuration [211]. The device demonstrated exceptional areal capacitance (11.24 F/cm^2 at 5 mA/cm^2), an impressive energy density of 0.62 mWh/cm^2, and a power density of 8 mW/cm^2. Even after 6000 charge/discharge cycles, the ASC retained 72.2% of its initial capacitance, highlighting excellent cyclic stability. The superior electrochemical behavior was attributed to the composite's tremella-like structure and the synergistic interaction between CuCo LDH and Ni_3S_2, which enhanced charge transfer and ion diffusion pathways.

In the application section, the NFMM composite was used as a positive electrode in an ASC, paired with AC as the negative electrode and 6 M KOH as the electrolyte [161]. The device exhibited battery-type behavior with a remarkable specific capacitance of 2079.6 F/g at 1 A/g, and retained 85% of its capacitance after 5000 cycles. The assembled NFMM//AC ASC delivered a high energy density of 67.3 Wh/kg at a power density of 750.9 W/kg, and maintained 89% capacitance retention after 5000 cycles, showcasing excellent cycling stability. The synergistic interaction between NiFe-LDH, $MnCO_3$, and MXene enhanced conductivity, charge storage, and structural stability, affirming NFMM's suitability for high-performance supercapacitor applications. The $MnCo_2O_4$@FeCoNi-LDH/CC was used as the positive electrode and AC/CC as the negative electrode in an all-solid-state ASC device, with a PVA/KOH gel as the electrolyte [162]. The device exhibited exceptional electrochemical performance with a high specific capacitance of 145.2 F/g at 1 A/g, an energy density of 51.66 Wh/kg at a power density of 890.81 W/kg, and outstanding cyclic stability, retaining 90.3% capacitance after 10,000 charge/discharge cycles. Figure 21 demonstrates the electrochemical performance and practical application of the MnCo@FCN-LDH/CC//AC/CC ASC device, including stability, energy storage capability, and its ability to power an LED. These results highlight the material's potential in flexible and high-energy-density supercapacitor applications.

The vacancy-rich NiCo-LDH (designated as AL-LDH) was employed as the positive electrode in a magnesium-ion HSC (Mg-HSC), with VS_2 as the negative electrode and $MgSO_4$ as the electrolyte [212]. The AL-LDH electrode delivered a specific capacitance of 47.70 mAh/g ($\approx$238.5 F/g) at 1 A/g in a two-electrode setup. The device achieved a high energy density of 48.44 Wh/kg and a power density of 937.49 W/kg, demonstrating excellent rate capability by retaining 4 mAh/g at 20 A/g. After 500 cycles, the device retained ~42% of its capacity, indicating moderate stability. These results highlight the effectiveness of defect engineering (via H vacancy formation) in boosting charge storage and delivery performance in LDH-based supercapacitor systems. The electrochemical performance of NiAl-LDH was explored by constructing an ASC with NiAl-LDH@CC as the positive electrode and AC cloth as the negative electrode, using 1 M KOH as the electrolyte [214]. The ASC exhibited excellent pseudocapacitive behavior with a high specific capacitance of 310 mF/cm^2, corresponding to an energy density of 51.67 μWh/cm^2 and a power density of 913.84 μW/cm^2 at a current density of 0.2 mA/cm^2, maintaining a stable operating voltage of 1.6 V. The device demonstrated outstanding long-term durability with 90% capacitance retention over 10,000 cycles and an average coulombic efficiency of 98.9%, emphasizing the suitability of NiAl-LDH@CC for high-performance, flexible energy storage systems. NiAl LDH was employed as the positive electrode in an ASC configuration, paired with AC or graphene as the negative electrode [167]. The ASC devices demonstrated excellent performance, such as a specific capacitance of 214.4 F/g, energy density of up to 76.23 Wh/kg, and power density of 1127.03 W/kg, depending on the configuration. These devices maintained outstanding cycling stability, retaining 91.4% to 95.84% of their original capacitance after 5000–10,000 cycles at high current densities, showcasing the LDH's potential for long-term energy storage applications.

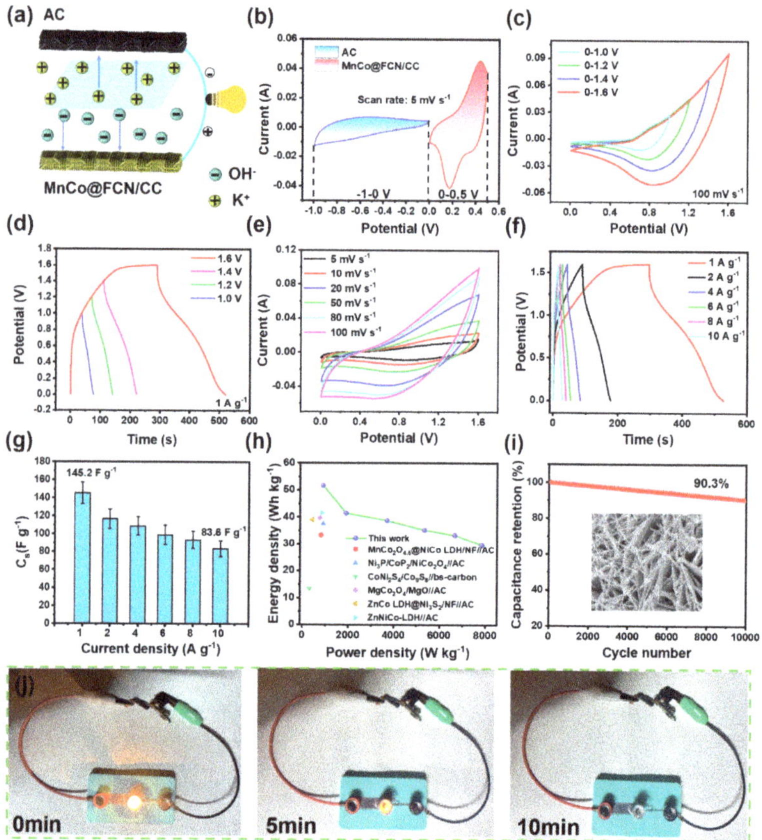

Figure 21. (**a**) Illustration of the MnCo@FCN-LDH/CC//AC/CC ASC setup. (**b**) CV comparison of AC and MnCo@FCN-LDH/CC electrodes at 5 mV/s. (**c**) CV curves of the ASC device at various voltages from 0–1.0 V to 0–1.6 V at 100 mV/s. (**d**) GCD curves at 1 A/g under the same voltage range. (**e**,**f**) CV and GCD performance of the device at 0–1.6 V. (**g**) Specific capacitance at different current densities. (**h**) Ragone plot comparing energy and power density with previous reports. (**i**) Cycling stability over 10,000 cycles at 1 A/g. (**j**) Digital photographs showing the device powering an LED for different durations. Reproduced with permission [162].

The optimized MP2 material, phosphorized MoB@NiCo-LDH with phosphorus vacancies, was employed as the positive electrode in an ASC device, with AC used as the negative electrode and 3 M KOH as the electrolyte [168]. The device exhibited outstanding electrochemical performance, including a high specific capacitance of 112.52 F/g at 1 A/g (two-electrode configuration), a maximum energy density of 39.91 Wh/kg at a power density of 948.25 W/kg, and an excellent power density of 7876.76 W/kg at 20.62 Wh/kg.

Furthermore, the device maintained 78.76% of its capacitance after 5000 cycles, demonstrating excellent cycling stability. These results highlight the synergistic effect of heterojunction formation and phosphorus vacancy engineering in enhancing charge transport kinetics and redox reactivity, thereby boosting the practical application potential of LDH-based electrodes in advanced energy storage systems. The NCS@CA heterojunction was employed as the positive electrode in a light-assisted ASC, with AC as the negative electrode and 3 M KOH as the electrolyte [236]. Under illumination, the device exhibited a remarkable specific capacitance of 112.2 F/g, a high energy density of 35.1 Wh/kg at a power density of 751.2 W/kg, and maintained 97.7% of its initial capacitance after 10,000 cycles, demonstrating excellent cyclic stability. Due to the semiconductor nature of NCS and LDH, the synergistic photoelectrochemical effect further enhanced charge separation and electrochemical kinetics, leading to substantial performance improvement over non-illuminated conditions.

In supercapacitor applications, the synthesized Te-NiMn LDH was employed as the positive electrode in a hybrid device, with AC as the negative electrode [169]. The resulting Te-NiMn LDH/C-MXene/NF//AC device achieved an impressive specific capacitance of 202.6 F/g at 2 A/g in a two-electrode configuration, along with a high energy density of 52.3 Wh/kg and a maximum power density of 6452 W/kg. Long-term cycling stability tests demonstrated excellent performance, with 77.3% capacitance retention after 10,000 cycles. The electrochemical performance of the Te-NiMn LDH/C-MXene/NF//AC device is illustrated in Figure 22. As shown in Figure 22a, the CV curves reveal that the Te-NiMn LDH/C-MXene/NF electrode exhibits a significantly higher current response than the AC electrode, indicating better electrochemical activity. Figure 22b presents CV measurements under different voltage ranges, showing that the area under the curve increases with higher potential windows, which reflects enhanced energy storage capacity. Figure 22c shows CV curves at various scan rates (5 to 50 mV/s) for a fixed voltage window of 1.5 V. The shapes of the curves remain consistent, suggesting good rate capability and reversibility. Figure 22d shows the GCD curves at different current densities, where the discharge time decreases with increasing current, typical of pseudocapacitive behavior. Long-term cycling performance is assessed in Figure 22e, with the device retaining 77.3% of its initial capacitance after 10,000 cycles at 5 A/g, indicating excellent stability. Finally, Figure 22f compares GCD curves before and after cycling. The slight change in shape further confirms the material's durability over extended use. These results highlight the synergistic enhancement due to CTAB-modified MXene conductivity and Te doping, which collectively facilitate improved redox kinetics, higher charge storage, and structural integrity over prolonged cycling.

The NC@CL nanosheet electrode was employed as the positive electrode in an ASC with AC as the negative electrode in a 2 M KOH electrolyte [141]. The NC@CL electrode delivered a high specific capacitance of 213 F/g at 0.2 A/g in the two-electrode device configuration, showcasing superior electrochemical performance due to its hierarchical nanostructure. The device achieved an energy density of 66.56 Wh/kg at a power density of 148.83 W/kg, maintaining 88.1% capacitance retention over 20,000 cycles, indicating excellent stability and reversibility. The enhanced performance is attributed to the synergistic interaction between NC and CL, large surface area, and efficient ion diffusion through the mesoporous framework.

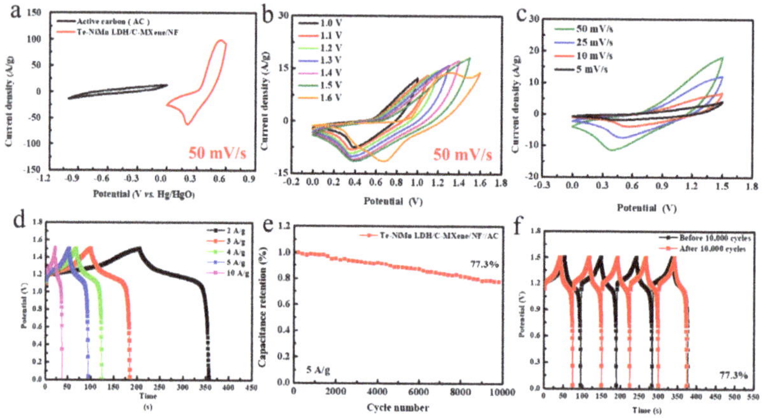

Figure 22. (**a**) CV comparison of Te-NiMn LDH/C-MXene/NF and AC electrodes at a scan rate of 50 mV/s. (**b**) CV profiles of Te-NiMn LDH/C-MXene/NF//AC device recorded under varying potential windows. (**c**) CV curves at different scan rates within a fixed voltage range of 1.5 V. (**d**) GCD curves at multiple current densities. (**e**) Cycling stability test showing capacitance retention after 10,000 cycles at 5 A/g. (**f**) GCD comparison before and after 10,000 charge/discharge cycles to evaluate long-term durability. Reproduced with permission [169].

The synthesized CF/CRBI-NiCo-LDH was employed as the positive electrode, while AC-coated carbon felt (CF/AC) was the negative electrode in an ASC [215]. The device utilized a cellulose separator and 6 M KOH aqueous electrolyte. In a two-electrode configuration, the supercapacitor achieved an impressive specific capacitance of 167.85 F/g at 0.5 A/g, with a high energy density of 57.2 Wh/kg and power density of 820 W/kg. Furthermore, the device retained 89% of its capacitance after 3000 cycles, highlighting excellent cycling stability and mechanical integrity. Figure 23a illustrates the coral reef-inspired structure of the CF/CRBI-NiCo-LDH electrode, promoting strong binding and fast redox transitions. CV curves (Figure 23b) confirm pseudocapacitive behavior with good rate capability, while GCD profiles (Figure 23c) show high coulombic efficiency across current densities. The specific capacitance trend (Figure 23d) indicates excellent rate performance, and the Ragone plot (Figure 23e) highlights the device's superior energy and power densities. The vertically grown nanoflake structure enhanced the charge storage through efficient ion transport and minimal resistance, confirmed by EIS and CV.

In the application as a supercapacitor, NiCo-LDH@TAC600-0 composite was used as the positive electrode, and tannin-derived AC (TAC600-2) served as the negative electrode in an ASC configuration with 6 M KOH as the electrolyte [237]. The device achieved a specific capacitance of 1250 F/g (three-electrode) and excellent performance in a two-electrode system, delivering a maximum energy density of 30.8 Wh/kg at a power density of 800 W/kg. Moreover, it demonstrated a stable cycle life with 72.5% capacitance retention after 5000 cycles at a high current density of 10 A/g. The sea urchin-like morphology contributed to a high rate of performance (82.8%) and facilitated fast ion/electron transport, validating its effectiveness in high-performance energy storage devices. In ASC configuration, CoNi LDH and its sulfurized derivatives were used as the positive electrode, while commercial AC was the negative electrode [235]. Figure 24a presents the synthesis and performance of the CoNiS-50-based electrode, highlighting its formation via sequential cation and anion exchange reactions to produce a flower-like CoNiS nanostructure. Among the synthesized materials, CoNiS-50 exhibited the most promising electrochemical performance, achieving a high specific capacitance of 150 F/g in the two-electrode configuration. The device delivered a maximum energy density of 37.8 Wh/kg at a power density of

750 W/kg. Furthermore, as shown in Figure 24a, the CoNiS-50-based ASC demonstrated excellent cycling stability, retaining 92.2% of its initial capacitance after 10,000 cycles at 10 A/g, attributed to the stabilized microstructure facilitated by residual $Mg(OH)_2$ and improved intrinsic conductivity from sulfur doping. Jia et al. [273] engineered NiCo-LDH/$NiCo_2S_4$ nanotube arrays modified with BPQDs, achieving an ultra-high specific capacitance of 2938.2 F/g, energy density of 133.7 Wh/kg at 800 W/kg, and 76.5% retention after 10,000 cycles. Figure 24b provides a visual overview of the energy storage mechanism. It outlines how charge is stored and transferred during electrochemical reactions, typically involving a combination of surface adsorption, redox reactions, and ion diffusion. This mechanism plays a key role in determining the overall performance of the electrode material, including its capacity, rate capability, and cycling stability. Figure 24c presents the design of the NiCo-LDH/$NiCo_2S_4$/BPQD//AC ASC. The electrochemical behavior of each electrode is shown in Figure 24d, where the CV curves confirm that the positive electrode has significantly higher redox activity than the AC negative electrode. Figure 24e displays CV curves at various voltage windows, indicating the device can operate stably up to 1.7 V. Figure 24f shows how the CV shape changes with increasing scan rates, suggesting good rate capability. The GCD curves in Figure 24g confirm that the device delivers longer discharge times at lower current densities. Finally, Figure 24h shows that the specific capacitance decreases as the current density increases, which is a typical trend in supercapacitor behavior.

Khalafallah et al. [213] developed Zn/CoLa LDH nanoarray-on-nanoarray electrodes using electrodeposition, achieving a specific capacitance of 269.4 F/g, an energy density of 59.9 Wh/kg at 800 W/kg, and a stability of 86.4% over 12,000 cycles in an ASC device. Liu et al. [162] synthesized $MnCo_2O_4$@FeCoNi-LDH on CC, achieving a high specific capacitance of 2235 F/g, an energy density of 51.66 Wh/kg at 890.81 W/kg, and 92.9% retention after 10,000 cycles in an ASC device. Luo et al. [274] prepared N-doped ZnNi-LDH via one-step hydrothermal synthesis, demonstrating a specific capacitance of 1265.7 C/g (~351.6 F/g), energy density of 64.6 Wh/kg at 850 W/kg, and excellent long-term cycle life when used with AC. Wei et al. [275] developed BPQD-embedded NiCoCu-LDH with a porous structure, reaching a specific capacitance of 1810.6 F/g, an outstanding energy density of 202.2 Wh/kg at 800 W/kg, and retained 81.4% capacity after 10,000 cycles. Moradi et al. [276] created a partially sulfurized CoMn-LDH/CoMn-S heterostructure, showing a capacity of 792.4 C/g (~220 F/g), energy density of 82.63 Wh/kg, power density up to 10,998 W/kg, and 94% stability over 6000 cycles. Wang et al. [277] introduced Ce into NiCo LDH@PBAs multi-core/shell structure, yielding a specific capacitance of 1847 F/g, energy density of 99.8 Wh/kg at 1283 W/kg, and 95.7% stability over 10,000 cycles.

Optimization strategies for LDH cathode materials in supercapacitors involve several synergistic approaches. Firstly, element doping, such as scandium doping, has promoted a superlong activation process and enhanced cycling durability while modifying the intrinsic electronic properties [278]. Secondly, tailoring the synthesis conditions to control the nanoscale morphology, including the design of hollow and quantum-layered structures, significantly improves ion diffusion and charge storage capability [119,243]. Modifying the intercalated anions and integrating LDHs with conductive substrates further optimizes electron transfer and overall capacitance performance [119,279]. These optimization methods, thus, address the challenges of moderate energy density and rate capability by enhancing both the redox activity and structural stability of LDH cathodes, ensuring improved electrochemical outcomes for supercapacitor applications [243,279].

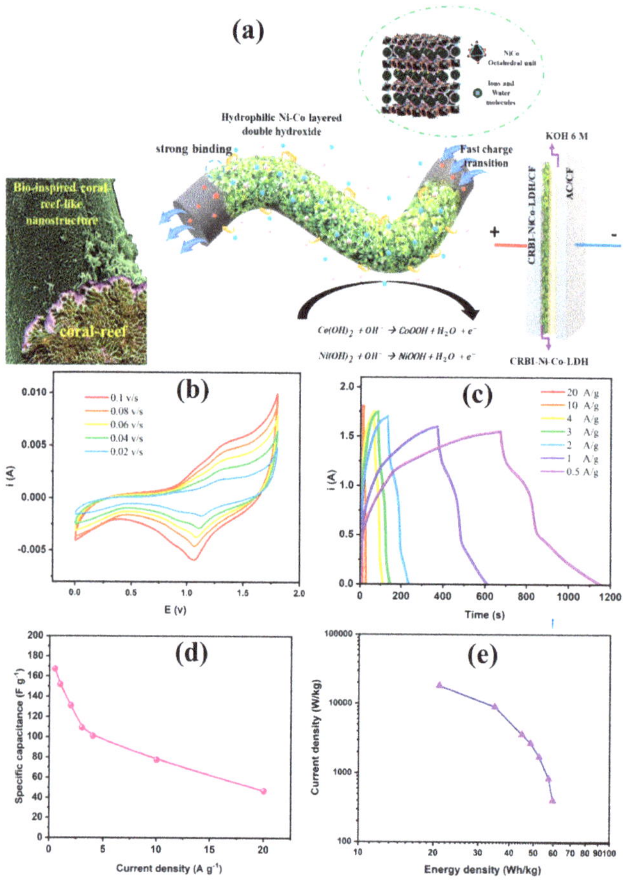

Figure 23. (**a**) Schematic illustration of the enhanced electrochemical charge storage mechanism in the electrosynthesized CF/CRBI-NiCo-LDH electrode, featuring a coral reef-inspired nanostructure that promotes strong binding, fast charge transfer, and redox reactions of Ni and Co hydroxides. (**b**) CV curves and (**c**) GCD profiles of the CF/CRBI-NiCo-LDH||CF/AC HSC at various scan rates and current densities. (**d**) Specific capacitance as a function of current density, and (**e**) Ragone plot comparing energy and power densities of the hybrid device. Reproduced with permission [215].

LDHs have emerged as highly effective electrode materials for supercapacitors, owing to their unique two-dimensional structure, tunable composition, and abundance of electroactive redox sites that support both pseudocapacitive and electric double-layer charge storage mechanisms. Their exceptional electrochemical behavior is attributed to the synergistic effects of compositional tailoring, such as metal-ion doping and anion intercalation, and advanced structural engineering, including nanosheet design, hierarchical porosity, and vacancy modulation. Integration with conductive matrices like carbon nanotubes, graphene, MXenes, and conductive polymers has been pivotal in mitigating LDHs' intrinsic low conductivity while enhancing rate capability, specific capacitance, and cycling durability. Particularly in asymmetric and hybrid supercapacitor configurations, LDHs paired with carbon-based or sulfide materials have demonstrated remarkable energy and power densities, extended cycle life, and mechanical robustness. Recent innovations such as heterostructures, phosphorization, photoelectrochemical enhancement, and defect engineering have pushed LDH-based devices toward practical viability in flexible, wearable, and high-performance energy storage systems. The continued development of LDH-based composites is central to advancing next-generation supercapacitors with superior energy density, fast charge/discharge kinetics, and long-term stability. A detailed summary of LDH-based materials applied in supercapacitor devices, highlighting their fabrication methods, device architecture, and electrochemical performance, is provided in Table 2.

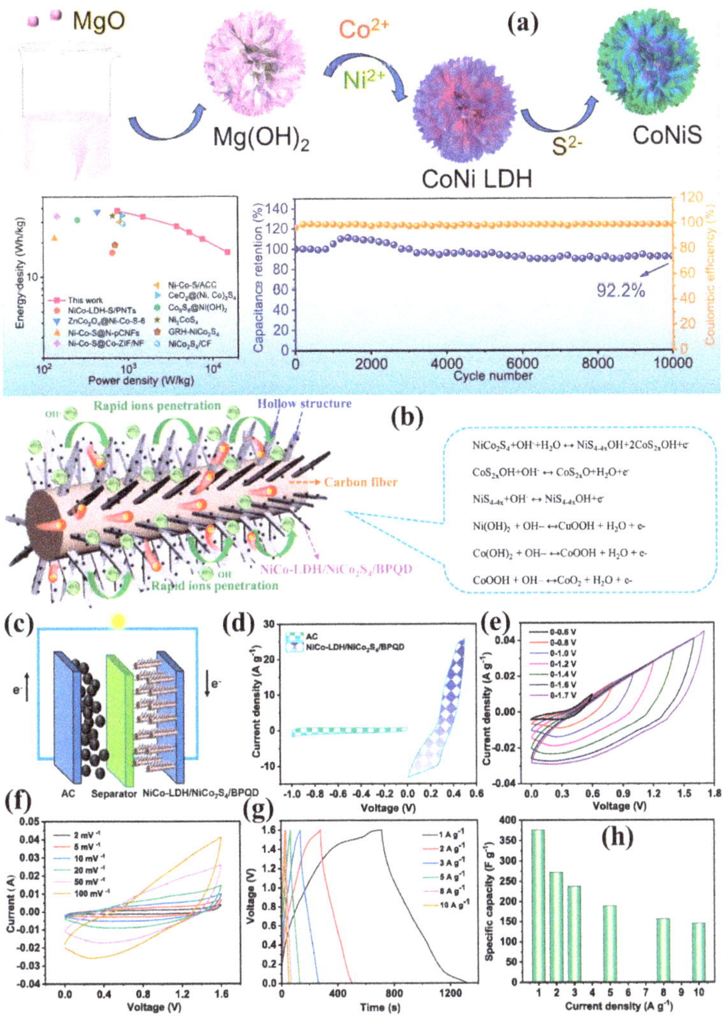

Figure 24. (**a**) An illustration shows the stepwise synthesis of CoNiS via combined cation and anion exchange, along with the energy and power density comparison of the CoNiS-50-based ASC and its cycling performance at 10 A/g. Reproduced with permission [235]. (**b**) Schematic representation illustrating the energy storage mechanism involved in the electrochemical process. (**c**) Schematic representation of the assembled NiCo-LDH/$NiCo_2S_4$/BPQD//AC ASC device. (**d**) CV curves of the NiCo-LDH/$NiCo_2S_4$/BPQD positive electrode and AC negative electrode measured at 2 mV/s. (**e**) CV profiles of the full device at varying voltage windows recorded at 100 mV/s. (**f**) CV curves at different scan rates ranging from 2 to 100 mV/s. (**g**) GCD curves at different current densities. (**h**) Specific capacitance values derived from the GCD results at various current densities. Reproduced with permission [273].

Table 2. Overview of LDH-based supercapacitor electrode materials with corresponding synthesis techniques, electrode configurations, and electrochemical performance metrics.

LDH Synthesis Procedure	Positive Electrode	Negative Electrode	Electrolyte	Specific Capacitance (F/g)	Energy Density (Wh/kg)	Power Density (W/kg)	Stability (%, No. of Cycles)	Ref.
Hydrothermal synthesis	NiCoLDH-1@JAC-2	JAC-2	PVA/KOH	750	100	250	95%, 10,000	[89]
Hydrothermal synthesis	NC@CL nanosheets	AC	1 M KOH	213	66.56	148.83	88%, 20,000	[141]
Hydrothermal synthesis	NiCo-LDH	AC	PVA/KOH	400.2 C/g	51.59 μWh/cm^2	1.125 mW/cm^2	70%, 10,000	[119]
Hydrothermal synthesis	Ni–Co LDH/NF (NCL3)	AC	1 M KOH	107	38	1702	76%, 2000	[151]
Hydrothermal + TA etching + sulfurization	S-NCCO (NiCo-LDH on S-ZIF-67)	AC	PVA/KOH	295	92.3	750	87%, 5000	[160]
Hydrothermal	NiFe-LDH/$MnCO_3$/MXene	AC	6 M KOH	215	67.3	750.9	89%, 5000	[161]
Hydrothermal synthesis	$MnCo_2O_4$@FeCoNi-LDH/CC	AC/CC	6 M KOH	145.2	51.66	890.81	90.3%, 10,000	[162]
Electrodeposition	NiAl LDH-rGO	Fe_2O_3-rGO	2 M KOH	214.4	76.23	800	95%, 5000	[167]
Hydrothermal synthesis + Phosphorization + $NaBH_4$ reduction	MP2 (MoB@NiCoP)	AC	3 M KOH	112.52	39.91	948.25	78.76%, 5000	[168]
Hydrothermal + Te soaking	Te-NiMn LDH/C-MXene/NF	AC	6 M KOH	202.6	52.3	6452	77.3%, 10,000	[169]
Two-step electrodeposition	CuCo LDH@Ni_3S_2	AC	3 M KOH	11.24 F/cm^2	0.62 mWh/cm^2	8 mW/cm^2	72.2%, 6000	[211]
Electrodeposition + Electrochemical Activation	Hv-rich NiCo-LDH	VS_2	1 M KOH	238.5 F/g	48.44	937.49	42%, 500	[212]
Electrodeposition of CoLa LDH nanoarrays followed by Zn^{2+} doping	Zn/CoLa LDH	AC	3 M KOH	269.4	59.9	800	86.4%, 12,000	[213]
Electrochemical deposition	NiAl-LDH@CC	CC	1 M KOH	310	51.67	913.84	90%, 10,000	[214]
Electrochemical deposition	CF/CRBI-NiCo-LDH	CF/AC	6 M KOH	167.85	57.2	820	89%, 3000	[215]
Solvothermal synthesis	CoNi LDH nanoflowers	AC	2 M KOH	768.3 C/g @ 1 A/g	37.1	748.0	93.7%, 4000	[232]
Solvothermal	NiCo-MOF@LDH-2 nanosheets	AC	2 M KOH	1873.9	49.8	422.4	83%, 10,000	[233]
Ion exchange	NiCo-LDH/ACC	NiCo-LDH/ACC	3 M KOH	876 mF/cm^2	0.352 mWh/cm^2	559.5 mW/cm^3	94.9%, 15,000	[234]
Cation/anion exchange	CoNiS-50	AC	6 M KOH	150	37.8	750	92.2%, 10,000	[235]
Two-step solvothermal	NCS@CoAl-LDH	AC	3 M KOH	112.2	35.1	751.2	97.7%, 10,000	[236]
Microwave hydrothermal synthesis	NiCo-LDH@TAC600-0	TAC600-2	6 M KOH	1250 (3-electrode)	30.8	800	72.5%, 5000	[237]

Table 2. *Cont.*

LDH Synthesis Procedure	Positive Electrode	Negative Electrode	Electrolyte	Specific Capacitance (F/g)	Energy Density (Wh/kg)	Power Density (W/kg)	Stability (%, No. of Cycles)	Ref.
Solvothermal synthesis followed by BPQD modification and hydrothermal sulfidation	NiCo-LDH/NiCo2S4/BPQD	AC	6 M KOH	376.1	133.7	800	76.5%, 10,000	[273]
Hydrothermal synthesis using 2-methylimidazole for N-doping	N-doped ZnNi-LDH	AC	3 M KOH	351.6	64.6	850	85.6%, 10,000	[274]
Precipitation and hydrothermal method with BPQD anchoring on CC	NiCoCu-LDH/BPQD	AC	6 M KOH	568.7	202.2	800	81.4%, 10,000	[275]
Electrochemical deposition followed by partial hydrothermal sulfidation	CoMn-LDH/CoMn-S	AC	3 M KOH	201	82.63	985	94%, 6000	[276]
Core/shell formation using Ce bridging between NiCo LDH and PBAs	NiCo-Ce@PBAs	AC	1 M KOH	222.3	99.8	1283	95.7%, 10,000	[277]

6. LDHs in Electrochemical Hydrogen Production

Hydrogen is a key clean energy carrier in the shift toward sustainable, low-carbon systems due to its high energy density, zero-emission profile, and broad applicability across sectors [280]. Efficient and affordable hydrogen production, particularly via water electrolysis, relies on effective catalysts for the HER and OER [281]. While noble metals like Pt, RuO_2, and IrO_2 offer excellent performance, their cost and scarcity limit widespread use [282]. LDHs have emerged as promising alternatives, thanks to their tunable brucite-like layered structure, high surface area, and rich active sites [283]. Their general formula, $[M^{2+}{}_{1-x}M^{3+}{}_{x}(OH)_2]^{x+}(A^{n-})_{x/n}\cdot yH_2O$, reflects their compositional flexibility, allowing precise control over layer charge, interlayer spacing, and anion content. LDHs also exhibit favorable redox behavior and surface characteristics, while modifications such as metal doping or hybridization with conductive materials can overcome their low intrinsic conductivity [284]. These properties make LDHs strong candidates for next-generation electrocatalysts in hydrogen production.

Pt-based catalysts are widely regarded as the benchmark for HER due to their exceptionally low overpotentials and high exchange current densities across a broad pH range. However, advanced LDH-based catalysts, particularly NiFe-LDH and CoFe-LDH, have emerged as competitive alternatives, especially under alkaline conditions. These LDHs can achieve overpotentials within 50–100 mV of Pt/C at a current density of 10 mA/cm^2, showcasing promising catalytic efficiency at a significantly reduced material cost [20,285,286]. In terms of stability, LDH catalysts exhibit excellent durability in alkaline electrolytes, maintaining over 90% of their initial performance after extended testing periods exceeding 50–100 h, whereas Pt-based catalysts, although stable in acidic environments, are prone to degradation in alkaline media. Furthermore, the use of earth-abundant and low-cost transition metals in LDHs provides a major economic advantage over expensive and scarce platinum. Nevertheless, LDHs still face limitations such as relatively lower electrical conductivity and reduced HER kinetics in acidic media, which currently limit their direct application in proton exchange membrane (PEM) electrolyzers [287]. Various strategies, including hybridization with conductive materials and defect engineering, are actively being explored to overcome these challenges [288].

6.1. Fundamentals of Electrochemical Hydrogen Production

Hydrogen production via water electrolysis involves passing an electric current through electrodes immersed in an electrolyte, triggering two key half-reactions: the HER at the cathode and the OER at the anode. Thermodynamically, water splitting requires a minimum of 1.23 V at standard conditions (25 °C, 1 atm), but in practice, a higher voltage (typically 1.8–2.0 V) is needed due to kinetic barriers and system resistances [289,290]. This additional energy input is known as the overpotential (η), caused by factors like charge transfer resistance, bubble formation, and ion diffusion limitations. Lowering η is critical for enhancing efficiency, often achieved by using advanced electrocatalysts that optimize interactions with H^+ or H_2O molecules. HER is a multi-step, pH-dependent process influenced by electrolyte type. In acidic media, H^+ ions are reduced via the Volmer (adsorption), Heyrovsky (electrochemical desorption), or Tafel (recombination) steps. In basic conditions, water molecules act as proton sources, undergoing similar steps but with a higher energy barrier due to stronger H–O bonds [291]. The general HER mechanism is described by Equations (1)–(6) and illustrated in Figure 25.

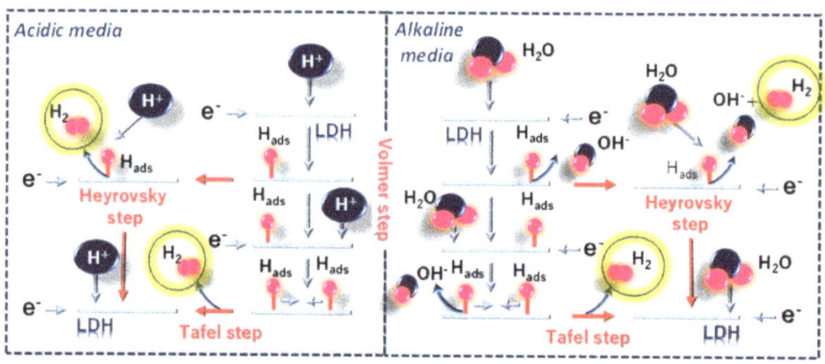

Figure 25. Schematic illustration of Volmer, Tafel, and Heyrovsky steps for HER in acidic and alkaline media.

Acidic HER:

$$H^+ + e^- \rightarrow H^* \text{ (Volmer)} \tag{1}$$

$$H^* + H^+ + e^- \rightarrow H_2 \text{ (Heyrovsky)} \tag{2}$$

$$H^* + H^* \rightarrow H_2 \text{ (Tafel)} \tag{3}$$

Alkaline HER:

$$H_2O + e^- \rightarrow H^* + OH^- \text{ (Volmer)} \tag{4}$$

$$H^* + H_2O + e^- \rightarrow H_2 + OH^- \text{ (Heyrovsky)} \tag{5}$$

$$H^* + H^* \rightarrow H_2 \text{ (Tafel)} \tag{6}$$

Due to slower kinetics in alkaline media, developing efficient electrocatalysts, such as composition/doping/interfacial engineering, etc., in LDHs is vital to lower overpotentials and enhance reaction rates, especially under non-acidic conditions [284,292]. Consequently, extensive research has turned toward low-cost, earth-abundant alternatives. Strategies such as carbon integration have improved conductivity and charge transfer, enhancing HER performance. Non-noble metal-based materials, including phosphides, sulfides, nitrides, borides, oxides, hydroxides, and especially LDHs, have gained prominence as bifunctional catalysts. Tailoring the composition, morphology, and electronic structure of LDHs through doping, defect engineering, and heterostructure construction has proven highly effective in enhancing their intrinsic activity and charge transfer efficiency for HER.

6.2. Mechanistic Role of LDHs Catalysts in Electrochemical HER

LDHs play a critical mechanistic role in HER due to their unique lamellar structure and tuneable composition. The brucite-like layers composed of divalent and trivalent metal cations create a high density of active sites and favorable charge environments that facilitate proton adsorption and electron transfer during HER. The metal composition significantly influences the catalytic behavior; for instance, incorporating transition metals like Ni, Co, or Fe enhances redox activity and electron conductivity, while synergistic interactions between metal centers modulate the hydrogen binding energy. Additionally, the interlayer anions and hydration layers contribute to ion transport and surface reactivity. HER performance of LDHs is strongly pH-dependent. In acidic media, LDHs often face stability challenges but can support HER through proton reduction if properly engineered. In neutral or alkaline environments, LDHs exhibit better structural durability, and the HER mechanism proceeds via water dissociation, making the Volmer step rate-limiting. Modifying the metal layers or coupling with conductive supports improves activity across all pH ranges. Among various electrocatalysts developed for water splitting, LDHs have gained considerable attention due to their tuneable composition, abundant active sites, and structural versatility [7]. While traditionally more active toward OER, several LDH systems have been tailored to exhibit excellent HER performance, especially in alkaline media [293,294]. Binary LDHs such as NiCo-LDH, NiMn-LDH, and CoMn-LDH have shown promising HER activity due to their redox flexibility and enhanced conductivity [295–297]. NiFe-LDH, though more renowned for OER, can be modified to support HER when integrated with conductive or catalytically active phases [298]. Ternary and doped LDHs, including NiCoFe-LDH, NiFeMn-LDH, and NiMo/NiFe-LDH, further enhance HER performance through synergistic electronic effects and optimized adsorption of intermediates [299–301]. Doping with elements like Cu, Mo, and W modulates the electronic structure, increases conductivity, and facilitates water dissociation [302–304]. Additionally, heterostructures such as NiFe-LDH@NiCoP, NiFe-LDH/NiS_2/VS_2, and LDH hybrids with MXene or graphene provide an ideal platform by coupling HER-active phases with conductive supports, promoting rapid charge transfer and improved stability [185,305,306]. These tailored LDH-based materials collectively demonstrate strong potential as noble metal-free HER electrocatalysts for efficient and scalable hydrogen production.

Another effective strategy involves combining LDHs with carbon-based materials such as graphene, CNTs, or reduced graphene oxide through the intercalation/exfoliation method. These LDH/carbon hybrids not only improve electrical conductivity but also provide a high surface area scaffold that prevents layer restacking and promotes better dispersion of active sites. The conductive carbon matrix facilitates rapid charge transport and enhances overall electrocatalytic kinetics. Exfoliation and intercalation of bulk NiFe LDHs into ultrathin nanosheets have proven to be powerful strategies for improving HER activity, primarily by enlarging surface area, enhancing charge transport, and exposing more active sites. Several advanced techniques, such as ultrasonic exfoliation, laser irradiation, and solid-phase assembly, have been employed to tailor their structure and electrochemical properties. For instance, Yao et al. fabricated a NiFe LDH nanosheet–defective graphene hybrid (NiFe LDH-NS@DG10) [307]. The synthesis of NiFe LDH@DG nanocomposite is successfully achieved through a straightforward electrostatic assembly of positively charged exfoliated NiFe LDH nanosheets with negatively charged defective graphene

(Figure 26a). Structural analyses confirm the formation of a well-dispersed hybrid architecture. XRD and AFM results validate the effective exfoliation into single-layer nanosheets (Figure 26b–e), while TEM and SAED demonstrate the preservation of crystallinity and uniform anchoring on the graphene support (Figure 26f,g). The EDS and HAADF mapping further confirm the homogeneous distribution of key elements, underscoring the intimate coupling between the LDH and DG components, as depicted in Figure 26h. This layered heterostructure achieved outstanding bifunctional performance with HER and OER overpotentials of 115 mV and 210 mV, respectively, at 10 mA/cm^2. This approach allows the coupling of diverse catalytic materials, creating junctions that foster synergistic interactions and facilitate more efficient charge transport. Such tailored interfaces can significantly lower the energy barriers for intermediate species and improve their adsorption behavior, thereby addressing the performance limitations typically observed in single-phase catalysts.

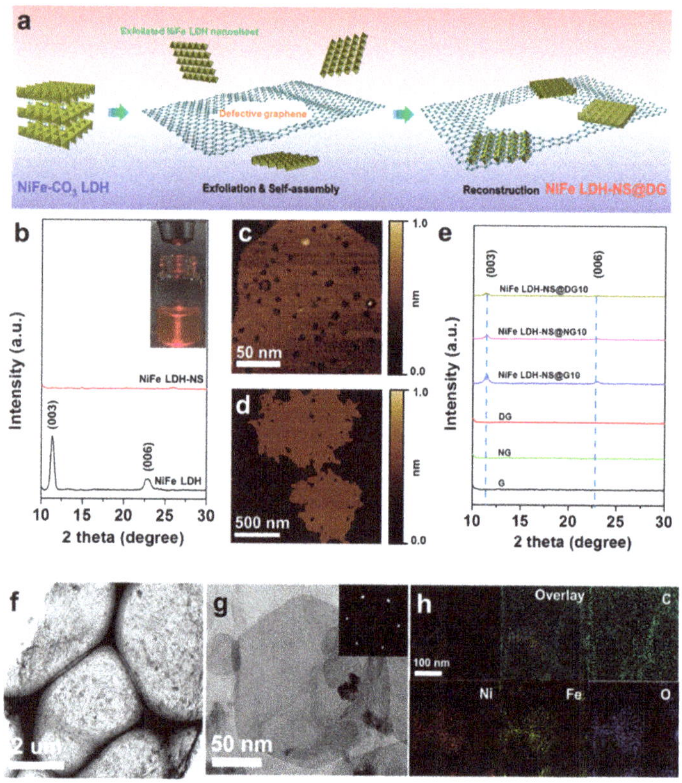

Figure 26. (**a**) Schematic depicting the synthesis pathway for the NiFe LDH-NS@DG hybrid, starting from bulk NiFe–CO_3 LDH, followed by exfoliation and integration with defective graphene, leading to the reconstructed nanocomposite. (**b**) XRD profiles comparing the bulk NiFe LDH and its exfoliated nanosheet form. (**c**,**d**) Atomic force microscopy (AFM) images showing the morphology of the exfoliated NiFe LDH nanosheets and defective graphene sheets, respectively. (**e**) XRD patterns of NiFe LDH nanosheets combined with various graphene types, including defective (DG), nitrogen-doped (NG), and pristine graphene (G). (**f**) Low-magnification TEM image displaying the overall morphology of the NiFe LDH-NS@DG10 composite. (**g**) High-resolution TEM highlighting the anchoring of NiFe LDH nanosheets on defective graphene, with the corresponding SAED pattern shown in the inset. (**h**) Elemental mapping via STEM-HAADF and EDS, confirming the uniform distribution of Ni, Fe, O, and C within the hybrid structure. Reproduced with permission [307].

LDH–metal composites, particularly those incorporating transition metals like Ni or Co, exhibit improved electron transfer and catalytic activity due to strong metal/metal interactions and enhanced redox behavior. These bimetallic or multimetallic systems often benefit from synergistic effects that optimize hydrogen adsorption and lower activation energy barriers. For example, Yao et al. developed a hierarchical heterostructured electrocatalyst, NiTe@CoFe LDH, by coupling conductive NiTe nanorods with amorphous CoFe-LDH via hydrothermal synthesis followed by electrodeposition [308]. This 3D architecture promotes rapid electron transfer, enhanced mass diffusion, and abundant active sites due to its large surface area and synergistic interfacial interactions. As a result, the catalyst exhibits excellent HER performance in alkaline media, requiring an overpotential of only 103 mV for HER at a current density of 10 mA/cm^2. The catalyst also demonstrates remarkable durability, maintaining stable performance for over 50 h. When used as both anode and cathode in a two-electrode configuration, the NiTe@CoFe LDH system achieves efficient overall water splitting at a low cell voltage of 1.56 V at 10 mA/cm^2, with admirable stability, indicating its strong potential for practical hydrogen production applications, as shown in Figure 27a–e. In addition, NiCo LDHs are promising electrocatalysts for hydrogen production due to their tunable redox properties, hydrophilicity, and structural flexibility. However, their HER performance is limited by poor conductivity and sluggish water dissociation kinetics. To address this, Xia et al. developed a hierarchical heterostructure comprising Cu-doped NiCo LDH grown on NiCo alloy (Cu-NiCo LDH/NiCo@CC) via a two-step electrodeposition process as shown in Figure 27f [285]. The NiCo alloy substrate enhanced conductivity, while Cu doping introduced oxygen vacancies and modulated the electronic structure. This optimized architecture delivered superior HER performance, achieving an overpotential of just 73 mV at 10 mA/cm^2 and a Tafel slope of 78.5 mV/dec in 1.0 M KOH. Electrochemical impedance spectroscopy showed a low charge transfer resistance of 1.48 Ω, and the catalyst also exhibited high surface capacitance (C_{dl} = 10.3 mF/cm^2) and excellent long-term stability. DFT calculations confirmed that Cu doping and interface effects enhanced water adsorption and charge transport, making this a promising strategy for efficient alkaline hydrogen evolution. Wang et al. developed a CoNiN@NiFe LDH hybrid electrocatalyst by electrodepositing NiFe-LDH onto NH_3-treated CoNiN grown on CC [309]. This structure combines the high conductivity of CoNiN with the catalytic activity of NiFe-LDH, enhancing charge transfer and active site exposure. In 1.0 M KOH, it delivered a low HER overpotential of 150 mV at 10 mA/cm^2, a Tafel slope of 169 mV/dec, and a charge/transfer resistance of 1.5 Ω, with excellent stability over 50 h. The synergy between components makes it a promising bifunctional catalyst for alkaline water splitting.

Furthermore, hybrid systems LDH with MOFs have shown great promise. MOFs offer a porous structure and high surface area that can be leveraged to host LDHs, creating hierarchical architectures with tailored pore channels and improved mass transport. For example, Huo et al. developed a novel strategy to fabricate monodispersed platinum (Pt) sites supported on NiFe-LDH via a targeted-anchoring and spontaneous-redox approach [286]. By engineering a MOF/LDH heterostructure, they introduced abundant defect sites that enabled the uniform dispersion and stabilization of Pt atoms. The synthesis of Pt@LDH-4h involves a stepwise strategy beginning with the formation of Fe-soc-MOF nanocubes, which serve as a structural scaffold, as shown in Figure 28a,b. Through urea-assisted hydrolysis in the presence of Ni^{2+} ions, a partial transformation occurs, generating a NiFe-LDH layer on the MOF surface and forming a MOF/LDH heterostructure rich in interfacial defects. These defects act as anchoring sites for Pt^{2+} ions introduced via K_2PtCl_4. A spontaneous redox reaction between Fe^{2+} in the MOF and Pt^{2+} leads to the in situ reduction of Pt to a lower valence state (~+0.96), resulting in the formation of monodispersed Pt sites, including both single atoms and nanoparticles (~4–5 nm). This process not only stabilizes the Pt species but also enhances electron transfer and active site exposure. The resulting Pt@LDH-4h catalyst demonstrated remarkable bifunctional activity for overall water splitting in alkaline media, achieving an impressively low overpotential of 58 mV for HER at 10 mA/cm^2. Among various LDH-based materials, NiAl-LDHs are attractive for their structural stability and tunable composition, and recent strategies have explored noble metal incorporation to enhance their catalytic performance. In this context, Kalusulingam et al. synthesized Pd-integrated NiAl-LDH nanosheets via a simple coprecipitation method, where Pd^{2+} was embedded into the brucite layer of NiAl-LDH to enhance electronic interactions and surface activity [310]. Pd-NiAl-LDH exhibited significantly improved HER performance in 1.0 M KOH, delivering an overpotential of 189 mV at 10 mA/cm^2 and a Tafel slope of 133.2 mV/dec, outperforming the pristine NiAl-LDH (258 mV, 137.4 mV/dec). The catalyst also showed a TOF of 40.37 s^{-1} at −0.4 V, indicating high intrinsic activity as depicted in Figure 29a–f. Moreover, the Pd-NiAl-LDH retained stability over 48 h of continuous HER operation, confirming its potential as a durable and efficient bifunctional catalyst for alkaline water electrolysis. The remarkable HER performance of Pd-NiAl-LDH in 1.0 M KOH is primarily attributed to its facilitation of the Volmer–Heyrovsky mechanism, enabling efficient hydrogen generation at reduced overpotentials, as shown in Figure 29g.

Transition metal compounds such as phosphides, oxides, and hydroxides are widely studied for water splitting, with phosphides like Ni_2P showing excellent HER activity due to their high conductivity and strong hydrogen affinity [311–313]. Coupling phosphides with LDHs enhances HER kinetics by improving water adsorption and proton transfer [185,314]. Ge et al. synthesized a Ni_2P/NiFe-LDH heterostructure featuring p–d orbital coupling P–Fe heterosites, which promoted O–H bond cleavage and improved charge transfer [315]. The catalyst achieved an HER overpotential of 230 mV at 100 mA/cm^2 and a Tafel slope of 47.43 mV/dec in 1.0 M KOH. DFT calculations revealed a significantly reduced Gibbs free energy (0.55 eV) for H_2O dissociation at the interface, confirming the synergistic effect of the heterostructure.

Zheng et al. developed a series of hybrid crystalline/amorphous electrocatalysts by decorating amorphous CoM LDHs (M = Mn, Fe, Ni, Cu, Zn) with crystalline CoMP nanostructures directly on nickel foam through partial dehydroxylation and phosphidation [316]. Among the variants, c-CoMnP/a-CoMn LDH/NF exhibited the best HER activity in 1.0 M KOH, requiring overpotentials of 170.3 mV and 254.9 mV to reach current densities of 100 and 500 mA/cm^2, respectively, with an onset potential of −0.134 V vs. RHE and a Tafel slope of 73.7 mV/dec. The enhanced performance was attributed to the synergistic effects between crystalline CoP domains and the amorphous LDH matrix, which increased active site density and improved charge transfer. The catalyst also demonstrated excellent stability over 60 h of continuous operation, positioning it as a promising candidate for high-rate alkaline hydrogen evolution.

Optimization of LDH cathode materials for hydrogen production focuses on tailoring composition, morphology, and surface properties to boost HER. Recent strategies include engineering 2D structures such as nanosheets or nanosheet arrays to increase the active surface area and thereby enhance charge transfer [317]. Additionally, compositional modifications, such as doping with transition metals or atomic platinum deposition, can tune the electronic structure, reduce reaction barriers, and promote efficient hydrogen adsorption [318]. Structural modulation via controlled annealing or anisotropic strain has also been shown to optimize phase evolution and lower water dissociation energy barriers, which further accelerates the HER kinetics [319,320]. Finally, surfactant-assisted exfoliation techniques help to overcome intrinsic conductivity limitations by exposing internal active sites and facilitating electron mobility [321]. These collective approaches contribute to the design of high-performance, stable, and cost-effective LDH cathodes for sustainable hydrogen production.

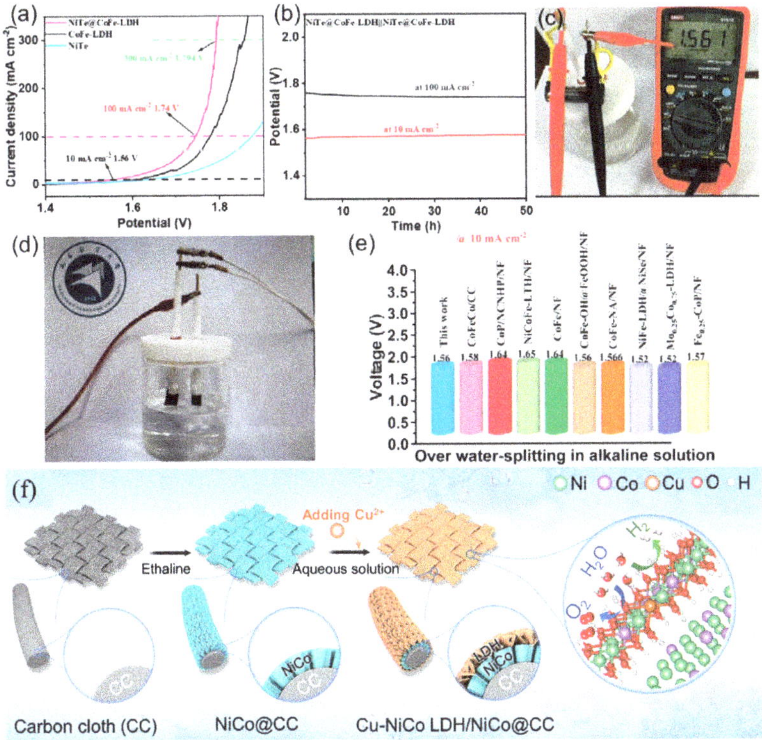

Figure 27. (**a**) Polarization curves of different electrocatalysts for overall water splitting recorded in 1.0 M KOH using a two-electrode setup at a scan rate of 1 mV/s. (**b**) Durability evaluation of NiTe@CoFe-LDH electrodes operated at current densities of 10 and 100 mA/cm^2 over a 50-h period. (**c**) Visual representation of the water-splitting device powered by a single battery, with a digital voltmeter showing the applied cell voltage. (**d**) Photograph of the assembled full-cell system under operation. (**e**) Comparative analysis of the cell voltages at 10 mA/cm^2 for NiTe@CoFe-LDH and other recently reported bifunctional catalysts in alkaline conditions. Reproduced with permission [308]. (**f**) Illustration of the synthetic strategy for constructing 3D hierarchical Cu-doped NiCo LDH nanosheets on NiCo alloy substrate and their deployment in overall water splitting applications. Reproduced with permission [285].

Integrating LDHs with MXenes significantly enhances catalytic performance by improving conductivity, preventing aggregation, and increasing exposure of active metal sites. This synergy accelerates gas evolution kinetics and boosts redox efficiency [306,322]. MXenes also offer corrosion resistance and mechanical reinforcement, improving overall catalyst durability. Though still emerging, LDH–MXene hybrids show great promise for water splitting applications. Xi et al. reported a CoFe-LDH-based catalyst synthesized via a partial in situ transformation of Co-MOF on a modified 3D MXene substrate (N-MXene), followed by the introduction of cationic vacancy defects to boost electrocatalytic activity [323]. The modified MXene provided a porous, conductive scaffold, while the vacancies tuned the electronic structure and enhanced active site exposure. The resulting hybrid, d-MLN (MOF-d-CoFe LDH@N-MXene), delivered excellent HER performance, requiring an overpotential of only 206 mV at 10 mA/cm^2 in 1 M KOH. It also showed a Tafel slope of 113 mV/dec, high electrochemical surface area (C_{dl} = 13.07 mF/cm^2), and charge transfer resistance as low as 1.72 Ω, confirming its fast kinetics and efficient electron transfer. Stability tests revealed minimal performance degradation over 60 h, highlighting its practical potential for sustainable hydrogen production. In one study, Shen et al. designed a 3D interwoven architecture combining $Ti_3C_2T_x$ MXene, reduced graphene oxide (RGO), and NiFe-LDH nanosheets via a hydrothermal co-assembly method. This structure, denoted as LDH (60%)/MX-RGO, featured a meso/macroporous network with a large BET surface area of 254.7 m^2/g, enhanced charge transfer, and strong electronic interactions among components. The optimized composite exhibited outstanding HER performance in 1 M KOH, with an overpotential of only 326 mV at 10 mA/cm^2 and a Tafel slope of 100 mV/dec. It also showed a high electrochemical double-layer capacitance (C_{dl} = 8.3 mF/cm^2) and a low charge/transfer resistance (2.7 Ω), indicating improved intrinsic activity and conductivity. The catalyst maintained excellent stability over 40 h of continuous operation and 2000 CV cycles, establishing its robustness and practical potential in alkaline HER systems. Table 3 presents a comparative overview of recent LDH-based HER electrocatalysts, highlighting the diversity in synthesis approaches and their corresponding electrocatalytic performances, including overpotential, stability, and Tafel slope.

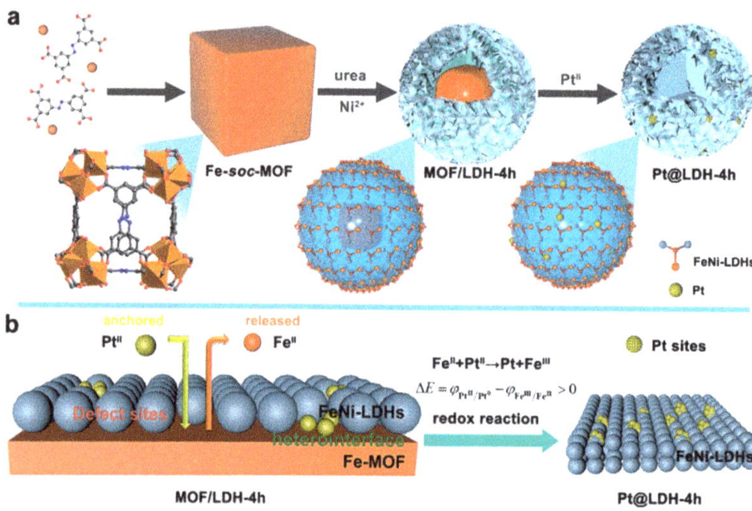

Figure 28. (**a**,**b**) Schematic representation of the strategic stepwise synthesis of Pt@LDH-4h catalysts tailored for efficient overall water splitting. Reproduced with permission [286].

Table 3. Summary of various LDH-based electrocatalysts for HER.

LDH Name	Synthesis Procedure	Electrolyte	HER Overpotential (η/mV, *j*/mA/cm^2)	Stability (Retention/%, Time/h)	Tafel Slope (mV/dec), FE (%)	Ref.
NiFe LDH-NS@DG10	Exfoliation method	1 M KOH	115, 10	~100%, 5.5 h	110, 97.3%	[307]
NiTe@CoFe-LDH	Hydrothermal and electrodeposition	1 M KOH	103, 10	~100%, 50 h	20, 100%	[308]
Cu-NiCo LDH/NiCo@CC	Electrodeposition process	1 M KOH	73, 10	. . ., 100 h	78.5, . . .	[285]
CoNiN@NiFe	Co-deposition	1 M KOH	150, 10	. . ., 60 h	167, 100%	[309]

Table 3. *Cont.*

LDH Name	Synthesis Procedure	Electrolyte	HER Overpotential (η/mV, j/mA/cm^2)	Stability (Retention/%, Time/h)	Tafel Slope (mV/dec), FE (%)	Ref.
Pt@LDH-4h	Hydrolysis reaction	1 M KOH	58, 10	~100%, 24 h	43.6, 95%	[286]
Pd-NiAl-LDH	Chemical precipitation	1 M KOH	189, 10	~100%, 48 h	133.2, . . .	[310]
Ni_2P/NiFe-LDH	Hydrothermal	1 M KOH	230, 100	~100%, 32 h	47.43, . . .	[315]
c-CoMnP/a-CoMn LDH/NF	Chemical and thermal route	1 M KOH	170, 100	99%, 60 h	73.7, . . .	[316]
MOF-d-CoFe LDH@N-MXene	Hydrothermal and electrophoretic deposition	1 M KOH	206, 10	. . ., 60 h	113, 93%	[323]
LDH(60%)/MX-RGO	Hydrothermal co-assembly	1 M KOH	326, 10	. . ., 40 h	100, . . .	[324]

LDH: Layered double hydroxide, HER: Hydrogen evolution reaction, FE: Faradaic efficiency

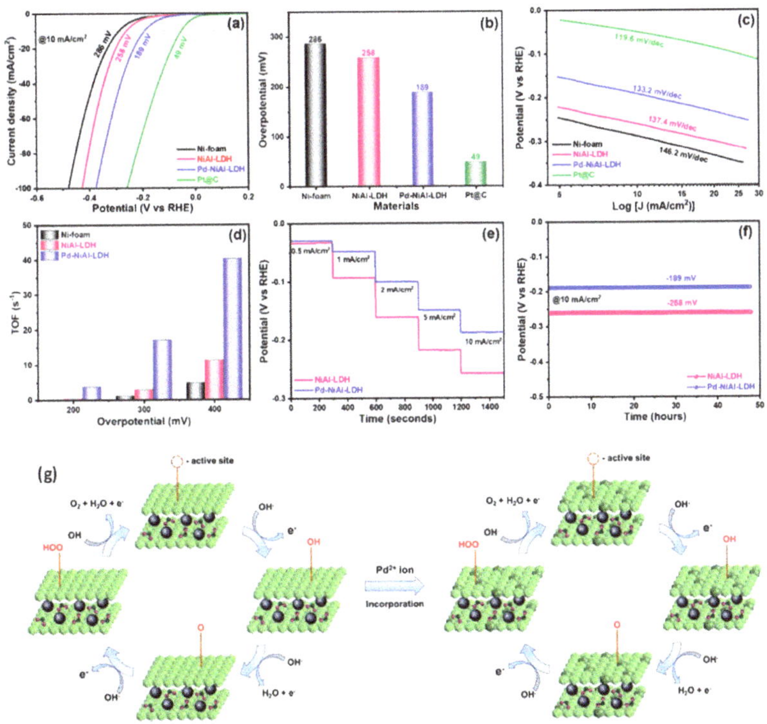

Figure 29. Evaluation of hydrogen evolution performance in 1.0 M KOH for NiAl-LDH, Pd-NiAl-LDH, Pt@C, and bare nickel foam: (**a**) Linear sweep voltammetry (LSV) profiles without iR compensation, (**b**) comparative HER overpotentials at 10 mA/cm^2, (**c**) Tafel plots indicating reaction kinetics, (**d**) TOF analysis, (**e**) stepwise chronopotentiometry tests across current densities from 0.5 to 10 mA/cm^2 with 300-s intervals for NiAl-LDH and Pd-NiAl-LDH, and (**f**) long-term stability test of Pd-NiAl-LDH at a constant current density of 10 mA/cm^2 over 48 h. (**g**) Illustration of the HER pathway involving two-electron transfer steps on NiAl-LDH and Pd-NiAl-LDH surfaces. Reproduced with permission [310].

6.3. Challenges in Catalytic Activity, Stability, and Practical Deployment

Despite their considerable promise, LDH-based electrocatalysts face several challenges that limit their widespread deployment in practical hydrogen production systems. A primary issue is their relatively modest intrinsic catalytic activity compared to noble metals, which stems from suboptimal hydrogen binding energies and sluggish electron transfer kinetics. While compositional tuning, doping, and heterostructure engineering have improved activity, achieving a balance between stability, conductivity, and catalytic efficiency remains complex. Durability under continuous operation is another major concern. Prolonged electrolysis can lead to structural disintegration, metal ion leaching, and the collapse of layered architectures, particularly under acidic or high-current conditions. These degradation mechanisms compromise long-term performance and hinder reliability. Scalability also presents significant hurdles. Many advanced LDH synthesis routes, such as hydrothermal growth, exfoliation, or electrodeposition, are challenging to scale due to batch limitations, material inconsistency, and the need for controlled environments. Moreover, integrating LDH materials into commercial electrolyzers and ensuring compatibility with system architectures and operational demands requires further development. To address these limitations, future efforts should focus on developing scalable and green synthesis methods that allow precise control over morphology and composition and enhancing interfacial engineering strategies, such as LDH–MXene, LDH–MOF, or LDH–graphene composites, to improve stability and charge transfer, in situ/operando characterization to uncover dynamic structural and electronic transformations under reaction conditions, and computational screening and machine learning approaches to accelerate the discovery of optimized compositions and architectures. Ultimately, advancing LDH-based electrocatalysts will require interdisciplinary strategies combining materials science, electrochemistry, and engineering design to bridge the gap from laboratory innovation to commercial application.

The utilization of LDHs in electrochemical hydrogen production presents several notable challenges related to their catalytic activity, stability, and practical deployment. This delves into these challenges, emphasizing the current understanding shaped by multiple academic explorations. One of the primary challenges in leveraging LDHs for electrochemical hydrogen production is their catalytic activity. Despite demonstrated potential, the efficiency of LDHs as electrocatalysts for HER remains limited when compared to noble metal counterparts. Transition metal-based LDHs exhibit favorable properties such as tunable composition and layered structures, yet often show inferior electrocatalytic efficiency. For instance, Wang et al. state that the catalytic performance is significantly influenced by the crystallographic phase and morphology of LDHs, where variations can lead to diverse electrochemical activities under identical conditions [314]. Luo et al. emphasize the necessity for adequately designed LDHs to provide more active sites while reducing reaction energy barriers [325]. Moreover, specific LDH compositions reveal limitations in their electrocatalytic performance; for instance, NiFe-based LDHs have been recognized as effective but still face challenges regarding the balance of active sites and reaction kinetics [326]. Optimizing the density of active sites while managing charge transfer efficiency is critical. Research by Liu et al. articulates that enhancing the uniform distribution of catalytic sites is vital for improving overall catalytic performance [327]. Ultimately, the ability of LDHs to function under diverse electrochemical conditions remains an area demanding further exploration and innovation.

In addition to catalytic activity, stability is a crucial concern regarding the performance of LDHs in practical applications. The intrinsic structure of LDHs is often susceptible to degradation under harsh electrochemical conditions, leading to a diminished operational lifetime [328]. For instance, during prolonged electrolysis, corrosion can occur, particularly in saline environments, resulting in the dissolution of components and loss of functionality. Research has demonstrated that specific compositions, such as $CoCO_3$/CoFe LDHs, show improved electrochemical stability, yet widespread stability challenges remain. Furthermore, Zhao highlights that ion leaching and structural collapse under continuous operational stress compromise the sustainable use of LDHs [329]. This instability limits their practical applications, especially in environments subjected to varying pH and current densities. Thus, integrating robust structural mechanisms capable of withstanding electrochemical stress while maintaining active sites is crucial for the successful deployment of LDHs in hydrogen evolution technologies.

In considering the practical deployment of LDHs in industry, several logistical and economic factors arise. The synthesis of LDHs must strike a balance between cost, scalability, and reproducibility. Current methods for producing LDHs, such as coprecipitation and hydrothermal techniques, may not be sufficiently scalable for industrial application, which is underscored by Chen et al., who point to the need for more efficient synthesis protocols that minimize costs and resource usage [330]. Additionally, the integration of LDHs into existing systems presents challenges regarding compatibility with associated materials, such as electrodes and supporting substrates [331]. Adjusting the physical and chemical properties of LDHs to facilitate optimal interaction with other components of electrochemical cells is critical [332]. The demand for multifunctional applications frames LDHs not merely as catalysts but as composite materials requiring structural and functional harmonization with other energy storage technologies. Moreover, environmental considerations, regulatory compliance, and societal acceptance play vital roles in deploying LDHs within commercial settings. As the push toward sustainable energy production intensifies, the environmental footprint of LDH processes must be carefully evaluated. Balancing lifecycle impacts while achieving functional deployment is an ongoing challenge faced by researchers and engineers [13].

Addressing the challenges of catalytic activity, stability, and deployment of LDHs in hydrogen production processes necessitates innovative strategies. Incorporating advanced synthesis methodologies can enhance the uniformity and availability of active sites within LDHs, as supported by various studies highlighting the effectiveness of nanosheet morphologies [314,333]. Additionally, engineering techniques such as doping and structural modification can enhance both the electrocatalytic performance and stability of LDHs [334]. Promising advancements include integrating LDHs with conductive substrates or hybrid materials that leverage the beneficial properties of both materials, thereby improving charge transfer and catalyst stability [89,335]. Achieving a harmonious integration of LDHs with materials such as MXenes or incorporating them into composite structures may offer pathways to overcome existing challenges [332]. Moreover, ongoing research focused on tailoring LDH compositions to optimize their electrochemical characteristics, stability, and economic viability is crucial. A deeper understanding of structure/performance relationships is fundamental to innovating LDH-based systems for hydrogen production applications. While LDHs present promising opportunities as electrocatalysts for hydrogen production, their utility is tempered by considerable challenges in catalytic activity, stability, and practical deployment. Addressing these concerns through innovative research and development is essential to realizing the potential of LDHs in contributing to sustainable hydrogen energy solutions.

7. Conclusions and Future Prospects

LDHs have emerged prominently as versatile and promising 2D materials for a wide array of energy applications, including batteries, supercapacitors, and hydrogen production. This review systematically outlines the foundational principles, synthesis methodologies, and application-specific advantages of LDHs, offering a comprehensive reference for researchers within this domain. In battery-related applications, LDHs demonstrate significant potential driven by their remarkable structural stability, chemically adjustable composition, and advantageous electrochemical properties. Specifically, in LIBs, LDHs have been shown to enhance electrode performance through stabilization of electrode/electrolyte interfaces, mitigation of volumetric expansion, and improvement in electrical conductivity. LSBs benefit substantially from the inherent capability of LDHs to adsorb polysulfides, effectively reducing the shuttle effect, thereby enhancing cycle stability and capacity retention. Similarly, NIBs leverage LDHs' expanded interlayer spacing, facilitating efficient diffusion of larger sodium ions, which directly addresses critical challenges in electrode kinetics and rate capability. Beyond conventional systems, LDHs hold significant promise in emerging battery technologies, including LIBs, CIBs, ZIBs, LSBs, and ZABs. They provide innovative solutions to longstanding challenges, such as corrosion resistance, dendritic growth prevention, and enhancement of sluggish ion diffusion processes.

In supercapacitor technologies, LDHs offer a compelling combination of high specific capacitance, excellent pseudocapacitive behavior, and robust electrochemical stability. Their layered structures provide abundant active sites and pathways conducive to rapid ion transport and electron mobility, positioning LDHs as exceptionally suitable electrode materials. Further optimization through doping and composite formation allows tailored improvements in energy density and power output. Nevertheless, considerable challenges remain, including their inherently low electrical conductivity and mechanical instability during extended cycling, necessitating innovative synthetic strategies or composite structures to overcome these limitations. Regarding catalytic hydrogen production, LDHs have exhibited substantial promise, especially in electrocatalytic water splitting processes. Their adjustable layer compositions and diverse elemental combinations enable precise tuning of catalytic properties, significantly boosting both HER and OER. Despite these advances, achieving industrial-level efficiency and durability poses substantial challenges, primarily due to insufficient intrinsic conductivity, limited specific surface areas of bulk forms, and operational durability under prolonged and harsh conditions.

Addressing these challenges requires strategic and innovative approaches. Future research should explicitly focus on precise control and reproducibility of LDH morphologies and electrochemical properties through meticulous optimization of synthesis parameters. Advanced synthesis techniques such as exfoliation into ultrathin nanosheets, integration with conductive substrates, or hybridization with complementary catalysts must be systematically developed and standardized to enhance material performance consistently. One of the most pressing challenges is the scalable and economically feasible production of LDHs. Current advanced synthesis methods, such as exfoliation, sol-gel processes, or hydrothermal treatments, are typically complex, expensive, and challenging to scale. Future research must prioritize the development of simplified, cost-effective, and scalable manufacturing techniques. Techniques such as continuous-flow reactors, microwave-assisted synthesis, and scalable hydrothermal reactors warrant intensive exploration due to their potential to deliver LDHs on a commercially viable scale. Integration into commercial energy systems demands attention to the practical aspects of LDHs, including long-term cycling stability in energy storage applications and durability under harsh catalytic conditions. Detailed investigations into degradation mechanisms, especially under realistic operational scenarios, will be crucial. For instance, comprehensive studies addressing the interfacial stability of LDHs within electrode structures under extensive cycling are necessary to design robust materials for commercial applications. Furthermore, understanding and optimizing complex interactions within composite structures incorporating LDHs is another critical area requiring in-depth study. This includes evaluating the interfacial chemistry between LDHs and conductive additives like carbon nanotubes, graphene, and conductive polymers.

Advanced analytical and characterization techniques, including in situ microscopy and spectroscopy, are needed to elucidate interfacial phenomena and interactions at the atomic scale, facilitating targeted optimization of composite performance.

To expedite and streamline the development of advanced LDH materials, computational modeling and simulation represent potent tools for predicting optimal material compositions and structures. Employing DFT, molecular dynamics, and machine learning algorithms can dramatically reduce experimental iterations, guide synthetic strategies, and facilitate the rapid design of optimized LDH-based materials. Exploration of novel and hybrid battery chemistries involving LDHs, such as hybrid supercapacitors and dual-ion batteries, could further expand their applications, potentially delivering high-performance and safer energy storage solutions. This requires systematic exploration and validation of LDHs within these novel battery architectures to assess performance, reliability, and safety comprehensively. In hydrogen production, future research directions must center on enhancing intrinsic catalytic activities, addressing conductivity and stability constraints. Hybrid electrocatalysts, combining LDHs with high-conductivity materials such as graphene, carbon nanotubes, or conductive polymers, present promising pathways for significant enhancements in electron transport and catalytic efficiency. Research must also focus on developing novel support structures and protective coatings to improve catalyst durability under industrially relevant operating conditions.

While LDHs have been extensively explored in conventional liquid electrolyte-based batteries, their application in solid-state batteries (SSBs) remains largely untapped, presenting a promising future direction. SSBs, offering enhanced safety, higher energy density, and broader operating temperature ranges, require materials with excellent ionic conductivity, chemical stability, and favorable interfacial properties. The unique layered structure of LDHs, combined with their tunable composition and ion exchange capability, suggests their potential use as solid electrolytes, interfacial layers, or electrode additives in SSBs. For instance, modifying LDHs to create highly conductive, defect-rich structures or composite materials could help facilitate ion transport across solid interfaces, a critical bottleneck in current SSB technologies. Moreover, their intrinsic ability to incorporate various metal cations could be exploited to engineer multifunctional interfaces that enhance both mechanical stability and electrochemical performance. Therefore, future research focused on tailoring the structure, conductivity, and interface compatibility of LDHs could open new pathways for their integration into solid-state battery systems, contributing significantly to the next generation of safe and high-performance energy storage devices.

Critically, researchers should adopt a pragmatic approach to LDH development, balancing theoretical performance enhancements against practical economic and industrial constraints. This includes comprehensive lifecycle assessments, economic viability studies, and scalability analyses to align research trajectories closely with industry requirements and commercial realities. LDHs exhibit immense potential in addressing pivotal challenges across energy storage and conversion technologies. Nevertheless, realizing their commercial and practical utility demands targeted research efforts addressing synthesis reproducibility, scalability, material stability, and interfacial interactions within composites. By leveraging advanced computational techniques, innovative synthesis methods, and rigorous practical assessments, LDHs can significantly impact future energy technologies. This review provides a robust foundation, clearly identifying critical areas requiring deeper exploration and establishing a structured path toward innovative, efficient, and sustainable LDH-based solutions.

Funding: This research received no external funding.

Acknowledgments: The authors acknowledge the support from the Graduate School of Energy Science, Kyoto University, Japan.

Conflicts of Interest: The authors declare no conflict of interest.

References

1. Chen, S.; Zhuo, Y.; Wang, X.; Li, S.; Lu, J.; Liu, D.; Pan, H.; Wang, Z. Advances of Layered Double Hydroxide Electrocatalysts for High-Current-Density Alkaline Water/Seawater Splitting. *Coord. Chem. Rev.* **2024**, *510*, 215832. [CrossRef]
2. Long, X.; Wang, Z.; Xiao, S.; An, Y.; Yang, S. Transition Metal Based Layered Double Hydroxides Tailored for Energy Conversion and Storage. *Mater. Today* **2016**, *19*, 213–226. [CrossRef]
3. Lin, Q.; Wang, L. Layered Double Hydroxides as Electrode Materials for Flexible Energy Storage Devices. *J. Semicond.* **2023**, *44*, 041601. [CrossRef]
4. Yu, W.; Deng, N.; Cheng, K.; Yan, J.; Cheng, B.; Kang, W. Advances in Preparation Methods and Mechanism Analysis of Layered Double Hydroxide for Lithium-Ion Batteries and Lithium-Sulfur Batteries. *J. Energy Chem.* **2021**, *58*, 472–499. [CrossRef]
5. Sahoo, D.P.; Das, K.K.; Mansingh, S.; Sultana, S.; Parida, K. Recent Progress in First Row Transition Metal Layered Double Hydroxide (LDH) Based Electrocatalysts towards Water Splitting: A Review with Insights on Synthesis. *Coord. Chem. Rev.* **2022**, *469*, 214666. [CrossRef]
6. Hu, X.; Zheng, W.; Wu, M.; He, Q.; Zhan, F.; Chen, L. Ternary Layered Double Hydroxide Cathode Materials for Electrochemical Energy Storage: A Review and Perspective. *Sustain. Energy Fuels* **2022**, *6*, 4551–4581. [CrossRef]
7. Zhou, D.; Li, P.; Lin, X.; McKinley, A.; Kuang, Y.; Liu, W.; Lin, W.F.; Sun, X.; Duan, X. Layered Double Hydroxide-Based Electrocatalysts for the Oxygen Evolution Reaction: Identification and Tailoring of Active Sites, and Superaerophobic Nanoarray Electrode Assembly. *Chem. Soc. Rev.* **2021**, *50*, 8790–8817. [CrossRef]
8. Jing, C.; Tao, S.; Fu, B.; Yao, L.; Ling, F.; Hu, X.; Zhang, Y. Layered Double Hydroxide-Based Nanomaterials for Supercapacitors and Batteries: Strategies and Mechanisms. *Prog. Mater. Sci.* **2024**, *150*, 101410. [CrossRef]
9. Zhang, D.; Guo, X.; Tong, X.; Chen, Y.; Duan, M.; Shi, J.; Jiang, C.; Hu, L.; Kong, Q.; Zhang, J. High-Performance Battery-Type Supercapacitor Based on Porous Biocarbon and Biocarbon Supported Ni–Co Layered Double Hydroxide. *J. Alloys Compd.* **2020**, *837*, 155529. [CrossRef]
10. Antil, B.; Olhan, S.; Vander Wal, R.L. Production of Graphitic Carbon from Renewable Lignocellulosic Biomass Source. *Minerals* **2025**, *15*, 262. [CrossRef]
11. Chaudhuri, H.; Yun, Y.S. A Critical Review on the Properties and Energy Storage Applications of Graphene Oxide/Layered Double Hydroxides and Graphene Oxide/MXenes. *J. Power Sources* **2023**, *564*, 232870. [CrossRef]
12. Tomboc, G.M.; Kim, J.; Wang, Y.; Son, Y.; Li, J.; Kim, J.Y.; Lee, K. Hybrid Layered Double Hydroxides as Multifunctional Nanomaterials for Overall Water Splitting and Supercapacitor Applications. *J. Mater. Chem. A Mater.* **2021**, *9*, 4528–4557. [CrossRef]
13. Fan, K.; Xu, P.; Li, Z.; Shao, M.; Duan, X. Layered Double Hydroxides: Next Promising Materials for Energy Storage and Conversion. *Next Mater.* **2023**, *1*, 100040. [CrossRef]
14. Sherryna, A.; Tahir, M. Recent Developments in Layered Double Hydroxide Structures with Their Role in Promoting Photocatalytic Hydrogen Production: A Comprehensive Review. *Int. J. Energy Res.* **2022**, *46*, 2093–2140. [CrossRef]
15. Mohanty, U.A.; Sahoo, D.P.; Paramanik, L.; Parida, K. A Critical Review on Layered Double Hydroxide (LDH)-Derived Functional Nanomaterials as Potential and Sustainable Photocatalysts. *Sustain. Energy Fuels* **2023**, *7*, 1145–1186. [CrossRef]
16. Zhao, Y.; Zhang, P.; Liang, J.; Xia, X.; Ren, L.; Song, L.; Liu, W.; Sun, X. Unlocking Layered Double Hydroxide as a High-Performance Cathode Material for Aqueous Zinc-Ion Batteries. *Adv. Mater.* **2022**, *34*, 2204320. [CrossRef] [PubMed]
17. Aziz, M.A.; Shah, S.S.; Mahnashi, Y.A.; Mahfoz, W.; Alzahrani, A.S.; Hakeem, A.S.; Shaikh, M.N. A High-Energy Asymmetric Supercapacitor Based on Tomato-Leaf-Derived Hierarchical Porous Activated Carbon and Electrochemically Deposited Polyaniline Electrodes for Battery-Free Heart-Pulse-Rate Monitoring. *Small* **2023**, *19*, 2300258. [CrossRef]
18. Jing, C.; Dong, B.; Zhang, Y. Chemical Modifications of Layered Double Hydroxides in the Supercapacitor. *Energy Environ. Mater.* **2020**, *3*, 346–379. [CrossRef]
19. Antil, B.; Elkasabi, Y.; Strahan, G.D.; Vander Wal, R.L. Development of Graphitic and Non-Graphitic Carbons Using Different Grade Biopitch Sources. *Carbon* **2025**, *232*, 119770. [CrossRef]
20. Yuan, Z.; Bak, S.M.; Li, P.; Jia, Y.; Zheng, L.; Zhou, Y.; Bai, L.; Hu, E.; Yang, X.Q.; Cai, Z.; et al. Activating Layered Double Hydroxide with Multivacancies by Memory Effect for Energy-Efficient Hydrogen Production at Neutral pH. *ACS Energy Lett.* **2019**, *4*, 1412–1418. [CrossRef]
21. Khorshidi, M.; Asadpour, S.; Sarmast, N.; Dinari, M. A Review of the Synthesis Methods, Properties, and Applications of Layered Double Hydroxides/Carbon Nanocomposites. *J. Mol. Liq.* **2022**, *348*, 118399. [CrossRef]

22. Riaz, S.; Rehman, A.; Akhter, Z.; Najam, T.; Hossain, I.; Karim, M.R.; Assiri, M.A.; Shah, S.S.A.; Nazir, M.A. Recent Advancement in Synthesis and Applications of Layered Double Hydroxides (LDHs) Composites. *Mater. Today Sustain.* **2024**, *27*, 100897. [CrossRef]
23. Jijoe, P.S.; Yashas, S.R.; Shivaraju, H.P. Fundamentals, Synthesis, Characterization and Environmental Applications of Layered Double Hydroxides: A Review. *Environ. Chem. Lett.* **2021**, *19*, 2643–2661. [CrossRef]
24. Ali Khan, A.; Tahir, M.; Khan, N. Recent Developments in Titanium Carbide (Ti_3C_2)-Based Layered Double Hydroxide (LDH) Nanocomposites for Energy Storage and Conversion Applications: A Minireview and Perspectives. *Energy Fuels* **2022**, *36*, 9821–9843. [CrossRef]
25. Parsapour, F.; Moradi, M.; Bahadoran, A. Metal-Organic Frameworks-Derived Layered Double Hydroxides: From Controllable Synthesis to Various Electrochemical Energy Storage/Conversion Applications. *Adv. Colloid. Interface Sci.* **2023**, *313*, 102865. [CrossRef] [PubMed]
26. Li, C.; Zhao, J.; Xie, L.; Wu, J.; Ren, Q.; Wang, Y.; Li, G. Surface-Adsorbed Carboxylate Ligands on Layered Double Hydroxides/Metal–Organic Frameworks Promote the Electrocatalytic Oxygen Evolution Reaction. *Angew. Chem.* **2021**, *133*, 18277–18285. [CrossRef]
27. Bukhtiyarova, M.V. A Review on Effect of Synthesis Conditions on the Formation of Layered Double Hydroxides. *J. Solid. State Chem.* **2019**, *269*, 494–506. [CrossRef]
28. Wang, Q.; Ohare, D. Recent Advances in the Synthesis and Application of Layered Double Hydroxide (LDH) Nanosheets. *Chem. Rev.* **2012**, *112*, 4124–4155. [CrossRef]
29. Li, F.; Zhang, L.; Evans, D.G.; Duan, X. Structure and Surface Chemistry of Manganese-Doped Copper-Based Mixed Metal Oxides Derived from Layered Double Hydroxides. *Colloids Surf. A Physicochem. Eng. Asp.* **2004**, *244*, 169–177. [CrossRef]
30. Xu, J.; Liu, X.; Zhou, Z.; Deng, L.; Liu, L.; Xu, M. Surface Defects Introduced by Metal Doping into Layered Double Hydroxide for CO_2 Photoreduction: The Effect of Metal Species in Light Absorption, Charge Transfer and CO_2 Reduction. *Chem. Eng. J.* **2022**, *442*, 136148. [CrossRef]
31. Forano, C. Environmental Remediation Involving Layered Double Hydroxides. In *Interface Science and Technology*; Elsevier: Amsterdam, The Netherlands, 2004; Volume 1, pp. 425–458.
32. Perera-Solis, D.D.; Pimlott, M.; Fidment, E.; Whiting, A.; Greenwell, H.C. Adding Value to Waste Minerals in a Circular Economy Framework: Ochre-Derived Layered Double Hydroxide Catalysts in Fatty Acid Ketonisation. *Minerals* **2019**, *9*, 681. [CrossRef]
33. Roy Chowdhury, P.; Medhi, H.; Bhattacharyya, K.G.; Mustansar Hussain, C. Layered Double Hydroxides Derived from Waste for Highly Efficient Electrocatalytic Water Splitting: Challenges and Implications towards Circular Economy Driven Green Energy. *Coord. Chem. Rev.* **2024**, *501*, 215547. [CrossRef]
34. Chowdhury, P.R.; Medhi, H.; Bhattacharyya, K.G.; Hussain, C.M. Recent Developments in Ni-Based Layered Double Hydroxides Extracted from Waste for Oxygen Evolution Reactions: A Review on the Pursuit for Sustainable Green Hydrogen. *Sustain. Energy Fuels* **2025**, *9*, 1447–1463. [CrossRef]
35. Jerome, M.P.; Alahmad, F.A.; Salem, M.T.; Tahir, M. Layered Double Hydroxide (LDH) Nanomaterials with Engineering Aspects for Photocatalytic CO_2 Conversion to Energy Efficient Fuels: Fundamentals, Recent Advances, and Challenges. *J. Environ. Chem. Eng.* **2022**, *10*, 108151. [CrossRef]
36. Singha Roy, A.; Kesavan Pillai, S.; Ray, S.S. Layered Double Hydroxides for Sustainable Agriculture and Environment: An Overview. *ACS Omega* **2022**, *7*, 20428–20440. [CrossRef]
37. Zubair, M.; Daud, M.; McKay, G.; Shehzad, F.; Al-Harthi, M.A. Recent Progress in Layered Double Hydroxides (LDH)-Containing Hybrids as Adsorbents for Water Remediation. *Appl. Clay Sci.* **2017**, *143*, 279–292. [CrossRef]
38. Li, L.; Soyhan, I.; Warszawik, E.; van Rijn, P. Layered Double Hydroxides: Recent Progress and Promising Perspectives Toward Biomedical Applications. *Adv. Sci.* **2024**, *11*, 2306035. [CrossRef]
39. Nemček, L.; Hagarová, I.; Matúš, P. Layered Double Hydroxides as Next-Generation Adsorbents for the Removal of Selenium from Water. *Appl. Sci.* **2024**, *14*, 8513. [CrossRef]
40. Altalhi, A.A.; Mohamed, E.A.; Negm, N.A. Recent Advances in Layered Double Hydroxide (LDH)-Based Materials: Fabrication, Modification Strategies, Characterization, Promising Environmental Catalytic Applications, and Prospective Aspects. *Energy Adv.* **2024**, *3*, 2136–2151. [CrossRef]
41. Li, C.; Jing, H.; Wu, Z.; Jiang, D. Layered Double Hydroxides for Photo(Electro)Catalytic Applications: A Mini Review. *Nanomaterials* **2022**, *12*, 3525. [CrossRef]
42. Gao, J.; Jin, B.; Shao, M. Layered Double Hydroxides Functionalization toward Rechargeable Batteries. *Particuology* **2024**, *91*, 138–154. [CrossRef]
43. Lu, L.; Zheng, Y.; Yang, R.; Kakimov, A.; Li, X. Recent Advances of Layered Double Hydroxides–Based Bifunctional Electrocatalysts for ORR and OER. *Mater. Today Chem.* **2021**, *21*, 100488. [CrossRef]
44. Gaur, A.; Sharma, J.; Lim, D.; Lee, H.I.; Han, H. Recent Advances in Electronic Structure Modifications of Layered Double Hydroxide (LDH) for the Water Splitting Application. *ChemCatChem* **2025**, *17*, e202401584. [CrossRef]

45. Lin, Z.; Li, X.; Zhang, H.; Xu, B.B.; Wasnik, P.; Li, H.; Singh, M.V.; Ma, Y.; Li, T.; Guo, Z. Research Progress of MXenes and Layered Double Hydroxides for Supercapacitors. *Inorg. Chem. Front.* **2023**, *10*, 4358–4392. [CrossRef]
46. Li, X.; Ren, J.; Sridhar, D.; Xu, B.B.; Algadi, H.; El-Bahy, Z.M.; Ma, Y.; Li, T.; Guo, Z. Progress of Layered Double Hydroxide-Based Materials for Supercapacitors. *Mater. Chem. Front.* **2023**, *7*, 1520–1561. [CrossRef]
47. Luo, F.; San, X.; Wang, Y.; Meng, D.; Tao, K. Layered Double Hydroxide-Based Electrode Materials Derived from Metal–Organic Frameworks: Synthesis and Applications in Supercapacitors. *Dalton Trans.* **2024**, *53*, 10403–10415. [CrossRef] [PubMed]
48. Jafari, M.; Ganjali, F.; Eivazzadeh-Keihan, R.; Maleki, A.; Geranmayeh, S. Recent Advances in Applications of Graphene-Layered Double Hydroxide Nanocomposites in Supercapacitors and Batteries. *FlatChem* **2024**, *45*, 100658. [CrossRef]
49. Aman, M.; Konduparty, P.; Sharma, S.; Srivastava, R.P.; Bhattacharyya, S.; Sharma, V.; Balani, K.; Jha, S.K.; Omar, S. Layered Double Hydroxide Based Composites for Energy Storage Applications: Insights into Supercapacitors and Batteries. *J. Energy Storage* **2025**, *116*, 116093. [CrossRef]
50. Fan, G.; Li, F.; Evans, D.G.; Duan, X. Catalytic Applications of Layered Double Hydroxides: Recent Advances and Perspectives. *Chem. Soc. Rev.* **2014**, *43*, 7040–7066. [CrossRef]
51. Cardinale, A.M.; Carbone, C.; Consani, S.; Fortunato, M.; Parodi, N. Layered Double Hydroxides for Remediation of Industrial Wastewater from a Galvanic Plant. *Crystals* **2020**, *10*, 443. [CrossRef]
52. Tsuneishi, T.; Sakamoto, H.; Hayashi, K.; Kawamura, G.; Muto, H.; Matsuda, A. Preparation of Hydroxide Ion Conductive KOH–Layered Double Hydroxide Electrolytes for an All-Solid-State Iron–Air Secondary Battery. *J. Asian Ceram. Soc.* **2014**, *2*, 165–168. [CrossRef]
53. Leung, D.W.J.; Laney, K.R.; Kenyon, P.; Rees, N.H.; Buffet, J.-C.; Chen, C.; O'Hare, D. Optimising the Acid–Base Ratio of Mg–Al Layered Double Oxides to Enhance CO_2 Capture Performance: The Critical Role of Calcination Conditions. *Dalton Trans.* **2024**, *53*, 6200–6206. [CrossRef] [PubMed]
54. Xu, S.-M.; Zhu, Y.-Q.; Chen, Z.-R.; Yang, J.-R.; Chen, X.; Yan, H. Theoretical Prediction of the Carrier Mobilities for $M^{II}{}_2M^{III}$-Cl-Layered Double Hydroxides in Three-Dimensional Directions. *J. Mater. Chem. C Mater.* **2022**, *10*, 9573–9585. [CrossRef]
55. Xu, J.; Deng, H.; Song, J.; Zhao, J.; Zhang, L.; Hou, W. Synthesis of Hierarchical Flower-like Mg_2Al-Cl Layered Double Hydroxide in a Surfactant-Free Reverse Microemulsion. *J. Colloid. Interface Sci.* **2017**, *505*, 816–823. [CrossRef] [PubMed]
56. Memon, J.; Sun, J.; Meng, D.; Ouyang, W.; Memon, M.A.; Huang, Y.; Yan, S.; Geng, J. Synthesis of Graphene/Ni–Al Layered Double Hydroxide Nanowires and Their Application as an Electrode Material for Supercapacitors. *J. Mater. Chem. A Mater.* **2014**, *2*, 5060–5067. [CrossRef]
57. Palapa, N.R.; Taher, T.; Mohadi, R.; Lesbani, A. Removal of Anionic Direct Dye Using Zn/Al, Zn/Fe and Zn/Cr Layered Double Hydroxides Toward Interlayer Distance. *Sci. Technol. Indones.* **2019**, *4*, 70–76. [CrossRef]
58. Ibrahimova, K.A.; Azizov, A.A.; Balayeva, O.O.; Alosmanov, R.M.; Abbasov, M.H. The sonochemical synthesis of PVA/Mg-Al-OH layered double hydroxide nanocomposite film. *Azerbaijan Chem. J.* **2023**, *1*, 141–148. [CrossRef]
59. Harada, K.; Nguyen, T.K.N.; Matsui, Y.; Fujii, K.; Grasset, F.; Ohashi, N.; Matsuda, M.; Uchikoshi, T. Observation of Stacking Faults and Photoluminescence of Laurate Ion Intercalated Zn/Al Layered Double Hydroxide. *Mater. Lett.* **2018**, *213*, 323–325. [CrossRef]
60. Kahl, M.; Golden, T.D. Corrosion Resistance of Electrochemically Synthesized Modified Zaccagnaite LDH-Type Films on Steel Substrates. *Materials* **2021**, *14*, 7389. [CrossRef]
61. Sun, X.; Neuperger, E.; Dey, S.K. Insights into the Synthesis of Layered Double Hydroxide (LDH) Nanoparticles: Part 1. Optimization and Controlled Synthesis of Chloride-Intercalated LDH. *J. Colloid. Interface Sci.* **2015**, *459*, 264–272. [CrossRef]
62. Geng, D.G. Removal of Nitrophenol from Water by Dodecyl Sulfate Modified Layered Double Hydroxides. *Asian J. Chem.* **2014**, *26*, 447–450. [CrossRef]
63. Mohadi, R.; Amri, A.; Badaruddin, M.; Ahmad, N.; Lesbani, A. Adsorption of Phenol from Aqueous Solution Using Zn/Al Layered Double Hydroxides-Cellulose Composite. *Sci. Technol. Indones.* **2023**, *8*, 123–128. [CrossRef]
64. Smalenskaite, A.; Sen, S.; Salak, A.N.; Ferreira, M.G.S.; Skaudzius, R.; Katelnikovas, A.; Kareiva, A. Sol-Gel Synthesis and Characterization of Non-Substituted and Europium- Substituted Layered Double Hydroxides $Mg_3/Al_{1-x}Eu_x$. *Curr. Inorg. Chem.* **2017**, *6*, 149–154. [CrossRef]
65. George, G.; Saravanakumar, M.P. Synthesising Methods of Layered Double Hydroxides and Its Use in the Fabrication of Dye Sensitised Solar Cell (DSSC): A Short Review. *IOP Conf. Ser. Mater. Sci. Eng.* **2017**, *263*, 032020. [CrossRef]
66. Yang, B.; Yang, Z.; Wang, R.; Feng, Z. Silver Nanoparticle Deposited Layered Double Hydroxide Nanosheets as a Novel and High-Performing Anode Material for Enhanced Ni–Zn Secondary Batteries. *J. Mater. Chem. A* **2014**, *2*, 785–791. [CrossRef]
67. Gong, M.; Li, Y.; Wang, H.; Liang, Y.; Wu, J.Z.; Zhou, J.; Wang, J.; Regier, T.; Wei, F.; Dai, H. An Advanced Ni–Fe Layered Double Hydroxide Electrocatalyst for Water Oxidation. *J. Am. Chem. Soc.* **2013**, *135*, 8452–8455. [CrossRef]
68. Shao, M.; Ning, F.; Zhao, J.; Wei, M.; Evans, D.G.; Duan, X. Hierarchical Layered Double Hydroxide Microspheres with Largely Enhanced Performance for Ethanol Electrooxidation. *Adv. Funct. Mater.* **2013**, *23*, 3513–3518. [CrossRef]

69. Jia, Z.; Wang, Y.; Qi, T. Hierarchical Ni–Fe Layered Double Hydroxide/MnO_2 Sphere Architecture as an Efficient Noble Metal-Free Electrocatalyst for Ethanol Electro-Oxidation in Alkaline Solution. *RSC Adv.* **2015**, *5*, 83314–83319. [CrossRef]
70. Gao, X.; Lei, L.; Chen, L.; Wang, Y.; He, L.; Lian, Y. Synthesis and Controlled Release of Vitamin C Intercalated Zn/Al Layered Double Hydroxide. *Asian J. Chem.* **2014**, *26*, 3471–3476. [CrossRef]
71. Iqbal, M.A.; Sun, L.; Fedel, M. Synthesis of Novel Cone-Shaped CaAl-LDH Directly on Aluminum Alloy by a Facile Urea Hydrolysis Method. *SN Appl. Sci.* **2019**, *1*, 1415. [CrossRef]
72. Liu, X.L. Chiral Recognition of Lysine by Cyclodextrin Intercalated Layered Double Hydroxides. *Asian J. Chem.* **2013**, *25*, 10277–10282. [CrossRef]
73. ABD MALEK, A.H.; YASIN, Y. Use of Layered Double Hydroxides to Remove Sunset Yellow FCF Dye from Aqueous Solution. *Chem. Sci. Trans.* **2012**, *1*, 194–200. [CrossRef]
74. Hanifa, Y.; Rahayu Palapa, N. Mg/Al Double Layer Hydroxides: Intercalation with $H_3[\alpha\text{-}PW_{12}O_{40}]\bullet nH_2O$. *Sci. Technol. Indones.* **2016**, *1*, 16–19. [CrossRef]
75. Rasouli, N. Application of a Novel, Efficient and Recyclable Photo Redox Catalyst (Zn–Al Layered Double Hydroxide/Eosin) for the Synthesis of Substituted Pyridine Derivatives under Visible Light Irradiation. *Appl. Organomet. Chem.* **2018**, *32*, e4585. [CrossRef]
76. Imron, M.; Said, M. Adsorption of Procion Red Using Layer Double Hydroxide Mg/Al. *Sci. Technol. Indones.* **2017**, *2*, 64–67. [CrossRef]
77. Mishra, G.; Dash, B.; Pandey, S. Layered Double Hydroxides: A Brief Review from Fundamentals to Application as Evolving Biomaterials. *Appl. Clay Sci.* **2018**, *153*, 172–186. [CrossRef]
78. Shin, J.; Kim, K.; Hong, J. Zn-Al Layered Double Hydroxide Thin Film Fabricated by the Sputtering Method and Aqueous Solution Treatment. *Coatings* **2020**, *10*, 669. [CrossRef]
79. Mattera, M.; Sorrenti, A.; De Gregorio Perpiñá, L.; Oestreicher, V.; Sevim, S.; Arteaga, O.; Chen, X.; Pané, S.; Abellán, G.; Puigmartí-Luis, J. "On-The-Fly" Synthesis of Self-Supported LDH Hollow Structures Through Controlled Microfluidic Reaction-Diffusion Conditions. *Small* **2024**, *20*, 2307621. [CrossRef]
80. Wang, C.; Xu, J.; Zhou, Z. A Mini-Review on CO_2 Photoreduction by MgAl-LDH Based Materials. *Energies* **2022**, *15*, 8117. [CrossRef]
81. Wang, L.; Yuan, Z.; Zhang, Y.; Guo, W.; Sun, X.; Duan, X. Sandwich Layered Double Hydroxides with Graphene Oxide for Enhanced Water Desalination. *Sci. China Mater.* **2022**, *65*, 803–810. [CrossRef]
82. Jaramillo-Hernández, C.; Oestreicher, V.; Mizrahi, M.; Abellán, G. Upscaling the Urea Method Synthesis of CoAl Layered Double Hydroxides. *Beilstein J. Nanotechnol.* **2023**, *14*, 927–938. [CrossRef] [PubMed]
83. Rosaiah, P.; Vadivel, S.; Prakash, N.G.; Dhananjaya, M.; Al-Asbahi, B.A.; Roy, S.; Chalapathi, U.; Park, S.-H. Fabrication of Porous Ni-Co LDH Nanocomposites as Efficient Electrodes for Supercapacitors. *Int. J. Energy Res.* **2023**, *2023*, 5793868. [CrossRef]
84. Cao, Y.; Zheng, D.; Li, X.; Lin, J.; Wang, C.; Dong, S.; Lin, C. Enhanced Corrosion Resistance of Superhydrophobic Layered Double Hydroxide Films with Long-Term Stability on Al Substrate. *ACS Appl. Mater. Interfaces* **2018**, *10*, 15150–15162. [CrossRef]
85. Kim, A.; Varga, I.; Adhikari, A.; Patel, R. Recent Advances in Layered Double Hydroxide-Based Electrochemical and Optical Sensors. *Nanomaterials* **2021**, *11*, 2809. [CrossRef] [PubMed]
86. Li, X.; Chen, F.; Zhao, B.; Zhang, S.; Zheng, X.; Wang, Y.; Jin, X.; Dai, C.; Wang, J.; Xie, J.; et al. Ultrafast Synthesis of Metal-Layered Hydroxides in a Dozen Seconds for High-Performance Aqueous Zn (Micro-) Battery. *Nanomicro Lett.* **2023**, *15*, 32. [CrossRef] [PubMed]
87. Xu, X.; Wang, J.; Zhou, A.; Dong, S.; Shi, K.; Li, B.; Han, J.; O'Hare, D. High-Efficiency CO_2 Separation Using Hybrid LDH-Polymer Membranes. *Nat. Commun.* **2021**, *12*, 3069. [CrossRef]
88. Belskaya, O.B.; Terekhova, E.N.; Gorbunova, O.V.; Muromtsev, I.V.; Trenikhin, M.V.; Salanov, A.N.; Likholobov, V.A. Synthesis of CuAl-LDHs by Co-Precipitation and Mechanochemical Methods and Selective Hydrogenation Catalysts Based on Them. *Inorganics* **2023**, *11*, 247. [CrossRef]
89. Shah, S.S.; Aziz, M.A.; Ali, M.; Hakeem, A.S.; Yamani, Z.H. Advanced High-Energy All-Solid-State Hybrid Supercapacitor with Nickel-Cobalt-Layered Double Hydroxide Nanoflowers Supported on Jute Stick-Derived Activated Carbon Nanosheets. *Small* **2024**, *20*, 2306665. [CrossRef]
90. Wu, F.; Liang, J.; Peng, Z.; Liu, B. Electrochemical Deposition and Characterization of Zn-Al Layered Double Hydroxides (LDHs) Films on Magnesium Alloy. *Appl. Surf. Sci.* **2014**, *313*, 834–840. [CrossRef]
91. Johnston, A.L.; Lester, E.; Williams, O.; Gomes, R.L. Understanding Layered Double Hydroxide Properties as Sorbent Materials for Removing Organic Pollutants from Environmental Waters. *J. Environ. Chem. Eng.* **2021**, *9*, 105197. [CrossRef]
92. Ibrahimova, K.A. The synthesis methods and applications of layered double hydroxides—A brief review. *NNC RK Bull.* **2022**, *4*, 16–29. [CrossRef]
93. Palapa, N.R.; Saria, Y.; Taher, T.; Mohadi, R.; Lesbani, A. Synthesis and Characterization of Zn/Al, Zn/Fe, and Zn/Cr Layered Double Hydroxides: Effect of M^{3+} Ions Toward Layer Formation. *Sci. Technol. Indones.* **2019**, *4*, 36–39. [CrossRef]

94. Huang, X.; Yang, X.; Li, G.; Ezeh, C.I.; Sun, C.; Snape, C. Hybrid Two-Step Preparation of Nanosized MgAl Layered Double Hydroxides for CO_2 Adsorption. In *Sorption in 2020s*; IntechOpen: Rijeka, Croatia, 2020.
95. G.Fouad, H.; Sayed Amin, A.; S.Ahmed, I.; A.Ali, A. Zinc-Aluminium Layered Double Hydroxides: Fabrication, Study and Adsorption Application for Removal Organic Dye from Aqueous Media. *Benha J. Appl. Sci.* **2022**, *7*, 53–61. [CrossRef]
96. Smalenskaite, A.; Kaba, M.M.; Grigoraviciute-Puroniene, I.; Mikoliunaite, L.; Zarkov, A.; Ramanauskas, R.; Morkan, I.A.; Kareiva, A. Sol–Gel Synthesis and Characterization of Coatings of Mg-Al Layered Double Hydroxides. *Materials* **2019**, *12*, 3738. [CrossRef]
97. Xu, Y.; Liu, T.; Li, Y.; Liu, Y.; Ge, F. Nanostructure Design and Catalytic Performance of Mo/ZnAl-LDH in Cationic Orchid X-BL Removal. *Materials* **2018**, *11*, 2390. [CrossRef]
98. Wijitwongwan, R.P.; Saothayanun, T.K.; Ogawa, M. Synthesis of NiFe Layered Double Hydroxides with Varied Layer Charge Densities: The Templating Effect of Dioctyl Sulfosuccinate. *Dalton Trans.* **2023**, *52*, 4692–4699. [CrossRef]
99. Zhang, J.-J.; Li, M.-Y.; Li, X.; Bao, W.-W.; Jin, C.-Q.; Feng, X.-H.; Liu, G.; Yang, C.-M.; Zhang, N.-N. Chromium-Modified Ultrathin CoFe LDH as High-Efficiency Electrode for Hydrogen Evolution Reaction. *Nanomaterials* **2022**, *12*, 1227. [CrossRef] [PubMed]
100. Cao, Y.; Zheng, D.; Luo, J.; Zhang, F.; Dong, S.; Pan, J.; Lin, C. Insight into the Fabrication of ZnAl Layered Double Hydroxides Intercalated with Organic Anions and Their Corrosion Protection of Steel Reinforced Concrete. *J. Electrochem. Soc.* **2019**, *166*, C617–C623. [CrossRef]
101. Fang, S.; Chen, K.; Yao, H.; Cao, Y.; Guo, S.; Wang, L.; Wang, Y.; Yu, S.; Wang, N. Preparation of Gallic Acid Intercalated Layered Double Hydroxide for Enhanced Corrosion Protection of Epoxy Coatings. *Coatings* **2023**, *13*, 128. [CrossRef]
102. Bani Hashemi, A.; Kasiri, G.; Glenneberg, J.; Langer, F.; Kun, R.; La Mantia, F. Electrochemical and Morphological Characterization of Zn−Al−Cu Layered Double Hydroxides as a Negative Electrode in Aqueous Zinc-Ion Batteries. *ChemElectroChem* **2018**, *5*, 2073–2079. [CrossRef]
103. Neethu, P.P.; Venkatachalam, G.; Venkatesha, N.J.; Joseph, D.; Sakthivel, A. Cobalt-Based Hydrotalcite: A Potential Non-Noble Metal-Based Heterogeneous Catalyst for Selective Hydrogenation of Aromatic Aldehydes. *Ind. Eng. Chem. Res.* **2023**, *62*, 4976–4986. [CrossRef]
104. Kitano, S.; Sato, Y.; Tagusari, R.; Zhu, R.; Kowalski, D.; Aoki, Y.; Habazaki, H. Facile Synthesis Approach of Bifunctional Co–Ni–Fe Oxyhydroxide and Spinel Oxide Composite Electrocatalysts from Hydroxide and Layered Double Hydroxide Composite Precursors. *RSC Adv.* **2023**, *13*, 10681–10692. [CrossRef]
105. Huang, L.; Megías-Sayago, C.; Bingre, R.; Zheng, Q.; Wang, Q.; Louis, B. Catalytic Performance of Layered Double Hydroxides (LDHs) Derived Materials in Gas-Solid and Liquid-Solid Phase Reactions. *ChemCatChem* **2019**, *11*, 3279–3286. [CrossRef]
106. Kovalenko, V.; Borysenko, A.; Kotok, V.; Nafeev, R.; Verbitskiy, V.; Melnyk, O. Determination of the Dependence of the Structure of Zn-Al Layered Double Hydroxides, as a Matrix for Functional Anions Intercalation, on Synthesis Conditions. *East.-Eur. J. Enterp. Technol.* **2022**, *1*, 12–20. [CrossRef]
107. Nyaba, L.; Munonde, T.S.; Mpupa, A.; Nomngongo, P.N. Magnetic Fe_3O_4@Mg/Al-Layered Double Hydroxide Adsorbent for Preconcentration of Trace Metals in Water Matrices. *Sci. Rep.* **2021**, *11*, 2302. [CrossRef] [PubMed]
108. Liu, J.; Yue, X.; Lu, X.; Guo, Y. Uptake Fluoride from Water by Starch Stabilized Layered Double Hydroxides. *Water* **2018**, *10*, 745. [CrossRef]
109. Yang, K.; Huang, Y.; Wang, P.; Tang, Y.; Zhu, Y.; Zhu, X.; Xu, Y.; Jiang, W.; Pan, L.; Li, Q.; et al. LDH Nanocrystal@amorphousness Core–Shell Structure Derived from LDH→LDO Transformation: Synergistically Enhanced Energy Stored for LIBs Anode. *Chem. Eng. J.* **2024**, *486*, 150416. [CrossRef]
110. Hwang, J.Y.; Park, H.; Kim, H.; Kansara, S.; Sun, Y.K. Advanced Cathodes for Practical Lithium-Sulfur Batteries. *Acc. Mater. Res.* **2025**, *6*, 245–258. [CrossRef]
111. Fortunato, M.; Sarapulova, A.; Schwarz, B.; Cardinale, A.M.; Dsoke, S. NiFe-NO_3 Layered Double Hydroxide as a Novel Anode for Sodium Ion Batteries. *Batter. Supercaps* **2024**, *8*, 2400451. [CrossRef]
112. Wu, Y.; Yuan, Q.; Wu, Z.; Zhao, Z.; Yin, Q.; Han, J. Interlayer-Spacing Regulation of NiFe LDH Nanosheets Cathode with High Rate Performance for Chloride Ion Battery. *J. Ind. Eng. Chem.* **2024**, *143*, 176–183. [CrossRef]
113. Li, J.; Li, M.; Xu, S.; Weng, H. Chloride Ion Batteries-Excellent Candidates for New Energy Storage Batteries Following Lithium-Ion Batteries. *Ionics* **2024**, *30*, 27–38. [CrossRef]
114. Song, Z.; Yin, Q.; Yang, S.; Miao, Y.; Wu, Y.; Li, Y.Z.; Ren, Y.; Sui, Y.; Qi, J.; Han, J. A High-Nickel Layered Double Hydroxides Cathode Boosting the Rate Capability for Chloride Ion Batteries with Ultralong Cycling Life. *Small* **2023**, *19*, 2302896. [CrossRef]
115. Chen, K.; Chen, C.; Long, J.; Zhou, G. MOF/LDH Cross Composites Derived Heterojunction Nanospheres as Highly Efficient Catalysts for ORR-OER and Rechargeable Zn-Air Batteries. *Appl. Surf. Sci.* **2024**, *657*, 159803. [CrossRef]
116. Allwyn, N.; Gokulnath, S.; Sathish, M. In-Situ Nanoarchitectonics of Fe/Co LDH over Cobalt-Enriched N-Doped Carbon Cookies as Facile Oxygen Redox Electrocatalysts for High-Rate Rechargeable Zinc–Air Batteries. *ACS Appl. Mater. Interfaces* **2024**, *16*, 20360–20374. [CrossRef] [PubMed]
117. Ogawa, M.; Asai, S. Hydrothermal Synthesis of Layered Double Hydroxide-Deoxycholate Intercalation Compounds. *Chem. Mater.* **2000**, *12*, 3253–3255. [CrossRef]

118. Jiang, F.; Xie, Y.; Zhang, H.; Zhang, L.; Gao, X.; Bai, H.; Yao, F.; Yue, H. Hierarchical Core-Shelled CoMo Layered Double Hydroxide@CuCo_2S_4 Nanowire Arrays/Nickel Foam for Advanced Hybrid Supercapacitors. *J. Colloid. Interface Sci.* **2025**, *677*, 150–157. [CrossRef]
119. Li, R.Y.; Shen, X.Y.; Li, J.; Zhao, D.P.; Zhao, R.D.; Wu, F.F. Enhanced Electrochemical Performance of NiCo-Layered Double Hydroxides: Optimal Synthesis Conditions and Supercapacitor Applications. *Adv. Sustain. Syst.* **2024**, *9*, 2400753. [CrossRef]
120. Zhi, P.X.; Guo, Q.L. Hydrothermal Synthesis of Layered Double Hydroxides (LDHs) from Mixed MgO and Al_2O_3: LDH Formation Mechanism. *Chem. Mater.* **2005**, *17*, 1055–1062. [CrossRef]
121. Ai, L.; Zhang, C.; Meng, L. Adsorption of Methyl Orange from Aqueous Solution on Hydrothermal Synthesized Mg-Al Layered Double Hydroxide. *J. Chem. Eng. Data* **2011**, *56*, 4217–4225. [CrossRef]
122. Knorpp, A.J.; Zawisza, A.; Huangfu, S.; Borzì, A.; Clark, A.H.; Kata, D.; Graule, T.; Stuer, M. Hydrothermal Synthesis of Multi-Cationic High-Entropy Layered Double Hydroxides. *RSC Adv.* **2022**, *12*, 26362–26371. [CrossRef]
123. Rybka, K.; Matusik, J. Exploring Synthesis Approaches, Properties and Applications of Layered Double Hydroxides Derived from Alternative Sources: A Comprehensive Review. *J. Clean. Prod.* **2025**, *498*, 145215. [CrossRef]
124. Sakai, M.; Imagawa, H.; Baba, N. Layered-Double-Hydroxide-Based Ni Catalyst for CO_2 Capture and Methanation. *Appl. Catal. A Gen.* **2022**, *647*, 118904. [CrossRef]
125. Liao, L.; Zhao, N.; Xia, Z. Hydrothermal Synthesis of Mg-Al Layered Double Hydroxides (LDHs) from Natural Brucite and $Al(OH)_3$. *Mater. Res. Bull.* **2012**, *47*, 3897–3901. [CrossRef]
126. Pan, Y.; Li, L.; Liu, D.; Wang, F.; Xia, M.; Wang, F. Macroscale Structures of Layered Double Hydroxides: A Review of Formation Strategies and Applications in Water Treatment. *J. Environ. Chem. Eng.* **2025**, *13*, 115284. [CrossRef]
127. Cheng, X.; Zhang, L.; Li, L.; Wu, H.; Zheng, J.; Sun, J.; Li, G. One-Step Hydrothermal Synthesis of Glucose-Induced Low Crystallinity NiCo-Based Layered Double Hydroxides for High-Performance Asymmetric Supercapacitors. *Chem. Eur. J.* **2024**, *31*, e202403439. [CrossRef]
128. Zhao, Y.; Xiao, F.; Jiao, Q. Hydrothermal Synthesis of Ni/Al Layered Double Hydroxide Nanorods. *J. Nanotechnol.* **2011**, *2011*, 646409. [CrossRef]
129. Zhang, F.; Du, N.; Song, S.; Liu, J.; Hou, W. Mechano-Hydrothermal Synthesis of Mg_2Al-NO_3 Layered Double Hydroxides. *J. Solid. State Chem.* **2013**, *206*, 45–50. [CrossRef]
130. Clark, I.; Dunne, P.W.; Gomes, R.L.; Lester, E. Continuous Hydrothermal Synthesis of Ca_2Al-NO_3 Layered Double Hydroxides: The Impact of Reactor Temperature, Pressure and NaOH Concentration on Crystal Characteristics. *J. Colloid. Interface Sci.* **2017**, *504*, 492–499. [CrossRef] [PubMed]
131. Wang, W.; Zhou, J.; Achari, G.; Yu, J.; Cai, W. Cr(VI) Removal from Aqueous Solutions by Hydrothermal Synthetic Layered Double Hydroxides: Adsorption Performance, Coexisting Anions and Regeneration Studies. *Colloids Surf. A Physicochem. Eng. Asp.* **2014**, *457*, 33–40. [CrossRef]
132. Tung, N.Q.; Van, D.T.C.; Thang, D.X.; An, N.T.K.; Trang, T.T.; Nhi, B.D.; Thao, N.P.; Son, L.T.; Huy, N.N.; Dung, N.T. Hydrothermal Synthesis of CuCoFe Layered Double Hydroxide and Its Performance in the Degradation of Antibiotics: Influencing Factors, Degradation Pathways, and Reaction Mechanism. *J. Environ. Chem. Eng.* **2023**, *11*, 110127. [CrossRef]
133. Prevot, V.; Caperaa, N.; Taviot-Guého, C.; Forano, C. Glycine-Assisted Hydrothermal Synthesis of NiAl-Layered Double Hydroxide Nanostructures. *Cryst. Growth Des.* **2009**, *9*, 3646–3654. [CrossRef]
134. Xiong, Q.; Li, X.; Zhou, M.; Chen, R.; Sun, C.; Zhou, Y.; Wang, S.; Qiu, X.; Song, M.; Wei, T. One-Pot Hydrothermal Synthesis of Flower-Shaped Zif-67@NiCo-LDH Heterostructure as Anode Materials for Lithium-Ion Batteries. *Ionics* **2023**, *29*, 1741–1749. [CrossRef]
135. Zhu, L.; Han, T.; Ding, Y.; Long, J.; Lin, X.; Liu, J. A Metal–Organic-Framework Derived $NiFe_2O_4$@NiCo-LDH Nanocube as High-Performance Lithium-Ion Battery Anode under Different Temperatures. *Appl. Surf. Sci.* **2022**, *599*, 153953. [CrossRef]
136. Luo, K.; Li, Y.; Yao, J.; Huang, B.; Zhu, Q.; Yang, J. Boosting the Lithium Storage Property of Nickel-Zinc Layered Double Hydroxides by Intercalation with Dodecyl Sulfate Anions. *Appl. Surf. Sci.* **2023**, *620*, 156850. [CrossRef]
137. Liu, Y.; Zhou, X.; Chen, P.; Cao, X.; Liu, D.; Wang, R. Properties of Hollow Yolk-Shell NiS_2/FeS_2@NC@NiFe LDH/FeO(OH) Nanoflower Microspheres as Anode Materials for Lithium-Ion Batteries. *J. Electroanal. Chem.* **2023**, *943*, 117606. [CrossRef]
138. Yang, T.; Zhang, Z.; Ma, P.; Li, H. Constructing 1D/2D NiCo-LDH Nanowire/MXene Composites for Efficient and Stable Lithium Storage. *Adv. Mater. Interfaces* **2024**, *11*, 2300962. [CrossRef]
139. Shah, S.S.; Aziz, M.A.; Rasool, P.I.; Mohmand, N.Z.K.; Khan, A.J.; Ullah, H.; Feng, X.; Oyama, M. Electrochemical Synergy and Future Prospects: Advancements and Challenges in MXene and MOFs Composites for Hybrid Supercapacitors. *Sustain. Mater. Technol.* **2024**, *39*, e00814. [CrossRef]
140. Yuan, Q.; Zhang, J.; Li, B.; Wu, Y.; Wu, Z.; Dou, Y.; Yin, Q.; Han, J. An MgAl Layered Double Hydroxide as a New Transition Metal-Free Anode for Lithium-Ion Batteries. *Chem. Commun.* **2023**, *59*, 13903–13906. [CrossRef]
141. Zhang, D.; Hua, Y.; Fu, X.; Cheng, C.; Kong, D.; Zhang, M.; Lei, B.; Liu, Z. Hierarchical Flower Array NiCoOOH@CoLa-LDH Nanosheets for High-Performance Supercapacitor and Alkaline Zn Battery. *Adv. Funct. Mater.* **2025**, *35*, e202414686. [CrossRef]

142. Yin, Q.; Song, Z.; Yang, S.; Zhao, Z.; Yuan, Q.; Qi, J.; Sui, Y.; Han, J. Heteroatoms Introduction via Temperature-Differential Doping Strategy Achieves Long-Cycle-Life NiCoMo-P LDH Cathode for Rechargeable Chloride-Ion Battery. *J. Electroanal. Chem.* **2023**, *951*, 117924. [CrossRef]
143. Zhao, G.; Deng, Z.; Wan, G.; Zhao, J.; Wang, G. Array Structured NiAl-Layered Double Hydroxides Grown on Graphene by Atomic Layer Deposition as Chloride-Ion Battery Cathode. *DeCarbon* **2025**, *8*, 100106. [CrossRef]
144. Yang, S.; Yin, Q.; Song, Z.; Xu, F.; Xie, Z.; Wu, Y.; Xu, S.; Li, Y.Z.; Zhao, D.; Xiao, B.; et al. Introducing High-Valence Molybdenum to Stimulate Lattice Oxygen in a NiCo LDH Cathode for Chloride Ion Batteries. *Mater. Horiz.* **2023**, *10*, 3429–3437. [CrossRef] [PubMed]
145. Wu, Z.; Wu, Y.; Yuan, Q.; Zhang, J.; Dou, Y.; Han, J. Aqueous Chloride-Ion Battery within a Neutral Electrolyte Based on a CoFe-Cl Layered Double Hydroxide Anode. *ACS Appl. Mater. Interfaces* **2023**, *15*, 38540–38549. [CrossRef]
146. Li, J.; Qin, Y.; Bai, Z.; Li, S.; Li, L.; Ouyang, B.; Kan, E.; Zhang, W. Investigating the Role of 3D Hierarchical Ni-CAT/NiFe-LDH/CNFs in Enhancing the Oxygen Evolution Reaction and Zn-Air Battery Performance. *Appl. Surf. Sci.* **2024**, *648*, 159080. [CrossRef]
147. Bhanuse, G.B.; Kumar, S.; Yu, C.C.; Fu, Y.P. Nanostructured $ZnCo_2O_4$@NiMn-LDH Electrodes for Supercapacitor and Zinc-Air Battery Application. *ACS Appl. Nano Mater.* **2024**, *7*, 13649–13663. [CrossRef]
148. Jia, Z.; Shang, J.; Xue, K.; Yang, X.; Wang, S.; Xu, C.; Wang, Q. N-doped Nanocarbon Inserted NiCo-LDH Nanoplates on NF with High OER/ORR Performances for Zinc-Air Battery. *ChemCatChem* **2023**, *15*, e202201469. [CrossRef]
149. Chen, J.; Liu, Z.; Liu, Y.; Cheng, Z.; Zhang, J.; Zhang, Q.; Li, Y.; Hou, C.; Li, K.; Wang, H. Efficient Bifunctional NiFe-LDH@Co_9S_8 Nanoflower Electrocatalysts Anchored with Pt Nanocrystal for Flexible Quasi-Solid Rechargeable Zinc Air Battery. *Energy Fuels* **2024**, *38*, 10264–10274. [CrossRef]
150. Liu, S.; Wan, R.; Lin, Z.; Liu, Z.; Liu, Y.; Tian, Y.; Qin, D.-D.; Tang, Z. Probing the Co Role in Promoting the OER and Zn–Air Battery Performance of NiFe-LDH: A Combined Experimental and Theoretical Study. *J. Mater. Chem. A Mater.* **2022**, *10*, 5244–5254. [CrossRef]
151. Dighe, P.S.; Redekar, R.S.; Tarwal, N.L.; Sarawade, P.B. Design and Development of the High-Performance Aqueous Asymmetric Supercapacitor Based on the Hydrothermally Grown Binder-Less Ni-Co LDH Nanosheets. *J. Energy Storage* **2024**, *88*, 111467. [CrossRef]
152. Chandra Sahoo, R.; Moolayadukkam, S.; Seok, J.H.; Lee, S.U.; Matte, H.S.S.R. Enhanced Charge Storage Capacity and High Rate Capabilities of Ni_2Co-Layered Double Hydroxides/Expanded-Graphite Composites as Anodes for Li-Ion Batteries. *J. Mater. Chem. A Mater.* **2023**, *11*, 7142–7151. [CrossRef]
153. Li, X.; Fortunato, M.; Cardinale, A.M.; Sarapulova, A.; Njel, C.; Dsoke, S. Electrochemical Study on Nickel Aluminum Layered Double Hydroxides as High-Performance Electrode Material for Lithium-Ion Batteries Based on Sodium Alginate Binder. *J. Solid. State Electrochem.* **2022**, *26*, 49–61. [CrossRef]
154. Du, P.; Xi, W.; Zhang, Y.; Wang, R.; Gong, Y.; He, B.; Wang, H.; Jin, J. Laser Induction of Hierarchically Micro/Nanostructured CoNi-LDH/C Composite for High-Performance Lithium-Sulfur Batteries. *J. Alloys Compd.* **2024**, *985*, 174043. [CrossRef]
155. Ye, Z.; Zhai, S.; Liu, R.; Liu, M.; Xu, Y.; Li, C.; Wang, X.; Mei, T. Sulfonate-Rich Polymer Intercalated LDH Artificial SEI Film to Enable High-Stability Li-S Batteries. *Chem. Eng. J.* **2024**, *479*, 147847. [CrossRef]
156. Zhen, M.; Meng, X.; Wang, X.; Zhang, Z.; Guo, S.-Q.; Hu, Z.; Shen, B. Layered Double Hydroxides-MXene Heterointerfaces with Abundant Anion Vacancies Expediting Sulfur Redox Kinetics for High-Performance Lithium–Sulfur Batteries. *Chem. Eng. J.* **2024**, *489*, 151285. [CrossRef]
157. Liu, Q.; Zhang, Y.; Zhou, Y.; Wang, M.; Li, R.; Yue, W. Layered Double Hydroxides Used as the Sulfur Hosts for Lithium-Sulfur Batteries and the Influence of Metal Composition on Their Performance. *J. Solid. State Electrochem.* **2023**, *27*, 797–807. [CrossRef]
158. Wang, H.Y.; Dai, Y.K.; Liao, K.M.; Deng, S.; Dai, G.P. Vertical Graphene Growth on LDH Nanosheets and Carbon Cloth Nanofibers with NiCo Nanoparticles as a Freestanding Host for High-Performance Lithium-Sulfur Batteries. *J. Phys. Chem. Lett.* **2025**, *16*, 1103–1113. [CrossRef]
159. Zhang, Y.; Zhang, Y.; Ma, L.; Yang, M.; Zhao, X. NiCr-Cl LDH/RGO Composite as Anode Material for Sodium-Ion Batteries. *J. Electron. Mater.* **2022**, *51*, 6067–6075. [CrossRef]
160. Li, B.; Li, J.; Li, P.; Zhang, N.; Shang, H.; Zhang, B. Sulfide Hollow ZIF-67 and NiCo-LDH Heterostructure for Superior Asymmetric Supercapacitors. *J. Alloys Compd.* **2025**, *1017*, 179128. [CrossRef]
161. He, L.; Cai, P.; Lai, H.; Lu, K.; Xu, Z.; Zeng, R.; Hao, C.; Wang, Z.; Gan, W. Sandwich-like NiFe-LDH/$MnCO_3$/MXene Ternary Nanocomposites Serve as Battery-Type Electrode for High-Performance Asymmetric Supercapacitor. *Chem. Eng. J.* **2025**, *504*, 159149. [CrossRef]
162. Liu, Y.; Wang, C.; Sun, H.; Duan, L.; Yang, Z.; Wang, X.; Liu, J. Construction of Flexible $MnCo_2O_4$@FeCoNi-LDH Electrode Materials with Nanoflower-like and Hierarchical Structure for High-Performance Asymmetric Supercapacitor. *J. Colloid. Interface Sci.* **2025**, *682*, 1051–1061. [CrossRef]

163. Tai, P.C.; Chung, R.J.; Wang, G.B.; Kongvarhodom, C.; Husain, S.; Yougbaré, S.; Chen, H.M.; Wu, Y.F.; Lin, L.Y. Polydopamine Derived Carbon Coated Cobalt Molybdenum Layered Double Hydroxide as Highly Stable Anode Materials of Sodium-Ion Battery. *J. Energy Storage* **2024**, *101*, 113961. [CrossRef]
164. Lin, Q.; Yuan, Z.; Wang, D.; Wei, W.; Wang, X.; Han, W.; Wang, L. Co-Co LDH-Derived $CoSe_2$ Anchored on N-Doped Carbon Nanospheres as High-Performance Anodes for Sodium-Ion Batteries. *Electrochim. Acta* **2022**, *432*, 141012. [CrossRef]
165. Ma, X.; Zhou, L.; Chen, T.; Sun, P.; Lv, X.; Yu, H.; Sun, X.; Leo Liu, T. High-Performance Aqueous Rechargeable NiCo//Zn Battery with Molybdate Anion Intercalated CoNi-LDH@CP Bilayered Cathode. *J. Colloid. Interface Sci.* **2024**, *658*, 728–738. [CrossRef]
166. Wang, C.; Han, L.; Yang, S.; Liu, Z.; Liu, M.; Li, B. Nanosheet-Structured ZnCo-LDH Microsphere as Active Material for Rechargeable Zinc Batteries. *J. Colloid. Interface Sci.* **2024**, *659*, 119–126. [CrossRef] [PubMed]
167. Zhang, Y.; Hu, S.; Li, C.; Yan, X.; Zhang, Y.; Yin, R.; Wei, Y.; Gao, K.; Gao, H. Advanced Strategies for Enhancing Electrochemical Performance of NiAl LDH Electrodes in Supercapacitors. *Coord. Chem. Rev.* **2025**, *531*, 216497. [CrossRef]
168. Jing, C.; Huang, L.; Tao, S.; Chen, Y.; Zhang, S.; Dong, W.; Ling, F.; Tang, X.; Li, Y.; Feng, L.; et al. Construction of MoB@LDH Heterojunction and Its Derivates through Phase and Interface Engineering for Advanced Supercapacitor Applications. *J. Colloid. Interface Sci.* **2024**, *660*, 10–20. [CrossRef]
169. Tai, P.-Y.; Sakthivel, M.; Peng, Y.-J.; Kubendhiran, S.; Lin, L.-Y.; Ho, K.-C. In Situ Growing Te-Doped Ni–Mn Layered Double Hydroxide on a Cetyltrimethylammonium Bromide-Modified MXene Conductive Layer for Binder-Free Supercapacitors. *ACS Appl. Energy Mater.* **2025**, *8*, 4122–4133. [CrossRef]
170. Han, X.; Li, J.; Lu, J.; Luo, S.; Wan, J.; Li, B.; Hu, C.; Cheng, X. High Mass-Loading NiCo-LDH Nanosheet Arrays Grown on Carbon Cloth by Electrodeposition for Excellent Electrochemical Energy Storage. *Nano Energy* **2021**, *86*, 106079. [CrossRef]
171. He, T.-L.; Zhang, X.-R.; Ma, Q.-L.; Hua, W.-M.; Xiong, H.-M. Positive Carbon Dots Induced Electrodeposition of NiCo-LDH Nanosheets for High-Performance Supercapacitors. *ACS Appl. Energy Mater.* **2024**, *7*, 5508–5516. [CrossRef]
172. Niu, Y.; Chang, L.; Huang, Y.; Sun, Q.; Lu, X.; Cheng, H. Electrochemical Constructing Versatile ZnAl-LDH Artificial Interface Layer Coated (002)-Textured Zn for Highly Reversible Zinc Anodes. *Chem. Eng. J.* **2024**, *499*, 155813. [CrossRef]
173. Lashgari, M.; Yamini, Y. Fiber-in-tube Solid-phase Microextraction of Caffeine as a Molecular Tracer in Wastewater by Electrochemically Deposited Layered Double Hydroxide. *J. Sep. Sci.* **2018**, *41*, 2393–2400. [CrossRef]
174. Seumo Tchekwagep, P.M.; Banks, C.E.; Crapnell, R.D.; Farsak, M.; Kardaş, G. Electrochemical Synthesis of NiCo Layered Double Hydroxides on Nickel-Coated Graphite for Water Splitting: Understanding the Electrochemical Experimental Parameters. *RSC Adv.* **2025**, *15*, 3969–3978. [CrossRef] [PubMed]
175. Cysewska, K.; Zając, M.; Łapiński, M.; Karczewski, J.; Rybarczyk, M.K.; Kamecki, B.; Jasiński, P.; Molin, S. The Effect of Cobalt Incorporation into Nickel–Iron Oxide/(oxy)Hydroxide Catalyst on Electrocatalytic Performance Toward Oxygen Evolution Reaction. *Energy Technol.* **2021**, *9*, 2100688. [CrossRef]
176. Wu, Q.; Li, F.; Sheng, H.; Qi, Y.; Yuan, J.; Bi, H.; Li, W.; Xie, E.; Lan, W. In Situ Fabrication of Hierarchical CuO@CoNi-LDH Composite Structures for High-Performance Supercapacitors. *ACS Appl. Mater. Interfaces* **2024**, *16*, 23241–23252. [CrossRef]
177. Kim, Y.; Jun, S.E.; Lee, G.; Nam, S.; Jang, H.W.; Park, S.H.; Kwon, K.C. Recent Advances in Water-Splitting Electrocatalysts Based on Electrodeposition. *Materials* **2023**, *16*, 3044. [CrossRef] [PubMed]
178. Pawar, S.A.; Yu, S.; Ju, E.; Seo, H.; Yeu, J.; Kim, J.; Patil, D.S.; Shin, J.C. Zinc Cobalt Layered Double Hydroxide Electrode for High-Performance Supercapacitor. *Appl. Sci. Converg. Technol.* **2019**, *28*, 164–168. [CrossRef]
179. Arbenin, A.Y.; Zemtsova, E.G.; Orekhov, E.V.; Sokolova, D.N.; Baburova, P.I.; Petrov, A.A.; Gaĭshun, V.E.; Smirnov, V.M. Features of Fabrication of Titanium Dioxide Based Coatings for Non-Lithographic Template Electrochemical Synthesis of Micron Metal Particle Arrays. *Gels* **2021**, *7*, 202. [CrossRef] [PubMed]
180. Guo, H.; Guan, Y.; Zhang, M.; Qin, J.; Guo, X.; Li, Z.; Zhang, B.; Tang, J. Dynamic Morphological Evolution of Co-Based Layered Double Hydroxide Nanosheets Investigated by In Situ Electrochemical-Atomic Force Microscopy. *J. Phys. Chem. C* **2023**, *127*, 5219–5229. [CrossRef]
181. Wang, T.; Wang, W.; Shao, W.; Bai, M.; Zhou, M.; Li, S.; Ma, T.; Ma, L.; Cheng, C.; Liu, X. Synthesis and Electronic Modulation of Nanostructured Layered Double Hydroxides for Efficient Electrochemical Oxygen Evolution. *ChemSusChem* **2021**, *14*, 5112–5134. [CrossRef]
182. Ramachandran, R.; Lan, Y.; Xu, Z.X.; Wang, F. Construction of NiCo-Layered Double Hydroxide Microspheres from Ni-MOFs for High-Performance Asymmetric Supercapacitors. *ACS Appl. Energy Mater.* **2020**, *3*, 6633–6643. [CrossRef]
183. Shahid, N.F.; Jamal, A.; Haq, G.; Javed, M.; Saifullah, M.; Anjum, M.A.R. Cobalt–Iron Layered Double Hydroxide Nanosheet-Wrapped Nitrogen-Doped Graphite Felt as an Oxygen-Evolving Electrode. *Energy Adv.* **2023**, *2*, 2109–2118. [CrossRef]
184. Ren, B.; Zhang, B.Y.; Zhang, H.; Jiang, X.; Yi, Q.; Xu, K.; Yu, H.; Hu, Z.; Jeerapan, I.; Ou, J.Z. CoNi Layered Double Hydroxide Nanosheets Vertically Grown on Electrodeposited Dendritic Copper Substrates for Supercapacitor Applications. *ACS Appl. Nano Mater.* **2022**, *5*, 2395–2404. [CrossRef]

185. Zhang, H.; Li, X.; Hähnel, A.; Naumann, V.; Lin, C.; Azimi, S.; Schweizer, S.L.; Maijenburg, A.W.; Wehrspohn, R.B. Bifunctional Heterostructure Assembly of NiFe LDH Nanosheets on NiCoP Nanowires for Highly Efficient and Stable Overall Water Splitting. *Adv. Funct. Mater.* **2018**, *28*, 1706847. [CrossRef]
186. Iqbal, M.A.; Sun, L.; LaChance, A.M.; Ding, H.; Fedel, M. In Situ Growth of a CaAl-NO_3^- -Layered Double Hydroxide Film Directly on an Aluminum Alloy for Corrosion Resistance. *Dalton Trans.* **2020**, *49*, 3956–3964. [CrossRef] [PubMed]
187. Gualandi, I.; Vlamidis, Y.; Mazzei, L.; Musella, E.; Giorgetti, M.; Christian, M.; Morandi, V.; Scavetta, E.; Tonelli, D. Ni/Al Layered Double Hydroxide and Carbon Nanomaterial Composites for Glucose Sensing. *ACS Appl. Nano Mater.* **2019**, *2*, 143–155. [CrossRef]
188. Alsaç, E.P.; Zhou, K.; Rong, W.; Salamon, S.; Landers, J.; Wende, H.; Smith, R.D.L. Identification of Non-Traditional Coordination Environments for Iron Ions in Nickel Hydroxide Lattices. *Energy Environ. Sci.* **2022**, *15*, 2638–2652. [CrossRef]
189. Shi, Z.; Yuan, Y.; Xiao, Q.; Li, Z.; Zhu, J. Carbonate Doped NiCo-LDH Modified with PANI for High Performance Asymmetric Supercapacitors. *CrystEngComm* **2022**, *24*, 3546–3555. [CrossRef]
190. Tan, Z.; Luo, X.; Wei, T.; Zhang, F.; Lv, Y.; Chen, L.; Shi, Y. In Situ Formation of NiAl-Layered Double Hydroxide with a Tunable Interlayer Spacing in a Confined Impinging Jet Microreactor. *Energy Fuels* **2020**, *34*, 8939–8946. [CrossRef]
191. Yan, Z.; Sun, H.; Chen, X.; Liu, H.; Zhao, Y.; Li, H.; Xie, W.; Cheng, F.; Chen, J. Anion Insertion Enhanced Electrodeposition of Robust Metal Hydroxide/Oxide Electrodes for Oxygen Evolution. *Nat. Commun.* **2018**, *9*, 2373. [CrossRef]
192. Ghadimi, A.M.; Ghasemi, S.; Omrani, A.; Mousavi, F. Nickel Cobalt LDH/Graphene Film on Nickel-Foam-Supported Ternary Transition Metal Oxides for Supercapacitor Applications. *Energy Fuels* **2023**, *37*, 3121–3133. [CrossRef]
193. Cysewska, K.; Rybarczyk, M.K.; Cempura, G.; Karczewski, J.; Łapiński, M.; Jasinski, P.; Molin, S. The Influence of the Electrodeposition Parameters on the Properties of Mn-Co-Based Nanofilms as Anode Materials for Alkaline Electrolysers. *Materials* **2020**, *13*, 2662. [CrossRef] [PubMed]
194. Kumar, M.; Singh, R.P. Transforming Prussian Blue Analogues: The Future-Centric Pathway to Rise of Thin Layered Double Hydroxides for Superior Water Oxidation: An Overview. *Int. J. Multidiscip. Res.* **2024**, *6*, 240529213. [CrossRef]
195. Yang, Q.; Wang, Q.; Long, Y.; Wang, F.; Wu, L.; Pan, J.; Han, J.; Lei, Y.; Shi, W.; Song, S. In Situ Formation of Co_9S_8 Quantum Dots in MOF-Derived Ternary Metal Layered Double Hydroxide Nanoarrays for High-Performance Hybrid Supercapacitors. *Adv. Energy Mater.* **2020**, *10*, 1903193. [CrossRef]
196. Li, Z.; Han, F.; Li, C.; Jiao, X.; Chen, D. Multi-Anion Intercalated Layered Double Hydroxide Nanosheet-Assembled Hollow Nanoprisms with Improved Pseudocapacitive and Electrocatalytic Properties. *Chem. Asian J.* **2018**, *13*, 1129–1137. [CrossRef]
197. Li, S.; Mo, T.; Chen, L.; Zhang, F.; Jamil, S.; Lu, Y.; Cai, Q. Hierarchical MgAl-Layered Double Hydroxide Growth on Porous MgO Template for Pollution Removal. *Environ. Prog. Sustain. Energy* **2022**, *41*, e13907. [CrossRef]
198. Wei, M.; Huang, Q.; Zhou, Y.; Peng, Z.; Chu, W. Ultrathin Nanosheets of Cobalt-Nickel Hydroxides Hetero-Structure via Electrodeposition and Precursor Adjustment with Excellent Performance for Supercapacitor. *J. Energy Chem.* **2018**, *27*, 591–599. [CrossRef]
199. Chavan, H.S.; Lee, C.H.; Inamdar, A.I.; Han, J.; Park, S.; Cho, S.; Shreshta, N.K.; Lee, S.U.; Hou, B.; Im, H.; et al. Designing and Tuning the Electronic Structure of Nickel–Vanadium Layered Double Hydroxides for Highly Efficient Oxygen Evolution Electrocatalysis. *ACS Catal.* **2022**, *12*, 3821–3831. [CrossRef]
200. Zhai, P.; Xia, M.; Wu, Y.; Zhang, G.; Gao, J.; Zhang, B.; Cao, S.; Zhang, Y.; Li, Z.; Fan, Z.; et al. Engineering Single-Atomic Ruthenium Catalytic Sites on Defective Nickel-Iron Layered Double Hydroxide for Overall Water Splitting. *Nat. Commun.* **2021**, *12*, 4587. [CrossRef]
201. Wang, G.; Liu, G.; Jin, Z. In Situ Coupled Nickel-Based Layered Double Hydroxides with MXene to Enhance Supercapacitor Performance. *J. Mater. Chem. C Mater.* **2023**, *11*, 10547–10561. [CrossRef]
202. Laipan, M.; Yu, J.; Zhu, R.; Zhu, J.; Smith, A.T.; He, H.; O'Hare, D.; Sun, L. Functionalized Layered Double Hydroxides for Innovative Applications. *Mater. Horiz.* **2020**, *7*, 715–745. [CrossRef]
203. Chen, Z.; Zhu, J.; Yang, S.; Wei, Z.; Wang, Y.; Chen, A.; Huang, Z.; Zhi, C. MXene Supported Electrodeposition Engineering of Layer Double Hydroxide for Alkaline Zinc Batteries. *Angew. Chem. Int. Ed.* **2024**, *63*, e202411443. [CrossRef]
204. Cheng, X.; Wu, D.; Gao, H.; Wang, Q.; Lv, P.; Yoon, S.S.; Wei, Q. A Mild, Configurable, Flexible CoNi-LDH(v)/Zn Battery Based on H-Vacancy-Induced Reversible Zn^{2+} Intercalation. *J. Energy Chem.* **2025**, *100*, 498–508. [CrossRef]
205. Inamdar, A.I.; Chavan, H.S.; Jo, Y.; Im, H.; Kim, H. Nonprecious Bimetallic NiFe-layered Hydroxide Nanosheets as a Catalyst for Highly Efficient Electrochemical Water Splitting. *Int. J. Energy Res.* **2021**, *45*, 16963–16972. [CrossRef]
206. Li, H.; Zhao, R.; Zhou, W.; Wang, L.; Li, W.; Zhao, D.; Chao, D. Trade-off between Zincophilicity and Zincophobicity: Toward Stable Zn-Based Aqueous Batteries. *JACS Au* **2023**, *3*, 2107–2116. [CrossRef]
207. Meng, J.; Song, Y.; Qin, Z.; Wang, Z.; Mu, X.; Wang, J.; Liu, X.X. Cobalt–Nickel Double Hydroxide toward Mild Aqueous Zinc-Ion Batteries. *Adv. Funct. Mater.* **2022**, *32*, 2204026. [CrossRef]

208. Dai, L.; Peng, S.; Wang, X.; Chen, B.; Wu, Y.; Xie, Q.; Ruan, Y. Three-Dimensional NiCoS Nanotubes@NiCo-LDH Nanosheets Core–Shell Heterostructure for High-Rate Capability Alkaline Zinc-Based Batteries. *RSC Adv.* **2024**, *14*, 7999–8006. [CrossRef] [PubMed]
209. Zhou, K.; Wang, S.; Zhong, G.; Chen, J.; Bao, Y.; Niu, L. Hierarchical Heterostructure Engineering of Layered Double Hydroxides on Nickel Sulfides Heteronanowire Arrays as Efficient Cathode for Alkaline Aqueous Zinc Batteries. *Small* **2022**, *18*, 2202799. [CrossRef]
210. Cai, X.; Jiang, T.; Wu, M. Confined Growth of NiFe LDH with Hierarchical Structures on Copper Nanowires for Long-Term Stable Rechargeable Zn-Air Batteries. *Appl. Surf. Sci.* **2022**, *577*, 151911. [CrossRef]
211. Zhou, P.; Ji, Y.; Zhang, B.; Zhang, P.; Zhang, S.; Wang, S. Rational Construction of Tremella-like CuCo LDH@Ni_3S_2 Nanocomposites as High-Performance Supercapacitor Electrode Materials. *J. Alloys Compd.* **2025**, *1021*, 179586. [CrossRef]
212. Xing, H.; Deng, X.; Wang, X. Alkaline Capacity Decay Induced Vacancy-Rich LDH for High-Performance Magnesium Ions Hybrid Supercapacitor. *J. Colloid. Interface Sci.* **2025**, *679*, 43–53. [CrossRef]
213. Khalafallah, D.; Ibrahim, M.A.; Hou, H.; Wang, J.; Liu, C.; Zhang, Q. Configuring Cations–Doped Cobalt Lanthanum LDH Nanoarray-on-Nanoarray Platforms for Supercapacitors. *Sustain. Mater. Technol.* **2025**, *43*, e01286. [CrossRef]
214. Duddi, R.; Shivani; Dhiman, S.; Kumar Singh, A.; Kamboj, N.; Kumar, S. Superior Cycle Stability and Enhanced Pseudocapacitive Performance of NiAl-LDH Decorated Carbon Cloth for Supercapacitor Application. *Appl. Surf. Sci.* **2025**, *679*, 161298. [CrossRef]
215. Mazloum-Ardakani, M.; Mozaffari, S.A.; Rasche, B.; Ebrahimi, F. Bio-Inspired Coral Reef-like NiCo-LDH Nanostructure Fabricated on Carbon Felt for High Performance Flexible Supercapacitor. *Sci. Rep.* **2025**, *15*, 10542. [CrossRef] [PubMed]
216. Klydziute, G.; Gliaudyte, L.; Sokol, D.; Vasiliauskiene, D.; Yang, T.C.-K.; Kareiva, A. Layered Double Hydroxides: Sol-Gel Synthesis, Characterization and Application. *Preprint* **2024**. [CrossRef]
217. Valente, J.S.; Lima, E.; Toledo-Antonio, J.A.; Cortes-Jacome, M.A.; Lartundo-Rojas, L.; Montiel, R.; Prince, J. Comprehending the Thermal Decomposition and Reconstruction Process of Sol−Gel MgAl Layered Double Hydroxides. *J. Phys. Chem. C* **2010**, *114*, 2089–2099. [CrossRef]
218. Mallakpour, S.; Dinari, M.; Behranvand, V. Ultrasonic-Assisted Synthesis and Characterization of Layered Double Hydroxides Intercalated with Bioactive N,N′-(Pyromellitoyl)-Bis-L-α-Amino Acids. *RSC Adv.* **2013**, *3*, 23303–23308. [CrossRef]
219. Sokol, D.; Vieira, D.E.L.; Zarkov, A.; Ferreira, M.G.S.; Beganskiene, A.; Rubanik, V.V.; Shilin, A.D.; Kareiva, A.; Salak, A.N. Sonication Accelerated Formation of Mg-Al-Phosphate Layered Double Hydroxide via Sol-Gel Prepared Mixed Metal Oxides. *Sci. Rep.* **2019**, *9*, 10419. [CrossRef]
220. Butenko, E.; Malyshev, A.; Kapustin, A. LDHs as Adsorbents of Phenol and Their Environmental Applications. *Am. J. Environ. Prot.* **2014**, *2*, 11–15. [CrossRef]
221. Mahjoubi, F.Z.; Khalidi, A.; Abdennouri, M.; Barka, N. Zn–Al Layered Double Hydroxides Intercalated with Carbonate, Nitrate, Chloride and Sulphate Ions: Synthesis, Characterisation and Dye Removal Properties. *J. Taibah Univ. Sci.* **2017**, *11*, 90–100. [CrossRef]
222. Shan, X.; Guo, Z.; Zhang, C.; Wang, W.; Zhao, L. Nickel Aerogel @ Ultra-Thin NiCo-LDH Nanosheets Integrated Freestanding Film as a Collaborative Adsorption and Accelerated Conversion Cathode to Improve the Rate Performance of Lithium Sulfur Batteries. *Chem. Eng. J.* **2024**, *488*, 151105. [CrossRef]
223. Yang, Y.; Ma, S.; Xia, M.; Guo, Y.; Zhang, Y.; Liu, L.; Zhou, C.; Chen, G.; Wang, X.; Wu, Q.; et al. Elaborately Converting Hierarchical NiCo–LDH to Rod–like LDH–Decorated MOF as Interlayer for High–Performance Lithium–Sulfur Battery. *Mater. Today Phys.* **2023**, *35*, 101112. [CrossRef]
224. Li, J.; Qiu, W.; Liu, X.; Zhang, Y.; Zhao, Y. NiCo-Layered Double Hydroxide to Composite with Sulfur as Cathodes for High-Performance Lithium-Sulfur Batteries. *ChemElectroChem* **2022**, *9*, e202101211. [CrossRef]
225. Chen, Q.; Zhou, J.; Zhu, Y.; Jin, C.; Zhang, J.; Wu, Y.; Tang, W. In Situ Transformation of LDH into $NiCo_2S_4$-NiS_2 Nano-Heterostructures on Hollow Carbon Boxes to Promote Sulfur Electrochemistry for High-Performance Lithium-Sulfur Batteries. *Energy Fuels* **2023**, *37*, 4711–4719. [CrossRef]
226. Li, M.; Li, Y.; Cu, Q.; Li, Y.; Li, H.; Li, Z.; Li, M.; Liao, H.; Li, G.; Li, G.; et al. Hollow and Hierarchical CuCo-LDH Nanocatalyst for Boosting Sulfur Electrochemistry in Li-S Batteries. *Energy Mater. Adv.* **2023**, *4*, 0032. [CrossRef]
227. Dong, H.; Qi, S.; Wang, L.; Chen, X.; Xiao, Y.; Wang, Y.; Sun, B.; Wang, G.; Chen, S. Conductive Polymer Coated Layered Double Hydroxide as a Novel Sulfur Reservoir for Flexible Lithium-Sulfur Batteries. *Small* **2023**, *19*, 2300843. [CrossRef]
228. Wu, Y.; Yang, S.; Liu, K.; Li, Q.; Wang, H.; Jiang, J. New Anode Materials for Sodium-Ion Batteries: Ni_xCa_{2-x}Al-Cl LDH and Co_yCa_{2-y}Al-Cl LDH Prepared from Dechlorination Wastes. *J. Energy Storage* **2024**, *101*, 113991. [CrossRef]
229. Li, Y.; Xu, C.; Li, D.; Zhang, Y.; Ni, G.; Xu, S.; Liu, B.; Huo, P. Sodium-Ion Batteries Enabled by CoSn-LDH Microspheres Anchored on Few-Layer Ti_3C_2 MXene. *ACS Appl. Nano Mater.* **2024**, *7*, 20598–20608. [CrossRef]
230. Yin, Q.; Wang, T.; Song, Z.; Yang, S.; Miao, Y.; Wu, Y.; Sui, Y.; Qi, J.; Li, Y.; Zhao, D.; et al. Computational High-Throughput Screening of Layered Double Hydroxides as Cathodes for Chloride Ion Batteries. *Chem. Eng. J.* **2023**, *459*, 141545. [CrossRef]

231. Zhang, G.; Xing, J.; Zhao, Y.; Yang, F. Hierarchical N,P Co-Doped Graphene Aerogels Framework Assembling Vertically Grown CoMn-LDH Nanosheets as Efficient Bifunctional Electrocatalyst for Rechargeable Zinc-Air Battery. *J. Colloid. Interface Sci.* **2021**, *590*, 476–486. [CrossRef]
232. Bao, E.; Ren, X.; Wang, Y.; Zhang, Z.; Luo, C.; Liu, X.; Xu, C.; Chen, H. Advanced Hybrid Supercapacitors Assembled with CoNi LDH Nanoflowers and Nanosheets as High-Performance Cathode Materials. *J. Energy Storage* **2024**, *82*, 110535. [CrossRef]
233. Wu, S.; Cai, D.; Tian, Z.; Guo, L.; Wang, Y. One-Step Synthesis of NiCo-MOF@LDH Hybrid Nanosheets for High-Performance Supercapacitor. *J. Energy Storage* **2024**, *89*, 111670. [CrossRef]
234. Owusu, K.A.; Wang, Z.; Saad, A.; Boakye, F.O.; Mushtaq, M.A.; Tahir, M.; Yasin, G.; Liu, D.; Peng, Z.; Cai, X. Room Temperature Synthesis of Vertically Aligned Amorphous Ultrathin NiCo-LDH Nanosheets Bifunctional Flexible Supercapacitor Electrodes. *Energy Environ. Mater.* **2024**, *7*, e12545. [CrossRef]
235. Zhang, Y.; Yao, Y.; Ye, H.; Chen, J.; Wan, L.; Du, C.; Zhang, Y.; Xie, M. Gradually Sulfurized Co-Ni LDH as Electrode for Highly-Stable Asymmetric Supercapacitor. *J. Energy Storage* **2024**, *91*, 112117. [CrossRef]
236. Hu, J.; Pan, Y.; Zhang, Q.; Dong, Z.; Han, S. Constructing Flower-Shaped $NiCo_2S_4$ @CoAl-LDH Heterojunction Nanosheets Exhibits Extraordinary Electrochemical Behavior for a Light-Assisted Asymmetric Supercapacitor. *Energy Fuels* **2024**, *38*, 6459–6470. [CrossRef]
237. Zhao, W.; Deng, J.; Li, M.; Du, G.; Fan, M.; Gao, H.; Yuan, Z. Rational Synthesis of Sea Urchin-like NiCo-LDH/Tannin Carbon Microsphere Composites Using Microwave Hydrothermal Technique for High-Performance Asymmetric Supercapacitor. *Adv. Compos. Hybrid. Mater.* **2025**, *8*, 215. [CrossRef]
238. Kovalenko, V.; Kotok, V.; Murashevych, B. Layered Double Hydroxides as the Unique Product of Target Ionic Construction for Energy, Chemical, Foods, Cosmetics, Medicine and Ecology Applications. *Chem. Rec.* **2024**, *24*, e202300260. [CrossRef] [PubMed]
239. Luo, Y.; Yang, Y.; Tian, Y.; Wu, Q.; Lin, W.-F.; Wen, M. Collaborative Reconstruction of FeOOH/FeNiCo-LDH Heterogeneous Nanosheets for Enhancing Anion Exchange Membrane Seawater Electrolysis. *J. Mater. Chem. A Mater.* **2025**, *13*, 7136–7148. [CrossRef]
240. Yang, L.; Lin, Q.; Guo, D.; Wu, L.; Guan, Z.; Jin, H.; Fang, G.; Chen, X.; Wang, S. Nitrogen Plasma Activates CoMn-Layered Double Hydroxides for Superior Electrochemical Oxygen Evolution. *Inorg. Chem.* **2023**, *62*, 17565–17574. [CrossRef]
241. Shankar Naik, S.; Theerthagiri, J.; Nogueira, F.S.; Lee, S.J.; Min, A.; Kim, G.-A.; Maia, G.; Pinto, L.M.C.; Choi, M.Y. Dual-Cation-Coordinated CoFe-Layered Double-Hydroxide Nanosheets Using the Pulsed Laser Ablation Technique for Efficient Electrochemical Water Splitting: Mechanistic Screening by In Situ/Operando Raman and Density Functional Theory Calculations. *ACS Catal.* **2023**, *13*, 1477–1491. [CrossRef]
242. Wu, K.; Shi, L.; Wang, Z.; Zhu, Y.; Tong, X.; He, W.; Wang, J.; Zheng, L.; Kang, Y.; Shan, W.; et al. A General Strategy to Generate Oxygen Vacancies in Bimetallic Layered Double Hydroxides for Water Oxidation. *Chem. Commun.* **2023**, *59*, 3138–3141. [CrossRef]
243. Zhou, J.; Wang, Y.; Chen, L.; Zhao, W.; Han, L. Precise Design and in Situ Synthesis of Hollow Co_9S_8 @CoNi-LDH Heterostructure for High-Performance Supercapacitors. *Dalton Trans.* **2023**, *52*, 12978–12987. [CrossRef] [PubMed]
244. Li, N.; Yang, T.; Huang, L.; Jiang, H.; Xiao, J.; Ma, X.; Lou, H.; Xie, C.; Yang, Y. Interfacial Coupling Engineering Boosting Electrocatalytic Performance of CoFe Layered Double Hydroxide Assembled on N-Doped Porous Carbon Nanosheets for Water Splitting and Flexible Zinc-Air Batteries. *ACS Appl. Mater. Interfaces* **2023**, *15*, 52530–52541. [CrossRef] [PubMed]
245. Jiang, H.; Ke, Q.; Qiu, X.; Chen, J.; Chen, P.; Wang, S.; Luo, X.; Rao, B. NiCo Layered Double Hydroxide Nanocages for High-Performance Asymmetric Supercapacitors. *Inorg. Chem. Front.* **2023**, *10*, 2154–2164. [CrossRef]
246. Zhai, H.; Wu, L.; Yu, L.; Li, L.; Wan, G.; Zhang, Y.; Yuan, X.; Wang, J.; Wang, G. Polypyrrole/NiFe-Layered Double Hydroxide Composite as an Anticorrosive Microwave Absorber. *J. Mater. Chem. C Mater.* **2024**, *12*, 11001–11011. [CrossRef]
247. Zhang, W.; Zhang, B.; Wang, Y.; Cao, X.; Wang, J.; Lu, W.; Guo, Y. Gelatin-Based Hydrogel Functionalized with Dopamine and Layered Double Hydroxide for Wound Healing. *Gels* **2024**, *10*, 318. [CrossRef] [PubMed]
248. Mohammadi Zardkhoshoui, A.; Arian, R.; Hosseiny Davarani, S.S. Tunable Construction of CuS Nanosheets@flower-like ZnCo-Layered Double Hydroxide Nanostructures for Hybrid Supercapacitors. *Ind. Chem. Mater.* **2023**, *1*, 443–457. [CrossRef]
249. Wang, L.; Li, J.; Meng, Q.; Xiao, M.; Liu, C.; Xing, W.; Zhu, J. Facilitating Active NiOOH Formation via Mo Doping towards High-Efficiency Oxygen Evolution. *Catal. Sci. Technol.* **2024**, *14*, 4166–4173. [CrossRef]
250. Lin, X.; Wang, Z.; Cao, S.; Hu, Y.; Liu, S.; Chen, X.; Chen, H.; Zhang, X.; Wei, S.; Xu, H.; et al. Bioinspired Trimesic Acid Anchored Electrocatalysts with Unique Static and Dynamic Compatibility for Enhanced Water Oxidation. *Nat. Commun.* **2023**, *14*, 6714. [CrossRef]
251. Das, J.P.; Nardekar, S.S.; Kesavan, D.; Bhunia, K.; Ravichandran, V.; Kim, S.-J. Unveiling the Effect of Growth Time on Bifunctional Layered Hydroxide Electrodes for High-Performance Energy Storage and Green Energy Conversion. *J. Mater. Chem. A Mater.* **2024**, *12*, 20179–20190. [CrossRef]
252. Chen, S.; Zheng, Z.; Li, Q.; Wan, H.; Chen, G.; Zhang, N.; Liu, X.; Ma, R. Boosting Electrocatalytic Water Oxidation of NiFe Layered Double Hydroxide via the Synergy of 3d–4f Electron Interaction and Citrate Intercalation. *J. Mater. Chem. A Mater.* **2023**, *11*, 1944–1953. [CrossRef]

253. Matsuda, K.; Okuda, A.; Kawashimo, M.; Fukuzaki, R.; Iio, N.; Tarutani, N.; Katagiri, K.; Inumaru, K. Molecular-Level Pictures of Chemical and Structural Transformations of Mg–Al Layered Double Hydroxide Crystals (Mg/Al = 2) at Elevated Temperatures. *J. Phys. Chem. C* **2023**, *127*, 12599–12605. [CrossRef]
254. Dillenburger, J.D.; Schulte, L.; Mahale, P.; Suleiman, M.; Mallouk, T.E. Anion-Dependent Structure, Dehydration, and Hydroxide Ion Conductivity of Magnesium Aluminum Layered Double Hydroxides. *Chem. Mater.* **2023**, *35*, 6437–6446. [CrossRef]
255. Fang, J.; Wen, D.; Zhang, Z.; Zheng, D.; Meng, A.; Su, Y.; Zhou, S. CoNiMn-Layered Double Hydroxide/$Ti_3C_2T_x$ Nanosheet Composites for Enhanced Pseudocapacitive Performance. *J. Phys. Conf. Ser.* **2023**, *2566*, 012004. [CrossRef]
256. Zou, L.; Sun, S.; Zhang, C.; Zhao, X. NiTi-Layered Double Hydroxide@Carbon Nanotube as a Cathode Material for Chloride-Ion Batteries. *Nanomaterials* **2023**, *13*, 2779. [CrossRef] [PubMed]
257. Bai, X.; Liu, Q.; Lu, Z.; Liu, J.; Chen, R.; Li, R.; Song, D.; Jing, X.; Liu, P.; Wang, J. Rational Design of Sandwiched Ni-Co Layered Double Hydroxides Hollow Nanocages/Graphene Derived from Metal-Organic Framework for Sustainable Energy Storage. *ACS Sustain. Chem. Eng.* **2017**, *5*, 9923–9934. [CrossRef]
258. Xu, J.; Tang, R.; Liu, M.; Xie, S.; Zhang, D.; Kong, X.; Jin, S.; Ji, H.; Zhang, T. Enhancing the Catalytic Activity of Layered Double Hydroxide Supported on Graphene for Lithium–Sulfur Redox Reactions. *Batteries* **2022**, *8*, 200. [CrossRef]
259. Hu, J.; Yuan, C.; Zhi, L.; Zhang, H.; Yuan, Z.; Li, X. In Situ Defect-Free Vertically Aligned Layered Double Hydroxide Composite Membrane for High Areal Capacity and Long-Cycle Zinc-Based Flow Battery. *Adv. Funct. Mater.* **2021**, *31*, 2102167. [CrossRef]
260. Zhang, D.; Cao, J.; Zhang, X.; Insin, N.; Wang, S.; Han, J.; Zhao, Y.; Qin, J.; Huang, Y. Inhibition of Manganese Dissolution in Mn_2O_3 Cathode with Controllable Ni^{2+} Incorporation for High-Performance Zinc Ion Battery. *Adv. Funct. Mater.* **2021**, *31*, 2009412. [CrossRef]
261. Yan, Z.; Wang, E.; Gao, J.; Yang, J.; Wu, C.; Jiang, L.; Zhu, M.; Sun, G. An Exceptionally Facile Synthesis of Highly Efficient Oxygen Evolution Electrodes for Zinc-Oxygen Batteries. *ChemElectroChem* **2017**, *4*, 2190–2195. [CrossRef]
262. Gao, Z.; Wang, J.; Li, Z.; Yang, W.; Wang, B.; Hou, M.; He, Y.; Liu, Q.; Mann, T.; Yang, P.; et al. Graphene Nanosheet/Ni^{2+}/Al^{3+} Layered Double-Hydroxide Composite as a Novel Electrode for a Supercapacitor. *Chem. Mater.* **2011**, *23*, 3509–3516. [CrossRef]
263. Zhang, D.; Cao, J.; Zhang, X.; Zeng, Z.; Insin, N.; Qin, J.; Huang, Y. Modification Strategies of Layered Double Hydroxides for Superior Supercapacitors. *Adv. Energy Sustain. Res.* **2022**, *3*, 2100183. [CrossRef]
264. Malak-Polaczyk, A.; Vix-Guterl, C.; Frackowiak, E. Carbon/Layered Double Hydroxide (LDH) Composites for Supercapacitor Application. *Energy Fuels* **2010**, *24*, 3346–3351. [CrossRef]
265. Zhang, C.; Zhu, Z.; Zhang, Y.; Shao, W.; Wu, D.; Peng, G.; Liu, Z. Introduction of Bimetallic Oxide-Modified Carbon Nanotubes for Boosting the Energy Storage Performance of NiCo-LDH Based in-Plane Micro-Supercapacitors on Paper. *Chem. Eng. J.* **2024**, *494*, 153242. [CrossRef]
266. Wang, J.; Ding, Q.; Bai, C.; Wang, F.; Sun, S.; Xu, Y.; Li, H. Synthesis of CNTs/CoNiFe-LDH Nanocomposite with High Specific Surface Area for Asymmetric Supercapacitor. *Nanomaterials* **2021**, *11*, 2155. [CrossRef] [PubMed]
267. Chen, H.; Hu, L.; Chen, M.; Yan, Y.; Wu, L. Nickel–Cobalt Layered Double Hydroxide Nanosheets for High-performance Supercapacitor Electrode Materials. *Adv. Funct. Mater.* **2014**, *24*, 934–942. [CrossRef]
268. Li, L.; Hui, K.S.; Hui, K.N.; Xia, Q.; Fu, J.; Cho, Y.-R. Facile Synthesis of NiAl Layered Double Hydroxide Nanoplates for High-Performance Asymmetric Supercapacitor. *J. Alloys Compd.* **2017**, *721*, 803–812. [CrossRef]
269. Shah, S.S. Biomass-Derived Carbon Materials for Advanced Metal-Ion Hybrid Supercapacitors: A Step Towards More Sustainable Energy. *Batteries* **2024**, *10*, 168. [CrossRef]
270. Jayababu, N.; Kim, D. CuCo LDHs Coated CuCoTe Honeycomb-Like Nanosheets as a Novel Anode Material for Hybrid Supercapacitors. *Small* **2021**, *17*, 2102369. [CrossRef]
271. Zhao, R.; Wang, M.; Zhao, D.; Li, H.; Wang, C.; Yin, L. Molecular-Level Heterostructures Assembled from Titanium Carbide MXene and Ni–Co–Al Layered Double-Hydroxide Nanosheets for All-Solid-State Flexible Asymmetric High-Energy Supercapacitors. *ACS Energy Lett.* **2018**, *3*, 132–140. [CrossRef]
272. Yang, J.; Yu, C.; Fan, X.; Qiu, J. 3D Architecture Materials Made of NiCoAl-LDH Nanoplates Coupled with NiCo-Carbonate Hydroxide Nanowires Grown on Flexible Graphite Paper for Asymmetric Supercapacitors. *Adv. Energy Mater.* **2014**, *4*, 1400761. [CrossRef]
273. Jia, H.; Wang, M.; Feng, M.; Li, G.; Li, L.; Liu, Y. Synergistic Enhancement of Supercapacitor Performance: Modish Designation of BPQD Modified NiCo-LDH/$NiCo_2S_4$ Hybrid Nanotube Arrays with Improved Conductivity and OH^- Adsorption. *Chem. Eng. J.* **2024**, *484*, 149591. [CrossRef]
274. Luo, C.-W.W.; Zhang, K.; Zeng, H.-Y.Y.; Yan, W.; Lv, S.-B.B.; Wu, G.-Z.Z. Engineering ZnNi-LDH with Improved Wettability by N-Direct-Doping for High-Performance Supercapacitor. *J. Alloys Compd.* **2025**, *1010*, 177406. [CrossRef]
275. Wei, X.; Li, M.; Li, L.; Feng, W.; Wu, H. Black Phosphorus Quantum Dots Embedded NiCoCu-LDH with Porous Structure for High-Performance Hybrid Supercapacitor. *Appl. Surf. Sci.* **2025**, *686*, 162203. [CrossRef]

276. Moradi, M.; Afkhami, A.; Madrakian, T.; Moazami, H.R.; Tirandaz, A. Partial Hydrothermal Sulfidation of Electrosynthesized Co-Mn Layered-Double-Hydroxide as an Active Material for Supercapacitor Applications. *J. Power Sources* **2025**, *629*, 235993. [CrossRef]
277. Wang, J.; Li, J.; Wang, F.; Zhang, X.; Wu, X.; Zhang, J.; Niu, S.; Chen, L. Introducing Ce into the Interface of NiCo LDH@PBAs to Build Multi Core–Shell Electrode with Superior Stability of Supercapacitor. *Chem. Eng. J.* **2024**, *488*, 150932. [CrossRef]
278. Hu, W.; Hu, B.; Wu, Z.; Zuo, M.; Song, Y.; Zheng, Q.; Shan, G.; Du, M. Exceeding 50 000 Cycle Durability of Layered Hydroxide-Based Hybrid Supercapacitor Through Scandium Doping-Induced Superlong Activation Process. *Small Struct.* **2024**, *5*, 2300574. [CrossRef]
279. Chen, R.X.; Cheng, L. Preparation of Flower-Like Layered Double Hydroxides Supercapacitor Cathode Based on Biomineralization Strategy. *Key Eng. Mater.* **2024**, *993*, 3–8. [CrossRef]
280. Bhandari, R.; Shah, R.R. Hydrogen as Energy Carrier: Techno-Economic Assessment of Decentralized Hydrogen Production in Germany. *Renew. Energy* **2021**, *177*, 915–931. [CrossRef]
281. Terlouw, T.; Bauer, C.; McKenna, R.; Mazzotti, M. Large-Scale Hydrogen Production via Water Electrolysis: A Techno-Economic and Environmental Assessment. *Energy Environ. Sci.* **2022**, *15*, 3583–3602. [CrossRef]
282. Haque, M.A.; Kawawaki, T.; Negishi, Y. Navigating Challenges and Possibilities for Improving Polymer Electrolyte Membrane Fuel Cells via Pt Electrocatalyst, Support and Ionomer Advancements. *Int. J. Hydrogen Energy* **2024**, *85*, 30–47. [CrossRef]
283. Zhang, B.; Ma, C.; Wang, J.; Zhang, X.; Tu, J. Layered Hydroxides as Electrocatalysts for Water Splitting. In *Metal Oxides and Related Solids for Electrocatalytic Water Splitting*; Elsevier: Amsterdam, The Netherlands, 2022; pp. 241–272, ISBN 9780323857352.
284. Ye, S.; Xu, Y.; Bai, X.; Liang, Z.; Liu, Q.; Wei, Q.; Yang, D.; Yang, W.; Gao, F.; Shi, Q. Efficient Bifunctional Water Splitting Catalysts Enabled by Crystalline-Amorphous Ni_xS_y@NiFe LDH Heterojunctions. *J. Mater. Chem. A Mater.* **2025**. [CrossRef]
285. Xia, X.; Wang, S.; Liu, D.; Wang, F.; Zhang, X.; Zhang, H.; Yu, X.; Pang, Z.; Li, G.; Chen, C.; et al. Electronic Modulation in Cu Doped NiCo LDH/NiCo Heterostructure for Highly Efficient Overall Water Splitting. *Small* **2024**, *20*, 2311182. [CrossRef]
286. Huo, J.M.; Ma, Z.L.; Wang, Y.; Cao, Y.J.; Jiang, Y.C.; Li, S.N.; Chen, Y.; Hu, M.C.; Zhai, Q.G. Monodispersed Pt Sites Supported on NiFe-LDH from Synchronous Anchoring and Reduction for High Efficiency Overall Water Splitting. *Small* **2023**, *19*, 2207044. [CrossRef] [PubMed]
287. Tucker-Quiñónez, A.M.; Rivadeneira-Mendoza, B.F.; Gorozabel-Mendoza, M.L.; Pérez-Almeida, I.B.; García-Guerrero, A.J.; Dueñas-Rivadeneira, A.A.; Yadav, K.K.; Zambrano-Intriago, L.A.; Rodríguez-Díaz, J.M. Challenges and Potential of Layered Double Hydroxides as Electrocatalytic Materials for Hydrogen Production from Water: A Review of Recent Advances and Applications. *Energy Nexus* **2025**, *17*, 100399. [CrossRef]
288. He, H.; Xiao, J.; Liu, Z.; Yang, B.; Wang, D.; Peng, X.; Zeng, L.; Li, Z.; Lei, L.; Qiu, M.; et al. Boosting the Hydrogen Evolution of Layered Double Hydroxide by Optimizing the Electronic Structure and Accelerating the Water Dissociation Kinetics. *Chem. Eng. J.* **2023**, *453*, 139751. [CrossRef]
289. Jia, Q.; Liu, E.; Jiao, L.; Li, J.; Mukerjee, S. Current Understandings of the Sluggish Kinetics of the Hydrogen Evolution and Oxidation Reactions in Base. *Curr. Opin. Electrochem.* **2018**, *12*, 209–217. [CrossRef]
290. Wang, S.; Lu, A.; Zhong, C.J. Hydrogen Production from Water Electrolysis: Role of Catalysts. *Nano Converg.* **2021**, *8*, 4. [CrossRef]
291. de Chialvo, M.R.G.; Chialvo, A.C. Hydrogen Evolution Reaction: Analysis of the Volmer-Heyrovsky-Tafel Mechanism with a Generalized Adsorption Model. *J. Electroanal. Chem.* **1994**, *372*, 209–223. [CrossRef]
292. You, H.; Wu, D.; Si, D.; Cao, M.; Sun, F.; Zhang, H.; Wang, H.M.; Liu, T.F.; Cao, R. Monolayer NiIr-Layered Double Hydroxide as a Long-Lived Efficient Oxygen Evolution Catalyst for Seawater Splitting. *J. Am. Chem. Soc.* **2022**, *144*, 9254–9263. [CrossRef]
293. Jadhav, H.S.; Roy, A.; Desalegan, B.Z.; Seo, J.G. An Advanced and Highly Efficient Ce Assisted NiFe-LDH Electrocatalyst for Overall Water Splitting. *Sustain. Energy Fuels* **2019**, *4*, 312–323. [CrossRef]
294. Shivakumar, P.; Monika, M.N.; Deepu, M.; Manjunatha Kumara, K.S.; Budagumpi, S.; Nagaraju, D.H. Amorphous and Crystalline Heterostructure $MoSe_2$/NiFe-LDH through p-n Junction Formation for Electrochemical Water Splitting-Synergistic Back Bonding Effect. *Int. J. Hydrogen Energy* **2024**, *65*, 74–82. [CrossRef]
295. Liu, G.; Huang, C.; Yang, Z.; Su, J.; Zhang, W. Ultrathin NiMn-LDH Nanosheet Structured Electrocatalyst for Enhanced Electrocatalytic Urea Oxidation. *Appl. Catal. A Gen.* **2021**, *614*, 118049. [CrossRef]
296. Cao, H.; Liu, B.; Bai, J.; Li, C.; Xu, G. Interfacial Engineering of Hierarchical Ultra-Thin NiCo-LDH Nanosheet Superstructures Nanofiber for Water Cracking Electrocatalysis. *J. Alloys Compd.* **2025**, *1010*, 178041. [CrossRef]
297. Yan, F.; Guo, D.; Kang, J.; Liu, L.; Zhu, C.; Gao, P.; Zhang, X.; Chen, Y. Fast Fabrication of Ultrathin CoMn LDH Nanoarray as Flexible Electrode for Water Oxidation. *Electrochim. Acta* **2018**, *283*, 755–763. [CrossRef]
298. Bodhankar, P.M.; Sarawade, P.B.; Singh, G.; Vinu, A.; Dhawale, D.S. Recent Advances in Highly Active Nanostructured NiFe LDH Catalyst for Electrochemical Water Splitting. *J. Mater. Chem. A Mater.* **2021**, *9*, 3180–3208. [CrossRef]
299. Fu, X.; Liao, H.; Zhang, Z.; Zheng, Y.; Lu, J.; Cheng, S.; Jiang, Y.; Gao, Y. Medium-Entropy Heterostructure of Crystalline NiCoFeP @ Amorphous NiCoFe-LDH for Industrial-Current Density and Ultrastable Overall Water Splitting. *Chem. Eng. J.* **2025**, *505*, 159520. [CrossRef]

300. Hong, C.; Ji, J.; Huang, J.; Zhang, Y.; Li, L. NiMo/NiFe-LDH Heterostructured Electrocatalyst for Hydrogen Production from Water Electrolysis. *Mater. Lett.* **2025**, *379*, 137664. [CrossRef]
301. Hou, Y.; Lu, Q.; Guo, Z.; Zhang, Y.; Zhang, Y.; Wan, X.; Lei, L. One-Step Electrodeposition Synthesis of Composites of NiFeMn Alloy and Its Hydroxide as an Efficient Bifunctional Water Splitting Catalyst. *J. Solid. State Chem.* **2025**, *348*, 125356. [CrossRef]
302. Ye, Z.; Wang, P.; Zhong, W.; Zheng, X.; Cai, W. Active-Site-Enriched Cu-Doped Ni-Fe Layered Double Hydroxide Nanosheets for Boosting the Oxygen Evolution Reaction. *New J. Chem.* **2023**, *47*, 9536–9539. [CrossRef]
303. Guo, J.; Wang, K.; Zhang, H.; Zhang, H. Enhanced Electrocatalytic Activity of Mo-Doped NiFe Layered Double Hydroxide Nanosheet Arrays for the Hydrogen Evolution Reaction. *ACS Appl. Nano Mater.* **2023**, *6*, 379–389. [CrossRef]
304. Ding, L.; Li, K.; Xie, Z.; Yang, G.; Yu, S.; Wang, W.; Cullen, D.A.; Yu, H.; Zhang, F.Y. W-Induced Morphological Modification of NiFe Layered Double Hydroxides as Efficient Electrocatalysts for Overall Water Splitting. *Electrochim. Acta* **2021**, *395*, 139199. [CrossRef]
305. Wang, T.; Zhang, X.; Yu, X.; Li, J.; Wang, K.; Niu, J. Interfacial Interaction in NiFe LDH/NiS_2/VS_2 for Enhanced Electrocatalytic Water Splitting. *Molecules* **2024**, *29*, 951. [CrossRef] [PubMed]
306. Udayakumar, A.; Dhandapani, P.; Ramasamy, S.; Angaiah, S. Layered Double Hydroxide (LDH)—MXene Nanocomposite for Electrocatalytic Water Splitting: Current Status and Perspective. *ES Energy Environ.* **2023**, *20*, 902. [CrossRef]
307. Jia, Y.; Zhang, L.; Gao, G.; Chen, H.; Wang, B.; Zhou, J.; Soo, M.T.; Hong, M.; Yan, X.; Qian, G.; et al. A Heterostructure Coupling of Exfoliated Ni–Fe Hydroxide Nanosheet and Defective Graphene as a Bifunctional Electrocatalyst for Overall Water Splitting. *Adv. Mater.* **2017**, *29*, 1700017. [CrossRef]
308. Yao, L.; Li, R.; Zhang, H.; Humayun, M.; Xu, X.; Fu, Y.; Nikiforov, A.; Wang, C. Interface Engineering of NiTe@CoFe LDH for Highly Efficient Overall Water-Splitting. *Int. J. Hydrogen Energy* **2022**, *47*, 32394–32404. [CrossRef]
309. Wang, J.; Lv, G.; Wang, C. A Highly Efficient and Robust Hybrid Structure of CoNiN@NiFe LDH for Overall Water Splitting by Accelerating Hydrogen Evolution Kinetics on NiFe LDH. *Appl. Surf. Sci.* **2021**, *570*, 151182. [CrossRef]
310. Kalusulingam, R.; Mariyaselvakumar, M.; Antonyraj, C.A.; Mathi, S.; Mikhailova, T.S.; Khubezhov, S.A.; Pankov, I.V.; Srinivasan, K.; Myasoedova, T.N. Highly Efficient Water Splitting with Pd-Integrated NiAl-LDH Nanosheets as Bifunctional Electrocatalysts. *Energy Fuels* **2023**, *37*, 13319–13330. [CrossRef]
311. Zhai, P.; Zhang, Y.; Wu, Y.; Gao, J.; Zhang, B.; Cao, S.; Zhang, Y.; Li, Z.; Sun, L.; Hou, J. Engineering Active Sites on Hierarchical Transition Bimetal Oxides/Sulfides Heterostructure Array Enabling Robust Overall Water Splitting. *Nat. Commun.* **2020**, *11*, 5462. [CrossRef]
312. Das, M.; Khan, Z.B.; Biswas, A.; Dey, R.S. Inter-Electronic Interaction between Ni and Mo in Electrodeposited Ni-Mo-P on 3D Copper Foam Enables Hydrogen Evolution Reaction at Low Overpotential. *Inorg. Chem.* **2022**, *61*, 18253–18259. [CrossRef]
313. Govind Rajan, A.; Carter, E.A. Microkinetic Model for pH- And Potential-Dependent Oxygen Evolution during Water Splitting on Fe-Doped β-NiOOH. *Energy Environ. Sci.* **2020**, *13*, 4962–4976. [CrossRef]
314. Wang, Y.; Zhang, M.; Liu, Y.; Zheng, Z.; Liu, B.; Chen, M.; Guan, G.; Yan, K. Recent Advances on Transition-Metal-Based Layered Double Hydroxides Nanosheets for Electrocatalytic Energy Conversion. *Adv. Sci.* **2023**, *10*, 2207519. [CrossRef] [PubMed]
315. Ge, Z.Q.; Li, J.; Zhang, H.J.; Liu, C.; Che, G.; Liu, Z.Q. p–d Orbitals Coupling Heterosites of Ni_2P/NiFe-LDH Interface Enable O–H Cleavage for Water Splitting. *Adv. Funct. Mater.* **2024**, *34*, 2411024. [CrossRef]
316. Zheng, K.; Ren, J.; Li, X.; Li, G.; Jiao, L.; Xu, C. Engineering Crystalline CoMP-Decorated (M = Mn, Fe, Ni, Cu, Zn) Amorphous CoM LDH for High-Rate Alkaline Water Splitting. *Chem. Eng. J.* **2022**, *441*, 136031. [CrossRef]
317. Feng, Y.; Smith, R.L.; Shen, F.; Qi, X. Structure-Oriented Electrochemical Synthesis of Layered Double Hydroxide Electrocatalytic Materials for 5-Hydroxymethylfurfural Oxidation. *ACS Sustain. Chem. Eng.* **2024**, *12*, 16905–16913. [CrossRef]
318. Yi, W.; Yu, R.; Jiang, H.; Wu, J.; Cheng, G.J. Unveiling the Synergistic Potential of Laser Chemical Solid-Phase Deposition of Atomic Platinum-Metal Layer on 2D Materials for Bifunctional Catalysts. *Adv. Funct. Mater.* **2024**, *34*, 2308575. [CrossRef]
319. Xiong, T.; Zhu, Z.; He, Y.; Balogun, M.-S.; Huang, Y. Phase Evolution on the Hydrogen Adsorption Kinetics of NiFe-Based Heterogeneous Catalysts for Efficient Water Electrolysis. *Small Methods* **2023**, *7*, 2201472. [CrossRef]
320. Zhai, Z.; Li, H.; Zhou, C.; Zheng, H.; Liu, Y.; Yan, W.; Zhang, J. Anisotropic Strain Boosted Hydrogen Evolution Reaction Activity of F-NiCoMo LDH for Overall Water Splitting. *J. Electrochem. Soc.* **2023**, *170*, 036509. [CrossRef]
321. Rosely, C.V.S.; John, H. Role of Surfactants on Electrocatalytic Activity of Co/Al Layered Double Hydroxides for Hydrogen and Oxygen Generation. *ChemCatChem* **2025**, *17*, 2401377. [CrossRef]
322. Wang, J.; Wei, X.; Song, W.; Shi, X.; Wang, X.; Zhong, W.; Wang, M.; Ju, J.; Tang, Y. Plasmonic Enhancement in Water Splitting Performance for NiFe Layered Double Hydroxide-N_{10}TC MXene Heterojunction. *ChemSusChem* **2021**, *14*, 1948–1954. [CrossRef]
323. Xi, C.; Ding, W.; Jiang, J.; Wang, Y.; Wang, S.; Zhang, Z.; Han, S. Unveiling the CoFe LDH Generated by Partial in Situ Conversion Strategy: Synergistic Effect of Modified MXene and Cationic Vacancies for Overall Water Splitting. *Fuel* **2024**, *370*, 131844. [CrossRef]
324. Shen, B.; Huang, H.; Jiang, Y.; Xue, Y.; He, H. 3D Interweaving MXene–Graphene Network–Confined Ni–Fe Layered Double Hydroxide Nanosheets for Enhanced Hydrogen Evolution. *Electrochim. Acta* **2022**, *407*, 139913. [CrossRef]

325. Luo, H.; Liu, C.; Chen, N.; Xue, K.; Yu, L.; Zhu, H.; Zhang, Y. Metal-Organic Frameworks Derived NiFe-Hydroxides for Efficient Electrocatalytic Hydrogen Evolution. *Adv. Comput. Mater. Sci. Res.* **2024**, *1*, 192. [CrossRef]
326. Lee, E.; Jeong, S.; Jeong, Y.; Kim, B.; Lee, K. Nanoscale-Confined Synthesis of 2D Metal Compounds for Electrochemical Applications. *Small Methods* **2025**, *9*, 2301782. [CrossRef] [PubMed]
327. Liu, J.; Zheng, K.; Hua, Y.; Peng, L.; Meng, X.; Zhang, C.; Su, R.; Tao, X.; Hu, H.; Guo, Y.; et al. LDHs and Their Derivatives for Electrochemical Energy Storage and Conversion Systems: Design, Insights and Applications. *ChemElectroChem* **2024**, *11*, e202400156. [CrossRef]
328. Yao, Y.; Zou, C.; Sun, S.; Guo, Y.; Hong, S.; Cai, Z.; Yang, C.; Zhuang, W.; Luo, F.; Hamdy, M.S.; et al. Ultrastable Seawater Oxidation at Ampere-level Current Densities with Corrosion-resistant $CoCO_3$/CoFe Layered Double Hydroxide Electrocatalyst. *Small* **2025**, *21*, 2409627. [CrossRef]
329. Ding, L.; Shen, Z.; Pan, H.; Chen, Y.; Hu, Y.; Zhao, G.; Hai, G.; Huang, X. Regulating Intermediate Adsorption and Promoting Charge Transfer of CoCr-LDHs by Ce Doping for Enhancing Electrooxidation of 5-Hydroxymethylfurfural. *Small* **2025**, *21*, 2409343. [CrossRef]
330. Chen, Y.; Zhang, Y.; Bai, X.; Zhao, J.; Yang, L.; Wang, X.; Wu, Q.; Hu, Z. Interlayer Engineering of Layered Double Hydroxides for Advanced Energy Storage and Conversion. *FlatChem* **2024**, *48*, 100775. [CrossRef]
331. Wang, Y.; Ying, Z.; Gao, Y.; Shi, L. Layered Double Hydroxide Nanosheets: Synthesis Strategies and Applications in the Field of Energy Conversion. *Chem. Eur. J.* **2024**, *30*, e202303025. [CrossRef]
332. Zhou, X.; Hou, Z.; Zhang, H.-Y.; Yu, J. Synergistic Coupling of NiVAl-Layered Double Hydroxide with Few-Layered $Ti_3C_2T_x$–MXene Nanosheets for Superior Asymmetric Supercapacitor Performance. *J. Mater. Chem. C Mater.* **2023**, *11*, 15571–15580. [CrossRef]
333. Jaramillo Hernández, C.; Oestreicher, V.; Mizrahi, M.; Abellán, G. An In-Situ Study of the Transformation of Hybrid Layered Hydroxides into Metallic Nanocomposites with Interest in Energy Storage Systems. In Proceedings of the ECS Meeting Abstracts 2023, Boston, MA, USA, 28 May–2 June 2023; Volume MA2023-01, p. 1322. [CrossRef]
334. Sun, X.; Chen, H.; Li, Y.; Zeng, D.; Qiu, P.; Zeng, H.; Ji, X.; Chen, L.; Shi, X. Efficient Harvesting Waste Heat by Zn-Ion Battery Under Thermally Regenerative Electrochemical Cycles. *Adv. Mater.* **2025**, *37*, 2418482. [CrossRef]
335. Abu Nayem, S.M.; Hardianto, Y.P.; Shuaibu, A.D.; Shah, S.S.; Islam, S.; Jafar Mazumder, M.A.; Aziz, M.A.; Saleh Ahammad, A.J. Biomass Derived Amino Acid Assisted Synthesis of FeNi Layered Double Hydroxide for Efficient Oxygen Evolution Reaction. *Inorg. Chem. Commun.* **2025**, *171*, 113574. [CrossRef]

MDPI AG
Grosspeteranlage 5
4052 Basel
Switzerland
Tel.: +41 61 683 77 34

Batteries Editorial Office
E-mail: batteries@mdpi.com
www.mdpi.com/journal/batteries

www.ingramcontent.com/pod-product-compliance
Lightning Source LLC
Chambersburg PA
CBHW061845210926
51850CB00006B/977
9783725861958

Soil Mechanical Systems and Related Farming Machinery

Editors

Ju-Seok Nam
Yongjin Cho

MDPI

Basel • Beijing • Wuhan • Barcelona • Belgrade • Novi Sad • Cluj • Manchester

Editors
Ju-Seok Nam
Kangwon National
University
Chuncheon
Republic of Korea

Yongjin Cho
Jeonbuk National University
Jeonju
Republic of Korea

Editorial Office
MDPI AG
Grosspeteranlage 5
4052 Basel, Switzerland

This is a reprint of articles from the Special Issue published online in the open access journal *Agriculture* (ISSN 2077-0472) (available at: https://www.mdpi.com/journal/agriculture/special_issues/HJ176P7057).

For citation purposes, cite each article independently as indicated on the article page online and as indicated below:

Lastname, A.A.; Lastname, B.B. Article Title. *Journal Name* **Year**, *Volume Number*, Page Range.

ISBN 978-3-7258-2349-9 (Hbk)
ISBN 978-3-7258-2350-5 (PDF)
doi.org/10.3390/books978-3-7258-2350-5

Contents

About the Editors

Ju-Seok Nam

Ju-Seok Nam received his B.S., M.S., and Ph.D. in Biosystems Engineering from Seoul National University, the Republic of Korea, in 2007, 2009 and 2014, respectively. He has been a Professor in the Department of Biosystems Engineering at Kangwon National University and a Principal Investigator of the Agricultural Field Machine Laboratory since 2016. His interests relate to the areas of design and reliability tests for agricultural machinery, simulation studies to prevent the overturning and rollover of agricultural machinery, agricultural ergonomics to reduce the work stress of farmers, and cultivating models to increase the mechanization rate of agricultural work. He won the Best Professor Award from Kangwon National University in 2021 and has published 24 papers in international journals in the past three years.

Yongjin Cho

Yongjin Cho received his B.S., M.S., and Ph.D. in Bio-Mechatronics Engineering from Sungkyunkwan University, the Republic of Korea, in 2006, 2008, and 2013, respectively. Since 2018, he has been a Professor at the Department of Bioindustrial Machinery Engineering at Jeonbuk National University. His research focuses on precision agriculture using smart agricultural machinery, including agriculture sensors, automation package technology for farming, and the development of a 110kW tractor for large agricultural applications. He also works on developing models to increase the mechanization rate of various agricultural tasks. In 2018, he received the Superior Paper Award from the ASABE (the American Society of Agricultural and Biological Engineers), and since 2020, and he has published 15 papers in international journals.

Preface

Farming machinery has a long history of development. However, the rate of farming mechanization is still low in many countries, and technological advances are needed to develop optimal machinery for specific agricultural work. This requires an in-depth understanding of both soil and mechanical systems as most farming machinery work on soil. This reprint aims to expand knowledge about soil and mechanical systems by introducing the latest related research.

The studies presented in this reprint include research on advanced machinery technology, soil-machine interactions, and soil characteristics. It contains original and advanced research with novel analytical and experimental methods. Most farming machine applications such as agricultural tractors for multi-purposed works, rotavators for soil preparation, planters and transplanters for planting crops, pot-seeding machines for seeding in pot trays, and machine vision systems for crop phenotyping information were covered in this reprint. It also includes research on soil machine interactions pertaining to traction force, load factor, tillage workload, and soil compaction as well as soil characteristics regarding fertilization effect and soil particle size distribution.

We believe that the research published in this reprint will be helpful to related researchers and contribute to expanding knowledge in this field. We appreciate all of the authors and reviewers for publishing their significant research results and providing excellent suggestions to improve the quality of this reprint.

Ju-Seok Nam and Yongjin Cho
Editors

agriculture

Editorial

Soil Mechanical Systems and Related Farming Machinery

Yongjin Cho [1] and Ju-Seok Nam [2,*]

1 Department of Bioindustrial Machinery Engineering, Jeonbuk National University, Jeonju 54896, Republic of Korea; choyj@jbnu.ac.kr
2 Department of Biosystems Engineering, Kangwon National University, 1 Kangwondaehak-gil, Chuncheon 24341, Republic of Korea
* Correspondence: njsg1218@kangwon.ac.kr

The mechanization of agricultural work has contributed significantly to the improvement of agricultural productivity and reduced production costs. Recently, its importance has been highlighted owing to the shortage of agricultural labor, especially in developing countries where urbanization is progressing rapidly [1,2]. Since the beginning of mechanization, various types of farming machinery related to soil preparation, sowing, fertilization, pest control, harvesting, and post-harvesting have been developed. In addition, customized farming machines suitable for the cultivation type and soil characteristics of each country and region have been developed [3]. Unlike other industrial machinery, farming machinery targets living organisms and operates on the soil; therefore, it should be designed considering its interaction with the soil. Thus, it is essential to understand the characteristics of both the soil and mechanical systems to optimize the design of farming machinery. This Special Issue covers the design, analysis, and testing of all types of farming machinery, with an in-depth understanding of soil and mechanical systems.

This Special Issue contains 14 papers, with 12 research articles, one communication article, and one technical note. The authors are from five countries: Indonesia, Iran, Poland, South Korea, and Ukraine. In terms of machines, agricultural tractors, rotavators, garlic planters, vegetable transplanters, pot-seeding machines, and machine vision systems were included. Design, measurement, simulation, and theoretical analyses were conducted to develop advanced farming machinery techniques and to extend the understanding of the interaction between soil and mechanical systems in farming machine applications. The papers in this Special Issue can be grouped into three categories: advanced farming machinery techniques, soil–machine interaction research, and soil property research.

The first category includes advanced farming machine techniques. This includes the development of novel or improved mechanical systems, advanced measurement and simulation methods, and the investigation of the characteristics of farming machines using theoretical or experimental methods. Eight papers were included in this category, half of which were related to agricultural tractors.

Kim et al. presented a modified method to improve the accuracy of a three-point hitch dynamometer, also known as a six-component load cell, for measuring the precise traction force of agricultural tractors [4]. They suggested modified force and moment equations that could consider the geometry and tilt of the dynamometer under real installation conditions. The accuracy of the modified method was verified through an actual field test. This study is important because accurate knowledge of the operating force is crucial for the application of smart farming technology that requires the precise control of attached implements, as well as for the optimal design and efficient use of agricultural tractors.

Unlike Kim et al. [4], who focused on measuring the traction force, Lebedev et al. analyzed the effect of a coupling weight to increase the traction efficiency of agricultural tractors by reducing the uneven distribution of vertical reactions between the wheels [5]. They employed theoretical and comparative analyses by incorporating existing scientific findings related to agricultural tractors operating in traction mode. This study revealed

Citation: Cho, Y.; Nam, J.-S. Soil Mechanical Systems and Related Farming Machinery. *Agriculture* **2024**, *14*, 1661. https://doi.org/10.3390/agriculture14091661

Received: 23 July 2024
Accepted: 26 July 2024
Published: 23 September 2024

that agricultural tractors with a center of mass offset to the front or rear axles had the highest probability of equal distribution of vertical reactions between the wheels of one axle, and agricultural tractors with a center of mass in the middle between the axles had the lowest probability. In addition, this study found that agricultural tractors can operate with maximum traction efficiency when the front and rear axles are locked during plowing work, with an uneven load distribution on the sides. Although further verification is required under conditions with various installed implements, this study provides good insights into the effect of an agricultural tractor's coupling weight and center of mass on traction performance.

Lim et al. developed a finite element (FE)-based simulation model of an agricultural tractor's frame-type rollover protective structure (ROPS) to replace a real ROPS test with a virtual simulation [6]. This study presents methods for developing 3D models of ROPS and setting the load and boundary conditions to accurately simulate the force and deformation applied to the ROPS. The accuracy of the proposed method was verified using actual test results. Virtual simulation technology for the ROPS of agricultural tractors can make significant savings in time and cost, and is also very important in terms of safety. Many developed countries are conducting research to develop this technology, and this study provides important technical information.

Ahn et al. investigated power loss in the dual-clutch transmission (DCT) of agricultural tractors using theoretical and experimental approaches considering the oil circulation effect [7]. Power losses in all of the components of the DCT, such as gears, shafts, bearings, clutches, and brakes, were calculated using the ISO standard and viscous fluid theory, and the theoretical analysis result was verified through an experiment using a three-axis dynamometer system. This study revealed that the oil level of the transmission system, oil circulation conditions, and operating characteristics of the transmission components, such as rotation speed, are important factors for power loss and transmission efficiency. This study has important implications as the interest in DCT and energy conservation is increasing worldwide.

Cho and Yang developed a real-time, low-cost, and high-throughput plant phenotyping system for commercial plant factories that can determine crop growth information, such as fresh weight, leaf length, leaf width, and number of leaves [8]. The developed system is composed of a low-cost phenotype sensor network with an integrated Raspberry Pi board and a camera module that allows easy modification depending on the crop growth environment. The applicability of the system is validated through experiments on Batavia lettuce cultivated in a plant factory. The cost-effective and highly productive characteristics of the developed system can be distinguished from those of existing phenotyping systems. The ripple effect of this research is significant, as it can be applied to all types of crop cultivation facilities, including smart farms.

Lim et al. studied garlic planters [9]. This study presented a simulation method using the discrete element method (DEM) to optimize the bucket size of the finger-type metering device. For highly accurate simulation results, techniques for creating 3D models of garlic cloves and buckets, deriving the physical and mechanical properties of garlic cloves, and setting the contact conditions between the garlic cloves and bucket were presented. The simulation method presented in this study was validated through experiments using an actual garlic clove and bucket. As a result of this study, the authors were able to obtain the optimal bucket size with a plant rate of 97.5%. Recently, DEM have been widely used in simulations of the interaction effects between farming machinery and soil or agricultural products, and this study provides important information for correctly applying DEM.

In countries such as Korea and Japan, where the per capita cultivation area is not large, vegetable transplantation is often performed using semi-automatic vegetable transplanters. Markumningsih et al. investigated the power consumption, static safety factor, and fatigue life characteristics of cam- and four-bar link-type transplanters during operation [10]. Each characteristic was measured and analyzed through extensive field tests using numerous sensors attached to the transplanter components. This study expands the understanding of the operating characteristics of semi-automatic vegetable transplanters, as well as the measurement and analysis techniques for the consumed power and static and dynamic safety of farming machines.

A mechanical pot-seeding machine was used to sow seeds into pot trays for transplanted crops. Plastic pot trays are currently being used, but are known to cause environmental problems. Hwang et al. developed an auto-feeding device for a mechanical pot-seeding machine applicable to paper-based biodegradable pot trays [11]. An auto-feeding device is used to supply pot trays individually in the first part of the mechanical pot-seeding machine, and a design appropriate for the physical and mechanical properties of the pot tray is essential. The authors designed and manufactured an auto-feeding device using mechanical and kinematic theories after measuring the properties of biodegradable pot trays. The performance of the developed device was evaluated through an actual test. The results of this study can be applied to technology for future environmental conservation.

The second category is soil–machine interaction research. It deals with the interaction effect between soil and mechanical systems, including the performance of machines that vary depending on the soil properties or the properties of soil that vary depending on the machine characteristics. Four studies were included in this category, all of which were related to agricultural tractors.

Bae et al. developed a traction force prediction model for agricultural tractors as part of a simulation study using DEM [12]. They presented a technique for modeling virtual soil environments in a simulation based on measured soil properties to obtain accurate simulation results. The traction force of an agricultural tractor during moldboard plow operations was simulated by considering the interaction between the soil and the plow blade. The simulation results were validated through an actual field test, which revealed higher accuracy than the existing empirical equation. Although the authors noted that further research is required to improve the prediction accuracy, this study provides good insight for simulating the interaction between soil and mechanical systems using DEM. This study can also be used as a reference for future digital twin modeling techniques for agricultural machinery.

Emission regulations for agricultural tractors are becoming stricter worldwide as concerns about environmental pollution increase. The load factor (LF) of agricultural tractors, which is the ratio of the actual engine power to the rated engine power, is an important indicator of air pollutant emissions. To obtain a more realistic LF in agricultural tractors, Min et al. investigated the effects of soil physical properties on the LF through extensive experiments under various soil conditions [13]. The results confirmed that soil variables had a significant influence on the LF, although the impact varied depending on the soil type. The authors noted that further research is needed to collect LF data under a wider range of tractor operating and soil conditions to improve reliability. This study has important implications related to environmental regulations of agricultural tractors.

Tillage is an essential agricultural task for the preparation of proper soil environments for crop growth. It is usually conducted by agricultural tractors with implements such as plows and rotavators, and the interaction between the soil and implementation has a decisive impact on tillage performance. Kim et al. analyzed the effects of the tillage type and gear selection on the workload of agricultural tractor–implement systems during rotary tillage [14]. The soil properties and mechanical operating characteristics were simultaneously considered to derive conclusions through extensive field experiments. The authors concluded that both the soil and mechanical conditions significantly influenced the tillage workload. The quantitative measurement results of this study can be used as reference data for designing agricultural tractors and implements as well as for understanding the workload characteristics during tillage operations.

Soil compaction caused by the passage of agricultural tractors has long been a subject of interest because of its significant effect on crop growth. Shahgholi et al. investigated the effects of the tire parameters of agricultural tractors on soil compaction through experiments and mathematical analysis [15]. The authors revealed that the tire size, tire inflation pressure, and number of passages are significant influencing factors of soil compaction, and their effects vary depending on the tire type and soil depth. A narrow tire is more effective for soil compaction because its soil bulk density changes significantly more than that of a normal tire under the same external offset condition. This study provides insight into the interaction between soil compaction and the mechanical or operating characteristics of agricultural tractors.

The final subject area was soil property research. It focuses on the quantification of soil properties for agricultural applications, with two papers in this category.

In precision agriculture, which minimizes the amount of input resources and maximizes production, spot-specific fertilization or variable control based on soil and crop characteristics is essential. Shin et al. developed a map of expected fertilization rates for nitrogen (N) and phosphorus (P_2O_5) based on quantified soil properties in salt-affected paddy fields [16]. The predicted N fertilization rate for the entire field was within 10.0 g to 25.7 g for each lot, whereas that of P_2O_5 was in the range of 0.68 g to 8.46 g. This large deviation indicates significant savings in resources compared to fertilizing the same amount overall. The technique used in this study can be applied to any type of fertilizer for precision agriculture.

Soil particle size distribution has a significant impact on the mechanical characteristics of the soil and its interaction with farming machines. Standard sieves were used to determine the size distribution of relatively large particles. However, for very small particles (less than approximately 5 mm in diameter), a hydrometer test is needed to determine the size distribution, which requires a considerable amount of labor and time. Kim et al. proposed a new method to determine the size distribution of small particles using a machine vision system with a red–green–blue (RGB) camera to overcome the shortcomings of hydrometer tests [17]. The experimental results demonstrate the reliability and accuracy of the proposed system, with only a 2.3% deviation from the actual particle size distribution. This study presents a novel and promising technique for assessing soil particle size distribution, which can be an alternative to traditional methods.

The papers in this Special Issue of Soil Mechanical Systems and Related Farming Machinery contain excellent research results that reflect the latest trends in the field. We are confident that this Special Issue will inspire researchers in the field and serve as excellent research reference data. We sincerely thank all of the authors who submitted their papers to this Special Issue and the reviewers of these papers for their constructive comments and thoughtful suggestions.

Author Contributions: Conceptualization, Y.C. and J.-S.N.; investigation, Y.C.; writing—original draft preparation, Y.C.; writing—review and editing, J.-S.N.; and visualization, Y.C. and J.-S.N. All authors have read and agreed to the published version of the manuscript.

Funding: This work was supported by Korea Institute of Planning and Evaluation for Technology in Food, Agriculture, and Forestry (IPET) through Machinery Mechanization Technology Development Program for Field Farming Program funded by Ministry of Agriculture, Food, and Rural Affairs (MAFRA) (RS-2023-00236201).

Conflicts of Interest: The authors declare no conflicts of interest.

References

1. Kim, S.J.; Lee, H.S.; Hwang, S.J.; Kim, J.H.; Jang, M.K.; Nam, J.S. Development of Seeding Rate Monitoring System Applicable to a Mechanical Pot-Seeding Machine. *Agriculture* **2023**, *13*, 2000. [CrossRef]
2. Choudhary, V.; Machavaram, R. Need of Automation in Paddy Nurseries for Raising Paddy Seedlings in India: A Review. *J. Biosyst. Eng.* **2022**, *47*, 209–222. [CrossRef]
3. Bagherpour, H.; Talab, M.M.S. Design, Construction, and Evaluation of a Precision Vegetable Reaper to Use in Small Plots. *J. Biosyst. Eng.* **2024**, *49*, 41–51. [CrossRef]
4. Kim, H.G.; Lee, J.W.; Kim, S.C.; Oh, J.; Shim, S.B. Development of a Modified Method for Measuring the Actual Draft Force Using a Tractor-Attached Dynamometer. *Agriculture* **2024**, *14*, 544. [CrossRef]
5. Lebedev, A.; Shuliak, M.; Khalin, S.; Lebedev, S.; Szwedziak, K.; Lejman, K.; Niedbała, G.; Łusiak, T. Methodology for Assessing Tractor Traction Properties with Instability of Coupling Weight. *Agriculture* **2023**, *13*, 977. [CrossRef]

6. Lim, R.G.; Kim, W.S.; Do, Y.W.; Siddique, M.A.A.; Kim, Y.J. Performance Evaluation of a Virtual Test Model of the Frame-Type ROPS for Agricultural Tractors Using FEA. *Agriculture* **2023**, *13*, 2004. [CrossRef]
7. Ahn, H.J.; Park, Y.J.; Kim, S.C.; Choi, C. Theoretical Calculations and Experimental Studies of Power Loss in Dual-Clutch Transmission of Agricultural Tractors. *Agriculture* **2023**, *13*, 1225. [CrossRef]
8. Cho, W.J.; Yang, M. High-Throughput Plant Phenotyping System Using a Low-Cost Camera Network for Plant Factory. *Agriculture* **2023**, *13*, 1874. [CrossRef]
9. Im, D.; Lee, H.S.; Kim, J.H.; Moon, D.J.; Moon, T.I.; Yu, S.H.; Park, Y.J. Bucket Size Optimization for Metering Device in Garlic Planter Using Discrete Element Method. *Agriculture* **2023**, *13*, 1199. [CrossRef]
10. Markumningsih, S.; Hwang, S.J.; Kim, J.H.; Jang, M.K.; Shin, C.S.; Nam, J.S. Comparison of Consumed Power and Safety of Two Types of Semi-Automatic Vegetable Transplanter: Cam and Four-Bar Link. *Agriculture* **2023**, *13*, 588. [CrossRef]
11. Hwang, S.J.; Jang, M.K.; Nam, J.S. Development of Autofeeding Device Applicable to a Biodegradable Pot Tray. *Agriculture* **2022**, *12*, 2097. [CrossRef]
12. Bae, B.M.; Kim, Y.S.; Kim, W.S.; Kim, Y.J.; Lee, S.D.; Kim, T.J. The Development of a Draft Force Prediction Model for Agricultural Tractors Based on the Discrete Element Method in Loam and Clay Loam. *Agriculture* **2023**, *13*, 2205. [CrossRef]
13. Min, Y.S.; Kim, Y.S.; Lim, R.G.; Kim, T.J.; Kim, Y.J.; Kim, W.S. The Influence of Soil Physical Properties on the Load Factor for Agricultural Tractors in Different Paddy Fields. *Agriculture* **2023**, *13*, 2073. [CrossRef]
14. Kim, Y.S.; Bae, B.M.; Kim, W.S.; Kim, Y.J.; Lee, S.D.; Kim, T.J. Working Load Analysis of a 42 kW Class Agricultural Tractor According to Tillage Type and Gear Selection during Rotary Tillage Operation. *Agriculture* **2023**, *13*, 1556. [CrossRef]
15. Shahgholi, G.; Moinfar, A.; Khoramifar, A.; Maciej, S.; Szymanek, M. Investigating the Effect of Tractor's Tire Parameters on Soil Compaction Using Statistical and Adaptive Neuro-Fuzzy Inference System (ANFIS) Methods. *Agriculture* **2023**, *13*, 259. [CrossRef]
16. Shin, J.; Won, J.; Kim, S.M.; Kim, D.C.; Cho, Y. Fertilization Mapping Based on the Soil Properties of Paddy Fields in Korea. *Agriculture* **2023**, *13*, 2049. [CrossRef]
17. Kim, D.; Song, J.; Park, J. Quantifying Soil Particle Settlement Characteristics through Machine Vision Analysis Utilizing an RGB Camera. *Agriculture* **2023**, *13*, 1674. [CrossRef]

agriculture

Article

Development of a Modified Method for Measuring the Actual Draft Force Using a Tractor-Attached Dynamometer

Hyo-Geol Kim [1], Jin-Woong Lee [2], Su-Chul Kim [3], Jooseon Oh [4,5,*] and Sung-Bo Shim [6,7,8,*]

1 TYM Co., Ltd., Seoul 06053, Republic of Korea; zion.kim@tym.world
2 Specialized Machinery and Robotics Group, Korea Institute of Industrial Technology, Gimje 54325, Republic of Korea; reallybear@kitech.re.kr
3 Department of Industrial Machinery DX, Korea Institute of Machinery and Materials, Daejeon 34103, Republic of Korea; sckim@kimm.re.kr
4 Department of Convergence Biosystems Engineering, Chonnam National University, Gwangju 61186, Republic of Korea
5 BK21 Interdisciplinary Program in IT-Bio Convergence System, Chonnam National University, Gwangju 61186, Republic of Korea
6 Department of Bio-Industrial Machinery Engineering, Kyungpook National University, Daegu 41566, Republic of Korea
7 Smart Agri Innovation Center, Kyungpook National University, Daegu 41566, Republic of Korea
8 Upland Field Machinery Research Center, Kyungpook National University, Daegu 41566, Republic of Korea
* Correspondence: jooseon.oh@jnu.ac.kr (J.O.); sbs80@knu.ac.kr (S.-B.S.); Tel.: +82-62-530-2155 (J.O.); +82-53-950-5793 (S.-B.S.)

Abstract: In this study, crank-locker kinematic equations were used to analyze the three-point hitch behavior when the dynamometer was connected to the work machine. The dynamometer was statically tested with a hydraulic actuator, and the accuracy of the three-way force and the moment was confirmed to be 96–99%. The calibrated dynamometer was put to the test on a real farm field, and data were collected using a data acquisition system. Using the transport pitch correction equation, the collected data can be transformed into more realistic data. International standards were used to determine the point of connection between the tractor, dynamometer, and implement. The results of this study made it possible to accurately measure force and moment, which will have an important role in future agricultural technologies such as autonomous agricultural operation.

Keywords: three-point hitch; tractor-attached dynamometer; force–moment equation; force measurement accuracy

Citation: Kim, H.-G.; Lee, J.-W.; Kim, S.-C.; Oh, J.; Shim, S.-B. Development of a Modified Method for Measuring the Actual Draft Force Using a Tractor-Attached Dynamometer. *Agriculture* **2024**, *14*, 544. https://doi.org/10.3390/agriculture14040544

Academic Editors: Jin He and Caiyun Lu

Received: 25 October 2023
Revised: 15 March 2024
Accepted: 25 March 2024
Published: 29 March 2024

1. Introduction

Modern agriculture is highly specialized and mechanized, and the agricultural workforce is decreasing as rural populations decline and age. As a result, the use of a tractor that can be equipped with a variety of agricultural implements to increase the labor force is becoming more common [1].

The tractor generates forces in various directions during operation. The generated force affects operation performance, the life of the implement, and tractor parts [2]. The tractor interacts with the soil during operation, and the force is changed by various variables, such as the soil properties and the tractor's shift level.

The traction force generated by the tractor can be estimated from the Wismer–Luth, Brixius equations, etc., which are registered in the ASABE test specification [3,4]. However, since these estimation methods are empirical formulas rather than theory-based methods, the prediction result cannot be considered accurate. Furthermore, because variables such as the soil cone index (CI) cannot be used in real-time formulas, predicting the traction force in real time is inaccurate. As a result, measuring the force with a sensor such as a load cell or strain gage yields the most accurate traction results. To address these issues, dynamometers with sensors have recently been developed to measure the force generated during operation [4].

An adjustable three-point hitch dynamometer with a draft capacity of 50 kN to measure forces on the tractor and mounted implements were developed [5]. With this dynamometer, it was confirmed that all mounted tillage implements in categories II and III such as plows, cultivators, and harrows could be tested [6]. And a bi-axial direct-mount strain gage lower link system calibrated for coincident and perpendicular loads up to 10 kN for measuring tractor–implement forces were developed [7]. The strain gage test method was confirmed to be the most appropriate when medium-type equipment is used with a tractor. However, because all of the tests were conducted in a lab, field validation is required.

A mathematical model was developed to determine the forces and the moments at the different couplers between bodies of the tractor–dynamometer–implement system, taking into account the mechanical characteristics of the dynamometer as well as the inclination of the terrain on which the tractor is working [8]. This model allows you to evaluate the dynamometer's destructive effect on the forces acting on the tractor. However, the effect of the connection between the dynamometer and the power take-off (PTO) was not considered. To measure traction forces, a dynamometer made up of three telescopic beams connected to a central T-shaped box was developed [9]. A three-point hitch-attached dynamometer basic conceptual diagram was presented. Different strain gage sets can be attached and arranged in the bridge to measure only lateral or vertical forces. However, because it was developed for use in a specific region, the use of the device presented is bound to be limited. Additionally, there is a lack of evidence that the measured force was exact using dynamometer.

A six-component load cell was developed to measure the tractor's traction force, and the accuracy of the load measurement using a six-component load cell was validated using a static load test [10]. In addition, a field test was conducted by installing the developed six-component load cell on the multi-tasking agricultural implement and measuring the working load. A method was presented for acquiring and post-processing traction force data while performing tillage operations using a three-point hitch [11]. The developed system is capable of running a large number of carried and partially carried machines in the II and III categories. In two past studies, a research was conducted to measure forces and moments using a hitch dynamometer, but it cannot be considered that accurate forces and moments were analyzed because changes in size and angle due to installation of the dynamometer were not considered.

Many research cases support the use of a three-point hitch dynamometer to measure the force applied to a three-point hitch during agricultural work confirmed [12,13]. To increase the precision of force measurement, the three-point hitch dynamometer primarily employs a six-component load cell. However, there are only a few cases where the geometrical change caused by the installation of the dynamometer is considered. Furthermore, no research has been conducted on how to correct the actual data that change depending on the angle of the dynamometer.

In this study, a modified method was presented to increase the accuracy of traction force prediction for tractors equipped with dynamometers. Using the existing equations, a modified equation considering the geometry of the dynamometer was developed. The equation's accuracy has been demonstrated through static load tests and a field test. By correcting the error of the measurement data that may occur due to the installation of the dynamometer, a modified method that can obtain data close to the actual is presented.

Securing realistic traction force data enables accurate measurement of the traction force. And it will enable the development of algorithms for performing autonomous agricultural operation through accurate traction measurement.

2. Materials and Methods

2.1. Three-Point Hitch

As shown in Figure 1, the three-point hitch is the most representative method used to connect the tractor and various implements [14]. The three-point hitch consists of two lower links and one upper link, and there are other devices to drive these links. The lift arm rotates by hydraulic power and is connected to the lift rod. The lift rod is connected to the lower link to pull up the lower link. The upper link moves dependently as the lower link moves up and down. The upper link is adjustable in length, and the upper and lower links each have a link point attached to the tractor and a hitch point attached to the implement. ISO 730 [15] defines mast height as the height of the triangle formed by the three hitch points.

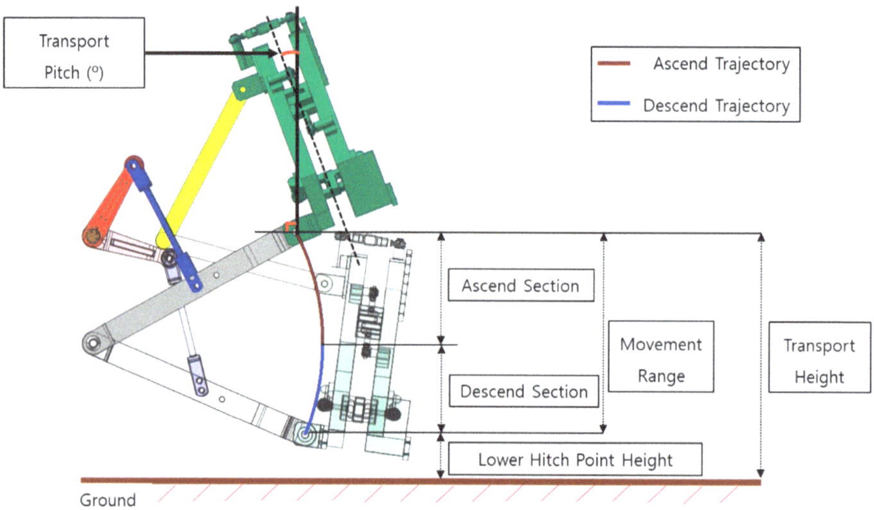

Figure 1. Element of three-point hitch.

When the three-point hitch ascends and descends, a height element with the ground and transport pitch of the implement occur. The transport pitch is defined as the angle between the working machine and the virtual vertical line. Transport pitch is determined by the range of use of the implement, which has a significant impact on load and traction. The transport pitch is measured when the minimum transport height is reached. The transport height is the vertical distance from the ground to the lower hitch point when the three-point hitch is ascended. The lower hitch point height is the vertical distance between the ground and the lower hitch point when the hitch descends. The movement range is actually the vertical displacement of the three-point hitch and is the transport height minus the lower hitch point height [16–19]. Transport pitch, transport height, lower hitch point height, and mast height are all specified by the ISO 730.

2.2. Three-Point Hitch-Type Dynamometer

2.2.1. Specification of Load Cell and DAQ

The dynamometer used in this study is a combination of triangular frame and one-axis load cell. The triangular shape was chosen for this study because it is known to be more structurally stable and easy to design. Because each link of the three-point hitch is connected one to one, the triangular geometry is relatively stable in terms of dynamics.

Single-axis load cells (CAS, SBA-2) are installed in 6 locations. Three load cells mounted in the tractor driving direction measure only the traction force. Two load cells mounted on the slope of a triangle measure longitudinal and vertical forces. Load cells mounted in the horizontal direction measure only longitudinal forces. The specifications of the load cell used in this study are shown in Table 1.

Table 1. Specification of the load cell used in this study.

Specification	Value
Max capacity (kgf)	2000
Rated output (mV/V)	3.0 ± 0.3
Zoro balance (mV/V)	0 ± 0.03
Combined error (%)	0.03
Repeatability (%)	0.01
Recommended excitation (V)	10
Maximum excitation (V)	15

Six load cells connected to the dynamometer are connected to the data acquisition system (DAQ) with a Wheatstone bridge. The data from each measured load cell are independently output and finally converted into target elements via equations. Output data are transferred to a PC via Ethernet and stored. The specifications of DAQ used for data acquisition are shown in Table 2. Each load cell is connected to the switch box by a connector and then from the switch box to the DAQ.

Table 2. Specification of DAQ used in this study.

Specification	Value
Model Name	Q.Brixx A108
Manufacturer	Gantner, Germany
Input Voltage (V)	Max 30
Input Current (mA)	Max 0.5
Upper Threshold (V)	>10
Lower Threshold (V)	<2
Analog Input Accuracy (%)	0.01–0.05
Repeatability (%)	0.003

2.2.2. Dynamometer Component Force and Moment

Because the dynamometer's load cell can only detect force in one direction, it must be converted to the needed factor using the combined load cell. The traction force is measured on F_a, F_b and F_c load cells mounted in the tractor's driving direction. Figure 2 shows a free body diagram of the load cell's three-direction force.

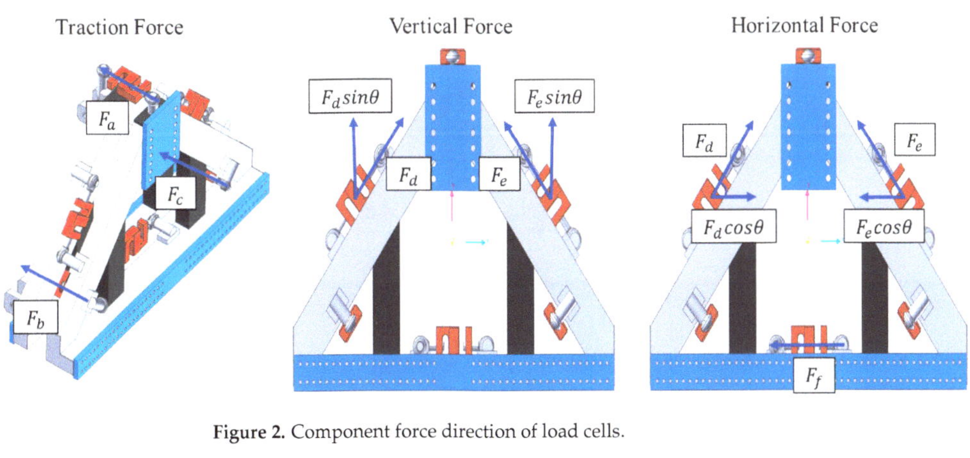

Figure 2. Component force direction of load cells.

Vertical force is calculated by multiplying the value measured by the F_d and F_e load cells mounted on the inclined plane of the triangle and the sine of the triangle. The longitudinal force is calculated by multiplying the values of the load cells F_d and F_e by the cosine of the triangle, and adding F_f which measures the longitudinal force.

The traction moment is calculated from the front of the dynamometer's geometric center. Forces F_d and F_e are converted to perpendicular components and recalculated as moments. Forces of F_a, F_b and F_c in the same moment, direction does not affect the traction moment. Force F_f is added to the moment in the tension direction.

The vertical direction moment is calculated based on the center of the dynamometer from above. F_a, F_f, which make up the center, and F_d and F_e, which are not in the plane, do not affect the vertical direction moment. Vertical moments are calculated simply with F_b and F_c, as well as a horizontal distance away from the center.

The horizontal direction moment is also calculated based on the side of the dynamometer from above. F_d, F_e, which make up the center, and F_f not in the plane, do not affect the horizontal direction moment. Horizontal moments are calculated only with F_a, F_b and F_c, and with a vertical distance away from the center.

Once the component force and moment are determined, the component force and moment may be calculated using the dynamometer shape and center point. P_h is the sum of the traction forces. P_v is the sum of the vertical forces, and P_s is the sum of the horizontal forces. M_h is the torsional moment acting in the traction direction? M_v is the torsional moment acting in the vertical direction, M_s is the torsional moment acting in the horizontal direction.

2.3. Tractor and Implement

The dynamometer was mounted on the 26 kW tractor and linked to the 2440 mm-long plows. Figure 3 shows the dynamometer and associated components attached to the tractor. The implement used in this study was category 1 type, and detailed information was omitted.

Figure 3. Dynamometer and instrumentation components.

2.4. Test

Static load tests and field tests were performed to compare the results measured by the dynamometer with those calculated by modified equation.

In the static load tests, the method was used to apply static load and moment through a hydraulic actuator to compare the input value with the actual input load. In field tests, the force was measured by performing the plow operation while maintaining the operating depth of the plow constant [20].

3. Results and Discussion

3.1. Modified Equation

The component force–moment equation for the triangular three-point hitch dynamometer was developed based on these considerations. Figure 4 summarizes the direction and magnitude of moments. The location information where the load cells are mounted on the dynamometer is shown in Figure 4. This is very important information regarding moment calculations. The location values in this study are as follows.

$$l_1 = 0.0506 \text{ m}$$

$$l_2 = 0.2878 \text{ m}$$

$$l_3 = 0.2461 \text{ m}$$

$$l_4 = 0.3081 \text{ m}$$

$$l_5 = 0.3233 \text{ m}$$

$$l_6 = 0.1938 \text{ m}$$

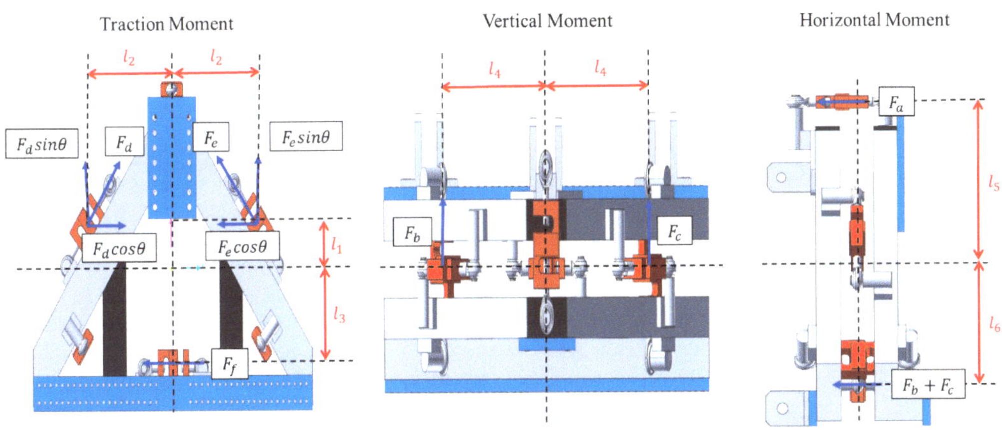

Figure 4. Moment direction of load cells.

Figure 5 shows the direction of each force and moment, and the dynamometer equation developed is represented by Equations (1)–(6). The coordinate system for calculation was set based on the end point of the dynamometer axis connected to the PTO axis. The tractor's forward direction was set to the x-axis, the left and right directions were set to the y-axis, and the vertical direction was set to the z-axis. This system is local coordinate system [21].

$$F_x = F_a + F_b + F_c \quad (1)$$

$$F_z = F_d sin\theta + F_e sin\theta \quad (2)$$

$$F_y = F_e cos\theta - F_d cos\theta + F_f \quad (3)$$

$$M_x = (F_d - F_e) \times [l_2 sin\theta + l_1 cos\theta] + l_3 \times F_f \quad (4)$$

$$M_z = (F_b - F_c) \times l_4 \quad (5)$$

$$M_y = F_a l_5 - (F_b + F_c) \times l_6 \quad (6)$$

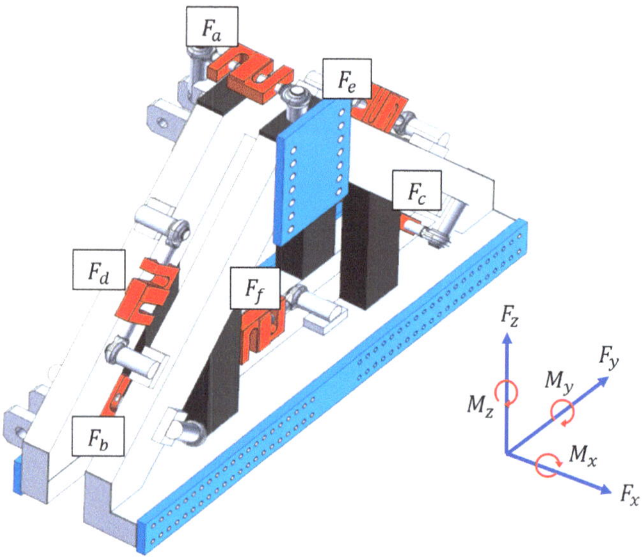

Figure 5. The force–moment direction of the dynamometer.

The dynamometer consists of a single-axis load cell, its value should be modified according to the pitch angle of the dynamometer. Since the dynamometer was initially connected vertically to the lower link, the transport pitch is the dynamometer's angle.

Let the x′y′z′ coordinate system move and rotate with the dynamometer such that the load cells are always in the y′z′ plane and with the same y′z′ coordinates (x′ coordinate of all load cells is zero since all load cells are in the y′z′ plane). Since the x′y′z′ coordinate system moves with the dynamometer, any forces and moments measured by the dynamometer load cells will be expressed in the x′y′z′ coordinate system by Equations (1)–(6), irrespective of the orientation of the dynamometer, even if it is upside down. The xyz coordinate system and x′y′z′ coordinate system can be simplified through Figure 6.

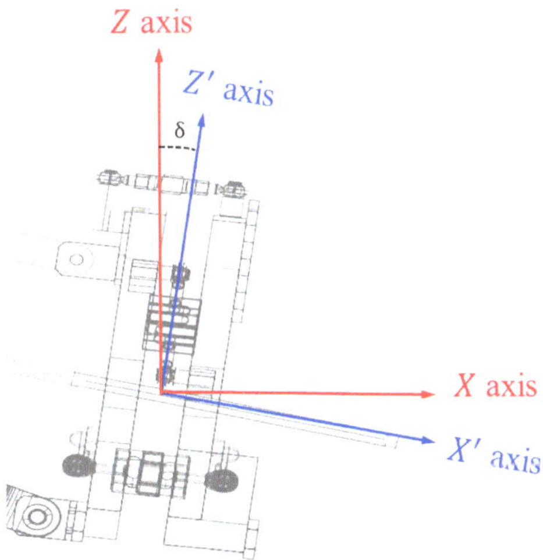

Figure 6. New coordinate system.

Transform these forces and moments in the x′y′z′ coordinate system at transport pitch, δ, to the global coordinate system, xyz. The transport pitch used in this study is -7.52° at lowest hitch point. All that is required to achieve this is a to pre-multiply the force and moment vectors by a standard 2-D coordinate transformation matrix.

Equation (7) rotates the x′y′z′ coordinate system about the y′ axis by an angle δ. Whether the signs for $sin\delta$ are positive or negative depends on whether the rotation angle is measured from x′y′z′ to xyz, or from xyz to x′y′z′ coordinate systems. The same rotation matrix can be used to transform the three moments, $M_{x'}$, $M_{y'}$ and $M_{z'}$ to M_x, M_y and M_z in the global xyz coordinate system. In fact, the three orthogonal forces and moments as measured by the dynamometer load cells, can be combined into a six dimensional load vector, and transformed from the x′y′z′ coordinate system by pre-multiplying by a single 6×6 partitioned transformation matrix.

$$\begin{bmatrix} F_x \\ F_y \\ F_z \end{bmatrix} = \begin{bmatrix} cos\delta & 0 & sin\delta \\ 0 & 1 & 0 \\ -sin\delta & 0 & cos\delta \end{bmatrix} \begin{bmatrix} F_{x'} \\ F_{y'} \\ F_{z'} \end{bmatrix} \quad (7)$$

3.2. Result of the Static Load Test

Tables 3 and 4 show the calculated force and moment when the static force and moment are applied. The moment is calculated by multiplying the vertical distance between the hydraulic actuator and the center of the dynamometer. And the error rate was calculated by comparing the calculated data and the input data.

Table 3. Calculated force and input force.

Specification	Calculated Force (kN)	Input Force (kN)	Error Rate (%)
Traction Force	14.709	14.551	1.1
Vertical Force	9.806	9.732	0.8
Lateral Force	4.903	5.032	2.6

Table 4. Calculated moment and input moment.

Specification	Calculated Moment (kN·m)	Input Moment (kN·m)	Error Rate (%)
Traction Moment	3.912	3.983	1.8
Vertical Moment	3.912	3.807	2.7
Lateral Moment	2.535	2.619	3.5

The component force test yielded an accuracy of 97.4% or higher in all regions, as well as satisfactory reliability. The component force tests have modest errors, which are predicted to be caused by manufacturing errors and misalignment when attaching the surface plates.

The error of the moment test is expected to be the measurement error and the moment arm length. The maximum accuracy of 98.2% and the minimum accuracy of 96.5% were calculated for the whole test, and the dynamometer is considered reliable even considering the error rate.

3.3. *Result of the Field Test*

To ensure that the dynamometer operates accurately in the field, the dynamometer was tested in the field. The test site is a test field located in Gimje-si, Korea. The force was measured by performing the plow operation while maintaining the operating depth of the plow constant. The operation velocity was set to 3 km/h, and the hitch height lever was set to the same value during all test conditions. During plow work, the power of the tractor was mainly consumed by the traction force, and the vertical force supporting the machine was second. Horizontal forces occurred insignificantly, and the effects of torsion were not considered in this study.

Figure 7 shows the traction force corrected by dynamometer installation, while Figure 8 shows the vertical force. When the implement descends due to the horizontal connection condition of the initial implementation, the transport pitch occurs in the negative direction, so the traction force decreases by the transport pitch sine value of the vertical force.

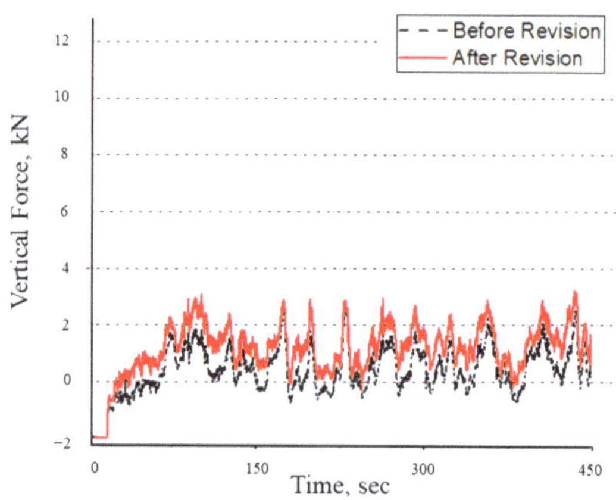

Figure 7. Traction data before and after revision.

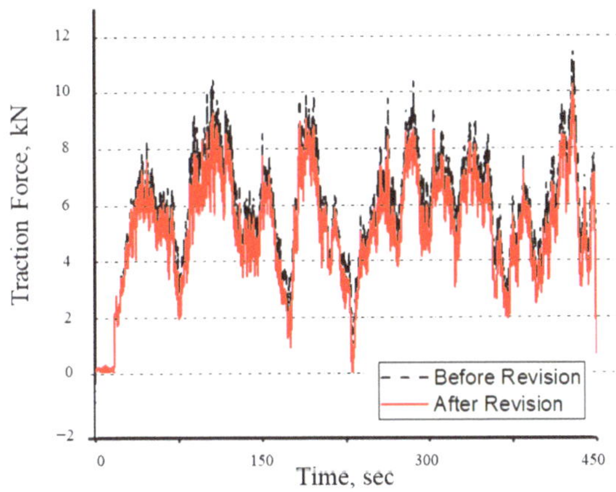

Figure 8. Vertical force data before and after revision.

The measured values of the traction force and vertical force itself are also changed by the transport pitch, but the transport pitch generated at the actual working depth is 0.994–0.99, and the difference value is insignificant, 0.06–1%, so there is virtually no difference.

When the transport pitch value occurs anticlockwise when the implement is lowered, the traction force increases in part with the vertical force and decreases in part with the traction force, resulting in a traction-vertical component force that differs from the initial measured force. The larger the transport pitch, the larger the difference between the correction value and the measured value is because the value of the traction force to be converted to the vertical force is greater than the vertical force to be converted to the traction force. When lowering the implement to the desired operation depth, the transport pitch affects this difference. In addition, the initial connection angle is thought to be important. As a result, net traction increased by approximately 11.4%.

As a result, the difference in vertical force is greater while operating with plows, cultivators, and sub-soilers, which demand a high traction force. When working with a rotavator or agricultural roller under conditions that demand little traction or have a high vertical force due to the weight of the implement itself, the difference in the traction force is expected to be greater.

Through this study, a method to accurately measure traction forces and moments was proposed. Static tests were performed in which known values were applied to the dynamometer, and the effectiveness of the method was proven as a result of low errors. It was confirmed that the results of applying this method to the field test showed a difference of 11.4% compared to the value measured using the existing method.

In order to realize open-field smart farms, which are in the spotlight as one of future agriculture, autonomous farming must be carried out. For autonomous agricultural operation, it is necessary to accurately measure the load applied to the three-point hitch during agricultural operation, and this study can contribute. In addition, it is necessary to correct the measurement results using a dynamometer and verify the effectiveness of this study through various field tests.

4. Conclusions

This study was conducted to design a three-point hitch dynamometer with an easy connection to the implement and to propose a calibration equation to obtain realistic data. The conclusions obtained in this study are as follows.

(1) The force–moment equation was developed considering the geometrical position of the load cell constituting the dynamometer and, as a result of static load testing using a hydraulic actuator, confirmed more than 97% accuracy in all sections.

(2) The plow work data were collected in the field with a dynamometer and verified in a static environment. A modified equation based on transport pitch was used to modify the measured data. The actual traction force is slightly less than the measured traction force, and the actual vertical force is 11.7% greater than the measured vertical force. Because the dynamometer is a single-axis load cell, the correction data using the transport pitch are thought to be closer to the actual data.
(3) Because traction is primarily used in plowing, a significant difference between the actual vertical force and the measured vertical force was confirmed because the traction force to be converted into vertical force was large. In contrast, in roller and landscape work, where vertical force is primarily used, the vertical force to be converted into the traction force is greater, so a significant difference between the actual traction force and the measured traction force is expected.
(4) This study can be used as a basis for open-field smart farm research. The core of open-field smart farm technology is autonomous agricultural operation, and for this, it is most important to accurately analyze the traction force required for agricultural operation. The significance of this study is that it enabled precise traction analysis for autonomous agricultural operation. In addition, it is expected that further research will be needed, such as real-time traction analysis and traction force prediction for real-time autonomous agricultural operation.

Author Contributions: Conceptualization and methodology, H.-G.K.; validation, J.O.; formal analysis, J.-W.L.; data curation, S.-C.K.; writing—original draft preparation, H.-G.K. and J.O.; writing—review and editing, J.O. and S.-B.S.; funding acquisition, S.-B.S. All authors have read and agreed to the published version of the manuscript.

Funding: This research was supported by Kyungpook National University Research Fund, 2021. This research was conducted with the support of the "Development of Safety Evaluation and Management Technology for High-Risk Construction Machinery (Project No. RS-2023-00244879)" implemented by the Ministry of Land, Infrastructure and Transport/Korea Agency for Infrastructure Technology Advancement and overseen by the Korea Construction Equipment Safety Institute.

Institutional Review Board Statement: Not applicable.

Data Availability Statement: Data is contained within the article.

Conflicts of Interest: Author Hyo-Geol Kim is employed by the company TYM Co., Ltd. The remaining authors declare that the research was conducted in the absence of any commercial or financial relationships that could be construed as a potential conflict of interest.

References

1. Koo, Y.M. PTO Torque and draft analyses of an integrated tractor-mounted implement for round ridge preparation. *J. Biosyst. Eng.* **2022**, *47*, 330–343. [CrossRef]
2. Beigi, M.; Ghazavi, M.A.; Ahmadi, I. Design and construction of load cell of a three point hitch dynamometer for tractor John Deere 3140. *J. Mod. Process. Manufaturing Prod.* **2014**, *3*, 47–58.
3. Wismer, R.D.; Luth, H.J. Off-road traction prediction for wheel vehicles. *Trans. ASAE* **1974**, *17*, 8–14. [CrossRef]
4. Alimardani, R.; Fazel, A.; Akram, A.; Mahmoudi, A.; Varnamkhsati, M.G. Design and Development of a three-point hitch dynamometer. *J. Agric. Technol.* **2008**, *4*, 37–52.
5. Brixius, W.W. *Traction Prediction Equations for Bias Ply Tires*; Transaction of the ASAE, 87-1622; American Society of Agricultural Engineers: St. Joseph, MI, USA, 1987.
6. Askari, M.; Komarizade, M.H.; Nikbakht, A.M.; Nobakht, N.; Teimourlou, R.F. A novel three-point hitch dynamometer to measure the draft requirement of mounted implements. *Res. Agric. Eng.* **2011**, *57*, 128–136. [CrossRef]
7. Khan, J.; Godwin, R.J.; Kilgour, J.; Blackmore, B.S. Design and calibration of a direct mounted strain gauged lower links system for measurement of tractor-implement force. *J. Eng. Appl. Sci.* **2006**, *1*, 22–25.
8. Roca, J.; Comellas, M.; Pijuan, J.; Nogués, M. Development of an easily adaptable three-point hitch dynamometer for agricultural tractors. Analysis of the disruptive effects on the measurements. *J. Soil Tillage* **2019**, *194*, 104323. [CrossRef]
9. Al-Jalil, H.F.; Khdair, A.; Mukahal, W. Design and performance of and adjustable three-point hitch dynamometer. *J. Soil Tillage* **2001**, *62*, 153–156. [CrossRef]
10. Kim, E.K.; Moon, S.G.; Oh, C.M.; Han, J.W. Working load measurement using 6-component load cell and fatigue damage analysis of composite working implement. *J. Agric. Life Environ. Sci.* **2017**, *29*, 225–236.

11. Kostic, M.; Rakic, D.Z.; Licen, H.H.; Malinovic, N.C. Design and construction of three point hitch device for measuring draft of tillage implement—Data acquisition and post processing analysis. *J. Food Agric. Environ.* **2014**, *12*, 1300–1307.
12. Kumari, A.; Raheman, H. Development of a Novel Draft Sensing Device with Lower Hitch Attachments for Tractor-Drawn Implements. *J. Biosyst. Eng.* **2024**, 1–9. [CrossRef]
13. Tewari, V.K.; Ravi, N.; Jha, K.R.; Ashok, A. Design and development of a three-Point-linkage dynamometer for tillage research. *Agric. Eng. Today* **2012**, *36*, 33–38.
14. *ASABE Standard S217*; Three-Point Free-Link Attachment for Hitching Implements to Agricultural Wheel Tractors. American Society of Agricultural and Biological Engineers: St. Joseph, MI, USA, 2001.
15. *ISO Standard 730*; Agricultural Wheeled Tractors—Rear-Mounted Three-Point Linkage—Part 1: Categories 1, 2, 3 and 4. ISO: Geneva, Switzerland, 2009.
16. *ISO Standard 11001-1*; Agricultural Wheeled Tractors and Implements—Three-Point Hitch Couplers—Part 1: U-Frame Coupler. ISO: Geneva, Switzerland, 2016.
17. *ISO Standard 11001-2*; Agricultural Wheeled Tractors and Implements—Three-Point Hitch Couplers—Part 2: A-Frame Coupler. ISO: Geneva, Switzerland, 2017.
18. *ISO Standard 11001-3*; Agricultural Wheeled Tractors and Implements—Three-Point Hitch Couplers—Part 3: Link coupler. ISO: Geneva, Switzerland, 1993.
19. *ISO Standard 11001-4*; Agricultural Wheeled Tractors and Implements—Three-Point Hitch Couplers—Part 4: Bar Coupler. ISO: Geneva, Switzerland, 1994.
20. Bentaher, H.; Hamza, E.; Kantchev, G.; Maalej, A.; Arnold, W. Three-point hitch-mechanism instrumentation for tillage power optimization. In Proceedings of the 14th Annual Conference of the Misr Society of Agricultural Engineering; 2006; Volume 22, pp. 923–936. Available online: https://citeseerx.ist.psu.edu/document?repid=rep1&type=pdf&doi=f470dc69970f41d2d6790174b64237c823e6ee35 (accessed on 14 March 2024).
21. *ISO Standard 5673-2*; Agricultural Tractors and Machinery—Power Take-Off Drive Shafts and Power-Input Connection—Part 2: Specification for Use of PTO Drive Shafts, and Position and Clearance of PTO Drive Line and PIC for Various Attachments. ISO: Geneva, Switzerland, 2017.

agriculture

MDPI

Article

The Development of a Draft Force Prediction Model for Agricultural Tractors Based on the Discrete Element Method in Loam and Clay Loam

Bo-Min Bae [1], Yeon-Soo Kim [1,*], Wan-Soo Kim [2,3], Yong-Joo Kim [4,*], Sang-Dae Lee [5] and Taek-Jin Kim [6]

1 Department of Bio-Industrial Machinery Engineering, Pusan National University, Miryang 50463, Republic of Korea; qhals806@pusan.ac.kr
2 Department of Bio-Industrial Machinery Engineering, Kyungpook National University, Daegu 41566, Republic of Korea; wansoo.kim@knu.ac.kr
3 Upland Field Machinery Research Center, Kyungpook National University, Daegu 41566, Republic of Korea
4 Department of Biosystems Machinery Engineering, Chungnam National University, Daejeon 34134, Republic of Korea
5 Smart Agricultural R&D Group, Korea Institute of Industrial Technology (KITECH), Gimje 54325, Republic of Korea; sdlee96@kitech.re.kr
6 Department of Drive System Team, TYM R&D Center, Iksan 54576, Republic of Korea; taek.kim@tym.world
* Correspondence: yskim23@pusan.ac.kr (Y.-S.K.); babina@cnu.ac.kr (Y.-J.K.); Tel.: +82-55-350-5422 (Y.-S.K.); +82-42-821-6716 (Y.-J.K.)

Abstract: In the field of agricultural machinery, various empirical field tests are conducted to measure design loads for the optimal design and implementation of tractors. However, conducting field tests is costly and time-consuming, with many constraints on weather and field soil conditions, and research utilizing simulations has been proposed as an alternative to overcome these shortcomings. The objective of this study is to develop a DEM-based draft force prediction model that reflects differences in soil properties. For this, soil property measurements were conducted in two fields (Field A in Daejeon, Republic of Korea, and Field B in Chuncheon, Republic of Korea). The measured properties were used as parameters for DEM-based particle modeling. For the interparticle contact model, the EEPA contact model was used to reflect the compressibility and stickiness of cohesive soils. To generate an environment similar to real soil, particle mass and surface energy were calibrated based on bulk density and shear torque. The soil property measurements showed that Field B had a higher shear strength and lower cone index and moisture content compared to Field A. The actual measured draft force was 19.47% higher in Field B than in Field A. In this study, this demonstrates the uncertainty in predicting draft force by correlating only one soil property and suggests the need for a comprehensive consideration of soil properties. The simulation results of the tillage operation demonstrated the accuracy of the predicted shedding force compared to the actual field experiment and the existing theoretical calculation method (ASABE D497.4). Compared to the measured draft force in the actual field test, the predictions were 86.75% accurate in Field A and 74.51% accurate in Field B, which is 84% more accurate in Field A and 37.32% more accurate in Field B than the theoretical calculation method. This result shows that load prediction should reflect the soil properties of the working environment, and is expected to be used as an indicator of soil–tool interaction for digital twin modeling processes in the research field of bio-industrial machinery.

Keywords: draft force; agricultural tractor; moldboard plow tillage; soil property

Citation: Bae, B.-M.; Kim, Y.-S.; Kim, W.-S.; Kim, Y.-J.; Lee, S.-D.; Kim, T.-J. The Development of a Draft Force Prediction Model for Agricultural Tractors Based on the Discrete Element Method in Loam and Clay Loam. *Agriculture* **2023**, *13*, 2205. https://doi.org/10.3390/agriculture13122205

Academic Editor: Kenshi Sakai

Received: 15 October 2023
Revised: 19 November 2023
Accepted: 24 November 2023
Published: 27 November 2023

1. Introduction

Among various agricultural operations, soil tillage operations such as moldboard plow generate the most soil resistance, which caused various dynamic stresses on the tractor. In addition, this is widely used as an indicator to evaluate the performance of agricultural machinery [1]. These tillage operations are performed in a wide range of soil environments and directly contribute to soil resistance and working load size. These factors act in combination and exhibit an atypical correlation with the working load, making it difficult to predict the working load in the soil environment [2]. Therefore, the design optimization of agricultural equipment typically relies on iterative prototyping and evaluation using soil bins and field experiments.

A number of studies have been performed on the various factors that affect the draft force of agricultural tractors, either through field experiments in soil bin test beds or in real agricultural environments. Raper [3] studied an analysis of minimum draft force generation as a function of subsoil geometry through soil bin tests, and the results showed that, the lesser the resistance of a bent shank compared to a straight shank, the more soil breakdown occurs. Kim et al. [4] conducted a study to analyze the effects of tillage depth and gear selection on the mechanical load and fuel efficiency of agricultural tractors during plow tillage. Typically, the working load increased as the travel speed increased, and there were different trends in production versus fuel usage depending on the type of agricultural machinery. Kim et al. [5] carried out a study for the progress of a real-time tillage depth measurement system. The development of an improved real-time tillage depth measurement system was to quantitatively confirm the increasing draft force with tillage depth. Development and design processes by the means of field testing have the advantage of reflecting actual field conditions, however they have a lot of disadvantages, including cost, time, the labor of building an instructional system, lower replicability, space constraints, and the effects of weather and soil conditions.

As an alternative to actual field experiments, simulation-based virtual field experiments can reduce the number of prototypes and accelerate design and product development by facilitating the creation of targeted work environments in any climate [6]. Virtual field experiments can be performed in two representative numerical methods. One method of virtual field experiments, the finite element method (FEM), is a stress analysis method that divides a structure into elements of virtually finite size and analyzes the structure as a collection of these elements. Analyses using FEM models use complex geometries as simple geometries, providing values that clearly indicate the location of high stresses or displacements. Fielke [7] carried out a soil–tool interaction analysis study of the cutting edges of tillage tools based on FEM. In Fielke's study, the effects of Poisson's ratio and cutting-edge geometry on tillage forces and soil flow were determined through FEM simulations. A study of soil–tool interaction was conducted by Ucgul et al. [8] by comparing discrete and finite element methods. The study showed that a better vertical force prediction was obtained using DEM software (EDEM 2.7^{TM}), whereas forward soil movements below the tillage depth were simulated more accurately using FEM. The prediction accuracy of the FEM simulation model was verified by soil bin experiments. A follow-up study analyzed the effects of the cutting angle and tool lift angle of the moldboard plow through FEM simulations using an elasto-plastic model [9,10]. However, numerical studies using FEM have the disadvantage that it is difficult to predict the flow of granular materials and is not suitable for modeling materials with large displacements, such as soil.

Another representative numerical analysis method, the discrete element method (DEM), has been used worldwide in agricultural machinery research because it can calculate the interaction of each granular material. In a study analyzing the analysis time and accuracy of fine-grained material sizes in relation to the computational power of the computer, Wang et al. [11] conducted a study on the effect of modeling particle size on the interaction between soil and subsoiler. The results, models with a radius of 4 to 7 mm, provided relatively accurate predictions. They also reported that, when modeling simulations were reduced from a radius of 5 mm to 3 mm, the solution time increased by about seven times. Obermayr et al. [12] carried out DEM simulations to forecast the draft force of a simple tool in a cohesionless powdery material, and validated the simulation consequences with small-scale laboratory tests, which showed that the draft force was predicted within a reasonable range despite the outdated particle modeling. Ucgul et al. [13] performed DEM modeling using a hysteretic spring contact model based on cohesion and adhesion. The results showed that the proposed contact model could be used to predict both tillage aeration and vertical tillage force for different speeds, operating depths, moisture contents, and compaction levels. Tekeste et al. [14] utilized DEM simulations to investigate the impact of tiller sweep–soil interaction, worn and hardened edges on soil–tool forces, and soil flow. Their findings revealed that carbide-treated worn sweeps exhibited comparable soil gradient forces and soil forward failure distances to those of new sweeps. However, most previous studies have been conducted in soil bin testing facilities under granular material or cohesionless soil conditions. [15] The soil within soil bin test facilities exhibits a lower cohesion compared to the natural paddy field, attributed to frequent indoor tillage operations; consequently, soil modeling in a soil bin test facility commonly employs a hysteretic spring contact model [16]. However, the DEM modeling of agricultural soils involves a nonlinear hysteretic spring model and employs the Edinburgh elasto-plastic adhesion (EEPA) contact model [17], which considers the compressible, adhesive, and nonlinear behavior of cohesive solids between particles and between particle shapes. Recently, a study predicted the working performance of agricultural machinery in response to DEM simulations of the properties of cohesive soil, a real agricultural soil. Kim et al. [18] conducted a study to predict the draft force with tillage depth by using the EEPA contact model to implement the modulus of compression and stickiness of cohesive soil in a DEM simulation program. In the same soil environment, draft force prediction studies have been conducted, but draft force prediction studies based on soil properties have not yet been conducted.

The specific objective of this study is to elucidate the soil–implement interaction by predicting the draft force in two soil working environments with different properties during moldboard plowing operations, and demonstrating the prediction accuracy by comparing the draft force predicted by traditional theoretical prediction equations and actual field experiment results.

2. Materials and Methods

2.1. Measurement of Field Soil Properties

To simulate the properties of the two fields and select them as modeling parameters, soil property measurement tests were performed respectively. Field A is located in Daejeon, Republic of Korea, and Field B is located in Chuncheon, Republic of Korea. The soil properties were calculated based on the uniformed grid (3 m $\times$ 3 m) sampling method in the field [19]. Each field was divided into 10 uniformed grids, each of which was sampled to measure the physical properties of the soil and vane shear and cone penetration tests to measure the mechanical properties. An actual field has different soil properties according to the depth [5]. Therefore, in this study, the target tillage depth of 20 cm was divided into several layers and the soil properties of each layer were measured. According to the depth of the blade of the ring-type vane shear tester (5 cm), the soil layer was divided into four layers: Layer 1 (0~5 cm), Layer 2 (5~10 cm), Layer 3 (10~15 cm), and Layer 4 (15~20 cm), and 10 times of soil sampling and shear vane tests were performed for each layer. Figure 1 displays the positions of the two examined fields and the soil property measurements obtained through uniform grid sampling.

For physical properties of the soil, soil sampling was conducted in both test fields, A and B, utilizing a 100 mL soil sampling tube (DIK-1801, Daiki Rika Kogyo Co., Ltd., Konosu, Japan) in conjunction with a soil sampling tool (DIK-1815, Daiki Rika Kogyo Co., Ltd., Konosu, Japan). To determine the soil moisture content, each soil sample underwent a 24 h drying process at 110 °C [20] using the oven drying method (SH-DO-100FGB, Samheung Energy, Sejong, Republic of Korea). The compaction of soil occurs due to the exclusion of air or water from void spaces, rearrangement of soil particles, and compaction of water and air within the voids. Soil texture is also an important indicator for soil modeling because it determines the degree to which soil particles rearrange and many other properties, such as water permeability [21]. The soil texture at the field site was analyzed using the USDA soil classification method [22] with a sieve shaker (HJ-4560, Heungjin, Gimpo,

Republic of Korea) [23]. The soil model's plasticity index value, a determinant for its plastic characteristics and compressibility [24], was assessed through the Atterberg limit test encompassing liquid and plastic limits [25].

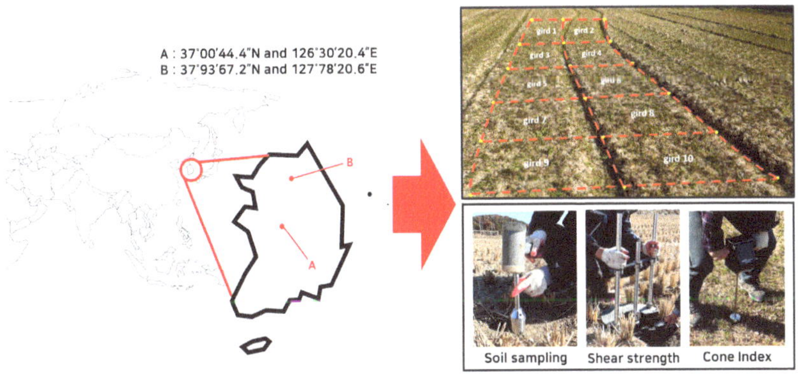

Figure 1. Location of test fields and uniformed grid sampling methods.

Shear strength and cone index are mechanical properties of soil that have a significant impact on working load [26,27]. These two properties were measured using a cone penetration tester (DIK-5532, Daiki Rika Kogyo Co., Ltd., Konosu, Japan) [28,29] and a ring-type vane shear tester (DIK-5503, Daiki Rika Kogyo Co., Ltd., Konosu, Japan). Shear strength was obtained from Equation (1) [30]:

$$\tau = \frac{3M}{2\pi({r_1}^3 - {r_2}^3)}, \quad (1)$$

where τ is the shear strength of the soil (Pa), M is the soil resistance torque that caused the vane shear test (Nm), nd r_1 and r_2 are the inner and the outer radii of the shear box (m).

2.2. Modeling Cohesive Soil with DEM

Modeling the virtual soil environment for simulation can be broadly divided into particle modeling and contact modeling. The modeling process proceeded as shown in Figure 2.

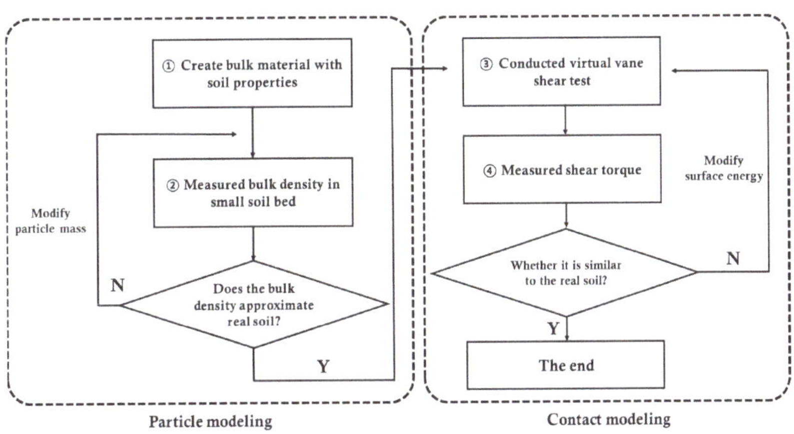

Figure 2. Modeling Procedures for Virtual Soil Environments.

The first step in conducting a DEM simulation is to model the particles. Modeling soil particles requires entering parameters such as solid density, Poisson's ratio, shear modulus, Young's modulus, and friction coefficient for both soil and steel. Table 1 shows the major particle model parameters in the DEM software (EDEM, 2022, Altair, Troy, MI, USA). The particle size of the soil was set based on past studies to reduce the simulation time without adversely affecting the prediction accuracy [18]. An increase in particle size leads to an increase in voids, so the process of calibrating the mass of the particles based on the actual measured bulk density was performed. The shear modulus and Young's modulus, which have a direct and significant impact on the draft force, were set based on measured values from field properties experiments. In prior work, Wang et al. [11] investigated subsoiler interaction with soil using DEM and they made a clear distinction between the friction coefficients of ploughed soil and soil beneath the subsoiler when selecting modeling parameters. In this study, the parameters used in Wang et al.'s study were selected to reflect the change in soil properties with depth. The restitution coefficients for soil–soil and soil–steel were selected from values widely used in DEM studies [31,32].

Table 1. Major particle model parameters in EDEM.

Properties	Value
Particle size of soil	Radii 5 mm
Particle mass of soil	Calibrated
Shear modulus	Measured
Young's modulus	Measured
Static friction coefficient of soil–soil	0.4
Rolling friction coefficient of plough layer	0.58
Rolling friction coefficient of hardpan layer	0.25
Static friction coefficient of soil–steel	0.24
Rolling friction coefficient of plough layer-steel	0.34
Rolling friction coefficient of hardpan layer-steel	0.14
Restitution coefficient of soil–soil	0.2
Restitution coefficient of soil–steel	0.3

The second step in conducting a DEM simulation is contact modeling. EDEM (EDEM, 2022, Altair, Troy, MI, USA) provides several contact models to reproduce various particle-to-particle contact behaviors. Soil with clay content exhibits a greater capacity for retaining water when compared to sandy soil [33]. This enhanced water retention capability creates a conducive environment for the flourishing of crops. Consequently, it is common to find clay-rich agricultural soils in Korea. This study employed the EEPA contact model to simulate the compressibility and stickiness of agricultural soil. The EEPA contact model is organized into various equations for numerical analysis. The vertical normal force of the EEPA model is given by Equation (2)

$$F_n = (f_{hys} + {f_n}^d)u, \tag{2}$$

where F_n is the total contact normal force, f_{hys} is the hysteretic spring stiffness, ${f_n}^d$ is the damping force, and u is the unit normal vector pointing to the center of the particle at the contact point.

$$f_{sys} = \begin{cases} f_0 + k_1\delta^n & if\ k_2(\delta^n - {\delta_p}^n) \geq k_1\delta^n \\ f_0 + k_2(\delta^n - {\delta_p}^n) & if\ k_1\delta^n > -k_{adh}\delta^x \\ f_0 - k_{adh}\delta^x & if\ -k_{adh}\delta^x > k_2(\delta^n - {\delta_p}^n) \end{cases} \tag{3}$$

where f_0 is the initial contact normal force, δ^n is the initial overlap, $-k_{adh}$ is the adhesive stiffness, and δ^x is the overlap due to adhesive force.

$$f_n{}^d = 2\sqrt{\frac{5}{6}}\beta_{NL}\sqrt{K_n m * v_n} \tag{4}$$

$$K_n = 2E * \sqrt{R * \delta^n} \tag{5}$$

$$\beta_{NL} = \frac{\ln e}{\sqrt{\ln e^2 + \pi^2}} \tag{6}$$

where β_{NL} is the coefficient of the normal dashpot in the nonlinear model, K_n is the Hertzian stiffness, $m*$ is the equivalent mass of the particle, and e is the user-defined restoration factor in the simulation.

Additionally, the contact tangential force (F_t) was obtained from Equation (7):

$$F_t = f_{ts} + f_{ds} \tag{7}$$

where f_{ts} is the tangential spring force and f_{td} is the tangential damping force.

$$f_{ts} = f_{ts(n-1)} + \Delta f_{ts} \tag{8}$$

$$\Delta f_{ts} = k_t\delta_t \tag{9}$$

$$k_t = \zeta_{tm}8G^*\sqrt{R^*\delta_n} \tag{10}$$

$$f_t^d = -2\sqrt{\frac{5}{6}}\beta_{NL}\sqrt{k_t m^*}\vec{v_t^r} \tag{11}$$

$$\beta_t = \sqrt{\frac{4m^*k_t}{1 + (\frac{\pi}{lne})^2}} \tag{12}$$

where $f_{ts(n-1)}$ is the tangential spring force at the previous time step, Δf_{ts} is the increment in the tangential force, k_t is the tangential normal stiffness, δ_t is the tangential displacement, ζ_{tm} is the tangential stiffness multiplier, G^* is the equivalent shear modulus, R^* is the equivalent radius, v_t is the relative tangential velocity between the two particles, and β_t is the linear tangential dashpot coefficient, which depend on the tangential stiffness.

$$\tau_i = -\mu_t f_{hys} R_i \omega_i \tag{13}$$

Finally, the rolling friction torque τ_i was obtained from Equation (9), μ_t is the rolling friction, R_i is the distance from the contact point to the center of the particle, and ω_i is the vector of the unit angular velocity of the particle at the contact point. Equations (2)–(13), used for particle force calculations in EDEM, show that the contact force, which consists of normal and tangential forces, is a function of the particle mass, particle overlap, particle size, Young's modulus, shear modulus, friction coefficient, and restitution coefficient.

Table 2 presents the key input parameters for the EEPA contact model. In this paper, the input parameters were determined by referring to a soil model with compressibility and adhesion from the Soil Starter Pack provided by DEM software (EDEM, 2022, Altair, Troy, MI, USA). The surface energy was calibrated with a virtual simulation using a ring-type vane shear tester to set the value. For cohesionless soils, the calibration is most often performed with a repose angle experiment. Cohesive soils sometimes exhibit non-flowing characteristics in certain areas. Hence, when calibrating cohesive soils, relying on vane shear test outcomes and comparisons proves more suitable than utilizing the repose angle for calibration. Repeated calibration through virtual vane shear tests was conducted for every soil layer to accurately anticipate the loads induced by tillage activities.

Table 2. Major contact model parameters in EDEM.

Properties	Value
Constant pull-off force	0
Surface energy	Calibrated
Tensile exponent	5
Tangential stiff multiplier	0.28571
Slope exponent	1.5
Contact plasticity ratio	0.75

2.3. *Design a Working Load Measurement System*

2.3.1. Tractor-Implement System

A 42 kW tractor (TX58, TYM, Iksan, Republic of Korea) was selected for this study considering the tillage load. Table 3 shows the specific specifications of the tractors utilized. Moldboard plows are commonly used in soil mechanization research due to their high working stability and soil resistance. In the field experiment, a six-row moldboard plow (WJSP-6S, Woongjin, Gimje, Republic of Korea) with a maximum working depth of 200 mm was used to perform the tillage test. The detailed specifications of the machine are presented in Table 4. The moldboard plow was reverse engineered in a 1:1 ratio using a 3D scanner for the virtual draft force prediction test.

Table 3. Specifications of the agricultural tractor used in this study.

Item		Specification
Company		TYM
Model		TX58
Wheel base (mm)		2155
Length × width × height (mm)		3695 × 1848 × 2560
Engine	Rated power (kW) Max torque (Nm)	42 @ 2000 rpm 211.8 @ 1600 rpm
Transmission gear selection		Main 4 stage/Sub 6 stage
PTO gear selection		1st 540 rpm 2nd 750 rpm 3rd 1000 rpm
Tire (inch)		Front 11.2–20 Rear 14.9–30
Maximum travel speed (km/h)		33.8

Table 4. Specifications of the moldboard plow used in this study.

Item	Specification
Product name	WJSP-6S
Manufacturing company	Woongjin
Type	Moldboard plow
Length × width × height (mm)	1930 × 1800 × 1235
Rake angle (deg)	30.76
Tillage width (mm)	270
Share length (mm)	360

Table 4. *Cont.*

Item	Specification
Share form	Pointed
Required power (kW)	40–52
Maximum tillage depth (mm)	Up to 200
Coulter type	Skid jig/Plain coulter with spring
Coulter diameters (mm)	340
Number of furrows	3
Required travel speed (km/h)	5–8

2.3.2. Working Load Measurement System

The load generated during tillage operations using a moldboard plow is strongly influenced by tillage depth and travel speed. For a comparative analysis of the effects of soil properties, it is requisite to verify the draft force at a constant tillage depth and travel speed with a device that accurately measures tillage depth and travel speed. The working load measurement system consists of a draft force measurement part, a tillage depth measurement part, and a real-time kinematic global positioning system (RTK-GPS) Mini Survey Antenna GPS1000, Swift Navigation, San Francisco, CA, USA). Figure 3 illustrates the load measurement system along with the installation positions of each measurement component. The part of draft force measurement consisted of a triangular jig with load cells (UU-T2, DACELL, Cheongju, Republic of Korea) to measure the draft force. The draft force (D) was obtained as the sum of the loads measured on the three load cells that measure tangential forces (F_A, F_B, F_C).

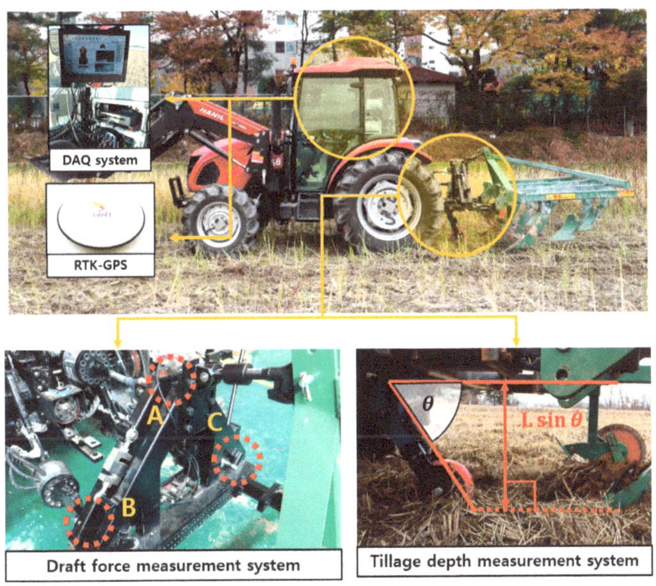

Figure 3. Configuration of field load measurement system.

$$D = F_A + F_B + F_C \quad (14)$$

The segment dedicated to tillage depth measurement facilitates the real-time tracking of the moldboard plow's vertical penetration depth that was achieved by affixing a custom-designed jig, housing an inclinometer (IS2MA090-U-BL, GEMACsensors, Chemnitz, Germany), at the base of the draft force measurement system [5].

$$d = L \sin\theta_1 - L \sin\theta_2 \tag{15}$$

where d is the tillage depth (cm), L is the length of the lower link of tillage depth measurement system (cm), θ_1 is the initial angle of the lower link measured using the inclination sensor when tillage depth is zero (deg), and θ_2 is the angle of the lower link measured using the inclination sensor when tillage depth occurs (deg).

The RTK-GPS is designed to attach to the tractor's center of gravity and measure the speed at which it is reduced by slip during tillage operations. The data acquisition system, Dewesoft X (Dewesoft 3X, Dewesoft, Trbovlje, Slovenia), was employed for concurrent field data measurements, operating at a sampling frequency of 1 kHz.

2.4. Field Experimental Design

The predicted draft force from the DEM simulation was compared to field tests to demonstrate its accuracy. Load measurement tests were performed on two fields in different locations to compare the draft force due to differences in soil properties. Field A was located at 37°00′44.4″ N and 126°30′20.4″ E and Field B was located at 37°93′67.2″ N and 127°78′20.6″ E. The tillage operation was performed by speeding the M3 gear (theoretical speed is 7.9 km/h) in four-wheel drive mode on a 100 m straight course with a moldboard plow. The measuring tractor's results from each field, including the travel speed, tillage depth, and draft force, were recorded across three drives and subsequently organized in ascending order based on tillage depth.

2.5. Theoretical Method for Draft Force Prediction

Typically, the validation of the predicted draft force from tillage operations involves comparing these values against those derived from traditional theoretical calculation methods [34]. For the theoretical calculations, ASAE Standard D.497.4 was used [35]. This method stands out as the simplest and most prevalent due to its reliance on soil texture details, tillage depth, implement geometry, and travel speed to make predictions.

$$D = F_i[A + BV + CV^2]\text{bd} \tag{16}$$

where F_i is the dimensionless soil texture adjustment with different values (if soil has fine texture F_i = 1, medium texture F_i = 0.7, and coarse texture F_i = 0.45), A, B, and C are machine-specific parameters determined by the implement type. In the case of moldboard flow, A is 652, B is 0, and C is 5.1.

3. Results

3.1. Result of Soil Properties Measurement

The results of the soil's physical and mechanical properties measured in the field according to the soil layer are presented in Table 5.

Atterberg limits testing showed that the plastic and liquid limits for Field A were 21.1 ± 1.55% and 34.59 ± 0.81%, respectively, and the plastic and liquid limits for Field B were 19.62 ± 1.22% and 32.31 ± 0.64%. The contact modeling parameters were selected based on the assumption that all soil layers were classified as plastic, which has a similar behavior to the compressible sticky model in the soil starter pack of DEM software (EDEM, 2022, Altair, Troy, MI, USA). A soil particle analysis revealed that Field A exhibited a loam texture composed of 40% sand, 48% silt, and 12% clay, while Field B demonstrated a clay loam texture comprising 40% sand, 28% silt, and 32% clay. The shear strength obtained based on the vane shear torque increased with depth, with Field A increasing by about 1.67 times to 25.19 kPa at Layer 1 and 42.25 kPa at Layer 4, and Field B increasing by about 1.43 times to 65.32 kPa at Layer 1 and 93.10 kPa at Layer 4. Identifying the hardpan layer's formation site is crucial, as tillage operations aim to disrupt this layer. Figure 4 shows the results of the cone penetration test for Field A and Field B. For the cone penetration test in Field A, the cone index is 491.5, 586.8, 706.4, and 2008.9 kPa, in the order of the layers. The cone index for field B is 548.96, 731.81, 824.75, and 1090.59 kPa, in the order of each layer. For both Field A and Field B, the cone index increases from Layer 3 to Layer 4 by 64.8% and 24.4%, respectively, indicating that the hardpan is located in Layer 4.

Table 5. Measured on-field soil properties according to soil layer.

Field	Properties	Soil Layer			
		Layer 1 (0~5 cm)	Layer 2 (5~10 cm)	Layer 3 (10~15 cm)	Layer 4 (15~20 cm)
A	Bulk density (g/cm^3)	1.50	1.60	1.72	1.90
	Moisture contents (%)	32.21	34.20	28.72	24.23
	Cone index (kPa)	465.48	547.92	636.96	1900.2
	Shear strength (kPa)	25.19	30.87	32.90	42.25
B	Bulk density (g/cm^3)	1.90	1.91	1.96	2.02
	Moisture contents (%)	26.56	23.81	22.49	20.24
	Cone index (kPa)	707.56	868.44	1063.16	2153.64
	Shear strength (kPa)	65.32	68.24	72.14	93.10

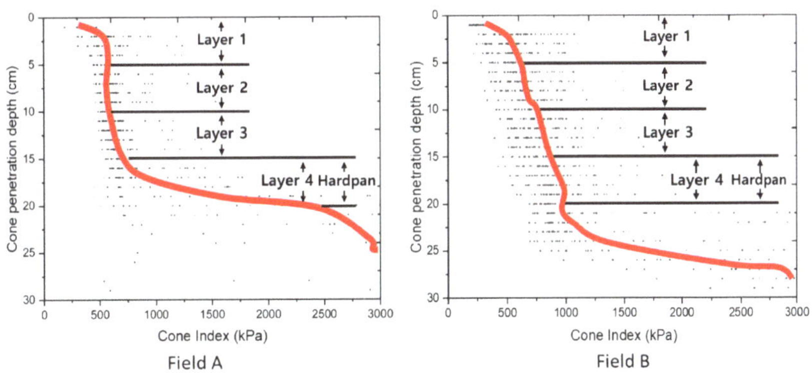

Figure 4. Results of cone penetration test.

3.2. Soil Modeling and Calibration Using DEM

Considering the computational power of the computer, a particle size of a 5 mm radius was selected for the soil modeling. The soil modeled using the discrete element method had more voids than the real soil due to the larger particle size. Therefore, the mass of the particles was calibrated based on the measured bulk density for a realistic draft force prediction. Soil particles were produced within a small soil bin (700 mm × 700 mm × 120 mm) and the total number of particles generated was 20,684. The calibration results are shown in Table 6.

Following the bulk density calibration, the interparticle surface energy was calibrated based on the shear torque obtained from the vane shear test. The vane shear tester was modeled as a CAD step file, as shown in Figure 5, and imported into DEM software (EDEM, 2022, Altair, Troy, MI, USA) to measure the shear torque generated by rotating the vane blades at the same angular speed as the field test. The generated particles were standardized with a radius of 5 mm, mirroring the simulation used for predicting draft force. The shear torque was measured by setting the vane shear tester to penetrate only the vane part to minimize the impact of vertical loading on the measurement. The surface energy values calibrated using shear torque are shown in Table 7. The shear strength error in Field A ranged from 8.51% to 54.9%, and the calibration process reduced the error by 0.16% to 0.89%. In the calibration process for Field A, the input surface energy values were

130, 150, 260, and 330 J/m^2 allocated to Layer 1, Layer 2, Layer 3, and Layer 4. For Field B, the error in the shear strength ranged from 9.4% to 39.53%, and the calibration process reduced the error from 0.52% to 1.35%. In the calibration of Field B, the input surface energy values allocated to Layer 1, Layer 2, Layer 3, and Layer 4 were 100, 300, 500, and 900 J/m^2 respectively.

Table 6. Calibration results of bulk density.

Field	Soil Layer	Procedure	Particle Mass (kg)	Bulk Density (kg/m^3)		Calibration Error (%)
				Simulated	Measured	
A	Layer 1	Initial	0.00144	1538.90	1496.58	2.83
		Calibrated	0.00138	1508.42		0.79
	Layer 2	Initial	0.00144	1559.09	1597.01	2.37
		Calibrated	0.00147	1597.44		0.03
	Layer 3	Initial	0.00144	1557.89	1715.29	9.18
		Calibrated	0.00159	1714.88		0.02
	Layer 4	Initial	0.00144	1565.49	1904.46	17.80
		Calibrated	0.00175	1904.81		0.02
B	Layer 1	Initial	0.00144	1269.91	1899.8	33.16
		Calibrated	0.00174	1901.43		0.09
	Layer 2	Initial	0.00144	1269.91	1905.64	33.36
		Calibrated	0.00175	1912.36		0.35
	Layer 3	Initial	0.00144	1905.69	1955.48	35.05
		Calibrated	0.00179	1956.07		0.03
	Layer 4	Initial	0.00144	1269.91	2021.54	37.18
		Calibrated	0.00185	2021.64		0.005

(a) (b)

Figure 5. Calibration of shear torque using vane shear test: (**a**) field experiments and (**b**) EDEM simulation.

3.3. *Prediction of Draft Force Based on DEM Simulations*

3.3.1. Results of Draft Force Measurements from Actual Field Tests

The actual field tests were conducted three times with the gear selection of the M3 for each field following the draft force measurement procedure. Table 8 shows travel speed and draft force for each tillage depth. The target tillage depth was 15–20 cm, and the average tillage depth was 16.28 cm in Field A and 16.9 cm in Field B. The actual travel speed measured through RTK-GPS was reduced by slip, showing Field A at 5.63 km/h and Field B at 5.43 km/h in gear M3. In order to determine the difference in draft force based on the soil properties, it was necessary to uniformize the tillage depth, which has a large impact on the draft force. Based on the tillage depth, the most data were recorded at 16–17 cm and the draft force was averaged at that depth. The results showed that the measured draft force was 12.38 kN in Field A and 14.79 kN in Field B.

Table 7. Calibration results for shear torque in small soil bin using field measurements values.

Field	Soil Layer	Procedure	Surface Energy (j/m^2)	Shearing Torque (Nm)		Calibration Error (%)
				Simulated	Measured	
A	Layer 1	Initial	100	4.73	5.17	8.51
		Calibrated	130	5.13		0.77
	Layer 2	Initial	100	5.11	6.33	19.27
		Calibrated	150	6.32		0.16
	Layer 3	Initial	100	3.81	6.75	43.56
		Calibrated	260	6.69		0.89
	Layer 4	Initial	100	3.91	8.67	54.9
		Calibrated	330	8.64		0.35
B	Layer 1	Initial	100	12.14	13.4	9.4
		Calibrated	100	13.55		1.19
	Layer 2	Initial	100	12.56	14	10.29
		Calibrated	300	13.84		1.11
	Layer 3	Initial	100	11.93	14.8	19.39
		Calibrated	500	15		1.35
	Layer 4	Initial	100	11.55	19.1	39.53
		Calibrated	900	19.2		0.52

Table 8. Draft force measurement results from actual field experiment.

Field A			Field B		
Tillage Depth (cm)	Travel Speed (km/h)	Draft Force (kN)	Tillage Depth (cm)	Travel Speed (km/h)	Draft Force (kN)
15	5.89 ± 0.35	11.61	15	5.83 ± 0.51	13.71
16	5.85 ± 0.31	12.11	16	5.68 ± 0.65	14.72
17	5.77 ± 0.33	12.64	17	5.47 ± 0.59	14.86
18	5.62 ± 0.34	13.29	18	5.61 ± 0.51	15.18
19	5.41 ± 0.39	13.53	19	4.99 ± 0.80	17.80
20	5.25 ± 0.48	14.30	20	4.97 ± 0.71	16.81

3.3.2. Creating a Virtual Soil Bed for Simulation

Figure 6 shows a large soil bed that represents the properties of the soil measured in this study. Considering the 1:1 ratio in full-scale moldboard plow geometry and the target tillage depth, the large soil bed size was set to 5000 mm $\times$ 2500 mm$\times$ 300 mm (length $\times$ width $\times$ depth) to minimize the interaction between the walls and the particles. The height of each soil layer in the virtual large soil bed was set to 50 mm for Layers 1, 2, and 3 and 150 mm for Layer 4. The total number of particles in the large soil bed was 4,004,939.

The simulation time was 3.2 s to ensure that the moldboard plow was sufficiently penetrated to represent a constant working load. The tillage depth in the simulation was set to 16.5 cm to compare with the actual field test conducted for the purpose of breaking the hardpan layer.

3.3.3. Accuracy Verification of DEM Simulation for Draft Force Prediction

Draft force is a force generated in the same axis as the working direction when the implement is towed by a tractor. To predict the draft force in this study, the total force generated in the Y-axis, the working direction, was analyzed through EDEM (EDEM, 2022, Altair, Troy, MI, USA). Then, the accuracy of the predicted draft force was demonstrated by comparing it to measured values from actual field experiments and to values obtained from the traditional theoretical method. As shown in Table 9, the simulation results show that Field A has a predicted draft force of 14.02 kN and Field B has a predicted draft force of 18.56 kN. The predicted draft forces obtained through the theoretical calculation method are 24.42 kN in Field A and 24.08 kN in Field B. Comparing the draft force measured in the actual field test to the draft force predicted by the simulation, the error for Field A is 13.25% and the error for Field B is 25.49%. The draft force predicted by the theoretical calculation method was 24.42 kN in Field A and 24.08 kN in Field B. Figure 7 shows a graph of the measured draft force from an actual field trial and predicted draft force based on simulation. Savitsky–Golay has the advantage of preserving the shape of the peaks when smoothing [36], so the graph of the draft force recorded in the form of a waveform was fitted with the Savitsky–Golay method. Figure 8 shows a comparison of Field A and Field B, which includes the actual measured draft force, the draft force predicted by the DEM simulation, and the values obtained using traditional theoretical methods. The actual measured draft force was 13.28% higher in Field B than in Field A, while the DEM simulation predicted that the draft force in Field B was 32.38% higher than that in Field A. Despite the errors, the increasing trend of the draft force could be demonstrated through the discrete element method.

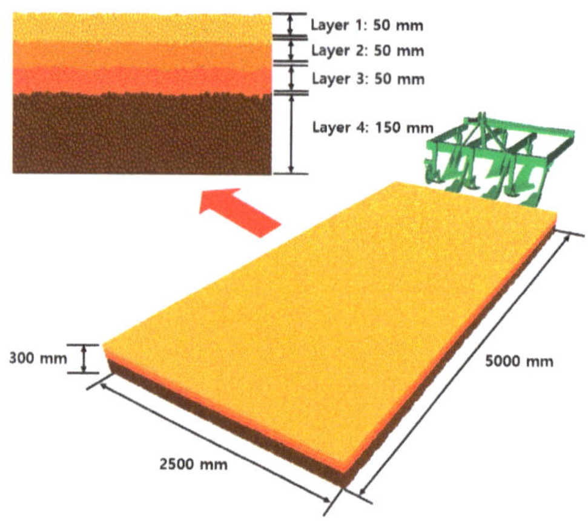

Figure 6. Virtual large soil bed made based on measured soil properties by soil layer.

Table 9. Result of draft force predict simulation.

Field	Tillage Depth (cm)	Travel Speed (km/h)	Draft Force (kN)		Predicted Error (%)
			Simulated	Measured	
A	16.5	5.63	14.02	12.38	13.25
B		5.43	18.56	14.79	25.49

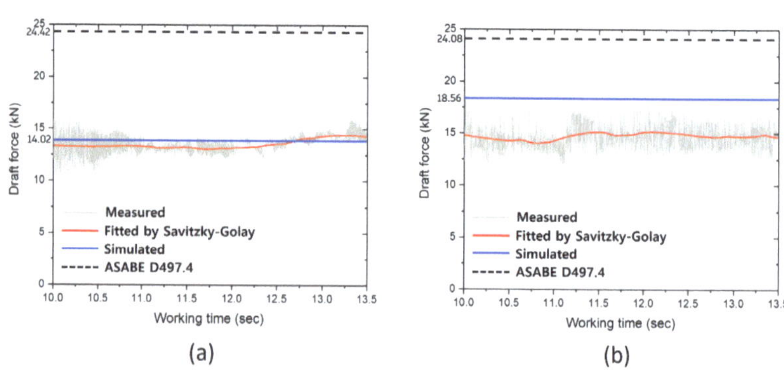

Figure 7. Comparison of draft force between actual field test and DEM simulation at 16.5 cm tillage depth. (**a**) Field A and (**b**) Field B.

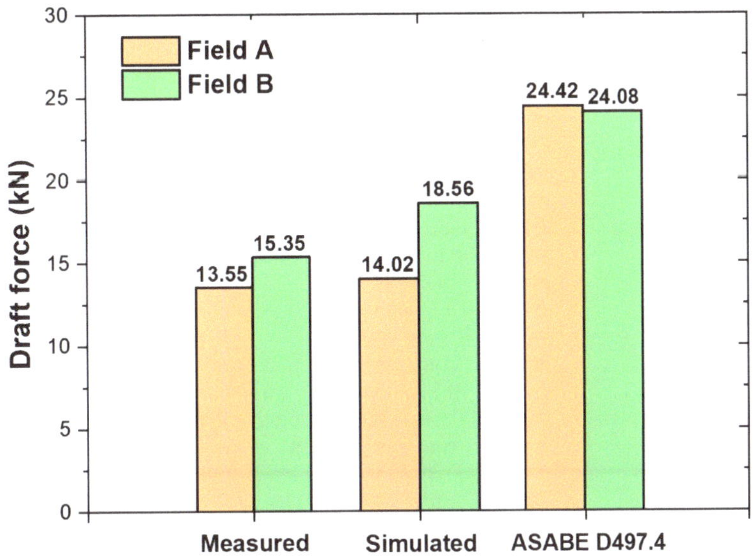

Figure 8. Comparison of draft force between Field A and Field B.

4. Discussion

In this study, a procedure for measuring soil properties was established, and property measurement tests were conducted at different depths of the soil in two cohesive fields with different properties. Measured properties include soil texture, moisture content, bulk density, shear strength, cone index, plastic limit, and liquid limit. First, the soil texture of the two fields was determined, with Field A as loam and Field B as clay loam. To determine the difference in draft force based on the properties of the working environment, the cone index of the soil was first compared. The cone index is one indicator of soil compaction, and past research has shown that the cone index is proportional to the workload of a tractor [26]. Compared to Field A, Field B was 1.52 times higher in Layer 1, 1.58 times higher in Layer 2, and 1.62 times higher in Layer 3, while in Layer 4, Field A was 1.84 times higher than Field B. In terms of shear strength, Field B measured higher than Field A in all soil layers: 2.59 times in Layer 1, 2.21 times in Layer 2, 2.19 times in Layer 3, and 2.2 times in Layer 4. For moisture content, Field A had a higher measured moisture content in all layers compared to Field B: 1.21 times at Layer 1, 1.44 times at Layer 2, 1.28 times at Layer 3, and 1.2 times at Layer 4. Rashidi et al. [37] reported that the draft force was the highest at 484 kgf in soil with 11.27% moisture content and the lowest was at 427 kgf in soil with 22.87% moisture content. In addition, S.A. Al-Suhaibani et al. [38] confirmed that the draft force decreases as the moisture content increases in soil at 5.08%, 5.14%, and 6.82%. Based on these past studies, for the two fields that were found to be plastic state by the Atterberg limit test, it can be assumed that the lower moisture content in Field B experienced more draft force. In the actual field test, the measured draft force was 12.38 kN in Field A and 14.79 kN in Field B, with Field B being 19.47% higher than Field A. Based on these results, the analysis of soil properties for draft force prediction should be treated comprehensively, and it is difficult to relate only one factor to draft force.

To reflect the properties of the measured soil in the particle modeling, two steps of calibration were performed. Since this process results in more voids than in real soil, the mass of the particles was calibrated based on the bulk density. As a result, the error was reduced to a range of 0.005–0.79%. In addition, unlike cohesionless soil, the cohesive soil modeling was calibrated for surface energy by performing a vane shear test rather than an angle of repose test. As a result, the virtual test closely reproduced the measured shear torque with an error from 0.16 to 1.35%. Despite calibrating the values of bulk density and shear torque to within 1.5% of each other, there was a prediction error of 13.25% for Field A and 25.49% for Field B. This indicates the need for more precise soil property measurements for modeling parameters that were not addressed, such as the coefficient of restitution and plasticity ratio. The fact that more prediction errors occurred in Field B compared to Field A is also an important discussion point. In this study, in the process of reflecting the measured soil properties into the particle modeling, a virtual vane shear test was performed to calibrate the surface energy based on the shear torque. As a result, Field A showed an error in shear torque of less than 1%, while Field B was calibrated with an error of up to 1.35%. It can be seen that an error in the calibration step led to more error in predicting the draft force.

The DEM-simulation-based predicted draft force was 14.02 kN in Field A and 18.56 kN in Field B, differences of 13.25% and 25.49%, respectively, when compared to the actual field test. Compared to the traditional theoretical calculation method, the DEM-simulation-based draft force prediction was 84% more accurate for Field A and 37.32% more accurate for Field B. Since the existing theoretical calculation method comprehensively classified soil texture into three categories (fine, medium, and coarse) and did not reflect other soil properties except soil texture, it could not show the difference in predicted draft force under the same type of implement and almost similar working conditions (tillage depth and travel speed, etc.). On the other hand, DEM-simulation-based draft force predictions were significant even under similar operating conditions (same tillage depth and similar travel speed), reflecting the soil properties that affect draft force.

5. Conclusions

In this study, a draft force prediction model using DEM was developed to verify the effect of soil properties on draft force through simulation. The soil textures of the two fields were loam and clay loam, and the difference in draft force due to the difference in the physical properties of the two fields could be seen. The developed prediction model was also verified by comparing the actual field experiment with the existing theoretical calculation method. Soil property measurements were conducted at each depth in two fields with different physical properties. The Atterberg limit test revealed that both soils were cohesive soil in a plastic state. Past studies have shown that, the higher the shear strength and cone index, and the lower the moisture content in a certain section, the higher the draft force. However, Field B had a higher shear strength, lower moisture content, and lower cone index compared to Field A. Therefore, it is difficult to relate to a single factor when predicting the draft force, and many factors must be considered together. The precision of the predicted draft force from DEM simulations was confirmed by juxtaposing it with the measured draft force obtained from real-field experiments. In Field A, the draft force prediction from DEM simulations demonstrated an accuracy rate of 86.75%, while, in Field B, it reached 74.51%. The process of calibrating the surface energy based on shear torque resulted in more errors in Field B than in Field A, which led to a larger error in the draft force prediction. In future research, it is recommended to further study the modeling of various agricultural soil environments to improve the accuracy of design load prediction for soil–machine systems. The developed DEM-based draft force prediction model is up to 84% more accurate than the existing theoretical method by ASABE D497.4. This demonstrated the importance of reflecting the physical and mechanical characteristics of the working environment when assessing the performance of agricultural machinery. Performance evaluations of soil tillage machines include not only the workload, but also the degree of soil crushing and the flow of soil to invert and properly drain the soil. Future research should focus on enhancing the initial prediction accuracy of workload by employing precise soil property measurements based on modal modeling parameters. Furthermore, it is anticipated that the acceleration of simulation-based design optimization will be achieved through the establishment of a comprehensive database encompassing diverse soil environments. This will enable a more thorough study of load predictions for various types of working machines beyond moldboard plow. For virtual simulation-based performance evaluations, it is important to obtain a database for defining various farming environments and analyzing loads. The results of this study are expected to be used as indicators of soil–tool interaction in future soil–tractor and soil–tire interaction studies for agricultural operations.

Author Contributions: Conceptualization, Y.-S.K. and Y.-J.K.; methodology, Y.-S.K. and W.-S.K.; software, B.-M.B. and T.-J.K.; validation, Y.-S.K. and B.-M.B.; formal analysis, Y.-S.K. and B.-M.B.; investigation, Y.-J.K. and S.-D.L.; data curation, Y.-S.K. and B.-M.B.; writing original, B.-M.B.; writing—review and editing, Y.-S.K. and W.-S.K.; supervision, Y.-J.K. and S.-D.L.; project administration, Y.-J.K. and S.-D.L.; funding acquisition, Y.-J.K. and S.-D.L. All authors have read and agreed to the published version of the manuscript.

Funding: This work was supported by the Industrial Strategic Technology Development Program (20014649, Development of intelligent technology and standard integrated controller for tractor attached implement) funded By the Ministry of Trade, Industry & Energy (MOTIE, Republic of Korea).

Institutional Review Board Statement: Not applicable.

Data Availability Statement: Data are contained within the article.

Conflicts of Interest: The authors declare no conflict of interest.

References

1. Mahatale, Y.V.; Tathod, D.V.; Chavan, V.K. Performance of Reversible Mold Board Plow. In *Emerging Technologies in Agricultural Engineering*; Apple Academic Press: Palm Bay, FL, USA, 2017; pp. 137–163.
2. Kim, W.-S.; Kim, Y.-J.; Park, S.-U.; Nam, K.-C.; Choi, C.-H. Analysis of Traction Performance for an Agricultural Tractor According to Soil Moisture Content during Plow Tillage. In Proceedings of the 2019 ASABE Annual International Meeting, Boston, MA, USA, 7–9 July 2019; pp. 2–7.
3. Raper, R.L. Force requirements and soil disruption of straight and bentleg subsoilers for conservation tillage systems. *Appl. Eng. Agric.* **2005**, *21*, 787–794. [CrossRef]
4. Kim, Y.S.; Kim, W.S.; Baek, S.Y.; Baek, S.M.; Kim, Y.J.; Lee, S.D.; Kim, Y.J. Analysis of Tillage Depth and Gear Selection for Mechanical Load and Fuel Efficiency of an Agricultural Tractor Using an Agricultural Field Measuring System. *Sensors* **2020**, *20*, 2450. [CrossRef] [PubMed]
5. Kim, Y.S.; Kim, T.J.; Kim, Y.J.; Lee, S.D.; Park, S.U.; Kim, W.S. Development of a Real-Time Tillage Depth Measurement System for Agricultural Tractors: Application to the Effect Analysis of Tillage Depth on Draft Force during Plow Tillage. *Sensors* **2020**, *20*, 912. [CrossRef] [PubMed]
6. Aikins, K.A.; Ucgul, M.; Barr, J.B.; Awuah, E.; Antille, D.L.; Jensen, T.A.; Desbiolles, J.M.A. Review of Discrete Element Method Simulations of Soil Tillage and Furrow Opening. *Agriculture* **2023**, *13*, 541. [CrossRef]
7. Fielke, J.M. Finite Element Modelling of the Interaction of the Cutting Edge of Tillage Implements with Soil. *J. Agric. Eng. Res.* **1999**, *74*, 91–101. [CrossRef]
8. Ucgul, M.; Saunders, C.; Fielke, J.M. Comparison of the Discrete Element and Finite Element Methods to Model the Interaction of Soil and Tool Cutting Edge. *Biosyst. Eng.* **2018**, *169*, 199–208. [CrossRef]
9. Bentaher, H.; Ibrahmi, A.; Hamza, E.; Hbaieb, M.; Kantchev, G.; Maalej, A.; Arnold, W. Finite Element Simulation of Moldboard–Soil Interaction. *Soil Tillage Res.* **2013**, *134*, 11–16. [CrossRef]
10. Ibrahmi, A.; Bentaher, H.; Hbaieb, M.; Maalej, A.; Mouazen, A.M. Study the Effect of Tool Geometry and Operational Conditions on Mouldboard Plough Forces and Energy Requirement: Part 1. Finite Element Simulation. *Comput. Electron. Agric.* **2015**, *117*, 258–267. [CrossRef]
11. Wang, X.; Zhang, S.; Pan, H.; Zheng, Z.; Huang, Y.; Zhu, R. Effect of Soil Particle Size on Soil-Subsoiler Interactions Using the Discrete Element Method Simulations. *Biosyst. Eng.* **2019**, *182*, 138–150. [CrossRef]
12. Obermayr, M.; Dressler, K.; Vrettos, C.; Eberhard, P. Prediction of Draft Forces in Cohesionless Soil with the Discrete Element Method. *J. Terramech.* **2011**, *48*, 347–358. [CrossRef]
13. Ucgul, M.; Fielke, J.M.; Saunders, C. Three-Dimensional Discrete Element Modelling (DEM) of Tillage: Accounting for Soil Cohesion and Adhesion. *Biosyst. Eng.* **2015**, *129*, 298–306. [CrossRef]
14. Tekeste, M.Z.; Balvanz, L.R.; Hatfield, J.L.; Ghorbani, S. Discrete Element Modeling of Cultivator Sweep-to-Soil Interaction: Worn and Hardened Edges Effects on Soil-Tool Forces and Soil Flow. *J. Terramech.* **2019**, *82*, 1–11. [CrossRef]
15. Bravo, E.L.; Tijskens, E.; Suárez, M.H.; Gonzalez Cueto, O.; Ramon, H. Prediction Model for Non-Inversion Soil Tillage Implemented on Discrete Element Method. *Comput. Electron. Agric.* **2014**, *106*, 120–127. [CrossRef]
16. DEM Solutions Ltd. *Theory Reference Guide, Version 2.6*; DEM Solutions Ltd.: Edinburgh, UK, 2014.

17. Morrisey, J.P.; Thakur, S.C.; Ooi, J.Y. EDEM Contact Model: Adhesive Elasto-Plastic Model. *Granul. Matter* **2014**, *16*, 383–400.
18. Kim, Y.S.; Siddique, M.A.A.; Kim, W.S.; Kim, Y.J.; Lee, S.D.; Lee, D.K.; Hwang, S.J.; Nam, J.S.; Park, S.U.; Lim, R.G. DEM Simulation for Draft Force Prediction of Moldboard Plow According to the Tillage Depth in Cohesive Soil. *Comput. Electron. Agric.* **2021**, *189*, 106368. [CrossRef]
19. Tan, K.H. *Soil Sampling, Preparation, and Analysis*; CRC Press: New York, NY, USA, 1995; ISBN 0-8247-9675-6.
20. O'Kelly, B.C.; Sivakumar, V. Water Content Determinations for Peat and Other Organic Soils Using the Oven-Drying Method. *Dry. Technol.* **2014**, *32*, 631–643. [CrossRef]
21. Klute, A.; Bradford, J.M.; Gupta, S.C. Compressibility. Methods Soil Anal. Part 1—Physical Mineral. Methods. *Soil Sci. Soc. Am. Am. Soc. Agron.* **1986**, 377–382. [CrossRef]
22. Ditzler, C.; Scheffe, K.; Monger, H.C. *USDA Handbook 18*; Government Printing Office: Washington, DC, USA, 2017.
23. Thien, S.J. A Flow Diagram for Teaching Texture-by-Feel Analysis. *J. Agron. Educ.* **1979**, *8*, 54–55. [CrossRef]
24. Ibrahim, N.M.; Rahim, N.L.; Amat, R.C.; Salehuddin, S.; Ariffin, N.A. Determination of Plasticity Index and Compression Index of Soil at Perlis. *APCBEE Procedia* **2012**, *4*, 94–98. [CrossRef]
25. *D4318-10*; Test Methods for Liquid Limit, Plastic Limit, and Plasticity Index of Soils (D4318-10), Report. ASTM International Standard: West Conshohocken, PA, USA, 2010.
26. Kim, W.S.; Kim, Y.J.; Baek, S.M.; Baek, S.Y.; Moon, S.P.; Lee, N.G.; Kim, T.J.; Siddique, A.A.; Jeon, H.H.; Kim, Y.S. Effect of the Cone Index on the Work Load of the Agricultural Tractor. *J. Drive Control* **2020**, *17*, 9–18. [CrossRef]
27. Benjamin, J.G.; Cruse, R.M. Tillage Effects on Shear Strength and Bulk Density of Soil Aggregates. *Soil Tillage Res.* **1987**, *9*, 255–263. [CrossRef]
28. *ASABE Standards EP542*; Procedure for Using and Reporting Data Obtained with the Soil Cone Penetrometer. ASAE: St. Joseph, MI, USA, 2009; Volume 1999.
29. *ASABE Standards S313.3*; Soil Cone Penetrometer. ASAE: St. Joseph, MI, USA, 2005; Volume 1999.
30. Srivastava, A.K.; Goering, C.E.; Rohrbach, R.P.; Buckmaster, D.R. *Engineering Principles of Agricultural Machines*, 2nd ed.; The American Society of Agricultural and Biological Engineers: St. Joseph, MI, USA, 2006.
31. Hang, C.; Gao, X.; Yuan, M.; Huang, Y.; Zhu, R. Discrete Element Simulations and Experiments of Soil Disturbance as Affected by the Tine Spacing of Subsoiler. *Biosyst. Eng.* **2018**, *168*, 73–82. [CrossRef]
32. Saunders, C.; Ucgul, M.; Godwin, R.J. Discrete Element Method (DEM) Simulation to Improve Performance of a Mouldboard Skimmer. *Soil Tillage Res.* **2021**, *205*, 104764. [CrossRef]
33. Bolotov, A.G.; Shein, E.V.; Makarychev, S.V. Water Retention Capacity of Soils in the Altai Region. *Eurasian Soil Sci.* **2019**, *52*, 187–192. [CrossRef]
34. Ahmadi, I. Development and Assessment of a Draft Force Calculator for Disk Plow Using the Laws of Classical Mechanics. *Soil Tillage Res.* **2016**, *163*, 32–40. [CrossRef]
35. Palmer, R.A. American Society of Agricultural Engineers. *Science* **1953**, *118*, 3. [CrossRef]
36. Savitzky, A.; Golay, M.J.E. Smoothing and Differentiation of Data. *Anal. Chem.* **1964**, *36*, 1627–1639. [CrossRef]
37. Ranjbar, I.; Rashidi, M.; Lehmali, H.F.; Mirmoradi, M.; Abdolalizadeh, E. Modeling of Double Action Disc Harrow Draft Force Based on Tillage Depth and Forward Speed. *Middle East J. Sci. Res.* **2013**, *17*, 1061–1067.
38. Al-Suhaibani, S.A.M.F.; Wahby, A.M.; Aboukarima, I.S.T. Study on the Effect of Soil Moisture Content and Plowing Speed on Draft Requirements of a Moldboard Plow. *J. Adv. Agric.* **2015**, *4*, 477–486.

agriculture

Article

Fertilization Mapping Based on the Soil Properties of Paddy Fields in Korea

Juwon Shin [1], Jinho Won [1], Seong-Min Kim [1,2], Dae-Cheol Kim [1,2] and Yongjin Cho [1,2,*]

[1] Department of Bioindustrial Machinery Engineering, Jeonbuk National University, Jeonju 54896, Republic of Korea; jw970429@naver.com (J.S.); wjh9446@jbnu.ac.kr (J.W.); smkim@jbnu.ac.kr (S.-M.K.); dckim12@jbnu.ac.kr (D.-C.K.)

[2] Institute of Agricultural Machinery ICT Convergence, Jeonbuk National University, Jeonju 54896, Republic of Korea

* Correspondence: choyj@jbnu.ac.kr

Abstract: The purpose of this study was to construct a map of expected fertilization rates for nitrogen (N) and phosphorus (P_2O_5) based on measurements of components in soil samples and to identify the spatial variabilities of four lots of a salt-affected paddy field in Korea. Four salt-affected paddy field lots in Korea were divided into 30 sectors for collecting soil samples. They were then analyzed for soil organic matter (SOM), silicon dioxide (SiO_2), total nitrogen (TN), and available phosphorus ($Av.P_2O_5$) in accordance with international standards. Expected fertilization rates of N and P_2O_5 were developed as prescription standards for the application of fertilizer to paddy fields. They were derived using a model of the fertilization rates of N and P_2O_5. To determine the presence of spatial correlation and continuity in the given fields, a spherical variogram was used. Based on the spherical model with the application of a regular kriging interpolation, maps of the contents of TN and $Av.P_2O_5$ as well as the expected fertilization rates of N and P_2O_5 at each sector of $1 \times 1\ m^2$ were developed. The expected fertilization rate of N at each sector appeared in the range of min. 10.0 g to max. 25.7 g, while that of P_2O_5 appeared in the range of min. 0.68 g to max. 8.46 g.

Keywords: fertilization rate; kriging; soil mapping; variogram

Citation: Shin, J.; Won, J.; Kim, S.-M.; Kim, D.-C.; Cho, Y. Fertilization Mapping Based on the Soil Properties of Paddy Fields in Korea. *Agriculture* **2023**, *13*, 2049. https://doi.org/10.3390/agriculture13112049

Academic Editor: Mojtaba Zeraatpisheh

Received: 4 October 2023
Revised: 19 October 2023
Accepted: 23 October 2023
Published: 26 October 2023

1. Introduction

Soil is regarded as an essential element for agriculture. It not only provides necessary nutrients and moisture needed for the growth of crops but also provides a stable foundation for the habitat. To produce agricultural products of good quality through an improvement in productivity and in the quality of crops, a sustainable management of soil is essential [1–3].

Fertilizer for soil is an important element that affects the growth of crops. It consists of nitrogen, phosphorus, potassium, and so on. Some farmers tend to exceed the level of reasonable fertilization rates needed for the growth of crops to maximize the yield of crops [4]. Conventional fertilization with excess manuring disturbs sustainable agriculture. An excess supply of nitrogen results in an exceeded reasonable level of nutrient needed for the growth of crops and reduces the resistance of crops against insect pests. Furthermore, it accelerates the acidification of the soil, leading to soil contamination [5]. An excess supply of phosphorus could disturb the absorption of nutrients by crops, reduce the level of the pH of the soil, and bring about eutrophication of the soil by creating an insoluble compound bound with either aluminum or iron. Excessive application of fertilizer could create various environmental problems, such as salt accumulation, overnutrition, accumulation of heavy metal and nitrate in the soil, and so on, resulting in adverse effects such as deteriorated growth of crops and consequential decrease in productivity [6]. Thus, the physicochemical properties of soils need to be considered for sustainable management of soil.

On the other hand, due to the geographical characteristics of Korea, the country is surrounded by the sea on three sides, and more than 66% of the country's land area is covered by mountains and lakes, leaving only about 26% of the land area available for agriculture. Therefore, reclamation projects were carried out to maximize land use and develop agricultural areas [7]. Reclamation refers to the conversion of water surfaces such as oceans, rivers, and lakes into arable land. About 81% of the reclaimed land in Korea is used for paddies, and about 21.9% of it is used as salt-affected paddies [8]. "Salt-affected paddy" refers to soil that has been recently reclaimed and has a high salt concentration. Compared to normal soils, salt-affected paddy soil has a lower fertility and a higher salt concentration, which reduces the availability of moisture to crops and can lead to an imbalance of nutrients needed for crop growth. In addition, salt-affected paddy soils have poorly developed soil structure, resulting in poor particle cohesion and low water-holding capacity [9]. Therefore, soil management technology that improves crop quality and yield while considering the chemical and physical properties of the soil is required [10], and sustainable agriculture and high-quality crop production are possible through proper fertilizer input [11].

Precision agriculture is a term describing the agricultural approach to maximize the level of productivity and quality by minimizing the input of resources such as water, fertilizer, manpower, and so on. It is also a term describing the technique to control the specified space of each site for the stable growth of crops [12]. Accordingly, a decision support system that takes spatial and temporal variabilities into account can be provided to keep reasonable levels of harvest and quality of crops. This could realize a safe production system in terms of the environment [13]. A technology that is capable of reducing uneven effects on crop growth is needed, utilizing the efficient processing of information of varying soil properties in the spaces of each lot of a field [14]. Variable rate technology has been introduced to solve such a problem. It could control the fertilization rate by considering spatial variabilities of corresponding factors in soils. To apply variable rate technology effectively, the identification of soil properties by considering spatial variabilities in the soil is essential [15].

To solve potential economic and environmental problems due to an excessive or deficient input of soil fertilizer, the development of a map of soils to design the application of variable rate technology is essential [16]. After developing a spatial statistics model of soil properties, a map of soil properties can be created by applying the spatial statistics model to a geographical database [17]. Rabi [18] has reported that unsampled data could be estimated using kriging interpolation in a variogram model developed based on sample data of soil collected at each site. Zhang and Kovacs [19] have developed a map of predicted soil components by employing sensor data applied to a spatial statistics model. Diacano and Castrignano [20] have applied variable rate technology to each location by employing a developed map of soil components. Aggelopoulou and Pateras [21] have developed a map of nitrogen content in soil through collected samples of soil, applied the fertilization of nitrogenous fertilizer to each lot, and reduced the fertilization rate of nitrogenous fertilizer by approximately 38%. Hong and Kim [22] have introduced variable rate technology to rice and reduced the fertilization rate by approximately 32% compared to conventional fertilization. The map of soil components can be utilized as an important tool for producing crops of uniform quality by taking spatial variability of soil into account.

Identification of soil composition is essential for soil mapping for variable rate technology. However, there is currently no active research on salt-affected paddies in Korea, and it is time to conduct research on salt-affected paddy soil for the realization of precise agriculture.

The purpose of this study is to analyze the soil composition of salt-affected paddies in Korea and to develop a variable rate technology application map for nitrogen and phosphate fertilizers. The specific objectives are as follows.

1. Collect soil samples from salt-affected paddies in South Korea and characterize the soil composition.
2. Derive expected nitrogen and phosphate rates based on the analyzed chemical composition.

3. Create a map of the total nitrogen and available phosphorus content of salt-affected paddies and a map of the expected nitrogen and phosphorus rates for variable rate technology through a variogram and kriging.

2. Materials and Methods

2.1. Analysis of Soil Components and Model of Fertilization Rate

The soil samples used in this study were collected from four places of a salt-affected paddy field (126°52′28″ E, 37°05′09″ N and 3 m elevation) in Suchon-ri, Jangan-myeon, Hwaseong-si, Gyeonggi-do Province in Korea during the period from 19 April 2022 to 23 April 2022 (Figure 1). Temperature and humidity in the ambient environment were 15.1 °C and 45%, respectively. The soil texture was silty clay. Silty clay is grayish in color, soft in texture, and abundant in organic matter, with a clay content of more than 50%, and is mainly distributed in ocean current areas. For the sampling of soils, the paddy field was divided into 30 sectors. The size of each sector was set as 11 × 9 m, 13 × 8 m, 13 × 9 m, and 13 × 9 m, respectively. A dedicated "Soil Sampler (Edelman Auger, Eijkelkamp, The Netherlands)" was used for the sampling of soils. A depth of 5 cm of surface soil was removed for sampling soils. A total of 120 soil samples were collected from the plough layer with a depth of 15 cm. Five samples were collected from each sector. Collected samples were mixed into a single lump to minimize potential deviations in the compositions of components in the soil sample.

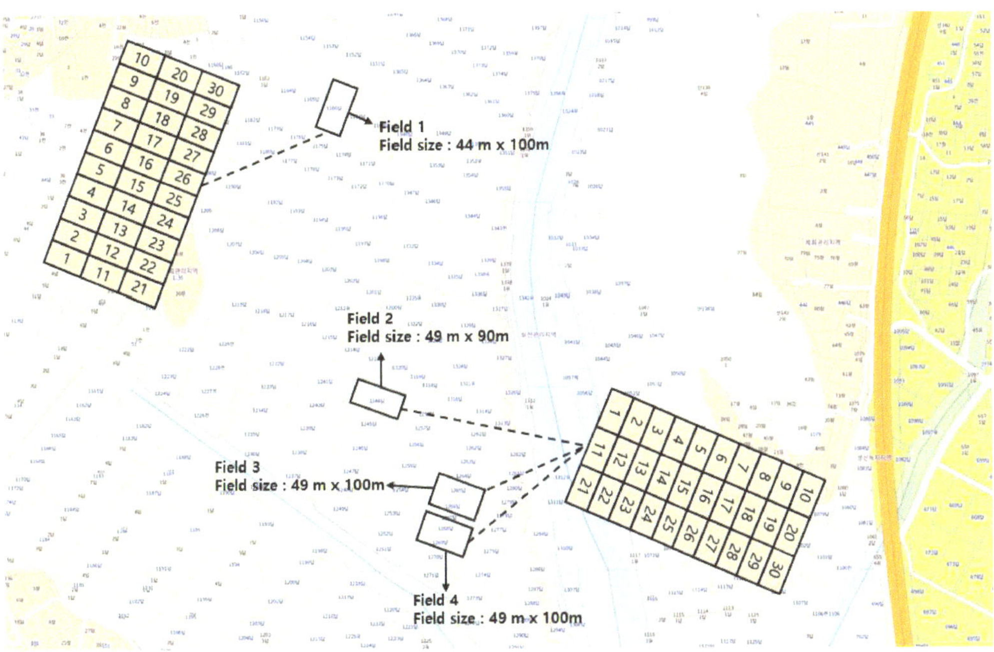

Figure 1. Location of soil sampling point in Hwaseong–si, Gyeonggi–do, Republic of Korea. All fields were divided into 30 sections. Each number was a soil sampling number.

The collected samples of the soil were forwarded to the "Center of Soil Verification" in the "Korea Agriculture Technology Promotion Agency" for analysis of the components in the soil samples. The "Manual of analysis procedures for comprehensive test lab [23]" and the "Method for Chemical Analysis of Soils [24]" of the "Rural Development Administration" in Korea were employed for analyses of soil organic matter (SOM), silicon dioxide (SiO_2), total nitrogen (TN), available phosphorus ($Av.P_2O_5$), and so on in the soil samples. The soil samples were pulverized finely and sifted through a 0.5 mm screen. The components of SOM, SiO_2, Av.P_2O_5, and TN in the soil samples were then analyzed by utilizing the Tyurin method, the method of stationary fresh water incubation, the Lancaster method, and the Dumas method, respectively.

The Tyurin method is an externally heated method of analysis, in which the soil sample is finely ground and sieved through a 0.5 mm sieve, and then 0.3 g of the homogenized soil sample and 10 mL of 0.4 N potassium dichromate sulfate-mixed solution are added to a 250 mL glass triangular flask. In our study, the mixture was then heated on a heating plate at 200 °C and boiled for 5 min from the beginning of bubbling, then removed from the heating plate; the potassium dichromate sulfate mixture was washed with distilled water, and about 5 mL of 85% phosphoric acid (H_3PO_4) and 5–6 drops of the indicator diphenylamine solution were added. Then, it was titrated with 0.2 M ferric ammonium sulfate solution, and the end point was measured when the color of the solution changed from black-brown to indigo-blue to green.

The fresh water constant temperature method was performed by placing 10 g of air- dried fine earth into a 100 mL centrifuge tube, shaking it in 60 mL of distilled water, removing the air, and sealing it. After that, it was placed in a thermostat at 40 °C for 7 days, and the supernatant was filtered through filter paper No. 6. Then, 10 mL of the filtered supernatant was placed in a test tube; 0.25 M HCl solution and 5 mL of ammonium molybdate solution were added; and 10 mL of sodium sulfite solution was added after standing for about 3 min. The treated supernatant was measured at an absorbance of 700 nm after about 10 min of incubation.

The Lancaster method was carried out by placing 5 g of air-dried fine earth in a triangular flask, dissolving 400 mL of HOAC and 300 mL of 10 M lactic acid in 6 L of distilled water, then adding 22.2 g of NH_4F, 133.3 g of ammonium sulfate, and 170 g of NaOH. Then, 20 mL of leachate adjusted to pH 4.25 by adding distilled water was added, shaken for 10 min, and filtered through No. 2 filter paper. Color development and determination were performed by the molybdenum (MO) method with ascorbic acid and the MO method with 1-amino-2-naphtol-4-sulfonic acid.

The Dumas method is a method for quantifying N_2O by reducing it to N_2 and measuring the volume of N_2 gas. The soil sample is oxidized by heating CuO to a high temperature of more than 600 °C; the resulting combustion gas is reduced to N_2 gas by contact with pure CO_2 and hot Cu, and CO is converted to CO_2 by contacting CuO again. The N_2–CO_2 mixture is passed through a nitrometer in a concentrated alkaline solution, and after the CO_2 is captured, the volume of N_2 gas is measured to quantify the nitrogen content.

The model of fertilization rate for N and P_2O_5, which was developed as a fertilization standard of rice [25], was also used for the analysis. The fertilization rate of N corresponds to the components of the soil of SOM and SiO_2. It can be calculated through Formula (1), whereas the fertilization rate of P_2O_5 corresponds to the soil component of Av.P_2O_5. It can be calculated by using Formula (2). Formulas (3) and (4) are transformations wherein a unit area of 10a is converted into 1 m^2 to be compatible with the scale of the lot.

$$N(kg/10a) = 9.05 - 0.108 \times OM + 0.020 \times SiO_2 \quad (1)$$

$$P_2O_5\ (kg/10a) = (100 - Av.P_2O_5) \times 0.1 \quad (2)$$

$$N(kg/m^2) = (9.05 - 0.108 \times OM + 0.020 \times SiO_2) \times 0.001 \quad (3)$$

$$P_2O_5\ (kg/m^2) = (100 - Av.P_2O_5) \times 0.1 \times 0.001 \quad (4)$$

where N is nitrogen (kg/10a); OM is organic matter (g/kg); SiO_2 is silicon dioxide (mg/kg); P_2O_5 is phosphorus (kg/10a); and Av.P_2O_5 is available phosphorus (100 mg/kg).

2.2. Spatial Statistics Model

A variogram generally uses measured data to predict unknown data, to determine the presence of correlation and continuity between data, and to estimate the interdependent distance between data [26,27]. "Correlation" and "variance" are measures to determine the presence of correlation between data. Autocorrelation represents a correlation between different variables at the point of measurement, given that only one datum is known at each point. Thus, it was regarded that it could be employed as a quantitative measure [28,29]. According to Lee and Jung [28], the autocovariance that expresses the quantitative measure of interrelationship between data points, which are distant from the point of specified data, is used. Autocovariance increases in accordance with decreasing distance between two data points. Therefore, it decreases in accordance with increasing distance between two data points [30]. In general, the semivariogram, which is half (1/2) of the variogram, is used. It provides the correlation length (range), threshold (sill), and nugget, which are necessary information for the kriging interpolation (Figure 2). Correlation length refers to the maximum distance of separation that represents the correlation between data points, whereas the threshold indicates the degree of variance of data. The nugget represents a constant value at the separation distance of 0. The selection of the model of the variogram is very important since the identified spatial correlation obtained through using the variogram would validate continuity in the interpolation of spatial data. Thus, the spherical model, which could be employed as an excellent one to determine the spatial variability and continuity of the soil, was used.

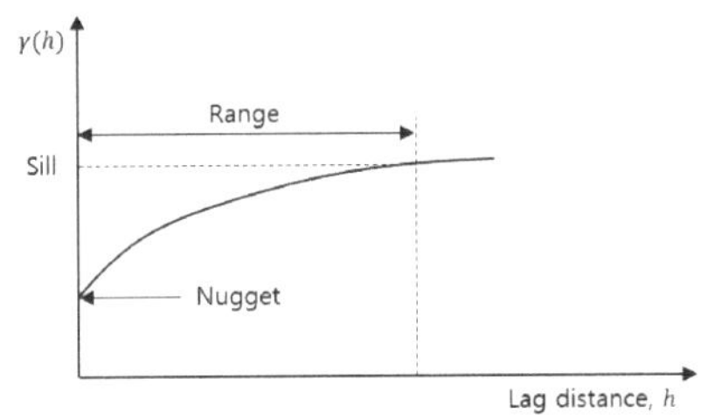

Figure 2. Sill, range, and nugget on variogram.

In this study, a regular kriging interpolation was applied to develop a soil map of the expected N and P_2O_5 fertilization amount suitable for salt-affected paddies, and the mentioned variogram and regular kriging interpolation were performed through Surfer ver. 22 (Golden Software, Golden, CO, USA).

Kriging interpolation renders expected differences between actual and estimated values to be 0 by using the BLUE (best linear unbiased estimator) of the minimum variance for the periphery or zone of given data in an arbitrary spatial space. Simultaneously, it renders dispersion of difference as minimized. A spherical model renders the tangential line at the point of 0 separation distance to intersect the threshold at the point of 2/3 of the correlation length, thereby enabling the derivation of threshold and correlation length.

3. Results and Discussion

3.1. Derivation of Expected Fertilization Rates Accordant with Soil Components

Table 1 presents the soil properties of collected soil samples from each field. Mean and standard deviation of the SOM of the entire field were 20.6 g/kg and 4.482 g/kg, respectively. Those of SiO_2 were 471.6 mg/kg and 72.80 mg/kg, respectively. Mean and standard deviation of Av.P_2O_5 were 44.24 mg/kg and 12.73 mg/kg, respectively. Those of TN were 0.12% and 0.02%, respectively. There were approximately less than 15% of differences compared to the average of the entire field in the statistics of SOM, SiO_2, and Av.P_2O_5. TN showed a difference of less than 20% compared to the average of the entire field. However, the SOM and Av.P_2O_5 in Field 2 and Field 4 exhibited respective values higher than the average of the entire field. In the case of SOM, its value in Field 2 was approximately 24% higher than the average of the entire field, while that of Field 4 was approximately 19% lower than the average of the entire field. With regard to Av.P_2O_5, the value of Field 2 was approximately 19% lower than the average of the entire field, while the value of Field 4 was approximately 37% higher than the average of the entire field. Such variability of the chemical components in the soils of each field was due to environmental factors comprising climate conditions, soil management approaches, and so on. According to Park [31], the variability in the chemical composition of soil could be created by various environmental factors such as the properties of specific soils, the size of the land area, and functions of time factors, consistent with the results of the analysis of soil components in the present study.

Table 1. Soil properties of collected soil samples.

		Min.	25%	Median	75%	Max.	Mean	* S.D.
All Fields (n = 120)	SOM [g/kg]	11.37	16.89	19.96	24.33	31.29	20.60	4.48
	SiO_2 [mg/kg]	300.88	414.60	466.04	524.30	637.40	471.60	72.80
	Total Nitrogen [%]	0.06	0.10	0.12	0.14	0.19	0.12	0.02
	Av.P_2O_5 [mg/kg]	15.39	35.20	41.84	52.35	75.43	44.24	12.73
Field 1 (n = 30)	SOM [g/kg]	11.37	16.74	21.98	23.68	29.91	20.90	4.38
	SiO_2 [mg/kg]	300.88	410.51	450.74	509.21	629.48	456.00	70.47
	Total Nitrogen [%]	0.06	0.10	0.13	0.15	0.19	0.13	0.03
	Av.P_2O_5 [mg/kg]	28.58	33.80	37.81	42.67	48.64	38.10	5.72
Field 2 (n = 30)	SOM [g/kg]	20.43	24.33	25.90	27.39	31.29	25.50	3.09
	SiO_2 [mg/kg]	327.74	387.23	421.58	481.50	565.71	437.40	68.03
	Total Nitrogen [%]	0.11	0.13	0.14	0.15	0.17	0.14	0.02
	Av.P_2O_5 [mg/kg]	15.39	29.31	35.61	41.74	48.78	35.70	9.63
Field 3 (n = 30)	SOM [g/kg]	15.25	17.21	19.05	21.08	24.74	19.20	2.47
	SiO_2 [mg/kg]	409.49	442.02	474.28	533.05	637.40	488.50	52.92
	Total Nitrogen [%]	0.9	0.10	0.11	0.12	0.15	0.11	0.01
	Av.P_2O_5 [mg/kg]	32.43	36.54	41.54	47.65	60.23	42.70	8.00
Field 4 (n = 30)	SOM [g/kg]	11.58	15.39	16.71	18.53	21.98	16.80	2.39
	SiO_2 [mg/kg]	375.14	438.22	505.41	571.92	696.36	504.60	80.42
	Total Nitrogen [%]	0.08	0.09	0.10	0.12	0.14	0.10	0.01
	Av.P_2O_5 [mg/kg]	48.36	54.78	58.53	64.75	75.43	60.60	9.06

* S.D. = Standard Deviation.

Fertilization rates for N and P_2O_5 were derived by using Formulas (3) and (4) (Table 2, Figures 3 and 4) to alleviate variability in soil composition and to obtain a reasonable fertilization rate for each field. According to the fertilization standards for crops in Korea, the optimal nitrogen fertilization rate is 11 to 18 kg/10a, and the optimal phosphorus fertilization rate is 4.5 kg/10a. Derivation of the expected fertilization rate of N for each sector used both SOM and SiO_2, while the derivation of the expected fertilization rate of P_2O_5 used Av.P_2O_5. Mean expected fertilization rates of N for Fields 1–4 were 15.9 kg/10a, 15.1 kg/10a, 16.8 kg/10a, and 17.4 kg/10a, respectively, while those of P_2O_5 were 6.1 kg/10a, 6.4 kg/10a, 5.7 kg/10a, and 4.3 kg/10a, respectively. The predicted fertilization rate of N exhibited a difference of less than 20% compared to the average of each field. The expected fertilization rate of N for Fields 3 and 4 should be set higher than those of Fields 1 and 2 (Figure 3). The difference in fertilization rate of P_2O_5 was insignificant except for that of Field 4.

Table 2. Means and standard deviations (SD) of expected fertilizer.

	All Fields		Field 1		Field 2		Field 3		Field 4	
	Mean (n = 120)	SD (n = 120)	Mean (n = 30)	SD (n = 30)	Mean (n = 30)	SD (n = 30)	Mean (n = 30)	SD (n = 30)	Mean (n = 30)	SD (n = 30)
Nitrogen fertilizer [kg/10a]	16.3	0.27	15.9	0.15	15.1	0.17	16.8	0.12	17.4	0.12
P_2O_5 fertilizer [kg/10a]	5.6	0.12	6.1	0.06	6.4	0.10	5.7	0.09	4.3	0.11

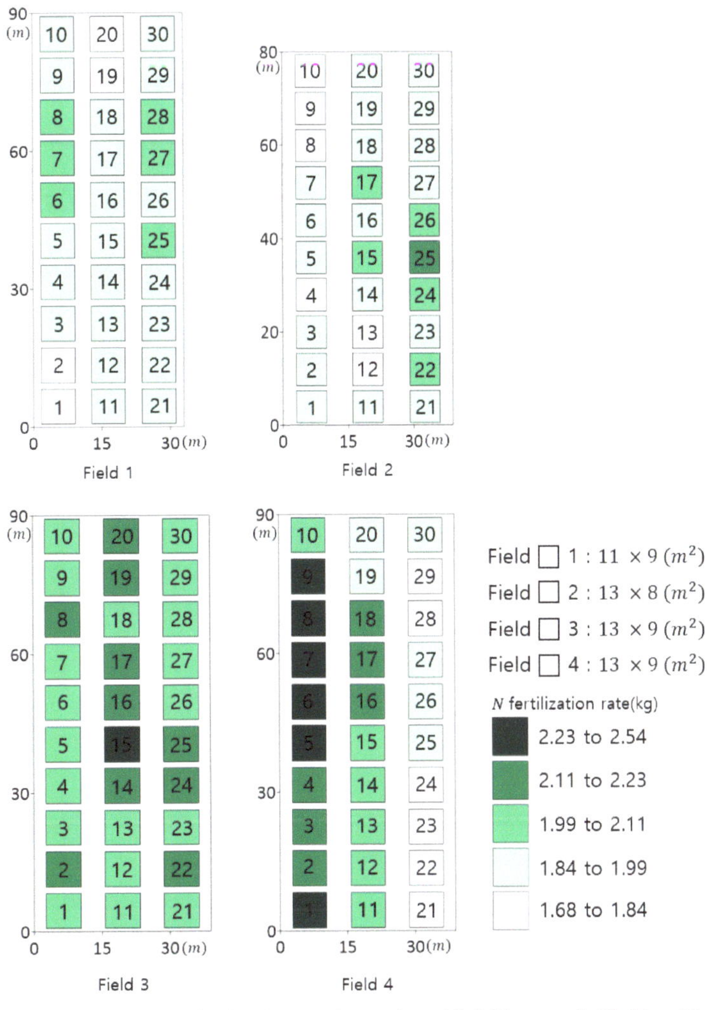

Figure 3. Expected N fertilization rate by section. All fields were divided into 30 sections. Each section size of field 1 to 4 was 11 × 9, 13 × 8, 13 × 9, and 13 × 9 m^2, respectively.

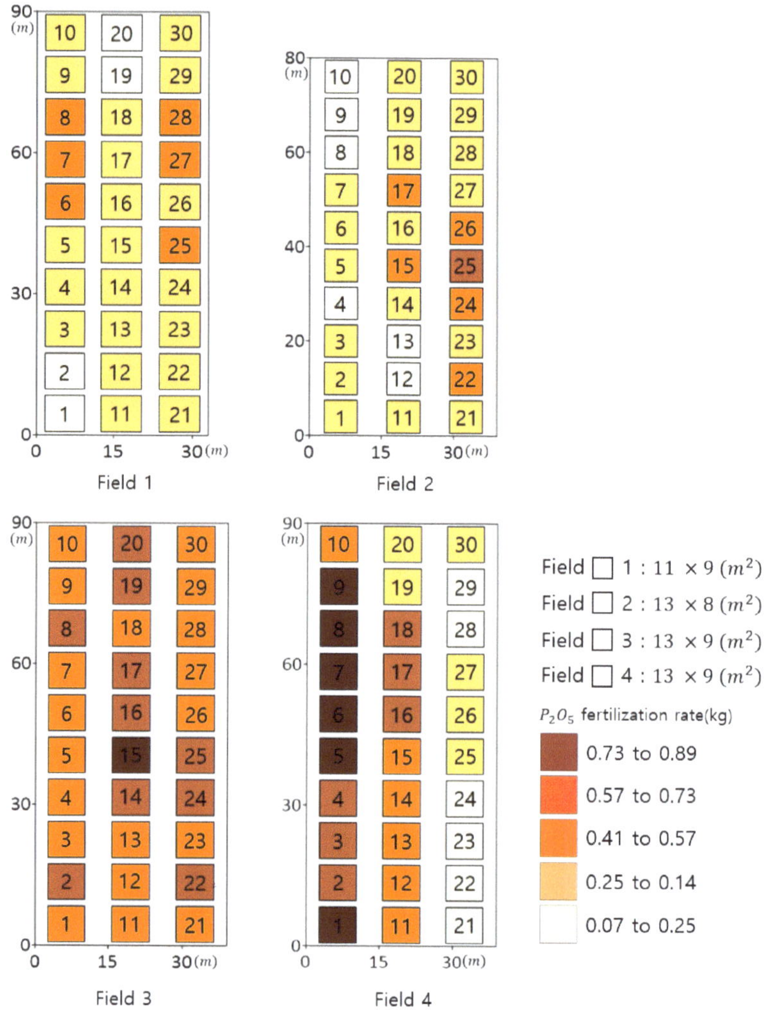

Figure 4. Expected P_2O_5 fertilization rate by section. All fields were divided into 30 sections. Each section size of field 1 to 4 was 11 × 9, 13 × 8, 13 × 9, and 13 × 9 m^2, respectively.

3.2. Spatial Statistics Analysis of Expected Fertilization Rate to Develop a Map of Variable Fertilization

The spherical model of the variogram was employed to construct a map of the variable fertilization of the expected fertilization rate of N and P_2O_5 for each field. The threshold (sill) and correlation length (range) of TN, Av.P_2O_5, N, and P_2O_5 are as presented in Table 3. Figure 5 represents the spherical model of the variogram of N and P_2O_5. The N fertilizer for Field 4 and the P_2O_5 fertilizer for Fields 2 and 3 exhibit a trend of larger spatial variability compared to those of other fields; the value of N fertilizer for Field 4 appeared as 43.26 m, which was higher than 20.43 m and 27.54 m for Fields 2 and 3. Contrarily, a higher value of the sill and a lower value of the range, compared to those of other fields, appeared in the cases of N fertilizer and Av.P_2O_5 fertilizer for Field 2. This was attributed to the lower spatial dependence of N fertilizer and Av.P_2O_5 fertilizer for Field 2 compared to the higher spatial variability in the given space.

Table 3. Sill and range for soil properties and expected fertilizer using variogram.

		Field 1		Field 2		Field 3		Field 4	
		Sill (m)	Range (m)	Sill (m)	Range (m)	Sill (m)	Range (m)	Sill (m)	Range (m)
Soil properties	TN in soil (%)	0.0051	276.60	0.0009	124.00	0.0006	220.00	0.0008	182.90
	Av.P_2O_5 in soil (mg/kg)	33.32	36.31	89.01	27.08	100	99.90	66.26	30.85
Expected fertilizer	N fertilizer (kg/Sector *)	0.0199	28.42	0.0220	20.43	0.0189	27.54	0.0614	43.26
	P_2O_5 fertilizer (kg/Sector *)	0.0033	36.31	0.0104	31.19	0.0112	73.68	0.0088	26.13

Sector * = Division of the entire field into 30 (Each sector size of fields 1–4 is 99 m^2, 104 m^2, 127 m^2, and 127 m^2).

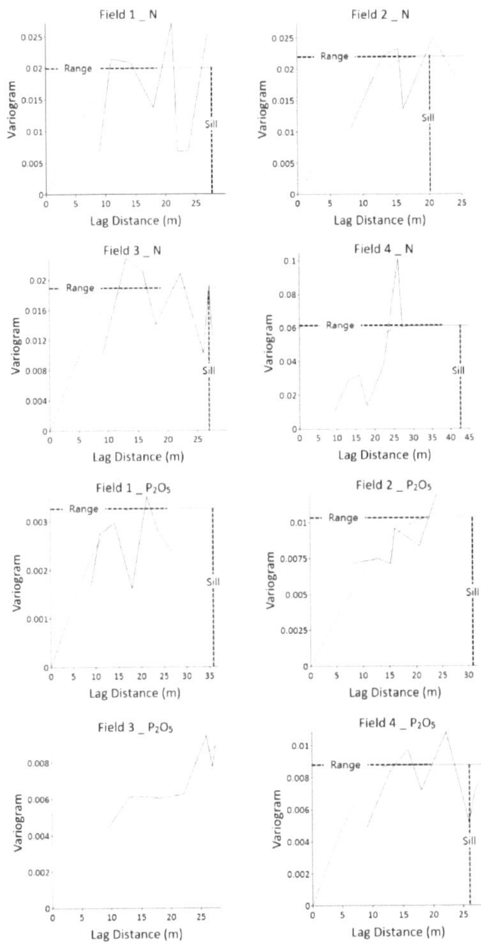

Figure 5. Variogram of N and P_2O_5 using spherical model. The blue and gray lines mean the spherical model and measured data, respectively.

3.3. *Map of Variable Fertilization Developed by the Application of Interpolation*

A map of soil components was developed by employing the regular kriging interpolation, the known spatial statistics technique, based on the spherical variogram model to obtain variable fertilization rates suitable for rice in the paddy fields of Korea. The soil map is divided into (1×1 m^2), taking into account the speed and working radius of the developing variable rate fertilizer applicator for accurate and precise variable rate fertilization. The expected fertilization map can show the appropriate fertilization amount for each compartment in the field (Figures 6 and 7). Figure 6 represents the map illustrating the predicted fertilization rate of TN (upper end of Figure 6) and N (lower end of Figure 6) in the soil for each field. Figure 7 represents the map illustrating the predicted fertilization rate of Av.P_2O_5 (upper end of Figure 7) and P_2O_5 (lower end of Figure 7) in soils for each field. In the map, the parts that correspond to less content or less predicted fertilization rate are painted in white color, and by way of contrast, the parts corresponding to more content or more predicted fertilization rate are painted in either green or orange color. In the case of the fertilization rate of N, the predicted fertilization rate per unit sector (1×1 m^2) appeared as 12 g to 26 g for Field 1 and Field 2 and 10.0 g to 21.7 g for Fields 3 and 4. With regard to the case of the fertilization rate of Av.P_2O_5, the predicted fertilization rate per unit sector (1×1 m^2) appeared as 5.15 g to 7.27 g for Field 1, 3.07 g to 8.46 g for Field 2, 3.16 g to 6.83 g for Field 3, and 0.68 g to 5.21 g for Field 4. The trends in the maps of the Av.P_2O_5 content and predicted fertilization rate of P_2O_5 in the soil for each field appeared oppositely (Figure 7). And the maps of the TN content and predicted fertilization rate of N in the soil for each field appeared with partially contrasting trends. This was attributed to the difference in the content levels of the SOM and SiO_2. It appears that with the utilization of the expected fertilization maps (N and P_2O_5) in the development of the variable rate fertilizer applicator, more precise variable rate fertilization will be achievable in the future. This is believed to enable environmentally friendly and sustainable agriculture, specifically, precision farming, by minimizing the input of fertilizer and increasing crop yields. Thus, it is presumable that the minimization of the input of fertilizer and the increase in production of crops would be enabled by utilizing the map of the predicted fertilization rate (of N and P_2O_5) in a way of realizing an environmentally friendly and sustainable agriculture, precision agriculture.

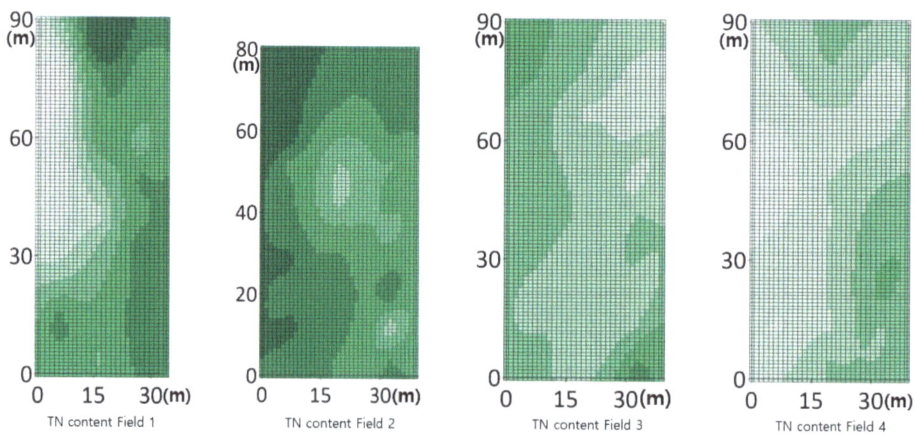

Figure 6. *Cont.*

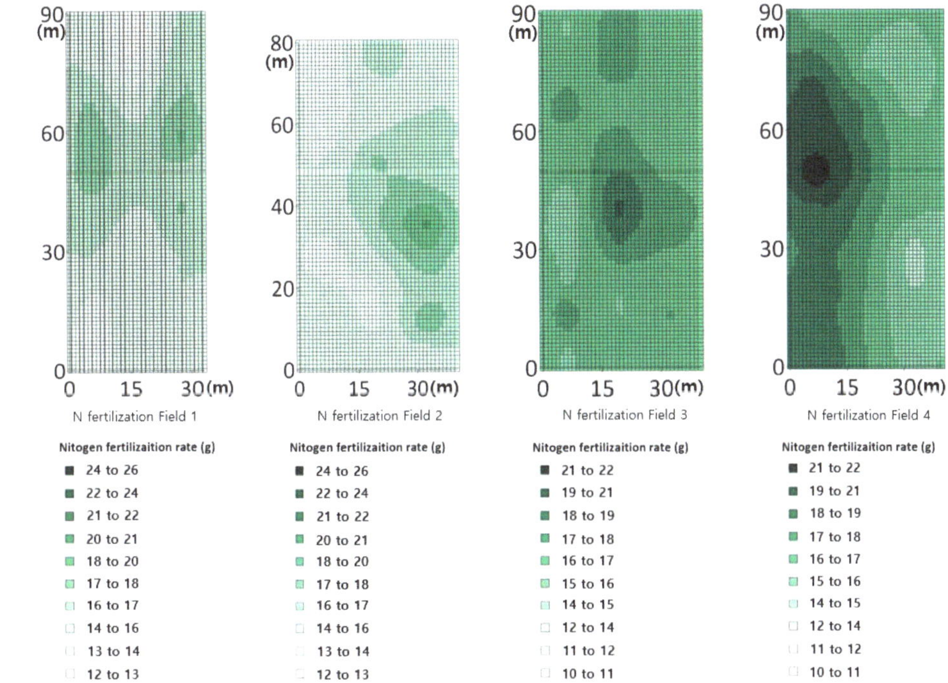

Figure 6. Mapping of TN content in soil and estimated N fertilization rate. Each square color represents the appropriate N fertilization rate for a 1 m^2 divided field.

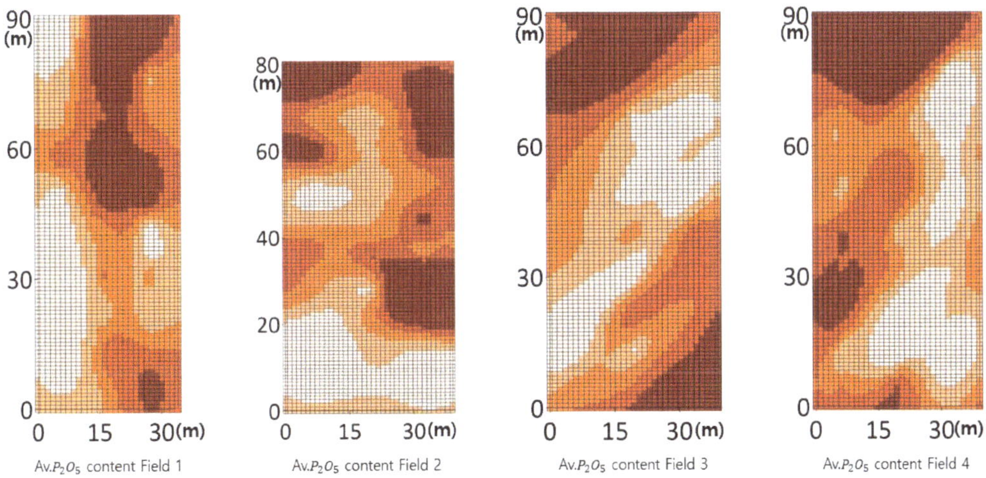

Figure 7. *Cont.*

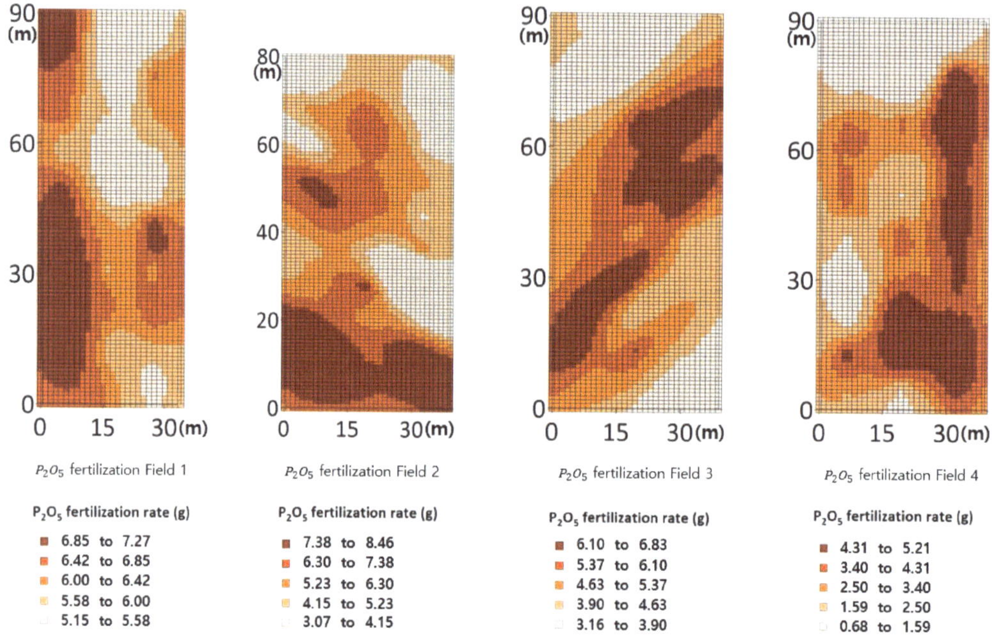

Figure 7. Mapping of Av.P_2O_5 content in soil and estimated P_2O_5 fertilization rate. Each square color represents the appropriate P_2O_5 fertilization rate for a 1 m^2 divided field.

4. Conclusions

The present paper intended to develop a map of the expected fertilization rate of N and P_2O_5 through the measurement of chemical components and the identification of spatial variability in soil samples collected from four lots of a paddy field in Korea. The paddy field was divided into 30 lots for the sampling of soils collected for the analysis. SOM, SiO_2, and Av.P_2O_5 manifested differences of less than 15% compared to the overall average of the field, while the TN exhibited a difference of less than 20% from the overall average of the field.

According to the prescription standard of fertilization for each crop based on chemical components comprising SOM, SiO_2, and Av.P_2O_5, the modeling for the expected fertilization rate of N and P_2O_5 was carried out, from which the predicted average of the fertilization rate of N and P_2O_5 per each lot appeared as 1.78 kg and 0.60 kg, respectively. The spherical model was employed to develop a map of soil components for the application of variable fertilization suitable for crops in Korean paddy fields. In addition, the map of the expected fertilization rate of N and P_2O_5 was developed through the application of regular kriging interpolation based on the spherical model for which the total field was divided into sectors of 1×1 m^2 to achieve precise variable fertilization. The predicted fertilization rate of N for each lot in the entire field appeared in the range from min. 10.0 g to max. 25.7 g, whereas that of P_2O_5 appeared in the range from min. 0.68 g to max. 8.46 g. The map of the predicted fertilization rate of N exhibited a trend partially contrasting with the map of the predicted fertilization rate of TN, whereas the map of the predicted fertilization rate of P_2O_5 manifested a fully contrasting trend with the map of Av.P_2O_5 content. This kind of map of the predicted fertilization rate can minimize the input of resources by enabling a reasonable application of N and P_2O_5 necessary for crops. The map of the expected fertilization rate of P_2O_5 and N developed in the present study needs to be validated through verifying the actual yield and evaluating the quality of rice resulting from its use.

Author Contributions: Methodology, S.-M.K.; Validation, D.-C.K.; Formal analysis, J.W.; Writing—original draft, J.S.; Writing—review and editing, Y.C. All authors have read and agreed to the published version of the manuscript.

Funding: This research was supported by the Rural Development Administration of Korea (PJ01710005) and the National Institute of Agriculture, Forestry, and Food Technology Planning and Evaluation (RS-2023-00236201).

Institutional Review Board Statement: Not applicable.

Data Availability Statement: Not applicable.

Conflicts of Interest: The authors declare no conflict of interest.

References

1. Kim, D. Development and Accuracy Evaluation of Field Soil Temperature Prediction Model by Depth Using Artificial Intelligence and Meteorological Parameters. Master's thesis, The Seoul National University, Seoul, Republic of Korea, 2002.
2. Passioura, J. Soil conditions and plant growth. *Plant Cell Environ.* **2002**, *25*, 311–318. [CrossRef] [PubMed]
3. Tahat, M.M.; Alananbeh, K.M.; Othman, Y.A.; Leskovar, D.I. Soil health and sustainable agriculture. *Sustainability* **2020**, *12*, 4859. [CrossRef]
4. Lemaire, G.; Gastal, F. N uptake and distribution in plant canopies. In *Diagnosis of the Nitrogen Status in Crops*; Springer: Berlin/Heidelberg, Germany, 1997; pp. 3–43.
5. Wei, W.; Yang, M.; Liu, Y.; Huang, H.; Ye, C.; Zheng, J.; Guo, C.; Hao, M.; He, X.; Zhu, S. Fertilizer N application rate impacts plant-soil feedback in a sanqi production system. *Sci. Total Environ.* **2018**, *633*, 796–807. [CrossRef] [PubMed]
6. Savci, S. An agricultural pollutant: Chemical fertilizer. *Int. J. Environ. Sci. Dev.* **2012**, *3*, 73. [CrossRef]
7. Jeong, K.S. Sihwa Construction Project. *J. Korean Water Resour. Assoc.* **1996**, *29*, 106–115.
8. Kim, H.; Kim, S.; Lee, D. The effects of irrigation water salinity levels on growth of rice by growth stage. *J. Korean Soc. Agric. Eng.* **2019**, *2019*, 73.
9. Lee, S.H.; Yoo, S.H.; Seol, S.I. Assessment of Salt Damage for Upland-Crops in Dae-Ho Reclaimed Soil. *J. Korean Environ. Agric.* **2000**, *19*, 358–363.
10. Monzon, J.P.; Calviño, P.; Sadras, V.O.; Zubiaurre, J.; Andrade, F.H. Precision agriculture based on crop physiological principles improves whole-farm yield and profit: A case study. *Eur. J. Agron.* **2018**, *99*, 62–71. [CrossRef]
11. Khuang, T.; Huan, T.; Hach, C. Study on fertilizer rates for getting maximum grain yield and profitability of rice production. *Omonrice* **2008**, *16*, 93–99.
12. Stafford, J.V. Essential technology for precision agriculture. In Proceedings of the Third International Conference on Precision Agriculture, Minneapolis, MN, USA, 23–26 June 1996; Wiley Online Library: New York, NY, USA; pp. 593–604.
13. Van Rees, H.; McClelland, T.; Hochman, Z.; Carberry, P.; Hunt, J.; Huth, N.; Holzworth, D. Leading farmers in South East Australia have closed the exploitable wheat yield gap: Prospects for further improvement. *Field Crops Res.* **2014**, *164*, 1–11. [CrossRef]
14. Cho, S.I.; Chang, Y.C.; Yeo, W.Y. Construction of a Spatioal Database for Realizing Precision Farming. *J. Biosyst. Eng.* **2002**, *27*, 161–172.
15. Sawyer, J.E. Concepts of variable rate technology with considerations for fertilizer application. *J. Prod. Agric.* **1994**, *7*, 195–201. [CrossRef]
16. Fu, W.; Tunney, H.; Zhang, C. Spatial variation of soil nutrients in a dairy farm and its implications for site-specific fertilizer application. *Soil Tillage Res.* **2010**, *106*, 185–193. [CrossRef]
17. Scull, P.; Franklin, J.; Chadwick, O.A.; McArthur, D. Predictive soil mapping: A review. *Prog. Phys. Geogr.* **2003**, *27*, 171–197. [CrossRef]
18. Sahoo, R.N. *Geostatistics in Geoinformatics for Managing Spatial Variability*; Indian Agricultural Research Institute: Pusa, India, 2003.
19. Zhang, C.; Kovacs, J.M. The application of small unmanned aerial systems for precision agriculture: A review. *Precis. Agric.* **2012**, *13*, 693–712. [CrossRef]
20. Diacono, M.; Castrignanò, A.; Troccoli, A.; De Benedetto, D.; Basso, B.; Rubino, P. Spatial and temporal variability of wheat grain yield and quality in a Mediterranean environment: A multivariate geostatistical approach. *Field Crops Res.* **2012**, *131*, 49–62. [CrossRef]
21. Aggelopoulou, K.; Pateras, D.; Fountas, S.; Gemtos, T.; Nanos, G. Soil spatial variability and site-specific fertilization maps in an apple orchard. *Precis. Agric.* **2011**, *12*, 118–129. [CrossRef]
22. Hong, Y.; Kim, G.; Kim, S.; Kang, T. Development of a soil sensor-based variable rate fertilizer applicator. In Proceedings of the KSAM 2013 Spring Conference, Jeju, Republic of Korea, 31 October–1 November 2013; Volume 18, pp. 139–140.
23. NAAS. *Manual of Analysis Procedures for Comprehensive Test Lab*; National Academy of Agricultural Science, Rural Development Administration: Suwon-si, Republic of Korea, 2017.
24. NAAS. *Methods of Soil Chemical Analysis*; National Academy of Agricultural Science, Rural Development Administration: Suwon-si, Republic of Korea, 2010.

25. NASS. *Fertilization Standard of Crops*; National Academy of Agricultural Science, Rural Development Administration: Suwon-si, Republic of Korea, 2017.
26. Ver Hoef, J.M.; Cressie, N. Multivariable spatial prediction. *Math. Geol.* **1993**, *25*, 219–240. [CrossRef]
27. Yang, F.-g.; Cao, S.-y.; Liu, X.-n.; Yang, K.-j. Design of groundwater level monitoring network with ordinary kriging. *J. Hydrodyn.* **2008**, *20*, 339–346. [CrossRef]
28. Lee, K.S.; Lee, D.H.; Jung, I.K.; Chung, S.O.; Sudduth, K.A. Sampling and calibration requirements for optical reflectance soil property sensors for Korean paddy soils. *J. Biosyst. Eng.* **2008**, *33*, 260–268.
29. Gringarten, E.; Deutsch, C.V. Teacher's aide variogram interpretation and modeling. *Math. Geol.* **2001**, *33*, 507–534. [CrossRef]
30. Hwang, S.H.; Ham, D.H. Evaluation of spatial downscaling methods for enhancement of spatial precipitation estimation. *J. Korean Soc. Hazard Mitig.* **2013**, *13*, 149–164. [CrossRef]
31. Park, N.-W. Integration of categorical data using multivariate kriging for spatial Interpolation of ground survey data. *Spat. Inf. Res.* **2011**, *19*, 81–89.

agriculture

Article

The Influence of Soil Physical Properties on the Load Factor for Agricultural Tractors in Different Paddy Fields

Yi-Seo Min [1], Yeon-Soo Kim [2], Ryu-Gap Lim [3], Taek-Jin Kim [4], Yong-Joo Kim [5,6,*] and Wan-Soo Kim [1,7,*]

1 Department of Bio-Industrial Machinery Engineering, Kyungpook National University, Daegu 41566, Republic of Korea; msg337@gmail.com
2 Department of Bio-Industrial Machinery Engineering, Pusan National University, Miryang 50463, Republic of Korea; yskim23@pusan.ac.kr
3 Department of Smart Agriculture, Korea Agriculture Technology Promotion Agency, Iksan 54667, Republic of Korea; limso@koat.or.kr
4 Department of Drive System Team, TYM R&D Center, Iksan 54576, Republic of Korea; taek.kim@tym.world
5 Department of Smart Agriculture Systems, Chungnam National University, Daejeon 34134, Republic of Korea
6 Department of Biosystems Machinery Engineering, Chungnam National University, Daejeon 34134, Republic of Korea
7 Upland Field Machinery Research Center, Kyungpook National University, Daegu 41566, Republic of Korea
* Correspondence: babina@cnu.ac.kr (Y.-J.K.); wansoo.kim@knu.ac.kr (W.-S.K.)

Abstract: The load factor (LF) of a tractor represents the ratio of actual engine power and rated engine power, and is an important indicator directly used in calculating national air pollutant emissions. Currently, in the Republic of Korea, a fixed value of 0.48 is used for the LF regardless of the working conditions, making it difficult to establish a reliable national air pollutant inventory. Since tractors perform work under soil conditions, soil physical properties directly affect the tractor LF. Therefore, it is expected that more accurate LF estimation will be possible by utilizing soil physical properties. This study was conducted to assess the impact of soil physical properties on the LF. Experimental data were collected in ten different soil conditions. Correlation analysis revealed that the LF exhibited strong correlations with SMC, soil texture, and CI, in that order. The coefficient of determination for the regression model developed using soil variables ranged from 0.678 to 0.926. The developed regression models generally showed higher accuracy when utilizing multiple soil variables, as compared to using a single soil variable. Therefore, an effective estimation of the LF through non-experimental methods can be achieved by measuring various soil properties.

Keywords: agricultural tractor; load factor; engine characteristic; soil physical properties; tillage operation

Citation: Min, Y.-S.; Kim, Y.-S.; Lim, R.-G.; Kim, T.-J.; Kim, Y.-J.; Kim, W.-S. The Influence of Soil Physical Properties on the Load Factor for Agricultural Tractors in Different Paddy Fields. *Agriculture* **2023**, *13*, 2073. https://doi.org/10.3390/agriculture13112073

Academic Editor: Jin He

Received: 27 September 2023
Revised: 24 October 2023
Accepted: 26 October 2023
Published: 29 October 2023

1. Introduction

Recently, there has been a significant amount of international interest in addressing environmental pollution problems [1]. The Korean government provides a clean air policy support system (CAPSS) for the management of national air pollutant emissions to analyze emissions and uses it to utilize national air conservation policies [2]. NRMS by non-road mobile machinery (NRMM), including agricultural machinery and construction machinery, is one of the key categories among the various sectors managed by CAPSS [3]. The Republic of Korea's NOx emissions from NRMMs were 311,748 tons in 2019, which is 28.7% of the country's annual NOx emissions [4].

Among the various machinery industries included in NRMM, agricultural machinery is essential for the production of high-efficiency and high-quality agricultural products. Tractors are among the most useful agricultural machinery. In 2019, the annual working area of tractors in the Republic of Korea was 21.7 ha/year, and the annual usage time was 139.9 h/year [5]. They can be specifically employed to perform multi-purpose agricultural tasks by attaching various pieces of working equipment [6].

In the Republic of Korea, national air pollutant emissions from agricultural machinery are calculated using the following formula [3,7], which takes into account the number of units owned, rated power, load factor (LF), and emission factor: Emission (g/y) = Number of machiney × Rated power (kW) × Load factor × Working hours (h/y) × Emission factor (g/kWh). LF is one of the most important factors in the above formula [8]. It is a quantitative indicator of the average power rating of the engine [9,10]. Currently, in the Republic of Korea, LF is used as a fixed value of 0.48 regardless of various conditions, such as the agricultural machinery type, model, power range, and operating conditions [11]. This greatly reduces the reliability of the national emissions inventory for agricultural machinery, and it is necessary to secure an LF generated under actual agricultural working conditions for an accurate representation.

In the domain of construction machinery, there have been studies aimed at quantifying the LF and emissions under actual operational conditions. Previous studies have indicated that the LF and emissions during actual operations are contingent on the specific type of construction equipment [12]. The conclusions of this study propose the potential for devising an emission factor model based on the LF. However, due to the influence of diverse emission control devices, the linear relationship between LFs and emission factors has proven elusive [13]. Barati and Shen (2016) developed an LF estimation model and an emission estimation model [14]. The LF model was developed using three operating parameters, including machine acceleration, road inclination, and machine speed, and the model demonstrated a high R^2 range of 0.973 to 0.986. And the emission model for CO_2, Co, HC, and NOx emission based on the LF exhibited a substantial R^2 range of 0.904 to 0.954. The conclusion of this study was that the LF and emissions have a strong linear relationship, and operational parameters have a high linear relationship with the LF, which suggests that emissions can be accurately estimated through operating parameters. As these operational parameters are crucial to the accuracy of the model, they should be selected taking into consideration the machine's characteristics.

Tractors are mainly operated in soil conditions, and the tractor's load (i.e., the LF) varies greatly depending on the soil conditions [15,16]. Numerous previous studies have indicated that the engine load differs depending on the soil physical properties. Inchebron et al. (2012) evaluated the traction performance of a tractor equipped with a moldboard plow at various tillage depths and SMC conditions [17]. It was demonstrated that elevated soil moisture content (SMC) led to increased rolling resistance and wheel slippage and decreased traction efficiency. Kim et al. (2021) analyzed the effect of SMC on tractor traction performance during moldboard plow operations [18]. The results of the study highlighted that the tractor's traction load varied in accordance with SMC levels. Rasool and Raheman (2018) conducted a study on improving the traction performance of mobile tractors according to the cone index (CI) [19]. The traction performance was assessed by towing a load tractor on the experimental tractor. It was identified that soils with a large CI increased the drawbar force and traction efficiency. Battiato and Diserens (2017) performed the simulation and validation of tractor traction performance under different soil texture conditions [20]. Traction tests were performed using a second tractor as a braking machine. As a result, the traction power of the tractor was found to be different in soils with different soil textures.

In previous studies, the emphasis was primarily on tractor wheel slip and traction force. Multiple soil variables were employed to describe the LF of the tractor. The soil physical properties have intricate interactions with the soil–tractor system, making it challenging to mathematically resolve these relationships. Consequently, numerous studies in terramechanics have been conducted based on empirical experimental values, primarily focusing on individual soil variables. While individual soil physical characteristics can influence the LF, employing a broader range of soil variables could provide a more comprehensive explanation of the engine load of the tractor [21,22]. Previous research indicates that SMC, CI, and soil texture are interrelated, and this relationship appears to be nonlinear [23]. In specific soil conditions, an increase in SMC can initially lead to an increase in CI, followed by a decrease, which is influenced by the soil's particle composition. These characteristics make it challenging to predict the impact of unaccounted soil conditions on the LF. Therefore, estimating tractor load based on various combinations of soil conditions can enhance a model's reliability. However, this approach has limitations due to the difficulty of obtaining soil and engine load data. Consequently, comprehensive studies that evaluate tractor LF based on diverse combinations of soil conditions are relatively scarce.

Research on the influence of soil physical properties on tractor LFs can aid in the appropriate estimation of the LF based on soil variables. Due to the highly diverse and challenging-to-control nature of soil conditions, it is impractical to experimentally evaluate the LF across all soil conditions. Estimating the LF using tractor specifications and soil variables could lead to more accurate exhaust gas emission calculations compared to applying a fixed value of 0.48. This may help to improve the reliability of emission data from tractors in the Republic of Korea.

Therefore, the aim of this study is to evaluate the effect of soil physical properties on the LF of tractors. This study is expected to provide useful information as a foundational study for estimating the tractor LF based on soil physical properties in the future. The specific objectives of this study are as follows: (1) the measurement of tractor engine data using a load measurement system, (2) the measurement of soil physical properties, (3) an evaluation of the influence of soil physical properties on the tractor LF, and (4) the development of an LF estimation model according to soil properties.

2. Materials and Methods

2.1. Experimental Equipment

2.1.1. Agricultural Tractor

In this study, a 78 kW agricultural tractor (S07, TYM, Gongju, Republic of Korea) was utilized. Table 1 details the specific tractor specifications utilized in this study. Figure 1 depicts the engine performance curve of the tractor.

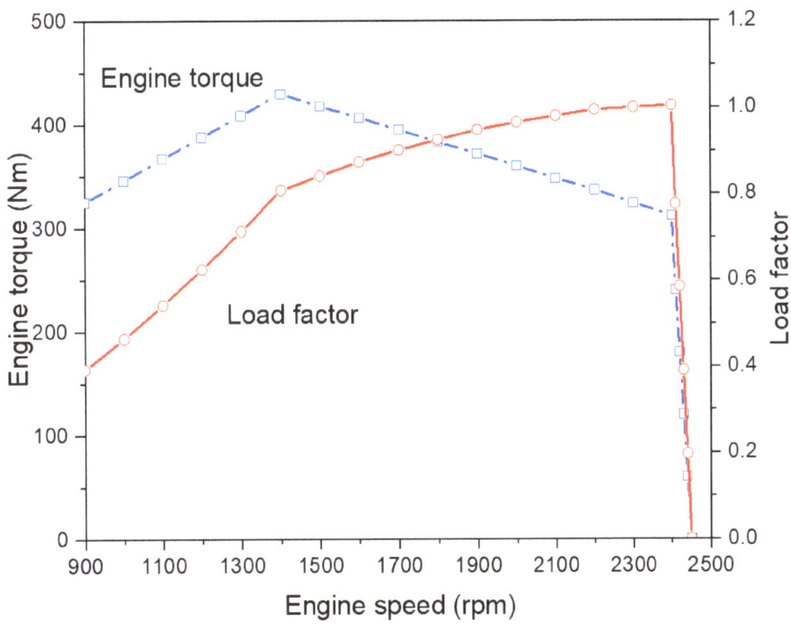

Figure 1. The engine performance curve of the tractor used in this study.

Table 1. Specification of the 78 kW agricultural tractor used in this study.

Item		Specifications
Dimensions (length × width × height) (mm)		4225 × 2140 × 2830
Empty weight (kg)		3985
Engine	Make	DOOSAN INFRACORE
	Type	In-line, 4-cycle
	Aspiration	Turbocharged and intercooler
	Rated torque (Nm)	324 @2300 rpm
	Rated power (kW)	78 @2300 rpm
	Max. torque (Nm)	430 @1400 rpm
	Torque rise (%)	31.3
	Maximum speed (rpm)	2450
	Bore × Stroke (mm)	98 × 113
	Total displacement (cc)	3409
	Compression ratio	17:1
	Dry weight (kg)	500
	Emission Compliance	TIER4-Final/EU STAGE IV
Transmission	Type	Power shuttle
	Gear stage (Forward/Reverse)	32/32

2.1.2. Measurement Equipment

In this study, a tractor equipped with a Tier-4 compliant electronic control unit was utilized. Engine torque and engine rotational speed during field operations were measured using a data measurement device (QuantumX MX840B, HBM, Darmstadt, Germany) through controller network area (CAN) communication. Among various soil physical properties, we selected three representative attributes: SMC, CI, and soil texture. SMC was measured through a soil moisture sensor (TDR350; Spectrum Technology, Aurora, IL, USA) with two rods, each 20 cm in length. CI was measured with a cone penetrometer (SC900; Spectrum Technology, Aurora, IL, USA). The average CI value, calculated from measurements obtained every 25 mm at a depth of 150 mm, was employed [24]. Table 2 demonstrates the specifications of the soil measurement equipment. Soil samples were collected at a depth of 150 mm. The soil particle proportion was determined using the hydrometer method of the Soil Environment Analysis Center of Chungnam National University, and soil texture was classified according to the USDA soil classification system.

Table 2. Specification of the soil measurement equipment used in this study.

Item	Specification
Soil moisture sensor	Measurement units: percentage of volumetric water content (VWC) Range: 0% VWC to saturation Accuracy: ±3.0% VWC
Cone penetrometer	Measurement units: cone index (kPa) Range: 0 to 45 cm, 0 to 7000 kPa Accuracy: ±1.25 cm, ±103 kPa

2.2. *Field Experiment*

2.2.1. Field Site

The field experiment was conducted at 10 sites in the Republic of Korea, as illustrated in Figure 2. All sites were paddy fields where rice was the primary crop, and only stubble remained at the time of the experiment. The experiments were conducted in March of the experimental year, just before the rice was transplanted, following the November harvest of the previous year. Information for each site is presented in Table 3.

Figure 2. Geospatial information of each experimental site.

Table 3. Information about the field experiment sites.

Sites	Field Size (m)	Location	Latitude	Longitude	Experimental Year
1	60 × 100	Seosan	36°46′44.4″ N	126°33′27.0″ E	2017
2	60 × 100	Seosan	36°46′46.2″ N	126°33′28.8″ E	2017
3	40 × 100	Cheongyang	36°30′35.3″ N	126°47′27.5″ E	2017
4	40 × 100	Cheongyang	36°30′35.9″ N	126°47′29.1″ E	2017
5	30 × 100	Anseong	36°56′42.7″ N	127°14′33.2″ E	2019
6	40 × 100	Anseong	36°56′37.8″ N	127°14′04.5″ E	2019
7	60 × 100	Dangjin	36°55′50.0″ N	126°37′57.5″ E	2019
8	40 × 100	Dangjin	36°56′04.1″ N	126°37′58.3″ E	2019
9	40 × 100	Dangjin	36°56′04.3″ N	126°37′59.8″ E	2019
10	40 × 100	Dangjin	36°56′04.8″ N	126°38′01.5″ E	2019

2.2.2. Soil Environment Measurement

Soil sampling can be carried out using various methods, including simple random sampling, systematic sampling, and stratified sampling. Samples are collected according to a regularized pattern in systematic sampling [25]. This method is often more accurate than simple random sampling because it guarantees uniform spatial coverage [26]. Therefore, in this study, soil samples were collected using a systematic sampling method. A total of ten soil samples were collected from each of the ten field sites, and SMC and CI were measured at the same points.

2.2.3. Field Experiment Conditions

In this study, the tractor implement used was an eight-blade moldboard plow (WJSP-8; Woongjin Machinery, Gimje, Republic of Korea). The dimensions of the moldboard plow were 2150 mm × 2800 mm × 1250 mm (length × width × height), with a weight of 790 kg. The tractor's driving gear stage was set to M3 low (7.09 km/h) during plow tillage [8]. The tillage depth was set to be in the range of 13 to 17 cm through the tractor's automatic tilling depth control system, and tillage work was performed with an average tillage depth of 15 cm.

2.3. Data Analysis

2.3.1. Load Factor Analysis

Engine torque and rotational speed data collected during tractor operations in the ten sites were analyzed to determine the LF. The engine power was calculated as in Equation (1), using time-based engine torque and rotational speed. Then, the average of engine power was calculated using time-based engine power. The analysis process of the engine power involved removing outliers using boxplot-based whisker analysis. In particular, this process was used to remove engine power values that were higher than the engine's rated power. The average of engine power data for each site was used in calculating the LF, as shown in Equation (2). One-way analysis of variance (ANOVA) was performed using IBM SPSS Statistics (SPSS 25, SPSS Inc., New York, NY, USA) to evaluate whether soil physical property and engine characteristic data from each site showed significant mean differences between groups.

$$EP = \frac{2\pi TN}{60{,}000}, \tag{1}$$

$$LF = \frac{EP_m}{EP_r}, \tag{2}$$

where EP represents the engine power, T is the engine torque, N is the engine rotational speed, LF is the load factor, EP_m is the measured engine power in real time, and EP_r is the rated engine power.

2.3.2. Correlation Analysis

A correlation analysis was conducted to analyze the impact of soil physical properties on the LF. The linearity between two variables was assessed using the Pearson correlation coefficient (r), as illustrated in Equation (3). The range of r values is from −1 to 1, where an absolute value of 1 indicates a perfect linear relationship and a value of 0 signifies no linear relationship. Each soil variable is considered to have a linear relationship with the LF when the absolute value of r is 0.7 or higher and the significance level of $p < 0.05$.

$$r = \frac{\sum_{i=1}^{n}(X_i - X)(Y_i - Y)}{\sqrt{\sum_{i=1}^{n}(X_i - X)^2 \sum_{i=1}^{n}(Y_i - X)^2}} \tag{3}$$

where X is the mean of sample group X, and Y is the mean of sample group Y.

2.3.3. Regression Analysis

The data collected from field experiments were used to develop regression models. Since the tractor and equipment conditions were consistent throughout the experiments, only soil variables were considered in the regression models. Considering the ease of data collection for each soil variable, regression models were constructed using both individual functions and combinations of functions. Equation (4) can be utilized in cases when only a soil moisture sensor is accessible. Equation (5) can be utilized in cases when only a soil strength sensor is accessible. Soil moisture sensors and soil strength sensors have been widely used as important tools for soil monitoring due to their affordability and ease of measurement. Equation (6) can be utilized when analyzing soil texture of an experimental field. The soil texture incorporates the proportions of sand, silt, and clay, but we only considered the sand proportion, which has the most significant impact on the LF. Using multiple soil texture variables had a detrimental effect on the model's performance, as stated in the paper. Typically, soil texture analysis involves several processes, such as sample collection, drying, and classification, making data measurement more challenging. Furthermore, in order to investigate whether the estimation performance of the model was enhanced by combining multiple soil variables, we employed all possible combinations of these variables (f_1, f_2, f_3, $f_1 + f_2$, $f_1 + f_3$, $f_2 + f_3$, $f_1 + f_2 + f_3$), as stated in the study.

$$LF = f_1(SMC) \tag{4}$$

where SMC is the soil moisture content.

$$LF = f_2(CI), \tag{5}$$

where CI is the cone index.

$$LF = f_3(S_a) \tag{6}$$

where S_a is the sand proportion.

The performance of the estimative model based on regression analysis was evaluated by referring to previous studies. For the assessment of model performance, four statistical metrics were chosen: the coefficient of determination (R^2), mean absolute percentage error (MAPE), root mean square error (RMSE), and relative deviation (RD) [27]. Each of these model performance metrics was calculated using Equations (7)–(10), based on the actual and estimated engine loads.

$$R^2 = \frac{\sum_{i=1}^{i=N}(y_i - y_a) - \sum_{i=1}^{i=N}(y_i - \hat{y}_i)}{\sum_{i=1}^{i=N}(y_i - y_a)}, \tag{7}$$

$$MAPE = \frac{1}{N}\sum_{i=1}^{i=N}\left|\frac{1}{y_i}(y_i - \hat{y}_i)\right| \times 100(\%), \tag{8}$$

$$RMSE = \sqrt{\frac{1}{N}\sum_{i=1}^{i=N}(\hat{y}_i - y_i)^2}, \tag{9}$$

$$RD = \frac{RMSE}{Mean} \times 100, \tag{10}$$

where y_a is the mean actual load factor, y_i is the actual load factor, and $\hat{y}_i$ is the estimated load factor.

3. Results

3.1. Soil Physical Properties

The collected CI and SMC from a total of ten sites are displayed in Figure 3. The CI and SMC exhibited large variations across ten measurements for each site. In terms of the CI, the largest variation was observed at site 9, ranging from 640.6 to 2085.0 kPa. As for the SMC, the most extensive range was observed at site 1, ranging from 24.2 to 41.0%. Additionally, in general, the CI exhibited a wider range than the SMC. The minimum value for CI was observed at site 4, measuring 236.6 kPa, while the maximum value was observed at site 9, measuring 2085.0 kPa. For SMC, the minimum value was noted at site 4, measuring 18.42%, while the maximum value was noted at site 8, measuring 45.90%.

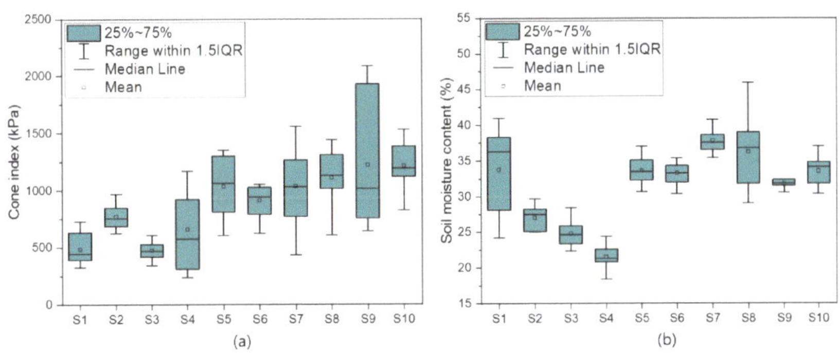

Figure 3. Results of box plot analysis of soil physical properties measured on field experiment sites (e.g., S1 refers to site 1): (**a**) cone index (CI) and (**b**) soil moisture content (SMC).

Figure 4 illustrates the results of the particle distribution proportion analysis for each particle type in the context of soil texture analysis. The sand proportion ranged from 24% to 85% across all sites, with Groups S1 to S4 exhibiting a sand proportion of over 70%. The silt proportion ranged from 13% to 57% across all sites, while Groups S5 to S10 displayed higher silt proportions, exceeding 35%. Clay ranged from 2% to 27%, with Groups S1 to S4 demonstrating a low clay proportion of less than 8%, and Groups S5 to S10 demonstrating a high clay proportion of over 18%. Sites with a high sand proportion exhibited relatively low CI values. These results are consistent with findings reported in previous studies [28].

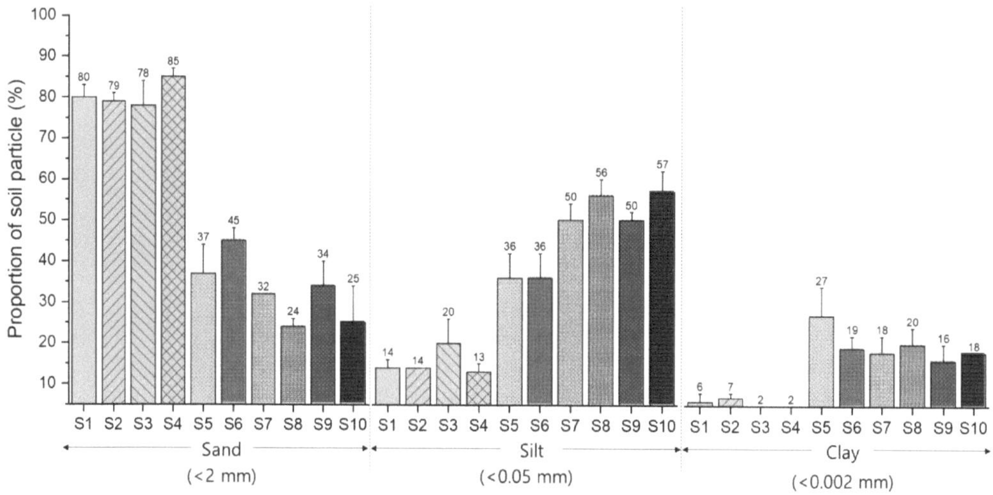

Figure 4. Results of sand, silt, and clay proportion analysis of soil particles by field experiment sites, where S# refers to the number of sites (i.e., S1 = Site 1).

Table 4 presents the analysis results of the soil physical properties of the ten sites used in this study. Soil textures were categorized as loamy sand in four sites, clay loam in one site, loam in two sites, and silt loam in three sites. An ANOVA analysis was conducted to determine whether there were significant differences in the means of soil physical properties data across different sites. The analysis results indicated statistically significant differences between sites, with the exception of a few sites with comparable values. Furthermore, in all cases except for the comparison between sites 8 and 10, each site exhibited statistically significant mean differences in at least one soil physical property. This suggests that data measurements were taken from sites with varying soil physical properties.

Table 4. Analysis results of averaged soil physical properties, including soil moisture content, cone index, soil particle proportions, and soil texture (N = 10).

Sites	SMC (%)	CI (kPa)	Soil Particle Proportions (%)			Soil Texture
			Sand (<2 mm)	Silt (<0.05 mm)	Clay (<0.002 mm)	
1	33.79 ^{b,c}	483 ^{e}	80.00 ^{b}	14.00 ^{e}	6.00 ^{d}	Loamy sand
2	27.01 ^{d}	768 ^{c,d}	79.00 ^{b}	14.00 ^{e}	7.00 ^{d}	Loamy sand
3	24.79 ^{d}	476 ^{e}	78.00 ^{b}	20.00 ^{d}	2.00 ^{e}	Loamy sand
4	21.55 ^{e}	656 ^{d,e}	85.00 ^{a}	13.00 ^{e}	2.00 ^{e}	Loamy sand
5	33.60 ^{b,c}	1034 ^{a,b,c}	37.00 ^{d}	36.00 ^{c}	27.00 ^{a}	Clay loam
6	33.27 ^{c}	910 ^{b,c,d}	45.00 ^{c}	36.00 ^{c}	19.00 ^{b}	Loam
7	37.84 ^{a}	1038 ^{a,b,c}	32.00 ^{e}	50.00 ^{b}	18.00 ^{b,c}	Loam
8	36.24 ^{a,b}	1111 ^{a,b}	24.00 ^{f}	56.00 ^{a}	20.00 ^{b}	Silt loam
9	31.77 ^{c}	1223 ^{a}	34.00 ^{e}	50.00 ^{b}	16.00 ^{c}	Silt loam
10	33.54 ^{b,c}	1212 ^{a}	25.00 ^{f}	57.00 ^{a}	18.00 ^{b,c}	Silt loam

Mean values within same column showing different superscripts are significantly different ($p < 0.05$). Duncan's least significant multiple-range test was employed to compare the means.

Figure 5 displays the results of mapping the 10 study sites onto the USDA soil texture classification triangle. Overall, sites 1–4 are predominantly situated within the loamy sand region, characterized by high sand proportions and low clay proportions. The remaining sites exhibited a balanced mixture of soil particles.

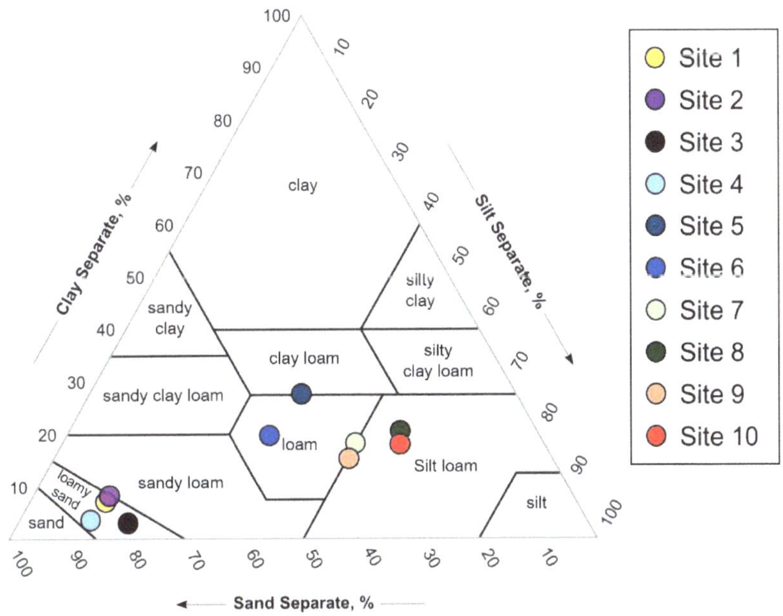

Figure 5. Classification results of the sampled soil by ten field experiment study sites mapped to the USDA soil texture triangle.

3.2. Engine Load Characteristics

The measurement results of the engine load characteristics at each site are depicted in Figure 6. Site 4 had the lowest average engine torque of 127.1 Nm, while site 7 had the highest average engine torque of 359.1 Nm. Site 6 had the lowest average engine rotational speed at 1967 rpm, while site 4 had the highest at 2450 rpm. The average engine power and average LF were lowest at site 4, measuring 32.60 kW and 0.418, respectively, and highest at site 10, measuring 78.33 kW and 1.004, respectively. The engine power and LF exhibited a similar box plot pattern to that of engine torque. Therefore, engine torque is estimated to have a greater effect on LF than engine rotation speed.

The statistical analysis results of the engine load characteristics are presented in Table 5. The average values of the engine load characteristics across all sites ranged as follows: engine rotation speed from 2148 to 2419 rpm, engine torque from 227.2 to 329.7 Nm, engine power from 57.52 to 76.76 kW, and LF from 0.737 to 0.984. According to the ANOVA results, both engine rotation speed and engine torque exhibited statistically significant mean differences across all sites. Conversely, there were no statistically significant differences observed in engine power and LF between sites 7 and 8 or between sites 5 and 9.

Figure 7 displays the results of mapping the average LF for each site to the engine performance curve. During soil tillage operations, tractors are generally operated at an engine rotational speed of 2400 rpm; however, when there is a demand for a higher load than the LF that can be output at this time, the engine rotational speed is reduced to output a higher LF [29]. Therefore, site 4, which has a low LF requirement, performs work in the range of 2400 rpm, whereas sites with a higher LF requirement have an engine operating point in a lower range of engine rotation speed.

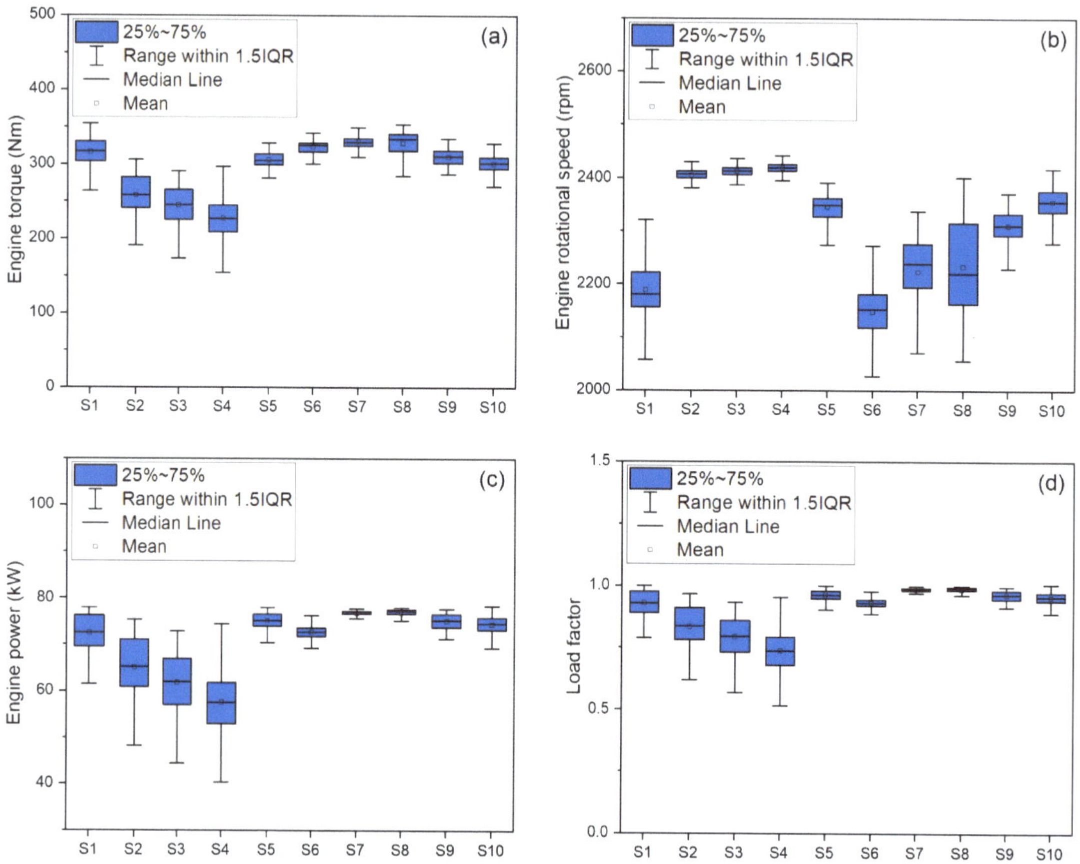

Figure 6. Results of box plot analysis of engine load measured on field experiment sites (e.g., S1 refers to site 1): (**a**) engine torque (ET), (**b**) engine rotation speed, (**c**) engine power (EP), and (**d**) load factor (LF).

Table 5. Results of engine load characteristics.

Sites	Engine Rotational Speed (rpm)	Engine Torque (Nm)	Engine Power (kW)	Load Factor
1	2189 ^{i}	316.4 ^{d}	72.47 ^{e}	0.929 ^{e}
2	2405 ^{c}	258.3 ^{h}	65.03 ^{f}	0.834 ^{f}
3	2413 ^{b}	244.6 ^{i}	61.77 ^{g}	0.792 ^{g}
4	2419 ^{a}	227.2 ^{j}	57.52 ^{h}	0.737 ^{h}
5	2345 ^{e}	305.8 ^{f}	75.07 ^{b}	0.962 ^{b}
6	2148 ^{j}	323.4 ^{c}	72.70 ^{d}	0.932 ^{d}
7	2224 ^{h}	329.7 ^{a}	76.71 ^{a}	0.983 ^{a}
8	2233 ^{g}	328.8 ^{b}	76.76 ^{a}	0.984 ^{a}
9	2310 ^{f}	310.3 ^{e}	75.05 ^{b}	0.962 ^{b}
10	2356 ^{d}	301.4 ^{g}	74.32 ^{c}	0.953 ^{c}

Means (±standard deviation) within same column showing different superscripts are significantly different ($p < 0.05$). Duncan's least significant multiple-range test was applied to compare the means.

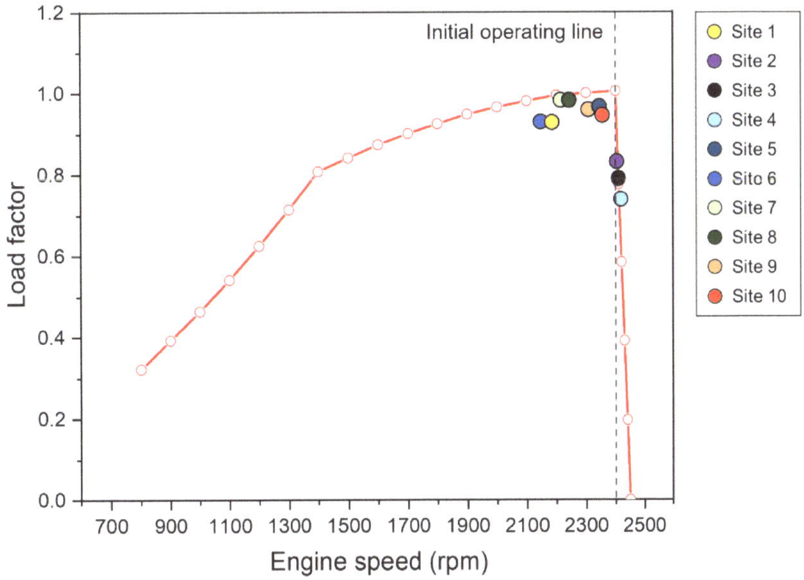

Figure 7. Load factor analysis results for each site on the engine performance curve.

3.3. *Effect of Soil Physical Properties on Engine Load*

Figure 8 presents the correlation matrix depicting the relationship between soil physical properties and engine load characteristics. The top portion of the matrix features bivariate scatterplots along with fitted lines, while the bottom section displays Pearson correlation coefficients. The engine speed exhibited a significance level of $p > 0.05$ between soil texture (sand, silt, and clay proportions) and the CI. Therefore, the correlation coefficients between engine speed and soil parameters are considered unreliable. In contrast, engine torque, engine power, and LF exhibited a significance level of $p < 0.05$ for all soil variables in the study. The LF exhibited correlation coefficients ranging from absolute values of 0.79 to 0.91 for soil parameters. Excluding engine power, the variables that had the most significant impact on the LF were engine torque (r = 0.96), SMC (r = 0.91), sand proportion (r = −0.85), and clay proportion (r = 0.84). Except for silt proportion and SMC, five of the soil variables mostly demonstrated a significance level of $p < 0.05$. The sand proportion and silt proportion displayed a very strong correlation coefficient of −0.98. Therefore, to address the issue of multicollinearity in the regression model, only sand proportion, one of the two soil variables with a higher correlation coefficient with load, was selected as the soil variable for developing the regression model.

Table 6 presents the developed regression models. The adjusted R^2 ranges from 0.638 to 0.902, indicating that each model can estimate the load rate with an accuracy ranging from 63.8% to 90.2%.

Models A–C represent single-variable regression models for soil physical properties. Model A had the highest R^2 at 0.824, while Model B had the lowest R^2 at 0.678. Model D used SMC and CI as soil variables. The combination of these two soil variables is significant because they can be relatively easily obtained through field sensors. However, Model D's adjusted R^2 of 0.800 was slightly lower than Model A's adjusted R^2 of 0.803, which solely used SMC as a variable. Thus, using both SMC and CI for load estimation may potentially reduce model accuracy. Models E and F utilized soil texture variables. Model E exhibited a higher adjusted R^2 value compared to Model A, while Model F outperformed Model B. Hence, incorporating the sand proportion variable can enhance the accuracy of the regression model. Model G incorporated all soil variables, displaying an adjusted R^2 value of 0.888, which was lower than Model E but higher than Model F. When comparing

Models E and G, the reason for the decrease in model accuracy can be attributed to the fact that the dependent variable is already well explained solely by the SMC variable. A high correlation between the SMC and CI variables can, in fact, decrease model accuracy. Therefore, it is crucial to choose a model with the appropriate number of variables.

Table 6. Regression model for estimating the load factor of the tractor.

Model	Source	Regression Model	R^2	R^2 Adj	S.E.
A	f_1	LF = 0.0147SMC + 0.4487	0.824	0.803	0.0388
B	f_2	LF = −0.000255CI + 1.1343	0.678	0.638	0.0525
C	f_3	LF = $-0.00292S_p + 1.0582$	0.718	0.683	0.0491
D	$f_1 + f_2$	LF = 0.0114SMC − 0.000076CI + 0.6168	0.844	0.800	0.0391
E	$f_1 + f_3$	LF = 0.00997SMC − $0.00147S_p$ + 0.6719	0.924	0.902	0.0274
F	$f_2 + f_3$	LF = −0.000126CI − $0.001804S_p$ + 1.1130	0.780	0.717	0.0465
G	$f_1 + f_2 + f_3$	LF = 0.0108SMC + 0.000030CI − $0.00162S_p$ + 0.6282	0.926	0.888	0.0291

Note: SMC = soil moisture content (%), CI = cone index (kPa), S_p = sand proportion (%).

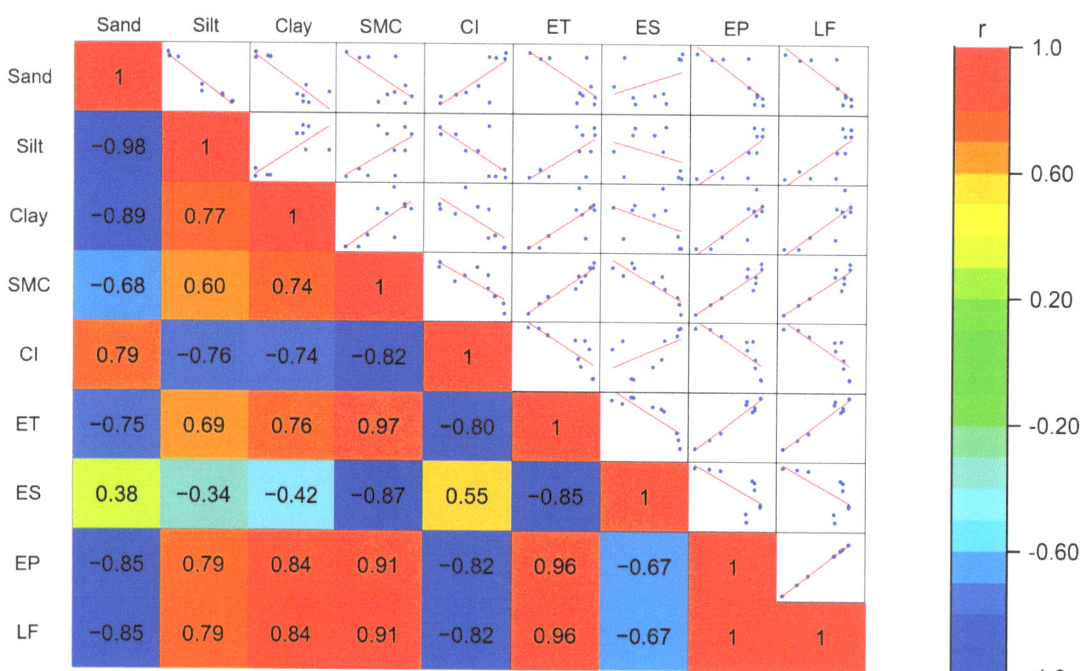

Note: SMC = soil moisture content, CI = cone index, ET = engine torque, ES = engine speed, EP = engine power, LF = load factor. The blue dots in the figure represent each data from 10 sites.

Figure 8. Results of correlation analysis of soil physical properties and engine load characteristics.

Table 7 displays the results of ANOVA analysis for each regression model, showcasing the degrees of freedom (Df), sum of squares (SS), and mean squares (MS). A higher F-value implies that the regression model better explains the variability in the dependent variable. Models E and A exhibited higher F-values of 42.278 and 37.381, respectively, indicating strong explanatory power. In contrast, Models F and B demonstrated lower F-values of 12.375 and 16.848, respectively, indicating relatively weaker explanatory capabilities. Consequently, the SMC is significantly more effective in explaining the load rate compared to the CI. Model G, which utilized all soil variables, yielded an F-value of 24.881. The *p*-value, calculated using the F-value and Df, indicates the probability that the model's dependent variable and independent variable have no significant correlation. All models demonstrated statistical significance with $p < 0.01$. Variance inflation factor (VIF) is used for diagnosing multicollinearity, and it is the reciprocal of tolerance. VIF values closer to 1 indicate that there is no correlation between independent variables. Typically, a VIF of 10 or higher suggests multicollinearity [30]. The VIF for each model ranged between 1.854 and 4.435. Therefore, all models did not exhibit multicollinearity issues.

Table 7. ANOVA results for each regression model.

Model		Degrees of Freedom (Df)	Sum of Squares (SS)	Mean Squares (MS)	F-Value	p-Value	Variable	Tolerance	Variance Inflation Factor (VIF)
A	Regression	1	0.0565	0.0565	37.381	0.000 *	SMC		
	Residual	8	0.0120	0.0015					
B	Regression	1	0.0464	0.0464	16.848	0.003 *	CI		
	Residual	8	0.0221	0.0028					
C	Regression	1	0.0492	0.0492	20.370	0.002 *	Sand		
	Residual	8	0.0193	0.0024					
D	Regression	2	0.0578	0.0289	18.964	0.001 *	SMC	0.331	3.201
	Residual	7	0.0107	0.0015			CI	0.331	3.201
E	Regression	2	0.0633	0.0316	42.278	0.000 *	SMC	0.539	1.854
	Residual	7	0.0052	0.0007			Sand	0.539	1.854
F	Regression	2	0.0534	0.0267	12.375	0.005 *	CI	0.370	2.703
	Residual	7	0.0151	0.0022			Sand	0.370	2.703
G	Regression	3	0.0634	0.0211	24.881	0.001 *	SMC	0.329	3.042
							CI	0.225	4.435
	Residual	6	0.0051	0.0008			Sand	0.367	2.722

* Significant at $p < 0.01$.

4. Discussion

The data obtained from previous studies were utilized to validate the regression models derived in this study [31]. The study evaluated the tractor's operational performance using the same tractor–moldboard plow combination. In contrast to this study, the moldboard plow had six blades and the tractor had a rated power of 42 kW. The soil physical properties were determined by measuring the SMC in the 150 to 200 mm layer and the CI as the average value in the 0 to 150 mm range, aligning with our data collection methodology for this study. The LF used measurements obtained under the condition of a working depth of 16 cm. The average LF for each soil physical characteristic reported in the previous studies is presented in Table 8.

Table 8. Average LF by soil physical properties for model validation [31].

Site	SMC (%)	CI (kPa)	Soil Texture			LF
			Sand (%)	Silt (%)	Clay (%)	
1	19.45	689.69	68	20	21	0.793
2	24.50	563.21	40	48	12	0.852
3	20.24	864.67	40	28	32	0.921

Table 9 illustrates the estimation value and analysis result of error for the model. The estimated average LF ranged from 0.734 to 0.956 at site 1, 0.808 to 0.991 at site 2, and 0.746 to 0.942 at site 3. MAPE, RMSE, and RD were used as an indicator to evaluate the model performance. The MAPE ranges of models A to G varied from 5.36% to 12.55%. With the exception of one case, Models D to G, which employed multiple soil variables, exhibited lower MAPEs compared to models A to C that utilized single soil variables. This suggests that employing multiple soil variables can enhance the estimative accuracy of the models. However, in the case of Model F, which used CI and sand proportion as variables, its MAPE was higher compared to the model C concerning the sand proportion. This underscores the fact that not all variables hold equal importance, and the choice of specific soil variables can significantly impact the model's performance. Furthermore, this emphasizes the conclusions drawn from the research in the Results section.

Table 9. Analysis of errors in validation data for each regression model.

	Items	Model A	Model B	Model C	Model D	Model E	Model F	Model G
Site 1	Average LF (Kim et al. [31])				0.793			
	Estimated LF	0.734	0.956	0.860	0.787	0.766	0.902	0.748
	Error (%)	7.43	20.58	8.47	0.80	3.44	13.79	5.64
Site 2	Average LF (Kim et al. [31])				0.852			
	Estimated LF	0.808	0.991	0.942	0.855	0.857	0.970	0.844
	Error (%)	5.20	16.21	10.47	0.26	0.57	13.77	1.00
Site 3	Average LF (Kim et al. [31])				0.921			
	Estimated LF	0.746	0.914	0.942	0.783	0.815	0.932	0.807
	Error (%)	19.09	0.84	2.19	15.02	11.58	1.11	12.43
Average	MAPE (%)	10.57	12.55	7.04	5.36	5.19	9.56	6.35
	RMSE	0.110	0.124	0.065	0.080	0.064	0.093	0.071
	RD (%)	12.87	14.44	7.66	9.35	7.44	10.85	8.31

The analysis result of RMSE and RD differed somewhat from MAPE. Model C demonstrated the second lowest values for RMSE and RD. This is the result of RMSE and RD being more responsive to outliers compared to MAPE. In practice, Models D, F, and G, which employed multiple soil variables, exhibited higher errors at specific sites. These results underscore the need for a more diverse dataset of LF data collected under various soil conditions to enhance the model's stability and applicability.

5. Conclusions

In this study, the effect of soil physical properties on LF of tractor engine was analyzed. A correlation analysis was performed between LF and five soil variables: SMC, CI, sand proportion, silt proportion, and clay proportion, and all soil variables were found to have a strong correlation with LF. However, the correlation coefficients between each soil variable and LF were different, indicating that the soil variables had different impacts. The LF estimation model was developed through a regression model using soil variables. The objective of developing this estimation model was to estimate the LF more accurately and compare the accuracy of models using a single soil variable and models using multiple soil variables. In many cases, models using multiple soil variables demonstrated higher accuracy compared to those using a single soil variable. Therefore, adding soil variables allows us to encompass the complexity and variability caused by soil conditions during actual tractor operations, leading to an improvement in model accuracy. However, in some instances, the addition of soil variables led to a decrease in model accuracy. This occurred due to the added soil variable having a relatively low correlation with LF, which increased the error of the model, or the increase in outlier values due to the high correlation between the added variables. Consequently, it was concluded that selecting an appropriate level of model complexity is crucial.

Kim et al.'s (2022) model validation was performed using soil physical properties and LF data reported in [31]. The MAPE of the models varied between 5.19 and 12.55%, with models using multiple soil variables tending to have lower errors. This supports the conclusions of this study. However, there was no correlation between the model's high R^2 value and the error in the verification results. Differences in tractor power, soil consolidation tools, range of soil conditions, etc., can affect the accuracy of the model. Therefore, LF data collected under a wider range of operating conditions and soil conditions are needed to improve the model's reliability and versatility. Research on data collection for these various conditions will continue to be conducted in future studies.

Author Contributions: Conceptualization, W.-S.K. and Y.-J.K.; methodology, W.-S.K.; software, Y.-S.M., Y.-S.K. and R.-G.L.; validation, Y.-S.K.; formal analysis, Y.-S.M. and T.-J.K.; investigation, Y.-S.M. and T.-J.K.; data curation, W.-S.K. and R.-G.L.; writing—original draft preparation, Y.-S.M. and W.-S.K.; writing—review and editing, Y.-S.M. and W.-S.K.; supervision, W.-S.K. and Y.-J.K.; project administration, Y.-J.K. and R.-G.L.; funding acquisition, Y.-J.K. All authors have read and agreed to the published version of the manuscript.

Funding: This work was carried out with the support of "Cooperative Research Program for Agri culture Science and Technology Development (Project No. PJ0156952022)" Rural Development Administration, Republic of Korea.

Institutional Review Board Statement: Not applicable.

Informed Consent Statement: Not applicable.

Data Availability Statement: The data presented in this study are available within the article.

Conflicts of Interest: The authors declare no conflict of interest.

References

1. Zhang, L.; Xu, M.; Chen, H.; Li, Y.; Chen, S. Globalization, green economy and environmental challenges: State of the art review for practical implications. *Front. Environ. Sci.* **2022**, *10*, 870271. [CrossRef]
2. National Institute of Environmental Research Center (NIER). *Standard Operations Procedure for the Construction of Supporting Data for National Air Pollutant Emissions*; NIER: Incheon, Republic of Korea, 2019; Available online: https://www.air.go.kr/article/view.do?boardId=8&articleId=89&boardId=8&menuId=49&currentPageNo=1 (accessed on 25 September 2023).
3. National Air Emission Inventory and Research Center (NAIR). *Handbook of Estimation Methods for National Air Pollutant Emissions V*; NAIR: Cheongju, Republic of Korea, 2022; Available online: https://www.air.go.kr/article/view.do?boardId=8&articleId=238&boardId=8&menuId=49&currentPageNo=1 (accessed on 25 September 2023).
4. National Air Emission Inventory and Research Center (NAIR). *2019 National Air Pollutant Emissions*; NAIR: Cheongju, Republic of Korea, 2022; Available online: https://www.air.go.kr/article/view.do?boardId=7&articleId=145&boardId=7&menuId=48&currentPageNo=1 (accessed on 25 September 2023).
5. National Institute of Agricultural Sciences (NIAS). *2019 Investigation of Agricultural Machinery Usage*; NIAS: Wanju, Republic of Korea, 2020.
6. Kim, W.S.; Lee, S.E.; Baek, S.M.; Baek, S.Y.; Jeon, H.H.; Kim, T.J.; Lim, R.G.; Choi, J.Y.; Kim, Y.J. Evaluation of exhaust emissions factor of agricultural tractors using portable emission measurement system (PEMS). *J. Drive Control* **2023**, *20*, 15–24.
7. Zhang, G.; Sandanayake, M.; Setunge, S.; Li, C.; Fang, J. Selection of emission factor standards for estimating emissions from diesel construction equipment in building construction in the Australian context. *J. Environ. Manag.* **2017**, *187*, 527–536. [CrossRef] [PubMed]
8. Lee, J.H.; Jeon, H.H.; Baek, S.Y.; Baek, S.M.; Kim, W.S.; Siddique, M.A.A.; Kim, Y.J. Analysis of Emissions of Agricultural Tractor according to Engine Load Factor during Tillage Operation. *J. Drive Control* **2022**, *19*, 54–61.
9. Tan, D.; Tan, J.; Peng, D.; Fu, M.; Zhang, H.; Yin, H.; Ding, Y. Study on real-world power-based emission factors from typical construction machinery. *Sci. Total Environ.* **2021**, *799*, 149436. [CrossRef] [PubMed]
10. Lee, S.E.; Kim, T.J.; Kim, Y.J.; Lim, R.G.; Kim, W.S. Analysis of Engine Load Factor for Agricultural Cultivator during Plow and Rotary Tillage Operation. *J. Drive Control* **2023**, *20*, 31–39.
11. Baek, S.M.; Kim, W.S.; Baek, S.Y.; Jeon, H.H.; Lee, D.H.; Kim, H.K.; Kim, Y.J. Analysis of engine load factor for a 78 kW class agricultural tractor according to agricultural operations. *J. Drive Control* **2022**, *19*, 16–25.
12. Lee, D.I.; Park, J.; Shin, M.; Lee, J.; Park, S. Characteristics of Real-World Gaseous Emissions from Construction Machinery. *Energies* **2022**, *15*, 9543. [CrossRef]
13. Shin, D.H.; Park, Y.S.; Yoo, C.; Park, S.H. Study on Real-Work NOx Emission Characteristics according to Load Factor of Excavator. *J. Drive Control* **2023**, *20*, 1–8.
14. Barati, K.; Shen, X. Operational level emissions modelling of on-road construction equipment through field data analysis. *Autom. Constr.* **2016**, *72*, 338–346. [CrossRef]
15. Koo, Y.M. PTO torque and draft analyses of an integrated tractor-mounted implement for round ridge preparation. *J. Biosyst. Eng.* **2022**, *47*, 330–343. [CrossRef]
16. Kim, J.T.; Im, D.; Cho, S.J.; Park, Y.J. A Study on the Prediction of Driving Performance of Agricultural Tractors Driving on Dry Sand. *J. Biosyst. Eng.* **2022**, *47*, 502–509. [CrossRef]
17. Inchebron, K.; Seyedi, S.M.; Tabatabaekoloor, R. Performance evaluation of a light tractor during plowing at different levels of depth and soil moisture content. *Int. Res. J. Appl. Basic Sci.* **2012**, *3*, 626–631.
18. Kim, W.S.; Kim, Y.J.; Park, S.U.; Kim, Y.S. Influence of soil moisture content on the traction performance of a 78-kW agricultural tractor during plow tillage. *Soil Tillage Res.* **2021**, *207*, 104851. [CrossRef]

19. Rasool, S.; Raheman, H. Improving the tractive performance of walking tractors using rubber tracks. *Biosyst. Eng.* **2018**, *167*, 51–62. [CrossRef]
20. Battiato, A.; Diserens, E. Tractor traction performance simulation on differently textured soils and validation: A basic study to make traction and energy requirements accessible to the practice. *Soil Tillage Res.* **2017**, *166*, 18–32. [CrossRef]
21. Han, G.; Kim, K.D.; Ahn, D.V.; Park, Y.J. Comparative Analysis of Tractor Ride Vibration According to Suspension System Configuration. *J. Biosyst. Eng.* **2023**, *48*, 69–78. [CrossRef]
22. Choudhary, S.; Upadhyay, G.; Patel, B.; Naresh; Jain, M. Energy requirements and tillage performance under different active tillage treatments in sandy loam soil. *J. Biosyst. Eng.* **2021**, *46*, 353–364. [CrossRef]
23. Ayers, P.D.; Perumpral, J.V. Moisture and density effect on cone index. *Trans. ASAE* **1982**, *25*, 1169–1172. [CrossRef]
24. *ASABE Standards EP542*; Procedure for Using and Reporting Data Obtained with the Soil Cone Penetrometer. American Society of Agricultural and Biological Engineers: St. Joseph, MO, USA, 2019.
25. Reicosky, D. *Managing Soil Health for Sustainable Agriculture Volume 2: Monitoring and Management*; Burleigh Dodds Science Publishing: Cambridge, UK, 2018.
26. Tan, K.H. *Soil Sampling, Preparation, and Analysis*; CRC Press: New York, NY, USA, 1995; ISBN 0-8247-9675-6.
27. Kim, W.S.; Lee, D.H.; Kim, Y.J.; Kim, Y.S.; Park, S.U. Estimation of Axle Torque for an Agricultural Tractor Using an Artificial Neural Network. *Sensors* **2021**, *21*, 1989. [CrossRef]
28. Kim, W.S.; Kim, Y.J.; Baek, S.Y.; Baek, S.M.; Kim, Y.S.; Choi, Y.; Kim, Y.K.; Choi, I.S. Traction performance evaluation of a 78-kW-class agricultural tractor using cone index map in a Korean paddy field. *J. Terramechanics* **2020**, *91*, 285–296. [CrossRef]
29. Chancellor, W.; Thai, N. Automatic control of tractor transmission ratio and engine speed. *Trans. ASAE* **1984**, *27*, 642–0646. [CrossRef]
30. Salmerón Gómez, R.; Rodríguez Sánchez, A.; García, C.G.; García Pérez, J. The VIF and MSE in Raise Regression. *Mathematics* **2020**, *8*, 605. [CrossRef]
31. Kim, Y.S.; Lee, S.D.; Baek, S.M.; Baek, S.Y.; Jeon, H.H.; Lee, J.H.; Kim, W.-S.; Shim, J.Y.; Kim, Y.J. Analysis of the effect of tillage depth on the working performance of tractor-moldboard plow system under various field environments. *Sensors* **2022**, *22*, 2750. [CrossRef]

agriculture

MDPI

Article

Performance Evaluation of a Virtual Test Model of the Frame-Type ROPS for Agricultural Tractors Using FEA

Ryu-Gap Lim [1], Wan-Soo Kim [2,3], Young-Woo Do [2], Md. Abu Ayub Siddique [4] and Yong-Joo Kim [4,5,6,*]

1 Department of Smart Agriculture, Korea Agriculture Technology Promotion Agency, Iksan 54667, Republic of Korea; limso@koat.or.kr
2 Department of Bio-Industrial Machinery Engineering, Kyungpook National University, Daegu 41566, Republic of Korea; wansoo.kim@knu.ac.kr (W.-S.K.); duddn3372@naver.com (Y.-W.D.)
3 Upland Field Machinery Research Center, Kyungpook National University, Daegu 41566, Republic of Korea
4 Department of Agricultural Machinery Engineering, Chungnam National University, Daejeon 34134, Republic of Korea; ayub@cnu.ac.kr
5 Department of Biosystems Machinery Engineering, Chungnam National University, Daejeon 34134, Republic of Korea
6 Department of Smart Agriculture Systems, Chungnam National University, Daejeon 34134, Republic of Korea
* Correspondence: babina@cnu.ac.kr

Abstract: In this study, a model of the frame-type ROPS (rollover protective structure) for an agricultural tractor was developed using FEA (finite-element analysis). Various boundary conditions were applied as input variables to replace the actual test of the ROPS with a virtual test. An optimization study was carried out based on the boundary conditions of the bolt, considering the ROPS part directly mounted on the tractor and the folding connection to the ROPS. The results of the virtual test were compared with those of the actual test, and the error was determined. The maximum error of the evaluation model was 7.0% for the force applied on the load and 9.6% for the corresponding ROPS deformation. All mounting bolts of the ROPS required modeling. In particular, we had to establish free boundary conditions for axial rotation to implement the folding connection. In addition, a simulation of the frame-type ROPS was conducted according to the mesh size. A convenient simulation time was obtained for a mesh size of 8~10 mm. Compared with the actual test, the accuracy was the highest with a mesh size of 6~8 mm.

Keywords: agricultural tractor; ROPS; virtual test; FEA

Citation: Lim, R.-G.; Kim, W.-S.; Do, Y.-W.; Siddique, M.A.A.; Kim, Y.-J. Performance Evaluation of a Virtual Test Model of the Frame-Type ROPS for Agricultural Tractors Using FEA. *Agriculture* **2023**, *13*, 2004. https://doi.org/10.3390/agriculture13102004

Academic Editor: Jin He

Received: 14 September 2023
Revised: 12 October 2023
Accepted: 13 October 2023
Published: 15 October 2023

1. Introduction

According to a survey on agricultural machinery accidents in Korea, tractor accidents due to a lack of safety measures account for 16.1% of all farming accidents, and overturning and fall accidents account for 43.3% of all tractor accidents [1]. According to a fatal accident investigation by the Ministry of Agriculture, Forestry and Fisheries of Japan conducted in 2005, about 70% of tractor fatal accidents were caused by tractor tipping or falls. This may be related to the fact that, in Japan, the installation rate of rollover protection structure (ROPS) is 69%, and only 48% of these frames are certified ROPSs, ensuring the required performance [2]. In 2002, approximately 52% of fatal accidents during tractor operation in the United States were caused by tipping, and the ROPS installation rate was 38% [3]. In Portugal, over the past 10 years, tractor accidents accounted for 79% of all deaths caused by farming accidents, and 38.6% of these accidents were caused by the overturning; the application rate of the ROPS for tractors was only 4.1% [4]. In Korea, the installation rate of the ROPS is 80.1%, but 15.6% of the ROPSs are then removed due to the conditions of the working environment [5]. It thus appears that, in most countries, tractor accidents are often fatal, which makes crucial the installation of the tractor ROPS with certified safety performance [6–8].

When a tractor overturns, the ROPS, which is intended to protect the driver, is subjected to external forces inducing crushing and deformation [9]. The tests performed to evaluate ROPS crushing and deformation resistance involve applying static loads to the tractor-installed ROPS to assess its efficiency in protecting the driver's safety zone. However, when changes are introduced in ROPS, even minor design changes such as thickness variations to improve its strength or changes in the location of bolt holes, the ROPS must be tested again and obtain a safety certification, which is time-consuming and expensive. For this reason, during the 2018 OECD (Organization for Economic Cooperation and Development) Tractor Technical Meeting, the OECD Tractor Division proposed to replace actual tests with virtual tests when assessing ROPS performance. Accordingly, the OECD Tractor Division is developing a virtual test for the ROPS. In technologically advanced countries, the importance of tractor ROPS virtual test technology is recognized through various studies.

Karakulak and Yetkin [10] argued that virtual testing can prevent wasting time and money in test repetitions whenever a minor change is introduced in a tractor ROPS design, and Blanco and Martin [11] confirmed that simulation tests required before ROPS certification tests are convenient in these respects [11]. Fabbri and Wards (2002) successfully simulated the test conditions defined in the OECD code to verify the effectiveness of FEA in reverse engineering of the tractor ROPS [12]. In Korea, Ha et al. [13] used finite-element analysis to design the ROPS that met the certification conditions and compared their results with those of an actual test, confirming that their method allowed satisfying the OECD code. Kim et al. [14] used the finite-element method to achieve the OECD certification of agricultural tractor cabin structures [14]. Additionally, Jang and Lee [15] obtained an optimal ROPS design using finite-element analysis, reducing the manufacturing time and cost and satisfying the performance requirements.

In this way, finite-element analysis can be applied in various conditions, allowing the optimization of various aspects such as object performance and reverse engineering [16–18]. It also has the great advantage of being able to repeatedly check load test results that cannot be repeated in reality. As mentioned earlier, the importance of ROPS performance certification is increasing, and at the same time, a simplification of the existing test methods is continuously required. One way to simplify the test methods is to introduce virtual testing, which has the advantages of being less expensive and time-consuming as well as more suitable for complex structure installations. However, a specific method for virtual testing has not been established yet, and most previous studies used empirical methods. In addition, as ROPS behavior is nonlinear, performance testing has to be conducted in conditions very similar to the real ones to increase accuracy; therefore, setting boundary conditions and constraints for the examined finite elements is very important.

The aim of this study was to develop a model for virtual testing as efficient as the performance test method recommended by the OECD tractor code and reflecting the structural characteristics of the ROPS. We propose a simplified model to validate the data obtained by virtual tests based on real-vehicle tests. We verified our model with data collected through repeated tests using a simplified ROPS model in previous research [19].

2. Materials and Methods

2.1. Experimental Design

In order to improve accuracy when performing finite-element analysis of the ROPS, it is important to select the most appropriate hardening model for the ROPS material. The stiffness of the model can be reduced by including soft deformable elements [20]. Since the shape of the joint portion of bolts is complex, it is difficult to establish it according to the joint conditions. That is why, depending on the purpose of the analysis, it is necessary to select a precise, practical, and simple model. The FE models can be developed for various conditions and can yield results similar to those obtained with actual tests only when suitable elements and constraints are set. Therefore, in this study, a virtual test was conducted by developing an FE model considering the characteristics of the tractor ROPS.

Rather than developing a new virtual test software, we verified it with an actual test using a dynamic commercial tool so that users can easily access it. The possibility of virtual testing was confirmed by modeling boundary conditions suitable for the frame-type ROPS and load plates considering actual tests. The procedure for developing the model is shown in Figure 1.

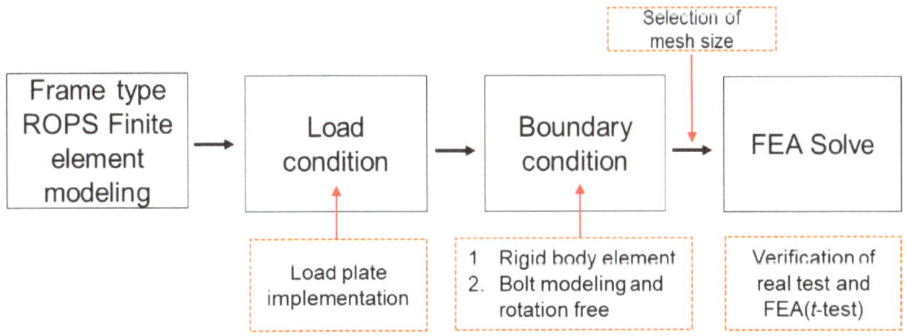

Figure 1. Procedure to develop the virtual test model for ROPS described in this study.

The required energy and forces for the ROPS performance test were calculated using the reference mass of an agricultural tractor of 1650 kg based on OECD code 4. The required energy and force were calculated as 2.310 kJ rear, 2.888 kJ side, and 0.578 kJ front and 33.0 kN crush, respectively. The loading sequence should be as follows: rear, primary crush, lateral, secondary crush, front. The crushing test was performed in both real and virtual tests. However, the crushing test was not analyzed in this study because it is a vertical load, not a concept of force direction and energy absorption.

2.2. Software

In general, various software packages are used for structural analysis. In this study, Marc ver. 2020 (MSC software Corp., Newport Beach, CA, USA), which is widely used for the structural analysis of the tractor ROPS in Korea, was chosen [21]. This is a general-purpose, commercial analysis software package that is compatible with various FE model files and has auto-remeshing and refinement functions useful for the analysis of the non-linear behavior of the ROPS in load tests. Marc has the advantage of being able to model the behavior according to nonlinear physical properties and environmental characteristics, considering the passage of time, by simply defining a contact model and providing a re-mesh function.

2.3. Simulation Conditions

2.3.1. FE Model for the Frame-Type ROPS

For a certification test of ROPS performance, the manufacturer must submit a drawing including detailed information, such as the tractor's basic characteristics, specifications for each ROPS component, the material used, and the type of assembly and mounting bolts, to an authorized testing agency. The submitted detailed drawings by the tractor manufacturer were used to create the initial FE model for virtual testing. In this study, a 3D model of each ROPS component was developed using Solidworks ver. 2019 (Dassault systems SolidWorks Corp., Waltham, MA, USA) based on a drawing of the tractor, as shown in Figure 2a. The main frame material of the ROPS model was square steel-pipe SPSR, and the folding and mounting portions were made of steel-plate SS400. The FE model used automatic meshing, and local corrections were made for mesh breaks and element shapes. In addition, each component was named and grouped to facilitate finding the requested information for the analysis. A grid was made based on the adaptive FE method for each part, and the shape of each part was accordingly presented. Its advantages allow us to predict and redesign the mesh breaks, shape, etc. for each assembly. Table 1 summarizes the mesh information for the nodes and elements of the entire ROPS and the geometric and material information for each part. The FE model of the ROPS included upper and lower frames, with 26,474 nodes and 25,951 elements. With respect to the shape, there were 24,654 square elements, 217 triangular elements, 792 hexagonal elements, and 276 octagonal elements. The FE model of the generated ROPS is shown in Figure 2b,c. The mainframe, steel plates, pins and mounting plates, and bolts are configured in the same way as the real thing. Therefore, it has the advantage of being able to analyze the FE model by various contact condition methods compared with the real one. In terms of geometry, the mainframe was a 3D shell with upper, middle, and lower parts, and the bolts for folding and mounting the ROPS were 3D solid. Based on previous research, as shown in Figure 3, the true stress–strain values of SS400 and SPSR were used in the simulation. As for the material properties of S45C bolts, the linear properties entered for automatic provision in the analysis software were used.

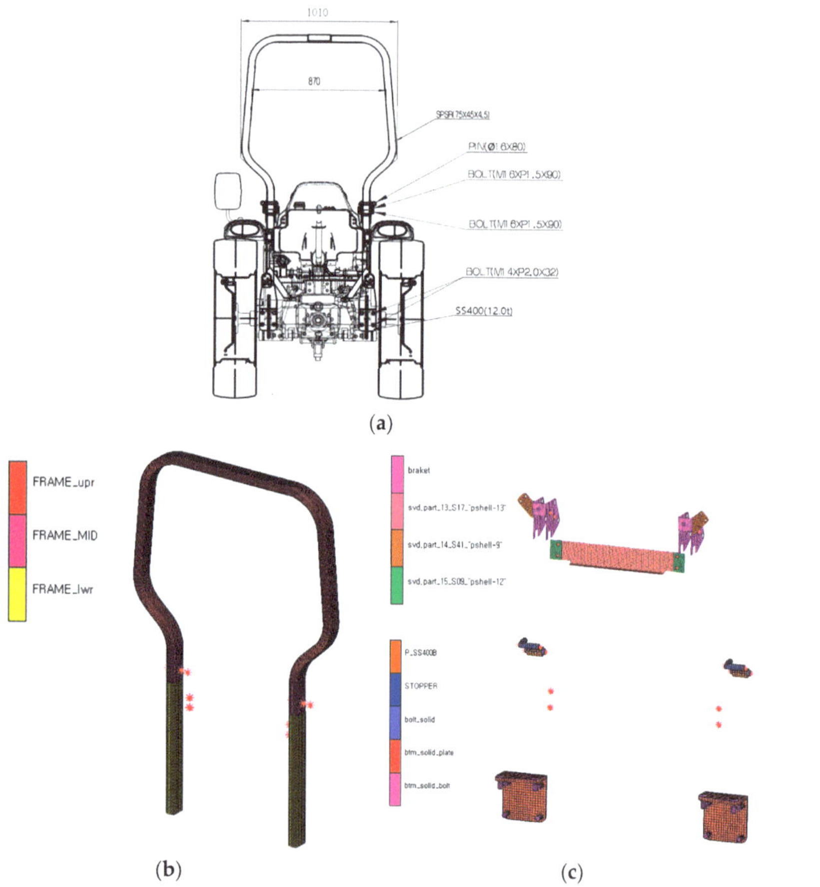

Figure 2. Collimation between the FE model and real model for frame-type ROPS: (**a**) drawing of frame-type ROPS; (**b**) frame of FE model; (**c**) folding part, plate, pin, and bolt of FE model. * Boundary conditions of contact area.

Table 1. Specifications of the FE model for ROPS proposed in this study.

Mesh		Geometric		
Components	**Specification (Number)**	**Components**	**Properties**	**No. of Elements**
Nodes	26,474	Frame upper part	3D Shell	14,565
		Frame mid part	3D Shell	1518
Elements	25,951	Frame lower part	3D Shell	5786
		Frame weld	3D Shell	142
Quad	24,654	Bolt case	3D Solid	144
		Stopper bolt	3D Solid	72
Tria	217	Bolt	3D Solid	652
		Bottom plate	3D Solid	2336
Hexa	792	Bottom bolt	3D Solid	1576
Octagon	276	Folding part bolt	3D thin walled	4
		Material Properties		
Components	**Mass Density (kg/m^3)**	**Elastic Modulus (GPa)**	**Poisson's Ratio**	**No. of Elements**
SPSR	7900	200	0.28	42,348
SS400	7900	200	0.28	8529
S45C Stopper	7850	205	0.29	76
S45C Bolt	7850	205	0.29	652

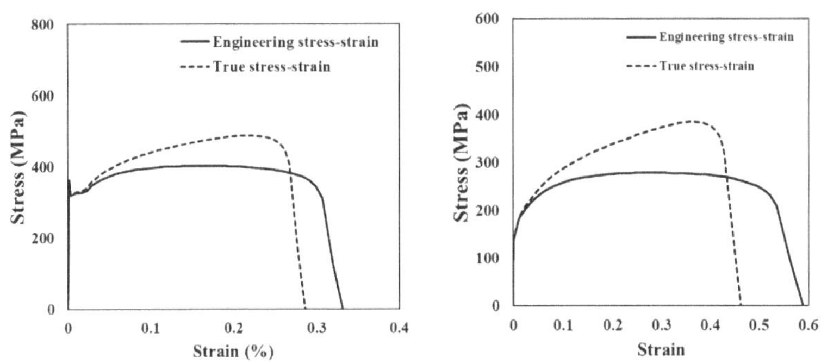

(a) (b)

Figure 3. True stress–strain of material used in this study. (**a**) SS400; (**b**) SPSR.

2.3.2. Load Conditions

In the simulation, the load was applied in the rear, side, and front of the ROPS, and the point contact condition was set at the same location as that in the actual test [22]. The evaluation of the ROPS was performed by applying loads with the energy required to simulate real conditions. However, there was no function available in the analysis program to estimate the required energy. Therefore, the representative deformation value for the required energy was selected based on the trail-error method and simulated the model. The load conditions including load direction and point in the ROPS FE model are shown in Figure 4. The ROPS test is a sequential analysis and should take into account the behavior changes caused by the previous test. In particular, the results of the ROPS test can be expressed in a strain–force graph obtained by determining strain energy, displacement, and force over time. The forces applied to the rear, side, and front of the ROPS are represented on the minus *Y* axis, minus *X* axis, and *Y*-axis, respectively.

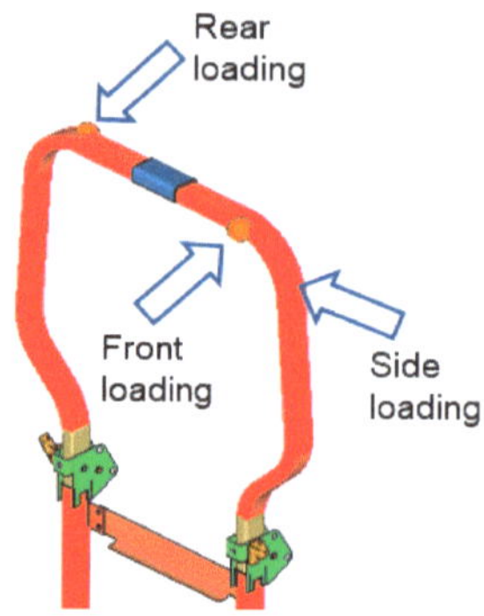

Figure 4. Loading conditions in the ROPS FE model.

In the actual test process, as shown in Figure 5a, a load plate of a certain size was attached to the front of the cylinder, and a load was applied. However, since the point of contact was set in the analysis, 2D plate elements were implemented by directly measuring the actual size of the load plate. The number of nodes for the central contact position of the plate element implemented in the analysis was 12,461 for the rear load and 26,475 for the front load, as shown in Figure 5b. As shown in Figure 6a, a load plate was installed to prevent the slipping and separation of the cylinder in order to apply the side load, and after measuring its size, a 2D plate element was introduced in the model, as shown in Figure 6b.

(**a**)

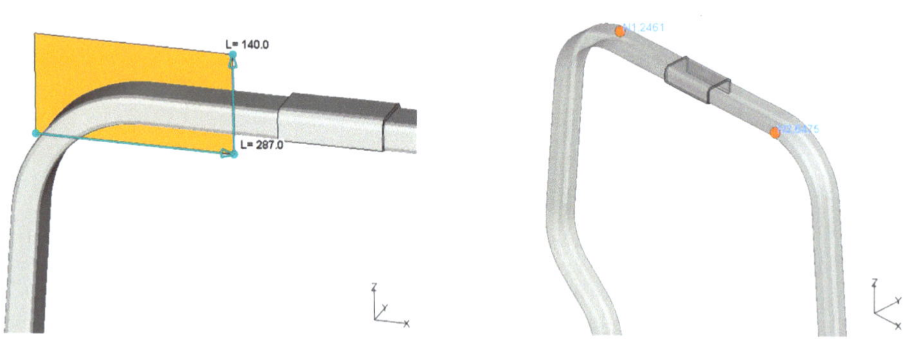

(**b**)

Figure 5. Load plate of the ROPS, for rear and front loading. (**a**) Actual plate; (**b**) 2D load plate size and node identification in the FE model.

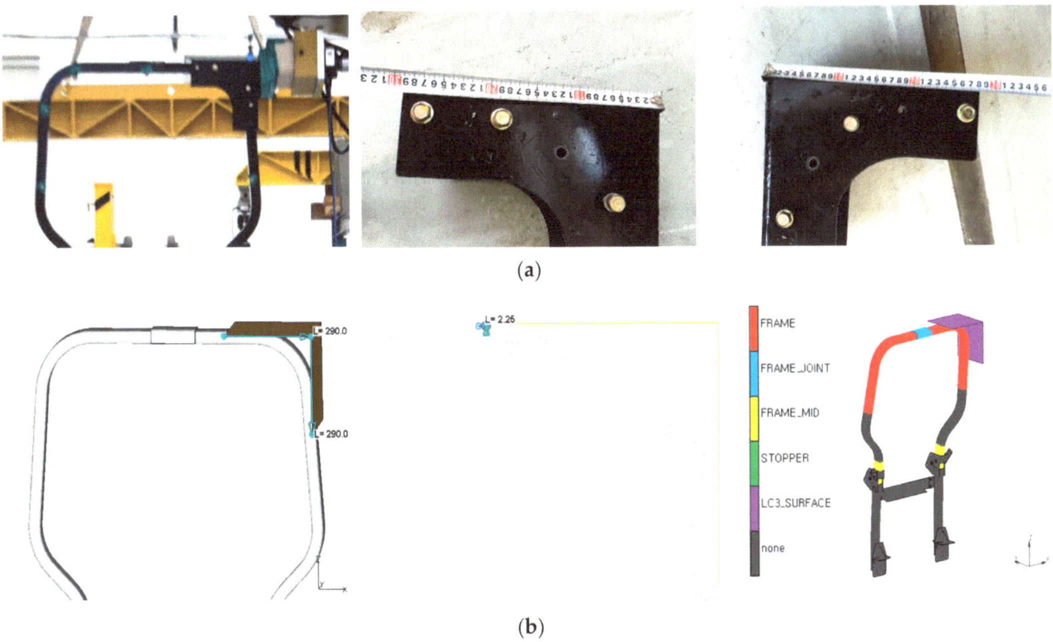

Figure 6. Load plate of the ROPS for side loading. (**a**) Actual; (**b**) 2D load plate size in the FE model.

2.3.3. Boundary Conditions

In the FE model, the mounting and folding parts of the ROPS consist of three-dimensional elements (3D shell, solid) connected to plate elements (2D plate). Therefore, when connecting solid elements and plate elements as well as beam elements in the normal direction of the plate elements, a hinge may be introduced. Therefore, a detailed analysis is required. The 3D solid elements have 3 degrees of freedom per node, while the shell plate elements have 6 degrees of freedom per node and generally are not stiff in the normal direction. The ROPS has a relatively complex shape in which solids and shell plates are geometrically connected. The boundary conditions regarded (1) the portion of the ROPS attached to the tractor and (2) the folding portion of the ROPS; the model was developed with two variables. As shown in Figure 7a, instead of 10 bolts for ROPS mounting, it was applied as a rigid body element (except for the bolt modeling process). For the second condition, we modeled the bolts as shown in Figure 7b and then constrained the displacement of the bolts on the x, y, and z axes on a rigid body, as shown in Figure 7c.

The ROPS upper frame can be folded for work convenience. As shown in Figure 8, it was difficult to model a behavior similar to the actual one when a load was applied because in the folding region, a solid element and a plate element were bound and had different degrees of freedom. Therefore, in this study, boundary conditions were set so that the FE model could obtain results that were as close to reality as possible. The boundary conditions were set as follows: (1) the bolts in the folding part were set to rigid restraints, (2) all fastening parts outside the bolts were connected, and (3) the fastening part inside the bolts was set free to rotate in the axial direction. In this study, boundary conditions were defined according to the Korean national standard [23]. This study focused on developing and verifying a virtual test model, and because too much content had to be covered to provide all mathematical equations, it was omitted, which can be confirmed in the standard. Table 2 shows the comparison between the existing methods and this study. The existing method, which did not consider the physical properties of the mounting bolt and the behavior of the connection, and the method of this study, which considered the physical properties of the modeled bolt and the contact condition of the connection, were compared. In addition, we compared it with the virtual test based on data from previous studies such as ROPS actual test, which did not exist until now.

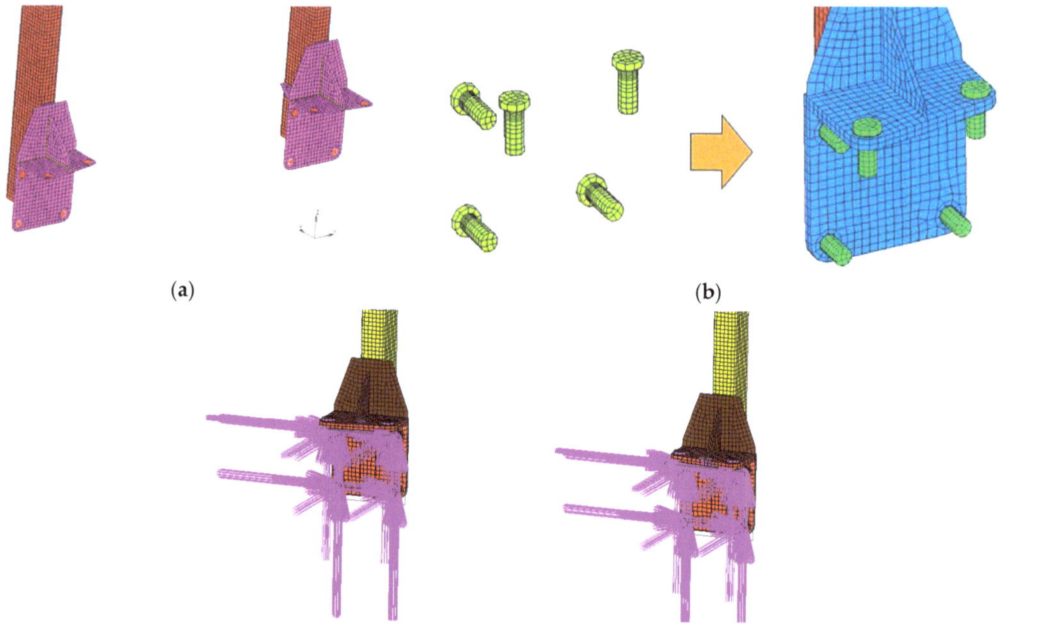

(a) (b)

(c)

Figure 7. Boundary conditions of the mounting parts. (**a**) Constraints for bolt hole; (**b**) bolt modeling; (**c**) constraints for modeled bolts.

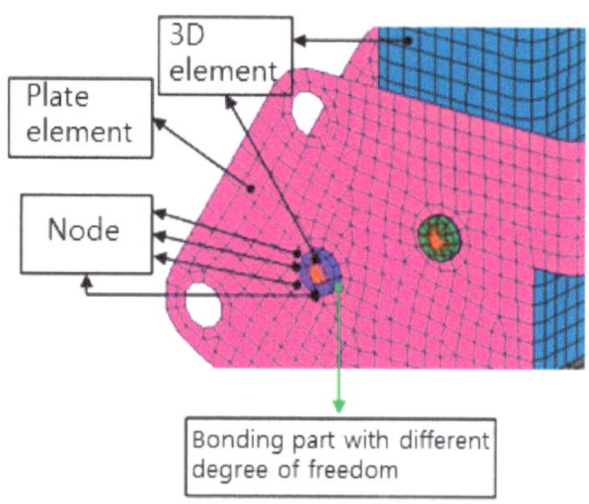

Figure 8. Model of the 3D shell and 2D plate of the folding portion of a tractor ROPS.

Table 2. Comparison of existing methods and this study.

Division	Existing Methods (FEA 1)	Study Methods (FEA 2)
ROPS mounting part	Rigid restraints	Bolts modeled Reflection physical property
ROPS folding part	Rigid restraints	Between plate element and 3D element was set rotation-free in the axial direction
Mesh size	Simple selection of large or small meshes	Apply mesh after performing simulation tests for each mesh size

2.3.4. Selection of the Mesh Size

If the same mesh is applied throughout the analysis, and deformation occurs in various parts during the structural analysis, some serious errors may occur, making it difficult to guarantee reliability. To reduce these errors, engineers often locally use fine meshes over the entire finite-element model. This is inefficient in terms of computing resources and time required for the analysis [24]. Also, the meshes of small size do not prevent excessive deformation in an unpredictable dynamics analysis. Therefore, it is important to find the optimal mesh size rather than simply choose a large or a small mesh size [25]. Therefore, in order to evaluate the effect of the mesh size on the FEA of the performance of the ROPS, five mesh sizes, i.e., 4, 6, 8, 10, and 15 mm, were selected using the re-meshing function, as shown in Figure 9. The optimal mesh size was selected by evaluating the difference between the actual test results and those of the FE model.

(a) (b) (c) (d) (e)

Figure 9. Mesh size of the ROPS in the FE model. Chosen mesh sizes: (**a**) 4 mm; (**b**) 6 mm; (**c**) 8 mm; (**d**) 10 mm; (**e**) 15 mm.

3. Results and Discussion

3.1. Performance of ROPS Simulation Test

3.1.1. Results by Boundary Conditions Rigid Body Element (FEA 1)

The automatic meshing with a mesh size of 6 mm was applied in the simulation. As a result of the simulation, we measured a force of 12.301 kN on the rear load, 19.515 kN on the side load, and 9.267 kN on the front load, and the corresponding values of maximum deformation were 253.9, 180.6, and 96.9 mm, respectively. Figure 10 shows a force–deformation graph that compares the obtained actual test values from a previous study with the virtual test values. It is evident that, in the simulation, the force sharply increased in each loading direction at the beginning, after the application of the load. This was not observed in the actual test. Because the folding part and the bolt mounting part of the ROPS were set to rigid constraints, a stress stiffening effect occurred. Therefore, it was measured to have higher rigidity and smaller deformation than the actual test.

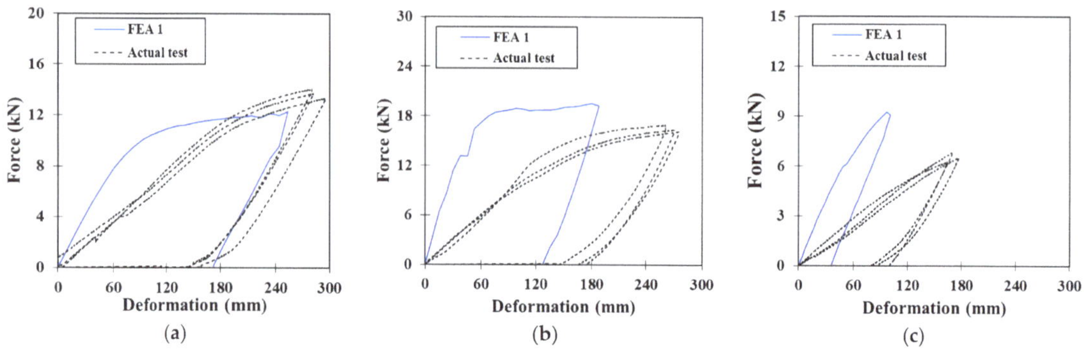

Figure 10. Force–deformation curve; results of FEA 1. (**a**) Rear loading, (**b**) side loading, and (**c**) front loading.

The errors and simulation results of FEA 1 are summarized in Table 3. The force and maximum deformation in the different loading directions were significantly different. In particular, the force error at the front load reached 43.4%. One-sample *t*-tests were performed comparing the mean force and the mean deformation measured in the actual test and in the simulation. As shown in Table 4, we found significant differences according to the *p* values at the confidence level of 95%.

Table 3. Results of the force and maximum deformation for ROPS FEA 1 by load direction.

Load Direction	Force (kN)			Maximum Deformation (mm)		
	Simulated (FEA 1)	Measured	Error (%)	Simulated (FEA 1)	Measured	Error (%)
Rear	12.301	13.681	10.1	253.9	285.9	11.2
Side	19.515	16.308	19.7	180.6	269.2	32.9
Front	9.267	6.461	43.4	96.9	170.2	43.1

Table 4. One-sample *t*-test on the results from FEA 1 and the actual analysis ($\alpha = 0.05$, two-tailed).

Test		N	Mean	SD	SE	t	df	*p*	Mean Difference
Actual rear/ FEA rear	Force	3	13.7	0.353	0.204	6.77	2.00	0.021 *	1.38
	Deformation	3	286	8.11	4.68	6.85	2.00	0.021 *	32.1
Actual side/ FEA side	Force	3	16.3	0.0402	0.0232	−138	2.00	0.001 *	−3.21
	Deformation	3	269	7.46	4.31	20.6	2.00	0.002 *	88.6
Actual front/ FEA front	Force	3	6.46	0.195	0.112	−25.0	2.00	0.002 *	−2.81
	Deformation	3	170	6.46	3.73	19.7	2.00	0.003 *	73.3

* $\alpha = 0.05$, *p* value < α; Therefore, there is a significant difference.

3.1.2. Results by Bolt Modeling and Free Axial Rotation (FEA 2)

External conditions such as load plates were implemented in the same way as in the actual test. In addition, the characteristics of the ROPS contact site were reflected in the

FEA boundary conditions. The test of the FE model was performed by setting the contact conditions of the mounting bolt and the folding steel plate. Figure 11 shows the FEA 2 stress distribution and deformation results of ROPS loading. A comparison between the simulation and the actual test is shown in the graphs in Figure 12. The graph obtained with the simulation appeared similar in shape to that obtained with the actual test. As before, a stress stiffening effect was seen at the beginning of the test. In fact, when compared with the actual test, the force at the beginning linearly increased as the deformation reached about 70 mm. But the errors were smaller overall. Table 5 shows the results for each load direction and the corresponding errors. We measured a force of 12.789 kN on the rear load, 17.135 kN on the side load, and 6.911 kN on the front load. The maximum deformation was 298.5 mm at the rear, 243.8 mm at the side, and 153.9 mm at the front. In summary, we found that the errors of load force and corresponding deformation were as follows: rear force 6.5%, deformation 4.4%; side force 5.1%, deformation 9.4%; front force 7.0%, deformation 9.6%. Overall, it showed significantly better performance than FEA 1. The results of one-sample *t*-tests comparing the actual and the FEA test results are shown in Table 6. We found significant differences only for the forward load deformation and the lateral load force and deformation, as indicated by the small *p*-value at a confidence level of 95%.

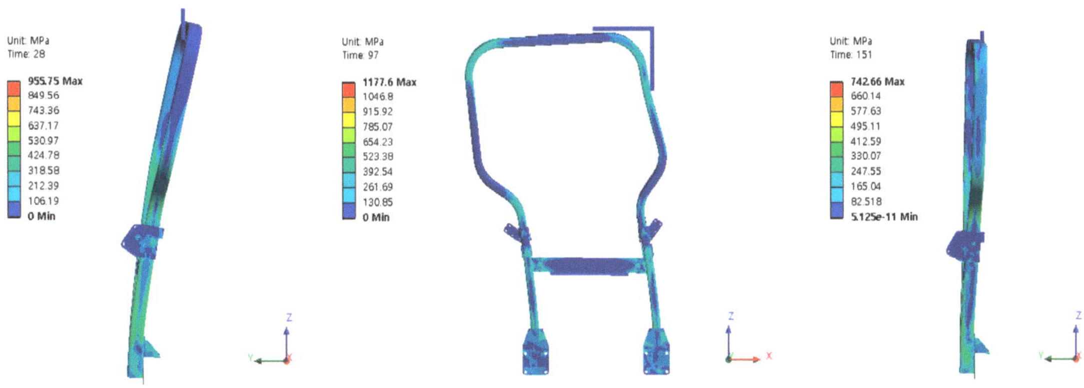

(a) (b) (c)

Figure 11. Evaluation of the distribution stress and deformation on ROPS according to the loading direction. (**a**) Rear loading; (**b**) side loading; and (**c**) front loading.

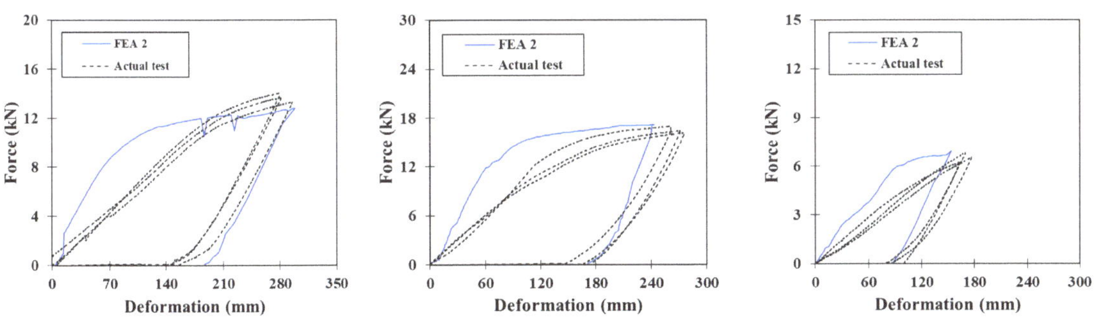

(a) (b) (c)

Figure 12. Force–deformation curve; FEA 2 results. (**a**) Rear loading, (**b**) side loading, and (**c**) front loading.

Table 5. Results of the force and maximum deformation for ROPS FEA 2 by load direction.

Load Direction	Force (kN)			Maximum Deformation (mm)		
	Simulated (FEA 2)	Measured	Error (%)	Simulated (FEA 2)	Measured	Error (%)
Rear	12.789	13.681	6.5	298.5	285.9	4.4
Side	17.135	16.308	5.1	243.8	269.2	9.4
Front	6.911	6.461	7.0	153.9	170.2	9.6

Table 6. One-sample *t*-test on the results from FEA 2 and the actual analysis (α = 0.05, two-tailed).

Test		N	Mean	SD	SE	t	df	*p*	Mean Difference
Actual rear/ FEM rear	Force	3	13.7	0.353	0.204	4.38	2.00	0.054	0.892
	Deformation	3	286	8.11	4.68	−2.68	2.00	0.116	−12.5
Actual side/ FEM side	Force	3	16.3	0.0402	0.0232	−35.7	2.00	0.001 *	−0.827
	Deformation	3	269	7.46	4.31	0.028	2.00	0.028 *	25.4
Actual front/ FEM front	Force	3	6.46	0.195	0.112	−4.01	2.00	0.057	−0.450
	Deformation	3	170	6.46	3.73	4.37	2.00	0.049 *	16.3

* α =0.05, *p* value < α; Therefore, there is a significant difference.

3.2. Simulation Test Results by Mesh Size

The mesh sizes in the ROPS FE model were set differently, and a test was performed on the rear load. The results showing the actual and FEA test results according to the mesh size are shown in Figure 13. The results are shown in Table 7. The time for solving the ROPS finite-element model was 21,285 s for an element size of 4 mm, 8315 s for an element size of 6 mm, 3383 s for an element size of 8 mm, 2540 s for an element size of 10 mm, and 4605 s for an element size of 15 mm. When the size of the element was 8 or 10 mm, the model was solved in a relatively short time, whereas when the element size was 4 mm, it took about 6 h to solve it. The largest error in the force value when comparing the results with those of the actual test was found for the element size of 10 mm and was 21.9%. In addition, the error in the force value was 19.3% and 18.8%, respectively, for the element sizes of 4 mm and 15 mm. Although the mesh size 6 mm showed the lowest error, this condition still required the second highest simulation time among the five mesh conditions. Therefore, the user must select the mesh size considering the appropriate simulation time and error. These results are similar to those presented in the previous research results of Kim and Jung (2012), who studied the relationship between element size, effective characteristics, and analysis time [26]. To replace a real-test ROPS performance, the virtual test is required to be highly accurate. It was observed that the lowest error in the virtual test compared with the actual test was for the mesh size of 6 mm.

Table 7. FEA results and errors for the ROPS FE model according to the mesh size.

Mesh Size (mm)	4	6	8	10	15
Simulation time (s)	21,285	8315	3383	2540	4605
Force (kN)	11.045	12.687	11.830	10.684	11.147
Error (%)	19.3	7.3	13.5	21.9	18.8

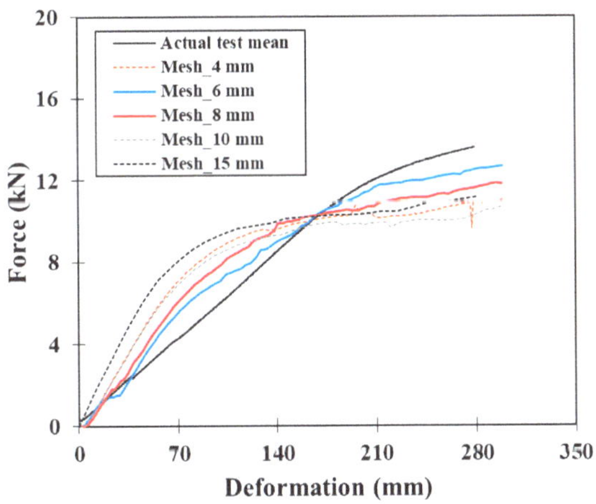

Figure 13. Comparison of the force–deformation curves obtained with the ROPS FE model and the actual test according to the mesh size.

4. Conclusions

The ROPS for the tractor driver's rollover safety must be subjected to load performance tests even when only minor design changes are introduced, such as changes in the position of bolts and in frame thickness. Tractors and the ROPS need to be examined and certified through actual testing, which is costly and requires manpower and money. Therefore, in this study, the characteristics of the tractor frame-type ROPS were considered to develop a simulation as a valid alternative to actual testing. In addition, an FE model was developed considering different boundary conditions. Simulations were performed, and the results were compared with those obtained with the actual test; the corresponding errors were then calculated. The main findings of this study were as follows:

(1) For the virtual test of the frame-type ROPS, the stress stiffening effect of the tractor and its mounted area must be considered. Therefore, all bolts should be modeled rather than using rigid body constraints. Additionally, the boundary condition of the bolt finite-element model must be set to rotation-free for the bolt itself.

(2) The size of the load plate in the actual performance test was taken into consideration and reflected in the load conditions. In addition, the boundary conditions of the ROPS part directly mounted on the tractor were simulated by applying bolt 3D modeling and free axial rotation rather than rigid body restraints. As a result, the slope of the force–deformation curve was improved. The measurement error with respect to the actual test measurement was reduced by up to 38.3%.

(3) A simulation was conducted to identify the mesh size suitable for the ROPS. As for the simulation time, when the mesh size was 8 or 10 mm, the simulation was relatively short. The error with respect to the measurement from the actual test was the smallest, i.e., at 7.3%, when the mesh size was 6 mm.

(4) We expect that the results of this study will soon allow the establishment of a secure virtual test technology for domestic ROPS. In addition, we expect that it will be possible to propose a domestic virtual test technology to the OECD tractor group. The manufacturers can thus verify ROPS design changes to improve ROPS quality through convenient virtual tests. This is expected to help ensure excellent ROPS quality by reducing costs and saving time.

In this study, the FE model for frame-type ROPS performance was proposed. The results of this study suggested the possibility of replacing real tests with virtual tests. In future research, in order to improve the accuracy of virtual testing, we would like to implement full virtual testing as a replacement for actual testing through advanced research such as individual research on the physical properties of materials.

Author Contributions: Conceptualization, R.-G.L. and Y.-J.K.; methodology, R.-G.L., Y.-J.K. and W.-S.K.; software, R.-G.L.; validation, R.-G.L. and W.-S.K.; formal analysis, R.-G.L. and W.-S.K.; investigation, R.-G.L., W.-S.K., Y.-W.D. and M.A.A.S.; data curation, R.-G.L.; writing—original draft preparation, R.-G.L., W.-S.K. and Y.-W.D.; writing—review and editing, R.-G.L., W.-S.K., Y.-W.D. and M.A.A.S.; supervision, R.-G.L. and Y.-J.K.; project administration, R.-G.L.; funding acquisition, R.-G.L. and Y.-J.K. All authors have read and agreed to the published version of the manuscript.

Funding: This work was supported by Korea Institute of Planning and Evaluation for Technology in Food, Agriculture and Forestry (IPET) through Technology Commercialization Support Program, funded by Ministry of Agriculture, Food and Rural Affairs (MAFRA) (821014-03 and 821053-03).

Institutional Review Board Statement: Not applicable.

Data Availability Statement: The data presented in this study are available within the article.

Conflicts of Interest: The authors declare no conflict of interest.

References

1. Rural Development Administration (RDA). *Current Status of Damage for Farmers Related to Agricultural Machinery*; RDA: Jeonju, Republic of Korea, 2019; pp. 20–38.
2. Muneki, T. Survey on the accident caused by agricultural machinery and its protective equipment. In Proceedings of the Japan-Korea Joint Seminar on the Safety of Agricultural Machinery, Saitama, Japan, 31 October 2006.
3. Rautiainen, R.H.; Reynolds, S.J. Mortality and morbidity in agriculture in the United States. *J. Agric. Saf. Health* **2002**, *8*, 259–276. [CrossRef] [PubMed]
4. Antunes, S.M.; Cordeiroa, C.; Teixeir, H.M. Analysis of fatal accidents with tractors in the centre of Portugal: Ten years analysis. *Forensic Sci. Int.* **2018**, *287*, 74–80. [CrossRef] [PubMed]
5. Kim, H.J.; Kim, K.W.; Choi, S.; Kim, J.S.; Kim, Y.Y.; Kim, J.O.; Kim, H.K.; Kwon, S.H. A Study on Improving the Tractor ROPS and Seatbelt use of Korean Farmers. *J. Biosyst. Eng.* **2010**, *35*, 294–301. [CrossRef]
6. Kim, J.H.; Hwang, S.J.; Nam, J.S. Simulation study on the safety of a fastening device of agricultural by-product collector. *J. Drive Control* **2023**, *20*, 42–49. [CrossRef]
7. Lee, D.; Yoo, H.J.; Shin, M.; Oh, J.; Shim, S.B. Analysis of overturning stability of small off-road vehicle. *J. Biosyst. Eng.* **2023**, *48*, 309–318. [CrossRef]
8. Iqbal, M.Z.; Islam, M.N.; Ali, M.; Kiraga, S.; Kim, Y.J.; Chung, S.O. Theoretical overturning analysis of a 2.6-kW two-row walking-type automatic pepper transplanter. *J. Biosyst. Eng.* **2022**, *47*, 79–91. [CrossRef]
9. Lim, R.G.; Kang, Y.S.; Kim, T.J. Measurement uncertainty calculation for improving test reliability of agricultural tractor ROPS test. *J. Drive Control* **2023**, *20*, 34–40. [CrossRef]
10. Karakulak, S.S.; Yetkin, E. Agricultural tractor cabin safety analysis and test correlation. *Int. J. Automot. Sci. Technol.* **2002**, *4*, 1–9. [CrossRef]
11. Blanco, D.; Martin, C.; Ortalde, A. Virtual ROPS and FOPS testing on agricultural tractors according to OECD standard code 4 and 10. In Proceedings of the 14th International LS-DYNA Users Conference, Detroit, MI, USA, 12–14 June 2016.
12. Fabbri, A.; Ward, S. Validation of a finite element program for the design of roll-over protective framed structures (ROPS) for agricultural tractors. *Biosyst. Eng.* **2002**, *81*, 287–296. [CrossRef]
13. Ha, C.W.; Kim, H.J.; Goo, N.S.; Kwon, Y.D. Finite element analysis of an agricultural cabin based on the OECD standard (code 4). *J. Biosyst. Eng.* **2003**, *28*, 431–436. [CrossRef]
14. Kim, H.J.; Goo, N.S.; Kwon, Y.D.; Ha, C.W.; Jung, H.K. Structural analysis of an agricultural tractor cabin for OECD certification. In Proceedings of the Korean Society of Mechanical Engineers Autumn Conference, Jeju, Republic of Korea, 27–29 June 2001.
15. Jang, H.; Lee, B. Optimal design of tractor cabin frame using design of experiment of taguch. *J. Korea Acad.-Ind. Coop. Soc.* **2015**, *16*, 7377–7384. [CrossRef]
16. Won, J.G.; Yoon, J.I.; Lee, H.A.; Chung, S.G.; Jeong, J.S. Simulation analysis on static safety of 55Hp-servo-based hydrostatic transmission. *J. Drive Control* **2022**, *19*, 34–42. [CrossRef]
17. Yu, Y.J.; An, Y.C.; Lee, K.H.; Park, J.H.; Lee, D.; Lee, C.H. Fatigue and severity analysis of drive axle parts according to forklift driving environment. *J. Drive Control* **2023**, *20*, 24–30. [CrossRef]
18. Kumar, N.; Tewari, V.K. Modification of reactive muffler in farm tractor to reduce noise level using finite element method. *J. Biosyst. Eng.* **2023**, *48*, 165–177. [CrossRef]
19. Lim, R.G.; Kang, Y.S.; Lee, D.H.; Kim, W.S.; Lee, J.H.; Kim, Y.J. Agricultural tractor roll over protective structure (ROPS) test using simplified ROPS model. *Korea J. Agric. Sci.* **2023**, *49*, 771–783. [CrossRef]
20. Agius, D.J.; Kourousis, K.I.; Takla, M.; Subic, A. Enhanced non-linear material modelling for analysis and qualification of rollover protective structures. *Proc. Inst. Mech. Eng. Part D J. Automob. Eng.* **2015**, *230*, 1558–1568. [CrossRef]

21. Song, H.S.; Kim, Y.Y.; Lim, R.G. Determination of the properties of structural steel for agricultural tractor roll-over protective structure by tensile test and finite element analysis. *J. Korean Soc. Mech. Technol.* **2019**, *21*, 567–576. [CrossRef]
22. OECD (Organization for Economic Cooperation and Development). *OECD Standard Code for the Official Testing of Protective Structures on Agricultural and Forestry Tractors (Statics Test)*; Tractor Standard Code 4-OECD; OECD Headquarters: Paris, France, 2021.
23. *KS B 7950*; Tractors for Agriculture and Forestry-Two-Post Roll-Over Protective Structures-Virtual Statistic Test Method and Acceptance Conditions. Korea Standards Association: Seoul, Republic of Korea, 2022.
24. Yoon, C.Y. Efficient adaptive finite element mesh generation for dynamics. *Comput. Struct. Eng. Inst. Korea* **2013**, *26*, 385–392. [CrossRef]
25. Heo, J.H.; Kim, H.S. Bending moment calculation method and optimum element size for finite element analysis with continuum elements. *Comput. Struct. Eng. Inst. Korea* **2018**, *31*, 9–16. [CrossRef]
26. Kim, S.W.; Jung, H.C. Investigation on the effect of mesh density in FE analysis for prediction of mechanical behavior of a PCM. In Proceedings of the Korean Society for Technology of Plasticity Conference, Jeju, Republic of Korea, 10–11 May 2012.

agriculture

Article

High-Throughput Plant Phenotyping System Using a Low-Cost Camera Network for Plant Factory

Woo-Jae Cho [1,2] and Myongkyoon Yang [3,4,*]

1 Department of Bio-Industrial Machinery Engineering, College of Agriculture & Life Sciences, Gyeongsang National University, Jinju 52828, Republic of Korea; woojae56@gnu.ac.kr
2 Institute of Smart Farm, Gyeongsang National University, Jinju 52828, Republic of Korea
3 Department of Bioindustrial Machinery Engineering, Jeonbuk National University, Jeonju 54896, Republic of Korea
4 Institute of Agricultural Machinery & ICT Convergence, Jeonbuk National University, Jeonju 54896, Republic of Korea
* Correspondence: yangmk@jbnu.ac.kr

Abstract: Plant phenotyping has been widely studied as an effective and powerful tool for analyzing crop status and growth. However, the traditional phenotyping (i.e., manual) is time-consuming and laborious, and the various types of growing structures and limited room for systems hinder phenotyping on a large and high-throughput scale. In this study, a low-cost high-throughput phenotyping system that can be flexibly applied to diverse structures of growing beds with reliable spatial–temporal continuities was developed. The phenotyping system was composed of a low-cost phenotype sensor network with an integrated Raspberry Pi board and camera module. With the distributed camera sensors, the system can provide crop imagery information over the entire growing bed in real time. Furthermore, the modularized image-processing architecture supports the investigation of several phenotypic indices. The feasibility of the system was evaluated for Batavia lettuce grown under different light periods in a container-type plant factory. For the growing lettuces under different light periods, crop characteristics such as fresh weight, leaf length, leaf width, and leaf number were manually measured and compared with the phenotypic indices from the system. From the results, the system showed varying phenotypic features of lettuce for the entire growing period. In addition, the varied growth curves according to the different positions and light conditions confirmed that the developed system has potential to achieve many plant phenotypic scenarios at low cost and with spatial versatility. As such, it serves as a valuable development tool for researchers and cultivators interested in phenotyping.

Keywords: plant phenotyping; phenotypic index; low-cost system; camera network; online monitoring; plant factory

Citation: Cho, W.-J.; Yang, M. High-Throughput Plant Phenotyping System Using a Low-Cost Camera Network for Plant Factory. *Agriculture* **2023**, *13*, 1874. https://doi.org/10.3390/agriculture13101874

Academic Editor: Signe Marie Jensen

Received: 31 August 2023
Revised: 15 September 2023
Accepted: 20 September 2023
Published: 25 September 2023

1. Introduction

Stressful situations cause morphological, physiological, and biochemical changes in crops. Thus, research on crop phenomics has been conducted to measure and utilize these changes in order to increase productivity, improve quality, and reduce energy and resource consumption through optimal control [1]. In the early stage, a specific stimulus was provided in a controlled environment, such as a smart farm, and response analysis and growth modeling were performed. In particular, to obtain characteristic information about crops, direct measurement, observation, or destructive methods are used. However, although these methods are generally simple and reliable, they require considerable time and labor because of the small number of activities that can be performed at one time [2,3]. In addition, when destructive methods are applied, the sample is damaged, and subsequent analysis is difficult.

To overcome these difficulties, various analyses using non-destructive methods have been conducted. Data available for analysis include images, environmental (temperature, humidity, light intensity, and nutrients) values, and physical (length and weight) values to quantify the genotype and phenotype [4]. Among them, image-based methods are used to analyze the external characteristics or growth of crops using two- (2D) or three-dimensional (3D) images or to indirectly explain internal phenomena through spectral images [5]. As several samples can be easily and automatically obtained using a camera, the time and cost required to acquire feature information can be reduced [6]. In addition, a large amount of information can be obtained using several images, broadening the scope of the analysis. This development coincides with the creation of integrated and mass crop production facilities; moreover, with the development of information communication technology (ICT) convergence technology, high-throughput analysis has been developed into a high-speed, mass-phenotypic method with high efficiency [7]. A high-throughput phenotyping platform refers to a system that integrates the storage of crop data, data analysis, robotics, and decision making. Growing demand, along with the development of numerous types of applicable data and technologies, has led to the requirement of several research studies for this platform and improvement of standardization, experimental methods, and analysis [8–10].

The process of stably collecting and analyzing images is important for analyzing the overall growth of crops through images. The process of acquiring the image should place no additional stress on the crop, and there should be no other influences such as noise or color distribution changes on the image. Several cameras can be utilized to acquire images. In addition, automatically moving and saving the images according to the conditions in a specific storage space can be advantageous because shooting is conducted over several days depending on the growth process. A system with continuity in which the characteristics of crops are automatically analyzed using various image processing and machine learning algorithms for each image and where model development and updates are performed is considered ideal. Thus, the entire process from image acquisition to analysis must be organically connected to enable high-speed mass-phenotypic analysis. However, most of the previously developed systems are very expensive and use algorithms applicable only to a specific environment, hindering their broad application. For example, in a previous study, when examining the number and area of leaves in the growth chamber to investigate the development of Arabidopsis, there were differences in the absence of various application; in addition, the study was soil-based [11]. Similarly to this study, in a 2D-based phenotyping system using a fixed camera, automatic water supply adjustment and area and color characteristics were analyzed. This study was conducted in a soil environment and applied to relatively small crops [12]. In some studies, 3D images and spectroscopy were applied by performing growth monitoring through individual moving ports. However, a limited number of parameters were used, and there were difficulties in system configuration using built-in software [13,14].

A plant factory is a mass crop production facility that can provide a more uniform growth environment than the open field through artificial lighting and complex environmental control in an enclosed space for the year-round production of crops [15]. However, even within these plant factories, spatial variations exist in micro-environmental conditions, such as light, temperature, and humidity, resulting in growth differences [3,16]. Moreover, it is necessary to capture and analyze at minimum hundreds of plant images per day to follow the rapid and intensive progression of growth in plant factories [4,17]. If the analysis is not accomplished in a real-time manner, the data will be accumulated and delay the response to the crop status. In the context, high-speed mass phenotyping approach can be a practical solution to provide location-specific information on crops growing in the cultivation bed, thereby allowing precise environmental control and growth management in plant factories such as yield estimation, crop harvest, or disease treatment.

Methods for acquiring image information according to location include acquiring multiple images while moving one camera and acquiring multiple data at a time by arranging multiple cameras at a certain location. When using a single mobile camera system, it is possible to use a high-cost camera with good performance rather than purchasing several cameras. Good performance provides high resolution and facilitates obtaining images of depth sensors or other wavelength bands rather than simple colors. Because the image is acquired with one device, there is no need to consider the error of the hardware itself. However, this method is relatively expensive and has limitations in repeatability, data acquisition time, and movement section restrictions because the camera position must be moved manually unless an additional automation device is employed in a situation where multiple crops need to be photographed. For example, Jiang et al. [18] developed an on-the-go monitoring platform to measure crop-growth variation according to the location; however, this system could only be moved for the first applied bed structure, and the configuration of the system became more complicated as the cultivation area increased. Yeh et al. [19] created a moving-image monitoring system using a moving arm. Its effectiveness was verified only on a small vertical bed, and rapid analysis was difficult because the growth information was confirmed via image processing through post-processing. Rossi et al. [20] reported a low-cost sliding platform for high-resolution plant phenotyping, but the processing time of more than 1 h would limit the application of the system to the level of laboratory analysis. A greenhouse-based low-cost phenotyping platform reported by Yassue et al. [21] is noteworthy since it proved its effectiveness in a greenhouse-level environment. However, it would not be applicable for vertical plant factories due to height problems. Bari et al. [22] applied a moving-cart embedded camera to the rails for monitoring growing peas in greenhouse, but the large area of the greenhouse meant that it was necessary to analyze the image via post-processing after the collection of images.

By contrast, as the method using several inexpensive cameras can secure a large amount of simultaneous image data, it is easy to analyze changes in the same location without additional correction processes. In addition, it can be flexibly configured for plant factories with relatively large and diverse production structures. However, there is a non-uniformity of information depending on the type and quality of images and the use of various hardware; therefore, it is necessary to develop an image processing technique to overcome this problem. Although An et al. [23] adopted the calibration of multiple-camera array and utilized it for plant phenotyping, the post-data collection and -processing made it difficult to access and analyze data in real time. Thrash et al. [24] also proposed a low-cost wireless phenotyping system, but it was only developed for a single index and a horizontal space. Gang et al. [25] developed a novel convolutional neural network-based model that can estimate various morphologic indices for plants grown in greenhouses, but its feasibility was not validated in an integrated phenotyping platform.

Considering the aforementioned limitations, it is necessary to develop a low-cost phenotyping system capable of collecting and processing data in real time while being structurally more flexible for crop monitoring in vertical plant factories. Furthermore, for phenotyping modeling aimed at analyzing the behaviors of various crops, it is necessary to simplify the determination of new indices or code modifications for correlation analysis.

For the purpose, in this study, a multi-camera network using an inexpensive image sensor and processor was developed. It was designed to have flexible space applicability through a wireless, miniaturized camera node configuration, and to make it easy to collect and analyze various phenotypic indices through a modularized image processing algorithm structure. Based on this, a system capable of performing high-speed mass-phenotypic analysis was constructed. Specifically, the developed system continuously and stably collected crop growth images from camera devices installed at different locations, classified them according to location and time, and stored them in an independent space of the computer server. The stored images were configured to analyze diverse crop characteristics through various image processing algorithms and to output the results. Additionally, the analyzed crop characteristics were used as data for a growth model for a specific situation, and the model was updated according to the data generation. The feasibility of the developed system was evaluated by collecting 2D images of lettuces for the entire growth cycle at various locations. To verify the applicability of the obtained phenotypic indices, growth information of lettuces growing under two different photo-environmental conditions within the plant factory were analyzed and compared.

2. Materials and Methods

2.1. Plant Growth Conditions

The camera system could be applied to any leafy vegetables with low heights, which are usually cultivated in the commercial plant factories [26]. In this study, Batavia lettuce (*Lactuca sativa* L.) was cultivated for the growth experiment, which is a type of lettuce that originated in Europe. Lettuce is one of the most important leafy vegetables in plant factories because of its popularity, with the highest consumption and economic importance throughout the world [27,28]. Individuals with similar weight, size, and shape were selected by direct sowing, and a monitoring experiment was conducted on the growth of the entire cycle from the second day after planting.

Cultivation was conducted in a container environment system located in Yongin, Gyeonggi-do, Republic of Korea (Figure 1). This container, a system that has an environment independent of the outside environment, is suitable for conducting reaction experiments of crops to a target factor. Crops were grown independently in a sponge medium serving as artificial soil, and the necessary nutrients were provided through a closed nutrient film technique system where the electrical conductivity and pH values without a separate nutrient solution supply device. Generally, EC and pH values below or above the optimal range can impede water and nutrient uptake in crops, thereby inhibiting growth and potentially inducing diseases [29]. Therefore, the nutrient solution was manually managed to maintain an appropriate concentration by measuring the EC and pH values at 2-day intervals through standard sampling. During cultivation, the temperature inside the container plant factory was maintained at an average of 22.1 °C in the range of 21–23 °C, and the humidity was maintained at an average of 65% in the range of 50% to 80%. CO_2 was maintained at 400–700 ppm as a factor relating to plant respiration.

Figure 1. Crop-growth container system with controlled environment.

2.2. Photoperiod Condition

Photoperiod is one of the most important variables affecting the plants' growth process, thereby inspiring researchers or industries to explore the effective light/dark cycle for leafy vegetable productivity [30,31]. Specifically, it was reported that differing pulse width modulation of light would be an important factor in the photosynthesis. Chen and Yang [32] reported that intermittent light irradiation could enhance the biomass and taste of lettuces. Therefore, not only to evaluate the applicability of the developed system but also to confirm the effect of different light/dark cycles on the growth of lettuces, an experiment was conducted to establish a camera network for two cultivation beds and to check the growth difference by varying the photoperiod conditions for each zone. The control (con) group was irradiated with light for 16 h and maintained in the dark for 8 h; for the pulse (pul) group, light–dark conditions were set at intervals of 2 h and 1 h, respectively. Accordingly, the daily light irradiation time of the two groups was 16 h, and the total irradiation time and the total incident light amount were the same under the two photoperiod conditions. However, it was expected that improved growth would appear for pul group due to the different light/dark cycles. A total of 14 Batavia lettuces were cultivated for each group, and 4 lettuces were used for acquiring destructive test data (Figure 2). Specifically, 12 camera systems were installed for each control and experimental group to acquire images, while each camera was installed to be located in the center of each target crop. With regard to the extra two cameras for each group, they were used to validate the phenotypic analysis during the cultivation. Growth characteristics such as fresh weights of shoots, leaf length, leaf width, and leaf numbers were measured for four samples every week. Fresh weights of shoots of lettuces were measured using a digital scale (Si-234, Denver Instrument, Denver, CO, USA), and the length and the width of leaves were measured using a ruler. The measured values were used to analyze the differences in data and to compare means among the treatments, respectively.

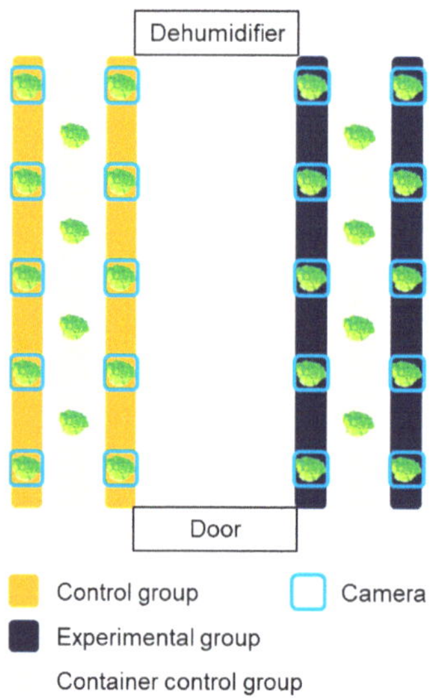

Figure 2. Experimental layout of growing bed and camera network installation.

2.3. Low-Cost Camera Network

For high-speed, mass-phenotypic analysis targeting plant factories, the narrow cultivation interval between crops and the restriction of height should be considered. Accordingly, the proposed system was configured in the form of a multi-point camera network that could freely attach and detach small camera modules and provide integrated access to information on each image collection node through a web connection. Moreover, the building cost of the camera network system is about 5% level compared to the on-the-go monitoring system by Jiang, Kim, and Cho [18] (Table 1). Although the cost will be increased as the number of camera nodes increases, it is still more reasonable than the on-the-go monitoring system when considering the scan area.

Table 1. Comparison of the camera network system with the on-the-go monitoring system by Jiang, Kim, and Cho [18].

Item	Camera Network System	On-the-Go System
Cost	~USD 50 per node	~USD 2200
Setup	Light and compact (~50 g L96 mm × W71 mm × D26 mm per node)	Bulky (~8 kg for conveyer belts, controllers, motors, and actuators)
Scan area (height of 0.45 m)	0.54 m × 0.4 m	0.94 m × 1.8 m
Plant scan time	<1 ms	~127 s
Ease of installation	Easy (due to the node's compact size and individual modularity)	Difficult (due to the complexity of conveyer belts and frames)
External equipment	Wi-Fi router	Linear actuators, a motor servo driver, and step motors

A Raspberry Pi 3B+ (Raspberry Pi Foundation, Cambridge, UK) was used as a low-cost microprocessor that can acquire images of crops for the experiment and transmit them to the server. It was selected because it is small, light, and inexpensive, has good compatibility with various equipment, and has good accessibility compared to other commercial products.

Various cameras, including a USB webcam, can be connected to the Raspberry Pi to take pictures. The Raspberry Pi Camera v2 (Raspberry Pi Foundation, UK) was selected and used to compare the stability, convenience, and image quality. This camera is composed of a module with a resolution of 3280 × 2464 and a performance of approximately eight megapixels. For the stability of data collection, the board and camera modules were configured in a one-to-one connection. The Raspberry Pi and camera module used in this study are not essential parts and can be replaced with other products depending on the experimental environment conditions.

Because the monitoring of crops is conducted remotely, the information obtained on the local Raspberry Pi has to be grasped remotely. Moreover, as mentioned above, an external system is needed to solve the problems associated with the use of multiple processors and limited local capacity. To this end, a computer acting as a server and multiple Raspberry Pi devices were connected to form a system to automatically import data. The computer acting as a server played an overall role in storing and analyzing data. The computer was configured with a Linux operating system environment of core i7-7700k CPU @4.2 GHz (Intel Corp., Santa Clara, CA, USA) and 32 GB RAM for continuity with Raspberry Pi.

Figure 3 shows the network diagram for system setup. This network diagram is an example for the photoperiod experiment of this study, and it is possible to add or remove a server, router, Raspberry Pi, etc., depending on the researcher's experimental environment. Several Raspberry Pi devices for shooting were connected to a single local router via Wi-Fi. A port forwarding method replaces the internal IP between the local router and the Raspberry Pi connected to the lower internal network and creates a new address that can be accessed remotely. Therefore, it was possible to remotely access the Raspberry Pi containing the image information for each location and transmit and receive real-time monitoring and image information. The external server was connected to several Raspberry Pi devices through the local router and used the secure shell (SSH) protocol to access and issue remote commands. Thus, the SSH class was implemented, and the image data were transmitted to the server at a specific period after creating a session using the port forwarding access address. In the server, there were folders designated for each Raspberry Pi, and images were saved in each folder with a name set using date and time information.

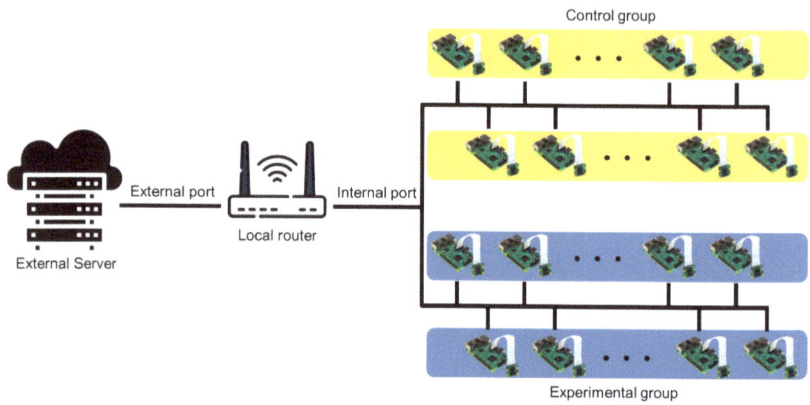

Figure 3. Diagram of network structure of the system.

2.4. Crop Image Collection

As described in the following passages, image data taken according to time conditions were saved in the Raspberry Pi and configured to be transmitted to the server at a specific point in time to enable continuous image shooting. When an image acquisition signal was transmitted to the Raspberry Pi according to the time condition (20 min intervals), a process for stabilizing the image sensor was performed, and the image shooting was performed after acquiring the local time required for phenotypic analysis. To stably acquire image data, the data transmission operation was interrupted during the acquisition process. Since it did not take much time to transmit one image, the data were transmitted when a certain number of images were accumulated in consideration of the system load in the transmission process. Accordingly, data collection to the server was performed once daily. The code for image acquisition was written using Python, and camera control was performed with a Python library called "PiCamera." The use of crontab provided a reduction in the system load, as the system was run only when necessary, rather than continuously.

As the images acquired have high resolution, one file has a size of approximately 4 MB. This is not a problem in short-term monitoring, but in the case of long-term experiments, there may be a problem of insufficient capacity inside the processor. Accordingly, flags for the number of consecutive shooting days, or the number of stored files can be configured to enable continuous monitoring by deleting previous files when the limit is exceeded. In addition, in the process of transmitting the image to the server, it was implemented to record and display logs such as connection errors, image errors, and transmission completion of the local Raspberry Pi to promptly respond to users.

The imaging system consisted of a top-view image acquisition system that could secure crop-growth images in container environmental conditions. The system was configured so that only one crop was placed in the center of one image. Camera modules were installed in the space between the LED lamps to obtain an accurate top-view image and to prevent light intensity interference. Each camera module was calibrated using a classic color checker (X-Rite, Grand Rapids, MI, USA) to compensate the spatial variability. Image acquisition was performed for additional 10 s to stabilize the shooting. Figure 4 shows the multi-camera network installed for multi-point growth image monitoring.

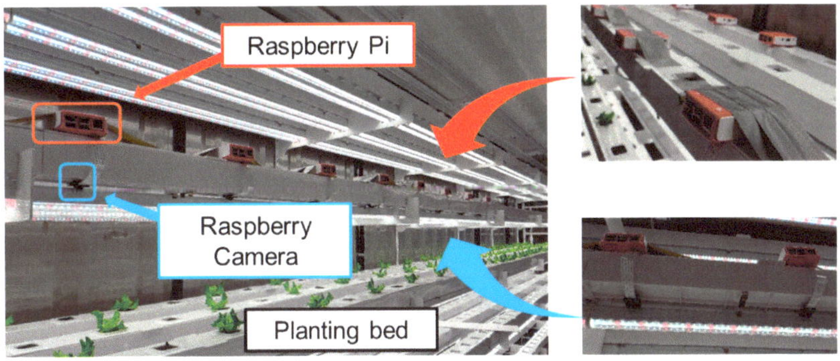

Figure 4. Views of multi-camera network-based high-throughput phenotyping system.

2.5. Phenotypic Index Processing Algorithms

An automatic extraction process of the phenotypic index required for analysis was performed on crop images stored in each folder of the server. Crop images acquired by our system consisting of a camera network contain various noises depending on the environmental conditions of the plant factory. Depending on the system configuration conditions, an image contains only one crop, but the image also includes several types of surrounding environments, such as planting bed, media, and other modules, as noise. The image processing steps to counter these issues are shown in Figure 5. Several algorithms were combined to extract the region of interest (ROI) corresponding to the crop parts required for analysis. ROI are initially extracted from the acquired images using the vegetation index and Otsu threshold method. Because small errors appear in these results, the final binary ROI was extracted through the intersection of colorimetric changes and small space filling methods. For the extracted ROI, the contour and pixel area were obtained, the center was found, the shortest and longest lengths were measured, and the average color information was identified. The extracted index formed a data frame with the structure of the crop, date, time, index 1, index 2,. . ., index n, which was stored in a comma-separated values (csv) format. It was also possible to check the image for specific results. The image processing was configured to be performed on the entire folder when a certain number of image data were added, including files that were not analyzed. The default value was set at daily intervals; thus, when the daily data transmission was completed, index extraction was performed only on new data. The newly created indicator data were stored cumulatively in the previously saved csv file to enable continuous analysis. For the storage and management of data, it is possible to consider a method of using a database rather than a simple csv format storage. In this study, a database was not considered because a simple configuration was sufficient. However, if the system grows in size, the use of a database would be necessary. OpenCV and PlantCV libraries were mainly used in the image processing.

For accurate analysis of crop image phenotypic indices, some aspects were considered on the images obtained. First, image distortion was considered [33]. Usually, in the case of an image containing several crops, errors occur in the center and edges of the image owing to the hardware characteristics of the camera lens. This phenomenon is called distortion. In the case of the configured system, only one crop was placed in the center of one camera to avoid distortion. Second, even by precisely positioning the camera and crop, physical skew occurs [34]. This can cause problems in image tracking because movement may occur as the crop grows; in addition, it depends on the characteristics of hydroponics. In this regard, an analysis method not affected by skew was used. Specifically, two methods were applied to avoid the effect of skew: first, one camera was placed for each crop so that the crop was located in the center of the image, thereby minimizing the effects of skew, which typically occurs in the corners of the image. Second, the projected area viewed from the top instead of the actual leaf area to extract phenotyping features because it could reduce skew interference caused by leaves growing in various directions. Lastly, because multiple cameras were used, the colors appeared slightly different for each camera, even when shooting the same subject. This was solved by performing a calibration process using a separate color panel as a hardware characteristic.

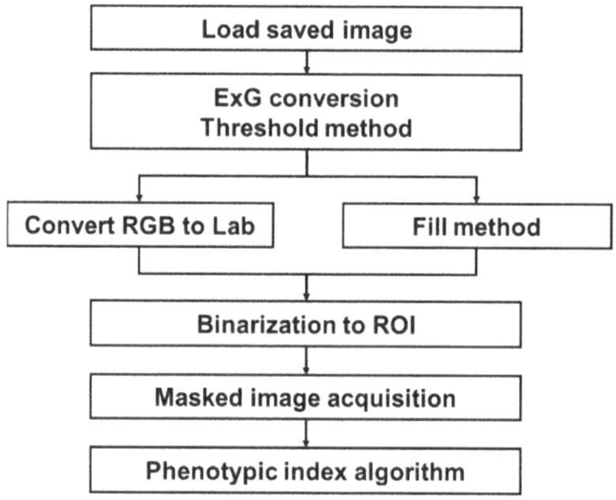

Figure 5. Overall flowchart for crop ROI extraction and phenotypic index.

The first step of the analysis of crops is the extraction of a ROI through background separation. If ROI extraction is not performed adequately, values calculated in the subsequent process become useless. Accordingly, several algorithms were combined and utilized. The target part was a crop, and the image to be analyzed was in the form of 2D red–green–blue (RGB). Thus, Excess Green (ExG), an image extraction method [35], was applied among the available vegetation indices [36]. As shown in Equation (1), R, G, and B color channels were used, normalized values divided by the maximum values (R_m, G_m, B_m) of each channel were employed, and the equation was derived by assigning weights to the green part.

$$\begin{gathered} R^* = \frac{R}{R_m},\ G^* = \frac{G}{G_m},\ B^* = \frac{B}{B_m} \\ r = \frac{R^*}{R^*+G^*+B^*},\ g = \frac{G^*}{R^*+G^*+B^*},\ b = \frac{B^*}{R^*+G^*+B^*} \\ ExG = 2g - r - b \end{gathered} \quad (1)$$

ExG uses only RGB colors and is an index affected by green; therefore, large and small noises could be included, as shown in Figure 6. Although the noise of small particles is also an issue, the artificial medium, which is usually used in plant factories, has a larger role in noises. The artificial medium is artificial soil constructed to enable crops to absorb and support the nutrient solution. In commercial plant factories, an empty artificial medium is sometimes placed to control the planting distance, but it could be recognized as a crop part in the image and included as noise. These noises cannot be distinguished by using ExG, so additional image processing is required.

In this study, the hydroponic system in plant factory was considered as the target field. Therefore, noise filtering for plant factory condition was established to extract the crop parts, excluding noise from the image extracted through the vegetation index. After converting the image masked with ExG to the Lab color system, the green-magenta and blue-yellow channels were extracted, and a threshold value determined based on the trial-and-error method was set to remove the noise. Since the lab color space reflects the uniformity of colors, thereby allowing more robustness for the illumination effect [37,38]. Relating to binarization for crop segmentation, by combining the two methods and performing an additional fill process on a small part, a sophisticated binary image of the ROI was extracted. The final target image was obtained by masking the original image on the binary image (Figure 7).

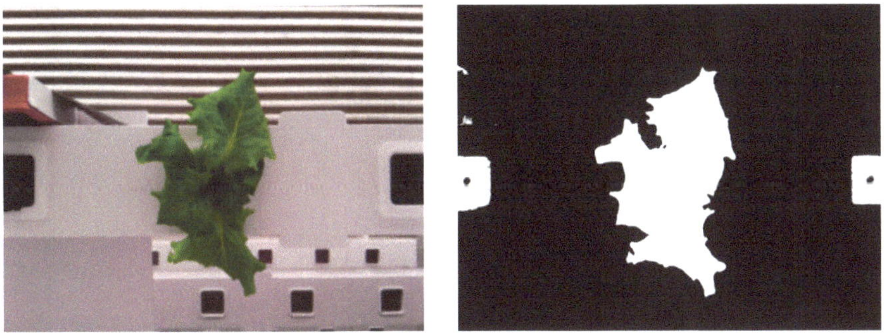

(A) (B)

Figure 6. Results of applying ExG conversion to crop images: (**A**) RGB and (**B**) ExG binary images.

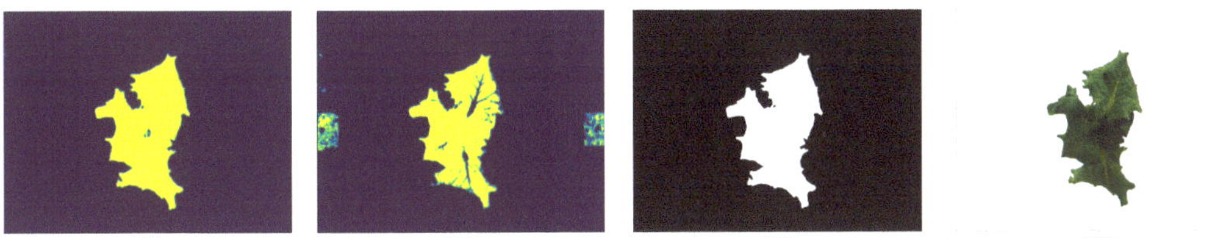

(A) (B) (C) (D)

Figure 7. Additional image processing to acquire images of the ROI: (**A**,**B**) noise removal through Lab color system, (**C**) binary image, and (**D**) mask image.

Although more sophisticated segmentation techniques such as a multilevel thresholding or combination of other color spaces such as hue–saturation–value (HSV), and hue–saturation–intensity (HSI) used for detecting the diseased leaf area could be used, the techniques require more processing time for one image, thereby inducing considerable delay for entire growing bed monitoring [37–39]. Therefore, in this study, the proposed image processing algorithm was simply constructed to achieve the real-time analysis of the morphological growth characteristics.

A process was performed to determine a phenotypic index that could be used for analysis with respect to the extracted ROI image. The center, contour, length, central axis, color space distribution, growth distortion, and color histogram were selected by referring to the useful phenotypic indices related to crop morphology in previous studies [19,40–42]. Table 2 shows a description of each indicator and the automatic acquisition algorithm.

All processes such as image acquisition, save, image processing, and phenotypic index extraction were modularized through classes and functions. Modularity facilitated the modification and addition of content. Although some parameters needed to be modified according to the experimental environment conditions, it was possible to easily change them according to the user environment through the modularization of the image processing algorithm. Since it was possible to remove only a specific part and add a sub-list of new functions and classes, it could be flexibly applied according to various experimental purposes and growing environments.

Although these various phenotypic indices can have meaning by themselves, they enable further analysis through correlation with actual crop characteristics or addition of an algorithm. To check this possibility, the manually measured values were used for analysis. In the process of conducting the experiment, crop characteristics such as fresh weight, leaf length, leaf width, and leaf number were manually measured and compared with the extracted phenotypic indices. After identifying the correlation, a trend line was drawn through a regression equation, and the coefficient of determination and root mean square error (RMSE) were obtained to verify the indices.

Table 2. Phenotypic index description and computation method.

Phenotypic Index	Description	Equation	Equation Number
 Center	• Obtained by using the moment value to find the coordinates of the center of mass p_{cm} • Reference point for radius and growth skew calculation	$m_{ji} = \sum_{x,y}\left(array(x,y)\cdot x^j \cdot y^i\right)$ $\overline{x} = \frac{m_{10}}{m_{00}}, \quad \overline{y} = \frac{m_{01}}{m_{00}}$	(2)
 Contour	• Point data acquired using the boundary of the binary mask image through ROI extraction • Used for projected leaf area calculation	$l_{contour} = \sum\left\lvert\overrightarrow{p_c{}^i - p_c{}^{i-1}}\right\rvert$ $p_c{}^i$: i-th point of the contour	(3)
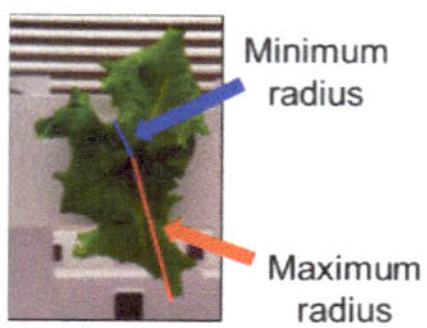 Radius (length)	• Minimum and maximum radii (length) from the central index to the contour using Euclidean distance • Available for estimation of leaf length and volume index	$l_{max\ radius} = max\left\{\left\lvert\overrightarrow{p_{cm} - p_c{}^i}\right\rvert\right\}$ $l_{min\ radius} = min\left\{\left\lvert\overrightarrow{p_{cm} - p_c{}^i}\right\rvert\right\}$ p_{cm}: center point	(4)
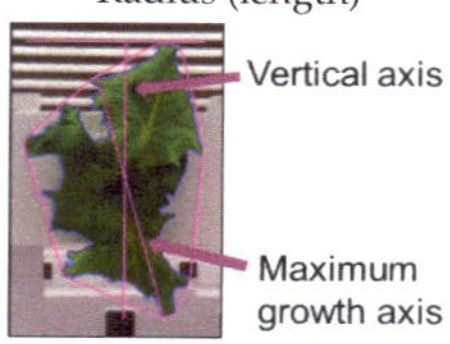 Central axis	• Vertical axis in space with respect to the center of the crop and the axis showing maximum growth • Alignment of the obtained crop image; evaluation of the crop growth behavior	$Axis_{max\ growth} = max\left\{\left\lvert\overrightarrow{p_c{}^i - p_c{}^j}\right\rvert\right\}$	(5)

Table 2. *Cont.*

Phenotypic Index	Description	Equation	Equation Number
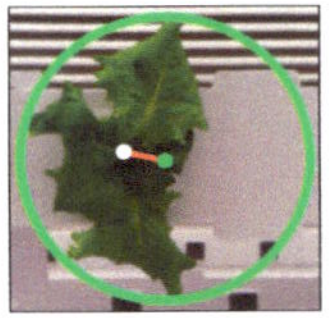 Growth skew	• Degree of deviation of the center of the crop from the center of both the circumscribed circle and the circumscribed rectangle with respect to the crop • Evaluation of the crop growth behavior	$l_{growth\ skew} = \mid \overrightarrow{p_{cm} - p_{cc}} \mid$	(6)
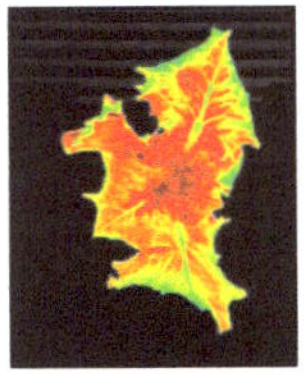 Color space distribution	• Degree of color intensity for the crop image (depth image: spatial map) • Available for estimation of crop height, volume index	-	-
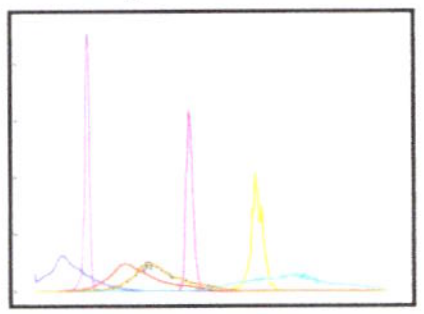 Color histogram	• Histogram for each color coordinate system of RGB, Lab, and HSV for the masked crop part • Color indices-based plant disease or stress evaluation	-	-

3. Results

3.1. Evaluation of Camera Network

It was confirmed that the developed system can easily identify and solve various problems that may occur in the process of image acquisition and processing through the log generated during the image transmission process. This detected several errors in the artificially generated data port, capturing data, and in the transmission process. Accordingly, it was possible to respond appropriately to the problems that arose, and the validity of subsequent analysis of the transmitted data could be secured.

The phenotypic index extraction algorithm was evaluated by performing the analysis on nine items, including center, contour, minimum radius, maximum radius, spatial vertical axis, maximum growth axis, color space distribution, growth distortion, and color histogram. As a result of performing automatic index extraction on a total of 35,000 image data, the total analysis time was approximately 2.43 s per image and included index values and image result output. The time required for analysis may vary depending on the number and type of indicators to be analyzed.

Figure 8 shows the analysis results using the phenotypic indices selected by the developed system for one crop. ROI extraction was performed through background separation, and the phenotypic index was automatically extracted according to a previously defined algorithm. The background was adequately removed using the vegetation index, color system conversion, and additional algorithms and by providing visualization information for all indices such as center, contour, and maximum/minimum radius. As a result of applying the original image and the phenotypic index, the entire growth cycle can be analyzed for approximately two weeks with an interval of two days for the image. This can provide a better understanding of the acquired information, such as the growth level of crops, growth status, and product quality. In addition, it is possible to access the saved analysis result and analyze it in the form of a data frame. In the developed system environment, the data frame can access each index value by using a "key" corresponding to the column, and it is possible to separate data for each crop by specifying a "value" condition of the key. Accordingly, it is possible to analyze individual crops to perform intensive analysis on crops of different growth by location.

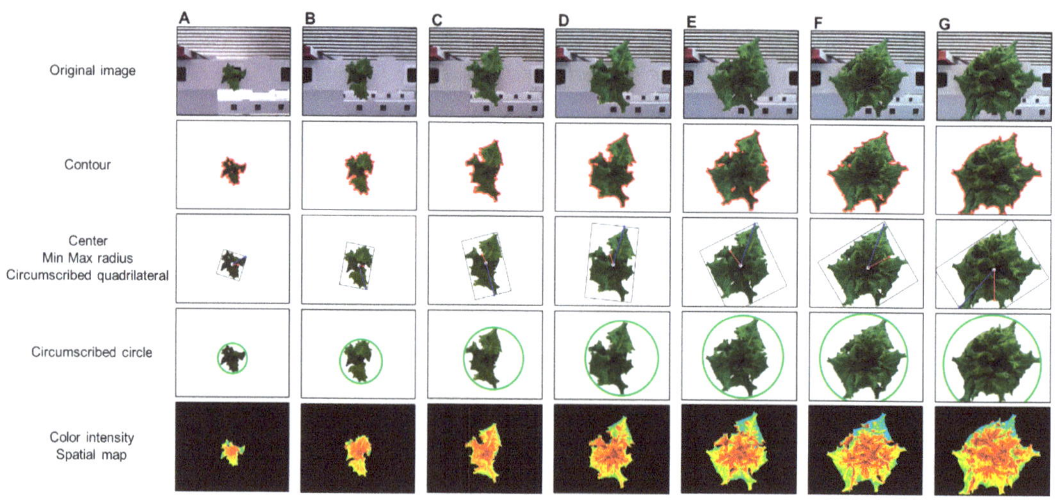

Figure 8. Changes in phenotypic indices according to crop growth: days (**A**) 1, (**B**) 3, (**C**) 5, (**D**) 7, (**E**) 9, (**F**) 11, and (**G**) 13.

3.2. Growth Analysis for Lettuces under Different Light Conditions

Figure 9 shows the area index changes for the pul group with a change in the photoperiod and for the "con group" with no change. In the pul group, as the light period during image acquisition and the dark period when acquisition was interrupted were short, the data appeared continuously, whereas the con group showed a step-like graph because the dark period was long. The two groups had the same duration of the entire light period; thus, the total number of images acquired in one day was the same, even though there were differences in the acquisition time. That is, the data of the two groups were compared for each day, not for the point value corresponding to the x-axis value of the graph.

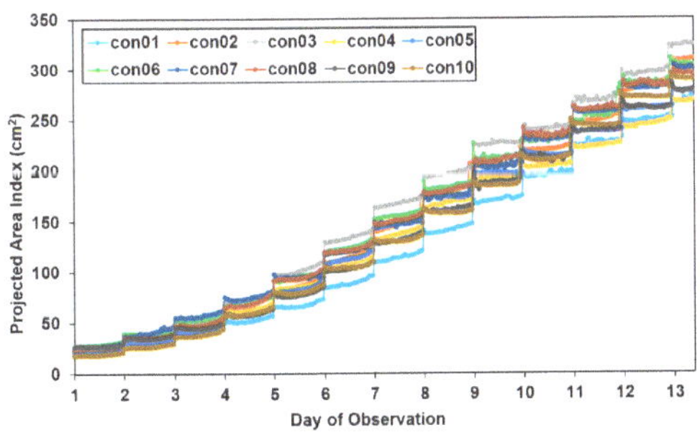

(A)

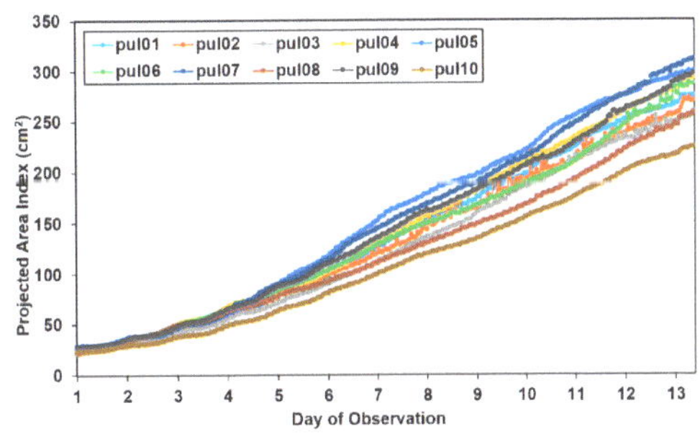

(B)

Figure 9. Growth curves for Batavia lettuces in different light periods: (**A**) control group (con) and (**B**) pulse group (pul).

In the growth curve, the growth pattern according to the growth location was different in both groups, indicating that there was spatial variation. Approximately 13 days after transplantation, the maximum area of lettuce was approximately 18% larger than the minimum area in the control group, whereas in the pulse group, it was 38% larger, confirming that the spatial variation was larger in the pulse group.

Among the ten phenotypic indices that can be extracted from our system, the indices that can be analyzed meaningfully in light condition experiments include contour, length, minimum radius, and maximum radius in addition to area. These phenotypic indices were closely analyzed to examine the growth difference according to the light period (Figure 10). Thus, a new data frame was created by calculating the average value and standard deviation of each indicator in the developed system interface, and it was expressed as a graph. In the graph, there was no significant difference in the minimum and maximum radius, but the growth was higher in the control group in both area and contour length.

Figure 10 shows the index change for one crop during the experimental period. As observed, it is possible to monitor the growth status, including the phenotypic indices selected for the crop growth of the entire period, using the developed system. However, based on the indices, additional considerations were identified in the analysis. In the case of the contour, it was possible to find a part showing a difference due to the complexity expressed in the growing process of the crop or the change in the crop shape according to the light/dark period. Accordingly, the maximum length from the center was measured by moving the position according to a specific period rather than a single point according to the passage of time. Although it was shown continuously on the graph, the actual measurement position changed significantly. The minimum length was considerably affected even by slight differences in these morphological characteristics; thus, this index showed the greatest change.

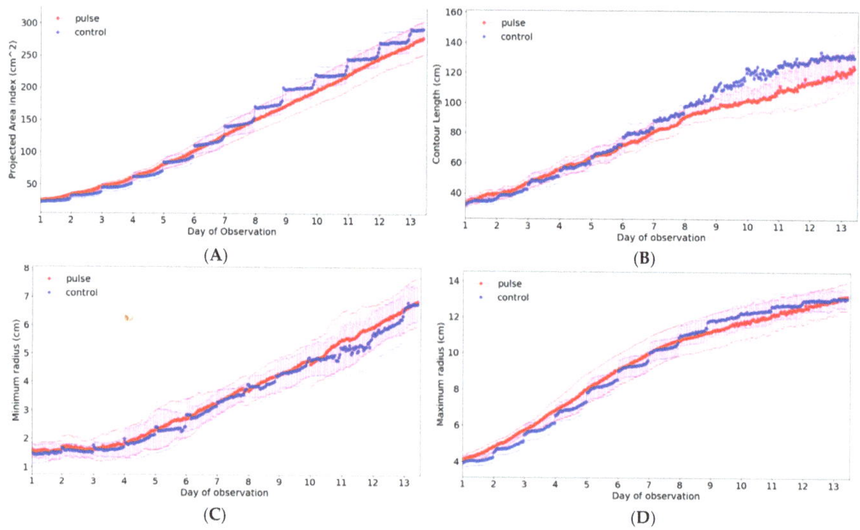

Figure 10. Analysis of phenotypic indices for Batavia lettuces in different light periods: (**A**) area, (**B**) contour length, (**C**) minimum radius, and (**D**) maximum length. The colored area indicates the standard deviations.

In order to verify the phenotypic index, the analysis was performed using the area index as a representative. Figure 11 shows the results of performing predictions using area index for the four characteristics of manually measured crops. Each regression formula was set to have the highest coefficient of determination. The area index showed the best correlation with fresh weight (R^2 = 0.9602 and RMSE = 2.23 g) (Figure 11A). In the early stages of small size, it was small with RMSE = 0.83 g, but it was confirmed that the error also increased as the size increased (Figure 11B–D). A comparison with the other three characteristics showed a low coefficient of determination compared to the fresh weight, and no specific characteristics were found.

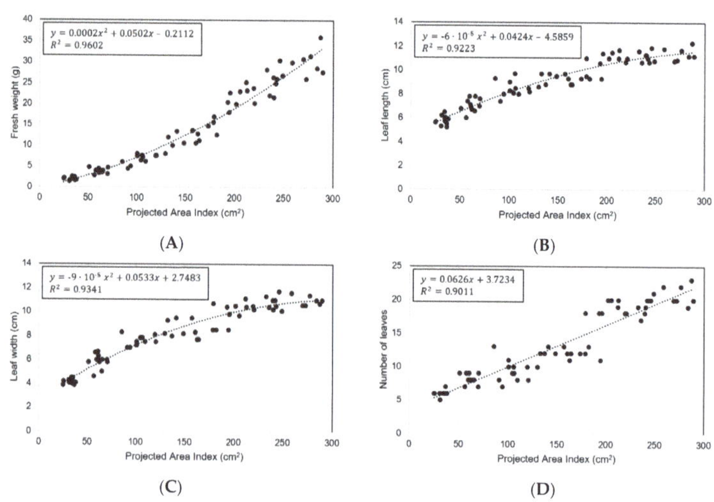

Figure 11. Predicting crop ground truth data through area index: (**A**) fresh weight, (**B**) leaf length, (**C**) leaf width, and (**D**) number of leaves.

4. Discussion

Among the phenotypic indices, the area that was the most intuitive index for crop growth was a part of the projected pixel points obtained from the top view. Although the area index would be different from the actual area because the index was calculated based on the top projection image without considering the lateral information, it can be used to provide spatial–temporal information of growing lettuces. Specifically, the site-specific growth monitoring might be feasible for the small number of plants because the variations in growth are displayed from the images for all growth periods [19,42]. In fact, this index enabled us to clearly identify both spatial variation in growth and differences in the photoperiods (Figures 9 and 10). In particular, looking at the control condition graph in Figure 10A, it can be seen that the y-axis value decreases at the beginning of the change from the dark period to the light period, and then it gradually increases thereafter. This is considered to be a phenomenon that occurs because the leaf area is expanded due to the horizontal decrease in the leaf during the dark period [31,43]. It can be seen from the graph that it decreases again as the light period begins, but it gradually increases again with plant growth. Meanwhile, another analysis was possible. Under the experimental conditions, when the light period was twice that of the dark stage, and considering the length of the x-axis expressed in the graph, it can be judged that the growth of the crop mainly occurred during the dark stage (Figure 10A). At the beginning of the experiment, the difference was less than two times, but on day 7, the difference in area was slightly larger. Initially, the control group showed similar or slightly smaller values, but 7 days after planting, it showed a larger value than the pulse group. In fact, even in the destruction data, the group with a change in the photoperiod had poor growth, and the final weight showed a difference of 13.89 g when comparing the average value based on the fresh weight of the above-ground part; thus, it was interpreted that this part was revealed by the area index. The initial hypothesis was that the efficiency could be increased by changing the photoperiod, but it was judged that the efficiency was rather low due to the short light period. The length of the contour was similar to that of the area, indicating a difference between the two groups. However, unlike the area index, where the difference gradually increased, the length of the contour showed an initial large difference, which gradually decreased (Figure 10B). As the crop grew, the edge of the leaf flattened, and the curvature seemed to disappear due to overlapping. Therefore, a different tendency from the continuously increasing area index was observed. In particular, a relatively fast saturation phenomenon was observed in the control group. The minimum and maximum radii from the center were also projected values, and the minimum length index was not suitable for analysis of the photoperiod (Figure 10C,D). This was because the minimum radius showed a large change depending on the growth of new leaves and the effect of the light period. The maximum radius showed a tendency similar to that of the area at the beginning, but with saturation at a relatively early period, and the two groups reached the same value in the second half of the experiment. Unlike the area and contour length indices that could increase in various directions on the image, the maximum radius index could not be measured beyond the measurable range; therefore, this point is shown in the graph. Moreover, researchers or growers can add new phenotypic indices based on the camera network system. Then, the system could be used for building a new phenotyping model by building a new model between the phenotypic indices and plant physiological behaviors such as nutrient deficiencies, diseases, or pests.

When referring to studies relating to fresh weight, it was found that the analysis using the area index extracted from this study has explanatory power, and it can be said that the low error shows the accuracy of the measurement (Figure 11A) [3,44]. Similarly, it was confirmed that the maximum radius index and the contour index had a high coefficient of determination in each leaf length and leaf width (Figure 11B,C). The results show the usefulness of the system for monitoring and determining the phenotypic indices of growing plants.

From the index, the growth of different light/dark cycle were observed and compared. The growth of the pul group was lower than that of the con group, indicating the longer light/dark cycle would be more beneficial for vegetative growth of lettuces. Specifically, lettuce biomass did not change unidirectionally with the extension or shortening of the light/dark cycle, but only specific light/dark cycle can strengthen root growth and water absorption of lettuces [32,45]. Therefore, it is important to find the optimal cycle of the illumination on growing lettuces. The results show the developed system would provide the continuous monitoring of growing lettuces, thereby allowing more effective and rapid analysis on such variables.

Table 3 shows the comparison of the high-throughput plant phenotyping system developed in this study with other existing systems. The most expensive system using a 3D camera array showed a highly accurate measurement performance with a slope of 1.01 and R^2 of 0.99 in the comparison with the actual values [46]. The automated growth measurement system using a conveyer also showed a highly linear predictability for the fresh weight, with a slope near 1.0 and R^2 of 0.95 [18]. Although the predictability of the developed system is lower than those systems (Figure 11), it is comparable to other systems with R^2 of 0.6–0.8 [20,21]. Considering the performance, a total cost of USD 1200 is quite reasonable. The system also can measure various phenotypic characteristics and has a high degree of freedom in analysis compared to other systems. Moreover, the use of other devices might be limited due to their high costs or spatial reasons in densely growing environment such as plant factories. In contrast, the proposed system has excellent space efficiency through a small-sized camera module and a network-based multi-connection protocol. Continuous monitoring of phenotypic indices relating to height, leaf area, volume and biomass are effective parameters that could allow farmers to conduct better fertilizer and water management as well as scheduling of their harvest [47,48]. In plant factories, leafy vegetables are usually harvested when the leaves reach full vegetative growth; thus, it is important to monitor the growth during the vegetative stage. Accordingly, in this paper, lettuce observation was continued from two days after transplantation until the end of the first harvest, and usable phenotypic indices were successfully presented.

Table 3. Comparison of the camera network system with the on-the-go monitoring system by Jiang, Kim and Cho [18].

System	Plant	No. of Plants	Phenotypic Characteristics	Code Availability	Cost	Ref.
Camera array HTPP *	Arabidopsis	1050	3	Commercial + Not open	~USD 43,200 (including camera)	[46]
Greenhouse-based HTPP *	Maize	756	3	Open	~USD 5000 (including camera)	[21]
An automated growth measurement system	Boston lettuce	45	4	Not open	~USD 2200 (including camera)	[18]
Sliding phenotyping platform	Maize, Tomato, Olive	1	8	Commercial	~USD 1300 (without camera)	[20]
Our system	Batavia lettuce	24	10	Open	~USD 1200 (including camera)	-

* HTPP: high-throughput plant phenotyping.

From the above results, it is demonstrated that the developed system can be applied to the real-time analysis of growth differences by group for various growth condition changes in plant factories, as well as spatial variation analyses according to the visualization of growth distribution. In future research, we will improve the measurement performance of the system by adding a side-view camera to collect information about crop height and areas that cannot be observed from the top view. In particular, we intend to apply a machine learning algorithm to the various phenotypic indices collected from the system to understand more complex crop-growth information and to develop a web-based interface, such that users can freely select and modify functions for each process.

5. Conclusions

In this study, we developed an analysis system for high-speed mass phenotyping based on a multi-point camera network using low-cost cameras and processor boards. To ensure its applicability to various cultivation structures and environments, a server computer was configured for several image collection modules to automatically import data from multiple points. Transmission stabilization conditions were applied for data collection stability, and a large amount of image information was transmitted from the local to the main server at daily intervals. When the transmission was completed, an ROI image was obtained through background separation, and 10 phenotypic indices were calculated using an image processing algorithm. The processing time of approximately 2.43 s per image confirmed the real-time analysis performance of the developed system. By applying the system to a commercial plant factory environment, it was possible to determine the growth distribution information and the influence of the growth environment for each individual based on various phenotypic indices through the data frame, confirming its feasible application. The projected area index-based estimation for fresh weight, leaf length, leaf width, and the number of leaves shows a high R^2 over than 0.9, indicating the system would be feasible for monitoring the growth information of growing lettuces. In addition, the high scalability of easily removing or adding programming parts through modularization of the calculation and analysis structure of the phenotypic index can be useful for growers or researchers who are concerned. In future research, we intend to introduce a side-view camera for crop height analysis and strengthen the field usability by adding machine learning algorithms. Overall, the developed system is expected to effectively help crop-growth management, as it can respond to changes in the dense plant factory cultivation environment and structural changes through low-cost camera installation and network connection configuration.

Author Contributions: Conceptualization, M.Y.; methodology, M.Y. and W.-J.C.; software, M.Y. and W.-J.C.; validation, M.Y.; formal analysis, M.Y. and W.-J.C.; writing—original draft preparation, M.Y. and W.-J.C.; writing—review and editing, M.Y. and W.-J.C.; visualization, M.Y. and W.-J.C.; supervision, M.Y.; project administration, M.Y.; funding acquisition, M.Y. All authors have read and agreed to the published version of the manuscript.

Funding: This work was supported by the National Research Foundation of Korea (NRF) grant funded by the Korea government (MSIT) (No. 2022R1C1C2005959), and the Technology development Program funded by the Ministry of SMEs and Startups (G21S325921401).

Institutional Review Board Statement: Not applicable.

Data Availability Statement: The data and code used in this study are available from the corresponding author upon request.

Acknowledgments: The authors would like to thank n.thing Inc. Republic of Korea for providing a container-type vertical farm.

Conflicts of Interest: The funders had no role in the design of the study; in the collection, analyses, or interpretation of data; in the writing of the manuscript; or in the decision to publish the results.

References

1. Li, L.; Zhang, Q.; Huang, D. A review of imaging techniques for plant phenotyping. *Sensors* **2014**, *14*, 20078–20111. [CrossRef] [PubMed]
2. Chen, D.; Neumann, K.; Friedel, S.; Kilian, B.; Chen, M.; Altmann, T.; Klukas, C. Dissecting the Phenotypic Components of Crop Plant Growth and Drought Responses Based on High-Throughput Image Analysis. *Plant Cell* **2014**, *26*, 4636–4655. [CrossRef]
3. Chen, W.-T.; Yeh, Y.-H.F.; Liu, T.-Y.; Lin, T.-T. An automated and continuous plant weight measurement system for plant factory. *Front. Plant Sci.* **2016**, *7*, 392. [CrossRef]
4. Fahlgren, N.; Gehan, M.A.; Baxter, I. Lights, camera, action: High-throughput plant phenotyping is ready for a close-up. *Curr. Opin. Plant Biol.* **2015**, *24*, 93–99. [CrossRef]
5. Kacira, M.; Ling, P. Design and development of an automated and Non–contact sensing system for continuous monitoring of plant health and growth. *Trans. ASABE* **2001**, *44*, 989. [CrossRef]
6. Sonnentag, O.; Hufkens, K.; Teshera-Sterne, C.; Young, A.M.; Friedl, M.; Braswell, B.H.; Milliman, T.; O'Keefe, J.; Richardson, A.D. Digital repeat photography for phenological research in forest ecosystems. *Agric. Meteorol.* **2012**, *152*, 159–177. [CrossRef]

7. Jiang, Y.; Li, C.; Paterson, A.H. High throughput phenotyping of cotton plant height using depth images under field conditions. *Comput. Electron. Agric.* **2016**, *130*, 57–68. [CrossRef]
8. Barker III, J.; Zhang, N.; Sharon, J.; Steeves, R.; Wang, X.; Wei, Y.; Poland, J. Development of a field-based high-throughput mobile phenotyping platform. *Comput. Electron. Agric.* **2016**, *122*, 74–85. [CrossRef]
9. Ge, Y.; Bai, G.; Stoerger, V.; Schnable, J.C. Temporal dynamics of maize plant growth, water use, and leaf water content using automated high throughput RGB and hyperspectral imaging. *Comput. Electron. Agric.* **2016**, *127*, 625–632. [CrossRef]
10. Naik, H.S.; Zhang, J.; Lofquist, A.; Assefa, T.; Sarkar, S.; Ackerman, D.; Singh, A.; Singh, A.K.; Ganapathysubramanian, B. A real-time phenotyping framework using machine learning for plant stress severity rating in soybean. *Plant Methods* **2017**, *13*, 23. [CrossRef]
11. Granier, C.; Aguirrezabal, L.; Chenu, K.; Cookson, S.J.; Dauzat, M.; Hamard, P.; Thioux, J.J.; Rolland, G.; Bouchier-Combaud, S.; Lebaudy, A. PHENOPSIS, an automated platform for reproducible phenotyping of plant responses to soil water deficit in Arabidopsis thaliana permitted the identification of an accession with low sensitivity to soil water deficit. *New Phytol.* **2006**, *169*, 623–635. [CrossRef] [PubMed]
12. Tisné, S.; Serrand, Y.; Bach, L.; Gilbault, E.; Ben Ameur, R.; Balasse, H.; Voisin, R.; Bouchez, D.; Durand-Tardif, M.; Guerche, P. Phenoscope: An automated large-scale phenotyping platform offering high spatial homogeneity. *Plant J.* **2013**, *74*, 534–544. [CrossRef] [PubMed]
13. Hartmann, A.; Czauderna, T.; Hoffmann, R.; Stein, N.; Schreiber, F. HTPheno: An image analysis pipeline for high-throughput plant phenotyping. *BMC Bioinform.* **2011**, *12*, 148. [CrossRef] [PubMed]
14. Shankar, T.; Malik, G.C.; Banerjee, M.; Dutta, S.; Praharaj, S.; Lalichetti, S.; Mohanty, S.; Bhattacharyay, D.; Maitra, S.; Gaber, A. Prediction of the effect of nutrients on plant parameters of rice by artificial neural network. *Agronomy* **2022**, *12*, 2123. [CrossRef]
15. Joo, H.-J.; Jeong, H.-Y. Growth analysis system for IT-based plant factory. *Multimed. Tools Appl.* **2017**, *76*, 17785–17799. [CrossRef]
16. Kang, W.H.; Zhang, F.; Lee, J.W.; Son, J.E. Improvement of canopy light distribution, photosynthesis, and growth of lettuce (Lactuca sativa L.) in plant factory conditions by using filters to diffuse light from LEDs. *Hortic. Sci. Technol.* **2016**, *34*, 84–93. [CrossRef]
17. Moriyuki, S.; Fukuda, H. High-Throughput Growth Prediction for Lactuca sativa L. Seedlings Using Chlorophyll Fluorescence in a Plant Factory with Artificial Lighting. *Front. Plant Sci.* **2016**, *7*, 394. [CrossRef]
18. Jiang, J.-s.; Kim, H.-J.; Cho, W.-J. On-the-go image processing system for spatial mapping of lettuce fresh weight in plant factory. *IFAC-Pap.* **2018**, *51*, 130–134. [CrossRef]
19. Yeh, Y.-H.F.; Lai, T.-C.; Liu, T.-Y.; Liu, C.-C.; Chung, W.-C.; Lin, T.-T. An automated growth measurement system for leafy vegetables. *Biosyst. Eng.* **2014**, *117*, 43–50. [CrossRef]
20. Rossi, R.; Leolini, C.; Costafreda-Aumedes, S.; Leolini, L.; Bindi, M.; Zaldei, A.; Moriondo, M. Performances Evaluation of a Low-Cost Platform for High-Resolution Plant Phenotyping. *Sensors* **2020**, *20*, 3150. [CrossRef]
21. Yassue, R.M.; Galli, G.; Borsato, R., Jr.; Cheng, H.; Morota, G.; Fritsche-Neto, R. A low-cost greenhouse-based high-throughput phenotyping platform for genetic studies: A case study in maize under inoculation with plant growth-promoting bacteria. *Plant Phenome J.* **2022**, *5*, e20043. [CrossRef]
22. Bari, M.A.A.; Fonseka, D.; Stenger, J.; Zitnick-Anderson, K.; Atanda, S.A.; Morales, M.; Worral, H.; Piche, L.; Kim, J.; Johnson, J. A greenhouse-based high-throughput phenotyping platform for identification and genetic dissection of resistance to Aphanomyces root rot in field pea. *Plant Phenome J.* **2023**, *6*, e20063. [CrossRef]
23. An, N.; Palmer, C.M.; Baker, R.L.; Markelz, R.J.C.; Ta, J.; Covington, M.F.; Maloof, J.N.; Welch, S.M.; Weinig, C. Plant high-throughput phenotyping using photogrammetry and imaging techniques to measure leaf length and rosette area. *Comput. Electron. Agric.* **2016**, *127*, 376–394. [CrossRef]
24. Thrash, T.; Lee, H.; Baker, R.L. A low-cost high-throughput phenotyping system for automatically quantifying foliar area and greenness. *Appl. Plant Sci.* **2022**, *10*, e11502. [CrossRef]
25. Gang, M.-S.; Kim, H.-J.; Kim, D.-W. Estimation of Greenhouse Lettuce Growth Indices Based on a Two-Stage CNN Using RGB-D Images. *Sensors* **2022**, *22*, 5499. [CrossRef]
26. Kozai, T. Current status of plant factories with artificial lighting (PFALs) and smart PFALs. In *Smart Plant Factory: The Next Generation Indoor Vertical Farms*; Springer: Singapore, 2018; pp. 3–13.
27. Coelho, A.F.S.; Gomes, É.P.; Sousa, A.d.P.; Glória, M.B.A. Effect of irrigation level on yield and bioactive amine content of American lettuce. *J. Sci. Food Agric.* **2005**, *85*, 1026–1032. [CrossRef]
28. Liu, Z.; Xu, Q. An Automatic Irrigation Control System for Soilless Culture of Lettuce. *Water* **2018**, *10*, 1692. [CrossRef]
29. Wortman, S.E. Crop physiological response to nutrient solution electrical conductivity and pH in an ebb-and-flow hydroponic system. *Sci. Hortic.* **2015**, *194*, 34–42. [CrossRef]
30. Shao, M.; Liu, W.; Zha, L.; Zhou, C.; Zhang, Y.; Li, B. Altering light–dark cycle at pre-harvest stage regulated growth, nutritional quality, and photosynthetic pigment content of hydroponic lettuce. *Acta Physiol. Plant.* **2021**, *43*, 9. [CrossRef]
31. Fricke, W. Night-time transpiration–favouring growth? *Trends Plant Sci.* **2019**, *24*, 311–317. [CrossRef]
32. Chen, X.-l.; Yang, Q.-c. Effects of intermittent light exposure with red and blue light emitting diodes on growth and carbohydrate accumulation of lettuce. *Sci. Hortic.* **2018**, *234*, 220–226. [CrossRef]
33. Ma, X.; Zhu, K.; Guan, H.; Feng, J.; Yu, S.; Liu, G. High-throughput phenotyping analysis of potted soybean plants using colorized depth images based on a proximal platform. *Remote Sens.* **2019**, *11*, 1085. [CrossRef]

34. Sinoquet, H.; Thanisawanyangkura, S.; Mabrouk, H.; Kasemsap, P. Characterization of the light environment in canopies using 3D digitising and image processing. *Ann. Bot.* **1998**, *82*, 203–212. [CrossRef]
35. Riehle, D.; Reiser, D.; Griepentrog, H.W. Robust index-based semantic plant/background segmentation for RGB-images. *Comput. Electron. Agric.* **2020**, *169*, 105201. [CrossRef]
36. Woebbecke, D.M.; Meyer, G.E.; Von Bargen, K.; Mortensen, D. Color indices for weed identification under various soil, residue, and lighting conditions. *Trans. ASAE* **1995**, *38*, 259–269. [CrossRef]
37. Hassanein, M.; Lari, Z.; El-Sheimy, N. A New Vegetation Segmentation Approach for Cropped Fields Based on Threshold Detection from Hue Histograms. *Sensors* **2018**, *18*, 1253. [CrossRef] [PubMed]
38. Jiao, Y.; Luo, R.; Li, Q.; Deng, X.; Yin, X.; Ruan, C.; Jia, W. Detection and Localization of Overlapped Fruits Application in an Apple Harvesting Robot. *Electronics* **2020**, *9*, 1023. [CrossRef]
39. Sibiya, M.; Sumbwanyambe, M. An Algorithm for Severity Estimation of Plant Leaf Diseases by the Use of Colour Threshold Image Segmentation and Fuzzy Logic Inference: A Proposed Algorithm to Update a "Leaf Doctor" Application. *AgriEngineering* **2019**, *1*, 205–219. [CrossRef]
40. Upadhyaya, H.D.; Ortiz, R.; Bramel, P.J.; Singh, S. Phenotypic diversity for morphological and agronomic characteristics in chickpea core collection. *Euphytica* **2002**, *123*, 333–342. [CrossRef]
41. Li, D.; Li, C.; Yao, Y.; Li, M.; Liu, L. Modern imaging techniques in plant nutrition analysis: A review. *Comput. Electron. Agric.* **2020**, *174*, 105459. [CrossRef]
42. Lee, U.; Chang, S.; Putra, G.A.; Kim, H.; Kim, D.H. An automated, high-throughput plant phenotyping system using machine learning-based plant segmentation and image analysis. *PLoS ONE* **2018**, *13*, e0196615. [CrossRef]
43. Zweifel, R.; Sterck, F.; Braun, S.; Buchmann, N.; Eugster, W.; Gessler, A.; Häni, M.; Peters, R.L.; Walthert, L.; Wilhelm, M.; et al. Why trees grow at night. *New Phytol.* **2021**, *231*, 2174–2185. [CrossRef]
44. Jung, D.-H.; Park, S.H.; Han, X.Z.; Kim, H.-J. Image processing methods for measurement of lettuce fresh weight. *J. Biosyst. Eng.* **2015**, *40*, 89–93. [CrossRef]
45. Chen, X.-l.; Li, Y.-l.; Wang, L.-c.; Yang, Q.-c.; Guo, W.-z. Responses of butter leaf lettuce to mixed red and blue light with extended light/dark cycle period. *Sci. Rep.* **2022**, *12*, 6924. [CrossRef] [PubMed]
46. An, N.; Welch, S.M.; Markelz, R.J.C.; Baker, R.L.; Palmer, C.M.; Ta, J.; Maloof, J.N.; Weinig, C. Quantifying time-series of leaf morphology using 2D and 3D photogrammetry methods for high-throughput plant phenotyping. *Comput. Electron. Agric.* **2017**, *135*, 222–232. [CrossRef]
47. Buelvas, R.M.; Adamchuk, V.I.; Leksono, E.; Tikasz, P.; Lefsrud, M.; Holoszkiewicz, J. Biomass estimation from canopy measurements for leafy vegetables based on ultrasonic and laser sensors. *Comput. Electron. Agric.* **2019**, *164*, 104896. [CrossRef]
48. Hu, Y.; Wang, L.; Xiang, L.; Wu, Q.; Jiang, H. Automatic Non-Destructive Growth Measurement of Leafy Vegetables Based on Kinect. *Sensors* **2018**, *18*, 806. [CrossRef] [PubMed]

agriculture

Article

Quantifying Soil Particle Settlement Characteristics through Machine Vision Analysis Utilizing an RGB Camera

Donggeun Kim [1], Jisu Song [2] and Jaesung Park [2,*]

[1] Graduate School of Agriculture, Kyoto University, Kyoto 606-8502, Japan; fiy3000@snu.ac.kr
[2] Department of Bio-Industrial Machinery Engineering, Pusan National University, Miryang 50463, Republic of Korea; jiisu9380@pusan.ac.kr
* Correspondence: jaesung.park@pusan.ac.kr

Abstract: Soil particle size distribution is a crucial factor in determining soil properties and classifying soil types. Traditional methods, such as hydrometer tests, have limitations in terms of time required, labor, and operator dependency. In this paper, we propose a novel approach to quantify soil particle size analysis using machine vision analysis with an RGB camera. The method aims to overcome the limitations of traditional techniques by providing an efficient and automated analysis of fine-grained soils. It utilizes a digital camera to capture the settling properties of soil particles, eliminating the need for a hydrometer. Experimental results demonstrate the effectiveness of the machine vision-based approach in accurately determining soil particle size distribution. The comparison between the proposed method and traditional hydrometer tests reveals strong agreement, with an average deviation of only 2.3% in particle size measurements. This validates the reliability and accuracy of the machine vision-based approach. The proposed machine vision-based analysis offers a promising alternative to traditional techniques for assessing soil particle size distribution. The experimental results highlight its potential to revolutionize soil particle size analysis, providing precise, efficient, and cost-effective analysis for fine-grained soils.

Keywords: soil particle size analysis; machine vision; RGB camera; settling characteristic; image analysis

Citation: Kim, D.; Song, J.; Park, J. Quantifying Soil Particle Settlement Characteristics through Machine Vision Analysis Utilizing an RGB Camera. *Agriculture* **2023**, *13*, 1674. https://doi.org/10.3390/agriculture13091674

Academic Editor: Ju-Seok Nam

Received: 4 July 2023
Revised: 21 August 2023
Accepted: 22 August 2023
Published: 24 August 2023

1. Introduction

Soil is composed of particles with varying sizes [1], and its properties vary depending on the particle size distribution. Soil texture is determined by its sand, silt, and clay content, and it is divided into coarse- and fine-grained soils based on the particle size distribution. These soil classifications are crucial because they facilitate the easy identification of general soil characteristics.

Particle size analysis separates soil by size to determine the particle size distribution [1]. Particle size analysis is one of the most basic and important soil property tests because soils can be classified according to their particle size distributions [2–4]. Using this method, the particle size distribution curve, which is the relationship curve of the percent finer (P) as a function of the particle size (D) of the soil, can be obtained. In addition to soil classification, the results of particle size analysis can be used for purposes such as index testing, profiling, and compliance [5]. They can also be used to estimate various soil properties such as permeability, shear strength, compressibility, conductivity, and consolidation [2,3].

In the particle size analysis tests, the particles with a diameter of 0.075 mm or more are analyzed via sieve analysis, and those with smaller diameters are analyzed using hydrometer testing [1–4,6–8]. Hydrometer testing is a test method for determining the particle size distribution of soil particles by measuring the change in density in a soil–water suspension over time as the soil particles settle in the suspension [2].

Hydrometer tests are widely used but have some limitations [9]. For example, they can be time- and labor-consuming [10] and exhibits operator and instrument dependency [3].

The hydrometer test involves multiple steps, including soil dispersion, sedimentation, and reading the hydrometer scale at specific time intervals. This makes it difficult to analyze a multitude of soil samples within a short period. Moreover, the accuracy of the hydrometer test depend on the skills and experience of the operator. Appropriate sample preparation, handling, and accurate hydrometer scale are critical for obtaining reliable results. Reading the hydrometer scale can be challenging because of several factors. The scale is often small and requires precise measurement. Variations in technique between different operators may introduce inconsistencies and affect the accuracy of the analysis. Despite these drawbacks, hydrometer testing is widely used for particle size analysis of fine-grained soils owing to its moderate cost, wide availability, and extensive existing references [5].

There are many alternatives to hydrometer testing, including laser diffraction (LD), X-ray absorption, and gamma-ray attenuation methods [8]. The aforementioned methods afford faster analysis and higher accuracy compared to hydrometer tests. However, they require specialized equipment for experimentation, which can be expensive to purchase, operate, and maintain. In particular, compared to experimental methods such as sieve analysis or pipette method, the test cost per sample is approximately 3 to 10 times more expensive, and the equipment cost is approximately 10 to 25 times more expensive [5]. As a result, these alternative methods have limitations that make them difficult to use except in specialized labs. It is also worth noting that hydrometer tests typically use 50 g of sample, whereas the aforementioned methods use much smaller amounts of sample; this can lead to problems concerning the representativeness of the test results [8]. In this regard, the sieve–hydrometer method is still widely used as the most representative method of soil particle size analysis.

Digital image analysis essentially entails analysis based on features found in a digital image of an object [11,12]. The process of analyzing digital images to quickly make a desired decision is referred to as machine vision. Digital image analysis is fast, inexpensive, and repeatable and has a wide range of applications [13–17]. As regards soils, digital image analysis is also being used to predict various soil properties such as water content [12,18–20], density [12,20], soil cracks [21,22], permeability [23,24], settling velocity [25,26] and strength [27,28].

Analyzing particle size distribution is also possible via digital image analysis. Soil particle size analysis through digital image analysis can be divided into two methods: measuring the size of individual particles in images and predicting the particle size distribution based on the features in the images. Measuring the size of individual particles is a widely used method for aggregates. Mora et al. [29] performed particle size analysis of coarse aggregates from 6.3 to 28 mm with high accuracy by digital image analysis. Ohm and Hryciw [16] developed a new image-based test called "sedimaging" to analyze particle sizes in the range of 0.075–2.0 mm, which has been typically performed by sieve analysis. Sudarsan et al. [30] characterized soil particle sizes using image analysis of microscope images. Bittelli et al. [31] conducted a comparative analysis of the pipette method, SediGraph method, LD method, and automated digital image analysis to determine the appropriate test method for particle size analysis of fine-graded soil and recommended the LD method as the standard method. Sun et al. [32] proposed a minimum image quality to obtain reliable results in image-based soil particle shape characterization. However, in the case of fine-grained soils with particle sizes of 0.075 mm or less, the size distribution is difficult to predict with typical digital image analysis because of the extremely small size of the particles and the difficulty of separating individual particles. As a result, the particle size analysis of fine-grained soils using digital image analysis is still limited. Thus, it is necessary to develop a particle size analysis method based on digital image analysis to analyze the particle size of fine-grained soil affordably and quickly.

Therefore, this study devised a machine vision-based analysis method to replace the hydrometer test to perform particle size analysis of fine-grained soil. The test method was developed by applying the principles of the hydrometer test. To reduce operator dependency and errors in the readings, an experimental method that does not use a hydrometer was proposed. A digital camera was used to analyze the settling properties of the soil, replacing the hydrometer. In particular, the analysis was performed based on digital images acquired using a conventional digital camera meant for general use. This enables efficient soil particle size analysis and affords the advantage of real-time analysis and automation.

This paper follows a structured approach to introduce a machine vision-based method for analyzing fine-grained soil particle sizes, aimed at replacing traditional hydrometer testing. It begins by outlining the rationale for the study and the development of the experimental technique, which eliminates the need for a hydrometer reading. The soil properties under investigation are detailed, and the methodology for particle size analysis, including sieve analysis and a particle size analyzer, is explained. The image acquisition process using a specially designed settling tank and digital camera is then elucidated. The core of the paper focuses on image analysis, establishing relationships between settling distance, particle size, and average image intensity, which replaces hydrometer readings. Results showcase the method's success in predicting particle size distribution across various soil types, followed by a discussion of practical considerations and limitations.

2. Hydrometer Test

The hydrometer test is a test method for obtaining the particle size distribution by measuring the density of soil–water suspension over time and obtaining the percent finer (P) by particle diameter (D) of the soil sample [7,33]. The hydrometer test calculates the diameter of a soil particle based on Stoke's law [3,4,34]. The settling velocity (v) of a particle having diameter D, according to Stoke's law, is as follows:

$$v = \frac{g(\rho_s - \rho_w)}{18\mu} D^2, \tag{1}$$

where g is the gravitational acceleration, ρ_s is the density of the soil particle, ρ_w is the density of water, and μ is the viscosity of water at T (°C). A particle having diameter D settles at a distance equal to the settling distance (L) in an elapsed time (t); therefore, $v = L/t$. Therefore, D can be determined as follows:

$$D = \sqrt{\frac{18\mu}{g(\rho_s - \rho_w)} \frac{L}{t}}. \tag{2}$$

L at time t is measured on the basis of the hydrometer reading (γ). When the hydrometer and measuring cylinder are tested as a pair, L and γ are in a linear relationship.

$$L = a\gamma + b. \tag{3}$$

Here, a and b are the slope and intercept, respectively. Assuming that at time $t = 0$, a particle of mass m_{s1} is uniformly suspended in a volume V, the mass of the particle contained is m_{s1}/V and the volume is $m_{s1}/(V\rho_s)$ in a unit volume of suspension. Thus, the volume and mass of water in a unit volume of suspension are $1 - m_{s1}/(V\rho_s)$, $\rho_w(1 - m_{s1}/(V\rho_s))$, respectively. Therefore, the density of the suspension (ρ) is as follows:

$$\rho = \frac{m_{s1}}{V} + \left(\rho_w - \frac{m_{s1}\rho_w}{V\rho_s}\right) = \rho_w + \frac{m_{s1}}{V}\left(\frac{\rho_s - \rho_w}{\rho_s}\right) \tag{4}$$

At time $t = t$, only particles having diameters smaller than D are present at depths smaller than L. Therefore, if the ratio of the mass of the particles having diameters smaller than D to the total mass is denoted $P(D)$, the density of the suspension at L is

$$\rho = \rho_w + \frac{m_{s1}}{V}\left(\frac{\rho_s - \rho_w}{\rho_s}\right)\frac{P(D)}{100} \tag{5}$$

In this case, ρ has the following relationship to the hydrometer reading (γ):

$$\frac{\rho_s - \rho_w}{\rho_w} = \gamma + C_m + F, \tag{6}$$

where C_m is meniscus correction and F is the temperature correction coefficient. Thus, using Equations (5) and (6), $P(D)$ can be calculated as follows:

$$P(D) = \frac{V}{m_{s1}} \frac{\rho_s}{\rho_s - \rho_w} (\gamma + C_m + F)\rho_w \times 100. \tag{7}$$

3. Materials and Methods

3.1. Soil Properties

Table 1 summarizes an overview of the soil types used in the present study and their corresponding sampling locations. The investigation focused on five forest soils, namely, B, R, GrB, DR, and Va. These are representative soils in the Korean forest soil group.

Table 1. Description and sampling location of soil samples.

Soil Group (Subgroup)	Symbol	Latitude	Longitude
Brown forest soils	B	37.309404	127.30986
Red & yellow forest soils (Red forest soils)	R	37.23417	126.80021
Gray brown forest soils	GrB	35.98825	127.63727
Dark red forest soils	DR	37.28844	126.84164
Volcanic ash forest soils	Va	33.414417	126.48421

Particle size analysis and soil property tests were carried out on a set of five soil samples. The tests included a specific gravity test, as well as liquid and plastic limit tests (LL and PL), along with a loss of ignition (LOI) test. These experiments focused on particles measuring 2 mm or smaller. Furthermore, particle size analysis was performed on particles measuring 0.075 mm or larger using sieve analysis, while a particle size analyzer Malvern Mastersizer 2000 (Malvern Panalytical Ltd., Worcestershire, UK) was employed to analyze the smaller particles. The resulting particle size distribution curve for each soil sample, which combines the outcomes from the sieve analysis and particle size analyzer, is shown in Figure 1.

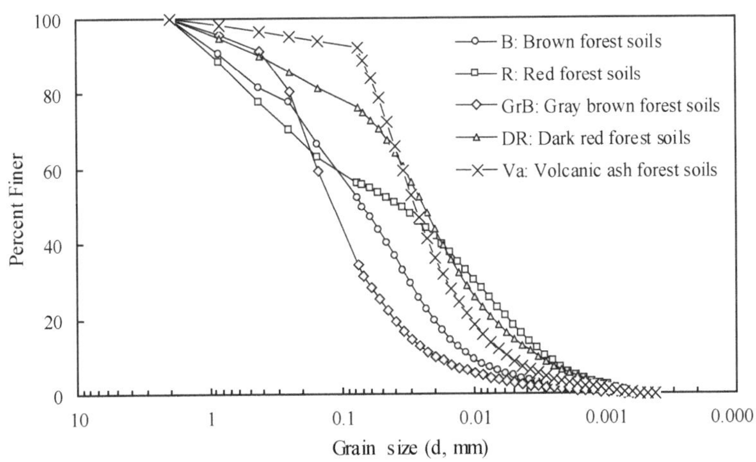

Figure 1. Particle size distribution curve of soil samples obtained using sieve analysis and a particle size analyzer.

Table 2 summarizes findings obtained from the particle size analysis and soil property tests. The specific gravity measurements for soil samples B, R, GrB, and DR were within the typical range for soils, while Va stood out with notably low specific gravity. Additionally, Va exhibited a significant LOI value of 32.6%, indicating a substantially higher organic matter content than the other soils examined. In terms of plasticity, B, GrB, and Va were classified as non-plastic, whereas R and DR showed low plasticity. Moving on to the particle size analysis results, GrB was identified as a coarse-grained soil, with 34.3% of particles passing through a 0.075 mm sieve, while the remaining were classified as fine-grained soils. Determining soil texture based on the sand, silt, and clay content proportions revealed that B, R, and GrB soils could be categorized as sandy loam, while DR and Va could be categorized as silt loam and silt, respectively.

Table 2. Result of soil property tests and particle size analysis.

Symbol	G_s *	LL *	PI *	LOI *	<0.075 mm (%)	Sand (%)	Silt (%)	Clay (%)	USDA *
B	2.53	N.P. *	N.P.	12.9	52.1	52.2	46.1	1.7	Sandy Loam
R	2.59	36.64	20.67	6.8	56.0	45.2	48.6	6.2	Sandy Loam
GrB	2.61	N.P.	N.P.	6.9	34.3	70.6	28.2	1.2	Sandy Loam
DR	2.57	28.78	11.79	9.0	76.3	26.4	67.8	5.8	Silt Loam
Va	2.07	N.P.	N.P.	32.6	92.0	14.5	82.2	3.3	Silt

* Gs: Specific gravity, LL: Liquid limit (%), PI: Plastic index, N.P.: Non plastic. LOI: Loss of ignition (%), USDA: US Department of Agriculture.

3.2. Image Acquisition during Settlement of Soil Particles

In this study, we aimed to predict the particle size distribution of fine-grained soil with a diameter of 0.075 mm or less based on digital image analysis. For this purpose, it is necessary to acquire a series of digital images of the soil–water suspension during settlement. The particle size analysis of fine-grained soils was performed using the hydrometer test as aforedescribed. The hydrometer test employs a measuring cylinder with an external scale for measuring the volume. In addition, measuring cylinders are typically made of glass, which can reflect light and result in low-quality images when used for capturing soil images. Hence, to overcome this issue, a settling tank was devised to replace the measuring cylinder, as illustrated in Figure 2.

(a)

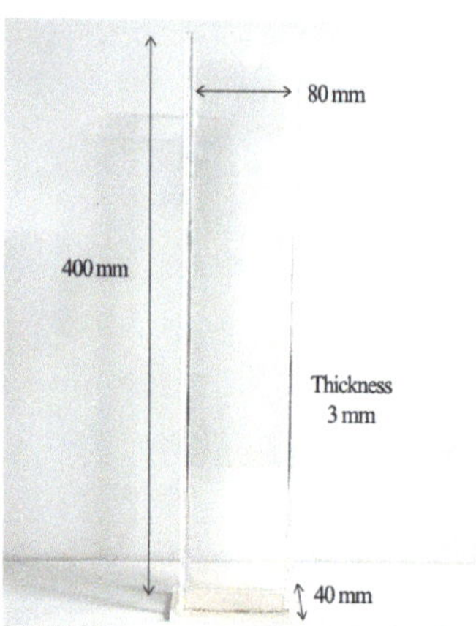

(b)

Figure 2. Settling tank for soil particle size analysis: (**a**) dimensions of settling tank; (**b**) experimental setup.

The settling tank used in the study has the following dimensions: a height of 400 mm, a width of 80 mm, and a depth of 40 mm. It is constructed using acrylic material having a thickness of 3 mm. Each soil–water suspension in the settling tank weighs 100 g, and the total volume of the suspension is 1000 mL. Before adding the soil to the tank, it was thoroughly dried. and only the portion that passed through the 2 mm sieve was used, consistent with the particle size analysis test.

To capture digital images of the soil–water suspension during the settling process, a Canon EOS 100d camera (Canon Inc., Tokyo, Japan) was used. The photographs were acquired from the front of the settling tank, and a white matte plate was positioned at the back of the tank to eliminate color distortion caused by the background. The photography sessions were conducted in an indoor studio to maintain consistent lighting conditions. The camera settings were configured as follows: a shutter speed of 0.125 s, an aperture value of 5.6, and an ISO setting of 200.

For automated continuous shooting, the camera's shooting time was controlled by a computer. Canon's EOS Utility program was employed for this purpose. The elapsed time from the initiation of the settling process to the moment the n-th photograph was taken was defined as T_n.

In the hydrometer test, readings from the hydrometer are recorded at specific time intervals: around 1, 2, 4, 15, 30, 60, 240, and 1440 min, with the option for more readings to enhance accuracy. In this study, photographs were acquired every 10 s within the first 20 min, followed by intervals of one minute until 1520 min, and then at ten-minute intervals up to 2980 min. From this collection of images, those taken within the initial 120 s, as well as at 3, 4, 6, 8, 12, 15, 30, 60, 120, 240, 480, 960, and 1440 min, were chosen for in-depth analysis. As a result, a total of 25 images were utilized for the image analysis and particle size analysis.

3.3. Image Analysis

In Figure 3, a specific region of interest (ROI) is highlighted within the digital image. The ROI corresponds to the area spanning the top to the bottom of the soil–water suspension in each settling tank. This area was designated as the ROI for analysis purposes. The dimensions of the ROI are 400 pixels in width and 2400 pixels in height.

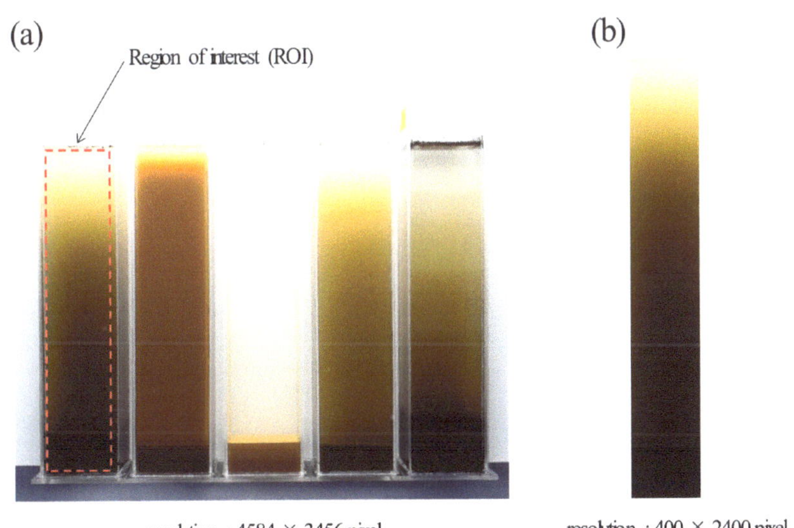

Figure 3. Region of interest in settling tank: (**a**) position of the region of interest; (**b**) cropped image of the region of interest.

Figure 4 illustrates the proposed concept of machine vision-based soil particle size analysis (MVSPSA). In the settling tank, the soil particles within the soil–water suspension gradually settle over time. As the particles settle, the color distribution within the suspension changes.

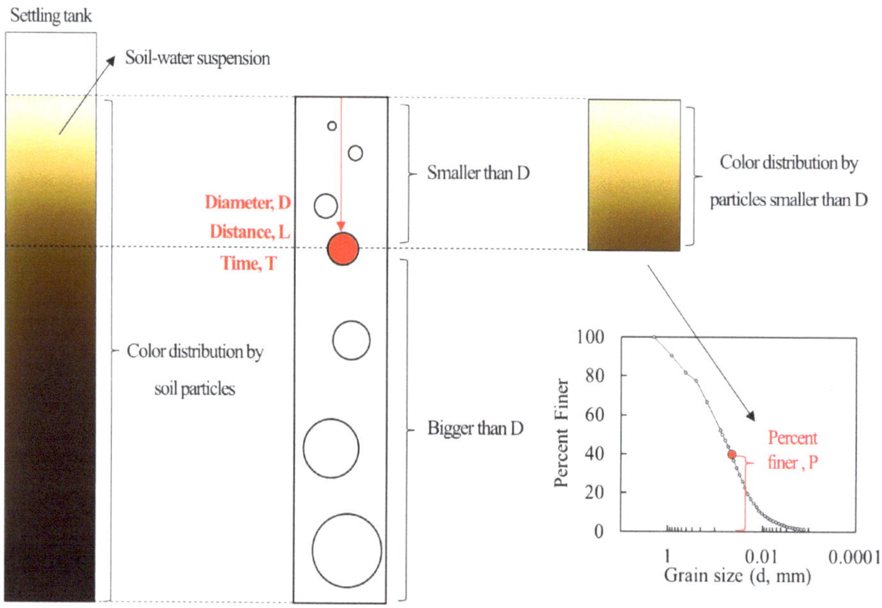

Figure 4. Conceptual diagram of machine vision-based soil particle size analysis.

The settling distance (L), at time T for a particle having a diameter D, can be determined using Equation (2). L is directly proportional to the square of D. Consequently, particles that have settled below the distance L have sizes larger than D, whereas particles above the distance L have sizes smaller than D. Consequently, the color distribution in the image, specifically within the range of L, is primarily influenced by particles having diameters smaller than D.

In the hydrometer test, the hydrometer reading (γ) at a particular value of L is measured using the hydrometer. This reading γ represents the density of the soil–water suspension, including particles having diameters smaller than D that have settled above L. According to Equation (7), γ is proportional to the percent finer (P). Thus, γ serves as a measure of the weight of particles with sizes smaller than D. In this study, instead of using a hydrometer, we sought to represent the weights of the particles smaller than D by using image features.

Figure 5 depicts a schematic diagram of the movement of a particle of diameter D over time and its color change in a soil–water suspension. When the end of settlement of a particle with diameter D is T_i, i images from T_1 to T_i each contain images L_1 to L_i. Each of these images represents the P values of particles smaller than D. Therefore, I_D, which is the average of the gray value of the i-th images ($I_{D,i}$) taken from T_1 to T_i, was calculated and used as a factor to predict the percent finer P of particles smaller than D.

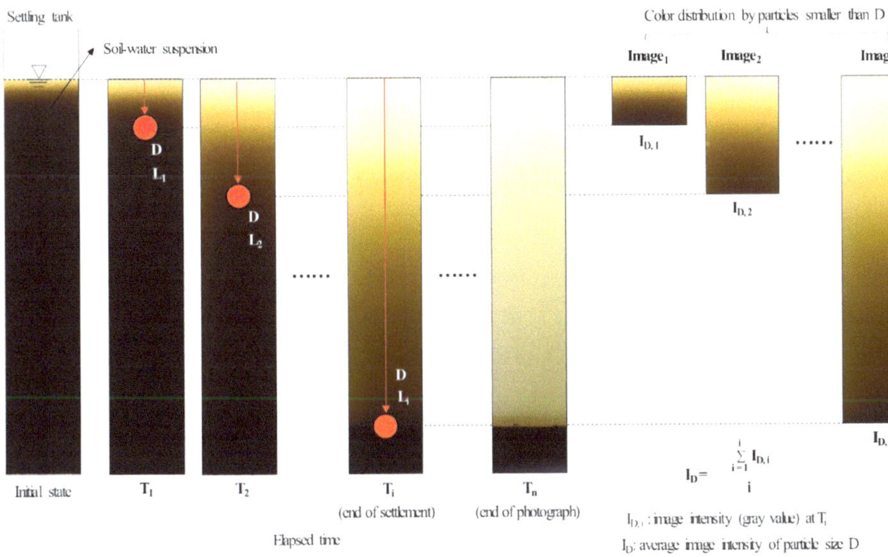

Figure 5. Calculation process of machine vision-based soil particle size analysis.

Python was utilized for performing image processing tasks, including crop and color extraction. Various libraries such as NumPy, pandas, pillow, OpenCV, and matplotlib, among others, were employed for this purpose. The ROI was extracted by cropping it from the original image. Subsequently, the ROI was converted into a grayscale image. To analyze the grayscale image, the average gray value was calculated from the top of the image up to a specific value of *L*.

Root mean squared error (RMSE) is used as a metric to evaluate the accuracy of image-based particle size analysis. RMSE is calculated as follows:

$$\text{RMSE} = \sqrt{\frac{1}{n}\sum_{i=1}^{n}(Predicted_i - Actual_i)^2} \quad (8)$$

where $Predicted_i$ is the predicted i-th percent finer and $Actual_i$ is the observed i-th percent finer. n is the number of data points.

4. Results

4.1. Image Analysis of Soil–Water Suspension during Settlement

Figure 6 shows the change in soil–water suspension and hydrometer reading during soil settlement over time. As time progresses, an observable trend emerges wherein the average color of the soil–water suspension gradually lightens. Specifically, in a soil–water suspension, the color tends to become lighter from the top downward. This phenomenon can be attributed to the gradual reduction in the concentration of suspended soil particles within the suspension. As settling occurs, the soil particles gradually settle down, leading to a decrease in their presence in the suspension. Consequently, the overall color of the suspension becomes lighter. Among the different soil types analyzed, Soil DR exhibited the most rapid color change, while Soil R exhibited the slowest. Within the initial 30-min period, most of the color change for Soil DR took place, with minimal additional color alteration thereafter.

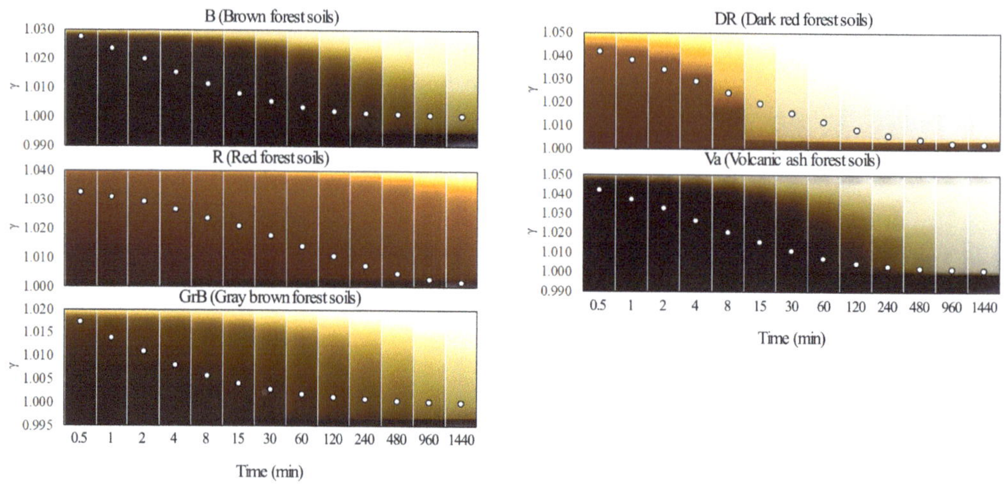

Figure 6. Changes in the hydrometer (γ) reading and color in soil–water suspension according to time.

The presence of dark areas at the top of the region of interest (ROI) can be attributed to the settling of organic matter that was initially present on the water surface. During the settling process, these organic particles tend to migrate downward and accumulate in the upper portion of the ROI, resulting in the observed dark areas. On the other hand, at the bottom of the settling tank, particles that have completed the settling process are observed. It's important to note that the volume of settled particles after the settling process varies for each soil type. This discrepancy arises due to differences in particle size distribution, organic content, and other soil-specific characteristics. These variations result in differences in the volume of settled particles among different soil samples.

Additionally, γ gradually decreases over time in the settling process. The initial γ values differ among the soils due to variations in fine content (0.075 mm sieve passage). However, as the settling process progresses, the hydrometer readings (γ) tend to converge towards a value close to 1. In hydrometer tests, γ is measured to calculate P, as expressed by Equation (7). Given this correlation between γ and particle sizes, it becomes feasible to explore whether certain image features of the soil change similarly to γ. By identifying image features that exhibit similar variations, the particle size distribution can be analyzed using image processing techniques.

Figure 7 shows the changes in gray value observed in a soil–water suspension as the settling process occurs. Over time, the gray value gradually increases, indicating a lightening of the soil–water suspension as the soil particles settle. The color change within the soil–water suspension varies across different elevations within ROI. Notably, the upper section of the soil–water suspension (0–10% range) exhibits significant color change initially but eventually converges towards a constant value after the completion of the settling process. In contrast, the lower section of the soil–water suspension (90–100% range) is where the sedimentation of the soil occurs. Consequently, the color value trend in this area differs from the rest of the ROI. For soil Va, it is worth noting that a relatively large volume of soil deposition occurs compared to the other soils. In fact, deposited soil is observed even within the 80–90% range of the suspension. As a result, subsequent analyses were performed within the 0–80% range of the ROI, excluding potential soil deposition, to ensure accurate color measurements in soil–water suspension.

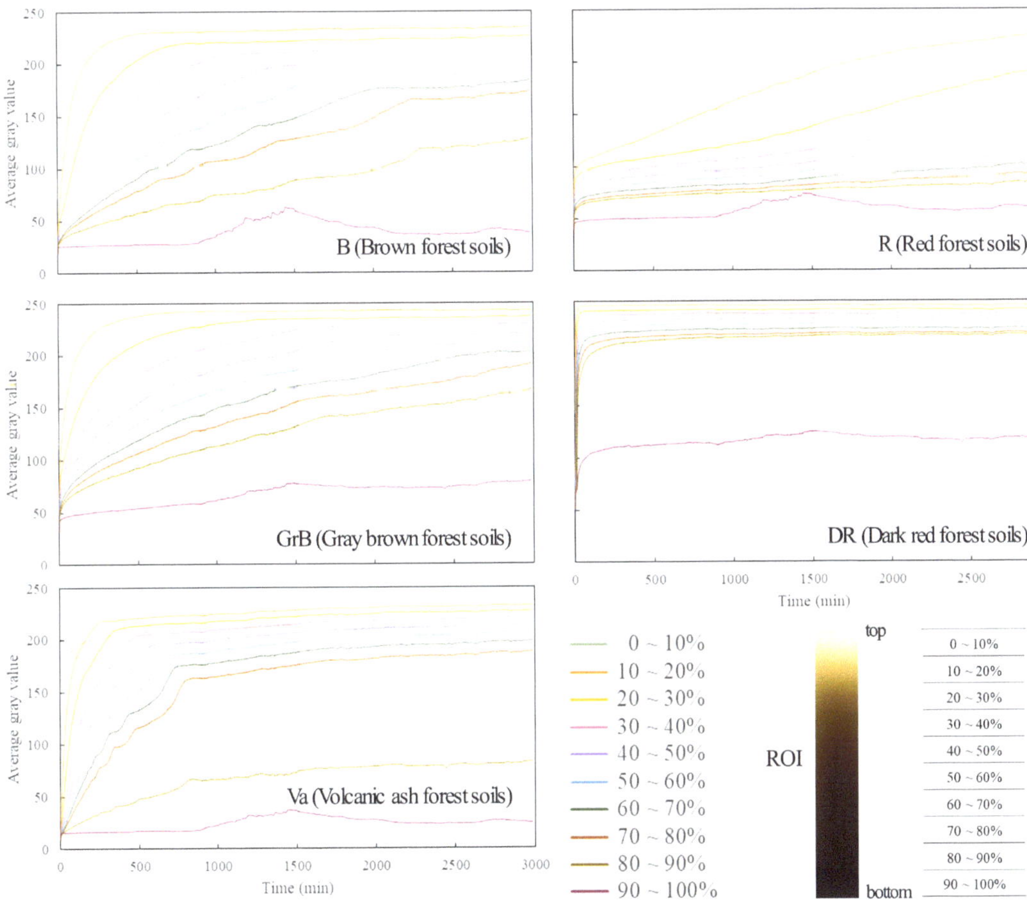

Figure 7. Changes of average gray value in soil–water suspension according to time.

4.2. Prediction of Particle Size Distribution by MVSPSA

Figure 8 shows the relationship between L and D for soil B. According to Equation (3), L is proportional to the square of D. Hence, larger diameter particles exhibit increasingly rapid settlement. Considering that the height of the soil–water suspension in the settling tank is 38.4 cm, it can be observed that a particle with a diameter of 0.075 mm takes approximately 1.6 min to completely settle. In contrast, a particle with a diameter of 0.038 mm takes approximately 6 min. Furthermore, a smaller particle with a diameter of 0.002 mm settles a distance of 23.9 cm after 1440 min, while a particle with a diameter of 0.001 mm settles a distance of 6.0 cm. Using this approach, the value of L, based on the D, can be calculated for each soil type. L is also influenced by the specific gravity (Gs) of the soil particles. This is because heavier particles tend to settle relatively faster. Consequently, soil types such as R, GrB, and DR, which have Gs approximately 1.5–3% higher than that of soil B, would exhibit increased values of L compared to soil B. In contrast, soil Va, with Gs approximately 18% smaller than that of soil B, shows a slower settling velocity compared to other soils, resulting in a decrease in L over the same period.

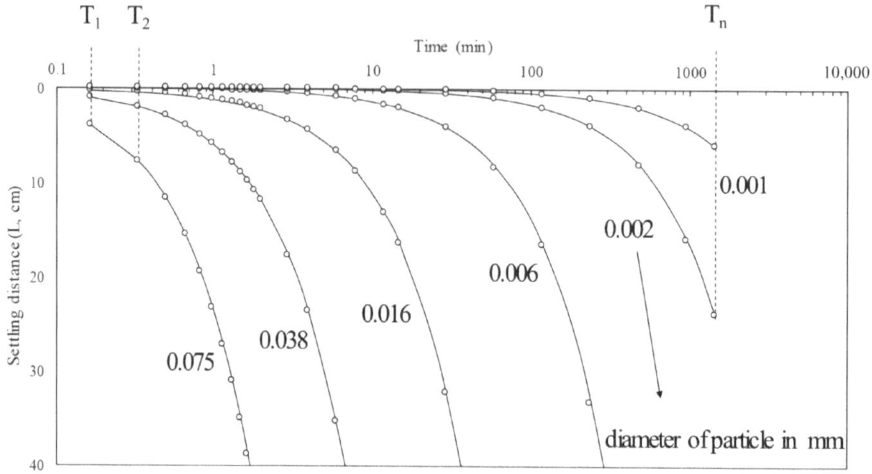

Figure 8. Relationship of settling distance (L) and time (T) for soil B.

Figure 9 illustrates the calculated values of I_D based on the particle diameter (D). In an 8-bit grayscale image, the pixel intensity or gray value is represented by a value ranging from 0 (black) to 255 (white). Consequently, when the soil content within the soil–water suspension is high, the corresponding gray value of the image tends to be closer to 0. Conversely, if there is a minimal amount of soil present, the gray value tends to be closer to 255. For ease of interpretation, the scale of gray values has been adjusted to range from 0 (white) to 1 (black) rather than the original 0 (black) to 255 (white). Therefore, an I_D value closer to 1 signifies a darker image with a higher concentration of soil particles, whereas an I_D value closer to 0 represents a lighter image with a lower concentration of soil particles. I_D shows a decreasing trend as the diameter of the particle decreases, similarly to γ. This indicates the possibility of replacing γ with I_D.

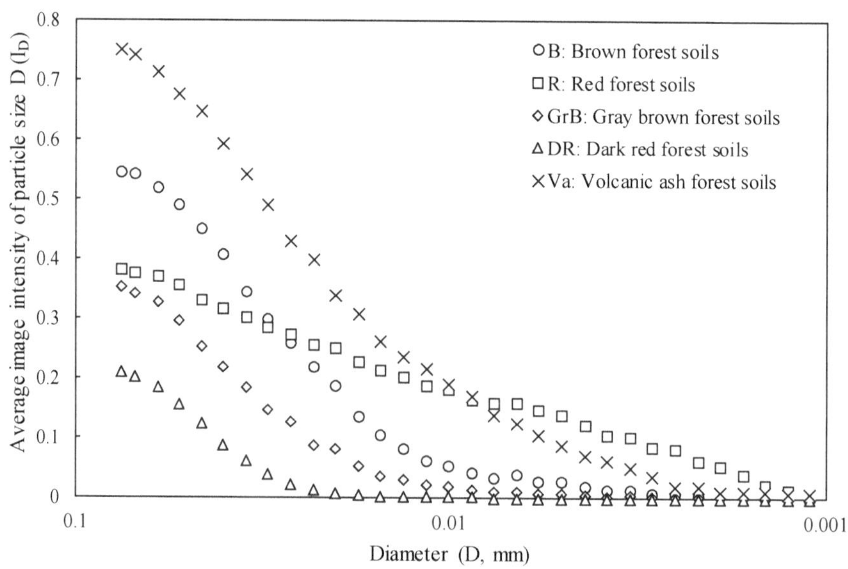

Figure 9. Calculation result of I_D for various diameter of soil particles.

Figure 10 shows the relationship between I_D and P. This indicates that P has an exponential tendency to increase with I_D. Depending on the value of I_D, P tended to increase exponentially. Consequently, it is inferred that by applying a suitable exponent to I_D, it is feasible to accurately predict P. However, it is important to note that Soil DR exhibited a distinct trend compared with the other four soil types. The relationship between I_D and P for Soil DR deviated from the expected exponential behavior observed in the remaining soils.

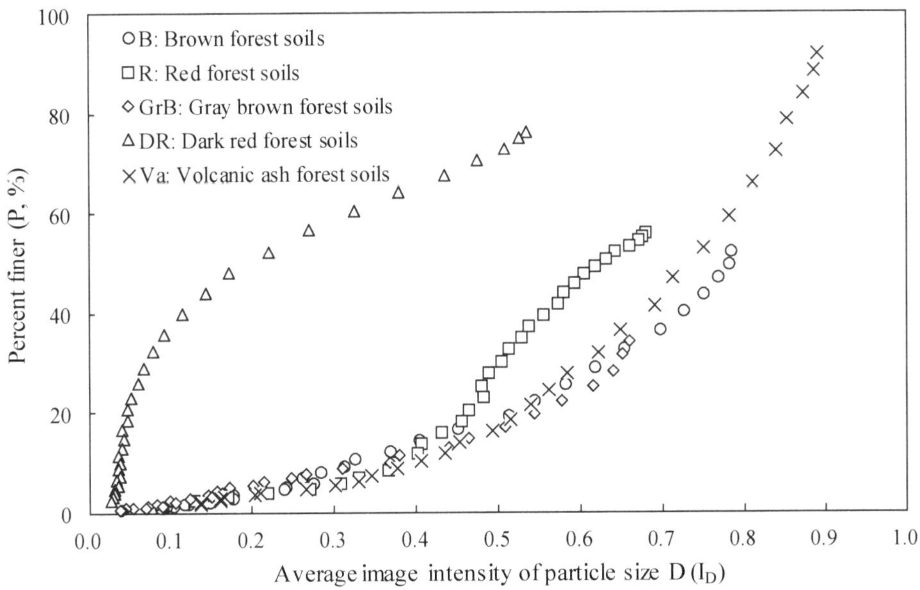

Figure 10. Relationship between average image intensity of particle size D and percent finer.

Considering the relationship between I_D and P shown in Figure 10, the prediction formula of P based on I_D is as follows:

$$P_D = {I_D}^m \times \frac{P_{0.075}}{I_{0.075}} \tag{9}$$

where m is exponent of I_D, $P_{0.075}$ is percent finer at $D = 0.075$ mm, $I_{0.075}$ is average image intensity of particle size = 0.075 mm. $P_{0.075}$ and $I_{0.075}$ were introduced to align the initial values of the image-based analysis results with the results of the sieve analysis. Depending on the value of m, the accuracy of the analysis results can vary. The optimal m for each soil was found to be different, but with the exception of DR, the other soils were found to be highly correlated when m ranged from 2 to 3. For Soil DR, the I_D did not exhibit an exponential relationship with P, resulting in a gradual decrease in correlation coefficient as m increased. On average, the highest correlation was found when m was 2.5. Therefore, it was decided to utilize m of 2.5 for predicting P using I_D across all soil types.

Figure 11 shows the results of the particle size analysis predicted using the proposed method. Except for Soil DR, the four remaining soil types yielded particle size distribution curves that closely matched the experimental results obtained from the laser particle size analyzer. Moreover, the root mean squared error (RMSE) of the prediction results, excluding Soil DR, demonstrated a high level of accuracy, ranging from 2.5 to 4.0%. This indicates that the proposed method provides reliable particle-size predictions for the analyzed soils, offering an effective alternative to hydrometer tests.

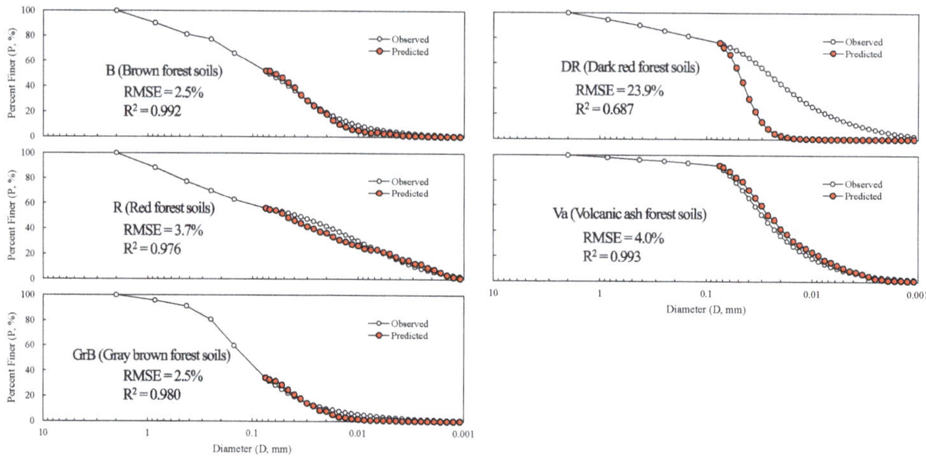

Figure 11. Particle size distribution curve predicted by machine vision-based soil particle analysis method for soil samples.

Figure 12 shows the USDA soil texture triangle depicting the sand, silt, and clay content calculated from the particle size analysis results. By utilizing the soil texture prediction derived from the image analysis, it becomes evident that the predicted soil texture closely corresponds to the soil texture determined from the actual particle size tests. This alignment between the predicted and calculated soil textures indicates the reliability and accuracy of the image analysis method for determining soil texture. The results confirm that the proposed approach successfully captures the essential characteristics of soil particles and enables the accurate prediction of soil texture based on particle size analysis.

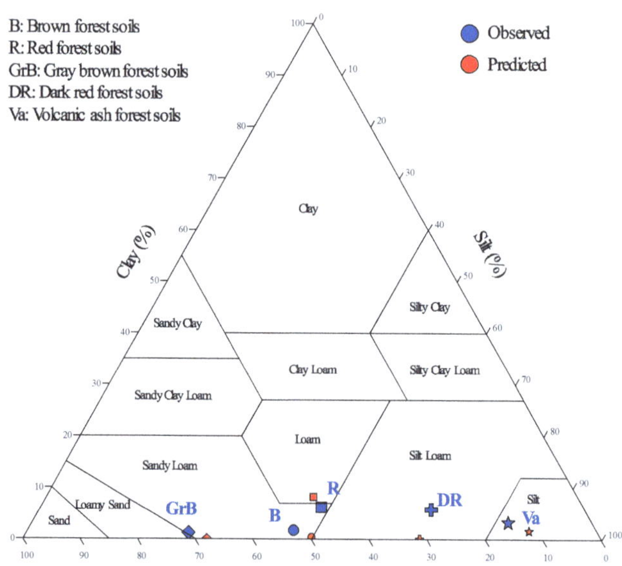

Figure 12. Soil texture triangle predicted by machine vision-based soil particle analysis method for soil samples.

5. Discussion

5.1. Practical Considerations and Limitations

There are important considerations when implementing the machine vision-based soil particle analysis method proposed in this study.

Firstly, the analytical method relies on the average color of the image. However, it should be noted that color values in digital images can vary based on camera settings and lighting conditions. This means that the analysis results for the same sample can differ due to external factors affecting image color. To address this issue, establishing standardized lighting and shooting conditions or developing appropriate measures for correction when variations in shooting conditions occur is crucial. By ensuring consistency in image acquisition, the reliability and comparability of the analysis results can be enhanced.

Additionally, the analysis of the DR sample was not conducted appropriately, leading to different results compared with the other soil samples (B, R, GrB, and Va). This discrepancy is attributed to the rapid color change observed in the DR images within a very short period. While the other soils exhibited gradual color changes over a duration of 1440 min (1 day), the DR sample underwent significant color change within the initial 30 min, followed by a minimal color change afterward. However, it should be considered that the sedimentation of particles in the DR sample may still be occurring during periods when color changes are not observed. This discrepancy could be due to inadequate lighting or camera settings, which may have hindered proper observation of the settling process of small particles. To address these issues and ensure accurate and reliable results, optimizing the experimental setup, lighting conditions, and camera settings is recommended to capture the full range of particle settlement. Additionally, further investigations and adjustments specific to the DR sample may be necessary to improve the analysis methodology for this particular soil type.

5.2. Advantages of Proposed Method

This study introduces a machine vision-based soil particle analysis method utilizing a standard digital camera, presenting it as a viable alternative to the hydrometer test. The method offers several notable advantages.

Firstly, it enables experimentation on many samples. In this study, a single test successfully predicted the particle size of five samples. Although the resolution of commercial cameras imposes limitations on sample size, future improvements in camera technology and test setup optimizations can increase the number of simultaneous tests.

Additionally, the proposed method offers simplicity compared to hydrometer testing. Unlike the labor-intensive process of ongoing measurements required in hydrometer testing, the test described in this study can be automated after the initial setup, enhancing efficiency and reducing manual labor.

Furthermore, advancements in camera technology can further enhance the results. While a common commercial digital camera was utilized in this study, specialized cameras such as NIR or hyperspectral cameras allow for the observation of wavelengths beyond RGB. This opens possibilities for analyzing samples like DR that were challenging to study in this particular research. However, a notable strength of this study lies in its reliance on commonly available cameras, ensuring wider accessibility and distribution.

The proposed machine vision-based sedimentation test offers advantages such as the ability to experiment on larger sample sizes, simplicity compared to hydrometer testing, and potential enhancements with the progression of camera technology. Notably, this method utilizes widely accessible cameras, making it practical for various applications.

For the automation of hydrometer testing, Murad et al. [35] presented an automated hydrometer testing system for about $70 using Time of Flight Distance (ToF). The system measures the distance between the hydrometer and the ToF sensor and converts it into a hydrometer reading. The study reported an R^2 of percent finer of 0.857 to 0.896 when comparing the results of the automated system to the results of the pipette method. In this study, R^2 at $m = 2.5$ was 0.976 to 0.993 for the soils except DR. The clay content of the soils in this study ranged from 1.2 to 6.2%, which is lower than the clay content of the soils used by Murad et al. [35], which ranged from 24.5 to 73.8%, which may explain the relatively high correlation. Compared to the above study, this study is considered to have the advantages of relatively simple system configuration and easy testing of a large number of samples. It is also significant that this study proposed a new method to analyze particle size by applying the principle of hydrometer test based on Stoke's law without using a hydrometer.

In particular, in this study, the percent finer was predicted based on the color change of the digital image, but in the further research, the hydrometer reading can be obtained directly by tracking the change of the position of the hydrometer in the digital image. In this regard, Souza et al. [36] applied machine vision technology to calibrate hydrometers that can be utilized in various industrial sites. Therefore, it is expected that machine vision-based hydrometer reading automation will be possible through further research.

6. Conclusions

This study focused on predicting the particle size distribution of fine-grained soil using digital image analysis. The study utilized a settling tank instead of a measuring cylinder to acquire digital images of soil–water suspensions during settlement. Image analysis techniques were applied to extract relevant features and predict the percent finer of particles smaller than a given diameter. The results showed a correlation between the average gray value of the images and the percent finer, indicating the potential of image-based analysis for particle size prediction.

The study investigated five forest soils, namely B, R, GrB, DR, and Va. The image analysis process involved cropping the region of interest (ROI) from the acquired images and calculating the average gray value up to a settling distance. The gray value decreased with decreasing particle diameter, similar to the hydrometer reading. The obtained gray values were then used to predict the percent finer of particles smaller than a given diameter.

The prediction results showed that the proposed method could closely align with the experimental results from a laser particle size analyzer for four out of five soil types. The RMSE of the prediction results ranged from 2.5% to 4.0%, indicating a high level of accuracy in particle size prediction.

Overall, this study demonstrates the potential of digital image analysis for predicting particle size distribution in fine-grained soils. The proposed method can provide a convenient and efficient alternative to traditional particle size analysis techniques, offering insights into soil properties and aiding in soil classification and characterization.

Author Contributions: Conceptualization, D.K. and J.P.; methodology, D.K. and J.P.; software, D.K. and J.S.; validation, D.K.; formal analysis, D.K. and J.S.; data curation, D.K. and J.P.; writing—original draft preparation, D.K. and J.S.; writing—review and editing, D.K., J.S. and J.P.; visualization, D.K. and J.S.; supervision, J.P.; funding acquisition, J.P. All authors have read and agreed to the published version of the manuscript.

Funding: This research was supported by the Basic Science Research Program of the National Research Foundation of Korea (NRF) and funded by the Ministry of Education (2021R1I1A1A01058373).

Institutional Review Board Statement: Not applicable.

Informed Consent Statement: Not applicable.

Data Availability Statement: Not applicable.

Conflicts of Interest: The authors declare no conflict of interest.

References

1. Kaur, A.; Fanourakis, G.C. Effect of sodium carbonate concentration in calgon on hydrometer analysis results. *Period. Polytech. Civ. Eng.* **2018**, *62*, 866–872. [CrossRef]
2. Ghalib, A.M.; Hryciw, R.D. Soil particle size distribution by mosaic imaging and watershed analysis. *J. Comput. Civ. Eng.* **1999**, *13*, 80–87. [CrossRef]

3. Faroughi, S.; Huber, C. A theoretical hydrodynamic modification on the soil texture analyses obtained from the hydrometer test. *Géotechnique* **2016**, *66*, 378–385. [CrossRef]
4. Kaur, A.; Fanourakis, G. Effect of hydrometer type on particle size distribution of fine grained soil. In Proceedings of the 1st Southern African Geotechnical Conference, Sun City, South Africa, 5–6 May 2016; pp. 1–8.
5. Abbireddy, C.O.; Clayton, C.R. A review of modern particle sizing methods. *Proc. Inst. Civ. Eng.-Geotech. Eng.* **2009**, *162*, 193–201. [CrossRef]
6. Wen, B.; Aydin, A.; Duzgoren-Aydin, N.S. A comparative study of particle size analyses by sieve-hydrometer and laser diffraction methods. *Geotech. Test. J.* **2002**, *25*, 434–442.
7. Al-Hadidi, M.T.; Al-Adhamii, R.; Ali, B. Estimate the ratio of total dissolved salts from the hydrometer test results. *Appl. Res. J. (Eng.)* **2015**, *1*, 444–450.
8. Kakuturu, S.P.; Kinzel, M.; Xiao, M. *Effects of Maximum Particle Size on the Results of Hydrometer Tests on Soils*; ASTM International: West Conshohocken, PA, USA, 2019.
9. Stott, P.; Monye, P.; Theron, E. Assessment of reliability of the hydrometer by examination of sediment. In Proceedings of the Proceedings of the first Southern African Geotechnical Conference, Sun City, South Africa, 5–6 May 2016; pp. 281–286.
10. Barman, U.; Choudhury, R.D. Soil texture classification using multi class support vector machine. *Inf. Process. Agric.* **2020**, *7*, 318–332. [CrossRef]
11. Ravikumar, R.; Arulmozhi, V. Digital image processing-a quick review. *Int. J. Intell. Comput. Technol. (IJICT)* **2019**, *2*, 11–19.
12. Kim, D.; Kim, T.; Jeon, J.; Son, Y. Soil-Surface-Image-Feature-Based Rapid Prediction of Soil Water Content and Bulk Density Using a Deep Neural Network. *Appl. Sci.* **2023**, *13*, 4430. [CrossRef]
13. O'Donnell, T.K.; Goyne, K.W.; Miles, R.J.; Baffaut, C.; Anderson, S.H.; Sudduth, K.A. Identification and quantification of soil redoximorphic features by digital image processing. *Geoderma* **2010**, *157*, 86–96. [CrossRef]
14. Prats-Montalbán, J.M.; de Juan, A.; Ferrer, A. Multivariate image analysis: A review with applications. *Chemom. Intell. Lab. Syst.* **2011**, *107*, 1–23. [CrossRef]
15. Chitradevi, B.; Srimathi, P. An overview on image processing techniques. *Int. J. Innov. Res. Comput. Commun. Eng.* **2014**, *2*, 6466–6472.
16. Ohm, H.-S.; Hryciw, R.D. Size distribution of coarse-grained soil by sedimaging. *J. Geotech. Geoenviron. Eng.* **2014**, *140*, 04013053. [CrossRef]
17. Ren, J.; Li, X.; Zhao, K. Quantitative analysis of relationships between crack characteristics and properties of soda-saline soils in Songnen Plain, China. *Chin. Geogr. Sci.* **2015**, *25*, 591–601. [CrossRef]
18. dos Santos, J.F.; Silva, H.R.; Pinto, F.A.; de Assis, I.R. Use of digital images to estimate soil moisture. *Rev. Bras. de Eng. Agrícola e Ambient.* **2016**, *20*, 1051–1056. [CrossRef]
19. Kim, D.; Son, Y.; Park, J.; Kim, T.; Jeon, J. Evaluation of calibration method for field application of UAV-based soil water content prediction equation. *Adv. Civ. Eng.* **2019**, *2019*, 2486216. [CrossRef]
20. Kim, D.; Kim, T.; Jeon, J.; Son, Y. Convolutional Neural Network-Based Soil Water Content and Density Prediction Model for Agricultural Land Using Soil Surface Images. *Appl. Sci.* **2023**, *13*, 2936. [CrossRef]
21. Peng, X.; Horn, R.; Peth, S.; Smucker, A. Quantification of soil shrinkage in 2D by digital image processing of soil surface. *Soil Tillage Res.* **2006**, *91*, 173–180. [CrossRef]
22. Abd El-Halim, A. Image processing technique to assess the use of sugarcane pith to mitigate clayey soil cracks: Laboratory experiment. *Soil Tillage Res.* **2017**, *169*, 138–145. [CrossRef]
23. Yan, L.; Meng, Q.-X.; Xu, W.-Y.; Wang, H.-L.; Zhang, Q.; Zhang, J.-C.; Wang, R.-B. A numerical method for analyzing the permeability of heterogeneous geomaterials based on digital image processing. *J. Zhejiang Univ. Sci. A* **2017**, *18*, 124–137. [CrossRef]
24. Bryk, M.; Kołodziej, B. Suitability of Image Analysis in Evaluating Air and Water Permeability of Soil. *Agronomy* **2021**, *11*, 1883. [CrossRef]
25. Smith, S.J.; Friedrichs, C.T. Image processing methods for in situ estimation of cohesive sediment floc size, settling velocity, and density. *Limnol. Oceanogr. Methods* **2015**, *13*, 250–264. [CrossRef]
26. Kim, D.; Son, Y.; Park, J. Prediction of settling velocity of nonspherical soil particles using digital image processing. *Adv. Civ. Eng.* **2018**, *2018*, 4647675. [CrossRef]
27. Wang, S.-N.; Xu, W.-Y.; Shi, C.; Zhang, Q. Numerical simulation of direct shear tests on mechanical properties of talus deposits based on self-adaptive PCNN digital image processing. *J. Cent. South Univ.* **2014**, *21*, 2904–2914. [CrossRef]
28. Li, H.-D.; Tang, C.-S.; Cheng, Q.; Li, S.-J.; Gong, X.-P.; Shi, B. Tensile strength of clayey soil and the strain analysis based on image processing techniques. *Eng. Geol.* **2019**, *253*, 137–148. [CrossRef]
29. Mora, C.; Kwan, A.; Chan, H. Particle size distribution analysis of coarse aggregate using digital image processing. *Cem. Concr. Res.* **1998**, *28*, 921–932. [CrossRef]
30. Sudarsan, B.; Ji, W.; Adamchuk, V.; Biswas, A. Characterizing soil particle sizes using wavelet analysis of microscope images. *Comput. Electron. Agric.* **2018**, *148*, 217–225. [CrossRef]
31. Bittelli, M.; Andrenelli, M.; Simonetti, G.; Pellegrini, S.; Artioli, G.; Piccoli, I.; Morari, F. Shall we abandon sedimentation methods for particle size analysis in soils? *Soil Tillage Res.* **2019**, *185*, 36–46. [CrossRef]

32. Sun, Q.; Zheng, J.; Coop, M.R.; Altuhafi, F.N. Minimum image quality for reliable optical characterizations of soil particle shapes. *Comput. Geotech.* **2019**, *114*, 103110. [CrossRef]
33. Baghban, A.; Bahadori, M.; Bahadori, A. Prediction of particle size distribution in hydrometer test analysis using LSSVM algorithm. In Proceedings of the Asia Pacific Confederation of Chemical Engineering Congress 2015: APCChE 2015, Incorporating CHEMECA 2015, Melbourne, Australia, 27 September–1 October 2015; pp. 1–13.
34. Özgan, E.; Öztürk, A. Genetic Algorithm Application for Hydrometer Tests. *Asian J. Chem.* **2009**, *21*, 6857–6868.
35. Murad, M.O.F.; Jones, E.J.; Minasny, B. Automated soil particle-size analysis using time of flight distance ranging sensor. *Soil Sci. Soc. Am. J.* **2020**, *84*, 690–699. [CrossRef]
36. De Souza, J.B.; Maio, M.F.; Belan, P.A. A Machine Vision for Automatic Hydrometers Calibration. *Int. J. Comput. Appl.* **2021**, *975*, 8887. [CrossRef]

agriculture

Article

Working Load Analysis of a 42 kW Class Agricultural Tractor According to Tillage Type and Gear Selection during Rotary Tillage Operation

Yeon-Soo Kim [1], Bo-Min Bae [1], Wan-Soo Kim [2], Yong-Joo Kim [3,*], Sang-Dae Lee [4] and Taek-Jin Kim [5,*]

1 Department of Bio-Industrial Machinery Engineering, Pusan National University, Miryang 50463, Republic of Korea; yskim23@pusan.ac.kr (Y.-S.K.); qhals806@pusan.ac.kr (B.-M.B.)
2 Department of Bio-Industrial Machinery Engineering, Kyungpook National University, Daegu 41566, Republic of Korea; wansoo.kim@knu.ac.kr
3 Department of Biosystems Machinery Engineering, Chungnam National University, Daejeon 34134, Republic of Korea
4 Smart Agricultural R&D Group, Korea Institute of Industrial Technology (KITECH), Gimje 54325, Republic of Korea; sdlee96@kitech.re.kr
5 Department of Drive System Team, TYM R&D Center, Iksan 54576, Republic of Korea
* Correspondence: babina@cnu.ac.kr (Y.-J.K.); taek.kim@tym.world (T.-J.K.); Tel.: +82-42-821-6716 (Y.-J.K.); +82-63-723-7753 (T.-J.K.)

Abstract: The objective of this study was to analyze the effect of tillage type (i.e., primary and secondary tillage) and gear selection (P1L2 to P1L4) on the working load of tractor–implement systems during rotary tillage. Soil properties change with depth, and differences in properties along the depth distribution, such as the location of formation of the hardpan layer, internal friction angle, and moisture content, affect the load of rotary tillage operations. Therefore, the physical properties of soil along the field depth distribution were measured to analyze the effect of tillage type and gear selection on workload in rotary tillage. In addition, a load measurement system equipped with PTO torque meter, axle torque meter, proximity sensor, and RTK-GPS were configured on the 42 kW agricultural tractor. The experimental results show that the combination of tillage type and gear selection has a wide-ranging effect on the tractor's workload and performance when the rotavator operated at the same tilling depth. Overall working load was higher by up to 14% (engine) and 29.1% (PTO shaft) in primary tillage compared to secondary tillage when the gear selection was the same. When the tillage type is the same, it was analyzed that the overall average torque increased by up to 35.9% (engine) and 33.9% (PTO shaft) in P1L4 compared to P1L2 according to gear selection. Based on load analysis results, it was found that the effect of gear selection (Engine: 4–14%, PTO: 12.1–28.6%) on engine and PTO loads was higher than that of tillage type (Engine: 31.6–35.1%, PTO: 31.9–32.8%), and the power requirement tended to decrease in secondary tillage. Therefore, working load should be considered according to the soil environment and tillage type when designing agricultural machinery system.

Keywords: working load; agricultural tractor; rotary tillage; soil property; tillage type

Citation: Kim, Y.-S.; Bae, B.-M.; Kim, W.-S.; Kim, Y.-J.; Lee, S.-D.; Kim, T.-J. Working Load Analysis of a 42 kW Class Agricultural Tractor According to Tillage Type and Gear Selection during Rotary Tillage Operation. *Agriculture* **2023**, *13*, 1556. https://doi.org/10.3390/agriculture13081556

Academic Editor: Jin He

Received: 28 June 2023
Revised: 1 August 2023
Accepted: 2 August 2023
Published: 3 August 2023

1. Introduction

In Korea, agricultural tractors have a working area of 18.9 ha per unit and an operation rate of 74.3%. Tractors not only simply pull attached implements, but also deliver the required power to drive attached implements through a power take-off (PTO) shaft [1]. In particular, agricultural power using PTO is mainly used for high-load agricultural work such as powered disc plow, rotary, and baler operation. Because tractors are subjected to large and irregular working load fluctuations depending on the type of agricultural work performed, it is necessary to analyze the largest fluctuation load on the tractor to ensure reliability [2,3].

For the optimal design of agricultural tractors and implements, it is necessary to measure and analyze the load data generated during agricultural work [4]. Basically, various load measurement tests were conducted for plow tillage. The initial working load analysis study was conducted mainly in the soil bin test bed to analyze soil–tool interaction. Raper [5] conducted a study related to the analysis of minimum draft force generation according to subsoiler geometry via a soil bin test. As a result, it was confirmed that the soil disruption was better with less resistance in the bentleg shanks than in the straight shanks. Other studies have conducted various types of soil–tool interaction study related to working conditions and load influence parameters on the basis of field experiments. Naderloo et al. [6] analyzed the tillage depth and travel speed effect on draft force using three types of plow treatment under clay loam soil conditions. Additionally, Moeenifar [7] conducted field experiments to analyze the influence of tillage depth and penetration angle on soil–thin blade interaction. In addition, Aday and Ramadhan [8] conducted draft force and distribution area analysis studies with respect to single-tine and double-tine conditions in the subsoiler. In another study, Kichler et al. [9] conducted field tests to analyze the effect of transmission gear selection on tractor performance and fuel cost as a function of the geometric shape of the subsoiler during deep tillage operations. The field test results showed that the productivity rate increased as the appropriateness of gear selection increased, while the fuel cost decreased. In another study, Kim et al. [10] analyzed the effect of soil water content on the traction performance of agricultural tractors during tillage operation. On the basis of the results of the test, it was concluded that overall tractor traction performance parameters such as axle torque, slip and traction increased proportionally with increasing soil water content. In addition, Kim et al. [11,12] conducted field tests to analyze the field environment (i.e., soil texture and soil properties) and the effect of tillage depth and gear selection on the working load of tractors during tillage operations.

In addition, some studies have conducted field load measurement tests for agricultural operations using PTO power. Lee et al. [13] conducted field tests to measure the PTO working load during rotary tillage and baler operation in Korean upland fields. In another study, Behera et al. [14] analyzed the effects of tillage depth and travel speed on PTO torque, draft force and fuel consumption in rota-cultivators. Additionally, Kim et al. [15,16] conducted a load analysis study on the effect of different gear combinations on the PTO working load during rotary tillage. In another study, Kim et al. [17] conducted field measurement testing of the PTO working load of a multi-purpose cultivator in order to identify the weak part of the PTO gear train.

Rotary tillage is used in both primary and secondary tillage operations. It is also useful for puddling the field before paddy transplantation. However, load analysis studies with respect to the soil working environments and tillage types used in primary and secondary methods have not been performed. Therefore, this study was conducted to evaluate the working load of 42 kW class agricultural tractor with respect to tillage type and gear selection during rotary tillage, which is the most widely used agricultural operation that uses tractor PTO power. The main purposes were to (1) develop a field load measurement system with an experimental design, (2) measure and analyze the working load data during rotary tillage, and (3) perform an analysis of the effect of tillage type and gear selection on tractor working load.

2. Materials and Methods

2.1. Measurement of Field Soil Properties

Basically, all working loads of agricultural machinery originate from the soil environment. Since the rotary work was performed in two fields with different physical properties, soil property measurement tests were performed. Figure 1 shows the soil sampling procedure of the field soil properties. Soil properties were measured using the uniform grid (3 m × 3 m) sampling method in the field [18]. The test field was divided into 10 uniform grid squared, and soil sampling was performed in each grid square to measure the physical properties of the soil, and in the case of the cone penetration test and vane shear test, which were performed to measure the engineering properties of the soil, 10 tests were performed in each grid, for a total of 100 tests.

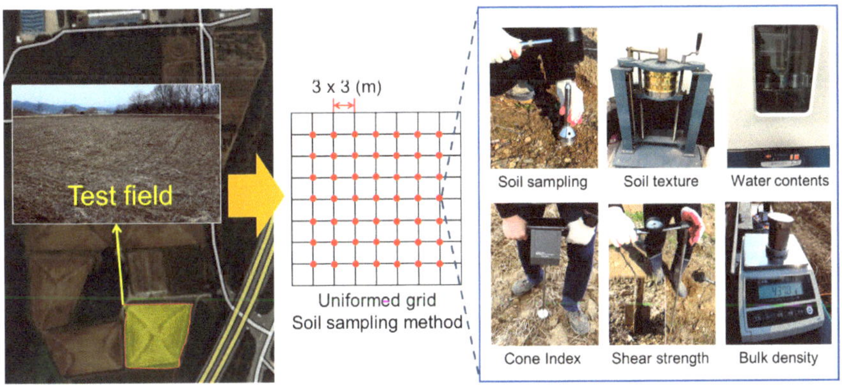

Figure 1. Sampling procedure of field soil property using uniformed grid method.

First, soil sampling was performed in each test field, A and B, using a 100 mL soil sampling tube (DIK-1801, Daiki Rika Kogyo Co., Ltd., Akagidai, Japan) and a soil sampling device (DIK-1815, Daiki Rika Kogyo Co., Ltd., Akagidai, Japan). To measure the soil water content, the soil specimen was dried at a temperature of 110 °C for 24 h [19] using the oven-drying method (SH-DO-100FGB, Samheung Energy, Sejong, Republic of Korea). The soil texture of field site was analyzed using the USDA soil classification method [20] using a sieve shaker (HJ-4560, Heungjin, Gimpo, Republic of Korea) [21].

Additionally, cone index and shear strength, which are the most important parameters among soil mechanical properties, and have the greatest impact on agricultural workload, were measured using a cone penetrometer (DIK-5532, Daiki Rika Kogyo Co., Ltd., Akagidai, Japan) [22,23] and a shear ring-type soil resistance meter (DIK-5503, Daiki Rika Kogyo Co., Ltd., Saitama, Japan), respectively.

In this study, the soil properties measured in this study were obtained using Equations (1)–(3) [24,25]:

$$W = \frac{100W_w}{W_s}, \quad (1)$$

where W is the soil water content (%), W_w is the water weight of the soil specimen, and W_s is the weight of solids in the soil specimen.

$$\gamma = \frac{W}{V}, \quad (2)$$

where γ is the bulk density (g/cm^3), W is the total weight of the soil specimen (kg), V is the total volume of the soil specimen (cm^3).

$$\tau = \frac{3M}{28\pi r^3}, \quad (3)$$

where τ is the shear strength of soil (kPa), M is the soil resistance torque that caused the on-field vane shear test (Nm), and r is the radius of shear box.

2.2. Tractor–Implement System

Figure 2a shows a rotavator (WJ185A, Woongjin Machinery, Gimje, Republic of Korea), often used in studies on soil–machine interactions to analyze working load using PTO power. The rotavator undertakes a high-load operation compared to other agricultural operations, such as plowing. Therefore, it is suitable for soil–tool interaction studies of agricultural machinery using PTO power. The rotavator has a maximum tillage depth of 200 mm, and the dimensions are 810 mm × 2020 mm × 1130 mm (length × width × height). In addition, it has 7 flanges and 42 tillage blades, with a working width of 1820 mm. The detailed specifications of the attachment implement are shown in Table 1.

(**a**) (**b**)

Figure 2. Tractor–implement system used in this study: (**a**) rotavator (WJ185A, Woongjin, Republic of Korea); (**b**) 42 kW class agricultural tractor (TX58, TYM, Gongju, Republic of Korea).

Table 1. Specifications of the rotavator.

Item	Specification
Company	Woongjin
Model	WJ185A
Required power (kW)	35~45
Length (mm) × width (mm) × height (mm)	810 × 2020 × 1130
Tillage width (mm)	1820
Weight (kg)	405
Max. tillage depth (mm)	200
Number of flanges	7
Number of blades	42

In this study, a 42 kW tractor (TX58, Tong Yang Moolsan, Gongju, Republic of Korea) was used, considering that the specifications of the rotavator indicate that a power source of 35~45 kW is required, as shown in Figure 2b. The overall dimensions of the agricultural tractor body were 3695 mm × 1848 mm × 2560 mm (length × width × height), with a rated power of 35.6 kW and a maximum torque of 211.8 Nm. The test tractor's total weight was 3894 kg, and with the inclusion of the DAQ system (229 kg) and a front loader (450 kg) (HIT400 L, Hanil Industry Co., Ltd., Gyeongsan, Republic of Korea). The agricultural tractor used in the field experiment was equipped with a mechanical transmission power shuttle, and through a combination of 4 main gears and 6 subgears, a total of 48 gears (24 forward gears, 24 reverse gears) could be shifted, depending the tilling work, and the maximum speed was 33.8 km/h. In addition, the PTO power train is a three-stage power train, and has rotational speeds of 540, 750, and 1000 rpm, respectively. The specifications of the agricultural tractor are shown in Table 2.

Table 2. Results of measured soil properties of test field according to soil depth.

Parameters	Soil Depth: 0 to 10 cm	Soil Depth: 10 to 20 cm
Water contents (WC, %)	20.6 ± 1.2	17.7± 1.1
Bulk density (γ, kg/m^3)	1571.7 ± 75.2	1830 ± 30.6
Cone index (CI, kPa)	801.8 ± 277.6	2065.9 ± 643.1
Shear strength (τ, kPa)	25.1 ± 7.2	64.3 ± 14.4

2.3. Working Load Measurement System

Figure 3 shows the overall measurement system of a 42 kW agricultural tractor for measuring working load during the rotary tillage. The DAQ (data acquisition) system was Dewesoft X (Dewesoft 3X, Dewesoft, Trbovlje, Slovenia), and the working load was measured simultaneously during rotary tillage at a sampling rate of 1 kHz. The working load measurement system used to analyze the effect of soil properties on the design load of the tractor–implement system during tillage operations is as follows:

$$V_{th} = \frac{\pi D_{rw} N_{rw} GR3.6}{60} \quad (4)$$

where V_{th} is the theoretical speed of the tractor in the working direction (km/h), D_{rw} is the diameter of the rear wheel axle (m), N_{rw} is the wheel axle rotational speed as calculated by proximity sensors (rpm), and GR is the gear ratio,

$$S = \left(\frac{V_{th} - V_a}{V_{th}}\right) \times 100 \quad (5)$$

where S is the slip ratio (%) and V_a is the travel speed as the actual working speed of the tractor, as measured by RKT-GPS (km/h).

Figure 3. Configuration of the field load measurement system.

The engine load measuring unit featured a 5 kNm strain gauge torque transducer mounted on the flex plate to measure the engine torque (T_e), which measured the torque data with non-contact sensor telemetry through an amplifier. The engine rotational speed (N_e) was measured by sampling at 100 Hz based on wireless CAN (Controller Area Network) communication and then synchronized with other working load data at 1 kHz using DAQ system. In addition, the engine power requirement can be obtained as follows:

$$P_e = \frac{2\pi T_e N_e}{60,000} \quad (6)$$

where P_e is the power requirement of the engine (kW), T_e is the engine torque (Nm) measured by the flex plate torque transducer, and N_e is the engine rotational speed (rpm) measured by controller area network (CAN) communication.

The torque of the wheel axle and the PTO shaft was measured using a flange-type torque transducer (Sensor telemetrie GmbH PCM16, MANNER, Spaichingen, Germany). The wheel axle torque and PTO torque data from each torque transducer were amplified by internal amplifiers. The torque range of the transducer was 15–30 kNm. In addition, a proximity sensor (CYGTS211B-PO2, Chen Yang Technologies GmbH & Co. KG, Finsing, Germany) was installed on the inner side of each wheel axle in order to measure wheel axle rotational speed. This method of measuring rotational speed has been used in previous studies [26,27]. An antenna integrating a torque transducer, a proximity sensor and an amplifier was installed inside and outside each wheel axle shaft to precisely measure the wheel axle load and the PTO axle load simultaneously. In the case of the proximity sensor, the measuring range was up to 20,000 rpm with a response frequency of 1–20 kHz. In addition, the power requirements of the agricultural tractor were analyzed in order to evaluate the working load of wheel axles and PTO shafts during rotary tillage operations, as follows:

$$P_W = \frac{2\pi T_W N_W}{60,000}, \quad (7)$$

where P_W is the power requirement of the wheel axle (kW), T_W is the wheel axle torque (Nm), as measured by the wheel torque meter, and N_W is the wheel axle rotational speed (rpm), as measured by the proximity sensor.

$$P_{PTO} = \frac{2\pi T_{pto} N_{pto}}{60,000}, \quad (8)$$

where P_{PTO} is the power requirement of the PTO shaft (kW), T_{PTO} is the PTO shaft torque (Nm), as measured by the PTO torque transducer, and N_{PTO} is the rotational speed of the PTO shaft (rpm), as measured by CAN communication.

To measure fuel consumption, an oval gear flowmeter (OG2-SS5-VHQ-B, Titan Enterprises, Sherborne, UK) was installed at the fuel injection inlet and outlet. Specific fuel consumption (SFC), which is often used as a performance indicator for fuel consumption, can be obtained using the following equation. The fuel cost was also calculated using the method described in another study [9].

$$FC = \frac{60(F_{in} - F_{out})}{0.83}, \quad (9)$$

$$SFC = \frac{FC}{P_e}, \quad (10)$$

$$PR = \frac{TS \times WW}{10}, \quad (11)$$

$$\text{Fuelcost} = FC \times \frac{1}{PR} \times FP, \quad (12)$$

where FC is the fuel consumption of the engine (kg/h), F_{in} is the flow rate (L/min) measured by the flow meter installed at the fuel inlet of the engine, F_{out} is the flow rate (L/min) measured by the flow meter installed at the fuel outlet of the engine, SFC is the specific fuel consumption (g/kWh), PR is the productivity rate (ha/h), TS is the travel speed measured by RTK-GPS (km/h), *WW* is the working width (m), fuel cost (USD/ha), and *FP* is the fuel price (1.31 USD/kg).

2.4. Field Experimental Design

In this study, field experiments were conducted to analyze the effect of tillage type and gear selection on the working load of agricultural tractors during rotary tillage. The field test site was located at 37°93′67.2″ N and 127°78′20.6″ E. After carrying out the soil property measurements using the equipment and method described in Section 2.1, rotary tillage was repeatedly performed using a tractor equipped with a field load measurement system.

The test was performed on the basis of the 3 × 2 split-plot design under three different gear selection (combinations of a PTO stage 1 and transmission gear stages L2 to L4 (i.e., P1L2, P1L3, and P1L4)) and two tillage types (primary and secondary tillage), as shown in Figure 4. Under no-load conditions, the first PTO gear stage had a PTO shaft rotational speed of 540 rpm, and the L2, L3, and L4 transmission gear stages had theoretical travel speeds of about 1.57, 2.25, and 3 km/h, respectively, at rated engine rotational speed. However, a load measurement test for rotary tillage was performed under full-throttle conditions in four-wheel-drive mode, taking into consideration the driving propensities of farmers. In addition, the plowing depth was set to about 10–12 cm, and the engine, axle, PTO, travel speed, and slip were measured three times with each test condition as the target measurement parameter.

(**a**) (**b**)

Figure 4. Load measurement test during rotary tillage: (**a**) primary tillage; (**b**) secondary tillage.

3. Results

3.1. Analysis of Soil Properties

In general, soil compaction in agricultural soils increases with the repeated use of heavy machinery such as tractors. Therefore, the distribution tendency of soil properties varies depending on the soil depth [28]. First, from the results of the cone penetration test analysis performed to see the distribution of mechanical properties of the soil, it was found that the cone index value rapidly increased at around 10 cm, as shown in Figure 5. Specifically, it showed an average value of 801.8 ± 277.6 kPa at a depth of about 0–10 cm and an average value of 2065.9 kPa at a depth of 10–20 cm. The average values at a depth of 10–20 cm were found to be 2.57 times higher than those measured at 0–10 cm.

Based on the results of the cone penetration test, the field soil properties were measured by dividing it into two depths. Table 3 shows the results of field measurement of soil quality divided into two depth sectors according to the soil depth, top soil (0–10 cm) and hardpan (10–20 cm). As a result, of soil analysis, the test field site was analyzed as loam (sand 46%, silt 34%, clay 20%). In general, it is known that the deeper the soil layer, the higher the soil compaction and the lower the porosity, and the lower the water content than the topsoil. This field test site also showed that the topsoil had an overall average soil water content of 20.6 ± 1.2%, which was 10% higher than that of the hardpan depth (17.7 ± 1.1%). As for the overall average bulk density result, the hardpan was analyzed to have an average of 1830 ± 30.6 kg/m^3, 16.4% higher than the top soil (1571.7 ± 75.2 kg/m^3), which showed similar bulk density results to the range of 1500–2100 kg/m^3 reported in studies using actual field experiments for tillage operations [29,30]. Additionally, as a comparison group, the bulk density of soil failure by each of the primary (rotary tillage) and secondary tillage (rotary tillage after plowing) was measured, and the results showed 1483.2 ± 10.9 kg/m^3 and 1431.2 ± 17.2 kg/m^3, respectively. Finally, in the case of shear strength, the average value of 64.3 ± 14.4 kPa in hardpan was 2.56 times significantly larger than top soil value (25.1 ± 7.2 kPa).

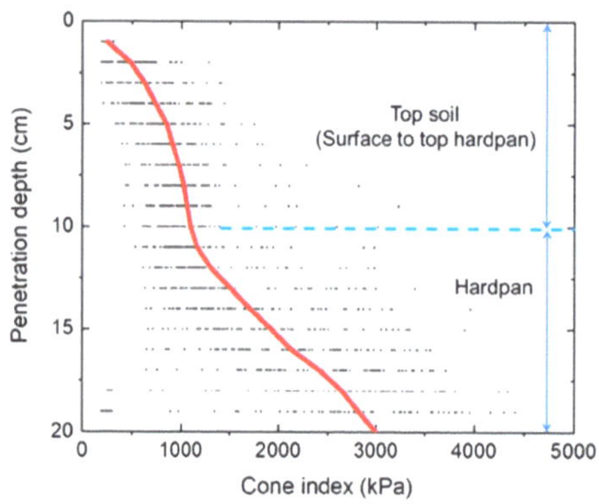

(a)

(b)

Figure 5. (**a**) Analysis of cone penetration test results; and (**b**) target tillage depth.

Table 3. Results of overall average travel speed and slip ratio according to driving conditions.

Driving Conditions	Theoretical Speed (km/h)	Travel Speed (km/h)	Slip Ratio (%)
Primary × P1L2	1.73 ± 0.03	1.7 ± 0.14	1.42 ± 5.31
Primary × P1L3	2.47 ± 0.04	2.44 ± 0.16	1.27 ± 6.21
Primary × P1L4	3.29 ± 0.06	3.26 ± 0.25	0.51 ± 6.54
Secondary × P1L2	1.73 ± 0.02	1.68 ± 0.08	2.43 ± 4.63
Secondary × P1L3	2.47 ± 0.02	2.42 ± 0.12	2.04 ± 4.81
Secondary × P1L4	3.29 ± 0.05	3.24 ± 0.17	1.43 ± 4.99

3.2. Travel Speed with Wheel Slippage

The overall average results of tillage depth and slip ratio are shown in Table 4. Since the travel speed was about 5–6% higher than the travel speed at the rated engine rotation speed in all gear stages, it was confirmed that the engine rpm was slightly higher than the rated speed. This was attributed to the fact that high loads such as plow tillage did not occur to such an extent that the engine load caused the engine rotational speed to drop below the rated engine rotational speed. In addition, the travel speed showed a very low slip ratio of 0.51~2.43%, considering the full-throttle condition of 2300 rpm. It was found that the harder the road surface (primary tillage) and the higher the travel speed (high transmission gear stage), the lower the overall average slip ratio, but the more frequent the negative slip based on larger data oscillation.

Table 4. Results of fuel efficiency by tillage type and gear selection.

Driving Conditions	FC [1] (kg/h)	SFC [2] (g/kWh)	PR [3] (ha/h)	Fuel Cost ($/ha)
Primary × P1L2	7.19	358.07	0.31	30.38
Primary × P1L3	7.56	336.10	0.44	22.51
Primary × P1L4	7.84	326.73	0.59	17.41
Secondary × P1L2	6.87	396.58	0.31	29.03
Secondary × P1L3	7.73	338.62	0.44	23.01
Secondary × P1L4	7.85	329.73	0.59	17.43

[1] FC: fuel consumption; [2] SFC: specific fuel consumption; [3] PR: productivity rate.

3.3. Engine Load

The overall measured results of engine torque are shown in Figure 6. In the case of engine rotation speed, only a difference of less than 1% was observed as a result of whether primary (within 0.83%) or secondary tillage (within 0.65%) conditions were employed under full-throttle (about 2300 rpm). In the case of engine torque, average engine torques of the working sections of 84.06 ± 7.24 Nm (P1L2), 95.67 ± 5.54 Nm (P1L3), and 111.55 ± 9.92 Nm (P1L4) were observed as a function of gear selection in primary tillage, and average engine torques of the working sections of 73.69 ± 6.86 Nm (P1L2), 91.88 ± 6.59 Nm (P1L3), and 100.15 ± 4.06 Nm (P1L4) were observed in secondary tillage. Under the same gear selection condition, the engine torque increased by 4.12% (P1L3) to 14.07% (P1L2) as a function of the tillage type. In addition, it was found that the engine torque increased by up to 32.7% (primary tillage) or 35.9% (secondary tillage) under the same tillage type. Basically, the overall average engine torque showed higher values in primary tillage under the same gear selection conditions, while there was a difference in engine torque of up to 51.37% as a function of the combination of tillage type and gear selection.

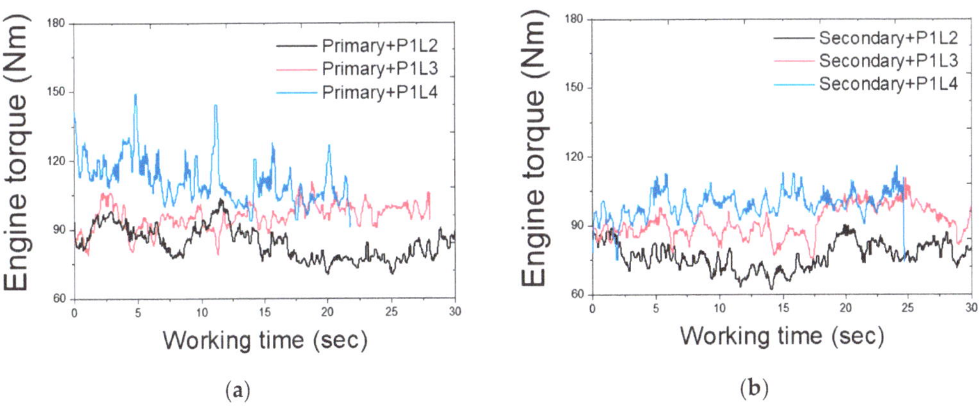

Figure 6. Results of engine torque data according to tillage type and gear selection during rotary tillage: (**a**) primary tillage; (**b**) secondary tillage.

3.4. PTO Shaft Load

The overall measured results of PTO shaft torque are shown in Figure 7. In the case of primary tillage, the overall average PTO shaft torque for the working section with respect to gear selection was 256.1 ± 36.8 Nm (P1L2 at 564.5 rpm), 286.2 ± 32.1 Nm (P1L3 at 563 rpm), and 343.1 ± 42.7 Nm (P1L4 at 559.8 rpm). In the case of secondary tillage, the results of PTO shaft torque were 200 ± 28.3 Nm (P1L2 at 565.4 rpm), 255 ± 29.7 Nm (P1L3 at 563.4 rpm), and 265.7 ± 22.2 Nm (P1L4 at 561.8 rpm). In the field test, the data measured for the secondary tillage method showed an average PTO torque value that was 10.1–22.6% lower than that of the primary tillage method due to the influence of the physical properties of soil, which changed rapidly with primary tillage, despite the rotary operation being performed at the same travel speed, gear selection, and tillage depth. This phenomenon of increasing PTO shaft torque with increasing forward speed and decreased PTO shaft torque for secondary tillage was also observed in a 2018 study analyzing performance using a combined offset disc harrow [31].

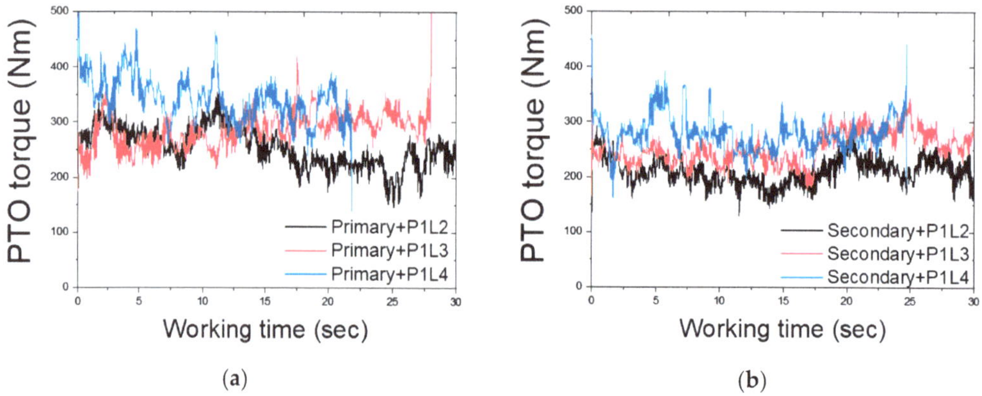

Figure 7. Results of PTO shaft torque as a function of tillage type and gear selection during rotary tillage: (**a**) primary tillage; (**b**) secondary tillage.

3.5. Axle Shaft Load

3.5.1. Front Wheel Axle

The overall measurement results for front wheel torque are shown in Figure 8. In the case of primary tillage, the overall average front axle torques for the working section according to each gear selection were 487.5 ± 124 Nm (P1L2 at 9.57 rpm), 566.6 ± 157 Nm (P1L3 at 13.73 rpm), and 541.2 ± 213.5 Nm (P1L4 at 18.16 rpm), respectively. In the case of secondary tillage, the results of front wheel axle torque were 823.9 ± 134.8 Nm (P1L2 at 9.73 rpm), 869.5 ± 173.1 Nm (P1L3 at 13.7 rpm), and 878.7 ± 108.7 Nm (P1L4 at 18.2 rpm), respectively. Overall, secondary tillage showed average values for each gear selection that were 1.55–1.69 times higher than those for primary tillage, which was attributed to the fact that the power distribution in the engine was affected by the higher PTO load in primary tillage. However, it was confirmed that the pitch angle at the center of gravity of the tractor continuously occurred due to the soil–tool interaction in primary tillage, resulting in unstable load transfer. The standard deviation of the tractor's torque was also higher by 23% at P1L4 than that for secondary tillage.

3.5.2. Rear Wheel Axle

The overall results measured for rear wheel torque are shown in Figure 9. Basically, the rear wheel axle load, like the front wheel axle torque, did not show a linear relationship according to gear selection. However, it was analyzed that the rear axle load under secondary tillage conditions was 21% (P1L2) to 49% (P1L3) higher for all gear selections than under primary tillage conditions. In the case of primary tillage, the overall average rear axle torque for the working section according to each gear selection was 555.9 ± 127 Nm (P1L2 at 6.41 rpm), 622 ± 167.8 Nm (P1L3 at 9.18 rpm), and 549.3 ± 213.1 Nm (P1L4 at 12.18 rpm), respectively. In the case of secondary tillage, the results of rear wheel axle torque were 673.7 ± 136.8 Nm (P1L2 at 6.42 rpm), 930.3 ± 195 Nm (P1L3 at 9.19 rpm), 772.7 ± 278.1 Nm (P1L4 at 12.2 rpm), respectively. Based on these results, it was also concluded that the power requirements distributed to the axles were relatively high, as the power requirements in the engine and PTO shaft were relatively reduced in the rear wheels as well.

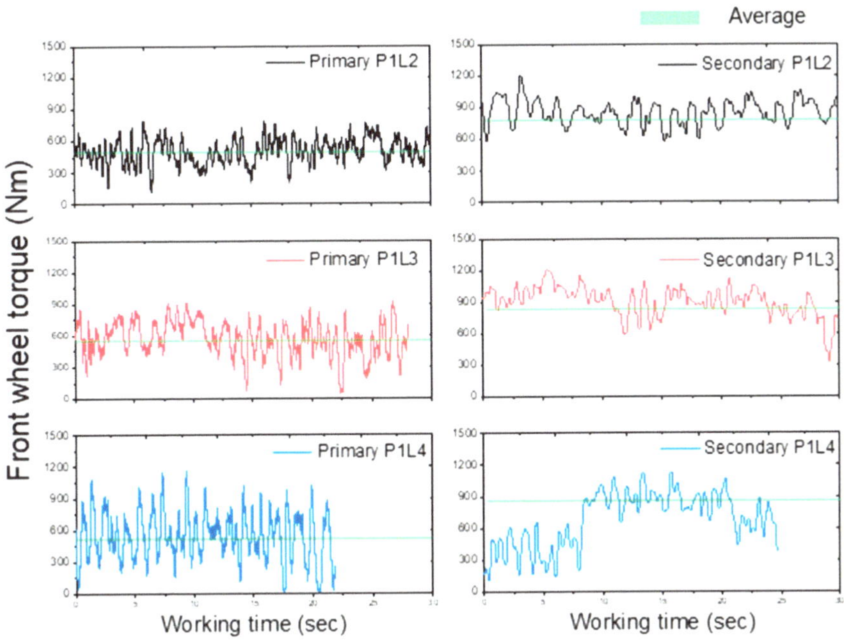

Figure 8. Results of front wheel axle torque data according to tillage type and gear selection during rotary tillage.

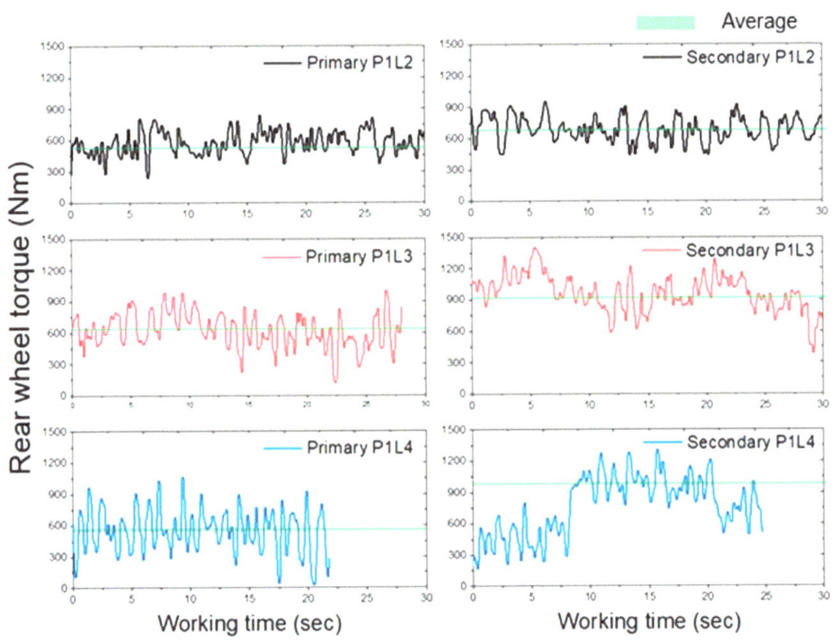

Figure 9. Results of rear wheel axle torque data according to tillage type and gear selection during rotary tillage.

3.6. Fuel Efficiency

The fuel efficiency of a 42 kW agricultural tractor depending on tillage type and gear selection is shown in Table 4. In terms of fuel usage, the FC when working with P1L4 was up to 1.09 times higher than the FC when working with P1L2. In terms of fuel efficiency, the SFC when working with P1L2 was up to 1.2 times higher than the SFC when working with P1L4. From a cost perspective, the fuel cost of working with P1L2 was up to 1.74 times higher than that when working with P1L4. In general, it has been reported that in order to obtain maximum fuel efficiency during tillage operations, it is required that the work be performed under the conditions of using the high torque ranges in the appropriate slip range [32]. Therefore, in the case of this test, where the slip rate did not change significantly when increasing the number of gears, the higher the torque value at a higher gear ratio, the better the fuel efficiency.

4. Discussion

In this study, the influence of soil–tool interaction (i.e., tillage type, gear selection, soil properties) during rotary plowing work was investigated. Based on a report on the effect of soil compaction on crop root growth from the USDA [33,34], test soil form an agricultural field was analyzed before the tillage operation in order to evaluate the degree to which the physical properties of the field affect (over 1630 kg/m^3) or limit crop root growth (over 1800 kg/m^3). The results of the test showed that the bulk density measured from 1571 to 1830 kg/m^3 decreased to a level of 1431 to 1480 kg/m^3. This was judged as indicating that tillage type directly affects the soil–tool interaction.

In general, there are reports suggesting that the rear axle load is 3 to 8 times greater than the front axle load during plowing [11,12]. However, unlike plowing, the load transfer increased toward the front wheel, and the front wheel axle torque showed a similar level to the rear wheel axle torque. In other words, the rotavator increases the soil thrust in the working direction, giving the same effect as pushing the vehicle forward. This was found to be the same as the effect of strengthening the soil–tyre interaction with an auxiliary iron wheel. In addition, for the same reason, it was found that negative slip, which momentarily produces a higher travel speed than the theoretical speed of the vehicle, partly and momentarily occurs in some sections.

Figure 10 shows the relative histogram results of power requirements as a function of tillage type and gear selection. On the basis of the results of power requirement analysis, it can be seen that engine and PTO power have tendencies that are different from the wheel axle power requirements. In the case of wheel axle power under the same gear selection conditions, it was found that the required power in secondary tillage was 21–81% higher than in primary tillage. The reason for this phenomenon is that primary tillage reduces the internal friction of the soil [35]. In another study by Kim et al. [36], it was reported that with increasing value of the slip ratio range, the axle torque also increased. Table 4 shows that the slip ratio increased more during secondary tillage than during primary tillage, overall. Therefore, it seems that the power requirement of the wheel axle increased more during secondary tillage than during primary tillage in this study. In the case of the rotary tillage performed in this study, it was found that the effect of tillage method was greater than the effect of gear selection for the axle load, which is in contrast to the engine and PTO load. Conversely, it was analyzed that the effect of gear selection (Engine power: 4–14%, PTO power: 12.1–28.6%) on engine and PTO loads was greater than that of tillage type (Engine power: 31.6–35.1%, PTO power: 31.9–32.8%), and the load tended to decrease in secondary tillage.

Compared to the test conditions in the specifications of the actual rotavator, considering that only up to 62.1% of the maximum engine capacity (42 kW) was used, it was deemed that this work could be performed in a harder and stickier environment than the loam soil in which this study was conducted. In addition, it was concluded that the work efficiency will increase more when rotary tillage is performed under the conditions of a tillage depth deeper than 12–13 cm, a wider work width (over 1820 mm), and a work speed faster than 3.24 km/h in the same working environment.

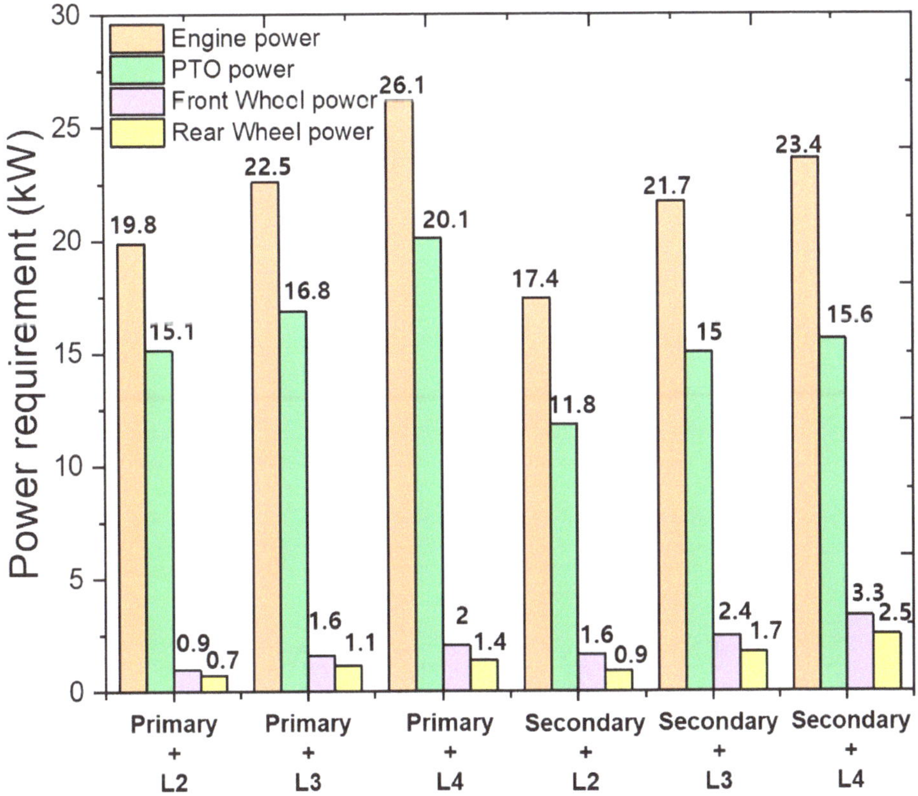

Figure 10. Analysis of power requirement as a function of tillage type and gear selection.

5. Conclusions

In this study, a field load measurement test was conducted to analyze the effect of tillage type (i.e., primary tillage or secondary tillage) and gear selection (i.e., P1L2, P1L3, or P1L4) on the tractor working load (i.e., engine load, wheel axle load, and PTO shaft load, as well as travel speed) during rotary tillage operations. The major conclusions obtained were as follows.

1. By measuring the soil properties according to depth, it was confirmed that the core index, shear strength, and water content, which are the main properties that affect the load of cultivation work, changed in accordance with the change in depth. In particular, soil bulk density was measured to decrease by 9–19% under rotary operation. Therefore, soil properties as a function of tillage type, which is a typical soil–tool interaction process, should be considered first when evaluating the performance of the operation of agricultural machinery or in field tests.
2. Overall average torque was higher by up to 14% (engine) and 29.1% (PTO shaft) in primary tillage than in secondary tillage when the selected gear was the same. When the tillage type was the same, it was found that the overall average torque increased as a result of gear selection by up to 35.9% (engine) and 33.9% (PTO shaft) in P1L4 compared to P1L2.

3. In the case of fuel efficiency, it was revealed that the effect of gear selection was greater than the effect of the tillage type. When working on loam field (soil water content of 17–20%, bulk density of 1571–1830 kg/m^3, and cone index of 801–2065 kPa), the most suitable gear for reducing fuel consumption was found to be P1L4.
4. In addition, based on the power requirement results, from the perspective of the machine, when working on loamy soil with a tillage depth of 10 to 13 cm, 66.7–77% of the power generated by the engine was consumed by the PTO shaft. This shows that the force applied to the implements during rotary operation is greater than the traction load.
5. In the case of the power requirements, the power required on the same tillage type increased with increasing gear ratio. During secondary tillage, the overall power required decreased due to changes in key soil properties such as bulk density, cone index, and vane shear torque. Therefore, it was judged that design modification is necessary in order to have a wide working width, and that in some cases, rotary tillage will be possible at a deeper tillage depth or at a higher travel speed. It is expected that performance evaluation and optimal design of soil operation machinery in various working environments will be possible through similar field verification procedures in future studies.

Recently, various agricultural work environments have been defined, and virtual agricultural performance evaluation or load prediction simulation studies are being conducted through modeling. It is important to secure a database for the definition of agricultural work environments and load analysis. The results of this study are expected to be of use in future studies for designing loads in research on modeling and the verification of agricultural machinery systems such as soil–tool and soil–tyre interaction during agricultural operation.

Author Contributions: Conceptualization, Y.-S.K. and T.-J.K.; methodology, Y.-S.K. and W.-S.K.; software, B.-M.B. and T.-J.K.; validation, Y.-S.K. and W.-S.K.; formal analysis, Y.-S.K. and W.-S.K.; investigation, Y.-J.K. and T.-J.K.; data curation, Y.-S.K. and B.-M.B.; writing—original draft preparation, Y.-S.K.; writing—review and editing, B.-M.B.; supervision, Y.-J.K. and S.-D.L.; project administration, Y.-J.K. and T.-J.K.; funding acquisition, Y.-J.K. All authors have read and agreed to the published version of the manuscript.

Funding: This work was supported by the Korea Institute of Planning and Evaluation for Technology in Food, Agriculture, Forestry (IPET) through the Eco-Friendly Power Source Application Agricultural Machinery Technology Development Program, funded by Ministry of Agriculture, Food and Rural Affairs (MAFRA) (322047-5), and it was also supported by the Korea Institute of Planning and Evaluation for Technology in Food, Agriculture, Forestry (IPET) through the Technology Commercialization Support Program, funded by the Ministry of Agriculture, Food and Rural Affairs (MAFRA) (821014-03).

Institutional Review Board Statement: Not applicable.

Data Availability Statement: Data sharing not applicable.

Conflicts of Interest: The authors declare no conflict of interest.

References

1. Kim, W.S.; Kim, Y.J.; Park, S.U.; Hong, S.J.; Kim, Y.S. Evaluation of PTO Severeness for 78 KW-Class Tractor According to Disk Plow Tillage and Rotary Tillage. *J. Drive Control* **2019**, *16*, 23–31. [CrossRef]
2. Lim, S.; Kwon, H.; Kang, Y.; Lee, P.; Kim, T.; Kim, Y.; Kim, Y. Power Analysis of a 3-KW Class Motor-Driven Multipurpose Walking-Type Transplanter. *J. Biosyst. Eng.* **2019**, *44*, 135–145. [CrossRef]
3. Kim, Y.J.; Chung, S.O.; Park, S.J.; Choi, C.H. Analysis of Power Requirement of Agricultural Tractor by Major Field Operation. *J. Biosyst. Eng.* **2011**, *36*, 79–88. [CrossRef]
4. Han, K.H.; Kim, K.U.; Wu, Y.G. Severness of Transmission Loads of Agricultural Tractor for Rotary Operations in Poorly Drained Paddy Field. *J. Biosyst. Eng.* **1999**, *24*, 293–300.
5. Raper, R.L. Force requirements and soil disruption of straight and bentleg subsoilers for conservation tillage systems. *Appl. Eng. Agric.* **2005**, *21*, 787–794. [CrossRef]

6. Naderloo, L.; Alimadani, R.; Akram, A.; Javadikia, P.; Zeinali Khanghah, H. Tillage Depth and Forward Speed Effects on Draft of Three Primary Tillage Implements in Clay Loam Soil. *J. Food Agric. Environ.* **2009**, *7*, 382–385.
7. Moeenifar, A.; Mousavi-Seyedi, S.R.; Kalantari, D. Influence of Tillage Depth, Penetration Angle and Forward Speed on the Soil/Thin-Blade Interaction Force. *Agric. Eng. Int. CIGR J.* **2014**, *16*, 69–74.
8. Aday, S.H.; Ramadhan, M.N. Comparison between the Draft Force Requirements and the Disturbed Area of a Single Tine, Parallel Double Tines and Partially Swerved Double Tines Subsoilers. *Soil Tillage Res.* **2019**, *191*, 238–244. [CrossRef]
9. Kichler, C.M.; Fulton, J.P.; Raper, R.L.; McDonald, T.P.; Zech, W.C. Effects of Transmission Gear Selection on Tractor Performance and Fuel Costs during Deep Tillage Operations. *Soil Tillage Res.* **2011**, *113*, 105–111. [CrossRef]
10. Kim, W.S.; Kim, Y.J.; Park, S.U.; Kim, Y.S. Influence of Soil Moisture Content on the Traction Performance of a 78-KW Agricultural Tractor during Plow Tillage. *Soil Tillage Res.* **2020**, *207*, 104851. [CrossRef]
11. Kim, Y.S.; Kim, W.S.; Baek, S.Y.; Baek, S.M.; Kim, Y.J.; Lee, S.D.; Kim, Y.J. Analysis of Tillage Depth and Gear Selection for Mechanical Load and Fuel Efficiency of an Agricultural Tractor Using an Agricultural Field Measuring System. *Sensors* **2020**, *20*, 2450. [CrossRef]
12. Kim, Y.S.; Lee, S.D.; Baek, S.M.; Baek, S.Y.; Jeon, H.H.; Lee, J.H.; Kim, W.S.; Shim, J.Y.; Kim, Y.J. Analysis of the Effect of Tillage Depth on the Working Performance of Tractor-Moldboard Plow System under Various Field Environments. *Sensors* **2022**, *22*, 2750. [CrossRef]
13. Lee, D.H.; Kim, Y.J.; Chung, S.O.; Choi, C.H.; Lee, K.H.; Shin, B.S. Analysis of the PTO Load of a 75 KW Agricultural Tractor during Rotary Tillage and Baler Operation in Korean Upland Fields. *J. Terramech.* **2015**, *60*, 75–83. [CrossRef]
14. Behera, A.; Raheman, H.; Thomas, E.V. A Comparative Study on Tillage Performance of Rota-Cultivator (a Passive—Active Combination Tillage Implement) with Rotavator (an Active Tillage Implement). *Soil Tillage Res.* **2021**, *207*, 104861. [CrossRef]
15. Kim, Y.J.; Chung, S.K.; Choi, C.H.; Lee, D.H. Evaluation of Tractor PTO Severeness during Rotary Tillage Operation. *J. Biosyst. Eng.* **2011**, *36*, 163–170. [CrossRef]
16. Kim, Y.J.; Chung, S.O.; Choi, C.H. Effects of Gear Selection of an Agricultural Tractor on Transmission and PTO Load during Rotary Tillage. *Soil Tillage Res.* **2013**, *134*, 90–96. [CrossRef]
17. Kim, Y.S.; Lee, P.U.; Kim, W.S.; Kwon, O.W.; Kim, C.W.; Lee, K.H.; Kim, Y.J. Strength Analysis of a PTO (Power Take-Off) Gear-Train of a Multi-Purpose Cultivator during a Rotary Ditching Operation. *Energies* **2019**, *12*, 1100. [CrossRef]
18. Tan, K.H. *Soil Sampling, Preparation, and Analysis*; CRC Press: New York, NY, USA, 1995; ISBN 0-8247-9675-6.
19. O'Kelly, B.C.; Sivakumar, V. Water Content Determinations for Peat and Other Organic Soils Using the Oven-Drying Method. *Dry. Technol.* **2014**, *32*, 631–643. [CrossRef]
20. Ditzler, C.; Scheffe, K.; Monger, H.C. *USDA Handbook 18*; Government Printing Office: Washington, DC, USA, 2017.
21. Thien, S.J. A Flow Diagram for Teaching Texture-by-Feel Analysis. *J. Agron. Educ.* **1979**, *8*, 54–55. [CrossRef]
22. *ASABE Standards EP542*; Procedure for Using and Reporting Data Obtained with the Soil Cone Penetrometer. ASAE, Ed.: St. Joseph, MI, USA, 2009; Volume 1999.
23. *ASABE Standards S313.3*; Soil Cone Penetrometer. ASAE, Ed.: St. Joseph, MI, USA, 2005; Volume 1999.
24. Park, J.G. *Bio-Production Machinery Engineering*, 1st ed.; CIR: Seoul, Republic of Korea, 2008.
25. Srivastava, A.K.; Goering, C.E.; Rohrbach, R.P.; Buckmaster, D.R. *Engineering Principles of Agricultural Machines*, 2nd ed.; American Society of Agricultural and Biological Engineers (ASABE): St. Joseph, MI, USA, 2006.
26. Nataraj, E.; Sarkar, P.; Raheman, H.; Upadhyay, G. Embedded Digital Display and Warning System of Velocity Ratio and Wheel Slip for Tractor Operated Active Tillage Implements. *J. Terramech.* **2021**, *97*, 35–43. [CrossRef]
27. Rasool, S.; Raheman, H.; Upadhyay, G. Development of an Instrumentation System for Evaluating the Tractive Performance of Walking Tractors. *Int. J. Curr. Microbiol. Appl. Sci.* **2017**, *6*, 759–770. [CrossRef]
28. Chung, S.O.; Sudduth, K.A. Soil Failure Models for Vertically Operating and Horizontally Operating Strength Sensors. *Trans. ASABE* **2006**, *49*, 851–863. [CrossRef]
29. Farid Eltom, A.E.; Ding, W.; Ding, Q.; Ali, A.B.B.; Eisa Adam, B. Effect of Trash Board on Moldboard Plough Performance at Low Speed and under Two Straw Conditions. *J. Terramech.* **2015**, *59*, 27–34. [CrossRef]
30. Kim, Y.S.; Kim, W.S.; Siddique, A.A.; Baek, S.Y.; Baek, S.M.; Cheon, S.H.; Lee, S.D.; Lee, K.H.; Hong, D.H.; Park, S.U.; et al. Power Transmission Efficiency Analysis of 42 KW Power Agricultural Tractor According to Tillage Depth during Moldboard Plowing. *Agronomy* **2020**, *10*, 1263. [CrossRef]
31. Upadhyay, G.; Raheman, H. Performance of Combined Offset Disc Harrow (Front Active and Rear Passive Set Configuration) in Soil Bin. *J. Terramech.* **2018**, *78*, 27–37. [CrossRef]
32. Grisso, R.; Pitman, R.; Perumpral, J.V.; Vaughan, D.; Roberson, G.T.; Hoy, R.M. *"Gear up and Throttle down" to Save Fuel*; Virginia Cooperative Extension Publication: Blacksburg, VA, USA, 2011; pp. 442–450.
33. Nyéki, A.; Milics, G.; Kovács, A.J.; Neményi, M. Effects of Soil Compaction on Cereal Yield: A Review. *Cereal Res. Commun.* **2017**, *45*, 1–22. [CrossRef]
34. USDA. *Soil Mechanics Level I. Module 3—USDA Textural Soil Classification. Study Guide*; USDA, Soil Convervation Service: Stillwater, OK, USA, 1987.

35. Park, J.G.; Lee, K.S.; Cho, S.C.; Noh, K.M.; Chung, S.O.; Chang, Y.C. Analysis of Variations in Mechanical Properties of Soil by Tillage Operations. *J. Biosyst. Eng.* **2007**, *32*, 215–222.
36. Kim, W.S.; Kim, Y.S.; Kim, Y.J. Development of Prediction Model for Axle Torque of Agricultural Tractors. *Trans. ASABE* **2020**, *63*, 1773–1786. [CrossRef]

agriculture

Article

Theoretical Calculations and Experimental Studies of Power Loss in Dual-Clutch Transmission of Agricultural Tractors

Hyoung-Jong Ahn [1,2], Young-Jun Park [2], Su-Chul Kim [3] and Chanho Choi [1,*]

1 Tractor Advanced Development Group, LS Mtron, Anyang 14118, Republic of Korea; hyoungjong.ahn@lsmtron.com
2 Department of Biosystems Engineering, Seoul National University, 1 Gwanak-ro, Gwanak-gu, Seoul 08826, Republic of Korea; yjpark95@snu.ac.kr
3 Department of Smart Industrial Machinery, Korea Institute of Machinery & Materials, Daejeon 34103, Republic of Korea; sckim@kimm.re.kr
* Correspondence: chance.choi@lsmtron.com; Tel.: +82-31-8045-0494

Abstract: Recent carbon neutrality policies have led to active research in the agricultural tractor sector to replace internal combustion engines, making it imperative to minimize power losses to improve efficiency. Dual-clutch transmissions (DCTs) have been employed in agricultural tractors primarily due to their short shift time and smooth shift feel. However, DCTs have a relatively large number of components and complex structures owing to spatial constraints, making it challenging to predict power losses. Therefore, to predict DCT power losses, this study defined oil churning by considering the structural characteristics and oil circulation and comparing and analyzing the theoretical calculation and test results of power losses at different oil levels. Power loss was calculated based on ISO standards and fluid viscosity theory, and tests were performed to verify. We calculated power losses based on the defined oil churning of a DCT in agricultural tractors and confirmed that their consistency in test results improved when reflecting the lubrication state, considering the structural features and oil circulation. In addition, the factors contributing to power loss under low- and high-speed conditions were analyzed by calculating the power loss for each component.

Keywords: dual-clutch transmission; power loss; agricultural tractor

Citation: Ahn, H.-J.; Park, Y.-J.; Kim, S.-C.; Choi, C. Theoretical Calculations and Experimental Studies of Power Loss in Dual-Clutch Transmission of Agricultural Tractors. *Agriculture* **2023**, *13*, 1225. https://doi.org/10.3390/agriculture13061225

Academic Editor: Kenshi Sakai

Received: 20 May 2023
Revised: 5 June 2023
Accepted: 9 June 2023
Published: 10 June 2023

1. Introduction

Agricultural tractors perform various farming tasks using the vehicle's rotational power or traction, such as plowing, transportation, and rotary work. Traditionally, diesel engines, particularly with manual transmissions, have been widely used owing to their high torque characteristics relative to engine output. Recently, demand for improved driver convenience and work efficiency, as well as technological pressure to replace fossil fuels in the agricultural tractor sector due to carbon neutrality policies, has been increasing.

Prominent alternative power sources for agricultural tractors being discussed include hydrogen fuel cells, batteries, and biodiesel. Research on small electrically driven tractors using batteries and motors, and studies exploring the technical and economic feasibility of small farms, are actively underway [1–3]. Other studies have proposed new energy management strategies for applying hybrid systems, comprising fuel cells and battery packs, to agricultural tractors [4,5]. Moreover, energy, economic, and environmental life-cycle assessment analyses have been conducted for applying biodiesel fuel to agricultural tractors [6]. However, alternative power sources generate less energy per unit volume and mass compared to diesel [7,8]. Consequently, when using alternative power sources, the continuous working time is reduced, making power losses in transmission systems an even more critical issue owing to their direct impact on production efficiency.

Furthermore, researchers are actively studying automatic transmission systems in response to market demand for improved driver convenience and work efficiency [9–12].

Automated transmissions (ATs) allow drivers to operate without a gear lever and can improve transmission productivity via the selection of the optimal gear ratio for farming tasks through control algorithms. Dual-clutch transmissions (DCTs), a type of AT, use two clutches responsible for even and odd gear steps in power transmission. Considering DCTs preselect the target gear shift stage and engage the clutch when shifting, they have a short shift time with no power interruption, resulting in a smooth shifting experience. However, due to the increased number of mechanical components needed to select even and odd gears—such as wet multi-plate clutches—and the complex supporting structures for components owing to spatial constraints within the transmission, power losses in the transmission must be considered.

Research on power loss in transmissions has been conducted primarily at the component level. Many studies have investigated the drag torque caused by oil viscosity between friction surfaces to calculate the power loss of wet multi-plate clutches [13,14]. The shear stress caused by oil viscosity between the ring and cone was considered for calculating the power loss of synchronizers, friction coefficients obtained through testing were proposed to predict bearing power loss [15,16]. In addition, research comparing analytical models and test results has been conducted to predict the power loss caused by oil churning in gears [17,18]. Moreover, researchers have conducted studies to calculate the power loss due to gear load transfer and conducted experimental verifications [19,20].

Power loss research on DCTs has primarily been conducted in the automotive sector, predicting power loss through theoretical predictions and experimental verifications [21,22]. Additionally, research considering power loss in transmissions has also been conducted in the agricultural tractor sector. The efficiency and energy consumption of tractor transmissions have been examined primarily in relation to operating conditions using vehicle tests [23,24]. Studies have also been conducted to predict transmission efficiency by examining the power loss generated by each component of agricultural DCTs through theoretical calculations [25].

Considering policy reasons driven by carbon neutrality requirements and consumer demand, a systematic study on the power transmission efficiency of agricultural transmissions is necessary. However, research on the efficiency of tractors has focused primarily on vehicle-level experimental verification, while no studies have examined the contribution of each component to the total power loss and compared theoretical calculations with actual test results. Moreover, the oil-churning state of the mechanical components must be defined to accurately predict the power loss of a transmission system. However, in transmissions such as the DCT—which has many components and circulates oil through hydraulic systems—oil levels change locally; hence, the structural characteristics must be considered. Moreover, studies on efficiency considering oil level changes due to the structural characteristics of agricultural DCTs remain insufficient.

Accordingly, this study compared the theoretical predictions of power loss in a DCT applied to agricultural tractors with power loss measurements obtained through laboratory tests. Considering the oil circulation for DCT lubrication, two power loss models, which considered the average oil level and oil circulation, respectively, were proposed, with theoretical predictions performed for each case. We used the ISO standard and viscosity fluid theory to calculate the power loss of components making up the transmission and constructed a measurement system using a 3-axis dynamometer to measure them. The power losses considering oil circulation and the average oil level were compared and the test results were analyzed.

2. Materials and Methods

2.1. DCTs for Agricultural Tractors

In this study, a 24-speed full-power shift DCT for 100-kW agricultural tractors was applied to 5-ton tractors (without attachments) and shifted by alternately engaging the left and right odd and even shaft clutches. This DCT comprises three parts—input, even, and odd shafts—which is relatively more components than other transmissions.

As shown in Figure 1, the target DCT comprises 64 shafts, 58 gears, 95 bearings, 5 clutches, 14 synchronizers, and 2 dog clutches, and includes 2 rear axle brakes and 1 parking brake within the transmission. However, the transmission system is assembled by dividing it into sections because it is challenging to arrange the numerous components that make up the DCT in a single case. Each section has a partition wall supported by housing and bearings, with empty spaces being left to allow free oil movement when designing the partition wall. However, the rear partition wall has a relatively thick wall and a narrow space to support the amplified load of the gear train, hindering oil circulation and isolating the space. Nonetheless, these spaces can be divided into three cells based on the partition walls isolating them.

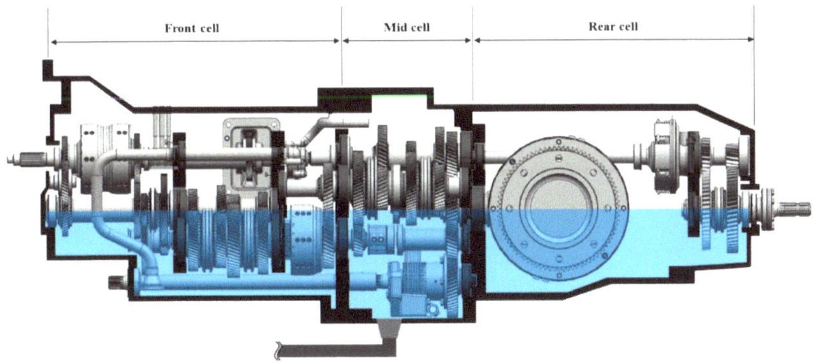

(**a**) Case I: Considering average oil level.

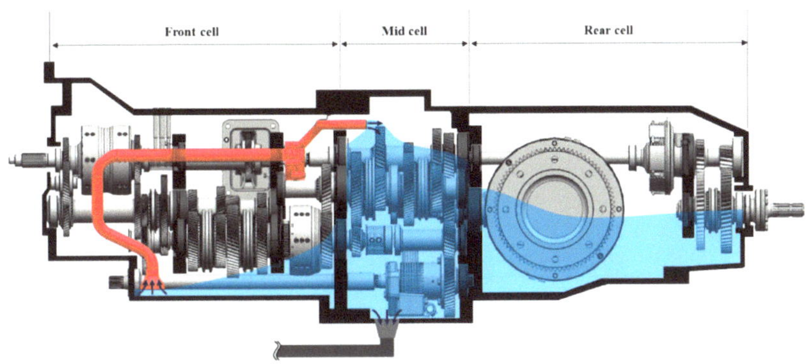

(**b**) Case II: Considering oil circulation.

Figure 1. Oil levels of a dual-clutch transmission (DCT) in full-power shift.

Figure 1 shows the structure and oil level of the DCT used in this study. The DCT is divided into three cells based on the partition walls supporting the rear part of the power transmission system: the front cell, where gears of the forward/reverse shift and main shift parts are arranged; the mid cell, where gears of the range shift and 4WD parts are arranged; and the rear cell, where the spiral bevel gear connected to the rear axle and PTO (power take-off) gears are arranged. As shown in Figure 1a, the transmission oil level is maintained at the height of the rear axle center.

The oil in agricultural tractors simultaneously performs the roles of lubrication for the components and hydraulic fluid for the hydraulic system. The transmission oil is churned by the immersed rotating components and circulated by the hydraulic system simultaneously. Therefore, the oil levels in each cell are dynamic and differ considerably. The DCT used in this study has an intake line at the bottom to supply oil to the hydraulic system. Moreover, to secure a stable oil supply flow when driving on a slope, forward/reverse and dual clutches are used, which forcibly supply oil from the front cell—where relatively more flow is discharged—to the mid cell through a pump. Figure 1b shows an oil-level application model reflecting the partition walls and oil circulation of the hydraulic system.

Theoretical power-loss predictions were performed for Case I, considering the average oil level of the transmission, and for Case II, considering the partition walls and hydraulic system, both verified through testing.

2.2. Power Loss Calculations

The ISO TR 14179-1 standard and drag torque due to viscous fluid were considered to calculate the power loss caused by the components making up the powertrain, with gears, bearings, shafts, clutches, and brakes considered for these calculations [26,27].

The total power loss can be expressed as the sum of fundamental, load-dependent, and speed-dependent power losses. The fundamental power loss is the power consumed to drive the tractor, a representative example being the power loss of the pump used to drive the hydraulic line. The measured fundamental power loss was 3.0 kW, applied equally to all gear steps. The load-dependent power loss occurs in the friction surfaces under load, resulting in a corresponding load-dependent power loss of gears and bearings. Speed-dependent power loss is the power loss due to motion resistance when components of the power transmission system rotate, implying that the power loss due to drag is caused by the oil applied for lubrication and cooling. For speed-dependent power loss, the power loss of clutches, bearings, gears, shafts, and brakes was considered. The total power loss (P_L) of the power transmission system can be calculated using Equation (1), with the factors considered for each power loss shown in Table 1.

$$P_L = \sum P_{HP} + \sum P_{GL} + \sum P_{GS} + \sum P_{BL} + \sum P_{BS} + \sum P_S + \sum P_{CL} + \sum P_{BR} \quad (1)$$

Table 1. Composition of the total power loss.

Power Loss	Component
Fundamental power loss	Hydraulic pump $[P_{HP}]$
Load-dependent power loss	Gear $[P_{GL}]$, Bearing $[P_{BL}]$
Speed-dependent power loss	Gear $[P_{GS}]$, Bearing $[P_{BS}]$, Clutch $[P_{CL}]$, Shaft $[P_S]$, Brake $[P_{BR}]$

2.2.1. Power Losses of the Gear and Shaft

The load-dependent power loss of the gear (P_{GL}) can be expressed as follows:

$$P_{GL} = \frac{f_m T_1 n_1 \cos^2\beta}{9549M} \quad (2)$$

where f_m denotes the mesh coefficient of friction (determined by the pitch line speed, lubricating oil viscosity, load size, and gear size), and M denotes the mesh mechanical advantage (a coefficient influenced by the gear size, pressure angle, and gear step).

The speed-dependent power loss of the gear (P_{GS}) and speed-dependent power loss of the shaft (P_S) can be expressed as the sum of power losses owing to the outer diameter (P_{GW1}), lateral part (P_{GW2}), and tooth surface (P_{GW3}). P_{GW1}, P_{GW2}, and P_{GW3} can be calculated using Equations (3)–(5), respectively, as follows:

$$P_{GW1} = \frac{7.37 f_g v n^3 D^{4.7} L}{A_g 10^{26}} \quad (3)$$

$$P_{GW2} = \frac{1.474 f_g \nu n^3 D^{5.7}}{A_g 10^{26}} \tag{4}$$

$$P_{GW3} = \frac{7.37 f_g \nu n^3 D^{4.7} F\left(\frac{R_f}{\sqrt{\tan\beta}}\right)}{A_g 10^{26}} \tag{5}$$

where R_f denotes the roughness factor (which is influenced by the transverse module).

2.2.2. Power Loss of the Bearing

The load-dependent power loss of the bearing (P_{BL}) can be expressed as follows:

$$P_{BL} = \frac{(M_1 + M_2)n}{9549} \tag{6}$$

where M_1 denotes the bearing-load-dependent torque (calculated using the load rating of the bearing, dynamic load, coefficient of friction, and bearing size) and M_2 denotes the cylindrical-roller-bearing axial-load-dependent moment (determined by the friction coefficient of the lubrication surface, axial bearing load, and bearing size).

The speed-dependent power loss of the bearing (P_{BS}) can be expressed as follows:

$$P_{BS} = \frac{(M_0 + M_3)n}{9549} \tag{7}$$

where M_0 denotes the load-independent frictional moment (influenced by the dynamic viscosity coefficient, dip factor, rotation speed, and size), and M_3 is the frictional moment of seals (a factor influenced by the size and type of the bearing).

2.2.3. Power Loss of the Clutch and Brake

Clutch drag torque occurs owing to the shear stress caused by the viscosity of the lubricating oil between the friction surfaces. The clutch power loss can be calculated by correcting the drag torque based on the supplied flow area, while the drag torque can be calculated using Equation (8) [27]. The oil is supplied from the shaft center, causing a ruptured section where the oil flow breaks into mist and steam. As a result, three flow sections are generated—T_{fa}, T_{ra}, and T_m; their relationships are expressed in Equations (9)–(11). The oil used was ISO VG 46; in addition, oil viscosity properties of 46 cSt at 40 °C and 6.7 cSt at 100 °C were considered. The viscosity of the lubricating oil in the mist state can be assumed to be 1/10 of the oil viscosity.

$$T_{CL} = (1 - r_{ag}) \cdot \left(T_{fa} + T_{ra} + T_m\right) \tag{8}$$

$$T_{fa} = \frac{\pi \mu \Delta\omega N}{2h} \cdot \left(r_c^{\,4} - r_i^{\,4}\right) \tag{9}$$

$$T_{ra} = \frac{\pi \mu \Delta\omega N}{h} \cdot \Phi \cdot \left(r_o^{\,2} - r_c^{\,2}\right) \tag{10}$$

$$T_m = \frac{2\pi \mu_{mist} \Delta\omega N}{h} \cdot \left[\frac{\left(r_o^{\,4} - r_c^{\,4}\right)}{4} - \frac{\Phi}{2} \cdot \left(r_o^{\,2} - r_c^{\,2}\right)\right] \tag{11}$$

The power loss of the clutch (P_{CL}) considering the drag torque of each section can be expressed as follows:

$$P_{CL} = T_{CL} \Delta n \frac{2\pi}{60 \times 1000} \tag{12}$$

Similarly, the power loss of the brake (P_B), which occurs owing to the shear stress caused by the fluid viscosity, can be expressed as follows:

$$P_B = (1 - r_{ag}) \frac{N\mu\alpha}{h} \left(\frac{2\pi n}{60}\right)^2 \left[\frac{r_o{}^4}{4} - \frac{r_i{}^4}{4}\right] \frac{1}{1000} \quad (13)$$

2.3. Power Loss Measurement of DCT

Figure 2 shows the system constructed for power loss measurement using a 3-axis dynamometer to control the rotation speed and torque. Sensors for measuring the rotation speed and torque are installed at the connection between the motor and the transmission to measure the input and output power. The input speed to the transmission is controlled at the input section, while the load applied to the transmission is controlled by controlling the torque at the output section. Additionally, a thermometer is installed inside the transmission system to monitor the oil temperature during the test. The test procedure is as follows. The speed of the input motor is slowly increased over 120 s from 0 to 2200 rpm. When the speed reaches 2200 rpm, the torque of the output motor is increased by 200 Nm to reach the target load level. While maintaining the target load for 1 min, the torque and rotation speed of the input and output sections are measured by the sensor. The oil temperature is maintained in the range of 50 to 90 °C, which is the normal operating temperature of the tractor. Motor and sensor specifications are shown in Table 2.

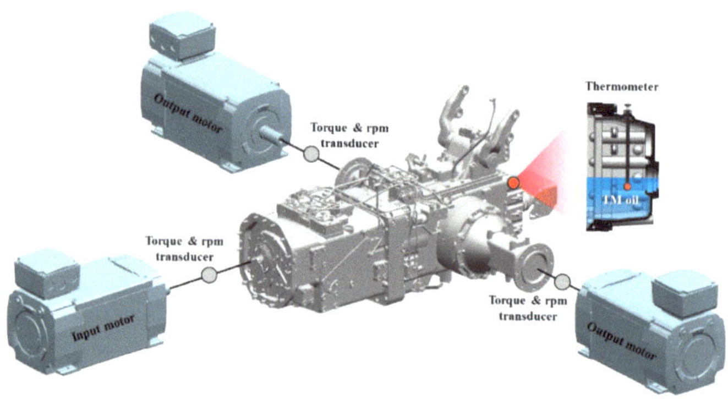

(**a**) layout of the 3-axis dynamometer system

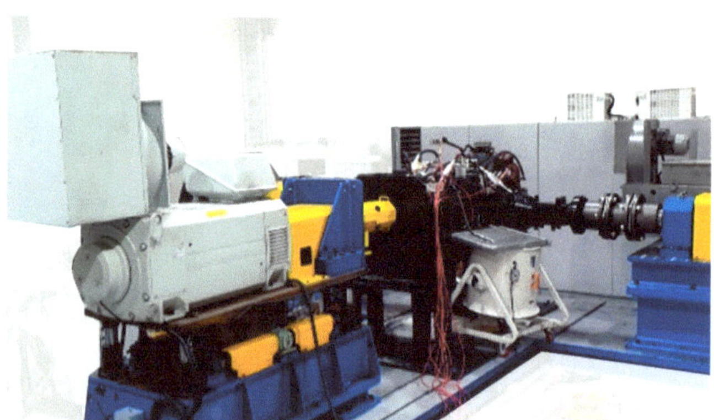

(**b**) photograph of the 3-axis dynamometer system

Figure 2. Power loss measurement system.

Table 2. Specifications of sensor and motor applied to 3-axis dynamometer.

Component	Specifications
Input motor	Rated voltage: 360 V, rated speed: 3090 rpm, rated power: 282 kW
Output motor	Rated voltage: 360 V, rated speed: 1800 rpm, rated power: 246 kW
Torque and rpm sensor	Nominal torque: 30 kNm, nominal rotational speed: 4000 rpm Magnetic rotational speed measuring system: 1024 pulses/revolution

For the test, the input rotation speed was set to the engine's rated rotation speed of 2200 rpm across all gear steps. The torque level applied to the transmission was determined separately for the two sections. In the max. traction force range, traction force was determined using the vehicle weight, with the engine's output not fully utilized. In the rated engine power range, the engine's rated output could be used, with the output torque of the rear axle decreasing as the vehicle speed increased. Figure 3 shows the test conditions, applying the wheel torque, vehicle speed, and input power at the rated rotation speed of the engine for the tractor used in this study. Tests were conducted in gear steps 2nd–24th, with the load conditions for each point shown in Figure 3.

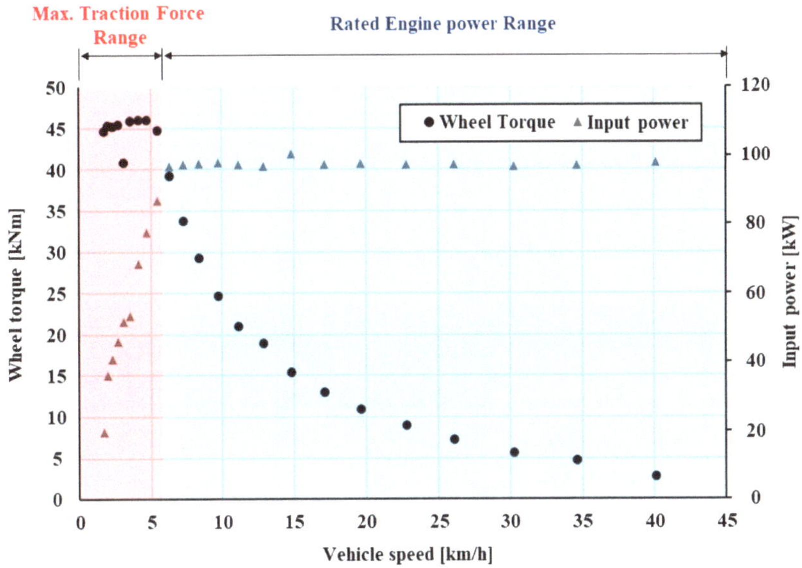

Figure 3. Input load conditions for experimental studies.

3. Results

3.1. Theoretical Power Loss Calculations

Figure 4 shows the theoretical calculation results of the power loss based on the equations presented in Section 2.1. The horizontal axis represents the vehicle speed at the engine's rated rotation speed for each gear step. For Case I and Case II, the total power loss can be calculated by adding the power loss of the DCT components, with power losses of less than 1 kW being combined and represented as a single item.

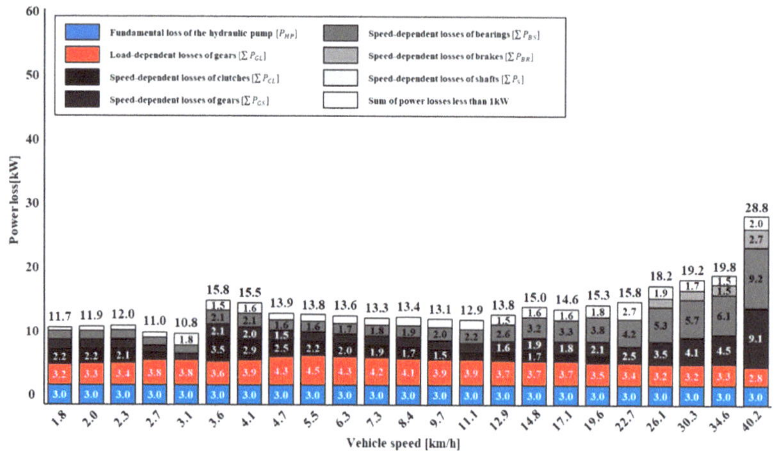

(**a**) Case I: Considering average oil level

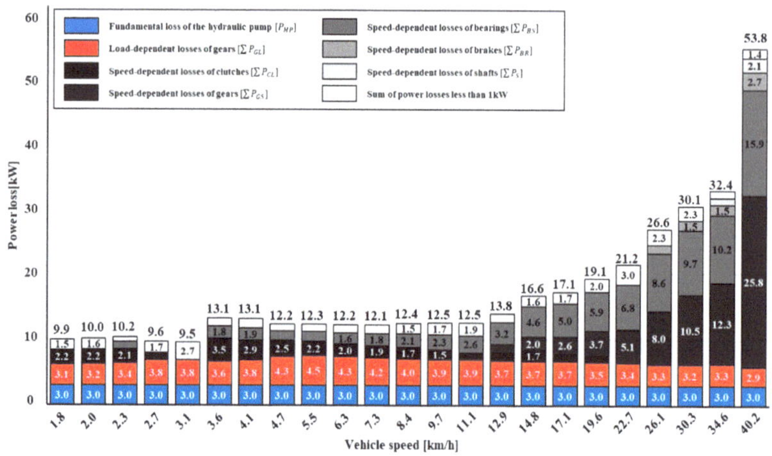

(**b**) Case II: Considering oil circulation

Figure 4. Theoretical calculation results of the power losses.

The components comprising the largest portion of overall power loss are the gears and bearings. Most of the load-dependent power loss occurs owing to the gears, the largest power loss occurring at 5.5 km/h for Cases I and II of 4.51 and 4.50 kW, respectively. As the vehicle speed increases at gear steps above 12 km/h, the speed-dependent power loss of the gears and bearings increases proportionally. Furthermore, the speed-dependent power loss of the gears and bearings is the highest at the top gear step of 40.2 km/h. In Case I, the speed-dependent power loss of the gears and bearings is 9.1 and 9.2 kW, respectively, while in Case II, it is 25.8 and 15.9 kW, respectively. The highest power loss occurs at the top speed of 40.2 km/h for both Cases I and II, with total power losses of 28.8 and 53.8 kW, respectively. This is because the speed-dependent power loss of each component is calculated to be higher in high-speed gear steps for Case II compared to Case I.

Figure 5 shows the power transmission efficiency calculated for Cases I and II. The power transmission efficiency for Cases I and II tends to increase as the vehicle speed increases in the max. traction force range (0–5.5 km/h) owing to the contribution of fundamental and load-dependent power losses relative to the total power. In the rated engine power range (5.5–40.2 km/h), the efficiency is maintained up to a speed of 11.1 km/h, but drops sharply as the speed increases in high-speed gear steps above 12.9 km/h. The maximum efficiency is 86.6% at 9.7 km/h for Case I and 87.6% at 7.3 km/h for Case II, while the minimum efficiency is calculated to be 70.4% for Case I and 44.7% for Case II at 40.2 km/h.

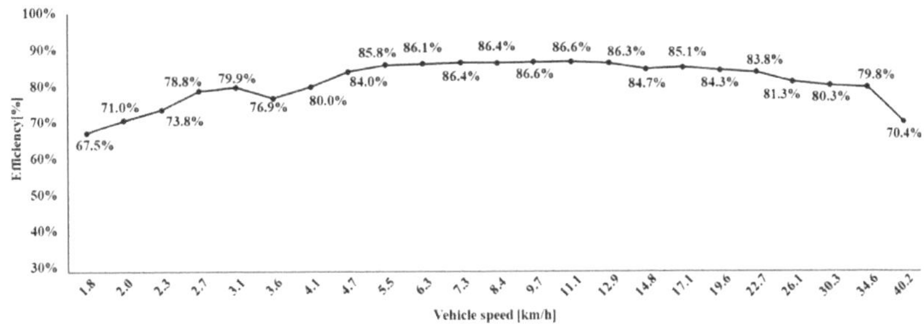

(**a**) Case I: Considering average oil level

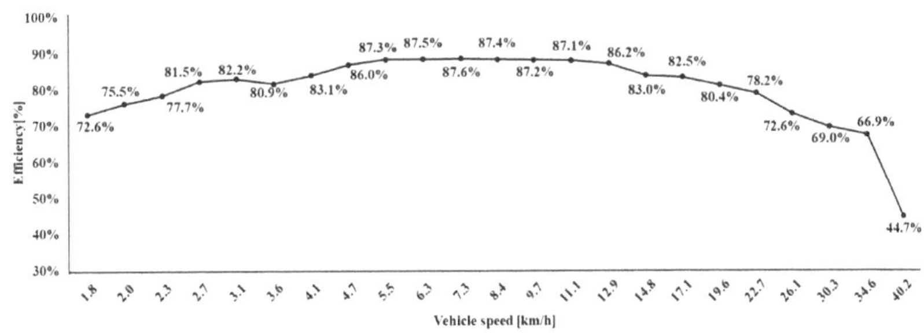

(**b**) Case II: Considering oil circulation

Figure 5. Power transmission efficiencies obtained by theoretical calculation.

3.2. Experimental Studies of Power Loss

Figure 6 shows the measurement results of power loss for each gear step of the DCT obtained through lab tests.

As the vehicle speed increases, the power loss of the transmission tends to increase. The power loss is lowest at 1.8 km/h (10.16 kW), and highest at 40.2 km/h (60.06 kW). Conversely, at lower gear steps with low vehicle speeds, the efficiency increases as the speed increases, it being highest at 5.5 km/h (83.2%). At gear steps with vehicle speeds above 5.5 km/h, the efficiency tends to decrease as the speed increases, the efficiency being the lowest at 40.2 km/h (37.2%).

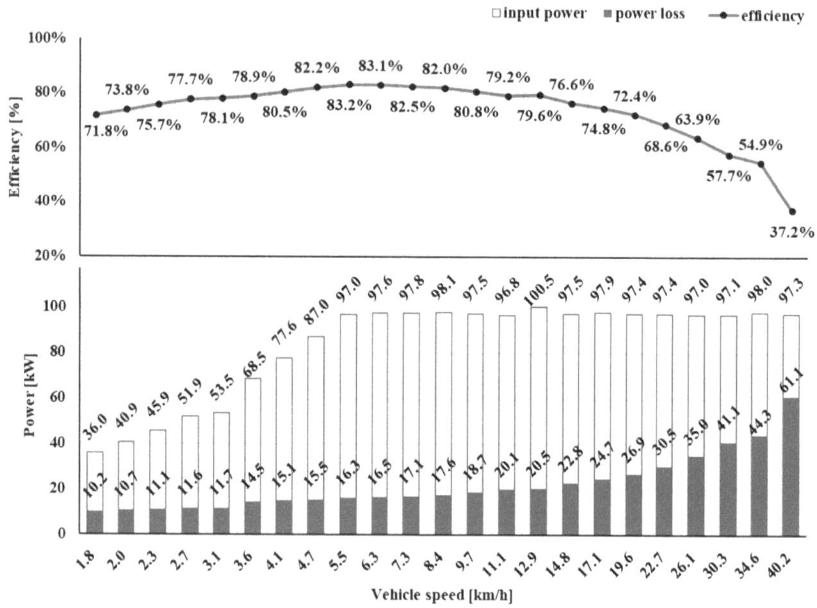

Figure 6. Power loss measurement results of the DCT.

4. Discussion

Comparison between Theoretical Calculations and Experimental Studies

Figure 7 compares the theoretical calculations and experimental studies of the power losses for Cases I and II. The tendency for power loss to increase as vehicle speed increased is the same for both Cases I and II, but the absolute level of power loss differs more as the speed increases.

As shown in Figure 7a, in Case I the theoretical calculations tend to be higher than the test results. However, as the speed increases in the high-speed gear steps, the measured power loss increases; at 40.2 km/h, the measured power loss is 60.06 kW, while the calculated total power loss is 28.80 kW, the error being as much as 31.26 kW.

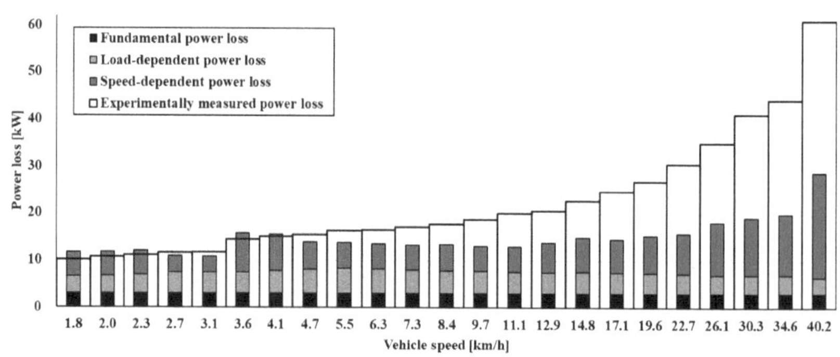

(a) Case I: Considering average oil level

Figure 7. *Cont.*

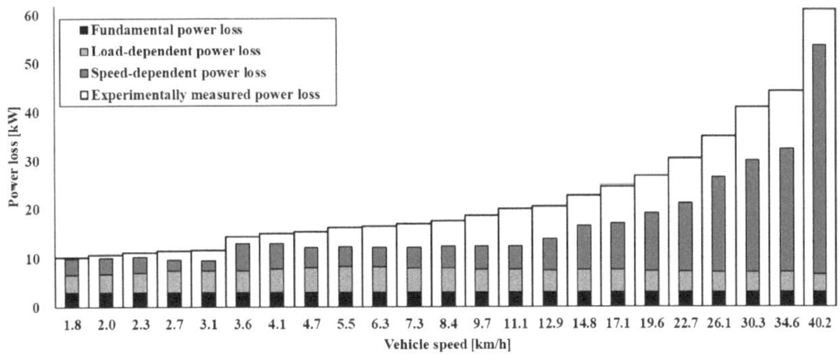

(**b**) Case II: Considering oil circulation

Figure 7. Comparison between calculations and measurements of the power loss.

As shown in Figure 7b, in Case II the load-dependent power loss is similar to that of Case I, where the speed-dependent power loss increases rapidly with the vehicle speed. The maximum power loss is 47.21 kW at 40.2 km/h, and the total power loss is 53.79 kW, exhibiting an error of 12.85 kW compared to the experimentally measured power loss of 60.06 kW. That is, the power loss error in high-speed gear steps is smaller than that of Case I.

To analyze the effects of the oil level on load-dependent and speed-dependent power losses, the theoretical calculation results for the power loss in Cases I and II were compared, as shown in Figure 8.

Figure 8a shows the comparison of load-dependent power losses for Cases I and II, which are similar regardless of the oil levels. The load-dependent power loss is the highest at 5.5 km/h, corresponding to the fastest gear step in the max. traction force range for agricultural tractors.

Figure 8b compares the speed-dependent power losses for Cases I and II. For both cases, the speed-dependent power loss tends to increase with vehicle speed. However, a local peak value is evident at 3.6 km/h owing to the considerable influence of temperature on the speed-dependent power loss. In the tests of 3.1 km/h and 3.6 km/h, the oil temperatures were 69.3 °C and 50.7 °C, respectively, showing a difference of 18.6 °C. As a result of calculating viscosity by the ASTM D341 method, the kinematic viscosity of the oil is 33.2 cSt and 63.61 cSt, respectively, for the tests of 3.1 km/h and 3.6 km/h. As the viscosity increases, it can be seen that the drag torque increases and results in a larger speed-dependent power loss.

According to the results shown in Figure 8b, at vehicle speeds below approximately 11.1 km/h, the power loss in Case I is greater; conversely, at speeds above 14.8 km/h, the power loss in Case II is greater. The differences in power loss between Cases I and II in each speed range are related to the structural features of the DCT used in the study.

The DCT used in the study comprises a gear train made up of 58 gears to implement 24 gear shift stages. The gears in the forward/reverse and main shift parts—including the dual clutches—are in the front cell, while gears in the range shift part are in the mid cell. The vehicle speed range for the primary work of agricultural tractors is defined by the range shift part, the transmission used in this study being divided into 0–4.1 km/h (low-speed range), 4.7–12.9 km/h (mid-speed range), and 14.8–40.2 km/h (high-speed range) depending on the applicable gear ratio. Consequently, the gears in the range shift part have larger gear ratio differences than other gears, the speed deviation of idle gears that do not transmit power also being greater.

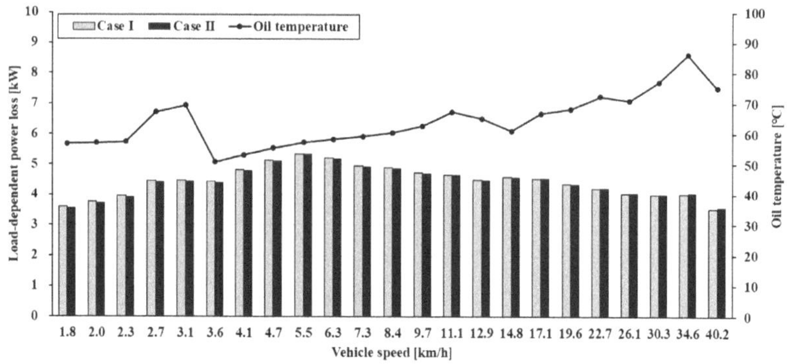

(**a**) Simulated load-dependent power loss.

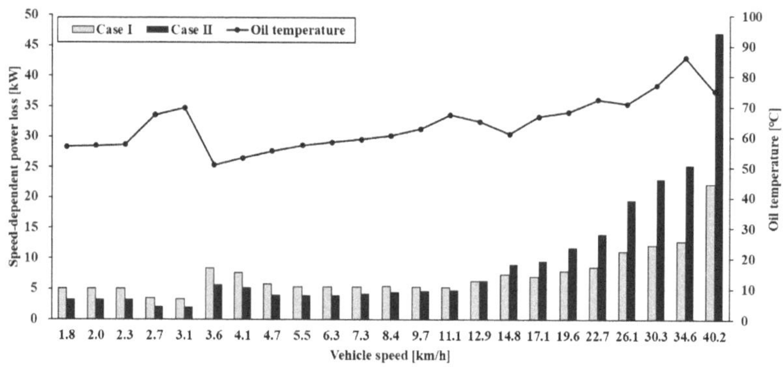

(**b**) Simulated speed-dependent power loss.

Figure 8. Power loss analysis results.

Figure 9 shows the average rotation speed and distribution of gears in the front and mid cells at the engine's rated rotation speed. The gears in the front cell—which are closer to the engine—determine the forward/reverse direction and perform the role of primary reduction in the engine's rotation speed. Consequently, their average rotation speed and deviation are similar. Therefore, the speed-dependent power loss of the gears in the front cell is not greatly affected by the vehicle speed.

By contrast, the rotation speed of the gears in the mid cell is similar to or lower than that of the gears in the front cell in the 1.8–12.9 km/h vehicle-speed range, the average rotation speed and deviation increasing sharply at speeds above 14.8 km/h compared to the gears in the front cell.

The rotation speed of the gears in the mid cell is more greatly influenced by driving speed than those in the front cell. Therefore, the speed-dependent power loss of the gears in the mid cell is affected by the driving speed, showing a similar tendency to the results shown in Figure 7b, where the speed-dependent power loss increases sharply at speeds above 14.8 km/h. Consequently, speed-dependent power loss has a greater impact on the components located in the front cell at low-speed gear steps and the mid cell at high-speed gear steps.

As in Case II of Figure 7b, when the oil level in the front cell is lowered and the oil level in the mid cell is raised, the speed-dependent power loss in the front cell decreases, reducing the total power loss at low speeds. Owing to the increased influence of speed-dependent power loss in the mid cell, the total power loss at high speeds increases. This trend is consistent with the test results, where power loss decreases at low speeds and increases sharply at high speeds.

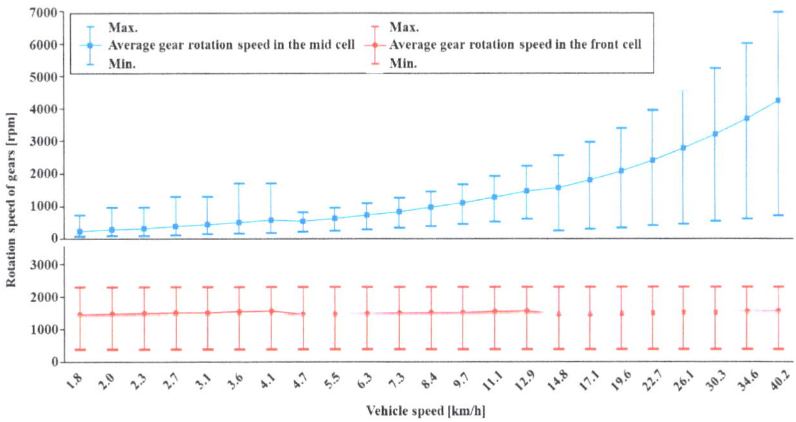

Figure 9. Rotation speed variation of each gear step.

Through this study, it was confirmed that the error between the theoretical calculation and the measured value of power loss was reduced by considering the oil level change caused by the oil circulation of the transmission. Therefore, in order to improve the efficiency of the transmission, the rotation speed of the internal components and the oil lubrication condition must be considered together. For example, gears with high rotation speeds should be located in positions that are not submerged in oil. Rotating components that are inevitably immersed in oil need to be designed with a small outer diameter and width to minimize power loss.

5. Conclusions

In this study, we calculated the power loss of an agricultural tractor DCT using the ISO standard and viscous fluid theory and measured the power loss of the DCT for each gear step using a 3-axis dynamometer measurement system. The calculated power loss and measurement results were compared and analyzed with the calculation results reflecting the oil levels considering the characteristics of the agricultural tractor DCT reviewed. The results of this study can be summarized as follows:

1. We calculated the power loss for each gear shift stage of a DCT applied to agricultural tractors using the ISO standard and the viscous fluid theory. The components accounting for the largest portion of the total power loss were the gears and bearings, with the efficiency calculated to be highest at 5.5 km/h, the fastest speed in the max. traction force range.
2. We confirmed that the theoretical calculation results of power loss in Case II were more consistent with the test results compared to Case I. Additionally, while the load-dependent power loss was not significantly affected by the oil level, the speed-dependent power loss was affected by the locally varying oil level owing to oil circulation.
3. The transmission components closer to the engine exhibited smaller speed deviations for each gear step, while the gears of the transmission components closer to the output shaft exhibited greater rotation speed and speed deviation between gears at higher gear steps. Owing to these driving characteristics, when the oil level in the cell closer to the output shaft rose, the power loss was lower at low-speed gear shift stages, while it increased sharply at high-speed gear steps.

4. Thus, we confirmed that, to predict the efficiency of agricultural transmissions, the oil level reflecting the transmission structure characteristics, oil circulation, and the operating characteristics of the transmission component—such as rotation speed and deviation—must be considered.

Author Contributions: Conceptualization, H.-J.A. and C.C.; methodology, H.-J.A. and S.-C.K.; investigation, Y.-J.P.; writing—original draft preparation, H.-J.A. and S.-C.K.; writing—review and editing, C.C. All authors have read and agreed to the published version of the manuscript.

Funding: This work was supported by the Technology Innovation Program (Development of common type dual-clutch transmission system and standardized control platform, No. 20011428), funded by the Ministry of Trade, Industry & Energy, Republic of Korea.

Institutional Review Board Statement: Not applicable.

Data Availability Statement: Not applicable.

Conflicts of Interest: The authors declare no conflict of interest.

Nomenclature

P_L	Total power loss, kW
P_{HP}	Fundamental power loss of the hydraulic pump, kW
P_{GL}	Load-dependent power loss of the gear, kW
P_{GS}	Speed-dependent power loss of the gear, kW
P_{BL}	Load-dependent power loss of the bearing, kW
P_{BS}	Speed-dependent power loss of the bearing, kW
P_S	Speed-dependent power loss of the shaft, kW
P_{CL}	Speed-dependent power loss of the clutch, kW
P_{BR}	Speed-dependent power loss of the brake, kW
f_m	Mesh coefficient of friction
M	Mesh mechanical advantage
T_1	Pinion torque, Nm
n_1	Pinion rotation speed, rpm
β	Operating helix angle/mean spiral angle, degrees
P_{GW1}	Gear windage and churning losses associated with smooth outside diameters, kW
P_{GW2}	Gear windage and churning losses associated with smooth sides of the disc, kW
P_{GW3}	Gear windage and churning losses associated with tooth surfaces, kW
f_g	Gear dip factor
v	Kinematic oil viscosity, m^2/s
n	Rotating speed, rpm
L	Length of element for gearing windage and churning, mm
A_g	Arrangement constant
F	Total face width, mm
R_f	Roughness factor for gear teeth
M_0	Load-independent frictional moment, Nm
M_1	Bearing load-dependent torque, Nm
M_2	Cylindrical roller bearing axial load-dependent moment, Nm
M_3	Frictional moment of seals, Nm
T_{CL}	Clutch torque loss, Nm
r_{ag}	Percentage value of the groove area to the friction material area
T_{fa}	Torque loss due to oil film in continuous section, Nm
T_{ra}	Torque loss due to oil film in ruptured section, Nm
T_m	Torque loss due to mist in ruptured section, Nm
r_c	Critical radius, m
r_i	Inner radius of the disk, m
r_o	Outer radius of the disk, m
N	Number of friction surfaces
μ	Fluid absolute viscosity, Pa·s

μ_{mis}	Absolute viscosity of mist, Pa·s
$\Delta\omega$	Difference in clutch rotation speed, rad/s
h	Clearance between plate and disc, m
Φ	Critical radius square, m^2
α	Angle of the area of brake caliper, rad

References

1. Ueka, Y.; Yamashita, J.; Sato, K.; Doi, Y. Study on the development of the electric tractor: Specifications and traveling and tilling performance of a prototype electric tractor. *Eng. Agric. Environ. Food* **2013**, *6*, 160–164. [CrossRef]
2. Vogt, H.H.; de Melo, R.R.; Daher, S.; Schmuelling, B.; Antunes, F.L.M.; dos Santos, P.A.; Albiero, D. Electric tractor system for family farming: Increased autonomy and economic feasibility for an energy transition. *J. Energy Storage* **2021**, *40*, 102744. [CrossRef]
3. Moreda, G.P.; Muñoz-García, M.A.; Barreiro, P.J. High voltage electrification of tractor and agricultural machinery—A review. *Energy Convers. Manag.* **2016**, *115*, 117–131. [CrossRef]
4. Martini, V.; Mocera, F.; Somà, A. Numerical Investigation of a Fuel Cell-Powered Agricultural Tractor. *Energies* **2022**, *15*, 8818. [CrossRef]
5. Yang, H.; Sun, Y.; Xia, C.; Zhang, H. Research on Energy Management Strategy of Fuel Cell Electric Tractor Based on Multi-Algorithm Fusion and Optimization. *Energies* **2022**, *15*, 6389. [CrossRef]
6. Hosseinzadeh-bandbafha, H.; Rafiee, S.; Ghobadian, B.; Mohammadi, P. Exergetic, economic, and environmental life cycle assessment analyses of a heavy-duty tractor diesel engine fueled with diesel–biodiesel-bioethanol blends. *Energy Convers. Manag.* **2021**, *241*, 114300. [CrossRef]
7. Madovi, O.; Hoffrichter, A.; Little, N.; Foster, S.N.; Isaac, R. Feasibility of hydrogen fuel cell technology for railway intercity services: A case study for the Piedmont in North Carolina. *Railw. Eng. Sci.* **2021**, *29*, 258–270. [CrossRef]
8. Bergthorson, J.M. Recyclable metal fuels for clean and compact zero-carbon power. *Prog. Energy Combust. Sci.* **2018**, *68*, 169–196. [CrossRef]
9. Kim, W.-S.; Kim, Y.-J.; Kim, Y.-S.; Baek, S.-M.; Baek, S.-Y.; Lee, D.-H.; Nam, K.-C.; Kim, T.-B.; Lee, H.-J. Development of Control System for Automated Manual Transmission of 45-kW Agricultural Tractor. *Appl. Sci.* **2020**, *10*, 2930. [CrossRef]
10. Kim, J.; Lee, H. Motor position control algorithm for an automated manual transmission of the agricultural tractor. *Proc. Inst. Mech. Eng. Part C J. Mech. Eng. Sci.* **2015**, *229*, 3341–3349. [CrossRef]
11. Brenninger, M.M. Fendt Vario CVT in Agricultural Tractors. *SAE Tech. Pap.* 2007. [CrossRef]
12. Tanelli, M.; Panzani, G.; Savaresi, S.M.; Pirola, C. Transmission control for power-shift agricultural tractors: Design and end-of-line automatic tuning. *Mechatronics* **2011**, *21*, 285–297. [CrossRef]
13. Pahlovy, S.A.; Mahmud, S.F.; Kubota, M.; Ogawa, M.; Takakura, N. Prediction of Drag Torque in a Disengaged Wet Clutch of Automatic Transmission by Analytical Modeling. *Tribol. Online* **2016**, *11*, 121–129. [CrossRef]
14. Yuan, Y.; Liu, E.A.; Hill, J.; Zou, Q. An Improved Hydrodynamic Model for Open Wet Transmission Clutches. *J. Fluids Eng.* **2007**, *129*, 333–337. [CrossRef]
15. Shen, Y.; Liu, Z.; Rinderknecht, S. Modelling of Power Losses of Transmission Synchronizers in Neutral Position. *SAE Tech. Pap.* 2018. [CrossRef]
16. Fernandes, C.M.; Marques, P.M.; Martins, R.C.; Seabra, J.H. Gearbox power loss. Part I: Losses in rolling bearings. *Tribol. Int.* **2015**, *88*, 298–308. [CrossRef]
17. Seetharaman, S.; Kahraman, A.; Moorhead, M.D.; Petry-Johnson, T.T. Oil Churning Power Losses of a Gear Pair: Experiments and Model Validation. *J. Tribol.* **2009**, *131*, 022202. [CrossRef]
18. Hu, X.; Jiang, Y.; Luo, C.; Feng, L.; Dai, Y. Churning power losses of a gearbox with spiral bevel geared transmission. *Tribol. Int.* **2019**, *129*, 398–406. [CrossRef]
19. Diab, Y.; Ville, F.; Velex, P. Prediction of Power Losses Due to Tooth Friction in Gears. *Tribol. Trans.* **2006**, *49*, 260–270. [CrossRef]
20. Fernandes, C.M.; Marques, P.M.; Martins, R.C.; Seabra, J.H. Gearbox power loss. Part II: Friction losses in gears. *Tribol. Int.* **2015**, *88*, 309–316. [CrossRef]
21. Vacca, F.; De Pinto, S.; Karci, A.E.H.; Gruber, P.; Viotto, F.; Cavallino, C.; Rossi, J.; Sorniotti, A. On the Energy Efficiency of Dual Clutch Transmissions and Automated Manual Transmissions. *Energies* **2017**, *10*, 1562. [CrossRef]
22. Habermehl, C.; Jacobs, G.; Neumann, S. A modeling method for gear transmission efficiency in transient operating conditions. *Mech. Mach. Theory* **2020**, 153. [CrossRef]
23. Strapasson, L.; Zimmermann, G.G.; Jasper, S.P.; Savi, D.; Sobenko, L.R. Energy efficiency of agricultural tractors equipped with continuously variable and full powershift transmission systems. *Eng. Agrícola* **2022**, *42*. [CrossRef]
24. Molari, G.; Sedoni, E. Experimental evaluation of power losses in a power-shift agricultural tractor transmission. *Biosyst. Eng.* **2008**, *100*, 177–183. [CrossRef]
25. Moon, S.P.; Moon, S.G.; Kim, J.S.; Sohn, J.H.; Kim, Y.J.; Kim, S.C. Transmission efficiency of dual-clutch transmission in agricultural tractors. *J. Drive Control* **2022**, *19*, 43–50.

26. *ISO/TR 14179-1*; Gears-Thermal Capacity-Part 1: Rating Gear Drives with Thermal Equilibrium at 95 °C Sump Temperature. International Standard: Geneva, Switzerland, 2001.
27. Park, K.; Kang, M.; Lee, J.; Son, W.C.; Harianto, J.; Kahraman, A. Development of an Analysis Program to Predict Efficiency of Automotive Power Transmission and Its Applications. *SAE Tech. Pap.* 2018. [CrossRef]

agriculture

Article

Bucket Size Optimization for Metering Device in Garlic Planter Using Discrete Element Method

Dongu Im [1], Ho-Seop Lee [2], Jae-Hyun Kim [1], Dong-Joo Moon [1], Tae-Ick Moon [3], Seung-Hwa Yu [4] and Young-Jun Park [1,2,5,*]

1 Department of Biosystems Engineering, College of Agriculture and Life Sciences, Seoul National University, 1 Gwanak-ro, Gwanak-gu, Seoul 08826, Republic of Korea; ehddn0112@snu.ac.kr (D.I.); kjh3tt@snu.ac.kr (J.-H.K.); 98dongdong@snu.ac.kr (D.-J.M.)

2 Research Institute of Agriculture and Life Sciences, College of Agriculture and Life Sciences, Seoul National University, 1 Gwanak-ro, Gwanak-gu, Seoul 08826, Republic of Korea; ghtjqdlek@snu.ac.kr

3 Taeung Farm Machinery, Hanseong-ro, Isan-myeon, Yeongju-si 36170, Republic of Korea; 01075010751@hanmail.net

4 Department of Agricultural Engineering, Upland Mechanization Team, National Institute of Agricultural Sciences, Nongsaengmyeong-ro, Deokjin-gu, Jeonju-si 54875, Republic of Korea; yush6210@korea.kr

5 Convergence Major in Global Smart Farm, College of Agriculture and Life Sciences, Seoul National University, 1 Gwanak-ro, Gwanak-gu, Seoul 08826, Republic of Korea

* Correspondence: yjpark95@snu.ac.kr

Abstract: In this study, the discrete element method was used to optimize the bucket size for the metering device in a garlic planter for enhancing the productivity of garlic farming according to the garlic size. Statistical information concerning the actual shape of garlic cloves was incorporated, and the mechanical properties of garlic were determined using the bulk density, sliding test, and repose angle test for enhancing the fidelity of the simulation model. The optimal bucket size achieving the target plant rate of 97.5% was determined using the developed discrete element model for the three garlic size groups. The linear search method was used for optimization, and batch simulation was performed to validate the optimized results and confirm the performance index of the metering device. A Gaussian distribution based on statistical information accounted for the various garlic sizes in each group. Finally, a metering test verified the reliability of the optimization technique. The differences between the simulation and test results were within 10% for all performance indices, including missing plant rate, multi-plant rate, and planting rate, indicating the high reliability of the analysis model. Subsequently, the larger garlic groups (Groups 2 and 3) exhibited metering performance close to the target plant rate.

Keywords: garlic planter; metering device; size optimization; discrete element method

Citation: Im, D.; Lee, H.-S.; Kim, J.-H.; Moon, D.-J.; Moon, T.-I.; Yu, S.-H.; Park, Y.-J. Bucket Size Optimization for Metering Device in Garlic Planter Using Discrete Element Method. *Agriculture* **2023**, *13*, 1199. https://doi.org/10.3390/agriculture13061199

Academic Editor: Mustafa Ucgul

Received: 15 May 2023
Revised: 1 June 2023
Accepted: 2 June 2023
Published: 5 June 2023

1. Introduction

The United Nations has identified zero hunger as one of the sustainable development goals to be achieved by 2030. Consequently, global interest in enhancing agricultural production rates has been growing [1]. However, in countries such as South Korea, the aging population and decreasing rural population pose challenges to these goals. One potential solution is the automation and mechanization of agriculture, which is particularly effective for labor-intensive crops. Garlic, a typical such crop, is a case in point; garlic planting is particularly strenuous and accounts for 15% of the total labor in garlic farming [2]. The mechanization of garlic planting could alleviate the manpower shortage, enhance the planting rate per hour, and increase overall production.

Garlic planters can be classified based on their power source and divided into self-propelled and tractor-attached types. They can also be categorized based on the posture of the garlic cloves to be planted, including upright planting, in which the root of the garlic cloves faces downwards, and lie-down planting, in which the garlic cloves are placed horizontally on the ground using free fall [3]. One planter component, the metering device, transfers garlic cloves from the hopper to the planting unit, which then plants them in the field based on the planting method. This is followed by soil covering and mulching. The metering device is a crucial component of the garlic planter. It plays a significant role in controlling the missing plant and multi-plant rates, key performance indicators of the garlic planter [4]. The missing plant rate refers to the frequency at which garlic cloves fail to be transferred to the planting unit, while the multi-plant rate indicates the frequency at which two or more garlic cloves are transferred to the planting unit simultaneously. Both performance indicators can cause significant losses for farmers and must be minimized. A high missing plant rate can result in a reduction in total production and/or additional labor required to replant in the missed area, while a high multi-plant rate leads to unnecessary waste of garlic cloves and can result in interference of the growth between several garlic cloves planted in a limited space, leading to reduced total production.

To reduce the rates of missing plants and multiple plants, an optimal design that considers the performance index of the metering device is necessary. It is imperative to anticipate the missing plant and multi-plant rates of the metering device during the development stage and supplement the design accordingly based on the results to achieve the optimal design of the metering device. Performance indicators of metering devices can be predicted through experimental methods using test devices and simulation methods. While performance prediction through experimentation offers the advantage of high accuracy when performed under the same conditions as the actual operation, it requires a significant amount of time and resources to build a test device, thereby limiting the production of samples that cover a wide range of design variables. Conversely, simulation-based methods offer high accuracy when using validated analysis models and, as there is no limit on the design range, can predict the performance of the metering device across a wide range of design parameters. Therefore, an analytical method that employs simulation is appropriate for optimal design beyond simple performance comparison.

To analyze the performance of garlic planters, various studies have been conducted utilizing both experimental and simulation methods. Kang et al. [5] conducted a yield comparison experiment as a testing method to determine the effects of planting angle and planting depth on garlic growth. The results confirmed that garlic yield was highest at planting angles of 45° or 90° and planting depths of 2–4 cm. Additionally, to improve the performance of the metering device, Kang et al. [3] experimentally determined the missing plant rate based on the bucket size. The results indicated that the bucket size of the metering device significantly affects the performance index. Zilpilwar et al. [6] determined the optimal cup size and operating speed of a cup-feed-type metering device through an experimental method similar to that in the aforementioned study. The authors considered the planting rate, planting interval, and clove damage rate. Zhang et al. [7] conducted one-factor and multifactor experiments to optimally design key components of a spoon-clamping type planter and optimize working parameters. As mentioned above, experimental studies offer high prediction accuracy using actual metering devices. However, these studies have limited applicability of predictive models depending on the sample range and expense of test equipment. Additionally, due to the restricted selection range of design variables, the scope for optimal design is limited.

In addressing the limitations of the experimental method, studies have been conducted using simulations to predict the performance of garlic planters. Im et al. [8] classified garlic samples into four groups based on size and developed a 3D computer-aided design (CAD) model using statistical results on garlic clove shape. The optimal bucket size of the metering device was then determined using the developed CAD model and multi-body dynamics model according to the size of the garlic cloves. Although the fidelity of the model was increased by incorporating the shape of actual garlic cloves in the analysis, validation of the determined optimal bucket size and simulation model was not performed. Guo et al. [9] analyzed the single seed rate, multiple seed rate, and missing seed rate of the metering device according to the adjustment size and operation speed using the discrete element method (DEM). This resulted in a single-seed metering performance qualification rate of over 80%. An actual field test was performed to validate the applicability of the optimal design, and the performance of the metering device met the required standards. While the authors derived the optimal design value using DEM, they did not consider the various sizes of garlic in the analysis model nor incorporate the actual shape of garlic cloves.

To minimize the missing seed rate of a finger clip plate garlic seed-metering device, Wang & Sun et al. [4] utilized the commercial software EDEM as a further investigation that executed optimal design using the DEM. The study scrutinized the missing seed and single seed rates based on the opening diameter of the device, operating speed, and the population number of garlic cloves. Moreover, a regression model was constructed via a Box–Behnken center combination test. However, the analysis model did not consider the actual shape of the target garlic, which was the garlic grown in Pizhou, and cloves were depicted as multi-spheres. The low accuracy of the resulting contact force calculation limits the validity of the study, similar to Guo et al. [9].

As another machine for the mechanization of garlic farming, Yu et al. [10] optimized the parameters for the root-cutting operation of the garlic harvester to minimize the root clearance rate and bulb damage. To achieve this optimization, they developed a mathematical model for assessing the root clearance rate and bulb damage.

Research has been conducted to enhance planting performance not only in garlic but also in other crops [11–17]. However, studies that utilized experimental approaches have limitations in design variables or operating conditions and are applicable only within specific sample ranges [3,5–7]. Previous studies that used simulation models executed optimal designs without design variable constraints; however, in some cases, the analysis model was not validated [8], and the actual shape of garlic cloves was not considered in the analysis model, constraining its reliability [10,11].

In this investigation, we aimed to surpass the limitations of prior research by incorporating the actual shape of garlic cloves into the discrete element model to determine the optimal bucket size of the metering device. We optimized by categorizing the garlic cloves into three groups, considering their size distribution. Furthermore, we conducted a metering simulation to validate the optimization results and validated the reliability of the optimization technique by comparing the outcomes with the actual test results. To incorporate the actual shape of garlic cloves, we adopted the statistical data proposed by Im et al. [8] into the discrete element model. We performed angle of repose and sliding tests to identify the mechanical parameters of the analysis model. Using the developed discrete element model, we determined the optimal bucket size to meet the target planting rate under the same operating conditions as the actual planter. We employed the sequential search method as the optimization technique and found the optimal solution by gradually reducing the bucket size. We conducted a metering simulation to validate that the determined bucket size satisfied the required performance index, we conducted a metering simulation and validated the optimization technique through a metering test. As a result of the metering simulation, the performance index of the required level was nearly achieved, and the reliability of the optimization technique was confirmed as the difference between the test result and simulation result in all performance indexes was within 10%.

Thus, in this study, we developed a discrete element model with high fidelity and executed the optimal design of the garlic planter metering device using the developed model. During actual metering operation, the optimization result of each group almost satisfied the target plant rate, and the proposed methodology can be used to optimize metering devices for other crops.

2. Materials and Methods

In this study, the DEM was employed to optimize the bucket size of the metering device of the garlic planter. To accurately represent the actual shape of garlic cloves in the analysis model, a 3D mesh model was created using the 3D CAD model for each of the three garlic size groups, which was previously developed by Im et al. [8]. Subsequently, the discrete element model parameters were determined through bulk density measurement, sliding test, and repose angle test. The optimal bucket size for each group was determined using the developed discrete element model, and a metering simulation was conducted to validate the optimal bucket size. Finally, an actual metering experiment was conducted to validate the optimization technique.

2.1. Metering Device Working Principle

The finger-type bucket metering device utilized in this study comprises the components illustrated in Figure 1. The finger-type bucket is affixed to the chain and derives its power from the sprocket. The lower sprocket is connected to the support wheel of the planter, and as the vehicle advances, it rotates in a clockwise direction owing to the frictional force on the ground. The bucket moves in unison with the rotation of the sprocket, collects the garlic cloves from the hopper in the carrying area, and transports them to the planting area. Throughout this process, the bucket size is determined based on the bucket guide setting, and according to the bucket size, any excess garlic cloves in the sorting area fall back into the hopper. The garlic that passes through the sorting area without falling rotates along the pitch diameter of the sprocket and undergoes a 180° phase shift as the bucket moves. As the phase shifts, gravity causes the garlic to fall onto the back of the finger-type bucket, which is one cycle ahead, and it is then planted through free fall while the planter traverses the planting area.

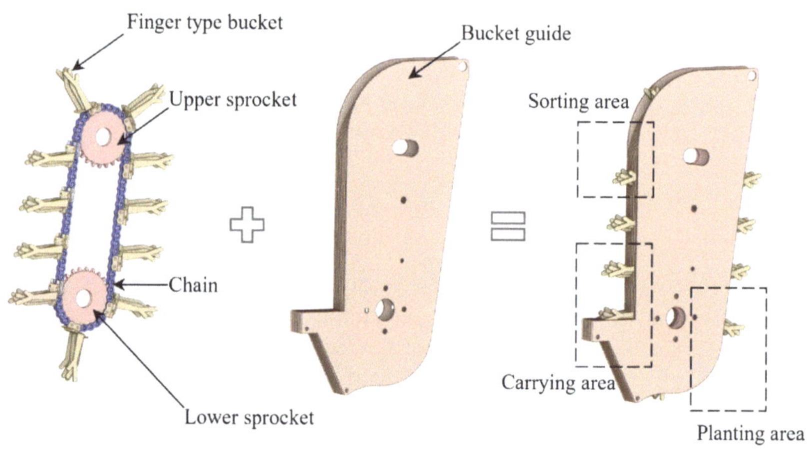

Figure 1. Configuration of metering device.

The bucket size varies based on the setting angle (α) of the bucket guide, as illustrated in Figure 2a, and the bucket guide rotates around the same rotation center as the lower sprocket, as shown in the figure. The position of the chain and sprocket does not change in accordance with the bucket guide settings, and the buckets affixed to the chain always follow the same trajectory. The bucket trajectory can be defined as the combination of the outermost points traced while the bucket moves, as indicated by the dotted line in Figure 2b. As the trajectory of the bucket is consistent, the bucket size (l) changes as the relative position of the guide changes. The bucket size can be defined as the minimum distance between the bucket trajectory and bucket guide, as shown in Figure 2b. When the center of gravity of the placed garlic cloves surpasses the bucket trajectory by a specific distance, the cloves fall under the influence of their weight.

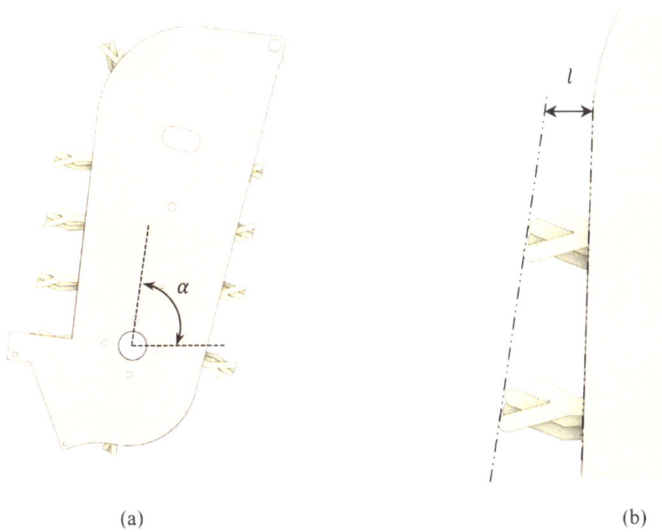

Figure 2. Mechanism of bucket size control in metering device. (**a**) Rotatable bucket guide. (**b**) Bucket size.

2.2. 3D Mesh Modeling

In this investigation, as the target garlic, we used garlic from the cold Uiseong region, identical to the sample used by Im et al. [8], and produced a 3D mesh model of garlic cloves using the 3D CAD model proposed in that study. Figure 3 displays samples of the actual garlic cloves of each group alongside the corresponding 3D CAD and mesh models for contact force calculation. The commercial software ANSYS Rocky can accommodate the actual particle shape in the analysis model by importing the mesh file (in STL format) to define the nodes required for contact force calculation (ANSYS Rocky, [18]). Excessively refined nodes demand a significant amount of computation and the analysis time can increase exponentially. Conversely, excessively coarse nodes can produce numerical errors during calculation. Therefore, selecting the appropriate number of nodes is critical. To ensure the accuracy and convergence of contact calculation, we utilized 3912 nodes for Group 1, 4730 nodes for Group 2, and 4497 nodes for Group 3 in the model, determined through trial-and-error procedures.

The 3D CAD model of garlic cloves used in this study comprises the elongated model, as presented in Figure 4a, and the average model, as shown in Figure 4b. To obtain statistical information about the shape of garlic cloves, Im et al. [8] measured the height and length of 160 garlic samples and developed a 3D CAD model to represent the shape of each group based on the statistical results. Similarly, we used an elongated model with an elongated length $(\mu_L + 2\sigma_L)$ and average height (μ_H) was used to determine the optimal bucket size, as was executed by Im et al. [8]. For the metering simulation, an average model with an average length (μ_L) and average height (μ_H) of the groups was employed. During metering work, mechanical vibration aligns the garlic cloves parallel to the finger-type bucket, and the length of the garlic cloves determines the location of their center of gravity. The length of the garlic clove model was set to $\mu_L + 2\sigma_L$, and the minimum bucket size allowing the clove to pass through the sorting area was determined to achieve the target planting rate of 97.5%. Further details are excluded as they are beyond the scope of this thesis. The average model presented in Figure 4b was utilized for the metering simulation described in Section 2.6. To validate that the optimal bucket size indeed satisfies the plant rate of 97.5%, we conducted a metering simulation. At this juncture, the average model was scaled up and down according to the Gaussian distribution to test the optimal bucket size for various garlic sizes.

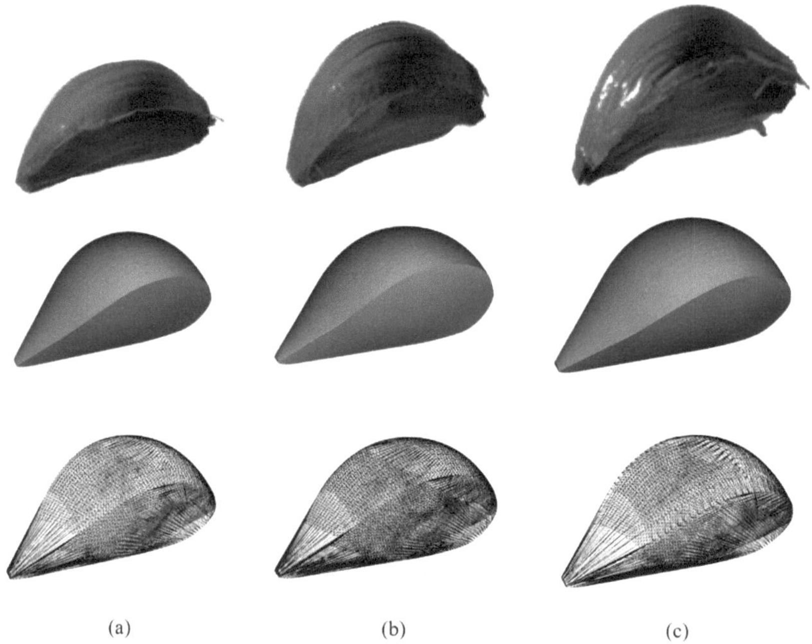

Figure 3. Contact calculation node for garlic clove. (**a**) Group 1. (**b**) Group 2. (**c**) Group 3.

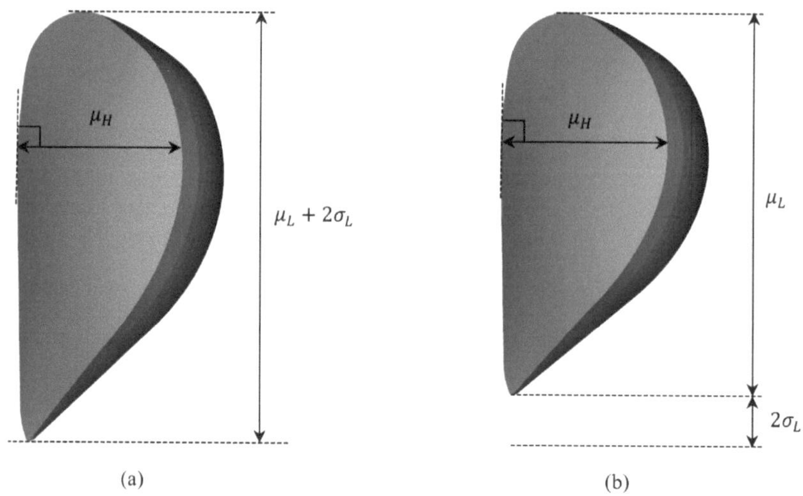

Figure 4. Garlic 3D models. (**a**) Elongated model. (**b**) Average model.

When constructing a discrete element model, it is necessary to define a mesh as a calculation node for the garlic cloves and the mechanical device with which they come into contact during planting. The garlic cloves come into contact with finger-type buckets and bucket guides, as shown in Figures 5 and 6. For this component, 8166 computational nodes were defined, as depicted in Figure 5b. Similarly, Figure 6a illustrates the 3D model of the bucket guide with 37,268 computational nodes, as shown in Figure 6b. The bucket guide consists of several plates, and contact with the garlic occurs in the anterior part, which is enlarged in Figure 6b. Therefore, a denser mesh was constructed in the contact area to efficiently calculate contact forces.

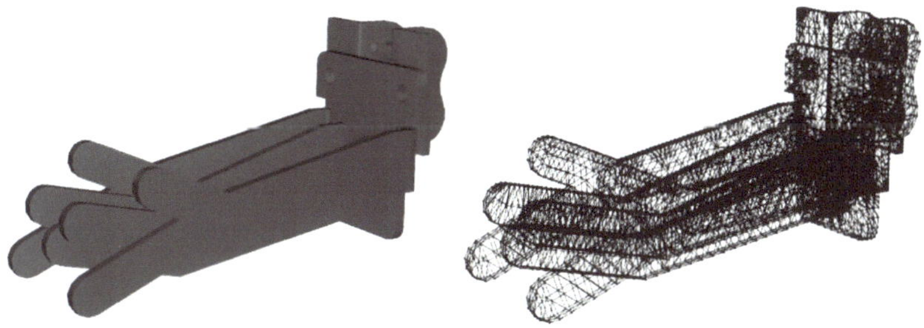

(a) (b)

Figure 5. Contact calculation node for the bucket. (**a**) Finger-type bucket. (**b**) Contact calculation node for bucket.

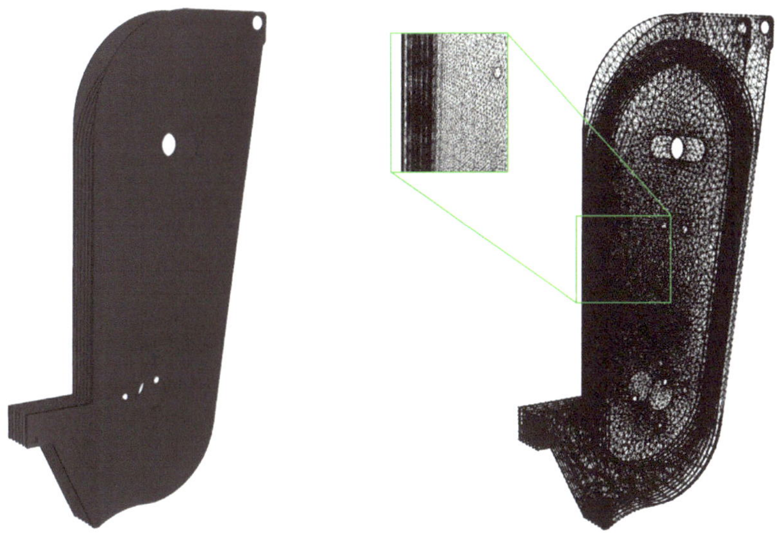

(a) (b)

Figure 6. Contact calculation node for the guide. (**a**) Bucket guide. (**b**) Contact calculation node for guide.

2.3. DEM Parameter Determination

It is essential to define discrete element parameters that precisely reflect the mechanical properties of actual garlic cloves to ensure the high accuracy of the discrete element model. These parameters include properties that reflect the mechanical characteristics of the particle itself, such as bulk density, Young's modulus, and Poisson's ratio, as well as parameters related to interactions with other mechanical devices, such as static friction coefficient and restitution coefficient. The parameters for the interaction between garlic and steel were determined to simulate the mechanical interaction with an actual steel device. We also determined the parameters for the interaction between acrylic and garlic to calibrate the repose angle. For visualization, we used an acrylic cylinder for the repose angle test and for the discrete element model for calibration. Table 1 summarizes the discrete element parameters and methods used to determine these parameters.

Table 1. DEM parameter determination.

Category	Property	Value	Unit	Determination
Garlic	Bulk density	605.45	kg/m^3	Directly measured
	Young's modulus	13.38	MPa	Park et al. [19]
	Poisson's ratio	0.16	-	Park et al. [19]
Garlic-garlic	Static friction coefficient	0.46	-	Repose angle calibration
	Tangential ratio	1	-	Default
	Restitution coefficient	0.50	-	Park et al. [19]
Acryl-garlic	Static friction coefficient	0.4	-	Sliding test
	Tangential ratio	1	-	Default
	Restitution coefficient	0.65	-	Yu et al. [20]
Steel-garlic	Static friction coefficient	0.31	-	Sliding test
	Tangential ratio	1	-	Default
	Restitution coefficient	0.65	-	Yu et al. [20]

The bulk density of garlic was directly measured using an acrylic cylinder, as depicted in Figure 7. In this study, the garlic cloves were used with their peels intact, resembling their natural state for planting. However, this preservation of the peels led to a notable disparity in true density. The presence of a gap between the garlic (white part) and the peel resulted in an overestimation of the clove's volume, consequently leading to imprecise measurements. By determining the bulk density of a specific set of garlic cloves, it becomes feasible to obtain a representative parameter more effortlessly. The values for Young's modulus and Poisson's ratio proposed by Park et al. [19] were employed. The value suggested by Yu et al. [20] for the restitution coefficient was used as an interaction parameter. The static friction coefficients and friction angle between steel–garlic, and acryl–garlic were determined via a sliding test, as shown in Figure 8. The garlic clove was placed on a plate made of the material to be measured, which was then slowly tilted; the angle at which the garlic clove began to move was considered the friction angle (ϕ) and the static friction coefficient was defined as tan (ϕ). Guo et al. [9] applied the same method to determine the static friction coefficient between garlic and steel plate. Finally, since fabricating a garlic plate is infeasible, the garlic–garlic static friction coefficient was determined through repose angle calibration.

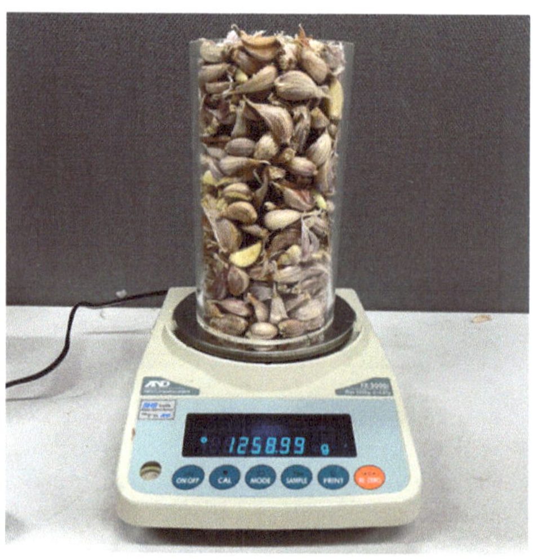

Figure 7. Bulk density measurement.

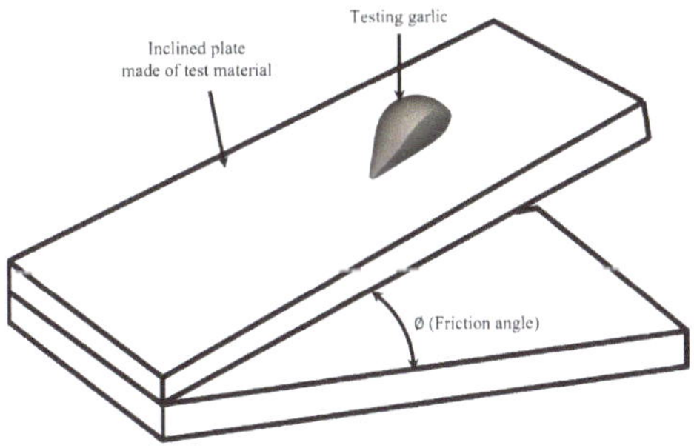

Figure 8. Sliding test for the determination of friction coefficient.

The repose angle calibration refers to the process of adjusting the garlic–garlic friction coefficient of the model such that the same value as the repose angle obtained through the actual test is obtained in the simulation result. For accurate calibration, fixed values should be used for the rest of the parameters except for the friction coefficient [21]. In this study, as shown in Table 1, calibration was performed after fixing the remaining values except for the garlic–garlic static friction coefficient. For the repose angle test, a cylinder lifting test (CLT) was performed, as shown in Figure 9 [22]. A total of eight repose angles were measured at 45° intervals in one test, and the reliability of the test results was improved by repeating the test three times. The measurement of the repose angle involved using a digital protractor attached to a steel plate. The steel plate was adjusted to be parallel to the pile of garlic cloves, and the angle of the steel plate was measured. When using large-sized garlic, the reproducibility of test results may not be secured owing to limited boundary conditions; hence, Group 1, which has the smallest size, was used. In addition, the CLT method can be affected by the lifting velocity according to the friction coefficient between the cylinder and the garlic [23]. Therefore, the dynamic effect was removed by performing CLT at a speed of 5 mm/s [22]. Consequently, an average of 30.53° and a standard deviation of 1.26° were obtained.

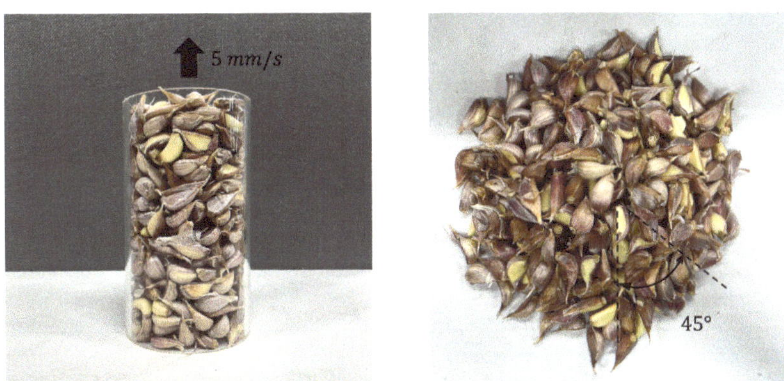

Figure 9. Repose angle test.

Figure 10a shows the actual repose angle test, and Figure 10b illustrates the response angle simulation for calibration purposes. An analysis was conducted by incrementally increasing the garlic–garlic friction coefficient by 0.1 from 0.3 to match the repose angle obtained from the test. Figure 11 shows the simulation results. Using data interpolation, it was confirmed that when the friction coefficient was 0.46, the angle of repose of 30.53° matched the test results.

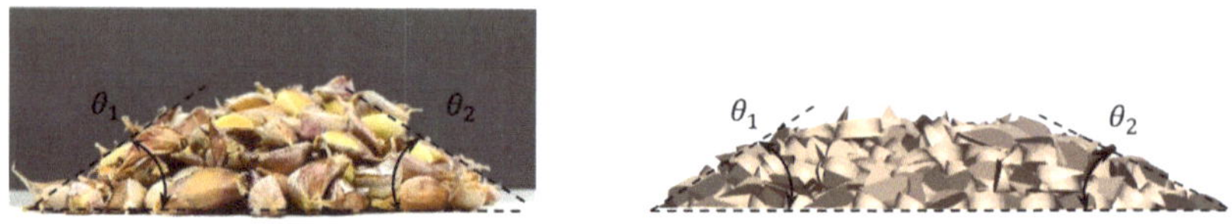

(a) (b)

Figure 10. DEM parameter calibration using repose angle test. (**a**) Repose angle (test). (**b**) Repose angle (simulation).

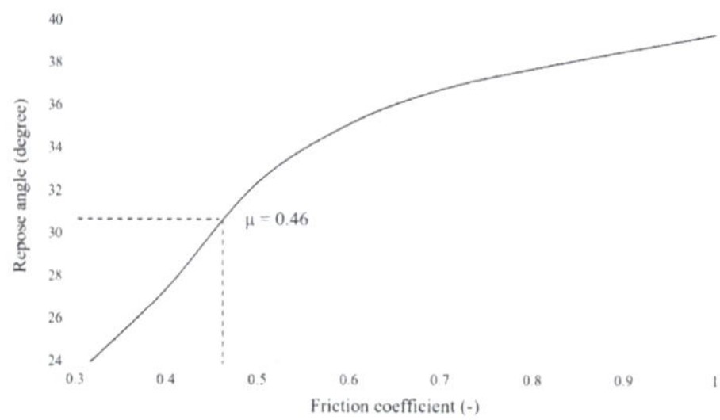

Figure 11. Repose angle according to the friction coefficient of garlic-garlic.

2.4. DEM Contact Model

2.4.1. Normal Force Model: Hysteretic Linear Spring Model

The study employed the hysteretic linear spring model proposed by Walton and Braun [23] to define contact in the normal direction of the discrete element model. This model is characterized by a rapid computation speed of plastic energy dissipation and is suitable as a model for compressible materials, such as garlic (ESSS Rocky [24]). The formula used to calculate the contact force in the normal direction is as follows.

$$F_n^t = \begin{cases} \min\left(K_{nl}s_n^t, F_n^{t-\Delta t} + K_{nu}\Delta s_n\right) \ if \ \Delta s_n \geq 0 \\ \max\left(F_n^{t-\Delta t} + K_{nu}\Delta s_n, \lambda K_{nl}s_n^t\right) \ if \ \Delta s_n < 0' \end{cases} \tag{1}$$

$$\Delta s_n = s_n^t - s_n^{t-\Delta t}, \tag{2}$$

where:

F_n^t is the normal elastic-plastic contact force at the current time t (N)
$F_n^{t-\Delta t}$ is the normal elastic-plastic contact force at the previous time $t - \Delta t$ (N)
Δs_n is the change in the contact normal overlap during the current time (mm)
K_{nl} is the loading contact stiffness (N/mm)
K_{nu} is the unloading contact stiffness (N/mm)
λ is a dimensionless small constant (its value in Rocky is 0.001) (-).

$$\frac{1}{K_{nl}} = \begin{cases} \frac{1}{K_{nl,p_1}} + \frac{1}{K_{nl,p_2}} \ for \ particle - particle \ contact \\ \frac{1}{K_{nl,p}} + \frac{1}{K_{nl,p}} \ for \ particle - boundary \ contact \end{cases} \tag{3}$$

$$K_{nu} = \frac{K_{nl}}{\varepsilon^2} \tag{4}$$

Here:

K_{nl,p_1} is the loading contact stiffness of particle one (N/mm)
K_{nl,p_2} is the loading contact stiffness of particle two (N/mm)
ε is the coefficient of restitution (-)

$$K_{nl,p} = E_p L \tag{5}$$

$$K_{nl,b} = E_b L \tag{6}$$

Here:
E_p is the particle material's bulk Young's modulus (Pa)
E_b is the boundary's bulk Young's modulus (Pa)
L is the particle size (mm)

2.4.2. Tangential Force Model: Linear Spring Coulomb Limit Model

Calculating the contact force in the tangential direction utilizes a linear spring Coulomb limit model. This model is categorized as an elastic-frictional model; the tangential force for purely elastic contact is computed using Equations (7) and (8), while the final tangential force uses Equation (9). This step is necessary as the tangential contact force must not surpass the Coulomb friction force.

$$F^t_{\tau,e} = F^{t-\Delta t}_{\tau} - K_\tau \Delta s_\tau \tag{7}$$

Here:
$F^t_{\tau,e}$ is the value of the tangential force at the current time (N)
$F^{t-\Delta t}_{\tau}$ is the value of the tangential force at the previous time (N)
Δs_τ is the tangential relative displacement of the particles during the timestep (mm)
K_τ is the tangential stiffness, which is defined as:

$$K_\tau = r_K K_{nl}, \tag{8}$$

where:
r_K is the tangential stiffness ratio (-).

$$F^t_\tau = \min\left(\left|F^t_{\tau,e}\right|, \mu F^t_n\right) \frac{F^t_{\tau,e}}{\left|F^t_{\tau,e}\right|}, \tag{9}$$

where:
μ is the friction coefficient (-).

2.5. Bucket Size Optimization Using DEM

A simulation model was created to optimize the bucket size of the metering device using the commercial software ANSYS Rocky for the DEM [18]. The bucket optimization method proposed by Im et al. [8] was employed to determine the optimal bucket size for each group. There exists a linear relationship between garlic size and optimal bucket size, and the sequential search method was used to optimize the design parameter by linearly decreasing the bucket size [8,25]. In the sorting process of the metering device, excess garlic cloves, which are more than one clove, are preferable to fall into the hopper via the guide. The rate at which the bucket was exhibited to be too small to contain a clove was defined as the missing plant rate. The target metering device is required to achieve a plant rate of 97.5% or more while simultaneously minimizing the multi-plant rate, as described in Equation (13). Although a bucket size larger than necessary is advantageous for achieving a plant rate of 97.5% or more, it may increase the multi-plant rate. Therefore, it is desirable to determine the smallest bucket size that satisfies the target plant rate.

$$Q = \left(1 - \frac{N_2}{N}\right) \times 100, \tag{10}$$

$$M = \frac{N_1}{N} \times 100, \tag{11}$$

$$C = \frac{N_2}{N} \times 100, \tag{12}$$

where:

Q is plant rate (%)
M is multi-plant rate (%)
C is missing plant rate (%)
N is the total number of trials (-)
N_1 is the number of metering trials with multiple cloves (-)
N_2 is the number of metering trials with empty bucket (-)

$$\begin{array}{c} minimize : M \\ subject\ to : \ Q \geq 97.5 \end{array} \quad (13)$$

The elongated model depicted in Figure 4a was utilized to achieve the target plant rate. The model was constructed by setting the length that determines the center of gravity of garlic cloves as $\mu_L + 2\sigma_L$. Assuming a normal distribution of garlic clove lengths, all cloves, except for the top 2.5% of the group, are successfully planted if they pass safely through the metering device [8]. Figure 12 illustrates the bucket size optimization simulation and displays the initial posture of the garlic clove for simulation purposes. The most unfavorable condition for the metering operation was selected as the initial location of the garlic clove to attain the target plant rate. The mechanical vibration and geometrical characteristics of the finger-type bucket ensure that garlic cloves are always positioned parallel to the bucket; when the center of gravity faces outward, it is disadvantageous for planting. Furthermore, due to the rolling phenomenon, when the round side of the garlic clove faces the bottom, the center of gravity moves inside the bucket trajectory, which can be beneficial for metering. Therefore, the flat side was arranged to face the bottom for the most unfavorable condition.

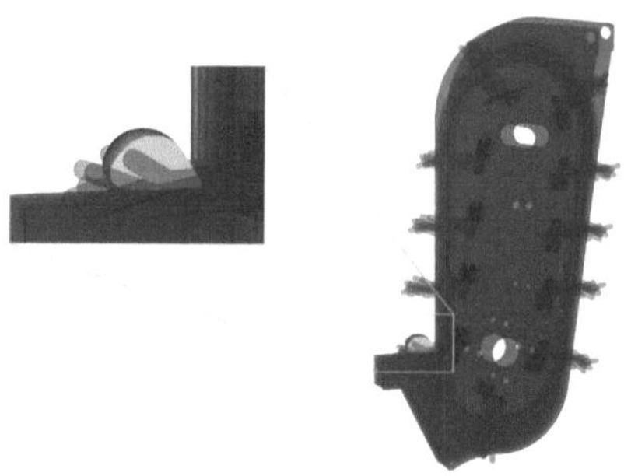

Figure 12. DEM simulation for the determination of optimal bucket size.

The dynamics model provided by ANSYS Rocky was utilized, and a rotational motion of 30 rpm was set on the lower sprocket, consistent with the actual operating conditions to establish the operation of the metering device. To carry out the sequential search method, the garlic clove model was positioned in the initial location in the most unfavorable orientation for planting, and multiple simulations were conducted while incrementally increasing the metering device guide angle (α) depicted in Figure 2a by 0.1°. The optimal bucket size was determined as the size immediately prior to the garlic clove of each group falling into the hopper via the guide. The design parameter interval was established to ensure applicability to the actual system as 0.1°, adjustable by the user.

2.6. Metering Simulation Using DEM

To validate the optimal bucket size of each group obtained through Section 2.5, a metering simulation was conducted using the DEM. The bucket size of the metering device was set to the optimum value obtained earlier, and the lower sprocket was rotated at 30 rpm, identical to the actual operating condition. Figure 13 depicts the metering simulation, demonstrating the sorting process that garlic cloves go through after being placed in the hopper. By incorporating the virtual sensor into the simulation model, as shown in Figure 13, the number of garlic cloves in the bucket that passed through the sorting area was analyzed in the time domain. The garlic clove model for metering simulation was based on statistical information on the shape of garlic cloves and used the average model illustrated in Figure 4b. Furthermore, for various sizes, garlic clove models were scaled up or down based on a Gaussian distribution and randomly placed in the hopper to perform metering simulations for various sizes. The software provides a scale distribution method to consider different sizes when the mean and standard deviation information is available [18]. The morphological statistical information of garlic cloves to define the Gaussian distribution was based on the values provided by Im et al. [8].

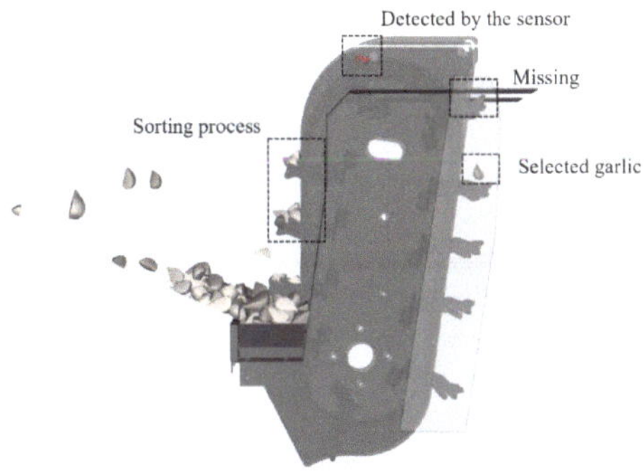

Figure 13. Metering simulation using DEM.

2.7. Metering Experiment

An actual metering experiment was conducted using a real metering device to validate the reliability of the discrete element analysis model and optimization technique. Figure 14a illustrates the actual garlic planter, while Figure 14b depicts the 1-row metering test machine. A boundary plate was installed on a 7-row metering device to create a 1-row metering test, establishing the same boundary conditions as the simulation model. Fifty garlic cloves were selected from each group, and a metering test was conducted with the sprocket rotating at 30 rpm, using the optimal bucket size as in the metering simulation. The test verified the plant rate (Q), multi-plant rate (M), and missing plant rate (C), and to ensure the reliability of the test results, it was repeated five times. The average value of each result was compared to the simulation result to validate the optimization methodology.

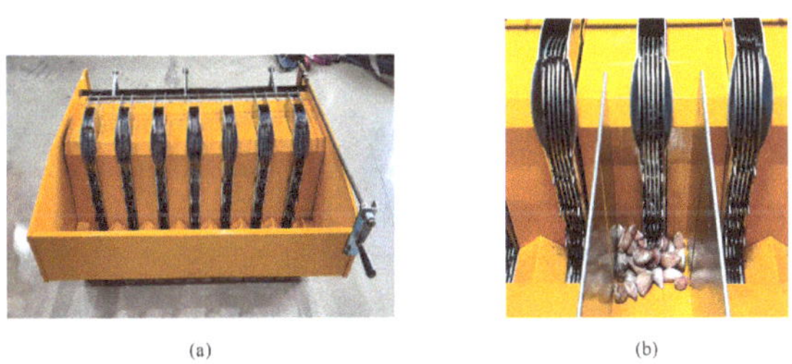

Figure 14. Metering experiment equipment. (**a**) Metering device. (**b**) 1-row metering test.

3. Results and Discussions

3.1. Bucket Size Optimization & Metering Simulation

This study aimed to optimize the bucket size of the metering device by initially positioning the garlic clove model under the most unfavorable conditions for metering. The optimal solution was then determined by gradually reducing the bucket size using the sequential search method. The design parameter interval was set at 0.1°, which the user can adjust to ensure system applicability. The optimization process revealed that the optimal bucket size increased with the size of the garlic clove, and it was found that Group 1 had an optimal bucket size of 16.11 mm, Group 2 had 22.75 mm, and Group 3 had 23.64 mm. Table 2 summarizes the results of the bucket size optimization. A batch simulation of the metering operation was conducted to validate the accuracy of optimization results. Once the optimal bucket size was obtained through the sequential search method, a metering simulation at the optimal setting was conducted for garlic of various sizes in each group, and the results are shown in Table 3. Some 40 metering operations were performed for Group 1, 37 for Group 2, and 43 for Group 3 to enhance the reliability of the metering performance results.

Table 2. Optimal bucket size for each group of garlic.

Group	Optimum Bucket Size l (mm)
Group 1	16.11
Group 2	22.75
Group 3	23.64

Table 3. Results of metering simulation.

Group	No. Trials	Plant Rate (Q)	Multi-Plant Rate (M)	Missing Plant Rate (C)
Group 1	40	92.50%	10.00%	7.50%
Group 2	37	97.30%	2.70%	2.70%
Group 3	43	97.70%	2.30%	2.30%

Except for Group 1, all groups were able to nearly achieve the target plant rate of 97.5%. Moreover, the multi-plant rate demonstrated a low level of under 3%. Group 1, being the smallest in size, exhibited the highest multi-plant rate at 10% and demonstrated poor metering performance. It is believed that the overlapping planting of more than one garlic clove occurred in the width direction of the finger-type bucket for Group 1, as shown in Figure 15. Conversely, in the case of Groups 2 and 3, this phenomenon did not occur because the thickness of the garlic clove was large, considering the bucket width. As such, they were deemed the most suitable groups for applying the metering device.

(a) (b)

Figure 15. Overlapped garlic cloves causing multi plant. (**a**) Front view. (**b**) Side view.

3.2. Experimental Result

To validate the bucket size optimization technique, an actual metering test was conducted, as depicted in Figure 14, and the results are illustrated in Table 4. The test was executed under conditions identical to the analysis model, including bucket size and rotational speed, to validate the optimization process using simulation. The test was repeated five times using 50 garlic cloves in each group to enhance the reliability of the test results. Average plant rates of 90.44% for Group 1, 97.97% for Group 2, and 94.95% for Group 3 were confirmed. It was observed that all groups except Group 1 came close to the target plant rate of 97.5%. Additionally, it was confirmed that Group 1 had a multi-plant rate of 1.47%, Group 2 had a rate of 2.85%, and Group 3 had a rate of 0.49%. These findings demonstrate significantly improved metering performance compared to the test results of Kang et al. [3].

Table 4. Results of the metering experiment.

Group	Test No	Plant Rate (Q)	Multi-Plant Rate (M)	Missing Plant Rate (C)
Group 1	1	89.29%	0.00%	10.71%
	2	89.09%	1.82%	10.91%
	3	90.91%	0.00%	9.09%
	4	89.29%	0.00%	10.71%
	5	94.00%	6.00%	6.00%
	Average	90.44%	1.47%	9.56%
Group 2	1	100.00%	2.04%	0.00%
	2	97.92%	6.25%	2.08%
	3	97.96%	4.08%	2.04%
	4	95.92%	2.04%	4.08%
	5	98.04%	0.00%	1.96%
	Average	97.97%	2.85%	2.03%
Group 3	1	97.56%	2.44%	2.44%
	2	97.62%	0.00%	2.38%
	3	93.18%	0.00%	6.82%
	4	93.18%	0.00%	6.82%
	5	93.18%	0.00%	6.82%
	Average	94.95%	0.49%	5.05%

Table 5 presents the simulation and test results for the plant rate, multi-plant rate, and missing plant rate. It was confirmed that the plant rate for both results had a maximum error of 2.8% for Group 3. The difference between the two results for the multi-plant rate was 8.5% for Group 1, 0.2% for Group 2, and 1.8% for Group 3. Lastly, for the missing plant rate, a relatively low level of error was confirmed, with a maximum error of 2.8% for Group 3. For Group 1, due to the relatively narrow width of cloves, multiple garlic cloves could be stacked in the bucket, as shown in Figure 15, and the metering performance was affected by both the width of the bucket and the length of the garlic cloves. The bucket size optimization technique applied in this study considered the length of garlic cloves as the only independent variable and not the effect of bucket width. Consequently, both simulation and test results displayed poor metering performance, and the reliability of the optimal solution was not fully guaranteed for Group 1. As a result, there was a relatively significant difference between the analysis and test results in Group 1 compared to those of Groups 2 and 3. Upon comparing the analysis and test results, it was determined that the analysis model predicted the metering performance of the actual metering device within 10% of all performance indicators, including plant rate, multi-plant rate, and missing plant rate. Particularly for Groups 2 and 3, the high reliability of the analysis model was confirmed, showing a difference in the metering performance of a maximum of 2.8% (plant rate of Group 3).

Table 5. Comparison between simulation and experiment results.

Group	Index	Simulation	Experiment	Difference
Group 1	Plant rate (Q)	92.50%	90.44%	2.10%
	Multi-plant rate (M)	10.00%	1.50%	8.50%
	Missing plant rate (C)	7.50%	9.60%	−2.10%
Group 2	Plant rate (Q)	97.30%	97.97%	−0.70%
	Multi-plant rate (M)	2.70%	2.90%	−0.20%
	Missing plant rate (C)	2.70%	2.00%	0.70%
Group 3	Plant rate (Q)	97.70%	94.95%	2.80%
	Multi-plant rate (M)	2.30%	0.50%	1.80%
	Missing plant rate (C)	2.30%	5.10%	−2.80%

4. Conclusions

In this investigation, the bucket size of the garlic planter metering device was optimized by utilizing the DEM, considering the size of various garlic cloves. To enhance the precision of the analysis model, the actual shape of the garlic cloves was incorporated using a 3D mesh, and the mechanical properties of garlic were found through various tests, such as the angle of repose test and sliding test. Subsequently, the optimal bucket size was determined for the three garlic size groups using the sequential search method based on the developed model. A metering simulation was carried out to validate the optimal value, and finally, a physical test was conducted under the same conditions to ensure the reliability of the optimization technique. The primary findings of the study are summarized as follows.

(1) Calibration was executed by conducting a repose angle test to establish the garlic-garlic friction coefficient. It was ascertained that the simulation model and the test produced the same repose angle when the garlic-garlic friction coefficient was 0.46.
(2) Bucket size optimization was carried out to achieve the target metering rate of 97.5%. Subsequently, it was determined that Group 1 had an optimal bucket size of 16.11 mm, Group 2 had that of 22.75 mm, and Group 3 had that of 23.64 mm.
(3) To validate the reliability of the determined optimal bucket size, a metering simulation was conducted, which showed that the plant rate was 90.44% for Group 1, 97.97% for Group 2, and 94.95% for Group 3. The results indicate that Groups 2 and 3 achieved a seeding rate close to the target value of 97.5%.
(4) To validate the reliability of the optimization methodology, an actual metering test was conducted under identical conditions. Upon comparing the metering performance, it was confirmed that all metering performances exhibited differences of less than 10%. Notably, in Groups 2 and 3, the difference in metering performance was within 2.8%, thus confirming the reliability of the optimization technique.
(5) Regarding Group 1, as depicted in Figure 15, the garlic clove widths were relatively narrow, leading to an overlap phenomenon. Consequently, the metering performance prediction exhibits low accuracy, and the metering performance is inferior to Groups 2 and 3.

The DEM was employed to confirm the optimal bucket size for each garlic group, and during actual metering work, the optimal bucket size resulted in nearly satisfactory target plant rates. Moreover, the proposed methodology enables the optimal design of various agricultural implements with similar mechanisms. Notably, this study solely considered the length of the finger-type bucket in the bucket optimization process, while the width was not taken into account. As a result, while Groups 2 and 3 met the target plant rate, Group 1 did not. Future studies will enhance the optimization completeness by incorporating the width as well as the bucket size.

Author Contributions: Conceptualization, D.I.; methodology, D.I.; software, J.-H.K. and D.-J.M.; validation, D.I. and H.-S.L.; resources, T.-I.M.; supervision, Y.-J.P.; project administration, S.-H.Y. All authors have read and agreed to the published version of the manuscript.

Funding: This work was carried out with the support of the "Cooperative Research Program for Agriculture Science & Technology Development (Project No. PJ016061)" Rural Development Administration, Republic of Korea.

Institutional Review Board Statement: Not available.

Data Availability Statement: No new data were created or analyzed in this study. Data sharing is not applicable to this article.

Acknowledgments: This work was carried out with the support of the "Cooperative Research Program for Agriculture Science & Technology Development (Project No. PJ016061)", Rural Development Administration, Republic of Korea.

Conflicts of Interest: The authors declare no conflict of interest.

References

1. Walsh, P.P.; Murphy, E.; Horan, D. The role of science, technology and innovation in the UN 2030 agenda. *Technol. Forecast. Soc. Change* **2020**, *154*, 119957.
2. Choi, D.K.; Cho, S.C.; Lee, C.S.; Park, S.H.; Kim, H.J.; Kang, T.K.; Kim, C.K.; Kwak, T.Y. Development of a Metering Device for the Garlic Planter. *Biosyst. Eng.* **2001**, *26*, 525–534.
3. Kang, T.G.; Lee, S.H.; Kim, J.G.; Kim, Y.K.; Choi, Y.; Choi, I.S.; Lim, J. A factor analysis of garlic metering device for development of a garlic planter. *J. Agric. Sci.* **2018**, *52*, 143–151.
4. Wang, H.; Sun, X.; Li, H.; Fu, J.; Zeng, X.; Xu, Y.; Wang, Y.; Liu, H.; Lü, Z. Design and parameter optimization of a finger clip plate garlic seed-metering device based on EDEM. *Agronomy* **2022**, *12*, 1543.
5. Kang, T.G.; Lee, S.H.; Choi, Y.; Kwon, Y.S.; Kim, T.H. Study on the optimum condition of mechanical sowing of a garlic. *J. Korea Acad.-Ind. Coop. Soc.* **2021**, *22*, 360–366.
6. Zilpilwar, S.; Yadav, R.; Gajjar, P.; Jakasania, R. Development and performance evaluation of metering mechanism for planting garlic cloves. *Agric. Res.* **2021**, *10*, 274–284.
7. Zhang, X.; Yi, S.; Tao, G.; Zhang, D.; Chong, J. Design and experimental study of spoon-clamping type garlic precision seeding device. *Wirel. Commun. Mob. Comput.* **2022**, *2022*, 5222651.
8. Im, D.; Kim, J.H.; Moon, D.J.; Moon, T.I.; Yu, S.H.; Park, Y.J. Optimization of bucket size of garlic metering device using multibody dynamics simulation. *Biosyst. Eng.* **2022**, *47*, 480–488.
9. Guo, H.; Cao, Y.; Song, W.; Zhang, J.; Wang, C.; Wang, C.; Zhu, L. Design and simulation of a garlic seed metering mechanism. *Agriculture* **2021**, *11*, 1239.
10. Yu, Z.; Yang, K.; Hu, Z.; Peng, B.; Gu, F.; Yang, L.; Yang, M. Parameter optimization and simulation analysis of floating root cutting mechanism for garlic harvester. Comput. *Electron. Agric.* **2023**, *204*, 107521.
11. Wang, J.; Tang, H.; Wang, J.; Li, X.; Huang, H. Optimization design and experiment on ripple surface type pickup finger of precision maize seed metering device. *Int. J. Agric. Biol.* **2017**, *10*, 61–71.
12. Kang, J.; Peng, Q.; Zhang, C.; Zhang, N.; Fang, H. Design and testing of a punching-on-film precision hole seeder for peanuts. *Trans. ASABE* **2020**, *63*, 1685–1696. [CrossRef]
13. Singh, R.C.; Singh, G.; Saraswat, D.C. Optimisation of design and operational parameters of a pneumatic seed metering device for planting cottonseeds. *Biosyst. Eng.* **2005**, *92*, 429–438. [CrossRef]
14. Wang, M.; Liu, Q.; Ou, Y.; Zou, X. Numerical simulation and verification of seed-filling performance of single-bud billet sugarcane seed-metering device based on EDEM. *Agriculture* **2022**, *12*, 983. [CrossRef]
15. Ryu, I.H.; Kim, K.U. Design of roller type metering device for precision planting. *Trans. ASABE* **1998**, *41*, 923. [CrossRef]
16. Han, D.; Zhang, D.; Jing, H.; Yang, L.; Cui, T.; Ding, Y.; Zhang, T. DEM-CFD coupling simulation and optimization of an inside-filling air-blowing maize precision seed-metering device. *Comput. Electron. Agric.* **2018**, *150*, 426–438. [CrossRef]
17. Sun, X.; Li, H.; Qi, X.; Nyambura, S.M.; Yin, J.; Ma, Y.; Wang, J. Performance parameters optimization of a three-row pneumatic precision metering device for brassica chinensis. *Agronomy* **2022**, *12*, 1011. [CrossRef]
18. ANSYS Rocky. 2023. Available online: https://www.ansys.com/products/fluids/ansys-rocky (accessed on 1 January 2023).
19. Park, D.; Lee, C.G.; Park, H.; Baek, S.H.; Rhee, J.Y. Discrete element method analysis of the impact forces on a garlic bulb by the roller of a garlic harvester. *Biosyst. Eng.* **2019**, *44*, 208–217. [CrossRef]
20. Yu, Z.; Hu, Z.; Peng, B.; Gu, F.; Yang, L.; Yang, M. Experimental determination of restitution coefficient of garlic bulb based on high-speed photography. *Int. J. Agric. Biol. Eng.* **2021**, *14*, 81–90. [CrossRef]
21. Chen, J.; Gao, R.; Liu, Y. Numerical study of particle morphology effect on the angle of repose for coarse assemblies using DEM. *Adv. Mater. Sci. Eng.* **2019**, *2019*, 8095267. [CrossRef]
22. Zeng, Y.W.; Jin, L.; Du, X.; Gao, R. Refined modeling and movement characteristics analyses of irregularly shaped particles. *Int. J. Numer. Anal. Methods* **2015**, *39*, 388–408. [CrossRef]
23. Walton, O.R.; Braun, R.L. Viscosity, granular-temperature, and stress calculations for shearing assemblies of inelastic, frictional disks. *J. Rheol.* **1986**, *30*, 949–980. [CrossRef]
24. ESSS Rocky, Release 2021 R2.2, DEM Technical Manual, ESSS Rocky DEM, S.R.L. 2021. Available online: https://rocky.esss.co/technical-library/ (accessed on 1 January 2023).
25. Kiefer, J. Sequential minimax search for a maximum. *Proc. Am. Math. Soc.* **1953**, *4*, 502–506. [CrossRef]

agriculture

MDPI

Communication

Methodology for Assessing Tractor Traction Properties with Instability of Coupling Weight

Anatoliy Lebedev [1], Mykhailo Shuliak [1], Stanislav Khalin [2], Sergei Lebedev [2], Katarzyna Szwedziak [3,*], Krzysztof Lejman [3], Gniewko Niedbała [4,*] and Tomasz Łusiak [5]

1 Agroengineering Department, Sumy National Agrarian University, 160 Herasyma Kondratieva Street, 4000 Sumy, Ukraine
2 Leonid Pogorilyy Ukrainian Scientific Research Institute, 08654 Doslidnitske, Ukraine
3 Institute of Agricultural Engineering, Wrocław University of Environmental and Life Sciences, 37b Chełmońskiego Street, 51-630 Wrocław, Poland
4 Department of Biosystems Engineering, Faculty of Environmental and Mechanical Engineering, Poznań University of Life Sciences, Wojska Polskiego 50, 60-627 Poznań, Poland
5 Department of Thermodynamics, Fluid Mechanics and Aviation Propulsion Systems, Lublin University of Technology, Nadbystrzycka 38 D, 20-618 Lublin, Poland
* Correspondence: katarzyna.szwedziak@upwr.edu.pl (K.S.); gniewko.niedbala@up.poznan.pl (G.N.)

Abstract: The purpose of the study is to increase the efficiency of using the tractor hitch weight in traction mode by reducing the uneven distribution of vertical reactions between the wheels. This work is grounded on a methodology that involves summarizing and analyzing established scientific findings related to the theory of tractors operating in traction mode. The analytical method and comparative analysis were employed to establish a scientific problem, define research objectives, and achieve the goal. The key principles of probability theory were applied in developing the empirical models of the tractor. The main provisions of the methodology for evaluating the traction properties of the tractor with the instability of the coupling weight were formulated. The method of evaluating the vertical reactions on the wheels of the tractor is substantiated, which is based on the measurement of the vertical reaction on one of the four wheels. It was proven that tractors with a center of mass offset to the front or rear axles have the greatest probability of equal distribution of vertical reactions between the wheels of one axle, and tractors with a center of mass in the middle between the axles have the lowest probability. It is theoretically substantiated and experimentally confirmed that when the tractor performs plowing work with uneven distribution of loads on the sides, its operation with maximum traction efficiency is ensured by blocking the front and rear axle drivers.

Keywords: tractor; traction properties; coupling weight; center of mass; vertical reactions; driving bridges; blocking

Citation: Lebedev, A.; Shuliak, M.; Khalin, S.; Lebedev, S.; Szwedziak, K.; Lejman, K.; Niedbała, G.; Łusiak, T. Methodology for Assessing Tractor Traction Properties with Instability of Coupling Weight. *Agriculture* **2023**, *13*, 977. https://doi.org/10.3390/agriculture13050977

Academic Editors: Ju-Seok Nam and Yongjin Cho

Received: 26 March 2023
Revised: 22 April 2023
Accepted: 26 April 2023
Published: 28 April 2023

1. Introduction

The classic approach to the tractor as an object of design and management [1] consists of imagining it as a traction machine that has one connection with the external environment (running gear). Such an idea leads to the setting of tasks for optimizing the properties of the tractor, including traction force and efficiency, movement resistance, load on the axles, etc. These tractor parameters are regulated during their tests in NTTL laboratories [2] in the United States of America and Germany [3].

When the tractor is working in traction mode, its connection with the soil is carried out through two channels (running gear, working implement), which lead to the instability of the traction force of the tractor and the resistance of the aggregated agricultural machine [4–6]. At the same time, the reactions between the wheels of the front and rear axles are assumed to be uniform, although due to different rolling resistance of the wheels, for example, when the tractor performs certain tasks, the reactions between the wheels are unevenly distributed [7–9]. The movement of the tractor during the execution of the technological process leads to longitudinal, lateral, and vertical oscillations of sprung and unsprung masses and tractor wheels. Longitudinal and lateral oscillations of the tractor lead to an increase in the dynamic tension of the soil due to skidding, which reduces its traction force, and vertical oscillations are the main cause of dynamic soil compaction [10,11]. At the same time, an uneven distribution of vertical reactions between the wheels of the tractor is noted, which affects the efficiency of using the coupling weight in traction mode [12,13]. The essential feature of the tractor's movement in the traction mode was not adequately covered in the technical literature when evaluating its support and traction properties.

The solution to the problem of the influence of a mobile machine's center of mass is considered in a number of foreign publications [14], which note the perspective of research in the direction of machine operation in traction mode. It is proposed to determine the energy parameters of the machine by differentiating its mass.

Practice presents science with the need to solve the problem of developing a methodology for assessing the uneven distribution of vertical reactions between the tractor wheels in traction mode.

The efficiency of using the towing weight of the tractor in traction mode is increased by reducing the uneven distribution of vertical reactions between the wheels.

The following tasks were set in this work:

- Justify the dependence of the tractor's traction properties on the use of the coupling weight of the driving axles.
- Estimate the distribution of vertical reactions between the wheels of one driving axle of a tractor.
- Justify the zone of the most likely reduction in the use of the towing weight of the four-wheel drive tractor.

2. Materials, Method and Results

This work is based on a methodology that summarizes and examines existing scientific findings on the theory of tractors in traction mode. The analytical method and comparative analysis were utilized to formulate the scientific problem, establish research objectives, and achieve the goal. The fundamental principles of probability theory were incorporated in developing the empirical tractor models.

Let us assume that the tractor, being on a flat horizontal surface, moves at a constant speed with the power take-off shaft turned off (Figure 1). Traction efficiency of the tractor can be expressed [1] by the following equation:

$$\eta_{\text{т}} = \frac{N_{\text{т}}}{N_e} = \frac{P_{\kappa p} \upsilon_{\text{д}}}{M_{\text{д}} n_{\text{д}}} \frac{716.2}{270} = 2.65 \frac{P_{\kappa p} \upsilon_{\text{д}}}{M_{\text{д}} n_{\text{д}}}, \tag{1}$$

where $N_{\text{т}}$—traction power of the tractor, kW; N_e—effective engine power, kW; $P_{\kappa p}$—traction force on the hook, kN; $\upsilon_{\text{д}}$—actual driving speed, km/h.; $M_{\text{д}}$—engine torque, kN•m; $n_{\text{д}}$—engine crankshaft rotation frequency (engine speed), r/min.

Let us imagine that the traction forces on the hook and the actual speed of movement, which are included in Equation (1) of the traction efficiency of the tractor, are functions of the dimensionless coefficients $\varphi_{1,2}$ of the use of the coupling weight of the driving axles in the following equation:

$$\varphi_{1,2} = \frac{P_{\kappa 1,2}}{Y_{1,2}}, \tag{2}$$

where $P_{\kappa 1,2}$—tangential traction force of the driving axle, kN; $Y_{1,2}$—soil reaction, numerically equal to the towed weight of the tractor, which is distributed to the drive axle ($Y_{1,2} = G_{1,2}$), kN. Here, and further, the index "1" refers to the front, and the index "2" refers to the rear axles.

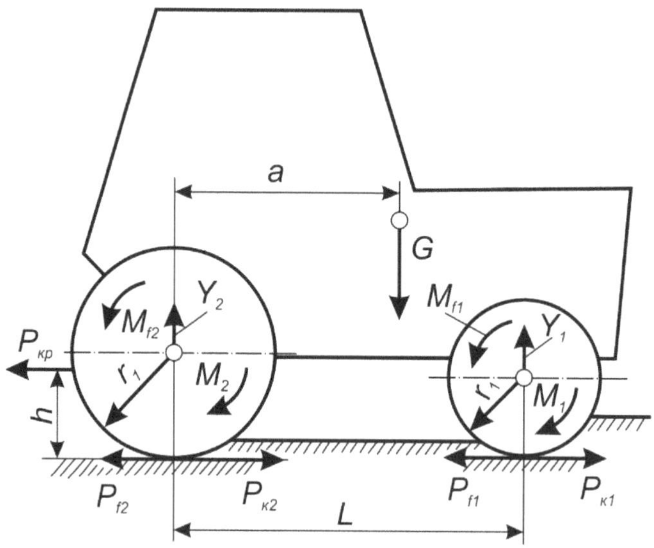

Figure 1. Diagram of the forces acting on the tractor.

To obtain the dependency $P_{кр} = f(\varphi_1, \varphi_2)$ we express $P_{кр} = P_{кр1} + P_{кр2} - P_{f1} - P_{f2}$; $P_{к1,2} = G_{1,2}\varphi_{1,2}$; $P_{f1,2} = G_{1,2}f_{1,2}$, where $P_{f1,2}$—force of resistance to rolling of driving axles; $f_{1,2}$—coefficient of resistance to rolling of driving axles.

Taking into account the redistribution of the coupling weight depending on the traction force, we obtain the following equation:

$$P_{кр} = \frac{G[A_1(\varphi_1 - f_1) + A_2(\varphi_2 - f_2)]}{A_3 + h(\varphi_1 - \varphi_2)}, \tag{3}$$

where $A_1 = a - f_2r_2$; $A_2 = L - a - f_1r_1$; $A_3 = L - h(\varphi_1 - f_2) + f_1r_1 - f_2r_2$; G—the weight of the tractor, kN; L—longitudinal base of the tractor, m; h—trailer point height, m; r_1r_2—rolling radii of the wheels, m.

Traction-supporting properties of the tractor depend mainly on its operating weight G_e, the conversion of which into traction force $P_{кр}$ is estimated by the coefficient of use of the towing weight $\varphi_{кр} = P_{кр}/G_e$. When moving a wheeled tractor on a flat agricultural background, with the coefficients of traction of the driving wheels with the soil $\varphi_{зч}$, rolling resistance f, and the share of weight falling on the driving wheels $\lambda_к$, the support and traction properties of a tractor with one driving axle are estimated according to the dependence $\varphi_{кр} = \varphi_{зч}\lambda_к - f$, and for all-wheel drive tractors—$\varphi_{кр} = \varphi_{зч} - f$. An analysis $\varphi_{кр}$ of tractors with one drive axle and four-wheel drive tractors shows that the greater the coefficient of traction in comparison with the coefficient of rolling resistance, the greater the tractor's margin of traction force.

In operating conditions, the traction force of the tractor varies from zero to the maximum value $P_{кр} = P_{крmax}$, which is determined by the traction properties at $G_e = const$.

Therefore, the coefficient $\varphi_{кр}$ changes from zero to $\varphi_{крmax}$. The nominal traction force corresponds to a certain value $\varphi_{крн}$. This statement was confirmed when the John Deere 8335 R tractor was moving in traction mode (Figure 2) [7,9,10,12].

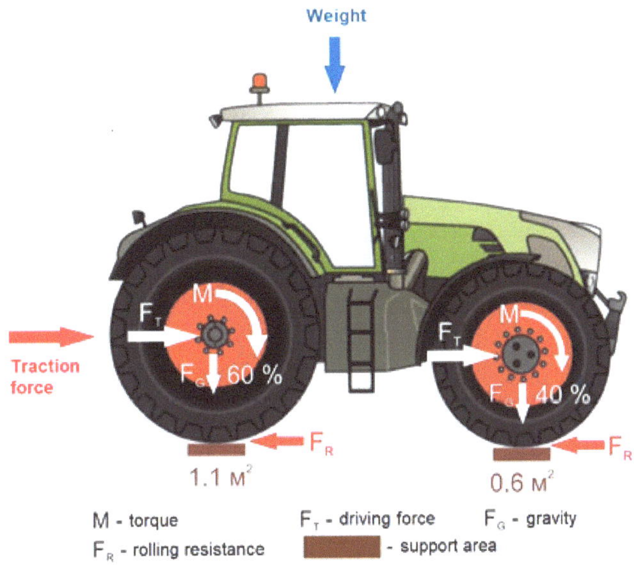

Figure 2. Diagram of the forces acting on the John Deere 8335 R tractor in traction mode.

Technical characteristics of the tractor John Deere 8335 R:
Nominal power of the engine, N_e, kW—246;
Nominal specific fuel consumption, g_e, g/kWh—224;
Engine torque reserve, %—41.4;
Traction power, N_e, kW—212;
Operating mass, m_e, kg—13,820;
Axle load, front/rear, kg—5528/8292;
Energy saturation, $E = N_e/m_e$—1.81;
Conditional traction efficiency, $\eta_T = N_T/N_e$—0.86.

Comparing the values of the loads acting on the front y_n and rear y_3 wheels of the tractor show that they do not remain constant. If the traction resistance line is parallel to the road surface, then the loads y_n and y_3 change due to the redistribution of the tractor's weight between the front and rear wheels. Reducing the load on the front wheels causes the same increase in the load on the rear wheels, and vice versa. The amount $y_n + y_3$ is equal to the weight of the tractor G_T. Instability in y_n and y_3 leads to an uneven distribution of vertical reactions between the wheels of the tractor, and a decrease in the efficiency of using its hitch weight in traction mode. At the same time, even when the center of mass of the tractor is located strictly in the longitudinal plane of symmetry, inequality of vertical reactions on the left and right wheels of the same axle is possible. This is due to the static uncertainty of the four-wheeled tractor as a spatial structure that has four supports (connections) and, accordingly, four vertical reactions of the connection. A well-founded methodology for solving this problem is based on the interrelationship of dimensionless (specific) forces on the wheels of the tractor as follows:

$$\gamma_{zi} = R_{z1}/G;\ \gamma_{z2} = b/L - \gamma_{zi};\ \gamma_{z3} = 0.5 - \gamma_{zi};\ \gamma_{z4} = \frac{a}{L} - 0.5 + \gamma_{zi}, \quad (4)$$

where R_{z1}—vertical reaction on the front wheel; G—total weight of the tractor; a, b, L—distances from the projection of the center of mass on the horizontal plane to the front and rear axles, the longitudinal wheel base of the tractor.

By specifying and measuring the vertical reaction on one of the four tractor wheels, the vertical reactions on the last three wheels can be accurately determined. Developed at the

Kharkiv branch of L. Pogorilyy UkrNDIPVT, the method for determining vertical reactions of tractor wheels involves determining two components: the position of the tractor frame in space and the vertical reaction to any of the four tractor wheels. In this case, the frame of the tractor is taken as a completely solid body and the vertical reaction is determined experimentally; the other three vertical reactions with the known weight of the tractor are determined analytically. For further modeling in this article, it was assumed that the tractor is installed on a perfectly flat surface, that is, the angles of inclination of the frame are equal to zero. In real operation, the position of the frame can be determined using acceleration sensors [15] and calculated analytically or using a digital gyroscope. At the stage of tractor design, it is advisable to imagine the vertical reaction on one of the wheels as a random variable, the distribution of which is subject to a normal law. Let us assume that such an independent random variable will be γ_{z2}, the changes of which are theoretically possible within the limits of $[0, b/L]$. With a decrease in γ_{zi} there is a decrease in γ_{z4}, which cannot take negative values, because this leads to a violation of the system equilibrium.

Thus, when $\gamma_{zi} = 0$ the following must be performed:

$$\gamma_{z4\min} = \frac{a}{L - 0.5} \geq 0 \text{ or } b/L \leq 0.5. \tag{5}$$

Therefore, when $b/L \leq 0$, the value γ_{zi} can vary within the following limits:

$$\frac{b}{L} - 0.5 \leq \gamma_{zi} \leq 0.5. \tag{6}$$

With the normal distribution law of vertical reactions on the wheels of the tractor, the density of the distribution of the specific load on one front wheel is estimated by the following equation:

$$f(\gamma_{zi}) = \left\{ \begin{array}{l} \frac{2394}{b/L}\mu_\sigma \exp\left[-\frac{1}{2}\left(\frac{6}{b/L}\gamma_{zi} - 3\right)^2\right] \text{at} \frac{b}{L} \leq 0.5; \\ \frac{2394}{1-b/L}\mu_\sigma \exp\left[-\frac{1}{2}\left(\frac{6}{1-b/L}\gamma_{zi} - \frac{3b/L}{1-b/L}\right)^2\right] \text{at} \frac{b}{L} > 0.5 \end{array} \right\} \tag{7}$$

where μ_σ—the selected scale of the root means square deviation, i.e., the value is conventionally taken as a unit during the analysis.

Figure 3 shows graphs of Function (7) at different positions of the tractor's center of mass (parameter b/L), taking into account the possible range of changes of γ_{zi}.

Analysis of the calculation results shows that tractors with an offset to the front or rear axle of the center of mass have the greatest probability of equal distribution of vertical reactions between the wheels of one axle. Tractors with a center of mass located in the middle between the axles have the lowest probability of equal distribution of vertical reactions at $b/L = 0.5$, and the largest—at $b/L = 0.9$ and $b/L = 0.1$.

If at $b/L = 0.1$, this probability is 0.4, then at $b/L = 0.5$, it is equal to 0.1.

When the tractor is working in the traction mode, the uneven distribution of vertical reactions leads to the deterioration of its traction–speed characteristics, since the total tangential traction force is determined by the wheel, which is in worse grip conditions. If we consider the specific vertical reactions on the wheels (the ratio of the corresponding reaction on the j-th wheel to the total weight of the tractor), taking into account their possible changes within the rms deviation, then the dependencies for their calculation are written in the following form:

$$\gamma_{z1} = m_{\gamma 1} + \sigma_{\gamma 1}; \; \gamma_{z2} = m_{\gamma 2} - \sigma_{\gamma 2}; \; \gamma_{z3} = m_{\gamma 3} - \sigma_{\gamma 3}; \; \gamma_{z4} = m_{\gamma 4} + \sigma_{\gamma 4}, \tag{8}$$

where $m_{\gamma 1}$, $m_{\gamma 2}$, $m_{\gamma 3}$, $m_{\gamma 4}$—mathematical expectation of specific vertical reactions on the corresponding wheels (1, 2—front wheels; 3, 4—rear wheels; 1, 3—left side wheels; 2,4—starboard wheels); $\sigma_{\gamma 1}$, $\sigma_{\gamma 2}$, $\sigma_{\gamma 3}$, $\sigma_{\gamma 4}$—root mean square deviations of the specific vertical reactions on the specified wheels.

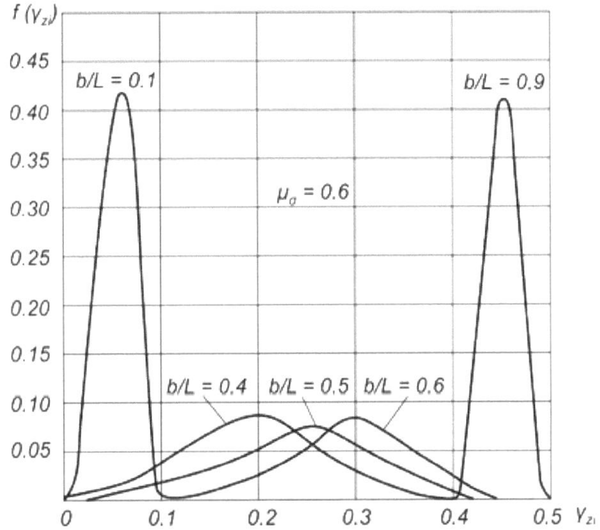

Figure 3. The density of the distribution of the specific load on one front wheel at the cut-off position of the tractor's center of mass.

It is obvious that

$$m_{\gamma 1} = m_{\gamma 1} = \frac{1}{2}\frac{b}{L};\ m_{\gamma 3} = m_{\gamma 4} = \frac{1}{2}\left(1 - \frac{b}{L}\right). \tag{9}$$

At $b/L \leq 0.5$ we have

$$\sigma_{\gamma} = \sigma_{\gamma 1} = \sigma_{\gamma 2} = \sigma_{\gamma 3} = \sigma_{\gamma 4} = \frac{1b}{6L}. \tag{10}$$

At $b/L > 0.5$ we have

$$\sigma_{\gamma} = \sigma_{\gamma 1} = \sigma_{\gamma 2} = \sigma_{\gamma 3} = \sigma_{\gamma 4} = \frac{1}{6}\left(1 - \frac{b}{L}\right). \tag{11}$$

The acceleration that occurs when the tractor accelerates in the absence of wheel slippage is determined by the wheel that is in worse grip conditions, that is, having a lower vertical load, as follows:

$$a = g\varphi\varphi_{KP}, \tag{12}$$

where g—free fall acceleration, g = 9.81 m/s^2; φ—coefficient of adhesion of the wheels to the road; φ_{KP}—the coefficient of use of the coupling weight in traction mode for a four-wheel drive tractor.

$$\varphi_{KP} = 2(\gamma_{z2} + \gamma_{z3}) = 2(m_{\gamma 2} + m_{\gamma 3} - 2\sigma_{\gamma}) = \begin{cases} 1 - \frac{2b}{3L}\ at\ \frac{b}{L} \leq 0.5 \\ \frac{1}{3} - \frac{2b}{3L}\ at\ \frac{b}{L} > 0.5 \end{cases} \tag{13}$$

Figure 4 presents a graph illustrating the dependence of φ_{KP} on the parameter b/L.

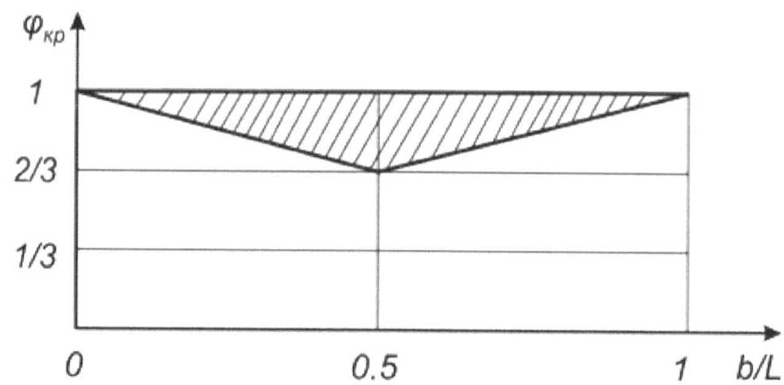

Figure 4. The zone of the most probable values of the reduction in the coefficient of use of the drawbar weight of a two-axle four-wheel drive tractor in traction mode (shaded).

Analysis of the graph in Figure 4 shows that the greatest decrease in the coupling weight coefficient of a two-axle four-wheel drive tractor in traction mode occurs when $\frac{b}{L} > 0.5$. This decrease can reach 33%.

In case of asymmetric application of the traction force on the tractor hook, for example, the resistance force of the plow, which is characterized by the distance a from the tractor axis to the point of application $P_{пл}$ and different rolling resistance of the wheels P_{fi}, the traction force required to overcome P_{fi} will be different for each wheel $P_{кi}$ (Figure 5).

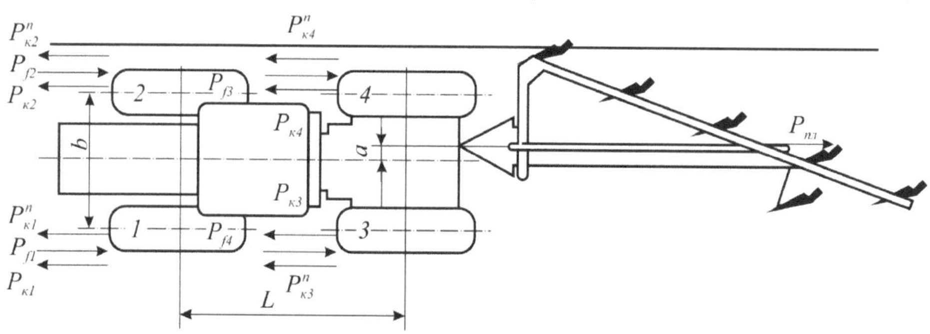

Figure 5. Diagram of the action of forces on the tractor's wheels during plowing operations.

To ensure uniform straight-line movement of the tractor, it is necessary to fulfill the condition $P_{кi} = P_{fi} + P^{n}_{кi}$, where $P^{n}_{кi}$—the traction force required to overcome the drag force of the plow.

For this case, the tangent of the traction force of the wheels is written in the following form:

$$\left.\begin{array}{l} P_{к1} = P_{пл}\frac{(b/2-a)}{b} + P_{к3} + \sum\left(P_{f1,3} + P^{n}_{к1,3}\right) \\ P_{к2} = P_{пл}\frac{(b/2-a)}{b} + P_{к4} + \sum\left(P_{f2,4} + P^{n}_{к2,4}\right) \\ P_{к3} = P_{пл}\frac{(b/2-a)}{b} + P_{к1} + \sum\left(P_{f1,3} + P^{n}_{к1,3}\right) \\ P_{к3} = P_{пл}\frac{(b/2-a)}{b} + P_{к2} + \sum\left(P_{f2,4} + P^{n}_{к2,4}\right) \end{array}\right\} \quad (14)$$

The analysis of this system of equations shows that at $P_{f1,3} \neq P_{f2,4}$ caused by different wheel rolling conditions when the tractor plows ($P_{f2} > P_{f4}$, $P_{f1} > P_{f3}$, $P_{f2} > P_{f1}$, $P_{f4} > P_{f3}$) has $P_{к2} > P_{к4}$, $P_{к1} > P_{к3}$, $P_{к2} > P_{к1}$, $P_{к4} > P_{к3}$.

In the Kharkiv branch of L. Pogorilyy UkrNDIPVT, in order to evaluate the dependence of the traction force of the tractor on the coupling weight with an uneven distribution of reactions between the axles, experimental studies of the HTZ-17021 tractor when aggregated with a PLN-5.35 plow were carried out. The results of these experimental studies are shown in Table 1.

Table 1. Torques (M_{Kp}) on the sun gears of the wheel gearboxes of the HTZ-17021 tractor with the PLN-5-35 plow (plowing winter wheat stubble to a depth of 25–27 cm).

y, m	a, m	M_{Kp}, Nm			
		M_{Kp1}	M_{Kp2}	M_{Kp3}	M_{Kp4}
0.150	0.36	947	2222	875	2150

y—the distance from the edge of the furrow to the outer edge of the wheel; a—asymmetry of traction load application; M_{Kp1}, M_{Kp2}, M_{Kp3}, M_{Kp4}—torques on the sun gears of the wheel gearboxes of the front left and right wheels, and the rear left and right wheels, respectively.

The analysis of this table shows that the front right wheel is the most loaded during plowing of the HTZ tractor, and the rear left wheel is the least loaded. Behind the sides of the tractor, the front wheels are 3–5% more loaded than the rear wheels. Let us take it for simplification that $P^n_{K1} = P^n_{K3}$ and $P^n_{K2} = P^n_{K4}$. In this case, the load on the wheels of the right P^n_Π and left $P^n_Л$ sides of the tractor, which is necessary to overcome the resistance of the plow, from the system of Equation (14), is written in the following form:

$$\left.\begin{aligned} P^n_\Pi &= P_{\Pi Л}\left(\tfrac{1}{2} + \tfrac{a}{b}\right), \\ P^n_Л &= P_{\Pi Л}\left(\tfrac{1}{2} - \tfrac{a}{b}\right) \end{aligned}\right\} \quad (15)$$

The asymmetrical connection of the plow, characterized by parameter a, leads to a greater load on the right side of the tractor compared to the left side. For a tractor of the HTZ type in plowing operations with a PLN-5-35 plow, this excess is 50–60%. In this case, to ensure stable rectilinear movement of the plow unit, it is necessary to fulfill the condition $M_2 \approx 1.5M_1$, $\omega_1 \approx 1.5\omega_2$.

For example, Figure 6 shows the universal characteristics of the traction efficiency of the HTZ-170 tractor when plowing winter wheat stubble with a PLN-5-35 plow to a depth of 25–27 cm (operating weight of the tractor 8.0 t, engine power 121.1 kW, speed 2.37 m/s).

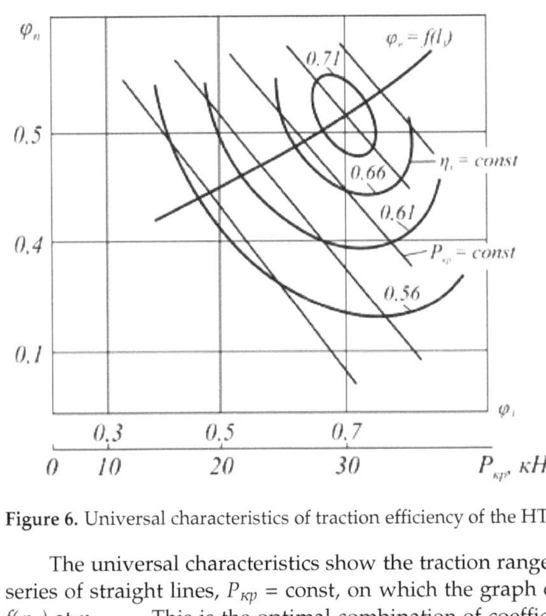

Figure 6. Universal characteristics of traction efficiency of the HTZ-170 tractor in plowing.

The universal characteristics show the traction range of the tractor in the form of a series of straight lines, P_{Kp} = const, on which the graph of the function is plotted $\varphi_\Pi = f(\varphi_3)$ at $\eta_{T\ max}$. This is the optimal combination of coefficients of the use of the driving axles' coupling weight, which leads to the operation of the tractor with maximum traction efficiency.

The analysis of the universal characteristics shows that $\eta_{\text{T}\ max} = 0.71$ of the KhTZ-17021 tractor during plowing will be in the case of blocked front and rear drive axles. All η_{T} points are located on the abscissa axis (because $\varphi_{\Pi} = 0$). At the same time, η_{T} on the universal characteristics will be below $\eta_{\text{T}\ max}$.

According to this characteristic, it is possible to estimate η_{T} when unlocking the differentials and simultaneously turning on the front and rear axles.

3. Discussion

The results of this research are mainly aimed at solving the scientific problem of increasing the efficiency of using the tractor hitch weight in traction mode by reducing the uneven distribution of vertical reactions between the wheels. New dependencies of the traction properties of the tractor on the use of its drawbar weight and a method of estimating the vertical reactions on the tractor's wheels from its drawbar weight are proposed. It was proven that the greatest efficiency of using the hitch weight of an all-wheel drive tractor in traction mode is achieved when its center of mass is located between the driving axles. At the same time, during plowing operations, the most energy-intensive technological process ensures its operation with maximum traction efficiency when the front and rear axle drivers are blocked.

Domestic and foreign scientists pay attention to the fact that the efficiency of tractors significantly depends on the position of its center of mass and the distribution of vertical reactions between the wheels. Domestic regulatory documents and methods of testing tractors according to the OECD Code 2 procedure in NTTL laboratories [2] in the United States of America and in Germany [3] do not provide for evaluating the efficiency of the tested tractors from the position of the center of its masses during the execution of the technological process. This is mainly a consequence of the lack of methods and instrumentation for evaluating the traction properties of the tractor during its unsteady movement.

The solution to this problem is possible using the method of partial accelerations [10], which is implemented at the standard L. Pogorilyy UkrNDIPVT, and the measuring and registration complex developed at the Kharkiv branch of L. Pogorilyy UkrNDIPVT, who has no analogues in the world. The presence of a solid underlying layer in the form of permafrost and the heavy precipitation typical of this period drastically reduce the load-bearing capacity of soils [16–20]. The increase in the tractor's weight under these conditions results in breaking of the upper layer of the soil down to permafrost, degradation of traction, and coupling properties and slippage, which, in turn, causes the increase in the man-made (technogenic) impact, and bad flotation even with small hook loads [21–24]. Numerous scientists worldwide have investigated this problem and proposed various solutions. One such study [2] examines the relationship between the load-carrying capacity and the speed of transport aggregate movement. The optimal ratio of these two factors helps to reduce the vertical load on the soil. The sustainable dead weight loading is also an efficient way to reduce the impact of the ground drive system on the soil when the tractor is in motion [25–28]. However, the dead weight loading increases metal consumption, fuel consumption, and the cost of the tractor. After analyzing both the foreign and domestic scientific literature, it was found that the established techniques used to enhance the traction and coupling properties of wheeled tractors [29,30] are not efficient when one needs to reduce the man-made (technogenic) impact of the ground drive system on the soil and simultaneously achieve sufficient traction properties of the machine-tractor units (MTA) or tractor-transport units (TTA). Therefore, the aim of this research is to determine new methods of improving the traction and coupling properties and reducing the anthropogenic (technogenic) impact on the arable soil of the drive system of a moving wheeled tractor, which are important components of increasing the productivity and efficiency of labor and mechanized work, as well as the rational use of natural resources and long-term conservation of soil horizons [11,31]. In their research, the authors of [16] determined that the optimal way to achieve this goal is to rationally distribute the weight of the coupling to the tractor propellers. For non-wheel drive modifications, the weight is redistributed from the non-driving front wheels to the driving rear wheels, with the front wheels pre-loaded to provide longitudinal stability if necessary.

4. Conclusions

Based on the results of this research, the methodology for evaluating the impact of the position of the tractor's center of mass on its traction properties is substantiated. The results of this research made it possible to form the main theoretical and scientific-practical conclusions:

1. The formulated methodology for evaluating the traction properties of a tractor with instability of the vertical reactions on its wheels. It was proven that when the tractor's center of mass is located in the longitudinal plane of symmetry, an uneven distribution of vertical reactions on the wheels of one axle of the tractor is possible.
2. The method of evaluating the vertical reactions on the wheels of the tractor, which is based on the measurement of the vertical reaction on one of the four wheels, is substantiated.
3. It was proven that tractors with the center of mass shifted to the front or rear axle have the greatest probability of equal distribution of the vertical reactions between the wheels of one axle, and tractors with the center of mass in the middle, between the axles, have the lowest probability.
4. It is theoretically substantiated and experimentally confirmed that when the tractor performs plowing work with an uneven distribution of the load on the sides, it is ensured that it works with maximum traction efficiency with the front and rear axles locked.
5. The issue of evaluating the influence of the tractor's center of mass on its traction properties when aggregated with mounted, trailed, and combined agricultural machines remains open. It is necessary to carry out theoretical and experimental research in this direction.

Author Contributions: Conceptualization, A.L., S.L. and S.K.; methodology, M.S. and T.Ł.; formal analysis, A.L., M.S., S.K., K.S. and T.Ł.; investigation, A.L., M.S., S.K., K.S. and T.Ł.; resources, A.L., S.L., M.S. and T.Ł.; writing—original draft preparation, A.L., M.S., S.K., K.S. and G.N.; writing—review and editing, K.S., K.L. and G.N.; visualization, K.S. and T.Ł.; supervision, K.L. and G.N.; project administration, K.S. and G.N.; funding acquisition, A.L., S.K. and K.S. All authors have read and agreed to the published version of the manuscript.

Funding: This research received no external funding.

Institutional Review Board Statement: Not applicable.

Data Availability Statement: Not applicable.

Conflicts of Interest: The authors declare no conflict of interest.

References

1. Guskov, V.V.; Velev, N.N.; Atamanov, Y.E.; Bocharov, N.F.; Ksenevich, I.P.; Solonsky, A.S. *Tractors: Theory: Manual for Students "Car & Tractor"*; Guskova, V.V., Ed.; Mashinostroenie: Moscow, Russia, 1988; 376p.
2. Nebraska Tractor Test Laboratory. Available online: https://tractortestlab.unl.edu/test-page-nttl (accessed on 16 April 2023).
3. DLG—Deutsche Landwirtschafts-Gesellschaft. Available online: https://www.dlg.org/de/landwirtschaft/tests (accessed on 16 April 2023).
4. Kalinin, E.; Shuliak, M.; Koliesnik, I. Optimization of machinery operation modes from the point of view of their dynamics. In Proceedings of the ICCPT 2019: Current Problems of Transport: Proceedings of the 1st International Scientific Conference, Ternopil, Ukraine, 28–29 May 2019; pp. 211–222.
5. Szwedziak, K.; Lusiak, T.; Grzywacz, Z.; Drozd, K. Numerical CFD Analysis of an Aerodynamic Head Cover of a Rotorcraft Motor. *Commun Sci. Lett. Univ. Zilina* **2018**, *20*, 42–47. [CrossRef]
6. Szwedziak, K.; Łusiak, T.; Bąbel, R.; Winiarski, P.; Podsędek, S.; Doležal, P.; Niedbała, G. Wind Tunnel Experiments on an Aircraft Model Fabricated Using a 3D Printing Technique. *J. Manuf. Mater. Process.* **2022**, *6*, 12. [CrossRef]

7. Lebedev, A.T.; Artjomov, N.P.; Grinenko, A.A. Traction efficiency of the tractor with uneven distribution of reactions between the wheels. In *Tractor Power Engineering in Crop Production*; Collection of Scientific Articles; KhNTUA: Kharkiv, Ukraine, 2003; pp. 49–56.
8. Fluck, R.C. Energy analysis for agricultural systems. In *Energy in Farm Production*; Elsevier: Amsterdam, The Netherlands, 2014; pp. 45–51.
9. Lebedev, S. The technical level of agricultural tractors on the market of Ukraine. *Tech. Technol. APK* **2014**, *11*, 8–12.
10. Lebedev, A.T.; Lebedev, S.A.; Korobko, A.I. *Qualimetry and Metrological of Tractors Testing*; Lebedev, A.T., Ed.; Miskdruk: Kharkiv, Ukraine, 2018; 394p.
11. Brouwer, F.M.; van Ittersum, M. (Eds.) *Environmental and Agricultural Modelling: Integrated Approaches for Policy Impact Assessment*; Springer: Dordrecht, The Netherlands, 2010; 322p.
12. Aleksandrov, E.E.; Volontsevych, D.O.; Karpenko, V.A.; Lebedev, A.T.; Peregon, V.A.; Samorodov, V.B.; Turenko, A.N. *Dynamics of Transport and Traction Wheeled and Tracked Vehicles*; KHNADU (KHADI): Kharkiv, Ukraine, 2001; 642p.
13. Maharjan, K.L.; Joshi, N.P. *Climate Change, Agriculture and Rural Livelihoods in Developing Countries*; Series: Advances in Asian Human-Environmental Research; Springer: Berlin/Heidelberg, Germany, 2013; 176p.
14. Hunt, D. *Farm Power and Machinery Management*; Wiley-Blackwell: Hoboken, NJ, USA, 2001; 368p.
15. Shuliak, M.; Klets, D.; Kalinin, Y.; Kholodov, A. Selecting a rational operation mode of mobile power unit using measuring and control complex. *CEUR Workshop Proc.* **2019**, *2387*, 141–151. Available online: https://ceur-ws.org/Vol-2387/20190141.pdf (accessed on 25 April 2023).
16. Kuznetsov, E.E. *The Use of the Multi-Axis Energy Devices of Class 1.4: Monograph*; FESAU: Blagoveschensk, Russia, 2013; 153p.
17. Kuznetsov, E.E.; Kuznetsova, O.A.; Polikutina, E.S. Energy consumption dependencies of an experimental machine-tractor aggregate equipped with a trailing weight distribution device. In Proceedings of the International Applied Research Conference, Samara, Russia, 11 April 2017; Current Issues in Science and Technology. pp. 26–29.
18. Kuznetsov, E.E.; Schitov, S.V. Hydro-Chain Regulating Device. Utility Model Patent No. 164794, 20 September 2016.
19. Kuznetsov, E.E.; Schitov, S.V.; Polikutina, E.S. Methodological Validation of the Design For Rational Trailing Weight Distribution Devices. AgroEcoInfo, no. 2. 2016. Available online: http://agroecoinfo.narod.ru/iournal/STATYI/2016/2/st209.doc (accessed on 25 April 2023).
20. Pentoś, K.; Mbah, J.T.; Pieczarka, K.; Niedbała, G.; Wojciechowski, T. Evaluation of Multiple Linear Regression and Machine Learning Approaches to Predict Soil Compaction and Shear Stress Based on Electrical Parameters. *Appl. Sci.* **2022**, *12*, 8791. [CrossRef]
21. Behnassi, M.; Syomiti Muteng'e, M.; Ramachandran, G.; Shelat, K.N. (Eds.) *Vulnerability of Agriculture, Water and Fisheries to Climate Change: Toward Sustainable Adaptation Strategies*; Springer: Dordrecht, The Netherlands, 2014; 336p.
22. Bulinski, J. Effect of Moisture Content in Soil on Its Compaction Caused by Multipie Running of Agricultural Vehicles. In *Annals of Warsaw University of Life Sciences*; Bulinski, J., Niemczyk, H., Eds.; SGGW: Warsaw, Poland, 2007; pp. 3–8.
23. Belyaev, V.I. Ecological Consequences of Conversion of Steppe to arable Land in Western Siberia. *Eur. Reg.* **2004**, *1*, 13–21.
24. Zakharov, N.S. Laws of change of quality of machines during use. In *Operation Efficiency Increase of Wheel and Track Machines under Severe Operating Conditions, Proceedings of the International Scientific and Technical Conference, Evanston, IL, USA, 25–27 July 1996*; Tyumen State Oil and Gas University: Tyumen, Russia, 1996; pp. 110–116.
25. Vorokhobin, A.V.; Vysotskaya, E.A.; Skuryatin, N.F. The reduction in the soil-packing impact caused by ground drive systems of the mobile energy devices. In *Modern Scientific and Practical Solutions for the 21st Century*; Materials of the International Applied Research Conference; OECD: Voronezh, Russia, 2016; pp. 15–19.
26. Selivanov, N.I. Improvement of Classification and Use of Energy-Packed Tractors. In *Krasnoyarsk State Agrarian University Bulletin*; Krasnoyarsk State Agrarian University: Krasnoyarsk Krai, Russia, 2016; Volume 4, pp. 113–119.
27. Selivanov, N.I.; Makeeva, Y.N. Dead Weight Loading of Wheeled Tractors during Soil Cultivation. In *Krasnoyarsk State Agrarian University Bulletin*; Krasnoyarsk State Agrarian University: Krasnoyarsk Krai, Russia, 2015; pp. 77–81.
28. Selivanov, N.I.; Makeeva, Y.N. Adaptation of wheeled tractors to soil cultivation technologies. *Mod. Issues Sci. Educ.* **2015**, 344.
29. Milan, J. *Advanced Transport Systems: Analysis, Modeling, and Evaluation of Performances*; Springer: Berlin/Heidelberg, Germany, 2014; 408p.
30. Mendes, A.; da Silva, E.L.S.; Santos, J.M.A. (Eds.) *Efficiency Measures in the Agricultural Sector: With Applications*; Springer Science & Business Media: Berlin/Heidelberg, Germany, 2012; 197p.
31. Lichtfouse, E. (Ed.) *Sustainable Agriculture Reviews*; Springer: Berlin/Heidelberg, Germany, 2015; Volume 15, p. 407.

agriculture

MDPI

Article

Comparison of Consumed Power and Safety of Two Types of Semi-Automatic Vegetable Transplanter: Cam and Four-Bar Link

Sri Markumningsih [1,2], Seok-Joon Hwang [1,3], Jeong-Hun Kim [1,3], Moon-Kyeong Jang [1,3], Chang-Seop Shin [4,*] and Ju-Seok Nam [1,3,*]

[1] Department of Biosystems Engineering, Kangwon National University, 1 Kangwondaehak-gil, Chuncheon 24341, Republic of Korea
[2] Department of Agricultural and Biosystems Engineering, Faculty of Agricultural Technology, Universitas Gadjah Mada, Jl. Flora No. 1, Bulaksumur, Yogyakarta 55281, Indonesia
[3] Interdisciplinary Program in Smart Agriculture, Kangwon National University, 1 Kangwondaehak-gil, Chuncheon 24341, Republic of Korea
[4] Department of Biosystems Engineering, Chungbuk National University, Chungdae-ro 1, Cheongju 28644, Republic of Korea
* Correspondence: shinchang7@cbnu.ac.kr (C.-S.S.); njsg1218@kangwon.ac.kr (J.-S.N.); Tel.: +82-43-261-2584 (C.-S.S.); +82-33-250-6497 (J.-S.N.)

Abstract: The consumed power and safety of cam and four-bar-link semi-automatic vegetable transplanters were analyzed and compared according to the engine speed and planting distances. A measurement system was constructed to obtain the torque, rotational speed, and strain at the corresponding locations of both transplanters. Field tests were conducted at three engine speeds and three planting distances for each type of transplanter. The torque and rotational speed data of the input shaft of the transplanting devices were used to calculate the power consumed during transplanting. The strain data were converted into stress values to calculate the static safety factor and fatigue life. The results show that the torque and consumed power of the cam transplanter were greater than those of the four-bar-link transplanter under similar operational conditions, owing to its rigid and heavier design. The consumed power increased as the engine speed increased for both types. The static safety factor and fatigue life exhibited different values depending on the measurement location with a sufficient safety margin. Although more skill is required in planting distance control owing to its manual adjustment, the four-bar-link type is more economical under similar operating conditions because of its smaller power requirement.

Keywords: cam type; consumed power; torque; vegetable transplanter; four-bar-link type

Citation: Markumningsih, S.; Hwang, S.-J.; Kim, J.-H.; Jang, M.-K.; Shin, C.-S.; Nam, J.-S. Comparison of Consumed Power and Safety of Two Types of Semi-Automatic Vegetable Transplanter: Cam and Four-Bar Link. *Agriculture* **2023**, *13*, 588. https://doi.org/10.3390/agriculture13030588

Academic Editor: Massimo Cecchini

Received: 10 February 2023
Revised: 23 February 2023
Accepted: 27 February 2023
Published: 28 February 2023

1. Introduction

A vegetable transplanter is an agricultural machine used to transplant vegetable seedlings from a nursery to a field [1,2]. There are two types of vegetable transplanters based on the seedling supply method: semi-automatic and fully automatic. For semi-automatic vegetable transplanters, the operator manually places the seedlings into the seedling cylinder. For fully automatic ones, the seedlings are automatically placed in the seedling cylinder using additional mechanical systems [3–5]. Fully automatic vegetable transplanters have high operational speeds with continuous operation, which saves labor. However, their working mechanism is more complicated and has special requirements, such as seedlings of a specific size and the use of special pot trays [6,7]. Although semi-automatic vegetable transplanters have limited speed, there are no restrictions on seedling size or pot tray type, and the working mechanism is more straightforward [8]. Both types are equally important, depending on the type of vegetable cultivation [9].

Vegetable transplanters are also classified based on the working mechanism of the transplanting device, which is an important component for planting seedlings in soil.

Four types of transplanting devices have been developed: wheel, rotary, four-bar-link, and cam types [10]. The most widely used are the cam and four-bar-link transplanters because of their relatively simple structures and ease of use [11]. The four-bar-link transplanting device uses four-bar-link mechanisms for planting seedlings, whereas the cam transplanter is characterized by a cam that opens and closes the hopper through contact with the bearing [10].

Several studies on both types of transplanting devices have been conducted, including analyses of working mechanisms and operational characteristics [12–14], modification of the transplanting device to improve performance through simulations [2,9], and safety analyses [11,15]. In contrast, research on the consumed power for both semi-automatic transplanting devices has yet to be conducted.

Research on the consumed power and load characteristics of agricultural machinery, such as tractors [16–22], potato harvesters [23], plows and rotavators [24–27], and other types of transplanters (including two-row riding fully automatic vegetable transplanters [28–30] and electric semi-automatic vegetable transplanters [31,32]), has been performed. Several studies on the consumed power and load characteristics of two-row riding fully automatic vegetable transplanters have been conducted, including one by Kim et al. [28], who measured the PTO (Power Take-Off) torque during field operation based on the planting distance. Kim et al. [29] analyzed the PTO shaft load with respect to the planting distance through load spectra and damage-level calculations. Then, Kim et al. [30] studied the PTO load spectrum under various working conditions, not only with various planting distances but also with different planting depths. Another type of vegetable transplanter studied for its power requirements is the electric semi-automatic vegetable transplanter. This machine is a modification of the conventional transplanting machine in which the power source, which was initially a conventional internal combustion engine, is changed into an electric motor. Lee et al. [31] measured the current to calculate and analyze the consumed power according to the planting distance and travel speed. Lim et al. [32] studied the power requirements of this machine based on the torque and rotational speed of the transplanting axle. Despite all the studies on the power consumed by various types of vegetable transplanters, none have been carried out for conventional semi-automatic cam- and four-bar-link transplanters. Therefore, comparative studies on the consumed power and safety of the two transplanting devices are needed for effective transplanting operations.

The purpose of this study was to analyze and compare the consumed power and safety of cam and four-bar-link semi-automatic vegetable transplanters according to engine speed and planting distance. The specific objectives were (1) to measure the torque, rotational speed, and strains on the corresponding spots of both transplanters, (2) to calculate the consumed power, static safety factor, and fatigue life from the measured data, and (3) to compare the torque and consumed power, static safety factor, and fatigue life for the two transplanters under similar operating conditions.

2. Materials and Methods

2.1. Test Equipment

Two types of vegetable transplanters were used in this study: cam and four-bar-link semi-automatic transplanters, which are mainly used on small farms. Figure 1 and Table 1 show the shapes and main specifications, respectively, of the cam semiautomatic vegetable transplanter. The shape and main specifications of the four-bar-link semi-automatic vegetable transplanter are shown in Figure 2 and Table 2, respectively. Both types of vegetable transplanters consist of engines, transmissions, control sections, wheels, seeding cylinders, transplanting devices, and molding wheels.

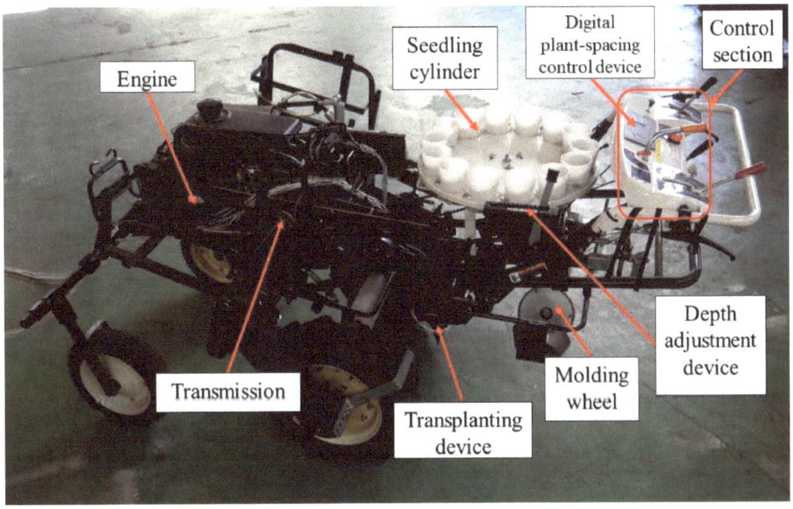

Figure 1. View of cam semi-automatic vegetable transplanter.

Table 1. Main specifications of the cam semi-automatic vegetable transplanter.

Items		Specification
Model		**KP-100KR**
Manufacturer, nation		KUBOTA, Japan
Length/width/height	mm	2150/1360/1130
Weight	kg	280
Engine	Rated power (kW)	2.6
	Rated speed (rpm)	1550
Wheel adjustment (inner/outer)	mm	750–1000/1100–1500
Planting distance	mm	350–900
Maximum working speed	m/s	0.57
Working efficiency	h/10a	1.5–2.5
Price	KRW	13,000,000

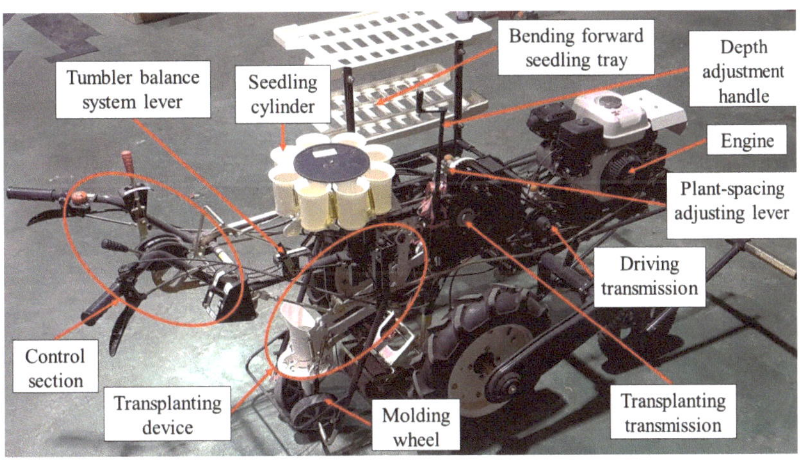

Figure 2. View of four-bar-link semi-automatic vegetable transplanter.

Table 2. Main specifications of the four-bar-link semi-automatic vegetable transplanter.

Items		Specification
Model		**KTP-30**
Length/width/height	mm	2125/1180/1510
Weight	kg	199
Engine	Rated power (kW)	3.4
	Rated speed (rpm)	1800
Wheel adjustment (inner/outer)	mm	360–750/1170–1550
Planting distance	mm	300–500
Maximum working speed	m/s	0.4
Working efficiency	h/10a	1.5–2.0
Price	KRW	4,900,000

The main difference between the two types is the design and working mechanism of the transplanting device. Figure 3 shows the transplanting device of cam and four-bar-link types. The four-bar-link type of transplanting device consists of several linkages and a transplanting hopper. In the cam type, the transplanting device consists of a frame, cam, bearing, crank, transplanting hopper, and some linkages. Additional differences exist in the transmission part and plant-spacing control device. The cam type has only one transmission to transmit the power from the engine to the wheel and transplanting device, while the four-bar-link type has (a) a driving transmission to transmit the engine power to the wheel and the planting unit and (b) a transplanting transmission to transmit the power received from the driving transmission to the crank of the four-bar-link mechanism of the transplanting device. The planting distance in the cam type is determined from the set value of the digital plant-spacing control device. The cam transplanter has a distance sensor that can measure the traveling distance of the transplanter. If the travel distance reaches the set planting distance in the digital plant-spacing control device, the transplanting device input shaft operates at a constant rotational speed so that it remains constant for various planting distances. In contrast, in the four-bar-link transplanter, the planting distance is determined by adjusting the number of shift stages of the transplanting transmission using a manual lever, which requires empirical skill to adjust the planting distance precisely. At a constant working speed, if the rotational speed of the input shaft of the transplanting device increases, the transplantation distance decreases.

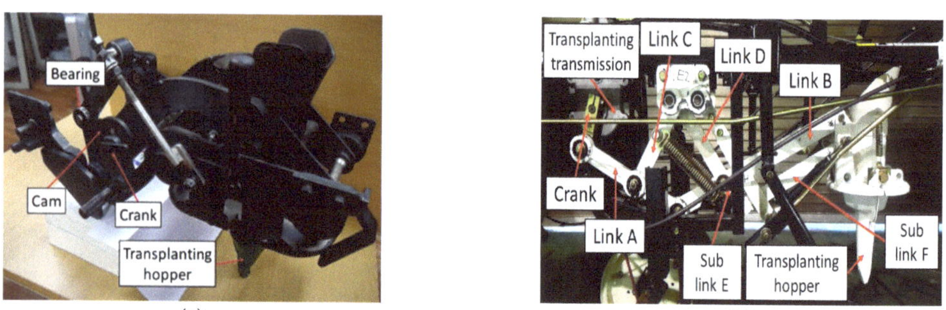

(**a**) (**b**)

Figure 3. Shapes of the transplanting devices of the semi-automatic vegetable transplanter types: (**a**) cam and (**b**) four-bar link.

In general, the working mechanism of a semi-automatic vegetable transplanter is as follows. The operator starts the engine and sets the desired planting distance digitally in the control section (for the cam type) or manually using the lever (for the four-bar-link type). The transplanter moves forward, and the transplanting device moves up and down. The operator places the seedlings manually in the seedling cylinder. The seedling cylinder is opened when the transplanting hopper of the transplanting device is in the top position to place the seedling into the hopper. Then, the transplanting device moves down, and the hopper is opened when it is in the bottom position to plant the seedling in the soil. The transplanting device moves up again, while the molding wheel covers both sides of the planted seedling with soil.

2.2. Work Conditions

Field tests were conducted by considering the primary operational conditions for vegetable transplanting. Semi-automatic vegetable transplanters usually work on ridges with soft tilled soil; therefore, the effect of soil strength can be ignored. The variables of the working conditions in this study were engine speed and planting distance. For the cam vegetable transplanter, the engine speeds were set to 1100, 1250, and 1550 rpm, and the planting distances were 0.35, 0.40, and 0.45 m; for the four-bar-link vegetable transplanter, the engine speeds were 1000, 1250, and 1500 rpm, and the planting distances were 0.35, 0.41, and 0.45 m. There are few differences in the engine speed and planting distance conditions between the cam and the four-bar-link types because fine adjustment is difficult. However, the difference is small and can be ignored when comparing the overall trend or tendency of consumed power and safety. The planting depth was set at 70 mm. The test was repeated three times for each working condition, and the data were analyzed using the average value. The experimental field was located in Sinbuk-eup, Chuncheon, Gangwon Province, in South Korea (37°56′24.0″ North and 127°46′59.1″ East with an altitude of 111.00 m.a.s.l.). The field consisted of ridges with a width and height of 0.6 and 0.3 m, respectively.

2.3. Measurement and Analysis

The torque and rotational speed of the input shaft of the transplanting device were derived to investigate the power consumed by the transplanting work of each type of vegetable transplanter. In addition, the operational stress of the transplanting device was measured for the safety analysis. The measurement system consisted of a torque sensor, an RPM sensor, strain sensors, a data acquisition device (DAQ), and a laptop. Figure 4 shows the instrument setup for the measurement system of the cam vegetable transplanter, whereas the measurement system of the four-bar-link type is shown in Figure 5. The signal from the sensors was transmitted to the DAQ through a telemetry system, and the data were stored on a laptop. The sampling frequency was set to 100 Hz after preliminary testing.

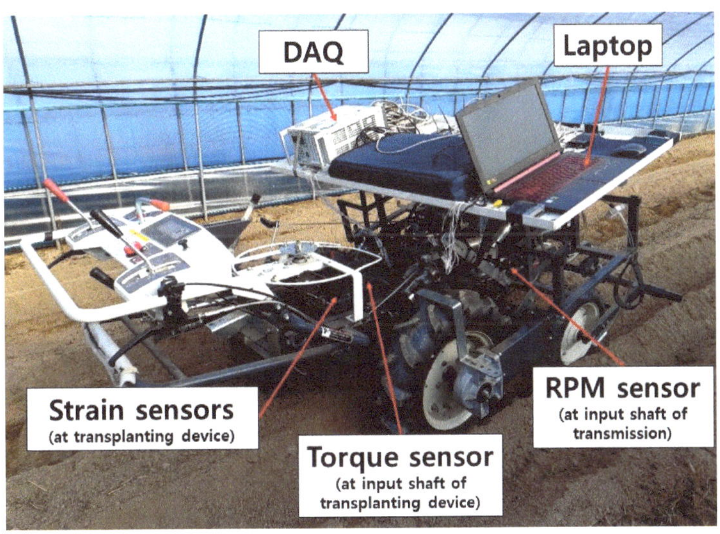

Figure 4. Measurement system for the cam vegetable transplanter.

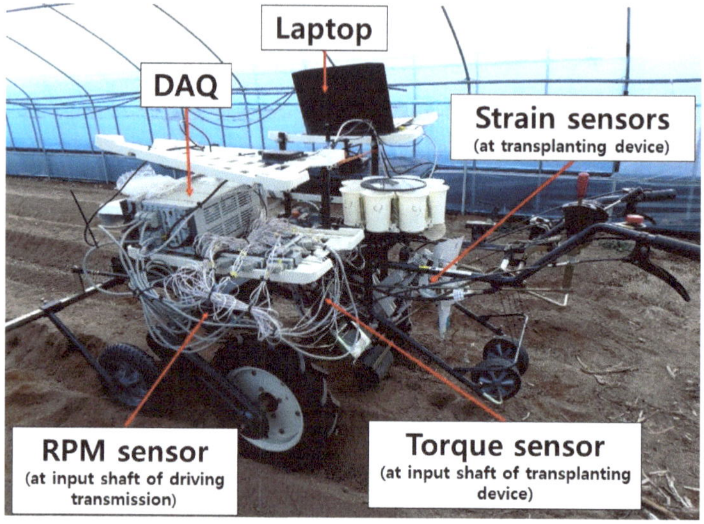

Figure 5. Measurement system for the four-bar-link vegetable transplanter.

Figures 6 and 7, respectively, show the locations of the torque sensor and rpm sensor for the cam and four-bar-link types. A torque sensor was installed at the input shaft of the transplanting device, and the rpm sensor was attached to the transmission input shaft. The rotational speed of the input shaft of the transplanting device was derived from the rotational speed of the transmission input shaft and the reduction gear ratio between the two shafts (Figure 8). The power consumed by the transplanting work was then calculated using Equation (1) [23].

$$P = \frac{2\pi \times T \times N}{60,000} \tag{1}$$

where P is the power consumed by the transplanting work (kW), T is the torque at the transplanting device input shaft (Nm), and N is the rotational speed of the transplanting device input shaft (rpm).

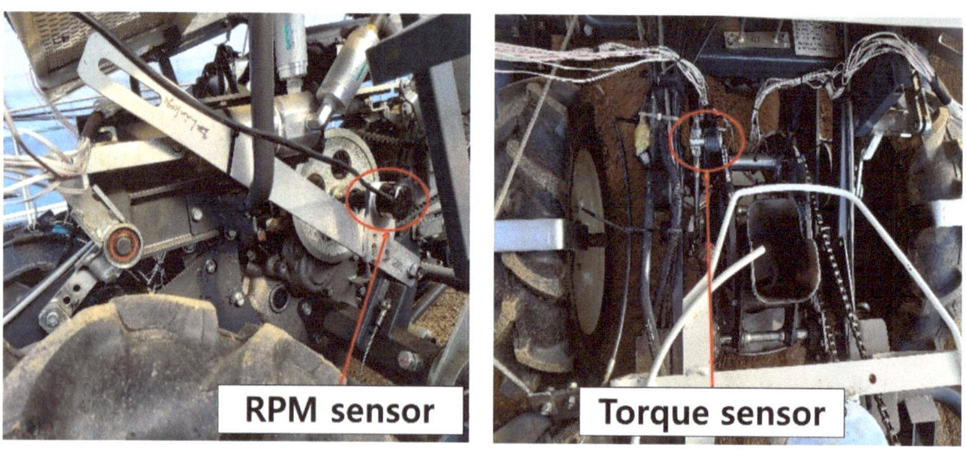

Figure 6. Location of torque and rpm sensors installed on cam vegetable transplanter.

Figure 7. Location of torque and rpm sensor installed on four-bar-link vegetable transplanter.

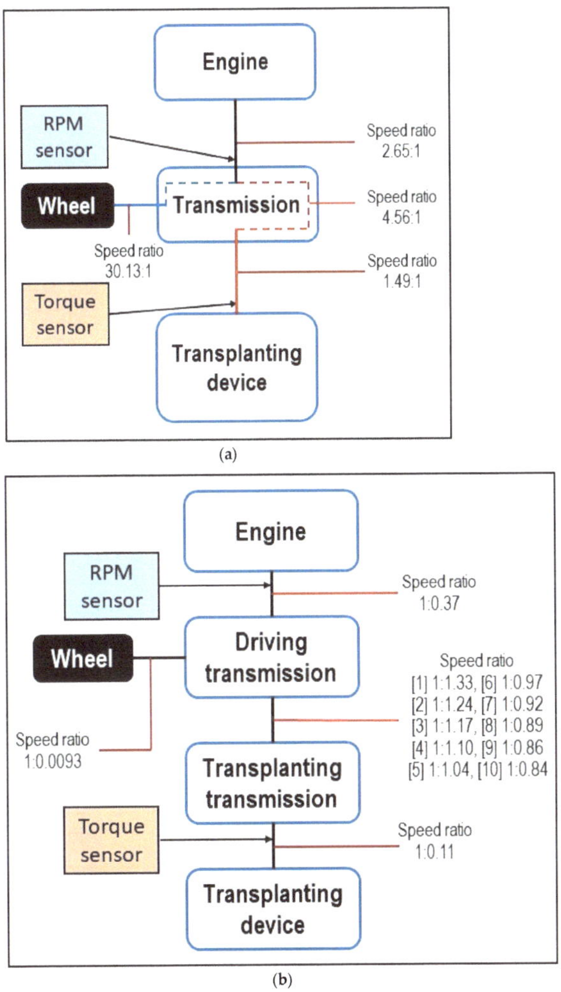

Figure 8. Reduction gear ratio of the transmission system: (**a**) cam and (**b**) four-bar-link types.

The strain sensors comprised 15 strain gauges attached to the main loading paths of the transplanting device to measure the stress during transplanting. Two types of strain gauges were used: a uniaxial strain gauge (KFGS-5-350-C1-11 L10M3R, KYOWA) to measure one-directional strain in the main loading direction, such as at linkages, and a triaxial strain gauge or rosette (KFGS-1-350-D17-11 L5M3S, KYOWA) to measure the strain in three directions in the main loading direction, such as at the surface of the transplanting hopper. The detailed attachment locations of the strain gauges are shown in Figure 9 and Table 3.

The measured strain data were converted into stress data for the safety analysis. Two types of safety analysis were conducted: static safety and dynamic safety (or fatigue life). For the static safety analysis, the static safety factor was calculated using the yield strength of the transplanting device material and the maximum stress that occurred during operation—see Equation (2). As for the maximum stresses, the maximum normal stress was used for uniaxial strain gauges, while the maximum von Mises stress was used for triaxial strain gauges. If the static safety factor is greater than 1.0, the design is considered statically safe. If the static safety factor is less than 1.0, this implies that the design is unsafe or that the part may fail.

$$SF = \frac{S_y}{\sigma_{max}} \quad (2)$$

where SF is the static safety factor, S_y is the yield strength of the transplanting device material (Pa), and σ_{max} is the maximum operational stress (Pa).

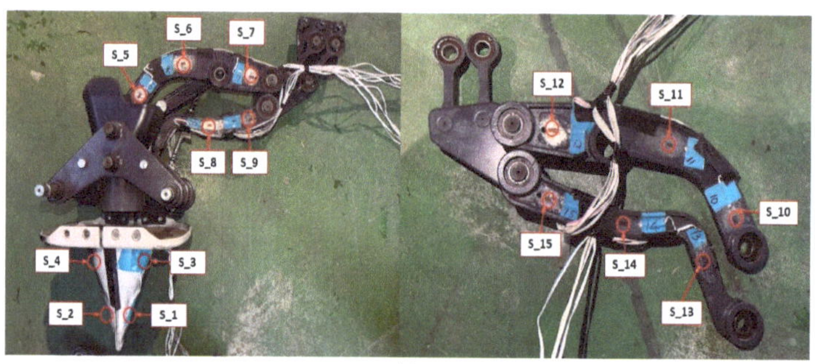

(**a**)

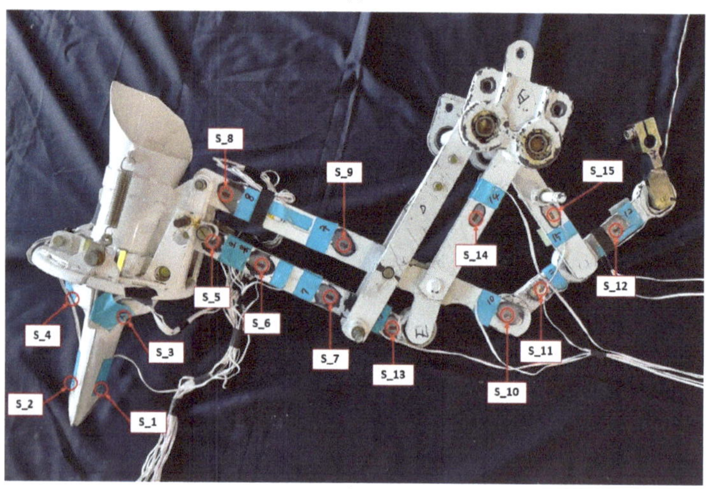

(**b**)

Figure 9. Location of the strain gauge on the transplanting device: (**a**) cam and (**b**) four-bar-link types.

Table 3. Detailed description of strain measurement location.

Strain Gauge Number	Locations	
	Cam Type	**Four-Bar-Link Type**
S_1	front lower side of transplanting hopper	front lower side of transplanting hopper
S_2	back lower side of transplanting hopper	back lower side of transplanting hopper
S_3	front upper side of transplanting hopper	front upper side of transplanting hopper
S_4	back upper side of transplanting hopper	back upper side of transplanting hopper
S_5	left side of left upper link	left side of sublink F
S_6	middle side of left upper link	middle side of sublink F
S_7	right side of left upper link	right side of sublink F
S_8	middle side of left bottom link	left side of link B
S_9	right side of left bottom link	right side of link B
S_10	right side of right upper link	right side of link B
S_11	middle side of right upper link	left side of link A
S_12	left side of right upper link	right side of link A
S_13	right side of right bottom link	right side of sublink F
S_14	middle side of right bottom link	middle side of link D
S_15	left side of right bottom link	middle side of link C

Fatigue life analysis is also required because the transplanting device is subjected to repeated cyclic loads when the vegetable transplanter is working. Figure 10 shows the fatigue life analysis procedure.

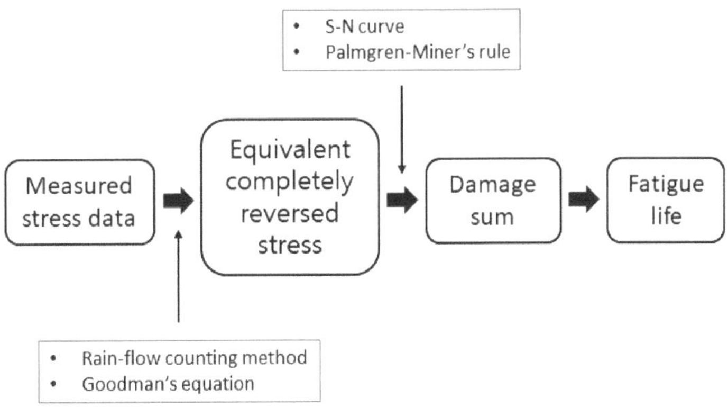

Figure 10. Fatigue life calculation procedure.

The measured stress was converted into equivalent, completely reversed stresses and their corresponding loading cycles by applying the rain-flow counting method and Goodman's equation [33,34]. The rain-flow counting method is used to obtain the loading cycles for each mean stress and stress amplitude from time-series stress data. Then mean stress and amplitude were used to derive the equivalent completely reversed stress using Goodman's equation—see Equation (3). The actual applied loading cycle (n_i) can be derived from the rain-flow counting. The life cycles (N_i) were determined from the S–N curve using the equivalent completely reversed stresses [35]. The ratio between the actual applied loading cycles and the life cycles is called "partial damage". The damage sum was calculated by summing the partial damages from all stresses that occurred during the operation using the Palmgren–Miner rule—see Equation (4) [36]. Fatigue failure occurs when the damage sum reaches 1.0 [37]. The damage sum and working time were used to calculate the fatigue life of the transplanting device, as shown in Equation (5). The 25.5 h of average annual usage time of vegetable transplanters was considered to determine the fatigue life in years [11].

$$\sigma_{eq} = \frac{S_u \sigma_a}{S_u - \sigma_m} \quad (3)$$

$$D = \sum_i \frac{n_i}{N_i} \quad (4)$$

$$L_f = \frac{1}{D} \times t \quad (5)$$

where σ_{eq} is the equivalent completely reversed stress (Pa), S_u is the ultimate strength (Pa), σ_a is the stress amplitude (Pa), σ_m is the mean stress (Pa), D is the cumulative damage sum, n_i is the number of applied loading cycles for the equivalent completely reversed stress i (cycles), N_i is the life cycles for the equivalent completely reversed stress i (cycles), L_f is the fatigue life (s), and t is the working time that generates the cumulative damage sum (s).

3. Results

3.1. Torque and Consumed Power

Figure 11 shows the time series torque data on the transplanting device input shaft according to the three engine speed conditions at a planting distance of 0.35 m, while Figure 12 shows the torque data according to the three planting distances at an engine speed of 1250 rpm. The shape and magnitude of the operational torque between the cam and the four-bar-link types are different because they have different designs and movement mechanisms. However, both have similarities in a regular fluctuation pattern because of the repeated up and down motions of the transplanting device.

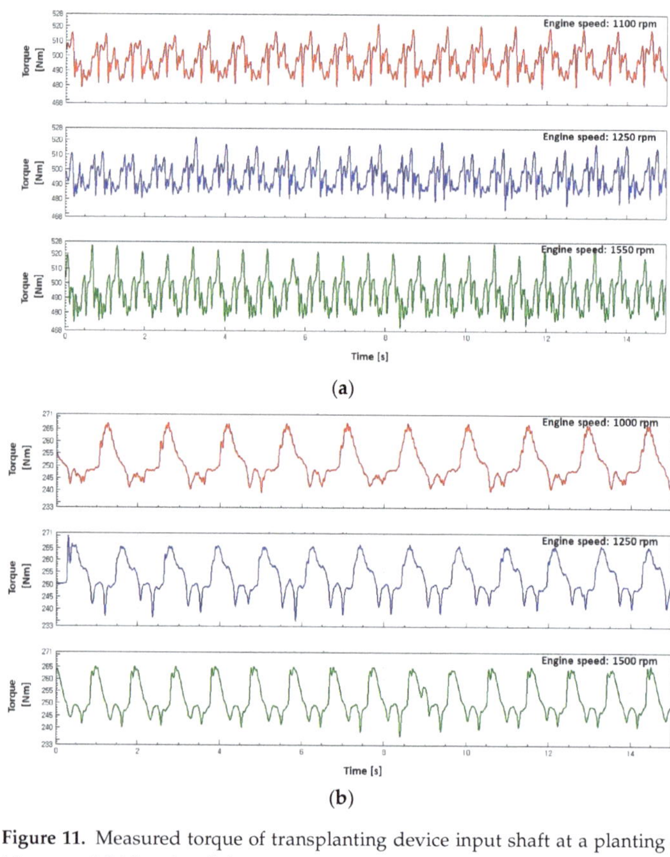

Figure 11. Measured torque of transplanting device input shaft at a planting distance of 0.35 m: (**a**) cam and (**b**) four-bar-link types.

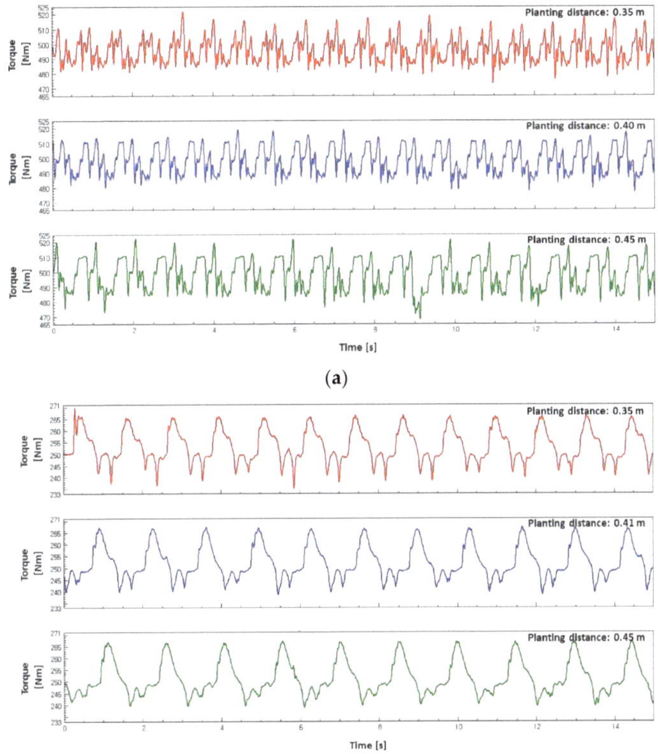

(**a**)

(**b**)

Figure 12. Measured torque of the transplanting device input shaft at an engine speed of 1250 rpm: (**a**) cam and (**b**) four-bar-link types.

Table 4 lists the average torque, rotational speed, and consumed power of the transplanting device input shaft of the two vegetable transplanters under different working conditions. Figure 13 shows a comparison of the average torque on the input shaft of the transplanting device. The cam type had a greater torque than the four-bar-link type under similar operating conditions. For the same type, the average torque showed no significant differences according to the engine speed and planting distance because the transplanting work was conducted under uniformly soft soil conditions.

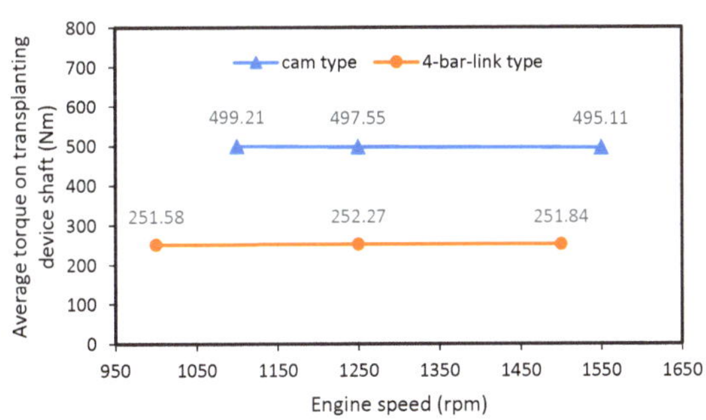

(a)

Figure 13. *Cont.*

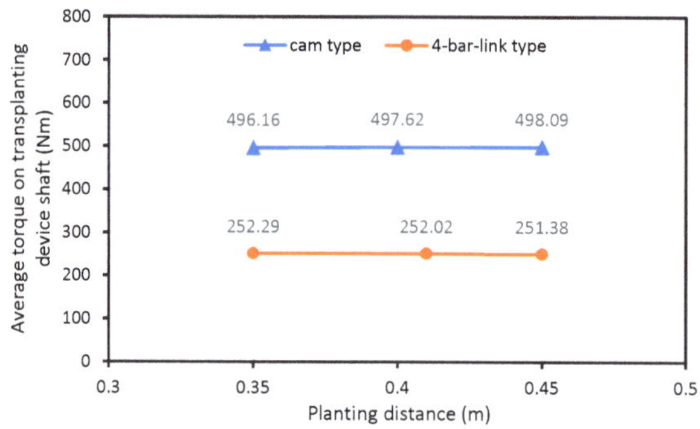

(**b**)

Figure 13. Comparison of torque on the transplanting device input shaft between cam and four-bar-link types according to working conditions: (**a**) engine speed and (**b**) planting distance.

Table 4. Torque, rotational speed, and consumed power of the transplanting device input shaft under different working conditions.

Type of Vegetable Transplanter	Engine Speed (rpm)	Planting Distance (m)	Torque of Transplanting Device Input Shaft (Nm)	Rotational Speed of Transplanting Device Input Shaft (rpm)	Consumed Power of the Transplanting Device Input Shaft (kW)
Cam	1100	0.35	497.55	66.3	3.45
		0.4	499.94	67.15	3.4
		0.45	500.15	63.44	3.32
	1250	0.35	496.1	73.64	3.83
		0.4	497.94	73.61	3.84
		0.45	498.61	73.56	3.84
	1550	0.35	494.83	89.94	4.66
		0.4	494.98	90.04	4.67
		0.45	495.51	90.23	4.68
Four-bar-link type	1000	0.35	251.66	41.24	1.09
		0.41	251.72	36.29	0.96
		0.45	251.36	31.83	0.84
	1250	0.35	253.2	50.78	1.35
		0.41	252.19	44.4	1.17
		0.45	251.41	42.38	1.12
	1500	0.35	252.01	63.02	1.66
		0.41	252.14	51.92	1.37
		0.45	251.37	48.01	1.26

Figure 14 shows a comparison of the consumed power of the transplanting device inputs for the two types. The consumed power based on engine speed is shown in Figure 14a, obtained from the average consumed power for three variations in planting distance. The consumed power according to the planting distance shown in Figure 14b is obtained from the average consumed power for three variations in engine speed. The consumed power of the cam type is almost three times greater than that of the four-bar-link type. The higher torque and consumed power of the cam type result from the rigid and heavier design of the transplanting device when compared to the four-bar-link type. The total weight of the cam type, including the transplanting device, is more than 1.4 times greater than that of the four-bar-link type, as shown in Tables 1 and 2.

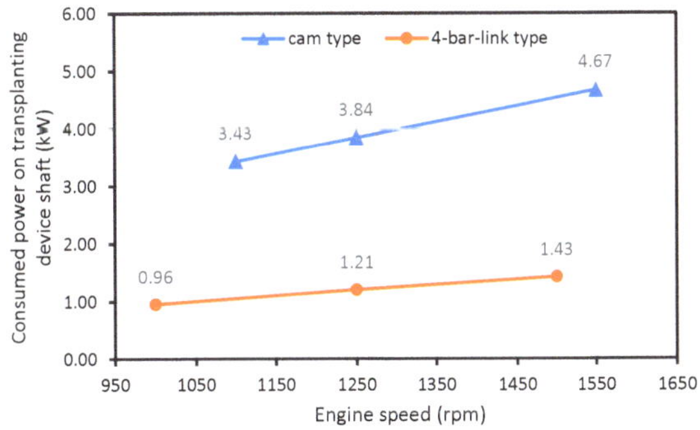

(a)

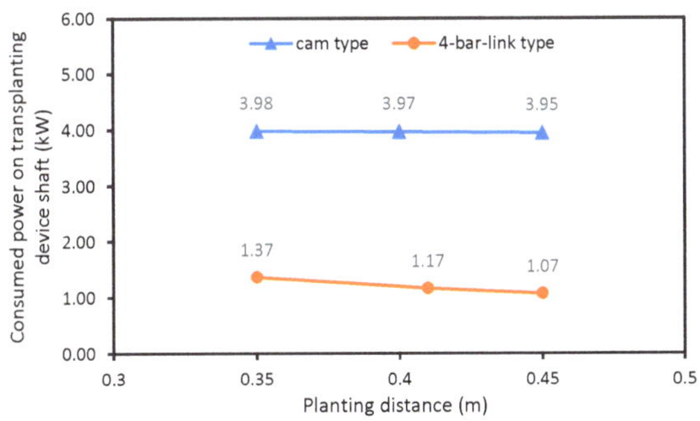

(b)

Figure 14. Comparison of consumed powers on transplanting device input shafts between cam and four-bar-link types according to working conditions: (**a**) engine speed and (**b**) planting distance.

The consumed power for both types showed a tendency to increase when the engine speed increased and planting distance decreased. The engine speed is directly proportional to the rotational speed of the input shaft of the transplanting device. The higher the engine speed, the higher the rotational speed of the transplanting device input shaft; thus, the consumed power also increases—see Figure 14a.

When the machine operates at the same engine speed, the difference in planting distance causes the period of one transplanting cycle to be different. The shorter the planting distance, the faster the period, and the higher the frequency. The mechanism for determining the planting distance of the cam transplanter uses a digital plant-spacing controller, which makes the rotational speed of the transplanting device input shaft constant for various planting distances. The torque of the shaft is similar according to the planting distance (Figure 13b). Therefore, the consumed power for cam transplanters based on planting distance showed no significant differences. However, the four-bar-link type determines the planting distance by adjusting the stages of the transplanting transmission, which makes the rotational speed of the input shaft of the transplanting device different for each planting distance. When the planting distance increases, the rotational speed decreases but the torque remains constant (Figure 13b); thus, the consumed power decreases—see Figure 14b.

3.2. Static Safety Factor

The static safety factors for each strain measurement spot in the transplanting devices are listed in Table 5. The results show that the static safety factor is greater than 1.0 for all locations and working conditions. This indicates that both types of vegetable transplanters are designed to be safe for static loads. The minimum static safety factor for the cam type was 3.35 at location S_4 at an engine speed of 1550 rpm and a planting distance of 0.35 m. For the four-bar-link type, it was 4.05 at location S_14 under an engine speed of 1500 rpm and a planting distance of 0.41 m. Location S_4 of the cam type, on the upper side of the hopper, has the minimum static safety factor because it receives a high reaction force when the hopper hits the soil and is subjected to a significant peak load. Location S_14 (link D) of the four-bar-link type is on the linkage connected to the body-fixing frame; therefore, it receives high reaction forces when the transplanting device is operated. In the four-bar-link type, the linkages are relatively thin and flexible compared with the cam type. Therefore, the location of the minimum static safety factor is one of the linkages of the four-bar-link type.

Figure 15 shows the tendency of static safety factor for both types. The four-bar-link type has a smaller static safety factor for many parts than the cam type. This is because of the thin and flexible linkage structure of the four-bar-link type.

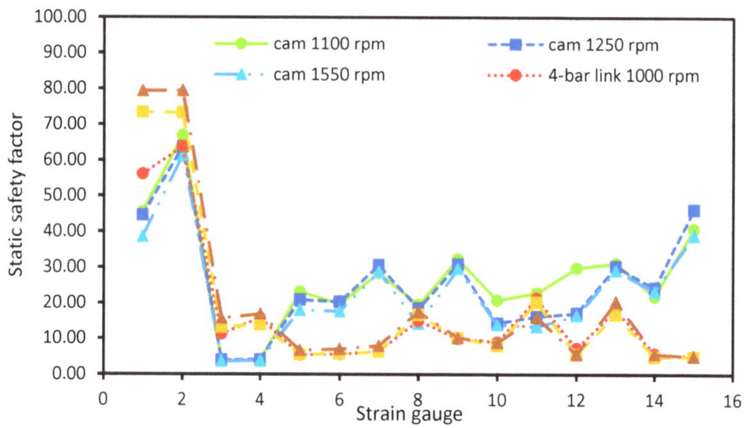

(**a**)

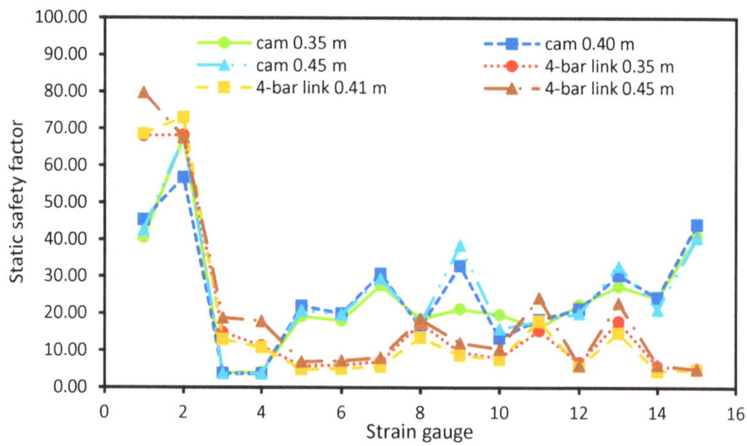

(**b**)

Figure 15. Comparison of static safety factor between cam and four-bar-link types on every strain gauge location in the transplanting device according to working conditions: (**a**) engine speed and (**b**) planting distance.

Table 5. Static safety factor on every spot of strain gauge in cam and four-bar-link types of vegetable transplanters under each working condition.

Type of Vegetable Transplanter	Engine Speed (rpm)	Planting Distance (m)	S_1	S_2	S_3	S_4	S_5	S_6	S_7	S_8	S_9	S_10	S_11	S_12	S_13	S_14	S_15
Cam Type	1100	0.35	39.21	73.85	4.41	4.18	23.25	20.07	28.3	22.29	20.84	20	19.51	28.29	28.22	20.43	38.96
		0.4	49.14	67.18	3.79	3.78	22.53	19.55	30.24	17.79	34.83	18.69	23.14	31.89	30.58	26.21	13.23
		0.45	47.99	59.81	3.84	3.37	23.81	19.7	26.58	18.54	41.35	23.74	25.88	26.44	34.74	18.96	40.26
	1250	0.35	45.94	68.86	3.83	4.89	17.69	17.24	26.08	17.47	15.79	21.02	14.07	18.46	27	30.45	47.71
		0.4	45.77	55.69	3.91	3.73	23.83	22.77	34.03	17.78	31.16	9.85	18.13	14.36	32.78	24.12	50.16
		0.45	41.75	66.64	4.08	3.6	21.15	21.03	31.62	20.12	45.41	12.16	16.56	18.85	31.09	18.31	40.62
	1550	0.35	36.31	60.12	3.7	3.35	16.62	16.87	28.26	15.66	27.09	18.04	14.86	20.33	26.79	21.51	36.53
		0.4	41.21	47.58	3.59	3.61	19.89	17.6	27.99	13.51	32.8	11.74	13.78	14.37	27.87	22.74	39.35
		0.45	38.12	76.39	3.52	4.28	17.56	18.44	30.02	13.32	28.9	11.74	11.36	14.82	32.75	25.83	40.83
Four-Bar-Link Type	1000	0.35	61.42	72.83	8.12	15.52	5.03	5.33	6.09	15.66	8.61	8.49	19.12	8.81	18.67	6.55	5.23
		0.41	60.7	62.83	8.28	13.65	4.41	4.67	5.26	10.83	7.54	8.17	16.97	6.79	13.98	4.29	4.95
		0.45	46.29	55.6	17.23	18.25	6.73	6.95	7.67	18.54	13.7	10.29	28.01	6.28	21.95	6.54	5.11
	1250	0.35	60.46	62.32	9.58	10.38	4.43	4.67	5.15	15.82	8.41	6.4	15.19	5.53	12.32	4.47	5.05
		0.41	88.82	67.85	14.3	14.82	5.67	5.83	6.33	16.78	10.27	8.2	22.91	5.56	17.58	5.02	5.02
		0.45	70.81	89.57	15.55	16.44	6.48	6.68	7.35	17.41	12.06	9.49	22.61	5.88	20.43	5.24	5.04
	1500	0.35	82.72	68.91	16.21	18.62	7.78	8.26	9.5	20.46	11.75	8.97	11.88	6.03	22.58	6.36	5.19
		0.41	69.72	75.16	9.89	10.46	4.63	4.76	5.17	12.94	8.37	6.42	13.94	4.89	12.72	4.05	4.95
		0.45	85.58	94.2	20.95	21.66	7.61	8.13	9.12	19.21	10.45	11.38	22.18	6	26.31	6.68	5.02

3.3. Fatigue Life

Table 6 lists the fatigue life data for every strain measurement point on the transplanting devices. The minimum fatigue life for the cam type was 95,603 years at location S_3 at an engine speed of 1550 rpm and a planting distance of 0.35 m. However, the minimum fatigue life for the four-bar-link type was 196,000 years at location S_14 under an engine speed of 1500 rpm and a planting distance of 0.41 m. Location S_3 on the cam type is on the upper side of the hopper and receives a large load when hitting the ground, so that these parts have the minimum fatigue life. However, the maximum stress value on S_3 is slightly lower than that at location S_4; therefore, there are differences in locations that have the minimum static safety factor and minimum fatigue life on the cam type. For the four-bar-link type, the minimum static safety factor and fatigue life occur at the same location (S_14).

Figure 16 shows the fatigue life tendency for both types. The cam type has more parts with a longer fatigue life than the four-bar-link type. However, for the transplanting device, the four-bar-link type exhibited a longer fatigue life. This is because the many linkages in the four-bar-link transplanting device can distribute the cyclic transplanting load more efficiently from the hopper to the linkages.

Table 6. Fatigue life (years) on every spot of strain gauge in cam and four-bar-link types of vegetable transplanters under each working condition.

Type of Vegetable Transplanter	Engine Speed (rpm)	Planting Distance (m)	S_1	S_2	S_3	S_4	S_5	S_6	S_7	S_8	S_9	S_10	S_11	S_12	S_13	S_14	S_15
Cam Type	1100	0.35	1.60E + 11	1.10E + 13	1.63E + 05	4.52E + 05	8.00E + 08	4.00E + 08	2.00E + 09	7.00E+08	5.80E + 10	3.00E + 09	4.00E + 08	5.00E + 09	2.00E + 09	7.90E + 09	1.70E + 10
		0.4	2.30E + 11	7.70E + 12	2.28E + 05	1.00E + 06	5.00E + 08	6.00E + 08	1.00E + 10	1.00E + 09	8.10E + 10	3.00E + 09	4.00E + 09	1.00E + 11	1.30E + 10	3.00E + 10	5.30E + 10
		0.45	1.30E + 11	6.30E + 12	1.66 E + 05	3.87E + 05	5.00E + 08	6.00E + 08	9.00E + 09	1.00E + 09	9.40E + 10	6.00E + 09	5.00E + 09	4.00E + 10	2.30E + 10	7.00E + 09	3.10E + 10
	1250	0.35	7.70E + 10	6.50E + 12	1.48 E + 05	8.69E + 05	2.00E + 08	5.00E + 08	4.00E + 09	6.00E + 08	5.70E + 09	5.00E + 09	2.00E + 08	3.00E + 09	4.60E + 09	5.40E + 09	3.40E + 10
		0.4	1.90E + 11	1.30E + 13	3.03 E + 05	4.41E + 05	1.00E + 09	1.00E + 09	3.00E + 10	2.00E + 09	1.10E + 11	1.00E + 08	1.00E + 09	1.00E + 09	2.60E + 10	1.10E + 10	3.70E + 10
		0.45	6.90E + 10	3.50E + 13	2.79 E + 05	7.00E + 05	8.00E + 08	8.00E + 08	2.00E + 10	2.00E + 09	1.80E + 11	6.00E + 08	5.00E + 08	3.00E + 09	1.70E + 10	4.10E + 09	3.40E + 10
	1550	0.35	4.30E + 10	1.90E + 12	9.56 E + 04	3.10E + 05	2.00E + 08	7.00E + 08	4.00E + 09	3.00E + 08	5.00E + 10	1.00E + 09	1.00E + 08	1.00E + 09	1.60E + 09	1.30E + 09	1.10E + 10
		0.4	5.30E + 10	2.10E + 12	9.72 E + 04	4.80E + 05	4.00E + 08	3.00E + 08	4.00E + 09	4.00E + 08	4.00E + 10	4.00E + 08	2.00E + 08	4.00E + 08	5.60E + 09	1.10E + 09	1.50E + 10
		0.45	5.40E + 10	2.90E + 12	2.08 E + 05	5.08E + 05	1.00E + 08	3.00E + 08	5.00E + 09	5.00E + 08	5.80E + 10	4.00E + 08	4.00E + 07	3.00E + 08	6.20E + 09	2.00E + 09	1.80E + 10
Four-Bar-Link Type	1000	0.35	3.60E + 13	1.30E + 14	2.00E + 09	9.00E + 10	1.00E + 07	2.00E + 07	5.00E + 07	9.00E + 08	4.00E + 07	4.00E + 07	2.00E + 09	8.00E + 07	1.30E + 10	1.30E + 07	1.25E + 06
		0.41	4.20E + 12	1.50E + 13	1.00E + 09	1.00E + 10	1.00E + 06	2.00E + 06	4.00E + 06	1.00E + 08	5.49E + 06	1.00E + 07	8.00E + 08	1.00E + 07	7.70E + 08	6.22E + 05	9.89E + 05
		0.45	1.30E + 15	5.20E + 15	2.00E + 12	2.00E + 12	5.00E + 08	7.00E + 08	9.00E + 08	1.00E + 07	1.50E + 09	2.00E + 08	1.00E + 10	4.00E + 07	1.20E + 11	2.20E + 08	1.13E + 06
	1250	0.35	5.90E + 12	3.50E + 12	5.00E + 09	4.00E + 09	7.79E + 05	1.00E + 06	2.00E + 06	2.00E + 08	5.49E + 06	7.00E + 06	2.00E + 08	1.00 E + 06	2.50E + 08	3.05E + 05	7.83E + 05
		0.41	3.20E + 13	5.20E + 14	3.00E + 10	2.00E + 10	5.00E + 06	6.00E + 06	1.00E + 07	5.00E + 08	1.80E + 07	2.00E + 07	1.00E + 09	2.00E + 06	2.10E + 09	6.48E + 05	9.36E + 05
		0.45	9.40E + 12	3.70E + 15	2.00E + 11	5.00E + 10	2.00E + 07	2.00E + 07	5.00E + 07	1.00E + 09	9.30E + 07	9.00E + 07	2.00E + 09	4.00E + 06	1.00E + 10	2.54E + 06	9.29E + 05
	1500	0.35	4.00E + 13	4.00E + 12	5.00E + 10	8.00E + 10	1.00E + 08	2.00E + 08	3.00E + 08	3.00E + 09	1.10E + 08	2.00E + 07	2.00E + 08	2.00E + 06	3.20E + 10	1.74E + 06	7.17E + 05
		0.41	4.50E + 12	3.70E + 12	9.00E + 09	3.00E + 09	1.00E + 06	1.00E + 06	3.00E + 06	1.00E + 08	4.51E + 06	7.00E + 06	2.00E + 08	1.00E + 06	2.60E + 08	1.96E + 05	6.85E + 05
		0.45	5.00E + 13	1.90E + 15	4.00E + 11	8.00E + 10	6.00E + 07	9.00E + 07	2.00E + 08	3.00E + 09	6.20E + 07	1.00E + 08	2.00E + 09	4.00E + 06	2.30E + 10	5.88E + 06	8.97E + 05

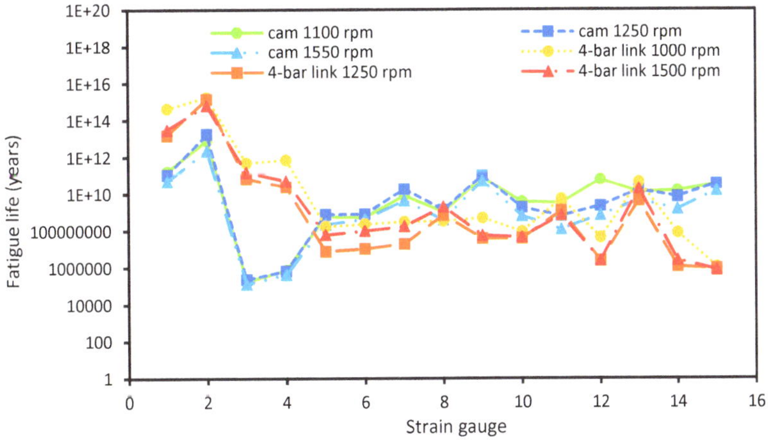

(a)

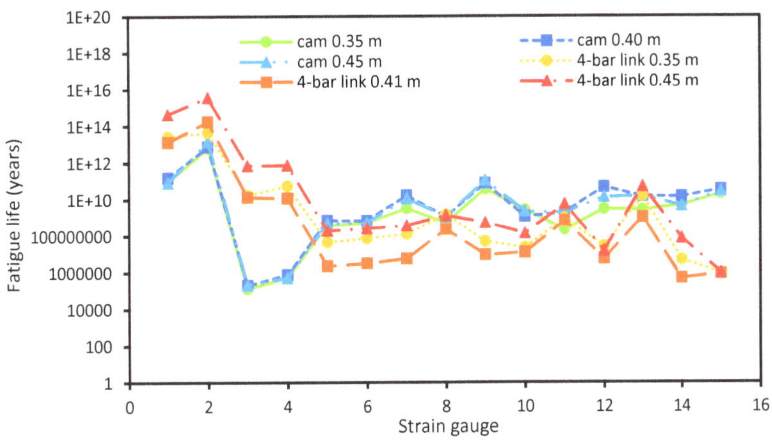

(b)

Figure 16. Comparison of fatigue life between cam and four-bar-link types on every strain gauge location on the transplanting device according to working conditions: (**a**) engine speed and (**b**) planting distance.

4. Conclusions

The torque, rotational speed, and strain were measured for cam and four-bar-link semi-automatic transplanting devices. The field test was conducted under various working conditions. The cam type has three engine speeds (1100, 1250, and 1550 rpm) and three planting distances (0.35, 0.40, and 0.45 m). The four-bar-link type has three engine speeds (1000, 1250, and 1500 rpm) and three planting distances (0.35, 0.41, and 0.45 m). The measured data were used to calculate and analyze the consumed power, static safety factor, and fatigue life of both types of transplanting devices.

The main results of this research are as follows:

1. Under similar operating conditions, the cam type had a greater torque and consumed more power than the four-bar-link type owing to its rigid and heavier design. The range of the consumed power on the transplanting device input shaft of the cam type was 3.32–4.68 kW, while it was 0.84–1.66 kW for the four-bar-link type. The

consumed power on the transplanting device input shaft increased when the engine speed increased and the planting distance decreased for both types.

2. The static safety factor was greater than 1.0 for both types at all measurement locations and under all working conditions. The minimum static safety factor for the cam type was 3.35 on the upper side of the transplanting hopper (S_4) at an engine speed of 1550 rpm and a planting distance of 0.35 m. For the four-bar-link type, the minimum static safety factor was 4.05 at one of the linkages (link D) at an engine speed of 1500 rpm and a planting distance of 0.41 m.
3. The minimum fatigue life for the cam type was 95,603 years at the upper side of the transplanting hopper (S_3) at an engine speed of 1550 rpm and a planting distance of 0.35 m. The minimum fatigue life for the four-bar-link type was 196,000 years at the same location with the minimum static safety factor (link D) at an engine speed of 1500 rpm and a planting distance of 0.41 m.
4. The rated work efficiencies of the cam and four-bar-link vegetable transplanters are similar. The cam type had a digital plant-spacing control device so that workers could work comfortably. However, considering the power consumption and price of the machine, the four-bar-link vegetable transplanter is highly economical.

Author Contributions: Data curation and formal analysis, S.M., S.-J.H., J.-H.K. and M.-K.J.; writing—original draft, S.M. and S.-J.H.; writing—review and editing, J.-S.N. and C.-S.S. All authors have read and agreed to the published version of the manuscript.

Funding: This work was supported by Korea Institute of Planning and Evaluation for Technology in Food, Agriculture and Forestry (IPET) through Agriculture, Food and Rural Affairs Convergence Technologies Program for Educating Creative Global Leader, funded by Ministry of Agriculture, Food and Rural Affairs (MAFRA) (320001-4), Republic of Korea. This research was also supported by Basic Science Research Program through the National Research Foundation of Korea (NRF) funded by the Ministry of Education (NRF-2020R1I1A3054353).

Institutional Review Board Statement: Not applicable.

Informed Consent Statement: Not applicable.

Data Availability Statement: Not applicable.

Conflicts of Interest: The authors declare no conflict of interest.

References

1. Kim, S.K.; Park, S.; Kwak, J.-H.; Choi, S.-K.; Chae, W.B.; Yang, E.Y.; Lee, M.J.; Jang, Y.; Seo, M.H.; Lee, S.H.; et al. Proper Plant Density for Mechanical Transplanting of Several Leafy Vegetables Under Korean Agricultural Condition. *J. Biosyst. Eng.* **2019**, *44*, 276–280. [CrossRef]
2. Hwang, S.-J.; Park, J.-H.; Lee, J.-Y.; Shim, S.-B.; Nam, J.-S. Optimization of Main Link Lengths of Transplanting Device of Semi-Automatic Vegetable Transplanter. *Agronomy.* **2020**, *10*, 1938. [CrossRef]
3. Khadatkar, A.; Mathur, S.M.; Gaikwad, B.B. Automation in Transplanting: A Smart Way of Vegetable Cultivation. *Curr. Sci.* **2018**, *115*, 1884–1892. [CrossRef]
4. Rasool, K.; Islam, N.; Ali, M.; Jang, B.-E.; Khan, N.A.; Chowdhury, M.; Chung, S.-O.; Kwon, H.-J. Onion transplanting mechanisms: A review. *Precis. Agric. Sci. Technol.* **2020**, *2*, 196. [CrossRef]
5. Kang, D.H.; Kim, D.E.; Lee, G.I.; Kim, Y.H.; Lee, H.J.; Min, Y.B. Development of a Vegetable Transplanting Robot. *J. Biosyst. Eng.* **2012**, *37*, 201–208. [CrossRef]
6. Moon, S.D.; Min, Y.B.; Park, J.C. Analysis of working capacity of a hand-fed transplanter. *J. Bio. Fac. Environ.* **1997**, *6*, 159–167. Available online: https://www.koreascience.or.kr/article/JAKO199711922405384.pdf (accessed on 18 December 2022).
7. Shao, Y.; Liu, Y.; Xuan, G.; Hu, Z.; Han, X.; Wang, Y.; Chen, B.; Wang, W. Design and Test of Multifunctional Vegetable Transplanting Machine. *IFAC-PapersOnLine* **2019**, *52*, 92–97. [CrossRef]
8. Kumar, G.V.P.; Raheman, H. Vegetable Transplanters for Use in Developing Countries—A Review. *Int. J. Veg. Sci.* **2008**, *14*, 232–255. [CrossRef]
9. Jo, J.S.; Okyere, F.G.; Jo, J.M.; Kim, H.T. A Study on improving the performance of the planting device of a vegetable trans-planter. *J. Biosyst. Eng.* **2018**, *43*, 202–210. [CrossRef]
10. Min, Y.B.; Kang, J.K.; Ryu, C.S. Development Onion Transplanter: Analysis a Transplanting Locus on The Type of Transplanting Devices for a Vegetable Transplanter. *J. Agric. Life Sci.* **2015**, *49*, 289–294. [CrossRef]

11. Sri, M.; Hwang, S.-J.; Nam, J.-S. Experimental Safety Analysis for Transplanting Device of the 4-Bar Link Type Semi-Automatic Vegetable Transplanter. *Agronomy* **2022**, *12*, 1890. [CrossRef]
12. Park, J.-H.; Hwang, S.-J.; Nam, J.-S. Operational Characteristics of a Domestic Commercial Semi-automatic Vegetable Transplanter. *J. Agric. Life Sci.* **2018**, *52*, 127–138. [CrossRef]
13. Park, J.-H.; Hwang, S.-J.; Nam, J.-S. Operational Characteristics of a Cam-type Vegetable Transplanter and Mechanism of a Transplanting Device. *J. Agric. Life Sci.* **2019**, *53*, 113–124. [CrossRef]
14. Iqbal, Z.; Islam, N.; Ali, M.; Kiraga, S.; Kim, Y.-J.; Chung, S.-O. Theoretical Overturning Analysis of a 2.6-kW Two-Row Walking-Type Automatic Pepper Transplanter. *J. Biosyst. Eng.* **2022**, *47*, 79–91. [CrossRef]
15. Sri, M.; Hwang, S.-J.; Nam, J.-S. Experimental safety analysis of transplanting device of the cam-type semi-automatic vegetable transplanter. *J. Terramechanics* **2022**, *103*, 19–32. [CrossRef]
16. Kim, Y.-J.; Chung, S.-O.; Park, S.-J.; Choi, C.-H. Analysis of Power Requirement of Agricultural Tractor by Major Field Operation. *J. Biosyst. Eng.* **2011**, *36*, 79–88. [CrossRef]
17. Lee, D.-H.; Kim, Y.-J.; Chung, S.-O.; Choi, C.-H.; Lee, K.-H.; Shin, B.-S. Analysis of the PTO load of a 75kW agricultural tractor during rotary tillage and baler operation in Korean upland fields. *J. Terramechanics* **2015**, *60*, 75–83. [CrossRef]
18. Choudhary, S.; Upadhyay, G.; Patel, B.; Naresh; Jain, M. Energy Requirements and Tillage Performance Under Different Active Tillage Treatments in Sandy Loam Soil. *J. Biosyst. Eng.* **2021**, *46*, 353–364. [CrossRef]
19. Koo, Y.M.; Kang, Y. Characteristics of Power and Fuel Use of a Tractor-Mounted Integrated Implement for Round Ridge Preparation. *J. Biosyst. Eng.* **2021**, *46*, 496–507. [CrossRef]
20. Swe, K.M.; Islam, N.; Chowdhury, M.; Ali, M.; Wing, S.; Jun, H.-J.; Lee, S.-H.; Chung, S.-O.; Kim, D.-G. Theoretical Analysis of Power Requirement of a Four-Row Tractor-Mounted Chinese Cabbage Collector. *J. Biosyst. Eng.* **2021**, *46*, 139–150. [CrossRef]
21. Shafaei, S.M.; Loghavi, M.; Kamgar, S. On the Reliability of Intelligent Fuzzy System for Multivariate Pattern Scrutinization of Power Consumption Efficiency of Mechanical Front Wheel Drive Tractor. *J. Biosyst. Eng.* **2021**, *46*, 1–15. [CrossRef]
22. Koo, Y.M. PTO Torque and Draft Analyses of an Integrated Tractor-Mounted Implement for Round Ridge Preparation. *J. Biosyst. Eng.* **2022**, *47*, 330–343. [CrossRef]
23. Lee, J.-Y.; Hwang, S.-J.; Nam, J.-S.; Kim, J.-G. Consumed-Power and Load Characteristics of Potato Harvesting Operation in Dry Field. *Korean Soc. Manuf. Process. Eng.* **2020**, *19*, 89–99. [CrossRef]
24. Nam, J.-S.; Kang, D.-S.; Kang, Y.-S.; Kim, K.-U.; Kim, D.-C. Comparison of Work Performance of Crank-type and Rotary-type Rotavators in Korean Farmland Conditions. *J. Biosyst. Eng.* **2012**, *37*, 140–147. [CrossRef]
25. Kim, M.-H.; Nam, J.-S.; Kim, D.-C. Comparison of Tillage and Loads Characteristics of Three Types of Rotavators: Rotary-type, Crank-type, and Plow-type. *J. Biosyst. Eng.* **2013**, *38*, 73–80. [CrossRef]
26. Kim, J.G.; Kim, Y.J.; Kim, J.H.; Shin, B.S.; Nam, J.S. Consumed-power and load characteristics of a tillage operation in an upland field in Republic of Korea. *J. Biosyst. Eng.* **2018**, *43*, 83–93. [CrossRef]
27. Lee, J.-Y.; Nam, J.-S. Load and Safety Analysis for Plow Operation in Dry Fields. *Korean Soc. Manuf. Process. Eng.* **2019**, *18*, 9–18. [CrossRef]
28. Kim, W.S.; Chung, S.O.; Choi, C.H.; Cho, J.S.; Choi, D.S.; Kim, Y.J.; Lee, S.D.; Hong, S.J.; Koo, S.M. Analysis of the PTO Torque of a Transplanter by Planting Condition. *J. Biosyst. Eng.* **2016**, *41*, 313–318. [CrossRef]
29. Kim, W.; Kim, Y.; Kim, Y.; Choi, C.; Inoue, E.; Okayasu, T. Analysis of the Load of a Transplanter PTO Shaft Based on the Planting Distance. *J. Fac. Agric. Kyushu Univ.* **2018**, *63*, 97–102. [CrossRef]
30. Kim, W.-S.; Kim, Y.-S.; Kim, T.-J.; Nam, K.-C.; Kim, T.-B.; Han, T.-H.; Im, R.-G.; Kim, Y.-H. Effects of planting distance and depth on PTO load spectrum of a small riding-type transplanter. *Int. J. Agric. Biol. Eng.* **2020**, *13*, 57–63. [CrossRef]
31. Lee, P.U.; So, J.H.; Nam, Y.S.; Choi, C.H.; Noh, H.S.; Shim, J.Y.; Hong, S.J. Power analysis of electric transplanter by planting distances. *Korean J. Agric. Sci.* **2018**, *45*, 290–297. [CrossRef]
32. Lim, S.-J.; Kwon, H.-J.; Kang, Y.-S.; Lee, P.-U.; Kim, T.-J.; Kim, Y.-J.; Kim, Y.-S. Power Analysis of a 3-kW Class Motor-Driven Multipurpose Walking-Type Transplanter. *J. Biosyst. Eng.* **2019**, *44*, 135–145. [CrossRef]
33. Downing, S.; Socie, D. Simple rainflow counting algorithms. *Int. J. Fatigue* **1982**, *4*, 31–40. [CrossRef]
34. Rychlik, I. A new definition of the rainflow cycle counting method. *Int. J. Fatigue* **1987**, *9*, 119–121. [CrossRef]
35. Juvinall, R.C.; Marshek, K.M. *Machine Component Design*, 5th ed.; John Wiley & Sons, Inc.: New York, NY, USA, 2020; pp. 338–340.
36. Paraforos, D.S.; Griepentrog, H.W.; Vougioukas, S.G.; Kortenbruck, D. Fatigue life assessment of a four-rotor swather based on rainflow cycle counting. *Biosyst. Eng.* **2014**, *127*, 1–10. [CrossRef]
37. Mattetti, M.; Molari, G.; Sereni, E. Damage evaluation of driving events for agricultural tractors. *Comput. Electron. Agric.* **2017**, *135*, 328–337. [CrossRef]

agriculture

MDPI

Article

Investigating the Effect of Tractor's Tire Parameters on Soil Compaction Using Statistical and Adaptive Neuro-Fuzzy Inference System (ANFIS) Methods

Gholamhossein Shahgholi [1,*], Abdolmajid Moinfar [1], Ali Khoramifar [1], Sprawka Maciej [2,*] and Mariusz Szymanek [2]

1 Department of Biosystems Engineering, University of Mohaghrgh Ardabili, Ardabil 56199-11367, Iran
2 Department of Agricultural, Forest and Transport Machinery, University of Life Sciences in Lublin, 20-950 Lublin, Poland
* Correspondence: shahgholi@uma.ac.ir (G.S.); maciej.sprawka@up.lublin.pl (S.M.)

Abstract: Many factors contribute to soil compaction. One of these factors is the pressure applied by tires and tillage tools. The aim of this study was to study soil compaction under two sizes of tractor tire, considering the effect of tire pressure and traffic on different depths of soil. Additionally, to predict soil density under the tire, an adaptive neuro-fuzzy inference system (ANFIS) was used. An ITM70 tractor equipped with a lister was used. Standard cylindrical cores were used and soil samples were taken at four depths of the soil inside the tire tracks. Tests were conducted based on a randomized complete-block design with three replications. We tested two types of narrow and normal tire using three inflation pressures, at traffic levels of 1, 3 and 5 passes and four depths of 10, 20, 30 and 40 cm. A grid partition structure and four types of membership function, namely triangular, trapezoid, Gaussian and General bell were used to model soil compaction. Analysis of variance showed that tire size was significant on soil density change, and also, the binary effect of tire size on depth and traffic were significant at 1%. The main effects of tire pressure, traffic and depth were significant on soil compaction at 1% level of significance for both tire types. The inputs of the ANFIS model included tire type, depth of soil, number of tire passes and tire inflation pressure. To evaluate the performance of the model, the relative error (ε) and the coefficient of explanation (R^2) were used, which were 1.05 and 0.9949, respectively. It was found that the narrow tire was more effective on soil compaction such that the narrow tire significantly increased soil density in the surface and subsurface layers.

Keywords: inflation pressure; bulk density; ANFIS; tire

Citation: Shahgholi, G.; Moinfar, A.; Khoramifar, A.; Maciej, S.; Szymanek, M. Investigating the Effect of Tractor's Tire Parameters on Soil Compaction Using Statistical and Adaptive Neuro-Fuzzy Inference System (ANFIS) Methods. *Agriculture* **2023**, *13*, 259. https://doi.org/10.3390/agriculture13020259

Academic Editors: Ju-Seok Nam and Yongjin Cho

Received: 29 December 2022
Revised: 16 January 2023
Accepted: 19 January 2023
Published: 20 January 2023

1. Introduction

Soil is a resource with a potentially rapid degradation rate and a very slow regeneration and formation process [1]. Thus, sustainable soil use is the only solution to global problems such as food security, energy and water demands, climate change and biodiversity [2]. In order to increase productivity in the agricultural sector, farms should be a large size, in which productivity can be increased by the use of more efficient machines [3]. For many years, the agricultural sector has tended to increase the size and weight of the tractor, which increases the risk of severe soil compaction [4].

One of the causes of soil compaction in mechanized agriculture is the passage of agricultural implements and tools. Soil compaction as a form of reducing the pore space between soil particles can significantly reduce soil production capacity [5]. According to data from seven countries in Europe and North America, a 14% drop in crop yield was reported in the first year after machine traffic [6]. Soil compaction is a form of physical degradation of soil that causes compacted soil particles to be closer to each other and the disappearance of voids in the soil, reducing porosity and soil permeability. This manifests as an increase in soil bulk density [7]. One method of assessing soil compaction is to measure its bulk density. When soil compaction occurs, the bulk density of the soil increases as the mass remains constant and the volume decreases. Density reduction decreases hydraulic conductivity and thus causes surface runoff and soil erosion by water. This induces forced currents in the soil pores, thus facilitating the transport of absorbable colloidal nutrients and pesticides to deeper horizons and groundwater, thereby reducing root growth and causing a loss of nitrogen and production of greenhouse gases through nitrogenization by anaerobic processes. As a result, soil compaction is one of the main causes of environmental and agricultural problems, and as a result causes significant economic damage to society and agriculture [8].

Increasing the load on the wheels of agricultural machinery is the most important cause of soil compaction created in deeper soil layers. The topsoil of arable lands is loosened every year by plowing operations, however, the subsoil often remains intact, which means that whenever the maximum load capacity is exceeded, additional soil compaction will occur deep in the soil. In addition, these effects are lasting, and the results of natural and artificial soil loosening are disappointing [9]. Excessive movement of agricultural machinery on the field and plowing at a constant depth gradually causes soil compaction, and this compaction restricts plant root growth and reduces crop yield [10]. To remove this hard layer created in the subsoil, which limits the longitudinal growth of roots and proper soil drainage, deep plowing is required, which greatly increases the required energy [11]. Another solution is to use energy-efficient combined (active-passive or passive-passive) tillage implements in place of conventional ones to reduce soil compaction, labor, and fuel cost as well as to save time in preparing the seedbed [12]. Another aspect affecting the protection of soil structure from compaction is the evolution of tires [13]. In this article we compare the effects of five tractor tires introduced between the 1970s and 2017 on soil stress and soil structure. Soil physical properties before and after one pass of each of the five tires were measured on undisturbed soil. The results of stress measurements showed the lowest mean normal stress at all depths with newer tires, both below the center of the tire and at the edge, even though the effect decreased with increasing depth. The results indicate that tire design can play an important role in reducing soil stress for a given wheel load. Tire design might then also help to reduce the compaction risk for larger wheel loads and other soil conditions which were tested in the study.

Maintaining proper ratios between solid, liquid and gas phases is of particular importance, because increasing the solid phase (i.e., reducing the liquid and gas phases) increases the mechanical strength of the soil and consequently reduces the growth and development of plant roots inside the soil. Reducing the development of root growth causes less water and nitrogen to be available to the plant. When the machine enters the field, the solid phase of the soil increases by a certain volume and the gas phase decreases. Soil density is one of the indicators of the destruction of the physical structure of the soil, which is defined as an increase in soil bulk density (solid phase mass to total volume), or a decrease in porosity [14]. Wheel load, tire type, and inflation pressure can increase soil bulk density and play an important role in soil compaction [15]. Almost all tires significantly increase soil compaction along the wheel track, while only some increase soil compaction near the track [15]. As the distance from the wheel path increases, a general decrease in soil density occurs, especially in the subsoil. Many researchers have reported that working with low-pressure tires can significantly reduce soil compaction and increase crop yield [16,17], while high tire pressure increases soil compaction [18]. Tire pressure control is an important factor in the control of and in soil compaction. A new tire pressure control system in forest machines was developed using a PressurePro solution. It was found that the installation of an automatic tire pressure control system leads to compaction reduction by 20%. Additionally, reduced tire pressure and its automatic control contribute to a minimal reduction in humus content and soil compaction over time [19]. Another researchers have investigated the effect of different tire inflation pressures of 100 and 200 kPa on selected physical soil properties by conducting field tests. The undisturbed samples were collected both in and between the tracks at depths of 0 to 0.5 m. The results indicated that fewer negative changes were found in the variant with a lower inflation pressure for all of the soil properties. Change in soil physical properties caused by the passage of the tractor were statistically significant for both tire pressures only at depths ranging from 0 to 0.1 m [20]. The effect of three types of wide, low-inflation pressure tires with similar dimensions on mean normal stress on the soil profile was also investigated. The results revealed a very limited effect of tire construction on mean normal stress. No differences were measured beneath the center, and differences at +0.3 m were found only at 0.2 m depth for the tires at the rear axle. The effect of low tire inflation pressure was limited to the upper parts of the soil profile for beneath the center of the tire, which was significant at depths of 0.2 m and 0.4 m. It was found that in order to reduce soil stress, tire design should allow for a large contact area and low inflation pressure [21].

Machine traffic plays an important role in soil compaction. Experiments have been carried out to study the effect of multiple passes of tillage operation on soil compaction beneath the tillage depth with three different tillage systems. The authors observed that active tillage machinery generally caused more compaction compared to passive tillage machinery for the same number of passes. However, active machinery required fewer vehicular passes [22]. In plowed soils, the first pass can dramatically increase soil compaction and cause soil degradation, while in firm soil and deep areas soil compaction occurs over time and with large numbers of tire passes [23]. Research has also shown that if the traffic exceeds 10 times, the benefits of using light tractors are lost and soil compaction is greatly increased. These studies have suggested that heavy tractors with less traffic should be used as much as possible because high traffic leads to compaction in the subsoil [24,25].

One method of assessing soil compaction is to measure its bulk density. When soil compaction occurs, the bulk density of the soil increases as the mass remains constant and the volume decreases. In this method, after determining soil bulk density at different depths and, based on the soil texture, measured data are compared with the standard values to determine the level of soil compaction. For example, soil compaction is considered significant for loamy clay soil when bulk density reaches more than 1.5 g/cm^3, and for sandy and loamy soil more than 1.6 g/cm^3 [26].

In recent years, the artificial intelligence approach has shown that it has great power to model and predict complex systems, thus it can be an alternative method to physical models that do not respond well to a large number of input variables [26]. Because agricultural systems such as soil are so complex, most researchers have focused on artificial intelligence to model different components of agricultural systems. The fuzzy neural adaptive inference system (ANFIS) is a multi-layered adaptive network consisting of the main elements and functions of fuzzy logic systems developed by Jang [27]. Each of the fuzzy systems and artificial neural networks (ANNs) has advantages and disadvantages. The fuzzy system is able to use human language and can use human experiences and experts while it is not able to learn [28]. In other words, the fuzzy system cannot be trained using observational data and it should be noted that this method does not provide good results in unpredictable conditions [29]. However, neural networks are self-learning using data sets. At the same time, neural networks are implicit and unable to use human language. They require a wide range of tested input and output data for successful modeling [30]. Although ANN is a powerful technique for modeling various problems in the real world, it has its own weaknesses. If the input data is vague or the uncertainty is relatively high, a fuzzy system such as ANFIS may be a better option [31].

In order to evaluate the potential of ANFIS in predicting the energy efficiency indicators of the drive wheel, experiments were performed in the soil reservoir. Input parameters included tire load, wheel speed and slip, each at three different levels. ANFIS was used with a combined method of descending gradient and least squares to find the optimum learning parameters using different membership functions (MFS). The results of this study showed high accuracy (MSE = 0.0166 and R^2 = 0.98) in prediction [31]. A fuzzy logic model was developed to describe soil fragmentation for seedbed preparation in a combination of primary and secondary tillage tools from subsoil, moldboard plow and disc. In this research, an intelligent model based on the Mardani fuzzy modeling principles approach was developed to predict soil fragmentation during tillage operations. Model inputs were: soil moisture, tractor speed and sampling depth. The fuzzy model consisted of 50 rules in which the three parameters of mean root mean square error (RMSE), relative error (e), and coefficient of determination (R^2) were used to evaluate fuzzy models. These parameters were calculated to be 0.17%, 3.95% and 0.99%, respectively. According to the results of this study, the fuzzy model can be used as a method to predict soil fragmentation during tillage operations [32]. In the study, the ANFIS model was used to predict soil compaction under tractor wheels. Model inputs included four factors: manure rate, number of passes, moisture and depth. The network structure was selected from the network segmentation type. The number of membership functions for each input was considered to be three and the membership functions of bell, triangle, trapezoid and Gaussian were tested for best performance. The results of this study showed that the ANFIS model has a higher accuracy in predicting soil density than the regression model [33].

The main objectives of this research are as follows:

1. Investigating the density created in different depths of the soil due to the passage of tractor wheels of different size;
2. Investigating the effect of inflation pressure changes of drive wheel and machine traffic on the soil density created at different depths;
3. Modelling the effect of tire size, tire pressure, and machine traffic on soil density at different depths using statistical and adaptive neuro-fuzzy inference system (ANFIS) methods

2. Materials and Methods

Field experiments were performed to measure the bulk density of the soil under two common and narrow radial tires. The first type of tire was a standard 8 layer tire with a size of 14–38. Three inflation pressures of 80, 120 and 160 kPa were used to evaluate the effect of tire pressure on soil compaction. The second type of tire was a 10 layer radial narrow tire with a size of 9.5–48. Inflation pressures of 240, 275 and 310 kPa were investigated. For each of these tire pressures, three traffics of 1, 3 and 5 were tested. In all tests, soil samples were obtained from four depths of 10, 20, 30 and 40 cm. The farm had been plowed already, and an ITM70 tractor equipped with a bedder-ridger on a three-point hitch was used for conducting tests. In this case, the weight measured on the rear axle was 2740 kg, and on the front axle 770 kg. In all experiments, the tractor speed was constant, equal to 3 km/h. The process of adjusting the desired speed was done in such a way that the tractor was moved to a distance of 30 m on the field to achieve a speed of 3 km/h with two gears and at engine speed of 1400 and 1600 rpm. the desired speed was obtained using a calibration diagram.

A hydrometric test (ASTM D-422) was used to determine the soil texture by determining the relative amount of sand, silt and clay. It was found that there were 43.92, 32.06 and 24.02% sand, soil and clay, respectively. According to the soil texture triangle, the soil texture was loamy. A standard proctor test was used to determine the soil critical moisture. According to Figure 1, a moisture level of 25% is the critical moisture content, and the highest value of compaction occurs at this humidity.

To measure the bulk density after the tire passes, standard cylinders were placed in groups of 3 at different depths of 10, 20, 30 and 40 cm measured from the center of the cylinders to the soil surface and in line with tire path direction in the center line of the tire track (Figure 2). In order that the cylinders were placed exactly at the desired height and in one direction inside the profile, a string was drawn from the middle of the profile and from the beginning to the end. Figure 3 shows how the cylinders were exerted inside the soil vertically in the direction of the string. Each group of cylinders was about 40 cm apart so that there were no problems when removing the cylinders. It should be noted that the soil surface appears loose in Figure 3; this is because the soil had been dug to place the cylinder at a predetermined depth.

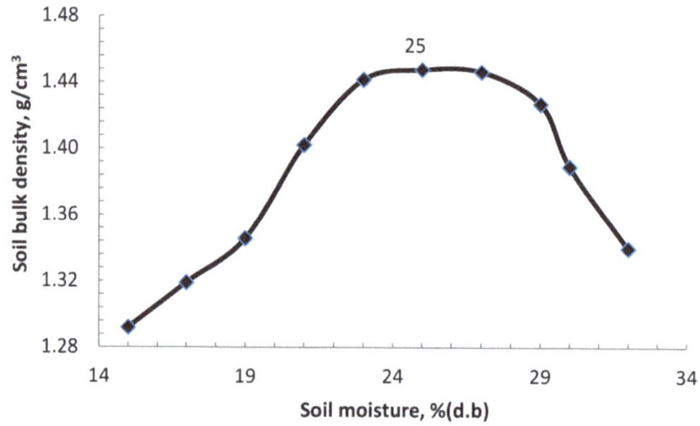

Figure 1. Determining the critical soil moisture using the proctor test.

Figure 2. Passing the tractor with two types of tires on the field.

Figure 3. Placement of the cylinders inside the soil profile.

To implement the fuzzy neural adaptive model, MATLAB 2014a software toolbox under Windows was used. According to the performed experiments, soil density was defined as system output, while tire type, tire inflation pressure, traffic and soil depth were independent inlet parameters (Figure 4). The data is usually divided into three categories: training data, validation data, and test data. The training data were used in the training process to calculate the gradient and optimize the parameters. One of the points that should be considered during the training is aiming to prevent the network from becoming specialized and the occurrence of over-fitting phenomena. For this purpose, a series of data was considered for validation during the training of the network. These data are actually part of the training data. In this way, in the regular intervals of the optimization process, the data obtained from the network were checked using the validation data. In this case, network training continued until the optimization error related to the evaluation data began to increase, and as soon as this error increased to a certain value or a certain number of repetition, network training was stopped. Finally, the test data was used to test the performance of the network after training. The test data was not used during network training, but it was used to compare the error rate. Accordingly, in this research, 70% of the data were considered for training, 15% for evaluation, and 15% for testing the network.

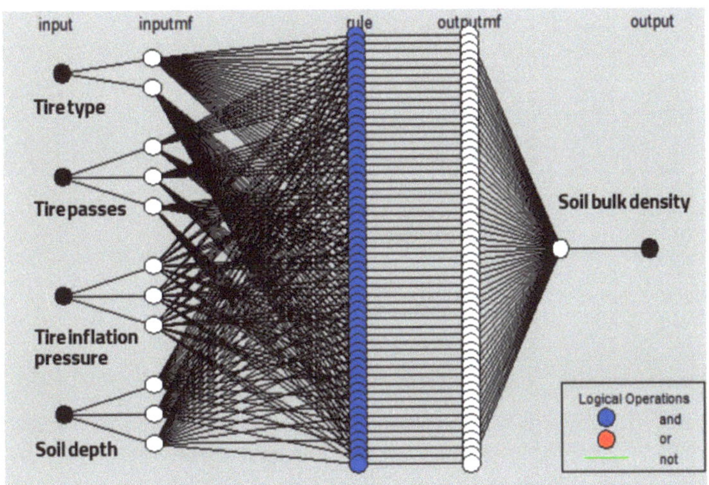

Figure 4. Structure topology of the neural-fuzzy adaptive fuzzy inference system (ANFIS).

A network separation structure was considered to create the network. Four commonly used membership functions, namely triangular, trapezoidal, Gaussian, and bell, were considered as membership functions to represent the inputs (Figure 5), and four different models were constructed using these membership functions.

The number of membership functions for each variable was considered proportional to the levels tested for that variable. Table 1 shows the membership functions for each variable.

The output membership function in this network is linear. A hybrid optimization method was used for network training. The number of rules created by the network was 54. Given that the number of rules created was very high, only a few of them are presented in Table 2.

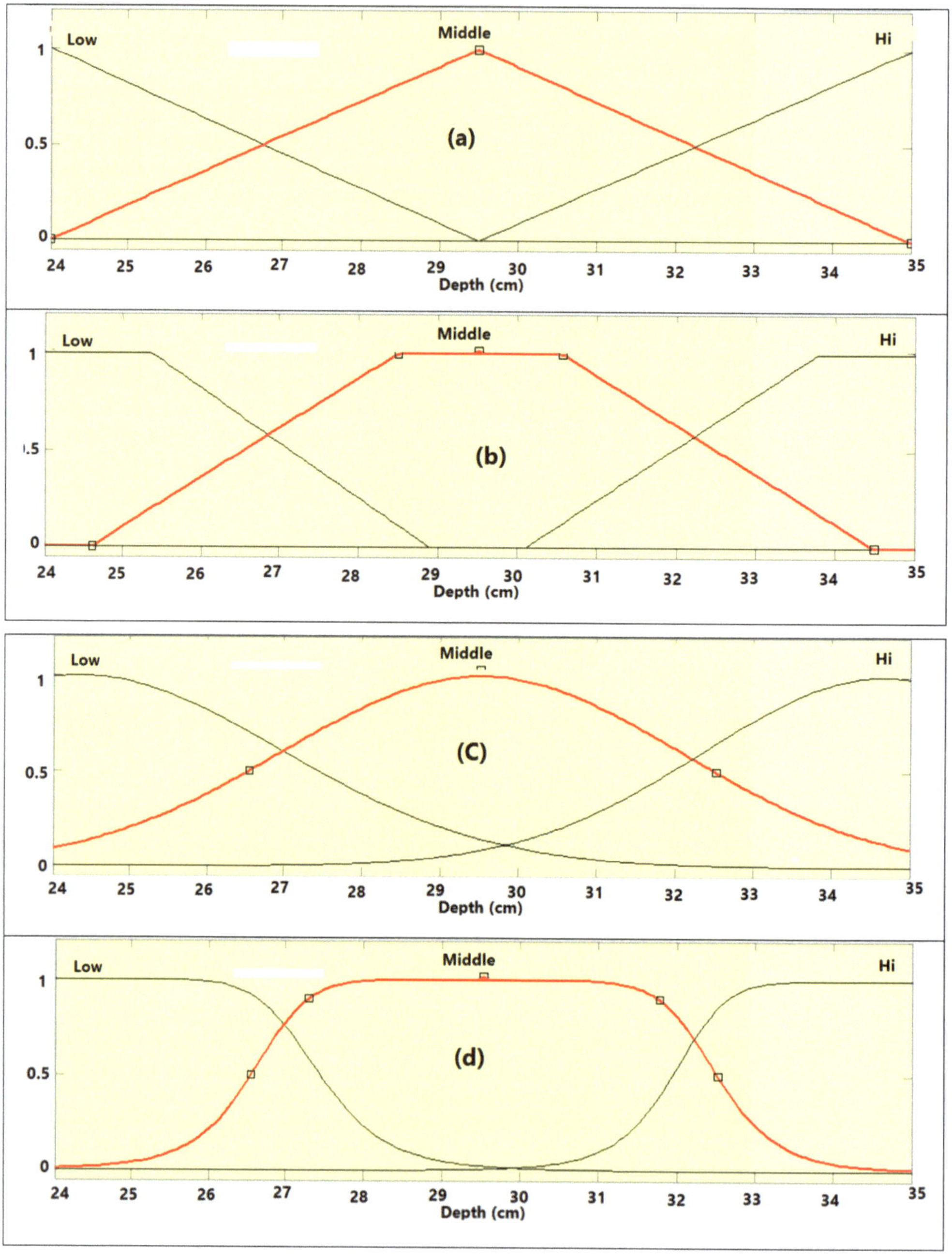

Figure 5. Membership functions used to represent inputs (**a**) Triangular, (**b**) Trapezoidal, (**c**) Gaussian, (**d**) Bell-shaped.

Table 1. Membership functions of input variables.

Input Variable	Membership Function		
Tire Type	**Narrow**		**Wide**
Traffic	max	ave	min
Tire pressure	high	normal	low
Depth	subsoil	mid soil	topsoil

Table 2. Some rules created in the ANFIS model.

	Input Parameters				Output Variable
Rules	**Tire Type(T)**	**Traffic (P)**	**Tire Inflation (IP)**	**Soil Depth (D)**	**Special Mass**
1	Wide	Min	Low	Topsoil	BD − 0.09035T + 0.06862P + 0.09035IP + 0.09035D + 0.09035
8	Wide	Min	High	Middle soil	BD = 0.01578T + 0.06565P + 0.07889IP + 0.03156D + 0.01578
16	Wide	Ave	High	Topsoil	BD = −0.00517T + 0.04506P − 0.02585IP −0.00517D − 0.00517
24	Wide	Max	Normal	Subsoil	BD = 0.00329T + 0.0256P + 0.009871IP + 0.009871D + 0.00329
32	Narrow	Min	Normal	Middle soil	BD = 0.05436T + 0.07231P + 0.08154IP + 0.05436D + 0.02718
40	Narrow	Ave	Normal	Topsoil	BD = −0.02717T+ 0.04873P − 0.04075IP − 0.01358D − 0.01358
48	Narrow	Max	Low	Subsoil	BD = 0.009582T + 0.02666P + 0.004791IP + 0.01437D + 0.004791
56	Narrow	Max	High	Subsoil	BD = 0.004228T + 0.02739P + 0.01057IP + 0.006341D + 0.002114

Finally, the performance of the models was evaluated based on the statistical criteria presented in Equations (1) and (2).

$$\mathrm{MAPE} = \frac{100\%}{N}\sum_{i=1}^{n}\frac{Y_{\mathrm{measured}} - Y_{\mathrm{predicted}}}{Y_{\mathrm{measured}}}, \tag{1}$$

$$R^2 = 1 - \frac{\sum_{i=1}^{N}\left(Y_{\mathrm{measured}} - Y_{\mathrm{predicted}}\right)^2}{\sum_{i=1}^{N}\left(\overline{Y}_{\mathrm{measured}} - Y_{\mathrm{predicted}}\right)^2}, \tag{2}$$

where:

MAPE: the absolute mean percentage of system error
R^2: is the coefficient of determination
N: the number of samples
$Y_{\mathrm{predicted}}$: the predicted values
Y_{measured}: the measured values
$\overline{Y}_{\mathrm{measured}}$: the average of measured values.

3. Results and Discussion

Table 3 shows the analysis of variance of soil bulk density for two types of ordinary and narrow tire. Data were analyzed in a 4 × 3 × 4 factorial design based on a randomized complete block design. The data coefficient of variation (CV) was 0.44%. Average initial soil bulk density was 1.017 g/cm^3, and based on data of initial bulk density it was uniform in all plots. In comparison with initial bulk density, statistical analysis showed that the main effects of tire type, traffic and depth on soil density change were significant at the level of 1% probability. The interactions of depth with traffic, depth with tire type, and traffic with tire type on soil bulk density were found to be significant at the 1% probability level. However, the combined interaction of depth, tire type and traffic on soil bulk density was not found to be significant at the 1% probability level.

Table 3. Analysis of variance of soil bulk density change due to tire type, traffic and depth.

Source of Variation	Df	Sum of Squares	Mean Squares	F
Soil depth	3	0.054	0.018	715.6 **
Traffic	2	0.015	0.008	302.1 **
Tire type	1	0.113	0.113	4490 **
Depth × pass	6	0.001	0.00016	5.5 **
Depth × tire type	3	0.002	0.00067	25.6 **
Pass × tire type	2	0.001	0.0005	13.5 **
Depth × pass × tire type	6	0	0	1.6 ^{ns}
Error	46	0.001	0	-

**: Highly significant, ns: Not significant.

3.1. The Binary Effect of Tire Type and Depth on Soil Bulk Density

The binary effect of tire type and depth on soil bulk density was evaluated using a Duncan test at 1% probability level, and is presented in Figure 6. According to the results, the maximum value of bulk density was 1.217 g/cm^3 for narrow tire at a depth of 10 cm, and the lowest value was 1.061 g/cm^3 which occurred for common tires at a depth of 40 cm. For both tires, a significant decrease in the soil bulk density was seen with increasing depth, but the intensity of these changes in the narrow tire and especially in shallow depths was greater than for the common tire. The bulk density of soil for the narrow tire, even at a depth of 40 cm, was greater than the bulk density of soil for a normal tire at a depth of 10 cm. The effect of tire size on stress applied to different soil layers was examined [34]. Their results showed that a narrow tire created much more stress in different layers of soil than a common tire, and as a result, it compacted the soil more. Stress for a narrow tire at a depth of 50 cm was higher than the stress of a common tire at depth of 30 cm.

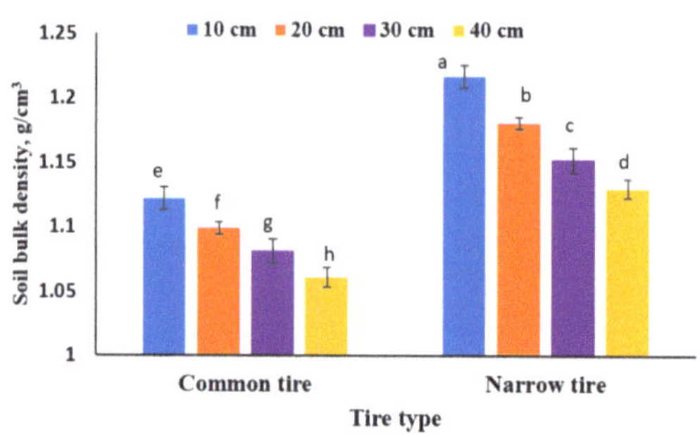

Figure 6. The main effect of tire type and depth on soil bulk density (different English alphabets shows significant difference between treatments).

3.2. The Binary Effect of Tire Type and Traffic on Soil Bulk Density

The results of comparing the mean interaction of tire type and depth on soil bulk density using a Duncan test at 1% probability level are presented in Figure 7. In both tires, the soil bulk density increased with increasing traffic. According to the results, the highest mean value of soil density was found for the narrow tire after five passes with a value of

1.185 g/cm^3 and the lowest for a normal tire after one pass with a value of 1.112 g/cm^3. It is noteworthy that the bulk density of soil for a narrow tire, even at one pass, is more than the bulk density of soil for a normal tire in five passes.

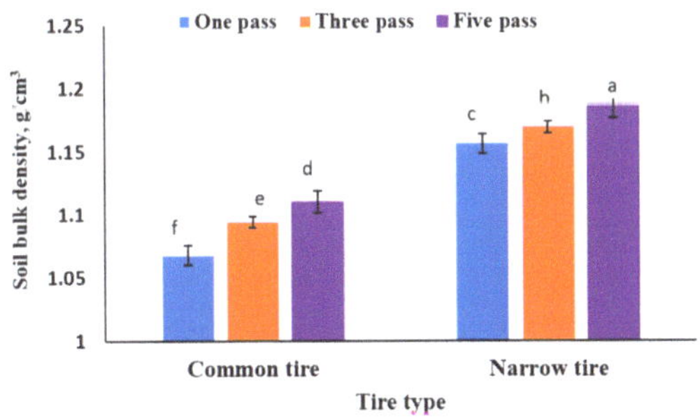

Figure 7. The main effect of tire type and traffic on soil bulk density (different English alphabets shows significant difference between treatments).

3.3. Results of Modeling Soil Bulk Density Using ANFIS

Four different models were developed to predict soil bulk density. Table 4 presents the structural parameters of the models along with their statistical criteria to evaluate their performance. According to the results presented in Table 4, it is clear that all models have a high ability ($R^2 \geq 0.99$ and ε (%) ≤ 2) to predict. The best model for predicting soil bulk density is a model that uses the bell membership function (Gbellmf) (R^2 = 0.9949 and ε (%) = 1.05).

Table 4. Specifications and evaluation results of adaptive fuzzy-neural inference system models (ANFIS).

Network Structure	Membership Function Type		Number of Membership Functions		Optimization Method	Test Result	
	Input	Output	Input	Epochs		Coefficient of Determination (R^2)	Relative Error (e)
Grid Partition	Triangular	Linear	2-3-3-3	20	hybrid	0.992	1.5
Grid Partition	Trapezoidal	Linear	2-3-3-3	20	hybrid	0.99	1.75
Grid Partition	Gaussian	Linear	2-3-3-3	20	hybrid	0.982	2.2
Grid Partition	Bell-shaped	Linear	2-3-3-3	20	hybrid	0.995	1.05

Table 5 presents the statistical characteristics of the stepwise regression model for predicting soil bulk density. The ANFIS models have poorer performance than the regression model due to a low coefficient determination of 0.96. However, regression models have valuable advantages, such a the fact that ANFIS models do not provide any information about the internal structure of the model and the relationships between independent and dependent variables, while the regression model directly deals with the impact of each factor and tries to provide a model in which the importance and impact of each factor is clear. According to Table 5 and the standard coefficients mentioned therein, type of tire, soil depth, inflation pressure, and finally the number of passes have the greatest impact on the bulk density.

Table 5. Statistical characteristics of stepwise regression model for predicting soil bulk density.

Model	Unstandardized Coefficients		Standardized Coefficients	t	Sig
	Beta	Std. Error	B		
Constant	0.98	0.006		154.528	<0.0001
Tire type	0.08	0.003	0.705	30.916	<0.0001
Depth	−0.002	0.001	−0.458	−20.091	<0.0001
Inflation pressure	0.031	0.002	0.444	19.452	<0.0001
Traffic	0.008	0.001	0.245	10.748	<0.0001

Figure 8 shows the relationship between measured and predicted values using ANFIS models under different operating conditions. In addition, the deviation between the measured values and the predicted values was calculated and plotted by ANFIS and multiple regression models. Figure 9 shows that the deviation distance of the predicted values of the ANFIS model (−0.011 to 0.0085) was much less than the deviation distance of the predicted values of the multiple regression model (−0.021 to 0.033).

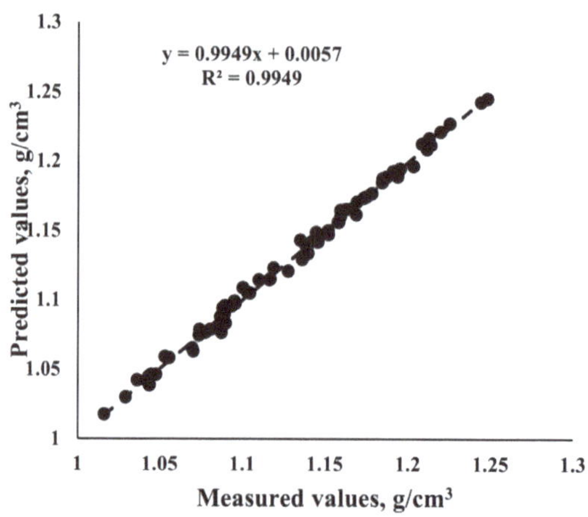

Figure 8. Relationship between measured and predicted values of soil bulk density.

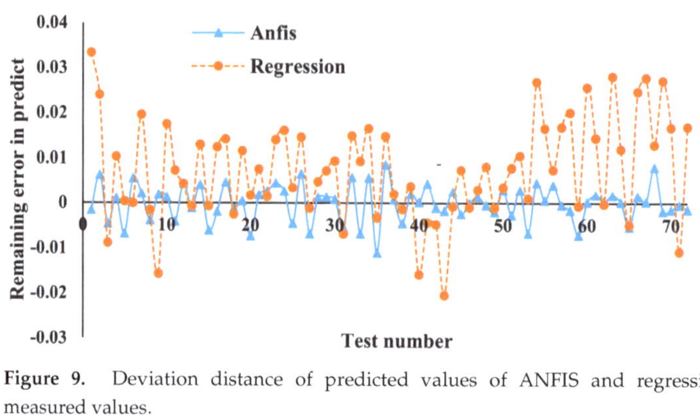

Figure 9. Deviation distance of predicted values of ANFIS and regression models from measured values.

Figure 10 shows the three-dimensional variation of soil bulk density with the interaction of tire type and soil depth. It indicates that with increasing depth, the bulk density of the soil decreased, indicating that the compaction created by the tire traffic of the tractor in the topsoil was higher than in the sub layers. It was found that at all depths the density change by the passage of common tire was less than the density created by the narrow tire. Such a low pressure at the soil surface and soil behavior was investigated using a flat tire [3].

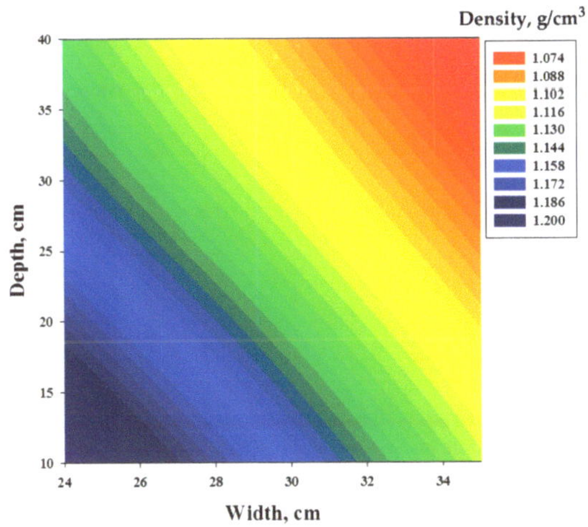

Figure 10. Binary effect of tire type and soil depth on the soil bulk density.

Figure 11 shows the binary effect of tire pressure and tire traffic on the soil bulk density. As the tire pressure decreases, the bulk density of the soil decreased. As the pressure decreased, the contact area between the tire and the soil increased, which reduces the pressure on the soil, resulting in decreased soil compaction and this was reported by other researchers [35–37]. With increasing traffic, the bulk density of the soil increased [33,38,39].

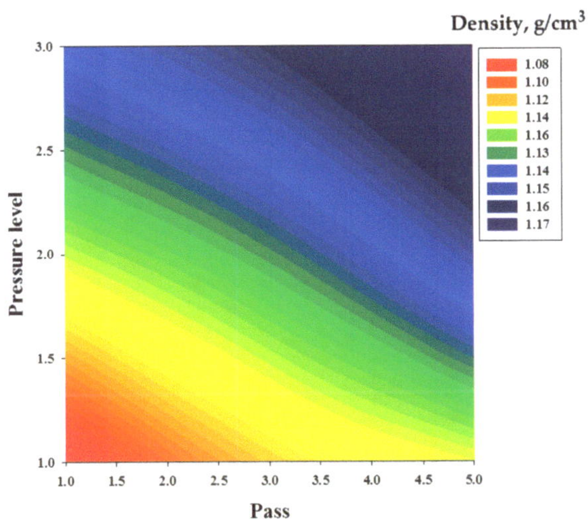

Figure 11. Binary effect tire inflation pressure and traffic on the soil bulk density.

Figure 12 illustrates with a surface plot of the interaction of tire type and traffic on the bulk density of the soil. With increasing traffic, when using a narrow tire, the soil bulk density increased sharply, but using a common tire was less effective on soil bulk density increment. This indicates that narrow tires are very destructive on soil structure.

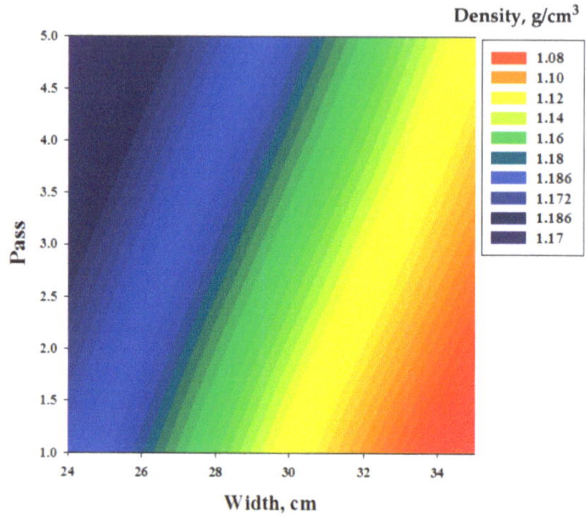

Figure 12. Tire type and traffic effect on the soil bulk density.

Figure 13 shows the interactive effect of tire type and tire inflation pressure on soil compaction. The soil bulk density changed strongly with increasing pressure when using a narrow tire, but less changes occurred when using a wide tire with increasing pressure.

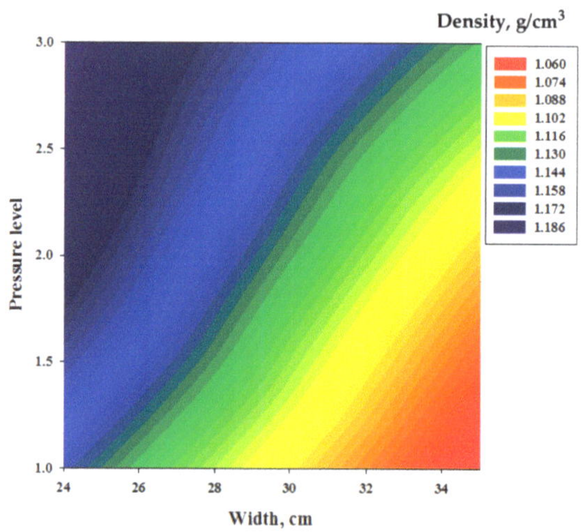

Figure 13. Three-dimensional surface curve of the main effect of tire type pressure and tire pressure on the soil bulk density of the soil.

4. Conclusions

1. The results showed that using a narrow tire increased the soil bulk density in different layers of soil much more in comparison with a common tire. A narrow tire compacted the soil more and increased the soil density, even at a depth of 50 cm, more than a common tire at a depth of 30 cm. Additionally, the soil bulk density of soil for a narrow tire at a depth of 40 cm was greater than that of a normal tire at a depth of 10 cm.
2. In both tires, the soil bulk density of the soil increased with increasing tire wheeling. According to the results, the highest mean value of soil density for narrow tire was after five passes, with a value of 1.185 g/cm^3. The lowest value was for a common

tire after one pass, with a value of 1.112 g/cm^3. It is noteworthy that the bulk density of soil for a narrow tire after one pass was more than the bulk density of soil for a common tire after five passes.

3. Using a narrow tire with tire wheeling increased the soil bulk density very sharply, but when using a wide tire, with increasing passes, fewer changes in the bulk density of the soil occurred. This indicates that using narrow tires should be considered to be very destructive on the soil structure. Additionally, with increasing tire pressure, and when using a narrow tire, the soil bulk density changed much more than when the common tire's pressure changed.

Author Contributions: Conceptualization, G.S. and A.M.; formal analysis, A.K., M.S.; investigation, A.M. and A.K.; methodology, G.S., A.M. and A.K.; resources, G.S., A.M. and A.K.; supervision, G.S.; visualization, G.S., A.M., A.K. and M.S.; writing—original draft, A.M. and A.K.; writing—review and editing, G.S., A.M., A.K., M.S. and S.M. All authors have read and agreed to the published version of the manuscript.

Funding: This research received no external funding.

Institutional Review Board Statement: Not applicable.

Data Availability Statement: Not applicable.

Conflicts of Interest: The authors declare no conflict of interest.

References

1. Van-Camp, L.; Bujarrabal, B.; Gentile, A.R.; Jones, R.J.; Montanarella, L.; Olazabal, C.; Selvaradjou, S.K. *Reports of the Technical Working Groups Established under the Thematic Strategy for Soil Protection*; Office for Official Publications of the European Communities: Luxembourg, 2004.
2. Lal, R. Soils and food sufficiency: A review. In *Sustainable Agriculture*; Springer: Dordrecht, The Netherlands, 2009; pp. 25–49.
3. Ansorge, D.; Godwin, R.J. The effect of tyres and a rubber track at high axle loads on soil compaction, Part 1: Single axle-studies. *Biosyst. Eng.* **2007**, *98*, 115–126. [CrossRef]
4. Arvidsson, J.; Westlin, H.; Keller, T.; Gilbertsson, M. Rubber track systems for conventional tractors—Effects on soil compaction and traction. *Soil Tillage Res.* **2011**, *117*, 103–109. [CrossRef]
5. Li, H.; Schindler, C. Analysis of soil compaction and tire mobility with finite element method. Proceedings of the Institution of Mechanical Engineers, Part K. *J. Multi-Body Dyn.* **2013**, *227*, 275–291.
6. Hakansson, I.; Reeder, R.C. Subsoil compaction by vehicles with high axle load-extent, persistence and crop response. *Soil Tillage Res.* **1994**, *29*, 277–304. [CrossRef]
7. Veronesi, F.; Corstanje, R.; Mayr, T. Mapping soil compaction in 3D with depth functions. *Soil Tillage Res.* **2012**, *124*, 111–118. [CrossRef]
8. Keller, T.; Lamandé, M.; Peth, S.; Berli, M.; Delenne, J.Y.; Baumgarten, W.; Or, D. An interdisciplinary approach towards improved understanding of soil deformation during compaction. *Soil Tillage Res.* **2013**, *128*, 61–80.
9. Voorhees, W.B. Long-term effect of subsoil compaction on yield of maize. *Adv. GeoEcol.* **2000**, *32*, 331–338.
10. Taylor, H.M.; Roberson, G.M.; Parker, J.J., Jr. Soil strength-root penetration relations for medium-to coarse-textured soil materials. *Soil Sci.* **1966**, *102*, 18–22. [CrossRef]
11. Barik, K.; Aksakal, E.L.; Islam, K.R.; Sari, S.; Angin, I. Spatial variability in soil compaction properties associated with field traffic operations. *Catena* **2014**, *120*, 122–133. [CrossRef]
12. Sarkar, P.; Upadhyay, G.; Raheman, H. Active-passive and passive-passive configurations of combined tillage implements for improved tillage and tractive performance: A review. *Span. J. Agric. Res.* **2021**, *19*, e02R01. [CrossRef]
13. Damme, L.; Stettler, M.; Pinet, F.; Vervaet, P.; Keller, T.; Munkholm, L.J.; Lamandé, M. The contribution of tyre-evolution to the reduction of soil compaction risks. In Proceedings of the 21th ISTRO Conference, Paris, France, 25–27 September 2018.
14. Hakansson, I. *Machinery-Induced Compaction of Arable Soils*; Incidence–Consequences–Counter-Measures; Reports from the Division of Soil Management; FAO: Québec City, QC, Canada, 2005; Volume 109.
15. Horn, R.; Way, T.; Rostek, J. Effect of repeated wheeling on stress/strain properties and ecological consequences in structured arable soils. *Rev. Cienc. Suelo Nutr. Veg.* **2001**, *1*, 34–40.
16. Boguzas, V.; Hakansson, I. Barley yield losses simulation under Lithuanian conditions using the Swedish soil compaction model. In Proceedings of the Conference on Sustainable Agriculture in Baltic States, Tartu, Estonia, 28–30 June 2001; Tartumaa Ltd.: Tartu, Estonia, 2001.
17. Ridge, R. Trends in sugar cane mechanization. *Int. Sugar J.* **2002**, *103*, 164–166.
18. Soane, B.D.; Dickson, J.W.; Campbell, D.J. Compaction by agricultural vehicles: A review III. Incidence and control of compaction in crop production. *Soil Tillage Res.* **1982**, *2*, 3–36. [CrossRef]

19. Kunickaya, O.; Hertz, E.; Kruchinin, I.; Tikhonov, E.; Ivanov, N.; Dolmatov, N.; Zorin, M.; Grigorev, I. Pressure control systems for tyre preservation in forestry machinery and forest soils. *Asian J. Water Environ. Pollut.* **2021**, *18*, 95–102. [CrossRef]
20. Šimečková, J.; Polcar, A.; Hammerová, A.; Votava, J.; Kumbár, V. Changes to the physical properties of the soil after the passage of an agricultural tractor. *Int. Agrophysics* **2021**, *35*, 97–105. [CrossRef]
21. Damme, L.; Stettler, M.; Pinet, F.; Vervaet, P.; Keller, T.; Munkholm, L.J.; Lamandé, M. Construction of modern wide, low-inflation pressure tyres per se does not affect soil stress. *Soil Tillage Res.* **2020**, *204*, 104798.
22. Upadhyay, G.; Raheman, H. Comparative analysis of tillage in sandy clay loam soil by free rolling and powered disc harrow. *Eng. Agric. Environ. Food* **2019**, *12*, 118–125. [CrossRef]
23. Chygarev, Y.; Lodyata, S. Research of tyre rigidity in terms of ecological safety of agricultural landscapes. In Proceedings of the VII Międzynarodowe Sympozjum Ekologiczne Aspekty Mechanizacji Nawożenia, Ochrony Roślin Uprawy Gleby i Zbioru Roślin Uprawnych, Warsaw, Poland, 18–19 September 2000; pp. 171–176.
24. Balbuena, R.H.; Terminiello, A.M.; Claverie, J.A.; Casado, J.P.; Marlats, R. Soil compaction by forestry harvester operation: Evolution of physical properties. *Rev. Bras. Eng. Agric. Ambient.* **2000**, *4*, 453–459. [CrossRef]
25. Seker, C.J.S.I.I.; Isildar, A.A. Effects of wheel traffic porosity and compaction of soil profile. *Turk. J. Agric. For.* **2000**, *24*, 71–78.
26. Abbaspour-Gilandeh, Y.; Rahimi-Ajdadi, F.; Asli-Ardeh, E.A.; Sharabiani, V.R.; Ardabili, M. Application of artificial neural network for predicting fuel consumption of tractor. In Proceedings of the International Agricultural Engineering Conference, Bangkok, Thailand, 7–10 December 2009.
27. Jang, J.S. ANFIS: Adaptive-network-based fuzzy inference system. *IEEE Trans. Syst. Man Cybern.* **1993**, *23*, 665–685. [CrossRef]
28. Buragohain, M.; Mahanta, C. A novel approach for ANFIS modelling based on full factorial design. *Appl. Soft Comput.* **2008**, *8*, 609–625. [CrossRef]
29. Dehnavi, A.; Aghdam, I.N.; Pradhan, B.; Varzandeh, M.H.M. A new hybrid model using step-wise weight assessment ratio analysis (SWARA) technique and adaptive neuro-fuzzy inference system (ANFIS) for regional landslide hazard assessment in Iran. *CATENA* **2015**, *135*, 122–148. [CrossRef]
30. Ertunc, H.M.; Hosoz, M. Comparative analysis of an evaporative condenser using artificial neural network and adaptive neuro-fuzzy inference system. *Int. J. Refrig.* **2008**, *31*, 1426–1436. [CrossRef]
31. Taghavifar, H.; Mardani, A. On the modeling of energy efficiency indices of agricultural tractor driving wheels applying adaptive neuro-fuzzy inference system. *J. Terramech.* **2014**, *56*, 37–47. [CrossRef]
32. Abbaspour-Gilandeh, Y.; Sedghi, R. Predicting soil fragmentation during tillage operation using fuzzy logic approach. *J. Terramech.* **2015**, *57*, 61–69. [CrossRef]
33. Ghadernejad, K.; Shahgholi, G.; Mardani, A.; Chiyaneh, H.G. Prediction effect of farmyard manure, multiple passes and moisture content on clay soil compaction using adaptive neuro-fuzzy inference system. *J. Terramech.* **2018**, *77*, 49–57. [CrossRef]
34. Lamandé, M.; Schjønning, P. Transmission of vertical stress in a real soil profile. Part II: Effect of tyre size, inflation pressure and wheel load. *Soil Tillage Res.* **2011**, *114*, 71–77. [CrossRef]
35. Mohsenimanesh, A.; Ward, S.M. On-the-move monitoring of soil-tire interaction on soft soil using wireless data acquisition. *Trans. ASABE* **2007**, *50*, 1919–1925. [CrossRef]
36. Mohsenimanesh, A.; Ward, S.M. Estimation of a three-dimensional tyre footprint using dynamic soil–tyre contact pressures. *J. Terramech.* **2010**, *47*, 415–421. [CrossRef]
37. Sohne, W. Fundamentals of pressure distribution and soil compaction under tractor tires. *Agric. Eng.* **1958**, *39*, 290.
38. Keller, T.; Défossez, P.; Weisskopf, P.; Arvidsson, J.; Richard, G. SoilFlex: A model for prediction of soil stresses and soil compaction due to agricultural field traffic including a synthesis of analytical approaches. *Soil Tillage Res.* **2007**, *93*, 391–411. [CrossRef]
39. Zink, A.; Fleige, H.; Horn, R. Load risks of subsoil compaction and depths of stress propagation in arable Luvisols. *Soil Sci. Soc. Am. J.* **2010**, *74*, 1733–1742. [CrossRef]

agriculture

MDPI

Technical Note

Development of Autofeeding Device Applicable to a Biodegradable Pot Tray

Seok-Joon Hwang [1,2], Moon-Kyeong Jang [1,2] and Ju-Seok Nam [1,2,*]

[1] Department of Biosystems Engineering, Kangwon National University, 1 Kangwondaehak-gil, Chuncheon 24341, Gangwon-do, Republic of Korea

[2] Interdisciplinary Program in Smart Agriculture, Kangwon National University, 1 Kangwondaehak-gil, Chuncheon 24341, Gangwon-do, Republic of Korea

[*] Correspondence: njsg1218@kangwon.ac.kr; Tel.: +82-33-250-6497

Abstract: In this study, a pot autofeeding device for a biodegradable pot tray was developed. The tensile strength and bending strength were measured to identify the physical properties of the biodegradable pot tray. As a result of the measurement, the tensile strength and bending strength of the biodegradable pot tray were 0.06 and 0.17 times smaller than those of the plastic pot tray. Therefore, a new type of pot tray extraction mechanism was designed, considering the physical properties, dimensions, and geometry of the biodegradable pot tray, and it was applied to the pot autofeeding device. The developed pot autofeeding device consists of a pot slot, pot-separating blades, pot holders, air cylinders, and a conveyor device. It can supply 240 pot trays per hour to the seeding process without deformation or damage to the biodegradable pot tray.

Keywords: autofeeding device; biodegradable; pot tray; separating blade; tensile strength

Citation: Hwang, S.-J.; Jang, M.-K.; Nam, J.-S. Development of Autofeeding Device Applicable to a Biodegradable Pot Tray. *Agriculture* **2022**, *12*, 2097. https://doi.org/10.3390/agriculture12122097

Academic Editor: Vadim Bolshev

Received: 15 November 2022
Accepted: 6 December 2022
Published: 7 December 2022

1. Introduction

As of 2019, 412,457 kilotons of paper and 183,031 kilotons of pulp were produced worldwide [1]. During the production of paper and pulp, various wastes are inevitably generated, including foreign matter, wastewater, and additives [2]. Paper sludge refers to the solid waste generated from the wastewater treatment process [3]. The amount of paper sludge generated accounts for approximately 50% of the total waste generated from the papermaking process [4]. Previously, paper sludge was treated using such methods as ocean dumping, landfill, and incineration. However, ocean dumping has been prohibited since 2014 owing to the London Convention and the Marine Pollution Prevention Act [5,6].

Under the influence of these policies, studies have been conducted to recycle paper sludge as a substitute for cement or to develop ecofriendly treatment measures [7–10]. In the agricultural sector, studies have been conducted to manufacture biodegradable pot trays that can replace the existing plastic pot trays using paper sludge [11,12]. When the hot pepper seedlings grown in the existing plastic pot trays and biodegradable pot trays were compared, there were no differences in plant length, leaf length, and the number of leaves, confirming the applicability of biodegradable pot trays [13,14]. Recently, studies have been conducted to improve the tensile and bursting strengths of biodegradable pot trays by mixing paper sludge with waste newspaper, wet strength agents, and water repellents [15,16]. Biodegradable pot trays can replace plastic pot trays, which are discarded through such methods as incineration after use, and they also have an advantage of being ecofriendly because they decompose in soil naturally [17]. In the case of transplanting using plastic pot trays, seedlings are removed from the pot trays and transplanted to the field. This extraction process may have a negative effect on the growth of crops because it can damage the roots of the seedlings [18]. In contrast, biodegradable pot trays can preserve the roots of seedlings because they are transplanted to the field together with the seedlings without an extraction process, and thus they are favorable for the growth of seedlings [16].

Owing to these benefits, biodegradable pot trays have been used for growing the seedlings of crops, such as tomatoes, cucumbers, lettuce, and hot peppers [19]. In this regard, there is a growing need for studies on the application of biodegradable pot trays to agricultural machines [18]. As a research case for applying biodegradable pot trays to agricultural machinery, a transplanter that can apply chain paper pots was developed [20,21]. Nambu and Tanimura (1992) developed the Ferris-type transplanting mechanism for biodegradable pot trays connected in a single line with applying the planting fingers to the transplanting device of the vegetable transplanter. The planting fingers gripped a cell of the biodegradable pot tray, then it rotated similarly to a Ferris wheel, and released the cells into the ground [20]. As a result of the performance evaluation, the work efficiency of the transplanter with the Ferris-type transplanting mechanism was 100 cells per minute [21]. Kumar and Raheman (2011) developed a transplanter that can transplant a biodegradable pot tray with multiple rows and columns. A supplied biodegradable pot tray was cut into multiple cells by circular blades with left–right movement. Then, each cell was dropped into the ground by the seedling drop tube [22]. Recently, a study has been conducted to shorten the time required for the cutting and extraction of biodegradable pot trays to less than one second for high-speed machine transplanting with circular blades [23].

On the other hand, there has been no prior study on the development of a pot-seeding machine capable of seeding in biodegradable pot trays. Pot-seeding machines significantly affect the increase in the mechanization rate of seeding and transplanting because they have high work efficiency and require less manpower, and pot trays that complete seeding using them are compatible with transplanters [24,25]. Therefore, studies on pot-seeding machines that can perform seeding in biodegradable pot trays are required.

A pot-seeding machine commonly used in agriculture consists of a pot autofeeding device, soil feeder, compressing device, seeding device, and covering soil feeder [26]. Among them, the pot autofeeding device is the important device which separates multi-stacked pot trays one by one and supplies them to the seeding process [27]. The mechanical operation of the pot autofeeding device continuously causes a load on pot trays during the pot tray extraction process. Biodegradable pot trays can be easily deformed and destroyed under load because their tensile strength (approximately 0.69 N/mm^2) is lower than that of polypropylene, which is used as the main material of existing pot trays (34 N/mm^2) [16,17,28]. Biodegradable pot trays are also exposed to continuous contraction and expansion in the manufacturing and storage processes because their materials are vulnerable to deformation by moisture. It is difficult to apply manufactured biodegradable pot trays to machinery because their uniformity and firmness are low [22]. To apply biodegradable pot trays to pot-seeding machines, the machines need to be designed so that biodegradable pot trays cannot be deformed or destroyed by minimizing the load caused by the mechanical operation. To this end, the physical properties of biodegradable pot trays must be considered in the design process of the pot autofeeding device.

This study is a preliminary study for the development of a pot-seeding machine that can be applied to biodegradable pot trays, and a pot autofeeding device for a biodegradable pot tray was developed. The tensile and bending strengths of the biodegradable pot tray were measured and reflected in the design of the pot autofeeding device. It is expected that the development of the pot autofeeding device will contribute to an improvement in the mechanization rate of seeding and the development of pot-seeding machines for biodegradable pot trays.

2. Materials and Methods

2.1. Biodegradable Pot Tray

Figure 1 shows the biodegradable pot tray. The biodegradable pot tray consists of paper sludge, a strength resin that increases the strength of paper, and a size agent for water resistance reinforcement. Their proportions are shown in Table 1 [16]. The total weight of the biodegradable pot tray is 227.17 g. As shown in Figure 2, the biodegradable pot tray has a characteristic of being bent by its own weight. The pot tray has a length of 440 mm, a width of 300 mm, and a height of 45 mm. Twelve cells are arranged in the lateral direction and eight cells in the longitudinal direction. Thus, the pot tray has a total of 96 cells. The shape of each cell is a truncated pyramid with square bases, and the lengths of upper base and lower base were 28.5 mm and 20.5 mm, respectively. At the bottom of the pot tray, there are empty spaces between the cells. The empty spaces are arranged at an interval of 36.5 mm.

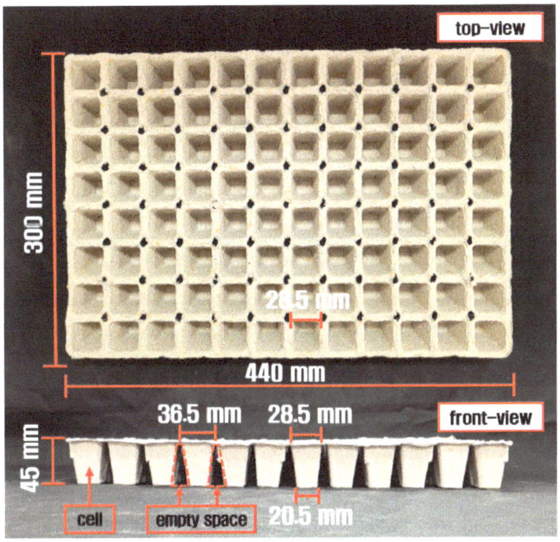

Figure 1. Dimensions of the biodegradable pot tray.

Table 1. Component ratio of the biodegradable pot tray ingredients.

Items	Ingredients of Biodegradable Pot Tray		
	Paper Sludge	**Strength Agent**	**Sizing Agent**
Component ratio	90.90%	4.55%	4.55%

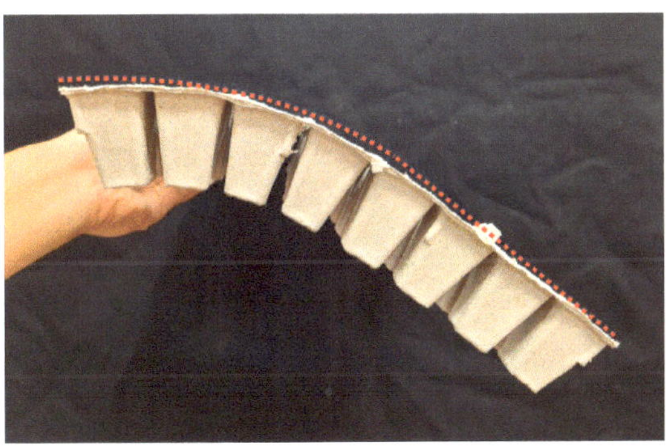

Figure 2. Deflection shape of biodegradable pot tray caused by its self-weight.

2.2. Characteristics of Current Pot Autofeeding Device

A pot-seeding machine commonly used in agriculture consists of a pot autofeeding device, soil feeder, compressing device, seeding device, and covering soil feeder (Figure 3). When multi-stacked pot trays are inserted into the autofeeding device, the pot trays are separated one by one by the pot tray extraction mechanism of the pot autofeeding device and transported to the soil feeder. The soil feeder fills each cell of the pot tray with soil. The compressing device then presses the soil filled in each cell into a conical shape to create a space for seeding. Subsequently, seeding is performed on the pressed soil by the seeding device. Finally, each cell of the pot tray is filled with covering soil by the covering soil feeder. The pot autofeeding device applied to a pot-seeding machine is an important device that significantly affects the improvement in the work efficiency of the pot-seeding machine because it automatically separates the inserted pot trays one by one and supplies them to the seeding process without requiring manpower. The commercial pot autofeeding devices have different pot tray extraction mechanisms. For representative pot tray extraction mechanisms, the hook-type, roller-type, and flat-type separators were applied. Each extraction mechanism operates as follows.

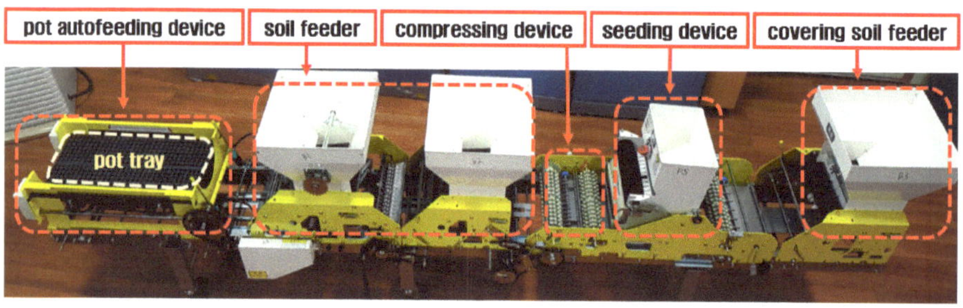

Figure 3. Picture of the mechanical pot tray seeding machine commonly used in agriculture.

A pot autofeeding device that extracts pot trays using the hook-type separator is shown in Figure 4. For this pot tray extraction mechanism, when stacked pot trays are placed in the pot autofeeding device, they are supported by pot holders. The hook-type separator is mounted on a cylindrical rod with rotational motion. Therefore, whenever the rod rotates, the hook-type separator also rotates. When the rotating hook-type separator contacts the top edge of the pot tray, it pulls down the pot tray in the direction of gravity to drop it. The separated pot tray is sent to the seeding process through the transport system.

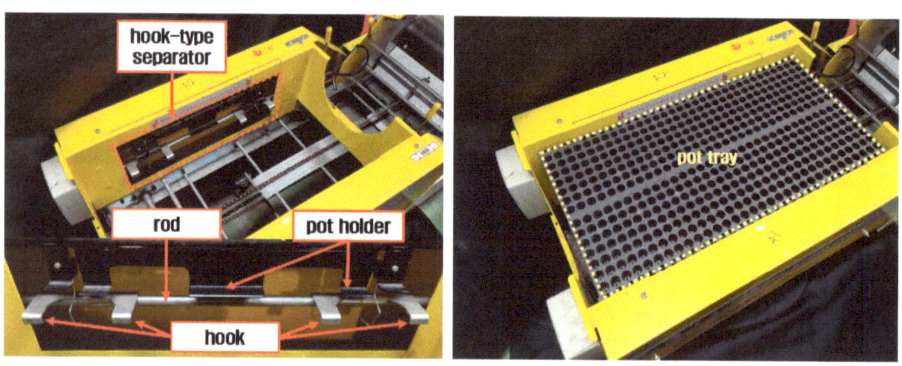

Figure 4. View of the pot autofeeding device with a hook-type separator applied.

Figure 5 shows the operation of the pot tray extraction mechanism using the flat-type separator. When stacked pot trays are placed in the pot autofeeding device, they are supported by pot holders. The flat-type separator can be moved in the vertical and horizontal directions because it is equipped with two air cylinders. The flat-type separator moves in the horizontal direction to contact the hollow between the pot trays and then moves in the vertical downward direction to separate the pot tray at the bottom. The separated pot tray is sent to the seeding process through the transport system.

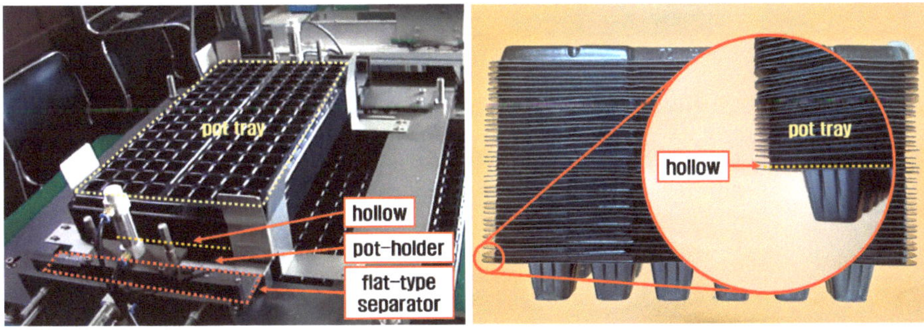

Figure 5. View of the pot autofeeding device with a flat-type separator applied.

Figure 6 shows the pot autofeeding device that extracts pot trays through the roller-type separator. For this pot tray extraction mechanism, when obliquely stacked pot trays are inserted into the pot autofeeding device, the top edge of the pot tray located at the bottom comes into contact with the roller-type separator. The roller-type separator then rotates to separate the pot tray and sends it to the seeding process.

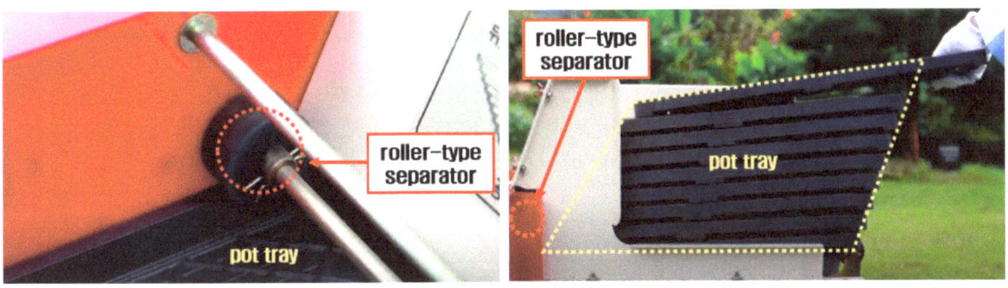

Figure 6. View of the pot autofeeding device with a roller-type separator applied.

2.3. Requirements for Pot Autofeeding Device

The following requirements must be considered for the design of a pot autofeeding device applicable to biodegradable pot trays.

1. To minimize the manpower required in the pot tray separation process and improve work efficiency, the pot autofeeding device must be designed so that multi-stacked biodegradable pot trays can be automatically separated one by one and sent to the seeding process.
2. The components of the pot autofeeding device must be designed considering the strength, dimensions, and geometry of biodegradable pot trays.
3. The components of the pot autofeeding device must be designed to prevent the bending of the biodegradable pot tray during the pot tray extraction process for the smooth operation of the pot autofeeding device.
4. The pot autofeeding device must be designed so that biodegradable pot trays cannot be deformed or destroyed by minimizing the load applied to the pot trays by the pot tray extraction mechanism.

As shown in Figures 4–6, plastic pot tray extraction methods that applied the hook-type, flat-type, and roller-type separators extract a pot tray by applying a load while the separator is in contact with the edge of the pot tray through mechanical behavior. Plastic pot trays are separated without deformation or destruction even if stress is concentrated on a small area, such as an edge. This is because the physical properties of the plastic material are considered, and an appropriate load is transferred to the edge of the pot tray within a range that does not exceed the tensile strength of the material.

However, biodegradable pot trays are unfavorable for application to machines because their firmness is lower than that of plastic materials, and thus, they are easily deformed and destroyed under a relatively small load [17,22,28].

Therefore, the applicability of the existing plastic pot tray extraction mechanisms must be examined to select a mechanism for the pot autofeeding device applicable to biodegradable pot trays. If the existing mechanisms are not applicable, a new-type pot tray extraction mechanism that can prevent deformation or destruction to biodegradable pot trays needs to be studied. For such a decision, the physical properties of biodegradable and plastic pot trays must be analyzed and compared first.

Accordingly, it is necessary to select a pot extraction mechanism that suitable with the biodegradable pot tray in order to develop a pot autofeeding device. In addition, the performance evaluation should be performed to confirm deformation or damage of the biodegradable pot tray using the developed pot autofeeding device.

2.4. Experimental Design

2.4.1. Tensile and Bending Strength

The tensile and bending strengths of biodegradable and plastic pot trays were measured. The tensile and bending strengths are important indicators of the physical properties of pot trays [16]. They were measured using a universal testing machine (AGS-50kNX, Scientific, Japan) (Figure 7 and Table 2). The measurements were obtained using three specimens for each experiment. The specimens were extracted from the outer walls of the pot trays (Figure 8). For tensile strength measurement, both ends of the specimen were fixed with two jigs and the load at the time of fracture was measured by pulling the specimen. As for the test conditions, the gauge length and tensile speed were 30 mm and 5 mm/min, respectively. The tensile strength was derived by dividing the measured load by the cross-sectional area of the specimen before measurement. The tensile strength of a pot tray can be determined through Equation (1) [29].

$$\sigma_t = \frac{W_{max}}{A_0} \quad (1)$$

where σ_t: tensile strength, N/mm^2; W_{max}: maximum tensile load, N and A_0: area of cross section, mm^2.

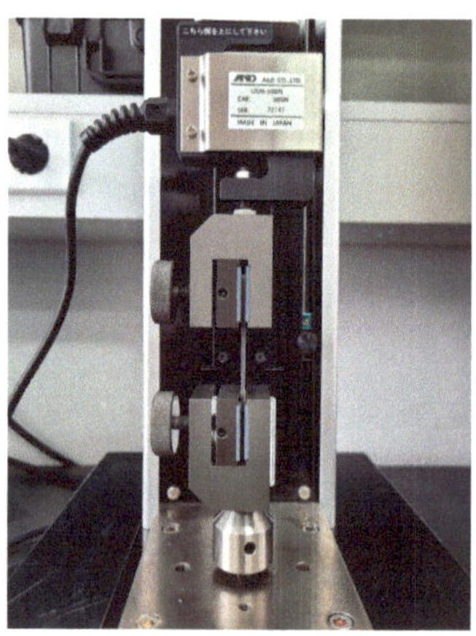

Figure 7. View of the universal testing machine used.

Table 2. Specifications of the universal testing machine.

Items	Specifications
Model/Company/Nation	AGS-50kNX/Scientific/Japan
Max. Load Capacity (kN)	50
Crosshead Speed Accuracy	0.1
Speed Range (mm/min)	0.001 to 1600
Max. Return Speed (mm/min)	2200

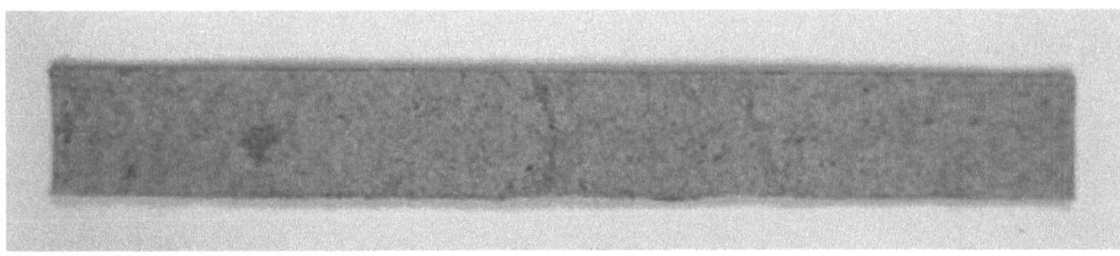

(**a**) biodegradable pot tray

(**b**) plastic pot tray

Figure 8. View of the specimen for measuring the tensile and bending stresses.

The bending strength was measured using the three-point bending method. While the specimen remained parallel to the ground, its bottom was supported with two jigs. The top of the specimen was pressed using a crosshead, and the generated load was measured. The distance between the jigs was 30 mm, and the crosshead speed was set to 2.8 mm/min. The bending strength of a pot tray can be determined through Equation (2) [30].

$$\sigma_b = \frac{3PL}{2BD^2} \quad (2)$$

where σ_b: bending strength, N/mm^2; P: maximum bending load, N; L: length between jigs, mm; B: width of the specimen, mm and D: depth of the specimen, mm.

2.4.2. Performance Evaluation

A pot extraction mechanism was selected for use with biodegradable pot trays, and a performance evaluation was performed using the pot autofeeding device developed by reflecting the selected pot extraction mechanism. The method of performance evaluation is as follows. The developed autofeeding device was operated with inserting the stacked biodegradable pot trays continuously. During the pot extraction process, it was confirmed that the components of the pot extraction mechanism were working normally. In addition, after the pot extraction process, the biodegradable pot trays supplied to the seeding process were collected, and then damage and deformation of the biodegradable pot trays were observed. The performance evaluation time was set to 1 h. The work efficiency of the pot autofeeding device was derived by counting the number of extracted biodegradable pot trays.

3. Results and Discussion

3.1. Result of Experiment

The results of the tensile strength and bending strength tests conducted on the biodegradable and plastic pot trays are as follows. The tensile strength was determined to be 12.66 $\pm$ 0.87 N/mm^2 for the plastic pot tray and 0.70 $\pm$ 0.03 N/mm^2 for the biodegradable pot tray (Table 3). The tensile strength of the biodegradable pot tray was 0.06 times that of the plastic pot tray. The bending strength was measured to be 22.26 $\pm$ 0.28 N/mm^2 for the plastic pot tray and 3.89 $\pm$ 0.54 N/mm^2 for the biodegradable pot tray (Table 4). The bending strength of the biodegradable pot tray was 0.17 times that of the plastic pot tray. It is judged that this difference occurred due to the characteristics of the main ingredient constituting the biodegradable pot tray and the plastic pot tray.

Table 3. Tensile strength of the biodegradable pot tray.

Items	Tensile Strength (N/mm^2)	
	Plastic Pot Tray	Biodegradable Pot Tray
1st Specimen	12.91	0.69
2nd Specimen	13.57	0.74
3rd Specimen	11.49	0.66
Average	12.66	0.70
Standard deviation	0.87	0.03

Table 4. Bending strength of the biodegradable pot tray.

Items	Bending Strength (N/mm^2)	
	Plastic Pot Tray	Biodegradable Pot Tray
1st Specimen	22.62	3.61
2nd Specimen	22.20	4.65
3rd Specimen	21.95	3.41
Average	22.26	3.89
Standard deviation	0.28	0.54

The results of the tensile and bending strength measurements show that the strength of the biodegradable pot tray is significantly lower than that of the plastic pot tray. Accordingly, it is judged that the hook-type, roller-type, and flat-type separators, which were developed for plastic pot trays, are not suitable for extracting biodegradable pot trays. Therefore, a pot autofeeding device with a new-type mechanism that can minimize deformation or damage to pot trays was developed considering the physical properties of the biodegradable pot tray.

3.2. Development of Pot Autofeeding Device

A pot autofeeding device with a new-type mechanism that can prevent deformation or destruction to the biodegradable pot tray in the pot tray extraction process was designed and fabricated considering the physical properties, dimensions, and geometry of the biodegradable pot tray.

3.2.1. Design of Main Components

The characteristics of each main component which constituting the pot autofeeding device are as follows.

1. Pot slot: the pot slot is the position where the biodegradable pot tray is inserted in the pot autofeeding device, and its geometry is shown in Figure 9. The length and width of the pot slot were designed to be 440 mm and 300 mm, respectively, considering the size of the biodegradable pot tray. The height of the pot slot is 240 mm, facilitating the input of 13 pot trays at a time.

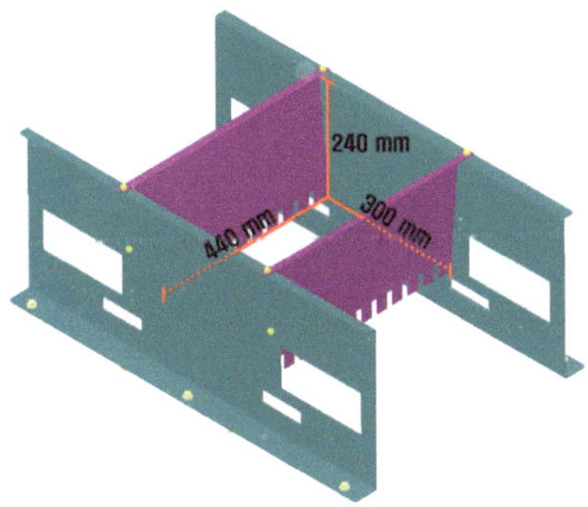

Figure 9. Shape of the pot slot.

2. Pot-separating blades: Figure 10 shows the geometry of the pot-separating blades. The blades are inserted into the spaces between the cells of the biodegradable pot tray to support or separate the pot tray. The biodegradable pot tray has a low bending strength, and thus, deflection occurs owing to its own weight. To prevent this, the number of pot-separating blades was maximized so that the entire pot tray could be supported. Therefore, 11 pot-separating blades were arranged at intervals of 36.5 mm to form an assembly. This interval is the same as the interval of the spaces located between the cells of the biodegradable pot tray. Each assembly is distinguished as either upper or lower pot-separating blades depending on its position. As the upper/lower pot-separating blades are located at both sides of the pot autofeeding device, 44 pot-separating blades were used in the pot autofeeding device. The thickness of the pot-separating blades was set to 4.5 mm, which was smaller than the width of the spaces located between the cells of the biodegradable pot tray (8 mm). Among the components of the pot autofeeding device, the pot-separating blades generate relatively large loads on the biodegradable pot tray during their operation. Such loads can be dispersed and geometry deformation can be minimized by designing the area in which the pot-separating blades are in contact with the biodegradable pot tray to be larger. The area can be increased by maximizing the length of the pot-separating blades. Considering that the width of the biodegradable pot tray is 300 mm, the length of the pot-separating blades was set to 150 mm. Therefore, the total length of contact between a pair of pot-separating blades and the biodegradable pot tray is 300 mm.

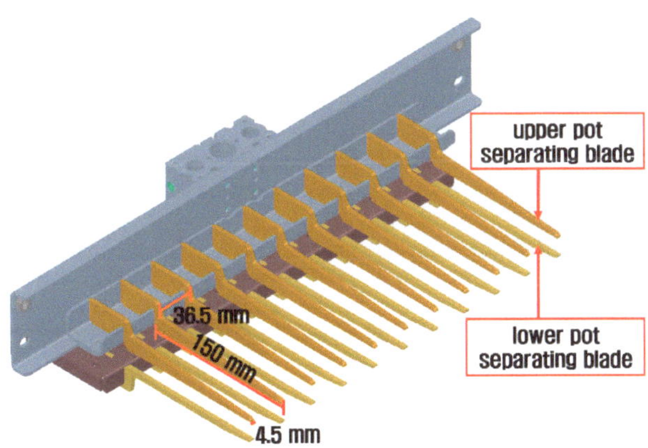

Figure 10. Shape of the pot-separating blade.

3. Pot holders: the pot holders support the inserted biodegradable pot tray, and their geometry is shown in Figure 11. The pot holders are located on both sides of the pot autofeeding device. Their width and height were set to 80 and 30 mm, respectively. Their length was set to 440 mm, which is the same as the length of the biodegradable pot tray.

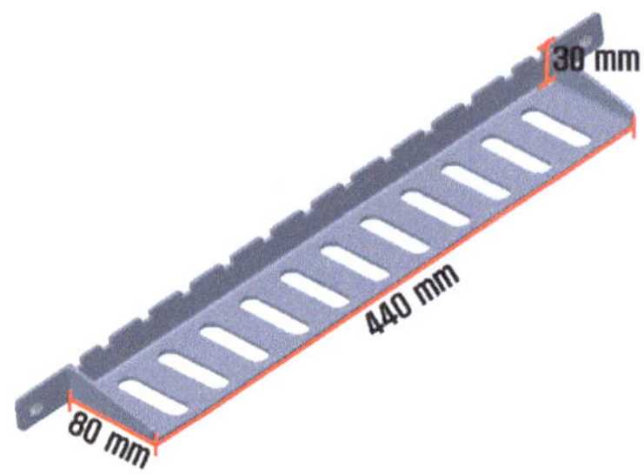

Figure 11. Shape of the pot holder.

4. Air cylinders: the air cylinders drive the pot-separating blades and pot holders, and their geometry is shown in Figure 12. Three types of air cylinders were used in the pot autofeeding device. Table 5 shows their specifications. The length of the pot-separating blades was set to 150 mm in Figure 10. During the pot tray extraction process, the blades need to insert 150 mm into the empty spaces of the biodegradable pot tray. Therefore, an air cylinder-ⓐ which has a stroke of 150 mm, was used for operating the pot-separating blades. The body of air cylinder-ⓐ is attached to the center of the frame outer wall, and the end of the piston rod is fixed to the pot-separating blades. When compressed air is supplied through position ①, the piston rod moves into the air cylinder body and pulls the pot-separating blades to the inside of the pot slot. Conversely, when compressed air is supplied through position ②, the piston rod protrudes and the pot-separating blades return to the outside of the pot slot. The width of the pot holders was set to 80 mm in Figure 11. The pot holders need to move 80 mm to support the biodegradable pot tray. Therefore, an air cylinder-ⓑ which has a stroke of 80 mm, was selected for operating the pot holders. The operation mechanism of air cylinder-ⓑ is the same as that of air cylinder-ⓐ. The body of air cylinder-ⓑ is attached on the frame outer wall, and the end of the piston rod of the air cylinder-ⓑ is fixed to the pot holders. Therefore, when compressed air is supplied to position ③, the pot holders move to the outside of the pot slot. When compressed air is supplied to position ④, they return to the inside of the pot slot. An air cylinder-ⓒ was designed to operate the lower pot-separating blade in Figure 10. The body of air cylinder-ⓒ and the end of the piston rod are fixed to the upper and lower pot-separating blades, respectively. At the beginning of the operation of the pot autofeeding device, compressed air is supplied to position ⑥, and the upper/lower pot-separating blades are in contact with each other. However, when compressed air is supplied to position ⑤, the lower pot-separating blade moves downward and is separated from the upper pot-separating blade. Air cylinder-ⓒ has a stroke of 20 mm. Therefore, the lower pot-separating blade moves 20 mm downward from the upper pot-separating blade. The proper pressure of all air cylinders was set to 0.6 bars.

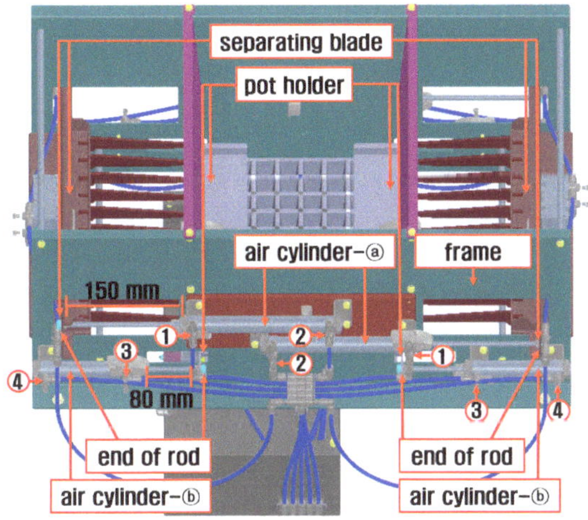

(**a**) side view

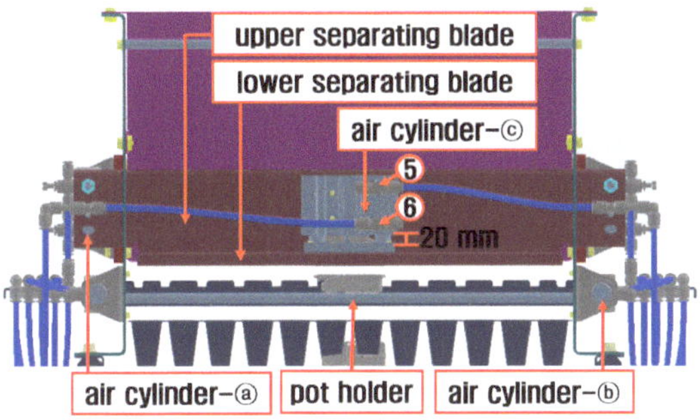

(**b**) front view

Figure 12. Shape of 3D model that the air cylinders applied to the pot autofeeding device.

Table 5. Specifications of the air cylinders used.

Items		Specifications		
Model/Company/Nation		ACS4-N LB20-S150 /KCC company/ South Korea	ACS4-N LB20-S80 /KCC company/ South Korea	ANGM 25-S20 /KCC company/ South Korea
Fluid type used		Air	Air	Air
Pressure (MPa)	Max.	1.0	1.0	1.0
	Min.	0.05	0.05	0.12
Piston speed (mm/s)	Max.	1000	1000	500
	Min.	750	750	50
Stroke (mm)		150	80	20
Inner diameter of rod (mm)		20	20	20

5. Conveyor device: The conveyor device transports the extracted biodegradable pot tray to the seeding process, and its geometry is shown in Figure 13. It has a length of 1800 mm and a height of 700 mm. Its width was set to 440 mm, which is the same as the length of the biodegradable pot tray. The conveyor device transports the biodegradable pot tray to the seeding process at a speed of 18.16 mm/s.

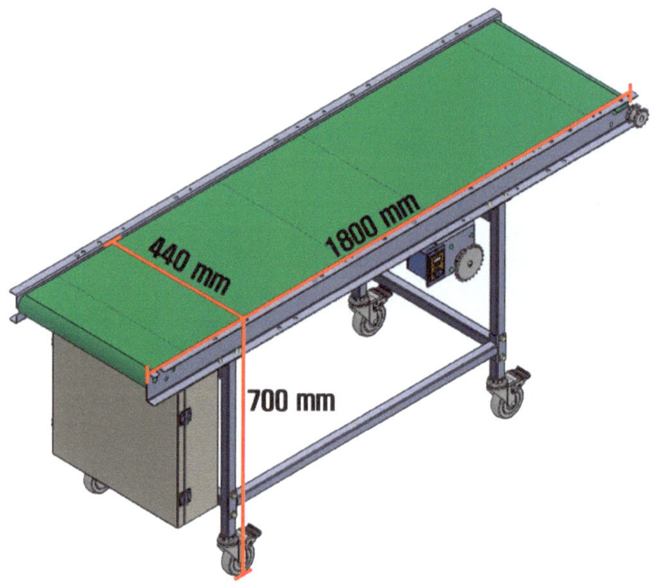

Figure 13. Shape of the conveyor device.

3.2.2. Operating Procedure of Developed Pot Autofeeding Device

Figure 14 shows the overall geometry of the pot autofeeding device that was designed considering the dimensions, geometry, and physical properties of the biodegradable pot tray. The main components of the pot autofeeding device are a pot slot, pot-separating blades, pot holders, air cylinders, and a conveyor device.

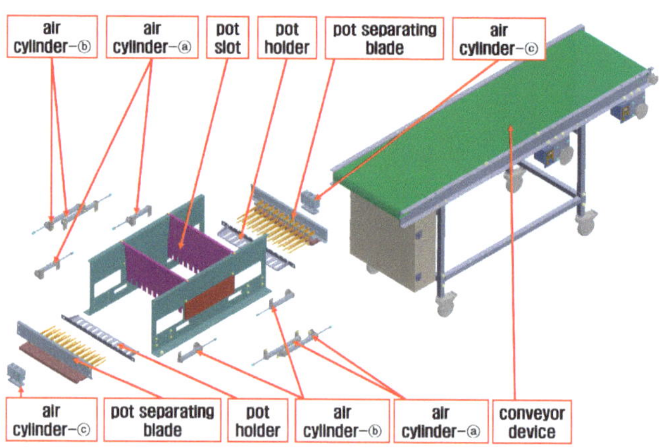

(**a**) 3D model

Figure 14. *Cont.*

(**b**) picture

Figure 14. Shape of the pot autofeeding device.

Figure 15 shows the operating procedure of the developed pot autofeeding device for the biodegradable pot tray. The operator places the biodegradable pot trays stacked in 13 layers into the pot slot. The lower part of the first biodegradable pot tray located at the bottom is supported by the pot holders located inside the pot slot (Figure 15a,b). Subsequently, air cylinder-ⓐ operates to insert the upper/lower pot-separating blades into the empty spaces between each cell of the second biodegradable pot tray (Figure 15c–e). In this instance, the upper pot-separating blade supports the lower part of the second biodegradable pot tray. Air cylinder-ⓑ operates to move the pot holders to the outside of the pot slot (Figure 15f). The first biodegradable pot tray does not drop vertically owing to the friction with the second biodegradable pot tray. Subsequently, air cylinder-ⓒ operates to move the lower pot-separating blade with the down-up motion (Figure 15g). The first biodegradable pot tray is separated from the second biodegradable pot tray as the lower pot-separating blade comes into contact with the upper part of the first biodegradable pot tray. The first biodegradable pot tray then drops onto the conveyor device to be transported to the seeding process (Figure 15h). Subsequently, the pot holders again move to the inside of the pot slot (Figure 15i), and the upper/lower pot-separating blades again move toward the outside of the pot slot (Figure 15j). Through this pot tray extraction mechanism, the biodegradable pot trays are sequentially extracted and transported to the seeding process one by one.

(**a**)

(**b**)

Figure 15. *Cont.*

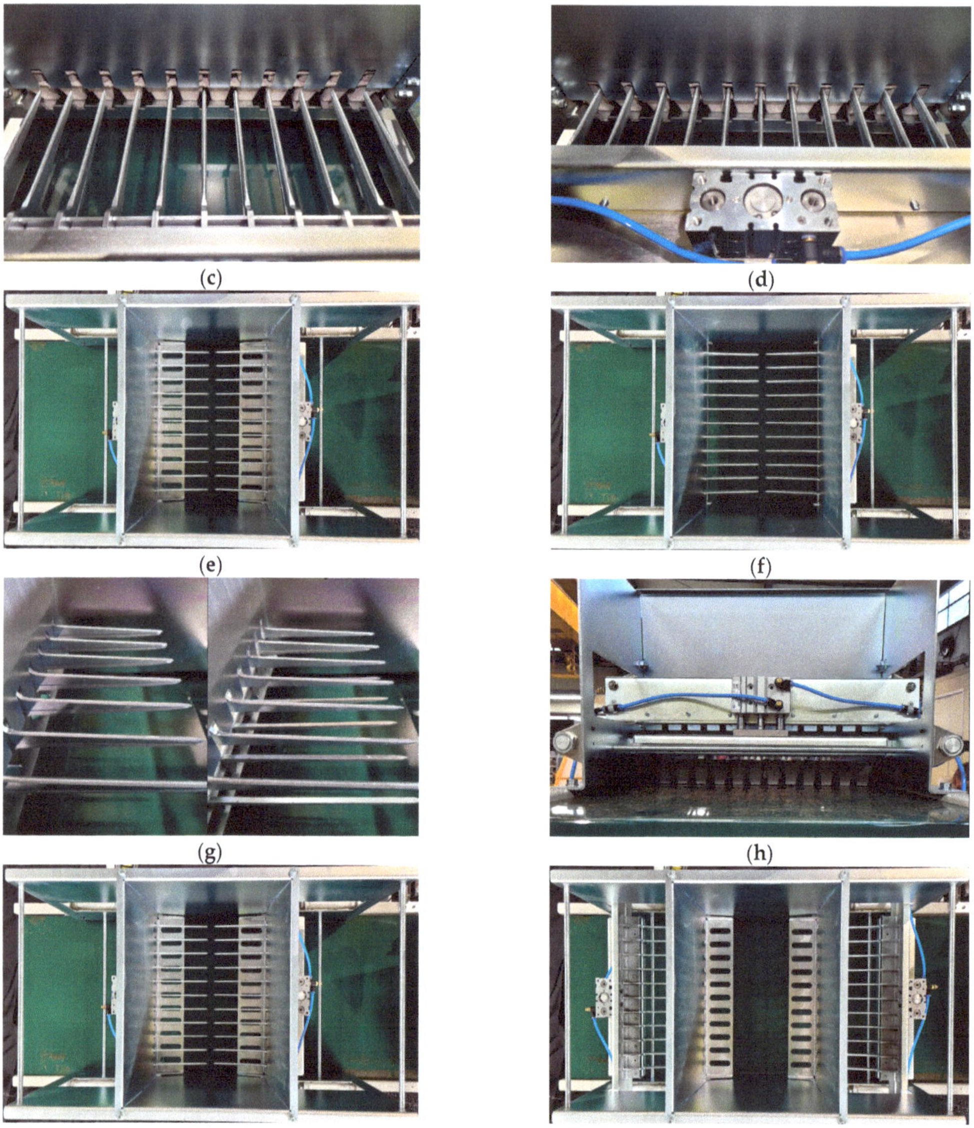

(c) (d) (e) (f) (g) (h) (i) (j)

Figure 15. Operating procedure of the pot autofeeding device for the biodegradable pot tray: (**a**) an initial condition; (**b**) supplying the biodegradable pot tray into the pot slot; (**c**) the shape of the pot separating blades before inserting; (**d**) inserting of pot separating blades into the spaces of biodegradable pot tray; (**e**) the shape of pot separating blades after inserting; (**f**) the pot holders move to the outside of the plot slot; (**g**) the lower pot-separating blade moves downward for separating the bottom biodegradable pot tray; (**h**) separated biodegradable pot tray drops onto the conveyor belt to be transported to the seeding process; (**i**) the pot holders return to the inside of the pot slot; (**j**) the pot-separating blades return toward the outside of the pot slot.

3.3. Result of Performance Evaluation

Figure 16 shows that conducting the performance evaluation of the pot autofeeding device. As a result of observing whether the pot autofeeding device operates normally under the condition that the biodegradable pot tray was inserted into the pot slot continuously, it was confirmed that the pot autofeeding device can automatically separate the stacked biodegradable pot trays one by one and send them to the seeding process without malfunction occurred. In addition, the result of counting the number of the biodegradable pot trays supplied for 1 h through the pot autofeeding device, it was observed that 240 biodegradable pot trays were supplied to the seeding process per hour.

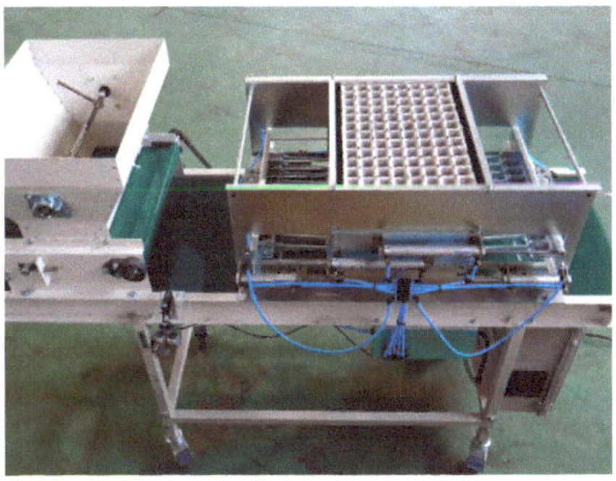

(**a**) initial condition

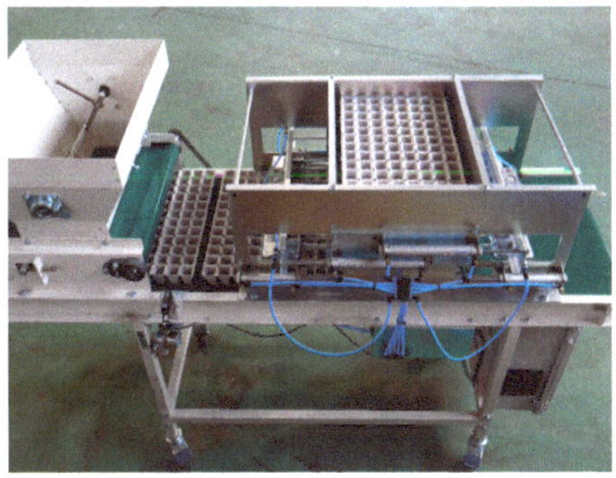

(**b**) operating condition

Figure 16. The picture of conduction the performance evaluation.

As a result of a complete enumeration survey regarding to the 240 biodegradable pot trays supplied to the seeding process, it was confirmed that no deformation or damage occurred in all biodegradable pot trays (Figure 17). This indicates that the new-type pot extraction mechanism is suitable for supplying the separating biodegradable pot trays. In particular, it was judged that the developed pot autofeeding device prevents the bending of the biodegradable pot tray due to its own weight in the pot extraction process by applying the pot separating blades, and it also prevents the deformation or destruction of the biodegradable pot tray by dispersing the load generated by the mechanical behavior of the components.

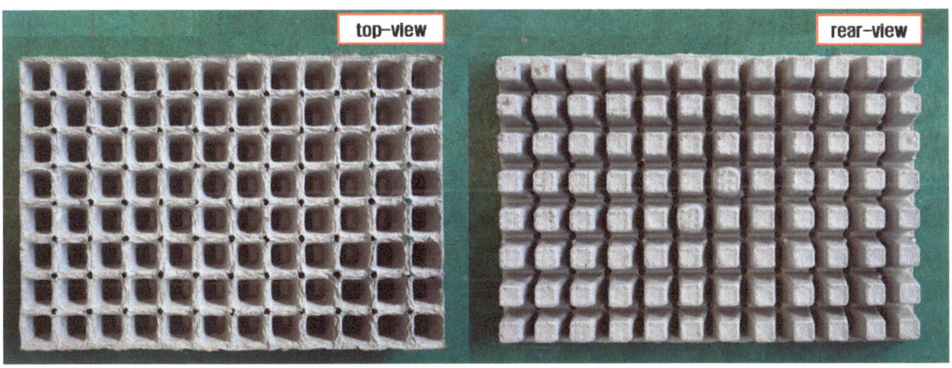

(**a**) before the pot extraction process

Figure 17. *Cont.*

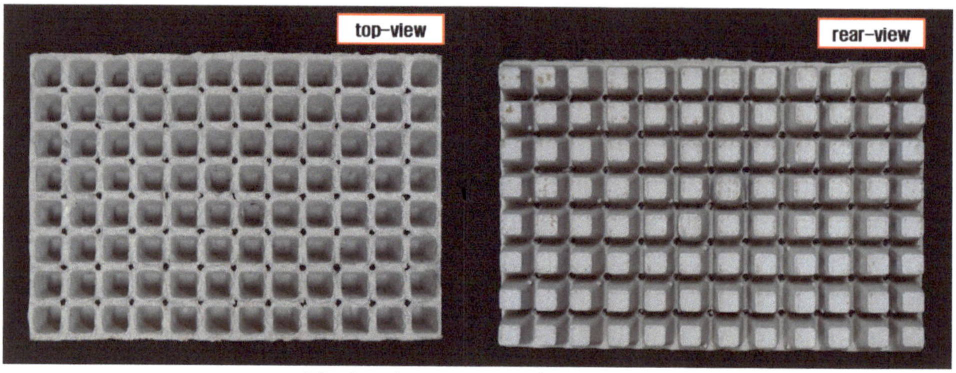

(b) after the pot extraction process

Figure 17. Picture of the biodegradable pot tray used for confirming the deformation and damage.

4. Conclusions

In this study, a pot autofeeding device for a biodegradable pot tray was developed considering the physical properties, dimensions, and geometry of the pot tray.

The pot tray extraction mechanisms commercialized for plastic pot trays were investigated, and their applicability to biodegradable pot trays was examined. To this end, specimens were secured from actual biodegradable and plastic pot trays, and their tensile and bending strengths were measured using a universal testing machine. The tensile and bending strengths of the biodegradable pot tray were derived as 0.70 ± 0.03 and 3.89 ± 0.54 N/mm^2, respectively, whereas those of the plastic pot tray were 12.66 ± 0.87 and 22.26 ± 0.28 N/mm^2, respectively. As a result of the measurement, the tensile strength and bending strength of the biodegradable pot tray were 0.06 and 0.17 times smaller than those of the plastic pot tray. It is judged that this difference occurred due to the material property of the biodegradable pot tray and the plastic port tray.

As the strength of the biodegradable pot tray was lower than that of the plastic pot tray, a new-type pot tray extraction mechanism was developed considering the physical properties, dimensions, and geometry of the biodegradable pot tray.

The developed pot autofeeding device consists of a pot slot, pot-separating blades, pot holders, air cylinders, and a conveyor device. The pot holders support the stacked biodegradable pot trays inserted into the pot slot. The pot separating blades insert into the stacked biodegradable pot trays and extract them one by one. Then, the separated biodegradable pot tray is transferred to the seeding process by a conveyor device. A performance evaluation was conducted to confirm that the components of the pot autofeeding device operate normally during the operation procedure. In addition, the biodegradable pot tray supplied to the seeding process was collected, and then damage and deformation occurring in the biodegradable pot tray were observed. As a result of observing whether the pot autofeeding device works normally, it was checked that the stacked biodegradable pot trays were extracted one by one automatically without malfunction occurred. The work efficiency of the pot autofeeding device was 240 pot trays per hour. As a result of a complete enumeration survey regarding to the 240 biodegradable pot trays supplied to the seeding process, it was confirmed that no deformation or damage occurred in all biodegradable pot trays. This indicates that the new-type pot extraction mechanism is suitable for supplying the separating biodegradable pot trays. Especially, it judged that the developed pot autofeeding device prevents the bending of the biodegradable pot tray due to its own weight in the pot extraction process by applying the pot separating blades, and it also prevents the deformation or destruction of the biodegradable pot tray by dispersing the load generated by the mechanical behavior of the components.

It is expected that the development of the pot autofeeding device for the biodegradable pot tray will reduce the manpower and labor load of seeding and improve the applicability of biodegradable pot trays.

Author Contributions: Investigation, S.-J.H., M.-K.J.; writing—original draft preparation, S.-J.H.; writing—review and editing, J.-S.N. All authors have read and agreed to the published version of the manuscript.

Funding: This work was supported by Korea Institute of Planning and Evaluation for Technology in Food, Agriculture and Forestry (IPET) through Agriculture, Food and Rural Affairs Convergence Technologies Program for Educating Creative Global Leader Program, funded by Ministry of Agriculture, Food and Rural Affairs (MAFRA)(716001-7).

Institutional Review Board Statement: Not applicable.

Data Availability Statement: Not applicable.

Conflicts of Interest: The authors declare no conflict of interest.

References

1. Korea Paper Association: Paper production by region. Available online: http://www.paper.or.kr/sub_5/5_1_3.php (accessed on 10 October 2022).
2. Skinner, H.J. Waste problems in the pulp and paper industry. *Ind. Eng. Chem.* **1939**, *31*, 1331–1335. [CrossRef]
3. Boni, M.R.; D'Aprile, L.; De Casa, G. Environmental quality of primary paper sludge. *J. Hazard. Mater.* **2004**, *108*, 125–128. [CrossRef]
4. Lee, Y.H.; Jo, H.M.; Kim, S.H.; Lee, J.Y. Fundamental study of the production of paper ash made from paper sludge I: Evaluation of the characteristics of paper ash made from paper sludge in a specialty paper mill. *J. Korea TAPPI* **2020**, *52*, 113–119. [CrossRef]
5. Ko, J.C.; Kim, S.H.; Park, Y.K.; Jeon, J.Y.; Kim, J.H.; Park, J.S. Pyrolysis characteristics of sludge discharged from paper mill process. *J. Korra* **2009**, *17*, 80–87.
6. Song, D.B.; Lim, K.H.; Jung, D.H.; Yoon, J.H. Analysis of drying performance and cost of large-capacity sewage sludge dryer. *J. Biosyst. Eng.* **2021**, *46*, 105–111. [CrossRef]
7. Takizawa, S.; Baba, Y.; Tada, C.; Fukuda, Y.; Nakai, Y. Pretreatment with rumen fluid improves methane production in the anaerobic digestion of paper sludge. *Waste Manag.* **2018**, *78*, 379–384. [CrossRef]
8. Bui, N.K.; Satomi, T.; Takahashi, H. Influence of industrial by-products and waste paper sludge ash on properties of recycled aggregate concrete. *J. Clean. Prod.* **2019**, *214*, 403–418. [CrossRef]
9. Chen, M.; Zheng, Y.; Zhou, X.; Li, L.; Wang, S.; Zhao, P.; Lu, L.; Cheng, X. Recycling of paper sludge powder for achieving sustainable and energy-saving building materials. *Constr. Build. Mater.* **2019**, *229*, 116874. [CrossRef]
10. Kang, I.K.; Oh, H.Y.; Han, D.Y.; Lim, Y.T.; Jang, M.M.; So, S.Y. An analysis of characteristics of asphalt concrete using paper sludge. *Conf. AIK* **2020**, *40*, 461.
11. Schrader, J.A.; Srinivasan, G.; Grewell, D.; McCabe, K.G.; Graves, W.R. Fertilizer effects of soy-plastic containers during crop production and transplant establishment. *Hortscience* **2013**, *48*, 724–731. [CrossRef]
12. Nandede, B.M.; Raheman, H.; Kumar, G.V.P. Standardization of potting mix and pot volume for the production of vegetable seedlings in paper pot. *J. Plant Nutr.* **2014**, *37*, 1214–1226. [CrossRef]
13. Seo, T.C.; An, S.W.; Jang, H.W.; Nam, C.W.; Chun, H.; Kim, Y.C.; Kang, T.K.; Lee, S.H. An approach to determine the good seedling quality of grafted tomatoes (*Solanum lycopersicum*) grown in cylindrical paper pot through the relation analysis between DQI and short-term relative growth rate. *J. Bio-Env. Control* **2018**, *27*, 302–311. [CrossRef]
14. Lee, J.H.; Lee, D.H.; Choi, G.M.; Kim, E.G.; Kim, T.H.; Kim, H.T. Analysis of plant guarding characteristics of biodegradable pot for transplant. *Proc. Conf. Korean Soc. Agric.* **2019**, *24*, 494–495.
15. Lee, J.Y.; Kim, C.H.; Lee, G.S.; Jo, H.S.; Nam, H.G.; Park, H.H.; Moon, S.O. Physico-chemical characteristics of biodegradable seedling pots made of paper mill sludges. *J. Korea TAPPI* **2014**, *46*, 9–18. [CrossRef]
16. Park, J.H.; Lee, G.H.; Lee, H.M.; Madhavi, B.G.K.; Arulmozhi, E.; Bhujel, A.; Moon, B.E.; Kim, H.T. Physical properties and plant growth according to the content of additives in the biodegradable pot for mechanical transplanter. *JALS* **2022**, *56*, 129–134. [CrossRef]
17. Park, H.S.; Son, K.Y.; Kang, J.R.; Seo, W.J.; Lee, S.J.; Lee, W.K. Study on properties of eco-friendly pot with biodegradable PLA/PBAT blend film. *J. Environ. Sci. Int.* **2015**, *24*, 1037–1043. [CrossRef]
18. Jang, Y.N.; An, S.W.; Chun, H.; Lee, H.J.; Wi, S.H. The growth of cucumber seedlings grown in paper pot trays affected by nutrient management during seedling period, seedling age, and night temperature after transplanting. *J. Bio-Env. Control* **2019**, *28*, 396–403. [CrossRef]
19. Seo, T.C.; An, S.W.; Kim, S.M.; Nam, C.W.; Chun, H.; Kim, Y.C.; Kang, T.K.; Kim, S.W.; Jang, K.S. Effect of the seedlings difference in cylindrical paper pot trays on initial root growth and yield of pepper. *J. Bio-Env. Control* **2017**, *26*, 368–377. [CrossRef]

20. Nambu, T.; Tanimura, M. Development of automatic transplanter using chain pot for vegetable crops. *Int. Symp. Transpl. Prod. Syst.* **1992**, *319*, 541–546.
21. Suggs, C.W.; Thomas, T.N.; Eddington, D.L.; Peel, H.B.; Seaboch, T.R.; Gore, J.W. Self feeding transplanter for tobacco and vegetable crops. *Appl. Eng. Agric.* **1987**, *3*, 148–152. [CrossRef]
22. Kumar, G.V.P.; Raheman, H. Development of a walk-behind type hand tractor powered vegetable transplanter for paper pot seedlings. *Biosyst. Eng.* **2011**, *110*, 189–197. [CrossRef]
23. Lee, M.H.; Moon, B.E.; Jo, J.M.; Choi, T.H.; Kim, H.T. Development of Automatic type transplanter for biodegradable seedling pot. *Proc. Conf. Korean Soc. Agric.* **2017**, *22*, 136.
24. Jun, H.J.; Yu, S.H.; Choi, I.S.; Kim, J.G.; Lee, S.H.; Kang, T.G.; Kim, Y.K.; Choi, Y. Study on compatibility between plug tray sowing, seedlings and transplanters for planting mechanization of sesame. *Proc. Conf. Korean Soc. Agric.* **2018**, *23*, 155.
25. Choudhary, V.; Machavaram, R. Need of automation in paddy nurseries for raising paddy seedlings in India: A Review. *J. Biosyst. Eng.* **2022**, *47*, 209–222. [CrossRef]
26. Hwang, S.J.; Lee, J.Y.; Nam, J.S. Irrigation System for a Roller-Type Onion Pot Seeding Machine. *Appl. Sci.* **2019**, *9*, 430. [CrossRef]
27. Hwang, S.J.; Nam, J.S. Development of automatic accumulating equipment for roller-type onion pot-seeding machine. *Appl. Sci.* **2019**, *9*, 2139. [CrossRef]
28. Juvinall, R.C.; Marshek, K.M. *Machine Component Design*, 5th ed.; WILEY (John Wiley & Sons): Hoboken, NJ, USA, 2015; pp. 338–340.
29. Malik, G.K.; Khuntia, A.; Mitra, J. Comparative effect of different plasticizers on barrier, mechanical, optical, and sorption properties of hydroxypropyl methylcellulose (HPMC)–based edible film. *J. Biosyst. Eng.* **2022**, *47*, 93–105. [CrossRef]
30. Sim, J.H.; Lee, J.B.; Hwang, S.S. Effect of glazing on the flexural strength of lithium disilicate glass ceramics. *Korean J. Dent. Mater.* **2019**, *46*, 185–194. [CrossRef]